THE EIGHT MOST COMMON ELEMENTS IN THE EARTH'S

Weight Percentage		Volume Percentage	
O	46.60	O	~94
Si	27.72	K	
Al	8.13	Na	
Fe	5.00	Ca	
Ca	3.63	Si	~6
Na	2.83	Al	in total
K	2.59	Fe	
Mg	2.09	Mg	
Total	98.59		

MOLECULAR WEIGHTS OF COMPONENTS COMMON IN SILICATES
(See also Table 4.2)

SiO_2	60.08	MnO	70.94
TiO_2	79.90	MgO	40.30
Al_2O_3	101.96	Na_2O	61.98
FeO	71.85	K_2O	94.20
Fe_2O_3	159.69	H_2O	18.02
CaO	56.08		

SOME OF THE MOST COMMON IONS, THEIR COORDINATION, AND RADII

(The number in brackets is **C.N.** = coordination number.) See Table 4.8 for a complete listing.

Ion	Coordination Number with Oxygen	Ionic Radius Å
O^{2-}		1.36 [3]
K^+	8-12	1.51 [8]–1.64 [12]
Na^+	8-6 } cubic to	1.18 [8]–1.02 [6]
Ca^{2+}	8-6 } octahedral	1.12 [8]–1.00 [6]
Mn^{2+}	6	0.83 [6]
Fe^{2+}	6	0.78 [6]
Mg^{2+}	6 } octahedral	0.72 [6]
Fe^{3+}	6	0.65 [6]
Ti^{4+}	6	0.61 [6]
Al^{3+}	6	0.54 [6]
Al^{3+}	4	0.39 [4]
Si^{4+}	4 } tetrahedral	0.26 [4]
P^{5+}	4	0.17 [4]
S^{6+}	4	0.12 [4]

Cubic coordination

Octahedral coordination

Tetrahedral coordination

MANUAL OF MINERALOGY

Cover

Optical photomicrograph of a naturally zoned single crystal of zircon ($ZrSiO_4$) from Sri Lanka, photographed between crossed polarizers. The total length of the crystal is about 6 mm, resulting in a magnification of about $70\times$ for the photograph. The different interference colors are due to variable UO_2 and ThO_2 contents of different layers in the crystal. Radioactive decay of U and Th causes radiation damage, which results in a lowering of the refractive index of zircon. The layering is presumed to be the result of the crystal's having grown in a magma chamber with differing UO_2 and ThO_2 contents in various parts of the magma. Transmission electron microscopy (TEM) reveals that some of the layers are crystalline and others are noncrystalline (aperiodic or metamict) due to alpha decay damage. Microfractures perpendicular to the layering (especially well-developed in parts of the crystal on the back cover) are the result of stress set up between layers, as a result of metamictization of some of the layers. Specimen courtesy of R. C. Ewing, University of New Mexico. Reference: B. C. Chakoumakos, T. Murakami, G. R. Lumpkin, and R. C. Ewing, 1987, Alpha-decay-induced fracturing in zircon: the transition from the crystalline to the metamict state, *Science*, v. 236, pp. 1556–1559.

Schematic representation of the zircon crystal illustrated on the cover.

Supplementary Materials

For laboratory and homework assignments: Klein, C., 1989, *Minerals and Rocks: Exercises in Crystallography, Mineralogy*, and *Hand Specimen Petrology*, John Wiley & Sons, Inc., New York, 402 pp. A *Solutions Manual* with completely worked answers to most of the assignments in *Minerals and Rocks: Exercises* is available to instructors who have adopted *Minerals and Rocks: Exercises* from John Wiley & Sons.

For classroom instruction: A 35-mm slide set that incorporates approximately 130 of the most important illustrations from the text will be available to instructors upon request from the publisher.

The *Manual of Mineralogy* was written by James D. Dana in 1848 and revised by him in 1857, in 1878 as the *Manual of Mineralogy and Lithology*, and in 1887 as the *Manual of Mineralogy and Petrography*. An edition number was given to some, but not all, reprintings of each revision. For example, the 1887 revision was reprinted in 1891 as the 10th Edition but in 1893, 1895, and 1900 as the 12th Edition.* Each subsequent revision, as *Dana's Manual of Mineralogy*, has been given an edition number as follows: *13*, 1912 and *14*, 1929 by William E. Ford. *15*, 1941; *16*, 1952; *17*, 1959; and *18*, 1971 by C. S. Hurlbut, Jr. *19*, 1977 by C. S. Hurlbut, Jr. and C. Klein. *20*, 1985 by C. Klein and C. S. Hurlbut, Jr.

The information regarding revisions by James D. Dana was supplied by Clifford J. Awald, Buffalo Museum of Science.

MANUAL OF MINERALOGY

(after JAMES D. DANA)

TWENTY-FIRST EDITION

Cornelis Klein
The University of New Mexico

Cornelius S. Hurlbut, Jr.
Harvard University

JOHN WILEY & SONS, INC.
New York   Chichester   Brisbane   Toronto   Singapore

Acquisitions Editor **Barry Harmon**
Marketing Manager **Catherine Faduska**
Production Supervisor **Deborah Herbert and P. M. Gordon Associates, Inc.**
Cover Designer **Marc Klein**
Manufacturing Manager **Andrea Price**
Illustration **Sigmund Malinowski**

This book was set in Optima by Waldman Graphics, Inc.
and printed and bound by Courier Stoughton.

Library of Congress Cataloging-in-Publication Data:

Klein, Cornelis, 1937–
    Manual of mineralogy / Cornelis Klein, Cornelius S. Hurlbut, Jr.
  —21st ed / after James D. Dana.
      p.   cm.
    Includes index.
    ISBN 0–471–57452–X
    1. Mineralogy.   I. Hurlbut, Cornelius Searle, 1909–
  II. Dana, James Dwight, 1813–1895.   III. Title.
  QE372.K54   1993
  549—dc20   92–42238
                                                    CIP

Printed in the United States of America

10 9 8 7 6 5 4 3 2

PREFACE

Mineralogy is an integral part of the undergraduate education of geologists and it may also be part of a curriculum in materials science. One or several courses in mineralogy are generally required in a geology or earth science undergraduate program. These courses are meant as prerequisites to upper-level courses in petrology, geochemistry, economic geology, or mineralogy itself. The undergraduate student's first course dealing with mineralogy must convey the concepts that underlie that science and must also address the descriptive aspects of the most common minerals. In most mineralogy courses the concepts and principles tend to be covered by the instructor in the lectures, whereas the descriptive aspects of the subject, such as the study of hand specimens of minerals and rocks, are generally treated in the laboratory. Because of this duality (concepts *versus* descriptive materials), instructors will have a personal preference in their emphasis of one category over the other. As such, some courses may strongly emphasize the internal order of crystals, the crystal chemistry and structural aspects of minerals, and the quantitative evaluation of physical properties such as color, magnetism, hardness, specific gravity, and optical parameters. On the other hand, other courses may immerse the students, mainly through extensive laboratory work, in the more descriptive and determinative aspects of mineralogy. This latter emphasis prepares the student well for hand specimen petrology, at the expense of good grounding in the more conceptual aspects of the science.

In this, the 21st edition of the *Manual of Mineralogy*, we have again striven to achieve a balance between concepts and principles on the one hand and the more systematic and descriptive treatment of mineralogy on the other. Such a goal, of a balanced presentation, requires the coverage of many subjects. Not

all of them can be treated in a one-year course, and much less so in a one-semester or one-quarter course. In this edition, Chapters 2 through 9 are concerned with concepts, principles, and techniques. Chapters 10 through 13 deal with systematic and descriptive mineralogy. Chapter 14 is an introduction to petrology, and Chapter 15 gives the student an introduction to the most common gems.

The 21st edition of the *Manual of Mineralogy* incorporates major revisions of and additions to the treatment of the subject given in the 20th edition. The present edition incorporates an almost completely new chapter on "Crystal Chemistry" (Chapter 4) and a major revision of "Mineral Chemistry" (incorporated into Chapter 5); it includes a new chapter on "Mineral Stability Diagrams" (Chapter 9) to prepare the student for the extensive use of such diagrams in the subsequent chapters (Chapters 10 through 14) on "Systematic Mineralogy" and "Mineral Assemblages"; it presents a more logical treatment of "Morphology" (Chapter 2) than the earlier edition, by the deemphasis of internal symmetry in Chapter 2; in Chapter 3, "Internal Order and Symmetry," it incorporates a newly developed, logical and well-illustrated sequence from one-, to two-, to three-dimensional order and symmetry content; and in the chapters on "Systematic Mineralogy" (Chapters 10 through 13) it eliminates reference to little used information (e.g., interfacial angles, axial ratios, and chemical data obtained by blowpipe) and provides updated information on important mineral localities. Throughout this new edition, there are abundant new illustrations to aid in the understanding of concepts discussed in the text.

The main unifying theme of this text, as in earlier editions, is that of "Crystal Chemistry"—the elucidation of the relationship between chemical composition, internal structure, and physical properties of crystalline matter. The material in the first eight chapters has been so arranged as to allow the instructor to select those chapters that he or she considers most essential. It will be impossible in most courses to cover the material in all of the chapters. For example, Chapters 7 and 8, on X-ray and optical mineralogy, respectively, will generally not play a large role in an introductory mineralogy sequence. These subjects are included, however, because in the subsequent treatment on "Systematic Mineralogy" each mineral species description includes a listing of diagnostic X-ray and optical properties. Similarly, Chapter 15, on gems, may not be part of the coverage in a specific course. However, if the student wishes to find out what X-ray and optical techniques are used for the determination of the various parameters, and what unique qualities distinguish gems from common minerals, the relevant discussion can be located in the appropriate chapters.

Chapter 1 provides a very brief introduction to the history of the science, to the sources of its literature, and to a few of the recent developments in instrumentation that allow for the chemical and structural evaluation of crystalline materials on very small scales. It also provides the student with a listing of those scientists who have made internationally recognized contributions to the fields of mineralogy, petrology, and geochemistry in the last five decades.

Chapter 2 deals with the external form and symmetry of crystals, whereas internal order is discussed in Chapter 3. The chapter is so arranged that the study of external form and the derivation of its inherent symmetry can be covered without reference to stereographic projection, if the instructor so desires. Stereographic projection is introduced on page 56, after which each of the crystal systems (and crystal classes) is treated systematically with stereograms. As in the 20th edition, the systematic treatment of crystal morphology begins with the crystal system with lowest symmetry and ends with the system with highest symmetry. Within each crystal system, however, the crystal classes are arranged in order of decreasing symmetry. Although this appears illogical with respect to the increasing symmetry of the crystal systems, this sequence of crystal classes is least cumbersome because the lower symmetry classes can be described as special cases of the highest symmetry class. Moreover, if time does not permit the development of all crystal classes, it is those of highest symmetry in each crystal system that are usually discussed.

Chapter 3 deals with order in one-, two-, and three-dimensional ordered patterns and provides an introduction to space groups. The logical sequence of discussion of the symmetry and translational aspects of one-, two-, and three-dimensional patterns has been much revised and improved in this edition. This should allow the student a better understanding of the unique symmetry and translation components of rows, of planar arrays (planar lattices and point groups), and of three-dimensional ordered arrays (including the development of Bravais lattices and their relationship to space groups). The ultimate goal of this chapter is to familiarize the student with the interrelation between external morphology, internal structure, and space group notation. Such a connection is worked out in considerable detail for three common minerals. A much expanded discussion of isostructuralism, polymorphism, polytypism, the metamict state, and structural complexities and defects follows the discussion of ideal order in three dimensions. Coverage of morphology (Chapter 2) precedes the treatment of internal order (Chapter 3). We chose this sequence because many students find it easier to locate symmetry in physical objects (e.g., crystal models) than in more abstract designs (such as patterns of motifs). However, this order need not be followed in a specific course, as neither chapter is a prerequisite for the other.

Chapter 4, on "Crystal Chemistry," gives a completely revised and much expanded treatment of the subject. For those students with a strong prior background in chemistry, it should provide a good review of the subject; for those with lesser chemical experience, it should provide an efficient and very readable introduction and overview. The coverage progresses from a discussion of the atom, the periodic table, the Bohr atomic model, the Schrödinger model, through quantum notation and electron configuration, to the ion, atomic and ionic radii, coordination, radius-ratio, chemical bonding mechanisms, and ends with an introduction to structure types, changes in structure type as a function of increasing depth in the Earth, and aspects of radioactivity.

Chapter 5, on "Mineral Chemistry," provides an introductory overview of commonly used chemical analytical techniques such as wet chemical analysis, atomic absorption spectroscopy, X-ray fluorescence and electron microprobe techniques, and optical spectrographic analysis. This is followed by a detailed discussion of compositional variation in minerals, as a result of various solid solution mechanisms. Exsolution is another major concept, before extensive discussion and examples of recalculations of mineral analyses and various schemes of graphic representation of chemical data.

Chapter 6, on "Physical Properties of Minerals," treats such properties as cleavage, hardness, density, color, and magnetism. In this chapter the sections on color and magnetism are completely rewritten and are quantitatively based on the crystal chemical concepts developed in Chapter 4.

Chapters 7 and 8, which deal specifically with X-ray crystallography and optical properties of crystals, are very similar to those in the previous edition.

Chapter 9, on "Mineral Stability Diagrams," is a completely new chapter that provides an introduction to a range of stability diagrams commonly used to illustrate mineral equilibria. It begins with a brief introduction to some basic aspects of thermodynamics, but the main part of the chapter is devoted to a basic understanding of one-, two-, and three- (or more) component diagrams, of diagrams for mineral reactions involving H_2O or CO_2, and Eh-pH representations. It is abundantly illustrated with examples of the stability relations of common rock-forming minerals, the majority of which are shown again in the subsequent Chapters (10 through 13) on "Systematic Mineralogy." This new Chapter 9 is designed specifically to provide the student with a basic understanding of mineral stability *before* the discussion of specific minerals (and/or groups) and the interpretive aspects of their occurrence.

Chapters 10 through 13 deal with the systematic descriptions of about 200 minerals, as well as their petrologic associations and the geographic locations of the most important mineral localities. The description of each mineral group, or major mineral series, is preceded by a synopsis of some of the highlights of its chemistry, crystal chemistry, and structure. Each of the approximately 200 descriptions contains a section on "Composition and Structure." Mineral structures are illustrated with the best available polyhedral and/ or ball-and-stick representations. Approximately 40 new illustrations (including structure representations, high resolution electron microscope—HRTEM—photographs, pictures of mineral specimens, and blackline drawings) have been added to these four chapters. Each of the mineral descriptions has been streamlined by elimination of data on interfacial angles, axial ratios, and blowpipe analysis results.

Chapter 14, on "Mineral Assemblages," presents a brief introduction to petrology providing a link between mineralogy and petrology. The material includes up-to-date classification schemes for igneous, sedimentary, and metamorphic rock types. Various graphic representations of assemblages in metamorphic rocks are also given.

Chapter 15 provides a brief overview of the most common gems and the techniques used in gemology. It contains four color plates of gem minerals and gems cut therefrom. The reason for the inclusion of this subject matter, which is normally absent from mineralogy texts, is as follows: Students who have mastered much of what is known as mineralogy generally are no better informed about gems than the layman. This is unfortunate but also unnecessary. Very few mineralogy courses, if any, will include much discussion of gems, let alone have a laboratory on gem materials. However, the student can read this chapter for his or her own edification, without specific reference to it in a course. Having read this chapter, the mineralogy student will be a well-informed customer in jewelry stores.

At the end of each chapter are selected references that will guide the student to the most relevant literature on the subject. The Determinative Tables (Appendix) and Mineral Index, all of which have been updated, are especially useful in the laboratory study and identification of unknown minerals.

In short, in writing the 21st edition of the *Manual of Mineralogy* our aim was for an introductory mineralogy book that would continue to serve two purposes: (1) to provide the student with an up-to-date understanding of basic concepts and principles in crystallography, crystal chemistry, chemistry, physical aspects of minerals, and introductory petrology concepts essential to the understanding of the genesis of minerals and rocks; and (2) to provide a reference for quick and unambiguous identification of the common minerals in the field and in the laboratory. This book is designed for both a beginning course in mineralogy and a permanent mineralogic reference. It is thus intended for students who will do further work in mineralogy as well as those for whom a single course will be their only exposure to the subject.

ACKNOWLEDGMENTS

We express our sincere appreciation to several colleagues who provided us with (1) reviews of the 20th edition of the *Manual of Mineralogy* and/or (2) reviews of the first six revised chapters of the present 21st edition:

Marc J. Defant, University of South Florida,
Malcolm D. Hill, Northeastern University,
David M. Jenkins, State University of New York at Binghamton,
Bruce M. Loeffler, Colorado College, and
Paul H. Ribbe, Virginia Polytechnic Institute and State University.

We also thank the following colleagues who reviewed specific sections of this text:

Andrew R. Campbell, New Mexico Institute of Mining and Technology,

Carl A. Francis, Harvard University,

John Husler, University of New Mexico,

Gregory J. McCarthy, North Dakota State University,

Henry O. A. Meyer, Purdue University,

William C. Metropolis, Harvard University,

Malcolm Ross, U.S. Geological Survey, Reston, Virginia, and

Crayton J. Yapp, University of New Mexico.

Many of the new photographs of high resolution transmission electron microscope (HRTEM) structures and of mineral specimens for this edition were generously provided by:

Jillian F. Banfield, University of Wisconsin,

Peter R. Buseck, Arizona State University,

Rodney C. Ewing, University of New Mexico,

Carl A. Francis, Harvard University,

Ermanno Galli, Instituto di Mineralogia e Petrologia, Modena, Italy,

Eugene A. Smelick, Princeton University,

David R. Veblen, Johns Hopkins University, and

LuMin Wang, University of New Mexico.

The word processing of most of the revisions in this edition was most efficiently and enthusiastically accomplished by Mabel T. Chavez of Santo Domingo Pueblo.

C. K. is grateful to his wife, Angela, for her continued support and forebearance.

Cornelis Klein
Cornelius S. Hurlbut, Jr.

CONTENTS

14. Mineral Assemblages: Introduction to Rock Types and Vein Mineralization 558

15. Gem Minerals 593

Appendix: Determinative Tables 613

Mineral Index 653
Subject Index 669

MANUAL OF MINERALOGY

CHAPTER 1

INTRODUCTION

Mineralogy is the study of naturally occurring, crystalline substances—minerals. Everyone has a certain familiarity with minerals for they are present in the rocks of the mountains, the sand of the sea beach, and the soil of the garden. Less familiar, but also composed of minerals, are meteorites and the lunar surface material. Gems are exceptionally beautiful and generally highly durable representatives of the mineral world. A knowledge of what minerals are, how they were formed, and where they occur is basic to an understanding of the materials largely responsible for our present technologic culture. For all inorganic articles of commerce, if not minerals themselves, are mineral in origin.

DEFINITION OF MINERAL

Although it is difficult to formulate a succinct definition for the word mineral, the following is generally accepted:

A mineral is a naturally occurring homogeneous solid with a definite (but generally not fixed) chemical composition and a highly ordered atomic arrangement. It is usually formed by inorganic processes.

A step-by-step analysis of this definition will aid in its understanding. The qualification *naturally oc-* *curring* distinguishes between substances formed by natural processes and those made in the laboratory. Industrial and research laboratories routinely produce synthetic equivalents of many naturally occurring materials including valuable gemstones such as emeralds, rubies, and diamonds. Since the beginning of the twentieth century, mineralogic studies have relied heavily on the results from synthetic systems in which the products are given the names of their naturally occurring counterparts. Such practice is generally accepted although at variance with the strict interpretation of *naturally occurring*. In this book *mineral* means a naturally occurring substance and its name will be qualified by *synthetic* if purposely produced by laboratory techniques. One might now ask how to refer to $CaCO_3$ (calcite) which sometimes forms in concentric layers in city water mains. The material is precipitated from water by natural processes but in a man-made system. Most mineralogists would refer to it by its mineral name, calcite, because humanity's part in its formation was inadvertent.

The definition further states that a mineral is a *homogeneous solid*. This means that it consists of a single, solid substance that cannot be physically subdivided into simpler chemical compounds. The determination of homogeneity is difficult because it is related to the scale on which it is defined. For example, a specimen that appears homogeneous to the naked eye may prove to be inhomogeneous, made up of several materials, when viewed with a microscope at high magnification. The qualification *solid* excludes gases and liquids. Thus H_2O as ice in a glacier is a mineral, but water is not. Likewise liquid mercury, found in some mercury deposits, must be excluded by a strict interpretation of the definition. However, in a classification of natural materials such substances that otherwise are like minerals in chemistry and occurrence are called *mineraloids* and fall in the domain of the mineralogist.

The statement that a mineral has a *definite chemical composition* implies it can be expressed by a specific chemical formula. For example, the chemical composition of quartz is expressed as SiO_2. Because quartz contains no chemical elements other than silicon and oxygen, its formula is definite. Quartz is, therefore, often referred to as a pure substance. Most minerals, however, do not have such well-defined compositions. Dolomite, $CaMg(CO_3)_2$, is not always a pure Ca-Mg-carbonate. It may contain considerable amounts of Fe and Mn in place of Mg. Because these amounts vary, the composition of dolomite is said to range between certain limits and is, therefore, *not fixed*. Such a compositional range may be expressed by a formula with the same atomic (or more realistically, ionic) ratios as pure $CaMg(CO_3)_2$ in which $Ca : Mg : CO_3 = 1 : 1 : 2$. This leads to a more general expression for dolomite as: $Ca(Mg,Fe,Mn)(CO_3)_2$.

A *highly ordered atomic arrangement* indicates an internal structural framework of atoms (or ions) arranged in a regular geometric pattern. Because this is the criterion of a crystalline solid, minerals are crystalline. Solids, such as glass, that lack an ordered atomic arrangement are called *amorphous*. Several natural solids are amorphous. Examples are: volcanic glass (which is not classified as a mineral because of its highly variable composition and lack of ordered atomic structure), limonite (a hydrous iron oxide), and allophane (a hydrous aluminum silicate); also several metamict minerals such as microlite, gadolinite, and allanite (in metamict minerals the original crystallinity has been destroyed, to various degrees, by radiation from radioactive elements present in the original structure). They, with the liquids water and mercury, which also lack internal order, are classified as mineraloids.

According to the traditional definition, a mineral is *formed by inorganic processes*. We prefer to preface this statement with *usually* and thus include in the realm of mineralogy those organically produced compounds that answer all the other requirements of a mineral. The outstanding example is the calcium car-bonate of mollusk shells. The shell of the oyster and the pearl that may be within it are composed in large part of aragonite, identical to the inorganically formed mineral.

Although several forms of $CaCO_3$ (calcite, aragonite, vaterite) and monohydrocalcite, $CaCO_3 \cdot H_2O$, are the most common biogenic minerals (meaning "mineral formed by organisms"), many other biogenic species have been recognized. Opal (an amorphous form of SiO_2), magnetite (Fe_3O_4), fluorite (CaF_2), several phosphates, some sulfates, Mn-oxides, and pyrite (FeS_2) as well as elemental sulfur are all examples of minerals that can be precipitated by organisms (see Lowenstam, 1981). The human body also produces essential minerals. Apatite, $Ca_5(PO_4)_3(OH)$, is the principal constituent of bones and teeth. The body can also produce concretions of mineral matter (calculi) in the urinary system. Such calculi are found to consist predominantly of calcium phosphates (such as hydroxylapatite, carbonate-apatite, and whitlockite), calcium oxalates that are very uncommon in the mineral world, and magnesium phosphates (see Gibson, 1974).

Petroleum and coal, however, frequently referred to as mineral fuels, are excluded; for, although naturally formed, they have neither a definite chemical composition nor an ordered atomic arrangement. However, in places coal beds have been subjected to high temperatures that have driven off the volatile hydrocarbons and crystallized the remaining carbon. This residue is the mineral, graphite.

HISTORY OF MINERALOGY

Although it is impossible in a few paragraphs to trace systematically the development of mineralogy, some of the highlights of its development can be singled out. The emergence of mineralogy as a science is relatively recent, but the practice of mineralogical arts is as old as human civilization. Natural pigments made of red hematite and black manganese oxide were used in cave paintings by early humans, and flint tools were prized possessions during the Stone Age. Tomb paintings in the Nile Valley executed nearly 5000 years ago show busy artificers weighing malachite and precious metals, smelting mineral ores, and making delicate gems of lapis lazuli and emerald. As the Stone Age gave way to the Bronze Age, other minerals were sought from which metals could be extracted.

We are indebted to the Greek philosopher Theophrastus (372–287 B.C.) for the first written work on minerals and to Pliny, who 400 years later recorded the mineralogical thought of his time. During the fol-

FIG. 1.1. Prospecting with a forked stick (A) and trenching (B) in the fifteenth century. (From Agricola, *De Re Metallica*, translated into English. Dover Publications, New York, 1950.)

lowing 1300 years, the few works that were published on minerals contained much lore and fable with little factual information. If one were to select a single event signaling the emergence of mineralogy as a science, it would be the publication in 1556 of *De Re Metallica* by the German physician Georgius Agricola. This work gives a detailed account of the mining practices of the time and includes the first factual account of minerals. The book was translated into English from the Latin in 1912[1] by the former President of the United States, Herbert Hoover, and his wife, Lou Henry Hoover. An illustration from this book is reproduced in Fig. 1.1. In 1669 an important contribution was made to crystallography by Nicolas Steno through his study of quartz crystals. (A portrait of Nicolas Steno is reproduced in Fig. 1.2.) He noted that despite their differences in origin, size, or habit, the angles between corresponding faces were constant (see Fig. 1.3). More than a century passed before the next major contributions were made. In 1780 Carangeot invented a device (contact goniometer) for the measurement of interfacial crystal angles (see

Fig. 1.5a). In 1783 Romé de l'Isle made angular measurements on crystals confirming Steno's work and formulated the law of the constancy of interfacial angles. The following year, 1784, René J. Haüy showed that crystals were built by stacking together tiny identical building blocks, which he called integral molecules (see Fig. 1.4). The concept of integral molecules survives almost in its original sense in the unit cells of modern crystallography. Later (1801) Haüy, through his study of hundreds of crystals, developed the theory of rational indices for crystal faces.

In the early nineteenth century rapid advances were made in the field of mineralogy. In 1809 Wollaston invented the reflecting goniometer, which permitted highly accurate and precise measurements of the positions of crystal faces. Where the contact goniometer had provided the necessary data for studies on crystal symmetry, the reflecting goniometer would provide extensive, highly accurate measurements on naturally occurring and artificial crystals (see Figs. 1.5c and d). These data made crystallography an exact science. Between 1779 and 1848 Berzelius, a Swedish chemist, and his students studied the chemistry of minerals and developed the principles of our present chemical classification of minerals.

[1]Published in 1950 by Dover Publications, Inc., New York.

FIG. 1.2. Portrait of Niels Stensen (Latinized to Nicolaus Steno). Steno was born in Copenhagen, Denmark, in 1638 and died in 1686. (From Scherz, G., *Steno, Geological Papers.* Odense University Press, 1969.)

In 1815 the French naturalist Cordier, whose legacy to mineralogy is honored in the name of the mineral *cordierite*, turned his microscope on crushed mineral fragments in water. He thereby initiated the "immersion method" which others, later in the century, developed into an important technique for the study of the optical properties of mineral fragments. The usefulness of the microscope in the study of minerals was greatly enhanced by the invention in 1828 by the Scotsman, William Nicol, of a polarizing device that permitted the systematic study of the behavior of light in crystalline substances. The polarizing microscope became, and still is, a powerful determinative tool in mineralogical studies. An early model is illustrated in Fig. 1.6. In the latter part of the nineteenth century Fedorov, Schoenflies, and Barlow, working independently, almost simultaneously developed theories for the internal symmetry and order within crystals which became the foundations for later work in X-ray crystallography.

The most far-reaching discovery of the twentieth century must be attributed to Max von Laue of the University of Munich. In 1912 in an experiment performed by Friedrich and Knipping at the suggestion of von Laue, it was demonstrated that crystals could diffract X-rays. Thus was proved for the first time the regular and ordered arrangement of atoms in crystalline material. Almost immediately X-ray diffraction

FIG. 1.3. Steno's drawings of various quartz and hematite crystals, illustrating the constancy of angles among crystals of different habits. (From Schafkranovski, J. J., 1971, *Die Kristallographischen Entdeckungen N. Stenens*, in *Steno as Geologist*. Odense University Press.)

FIG. 1.4. Illustration of the concept developed by R. J. Haüy (1743–1826) of tiny identical building blocks underlying the external form of crystals. In this figure the development of a dodecahedron of garnet is shown. (From Marr, G. M., 1970, *Geschichte der Kristallkunde*. Reprinted by Sandig, Walluf, Germany.)

FIG. 1.5. Examples of instruments used for the measurement of angles between crystal faces.
(a) Brass contact goniometer of the Carangeot type. This was used at Harvard University in 1797
(see Frondel, 1983). *(b)* A two-circle contact goniometer based on the 1896 design of Victor Goldschmidt. *(c)* The earliest reflecting one-circle goniometer as invented by W. H. Wollaston in 1809.
(From Tschermak, G. and Becke, F., 1921, *Lehrbuch der Mineralogie*. Hölder-Pichler-Tempsky,
Vienna.) *(d)* A two-circle reflecting goniometer as developed in the latter part of the nineteenth century. (From Groth, P., 1895, *Physikalische Krystallographie*, Leipzig.) Compare with Fig. 2.40.

FIG. 1.6. Polarizing microscope as available in the mid-nineteenth century. (From Tschermak, G. and Becke, F., 1921, *Lehrbuch der Mineralogie*. Hölder-Pichler-Tempsky, Vienna.) Compare with Fig. 8.10.

became a powerful method for the study of minerals and all other crystalline substances, and in 1914 the earliest crystal structure determinations were published by W. H. Bragg and W. L. Bragg in England. (Their photographs are given in Fig. 1.7.) Modern X-ray diffraction equipment with on-line, dedicated computers has made possible the relatively rapid determination of highly complex crystal structures. The advent of the electron microprobe in the early 1960s, for the study of the chemistry of minerals on a microscale, has provided yet another powerful tool that is now routinely used for the study of the chemistry of

FIG. 1.7. Portraits of *(a)* Sir William Henry Bragg (1862–1942) and *(b)* his son Sir William Lawrence Bragg (1890–1971). Father and son received the Nobel Prize for Physics in 1915. Both men are eminently known for their researches in the field of crystal structure by X-ray methods. (*a* from Godfrey Argent, London, photograph by Walter Stoneman; *b* from Times Newspapers, Ltd., London.)

FIG. 1.8. A computer-automated electron microprobe, with the electron beam column and X-ray spectrometers on the right; computer control and read-out systems are on the left. This is a Camebax SX 50, manufactured by Cameca SA, Courbevoie, France. (Courtesy of Cameca Instruments Inc., Stamford, Conn.)

FIG. 1.9. Schematic cross section through the electron optical column and X-ray spectrometer of an electron microprobe.

minerals, synthetic compounds, and glasses. Electron microprobes (see Figs. 1.8 and 1.9) can provide accurate, many-element analyses of solid materials in a grain size as small as about one micrometer (0.001 mm). The majority of mineral analyses now produced are made by electron microprobe not only because of the fine focus of the electron beam of the instrument but because the analyses can be made *in situ*—on specific mineral grains in polished sections and polished thin sections of rocks. This has eliminated the laborious process of mineral separation and concentration, which is a requirement for several other mineral analysis techniques (see Chapter 5 for further discussion of analytical techniques).

Since early 1970 another electron beam instrument, which can magnify the internal architecture of minerals many millions of times, has produced elegant and powerful visual images of atomic structures. This instrument, the transmission electron micro-scope, is illustrated in Fig. 1.10, and a schematic representation of the magnification process of an object into a final and highly enlarged image is shown in Fig. 1.11. The most visually instructive application of this technique is known as high-resolution transmission electron microscopy (HRTEM), which allows the study of crystalline materials at resolutions approaching the scale of atomic distances (see P. R. Buseck, 1983). The technique can produce projected two-dimensional images of three-dimensional crystalline structures. These images show that many minerals have infinitely extending, periodic (meaning: perfectly repeating) internal structural arrangements; an example of such a "perfect" structure is illustrated by the HRTEM image in Fig. 1.12 for the chemically complex mineral tourmaline. The HRTEM images have also shown that minerals may contain defects that are deviations from idealized ("perfect") structures (see Figs. 3.52 and 3.53).

FIG. 1.10. A transmission electron microscope, Philips System CM12. The far left-hand cabinet is the high-voltage generator; the tall vertical feature is the electron column with control panels on either side and a viewing screen at table-top level; the viewing screen is used for transmission electron microscope (TEM) imaging and for display of electron diffraction patterns. In the middle of the electron column and slightly to the right is the sample assembly holder. An energy dispersive X-ray detector (EDX) with liquid nitrogen dewar for cooling is affixed to the right of the column. The cabinet to the far right with the TV monitor is the analyzer (for chemical analysis). (Courtesy of Philips Electronic Instruments, Inc., Mahwah, N.J.)

The field of mineralogy now encompasses a very broad area of study that includes X-ray, electron, and neutron diffraction by minerals, mineral synthesis, crystal physics, the evaluation of the thermodynamic stability of minerals, petrography (the study of rocks and minerals in thin section), petrology (the study of rocks), experimental petrology, and aspects of metallurgy and ceramics. Because it is difficult to predict which contributions (made in relatively recent times) to the science of mineralogy will prove to be most enduring and important, we have listed in Table 1.1 the Roebling medalists and some of their major research contributions. This list illustrates the diversity of internationally recognized professional contributions, and it is a reasonable assumption that it includes those mineralogists whom future historians will regard as the mineralogical giants of our time.

The Roebling medal was established in 1937 by the Mineralogical Society of America in memory of Colonel Washington A. Roebling (1837–1926), who had made a generous financial gift to the society in 1926. Colonel Roebling, the designer of such well-known suspension bridges as those over the Niagara River at Niagara Falls, over the Allegheny River at Pittsburgh, over the Ohio River in Cincinnati, and the Brooklyn Bridge over the East River in New York City, had a deep, life-long interest in the study of minerals. The Roebling medal signifies the highest recognition of achievement American mineralogy can bestow on outstanding investigators in America or abroad. The presentation and acceptance speeches for each Roebling award can be found in *The American Mineralogist*.

In 1927, Roebling's son John donated his father's mineral collection of about 16,000 specimens to the National Museum of Natural History (Smithsonian Institution) in Washington, D.C. This collection, known as the Washington A. Roebling mineral collection, was undoubtedly one of the largest and finest private collections of its time. This acquisition, together with another collection of about 9100 specimens (the Canfield collection), made the Smithsonian mineral collection one of the best in the world (Roe, A., 1990, Washington A. Roebling, his life and his mineral collection, *Mineralogical Record*, v. 21, pp. 13–30).

FIG. 1.11. Schematic cross section through the column of a transmission electron microscope, showing the ray path for structural imaging. The four lenses are electromagnetic lenses.

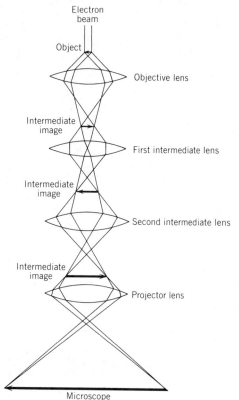

FIG. 1.12. High-resolution transmission electron microscope (HRTEM) image of the structure of tourmaline. (From Iijima, S., Cowley, J. M., and Donnay, G., 1973, High resolution electron microscopy of tourmaline crystals. *Tschermaks Mineralogische Petrographische Mitteilungen*, v. 20, pp. 216–224.) The white areas in the photograph correspond to regions of low electron density in the structure of tourmaline. The 6-fold pattern is the image of the 6-fold Si_6O_{18} ring in tourmaline (compare with Fig. 13.41). The bar scale represents 15 angstroms.

Table 1.1

RECIPIENTS OF THE ROEBLING MEDAL AWARDED BY THE MINERALOGICAL SOCIETY OF AMERICA

Award Year	Recipient and Institutional Affiliation	Examples of Outstanding Contributions to Mineralogy
1937	Charles Palache, Harvard University	goniometric studies in crystallography and mineral paragenesis; minerals of Franklin Furnace, N.J.
1938	Waldemar T. Schaller, U.S. Geological Survey	chemical mineralogy; crystallography and paragenesis of pegmatite minerals
1940	Leonard James Spencer, British Museum (Natural History)	mineral systematics; long-time editor of *Mineralogical Magazine* and *Mineralogical Abstracts*
1941	Esper S. Larsen, Jr., Harvard University	mineralogy and petrology; the determination of the optical properties of nonopaque minerals
1945	Edward H. Kraus, University of Michigan	occurrence and origin of minerals; crystallographic forms; development of apparatus for mineral testing
1946	Clarence S. Ross, U.S. Geological Survey	petrography and petrology, particularly clay minerals; microscopic techniques; geochemistry of ore deposits
1947	Paul Niggli, Technische Hochschule, Zurich	crystallography and structure of minerals; igneous and metamorphic rocks; minerals of the Swiss Alps
1948	William Lawrence Bragg, Cavendish Laboratory, University of Cambridge	crystal structure determinations by X-ray diffraction techniques; recipient (jointly with his father, William Henry Bragg) of the Nobel Prize for Physics in 1915
1949	Herbert E. Merwin, Geophysical Laboratory, Carnegie Institution	crystal optics; characterization of synthetic, transparent, and opaque phases
1950	Norman L. Bowen, Geophysical Laboratory, Carnegie Institution	application of experimental physico-chemical data and principles to petrological problems
1952	Frederick E. Wright, Geophysical Laboratory, Carnegie Institution	optical properties of minerals; design of improved petrographic microscope and test plates
1953	William F. Foshag, U.S. National Museum	characterization of new minerals; minerals of the United States and Mexico; gemology
1954	Cecil Edgar Tilley, University of Cambridge	application of physico-chemical principles to the study of mineral assemblages in metamorphic and igneous rocks
1955	Alexander N. Winchell, University of Wisconsin	evaluation of major rock-forming silicate groups; textbooks on optical mineralogy
1956	Arthur F. Buddington, Princeton University	quantitative petrologic studies of many rock types and of ore deposits
1957	Walter F. Hunt, University of Michigan	mineralogy; editor for 35 years of *The American Mineralogist*
1958	Martin J. Buerger, Massachusetts Institute of Technology	structural crystallography; development of single crystal X-ray techniques; textbooks in crystallography
1959	Felix L. Machatschki, University of Vienna	atomic arrangement of major silicate groups; atomic substitution; relation of crystal structure to paragenesis
1960	Thomas F. W. Barth, Oslo University	petrology and X-ray crystallography; petrogenetic relations of rocks in regions of Norway (e.g., the Oslo region) and North America
1961	Paul Ramdohr, University of Heidelberg	reflected light microscopy; mineralogy and genesis of ore deposits; major text on ore mineralogy

(continued)

Table 1.1
(continued)

Award Year	Recipient and Institutional Affiliation	Examples of Outstanding Contributions to Mineralogy
1962	John W. Gruner, University of Minnesota	X-ray crystallography of clay minerals; mineralogy and petrology of iron formations; uranium mineralization
1963	John Frank Schairer, Geophysical Laboratory, Carnegie Institution	experimental studies of silicate phase equilibria
1964	Clifford Frondel, Harvard University	structural mineralogy; characterization of many new minerals; mineralogy of uranium and thorium; co-author of *Dana's System*
1965	Adolph Pabst, University of California, Berkeley	X-ray crystallographic and structural mineralogy; evaluation of the metamict state and of sheet silicates
1966	Max H. Hey, British Museum (Natural History)	mineral chemistry of zeolites, the chlorite group, and meteorites; author of *Index of Mineral Species;* long-time editor of *Mineralogical Magazine*
1967	Linus Pauling, University of California, San Diego	crystal and molecular structures; quantum chemistry; theory of chemical bonding; author of *The Nature of the Chemical Bond;* recipient of the Nobel Prize in Chemistry in 1954 and the Nobel Peace Prize in 1962
1968	Tei-ichi Ito, University of Tokyo	structural crystallography; polymorphism; X-ray powder diffraction
1969	Fritz Laves, Technische Hochschule, Zurich	structure and paragenesis of feldspars; crystal chemistry of metallic compounds
1970	George W. Brindley, Pennsylvania State University	structural crystallography of layer silicates; clay mineralogy
1971	J. D. H. Donnay, Johns Hopkins University	crystallography; crystal optics; relationship of morphology and structure; twinning; author of *Crystal Data*
1972	Elburt F. Osborn, Bureau of Mines, U.S. Department of Interior	experimental petrology in rock-forming systems; crystallization and differentiation trends in magmas
1973	George Tunell, University of California, Santa Barbara	experimental investigations of oxidized ore minerals; physical-chemical evaluations of ore-forming processes
1974	Ralph E. Grim, University of Illinois, Champaign-Urbana	clay mineralogy; author of *Clay Mineralogy* and *Applied Clay Mineralogy*
1975	O. Frank Tuttle, Stanford University	experimental petrology; development of hydrothermal research equipment; experimental studies on the origin of granite
1975	Michael Fleischer, U.S. Geological Survey	trace element geochemistry; evaluation of new mineral species; long-time chairman of international commission on new minerals; long-time editor of section on mineralogy in *Chemical Abstracts*
1976	Carl W. Correns, University of Göttingen	physical-chemical evaluations of sedimentology; chemical and physical oceanography; mineralogy of sediments
1977	Raimond Castaing, University of Paris	inventor of the electron microprobe and pioneering work on theory of quantitative analysis; development of ion beam microprobe

Table 1.1
(continued)

Award Year	Recipient and Institutional Affiliation	Examples of Outstanding Contributions to Mineralogy
1978	James B. Thompson, Jr., Harvard University	theoretical evaluation of petrologic systems; thermodynamics of mineral systems; metamorphic reactions in pelitic schists; crystal chemistry of amphiboles
1979	William H. Taylor, Cavendish Laboratory, University of Cambridge	structural crystallography; structural features of feldspar, zeolite, and aluminosilicate minerals
1980	D. S. Korzhinskii, Academy of Sciences, Moscow	application of chemical thermodynamics to petrology; author of *Physicochemical Basis for the Analysis of the Paragenesis of Minerals* and of *Theory of Metasomatic Zoning*
1981	Robert M. Garrels, University of South Florida	theoretical studies of ore formation; chemical thermodynamics of mineral systems; phase diagrams for minerals at low temperature; chemical evolution of the ocean and atmosphere; co-author of *Solutions, Minerals and Equilibria*
1982	Joseph V. Smith, University of Chicago	structural crystallography of rock-forming minerals; zeolites and feldspars; lunar mineralogy and petrology; author of *Feldspars* (2 volumes)
1983	Hans P. Eugster, Johns Hopkins University	solid-fluid equilibria in hydrothermal systems; chemical sedimentation and water chemistry such as in active salt lakes
1984	Paul B. Barton, Jr., U.S. Geological Survey	petrology of ores; the chemistry and physics of the ore-forming process
1985	Francis J. Turner, University of California, Berkeley	metamorphic petrology
1986	Edwin Roedder, U.S. Geological Survey	fluid inclusions in minerals
1987	Gerald V. Gibbs, Virginia Polytechnic Institute and State University	mathematical foundations of crystallography; application of molecular orbital theory to chemical bonding
1988	Julian R. Goldsmith, University of Chicago	order–disorder in feldspars; phase equilibria of carbonates; oxygen-isotope fractionations among minerals
1989	Helen D. Megaw, University of Cambridge	X-ray crystal structure of feldspars; origin of ferroelectricity in oxides
1990	Sturges W. Bailey, University of Wisconsin	structural and crystal chemical studies of layer silicate minerals
1991	E-an Zen, U.S. Geological Survey	application of thermodynamics to petrology; the temperature-pressure regime of the Appalachian mountain belt; igneous thermobarometry
1992	Hatten S. Yoder, Jr., Geophysical Laboratory, Carnegie Institution	experimental petrology and its application to mineral paragenesis; study of the role of water in metamorphism and the petrogenesis of igneous rocks; author of *Generation of Basaltic Magma*
1993	Brian Mason, National Museum of Natural History (Smithsonian Institution)	characterization of a broad range of terrestrial and meteoritic mineralogy; author of *Principles of Geochemistry,* and of *Meteorites;* co-author of *Mineralogy*

ECONOMIC IMPORTANCE OF MINERALS

Since before historic time, minerals have played a major role in humanity's way of life and standard of living. With each successive century they have become increasingly important, and today we depend on them in countless ways—from the construction of skyscrapers to the manufacture of computers. Modern civilization depends on and necessitates the prodigious use of minerals. A few minerals such as talc, asbestos, and sulfur are used essentially as they come from the ground, but most are first processed to obtain a usable material. Some of the more familiar of these products are bricks, glass, cement, plaster, and a score of metals ranging from iron to gold. Metallic ores and industrial minerals are mined on every continent wherever specific minerals are sufficiently concentrated to be economically extracted.

According to the U.S. Bureau of Mines (as reported in *Geotimes*, 1989, v. 34, p. 19), "each year, every American requires 40,000 pounds of new minerals. At that level of consumption, the average newborn infant will need a lifetime supply of 795 pounds of lead (mainly for car batteries, solder, and electronic components); 757 pounds of zinc (as an alloy of copper to make brass, as protective coatings on steel, and as chemical compounds in rubber and paints); 1500 pounds of copper (mainly used in electrical motors, generators, communication equipment, and wiring); 3593 pounds of aluminum (for all sorts of things such as beverage cans, folding lawn chairs, and aircraft); 32,700 pounds of pig iron (for kitchen utensils, automobiles, ships, and large buildings); 28,213 pounds of salt (for cooking, highway de-icing, and detergents); and 1,238,101 pounds of stone, sand, gravel, and cement (for building roads, homes, schools, offices, and factories)." A nationwide commitment to recycling will lower many of the above estimates.

The location of mineable metal and industrial mineral deposits, and the study of the origin, size, and ore grade of these deposits is the domain of economic geologists. But a knowledge of the chemistry, occurrence, and physical properties of minerals is basic to pursuits in economic geology.

NAMING OF MINERALS

Minerals are most commonly classified on the basis of the presence of a major chemical component (an anion or anionic complex) into oxides, sulfides, silicates, carbonates, phosphates, and so forth. This is especially convenient because most minerals contain only one major anion. However, the naming of minerals is not based on such a logical chemical scheme.

The careful description and identification of minerals often require highly specialized techniques such as chemical analysis and measurement of physical properties, among which are the specific gravity, optical properties, and X-ray parameters that relate to the atomic structure of minerals. However, the names of minerals are not arrived at in an analogous scientific manner. Minerals may be given names on the basis of some physical property or chemical aspect, or they may be named after a locality, a public figure, a mineralogist, or almost any other subject considered appropriate. Some examples of mineral names and their derivations are as follows:

Albite ($NaAlSi_3O_8$) from the Latin *albus* (white), an allusion to its color.

Rhodonite ($MnSiO_3$) from the Greek *rhodon* (a rose), an allusion to its characteristically pink color.

Chromite ($FeCr_2O_4$) because of the presence of a large amount of chromium in the mineral.

Magnetite (Fe_3O_4) because of its magnetic properties.

Franklinite ($ZnFe_2O_4$) after a locality, Franklin, New Jersey, where it occurs as the dominant zinc mineral.

Sillimanite (Al_2SiO_5) after Professor Benjamin Silliman of Yale University (1779–1864).

An international committee, the Commission on New Minerals and New Mineral Names of the International Mineralogical Association, now reviews all new mineral descriptions and judges the appropriateness of new mineral names as well as the scientific characterization of newly discovered mineral species. The *Glossary of Minerals Species*, published in 1991 by Michael Fleischer (see end of this chapter for complete reference), lists the internationally recommended names for nearly 3500 mineral entries. This text will use the names given in that listing.

REFERENCES AND LITERATURE OF MINERALOGY

The first comprehensive book on mineralogy in English, *A System of Mineralogy*, was written by James D. Dana in 1837. Since then, through subsequent revisions, it has remained a standard reference work. The last complete edition (the sixth) was published in 1892, with supplements in 1899, 1909, and 1915. Parts of a seventh edition, known as *Dana's System of Mineralogy*, appeared as three separate volumes in 1944, 1951, and 1962. The first two volumes treat the nonsilicate minerals, and volume three deals with silica (quartz and its polymorphs). Additional volumes on silicates are in preparation. A somewhat more recent reference is the five-volume work, *Rock-Forming Minerals*, by W. A. Deer, R. A. Howie, and J. Zussman; similar volumes are now being published sequentially (see complete reference at the end of this chapter). The treatment of the physical properties of all minerals in *Dana's System* is exhaustive. The coverage in *Rock-Forming Minerals*, however, is more topical and expansive in the areas of chemistry, structure, and experimental studies, but is essentially confined to the rock-forming minerals. Another very useful reference on the chemistry and nomenclature of minerals is *Mineralogische Tabellen* (in German) by H. Strunz. An in-depth treatment of topical subjects in mineralogy is provided by *Reviews in Mineralogy*, vols. 1 to 27, published by the Mineralogical Society of America. Example titles are: *Orthosilicates* (vol. 5), *Pyroxenes* (vol. 7), *Amphiboles and Other Hydrous Pyriboles—Mineralogy* (vol. 8), *Micas* (vol. 13), *Hydrous Phyllosilicates (Exclusive of Micas)* (vol. 19), and so on.

The diverse literature of mineralogy is found in papers published in scientific journals all over the world. The most widely circulated mineralogical journals in the English-speaking world are the *American Mineralogist*, published by the Mineralogical Society of America, the *Canadian Mineralogist*, published by the Mineralogical Association of Canada, and the *Mineralogical Magazine*, published by the Mineralogical Society of Great Britain. *The Mineralogical Record*, first published in 1970, devotes itself to mineralogical subjects that are often of more general appeal to hobbyists or mineral collectors than those found in the other above-mentioned journals.

REFERENCES AND SUGGESTED READING

Standard Mineralogical Reference Works

Dana, J. D., *A System of Mineralogy*, 7th ed., vol. 1, 1944; vol. 2, 1951; vol. 3, 1962. John Wiley & Sons, New York. Rewritten by C. Palache, H. Berman, and C. Frondel.

Deer, W. A., Howie, R. A., and Zussman, J., 1962, *Rock-Forming Minerals*, 5 vols. John Wiley & Sons, New York. Three completely revised volumes (nos. 1A, 1B, and 2A) were published in 1982, 1986, and 1978, respectively.

Fleischer, M., 1991, *Glossary of Mineral Species*. The Mineralogical Record, Tucson, Ariz., 256 pp.

Nickel, E. H. and Nichols, M. C., 1991, *Mineral Reference Manual*. Van Nostrand Reinhold, New York, 250 pp.

Reviews in Mineralogy, 1974–1992, vols. 1 to 27. Mineralogical Society of America, Washington, D.C.

Strunz, H., 1970, *Mineralogische Tabellen*, 5th ed. Akademische Verlagsgesellschaft, Leipzig, 621 pp.

Historical Accounts

Agricola, G., *De Re Metallica*, translated from the first Latin edition of 1556 by Herbert C. Hoover and Lou H. Hoover in 1950. Dover Publications, New York, 638 pp.

Ford, W. E., 1918, The growth of mineralogy from 1818 to 1918. *American Journal of Science*, v. 46, pp. 240–254.

Frondel, C., 1983, An overview of crystallography in North America, in *Crystallography in North America*, D. McLachlan and J. P. Glusker, eds. American Crystallographic Association, New York, pp. 1–24.

Greene, F. C. and Burke, J. G., 1978, The science of minerals in the age of Jefferson. *American Philosophical Society*, vol. 68, pt. 4, 113 pp.

Hazen, R. M., 1984, Mineralogy: A historical review. *Journal of Geological Education*, vol. 32, pp. 288–298.

Knopf, A., 1941, Petrology. *Geological Society of America*, 50th anniversary vol., Geology 1888–1938, pp. 333–364.

Kraus, E. H., 1941, Mineralogy. *Geological Society of America*, 50th anniversary vol., Geology 1888–1938, pp. 307–332.

Pirsson, L. V., 1918, The rise of petrology as a science. *American Journal of Science*, v. 46, pp. 222–240.

Economic Geology

Brookins, D. G., 1990, *Mineral and Energy Resources: Occurrence, Exploration, and Environmental Impact*. Merrill Publishing Co., Columbus, Ohio, 389 pp.

Cameron, E. G., 1986, *At the Crossroads: The Mineral Problems of the United States*. John Wiley & Sons, New York, 320 pp.

Craig, J. R., Vaughn, D. J., and Skinner, B. J., 1988, *Resources of the Earth*. Prentice Hall, Englewood Cliffs, N. J., 395 pp.

Evans, A. M., 1987, *Introduction to Ore Geology*, 2nd ed. Blackwell Scientific Publications, London, 358 pp.

Guilbert, J. M. and Park, C. F., Jr., 1986, *The Geology of Ore Deposits*. W. H. Freeman, New York, 984 pp.

Lefond, S. J., ed., 1983, *Industrial Minerals and Rocks*, 5th ed. American Institute of Mining, Metallurgical and Petroleum Engineers, Baltimore, Maryland, 722 pp.

Skinner, B. J., 1989, Resources in the 21st century: Can supplies meet needs? *Episodes*, v. 12, pp. 267–275.

————, 1989, Mineral resources of North America, in Bally, A. W., and Palmer, A. R., eds., *The Geology of North America—An Overview*, v. A of *Geology of North America*. Geological Society of America, Boulder, Colo., pp. 575–583.

U.S. Bureau of Mines, various years, *Minerals Yearbook*. U.S. Government Printing Office, Washington, D.C.

Additional References

Buseck, P. R., 1983, Electon microscopy of minerals. *American Scientist*, v. 71, no. 2, pp. 175–185.

Gibson, R. I., 1974, Descriptive human pathological mineralogy. *American Mineralogist*, v. 59, pp. 1177–1182.

Lowenstam, H. A., 1981, Minerals formed by organisms. *Science*, v. 211, pp. 1126–1131.

CHAPTER 2

CRYSTALLOGRAPHY: EXTERNAL FORM

Minerals, with few exceptions, possess the internal, ordered arrangement that is characteristic of crystalline solids. When conditions are favorable, they may be bounded by smooth plane surfaces and assume regular geometric forms known as *crystals*. Today, most scientists use the term *crystal* to describe any solid with an ordered internal structure, regardless of whether it possesses external faces. Because bounding faces are mostly an accident of growth and their absence in no way changes the fundamental properties of a crystal, this usage is reasonable. Thus we may frame a broader definition of a crystal as a *homogeneous solid possessing long-range, three-dimensional internal order*. The study of crystalline solids and the principles that govern their growth, external shape, and internal structure is called *crystallography*. Although crystallography was originally developed as a branch of mineralogy, today it has become a separate science dealing not only with minerals but with all crystalline matter.

This chapter presents mainly those aspects of crystallography that relate to the *external form*, or *morphology*, of crystals. This subject is known as *morphological crystallography*. In our discussion of the external form of crystals we must, however, incorporate many of the findings that relate to the *internal order* of crystals because the principles that control the crystal's atomic structure also affect the shapes of the crystal faces and the angles between them. Aspects of the internal order and structure of crystals will be developed more fully in Chapter 3.

In this book, the general term *crystalline* is used to denote the ordered arrangement of atoms in the crystal structure. The term *crystal*, without a modifier, is generally used in the traditional sense of a regular geometric solid bounded by smooth plane surfaces. *Crystal* is also used in its broader sense with modifiers indicating perfection of development. Thus, a crystalline solid with well-formed faces is *euhedral*; if it has imperfectly developed faces, it is *subhedral*; and without faces, *anhedral*. These adjectives are derived from the Greek *hedron*, meaning face, and the Greek roots *eu*, meaning good, *an*, meaning without, and the Latin root *sub*, meaning somewhat.

Crystalline substances may occur in such fine-grained aggregates that their crystalline nature (or crystallinity) can be determined only with the aid of a microscope. These are designated as *microcrystalline*. If the aggregates are so fine that the individual crystallites cannot be resolved with the microscope but can be detected by X-ray diffraction techniques, they are referred to as *cryptocrystalline*. Although most substances, both natural and synthetic, are crystalline, some lack any ordered internal atomic arrangement and are called *amorphous*. Naturally occurring amorphous substances are designated as *mineraloids*.

CRYSTALLIZATION

Crystals are formed from solutions, melts, and vapors. The atoms in these disordered states have a random distribution, but with changing temperature, pressure, and concentration they may join in an ordered arrangement characteristic of the crystalline state.

As an example of crystallization from solution consider sodium chloride dissolved in water (see Fig. 2.1). If the water is allowed to evaporate, the solution contains more and more Na^+ and Cl^- per unit volume. Ultimately, the point is reached when the remaining water can no longer retain all the salt in solution, and solid salt begins to precipitate. If the evaporation of the water is very slow, the ions of sodium and chlorine will group themselves together to form one or a few crystals with characteristic shapes and often with a common orientation. If evaporation is rapid, many centers of crystallization will be set up, usually resulting in many small, randomly oriented crystals.

FIG. 2.1. Schematic representation of a *nucleus* of NaCl in an evaporating saline lake. The *nucleus*, as drawn, consists of only 125 regularly packed ions (Na^+ and Cl^-). Additional ions will accrete, in an ordered array, on the outside of the nucleus cube, thus allowing it to grow to a larger crystal. The scale used for ionic size is the angstrom unit ($1\text{Å} = 10^{-8}$ cm). A cubic crystal of NaCl, 1 cm along each edge, would contain approximately 10^{23} ions or atoms.

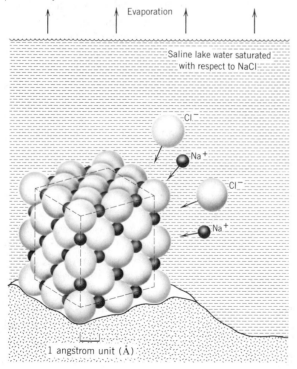

Crystals also form from solution by lowering the temperature or pressure. Hot water will dissolve sightly more salt, for instance, than cold water; and, if a hot solution is allowed to cool, a point may be reached where the solution becomes sufficiently concentrated that salt will crystallize. Again, the higher the pressure, the more salt water can hold in solution. Thus, lowering the pressure of a saturated solution will result in supersaturation and crystals will form. Therefore, in general, crystals may form from a solution by evaporation of the solvent, by lowering the temperature, or by decreasing the pressure.

A crystal is formed from a melt in much the same way as from a solution. The most familiar example of crystallization from fusion is the formation of ice crystals when water freezes. Although it is not ordinarily considered in this way, water is fused ice. When the temperature is lowered sufficiently, the H_2O molecules, which were free to move in any direction in the liquid state, become fixed and arrange themselves in a definite order to build up a solid, crystalline mass. The formation of igneous rocks from molten magmas, though more complicated, is similar to the freezing of water. In a magma many of the ions of the elements are in an uncombined state, although there is considerable cross-linking of ions and ionic groups. The crystal growth in a cooling magma is the result of two competing tendencies: (1) thermal vibrations that tend to destroy the nuclei of potential minerals, and (2) attractive forces that tend to aggregate atoms (and/or ions) into crystal structures. As the temperature falls, the effect of the first tendency diminishes, which allows the effect of the attractive forces to dominate.

Although crystallization from a vapor is less common than from a solution or a melt, the underlying principles are much the same. As the vapor is cooled, the dissociated atoms or molecules are brought closer together, eventually locking themselves into a crystalline solid. Familiar examples of this mode of crystallization are the formation of snowflakes from air laden with water vapor, and the formation of sulfur crystals at the base of fumaroles, or the neck of volcanoes.

Crystal Growth

Anyone who has collected minerals or has viewed mineralogical exhibits in museums, in rock shops, or in jeweler's display cases knows the aesthetic beauty and attraction of well-formed crystals. Most of these crystals are the result of chemical deposition from a solution (or a melt) in an open space, such as a vug or cavity in a rock formation.

The question that arises is, how do such well-formed crystals grow from small to larger ones? Or one might rephrase the same question in chemical terms, such as, how are the chemical building blocks (atoms, ions, or ionic clusters) incorporated into a well-ordered crystalline pattern?

Here we will briefly discuss some of the most basic aspects of crystal growth. The first stage in the growth of a crystal is that of *nucleation*, which implies that growth can commence only after a *nucleus* (or *seed*) has been formed. In most cases, the *nuclei* are the initial products of precipitation (in a water-rich environment) or of crystallization (as in a melt). The *nucleus* is the result of the coming together of various ions (in the solution or melt) to form the initial regular structural pattern of a crystalline solid. For example, in an evaporating salt lake, the conditions might be appropriate for the random precipitation of nuclei of NaCl. This means that Na^+ and Cl^- ions in the lake water are combining with each other in a regular cubic array of alternating ions of Na^+ and Cl^- as required by the structure of NaCl (halite, or rock salt; see Fig. 2.1). The formation of a single crystal of halite is generally preceded by the random formation of large numbers of potential nuclei. Most nuclei may not lead to further crystal growth, because in a saturated solution (saturated with respect to Na^+ and Cl^- ions) there is also a tendency for nuclei to go back into solution (to be redissolved). This is because these very small beginnings of an ordered structure have a very large surface area with respect to volume. A high surface area implies that there are many atoms (on the outer surface of the crystal) with unsatisfied chemical bonds (see Fig. 2.2). Such a crystal (or mineral grain)

with a high surface area is more soluble than a crystal (or mineral grain) with a high volume, in which most of the atoms are internal with completely satisfied chemical bonds.

If a nucleus is to survive, it must grow rapidly enough to reduce its surface energy (calculated as the ratio of surface area to volume) and thus its solubility. If a nucleus reaches a *critical size* through rapid deposition of further layers of ions, it will have a high chance of surviving as a larger crystal. An idealized picture of crystal growth is one of enlargement of the nucleus by the ordered accretion of additional ions to its outer surfaces (see Fig. 2.1). Such a picture, we now know, is too simple. The outer solid surface of a nucleus (or crystal) in contact with a saturated solution represents a surface of unsatisfied chemical bonds (see Fig. 2.2). The energy of such a surface is lowered when an atom attaches itself to it, and the amount of energy released by such an attachment depends on where such an attachment occurs. For example, in Fig. 2.2b the "step" location on the surface of the crystal is where it can lose energy through the attachment by additional ions. This is because in ionically bonded crystals, as in halite (NaCl), the energy of attachment is greatest at corners, intermediate at edges, and least in the middle of faces. The greater attraction of atoms to the corners of ionic crystals commonly leads to rapid growth in these directions. This may result in the formation of *dendrites* with branches radiating from a central core. In some types of crystals, which differ from NaCl because of their nonionic bonding, it is thought that atoms accrete on the outer surface as clumps of atoms. Such clumps, shown as blocks in Fig. 2.2c, provide the outer surface with

FIG. 2.2. (a) A section through a corner of an NaCl crystal showing well-bonded, closely packed ions in the internal part of the crystal and unsatisfied chemical bonds at the outer surfaces of the crystal (~ represents unsatisfied and = represents satisfied chemical bonds). (b) A crystal surface showing a submicroscopic step. Attachment of ions at the location of such a step lowers the energy of the crystal surface. This energy is the cumulative result of the unsatisfied bonds. (c) Submicroscopic clumps of atoms, shown as blocks, attached to the outer three surfaces of a crystal. Such blocks create steps for the attachment of new layers of ions on the outer surfaces of the crystal.

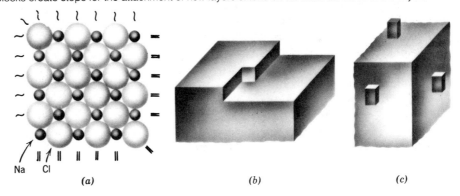

Na   Cl

(a)   (b)   (c)

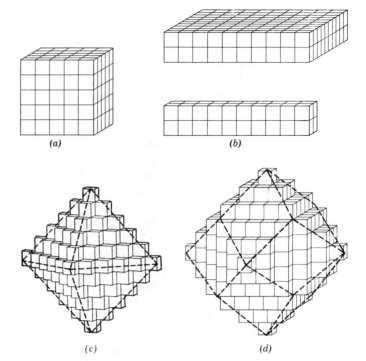

FIG. 2.3. Different external shapes produced by systematic stacking of cubic unit cells. *(a)* Perfect cube, *(b)* distorted cubes, *(c)* octahedron, and *(d)* dodecahedron. The octahedral and dodecahedral forms are the result of systematic additions of units along directions of accelerated growth. Compare with Fig. 1.4.

steps along which a new outer layer of crystal can be built up.

In this brief introduction to the subject of crystal growth we imply that the addition of ions (or atoms, or clumps of atoms) to the outside of a crystal occurs in a regular and continuous pattern. It is now known that this is generally not the case. Indeed, natural and synthetic crystals commonly contain *imperfections* and these will be discussed more fully in Chapter 3.

INTERNAL ORDER IN CRYSTALS

The internal order or crystal structure of a mineral can be thought of as a motif (or group of atoms) repeated on a lattice (which is a periodic array of points in space). The lattice comprises the *translation* component of the internal order, which will be discussed in Chapter 3. The motif, or group of atoms, has a certain symmetry that may be reflected in the crystal's external shape.

The first scientist to demonstrate that the external crystal form of a mineral (its morphology) is an expression of its internal order was René-Just Haüy (1743–1822; see Fig. 1.4). Figure 2.3 is an illustration of Haüy's concept of "integral molecules" that are regularly stacked together to achieve various commonly developed forms. Haüy coined the word *"molécule,"*

by which he meant the modern concept of unit cells. A unit cell is the smallest unit of a structure (or pattern) that can be indefinitely repeated to generate the whole structure (or to generate the complete print of a pattern; for further discussion see Chapter 3).

The *three-dimensional internal order* of a crystal can be considered as the periodic repetition of a *motif* (a unit of pattern) in such a way that the environment of and around each repeated motif is identical (see Chapter 3 for further discussion of ordered patterns). A simple and ordered arrangement of a motif in two dimensions is shown in Fig. 2.4 with a comma as the motif. In crystals the motifs may be molecules such as H_2O, anionic groups such as $(CO_3)^{2-}$, $(SiO_4)^{4-}$, or $(PO_4)^{3-}$, cations such as Ca^{2+}, Mg^{2+}, Fe^{2+}, atoms such as Cu, or combinations of anionic groups, ions, and/or atoms. Figure 2.5 illustrates the regular (ordered) arrangement of triangular $(CO_3)^{2-}$ groups and Ca^{2+} ions in the rhombohedral structural outline of a unit cell of calcite ($CaCO_3$).

FIG. 2.4. Two-dimensional order. The comma is the motif.

,   ,   ,   ,

,   ,   ,   ,

,   ,   ,   ,

FIG. 2.5. The atomic structure of calcite, CaCO₃. The outline of the unit cell is rhombohedral in shape. The locations of the carbon, calcium, and oxygen are shown. The carbonate group, $(CO_3)^{2-}$, has the shape of an equilateral triangle with carbon in its center and oxygen at the three corners. The calcium ions and carbonate groups are the motif units of the structure.

Carbon •
Calcium ●
Oxygen ◐

SYMMETRY ELEMENTS (WITHOUT TRANSLATION)

The motif used in Fig. 2.4, the comma, contains *no symmetry*. However, many motifs, such as those used to create two-dimensional printed patterns, do contain *symmetry*. The geometrical locus that aids in the visualization of the symmetry of an ordered arrangement is known as a *symmetry element*. *Rotation axes*, *mirror planes*, and *centers of symmetry* are examples of such symmetry elements. The act of rotation about an axis, the act of reflection by a mirror, and the act of inversion about a central point are collectively referred to as *symmetry operations*.

Rotation alone, through an angle (α), about an imaginary axis, generates another motif, or several other motifs. In Fig. 2.6a the angle α of 180° generates a pattern with two hands. A *rotation axis* is a line about

FIG. 2.6. *(a)* Generation of a pattern by rotation of a motif through an angle of 180°. *(b)* Right- and left-handed motifs related by a mirror reflection. *(c)* Motifs related by inversion through a center. *(d)* Motifs related to 180° rotation and subsequent inversion; this is known as roto-inversion (see also Fig. 2.12a).

which a motif unit may be rotated and repeat itself in appearance once or several times during a complete rotation (see Fig. 2.6a).

Rotational symmetry is generally expressed by any whole number (n) from 1 to infinity. The number n expresses the number of times a motif unit is repeated during a complete (360°) rotation. A rotational symmetry of $n = 1$ means that after a complete rotation of 360° about an axis, all aspects of an object (or figure) come into coincidence with themselves just once. The other limiting case of rotational symmetry is with a rotation axis of infinite order ($n = \infty$). An object possessing this kind of an axis may be made to coincide with itself by any angle of rotation, because the amount of rotation necessary is infinitely small. Figure 2.7 shows how a hollow cylinder is compatible with a rotational axis, parallel to the length of the cylinder, of $n = \infty$. It also shows that this same hollow cylinder with a single notch has rotational symmetry of $n = 1$.

FIG. 2.7. The two extremes of rotational symmetry ($n = \infty$ and $n = 1$) as shown by independent objects: (a) a perfect, hollow cylinder and (b) a hollow cylinder with a V-shaped notch at the top.

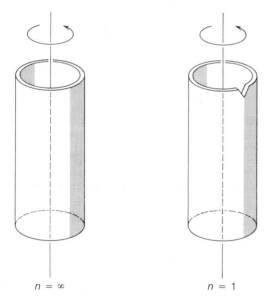

$n = \infty$ $n = 1$

Plan view:

(a)   (b)

We can fabricate objects, or draw shapes, with rotational symmetries that lie between the two extremes of $n = \infty$ and $n = 1$ (where n is a whole number and an aliquot part of 360°). We might, for example, decide to construct a 36-sided cannister in which each side is offset from its adjoining sides by 10°. Such an object would have rotational symmetry of $n = 36$. The rotational symmetries of some shapes and objects are illustrated in Fig. 2.8.

It should be noted, however, that until now we have concerned ourselves with the rotational symmetry of independent objects. We have not considered the fact that we may wish to arrange such objects into an ordered pattern. When the symmetry of the motif is constrained by the translational symmetry of the lattice only certain rotations are possible (see Chapter 3).

The types of rotation found in internally ordered crystals, and also expressed in their external shape (morphology) are *1-fold* ($\alpha = 360°$), *2-fold* ($\alpha = 180°$), *3-fold* ($\alpha = 120°$), *4-fold* ($\alpha = 90°$), and *6-fold* ($\alpha = 60°$). A 5-fold axis and 7- and higher-fold axes are not possible. This will be proved geometrically in Chapter 3 (see page 115), after lattice translation has been discussed. Intuitively, this becomes clear when one tries to completely cover a plane surface with a five-sided motif such as a pentagon, without mismatches and gaps (see Fig. 2.9a). On the other hand, Fig. 2.9b shows how hexagons can completely cover a surface. The geometric picture in Fig. 2.9a shows mismatches and gaps, which tend *not* to occur in the crystal structures of minerals. On an atomic scale, such gaps could represent unsatisfied chemical bonds or overstretched bonds between ions (or atoms), both of which are absent in ordered crystal structures.

The possible rotation axes are portrayed in Fig. 2.10 with the graphic symbols used to represent them. The number of duplications of the motif during a 360° rotation gives the rotation axis its name. For example, two equivalent units per 360° rotation are related by a *2-fold rotation axis*.

Rotation produces patterns in which the original motif and those generated from it are identical in orientation with respect to each other. In other words, the original motif and the newly generated one have the same "handedness." The original motif and those generated from it by rotation are, therefore, said to be *congruent*.

A *reflection* produces a mirror image across a mirror plane, *m* (Fig. 2.6b). In this case the generated motif has the opposite handedness of the original motif, and the two make an *enantiomorphic pair* (meaning that the motifs are related by a mirror and that they

(a)   (b)   (c)

FIG. 2.8. Objects showing elements of rotational symmetry only. *(a)* A child's pinwheel with a 4-fold axis (4) perpendicular to the face of the wheel. *(b)* A pattern of three lizards that are related to each other by a 3-fold rotation axis (3) perpendicular to the page. The 3-fold axis intersects the page at a point between the three heads. (Redrawn from plate 38 by M. C. Escher in *Fantasy and Symmetry; The Periodic Drawings of M. C. Escher*, 1965 by Caroline H. MacGillavry. Harry N. Abrams, Inc., Publishers, New York; copyright © 1939, M. C. Escher/Cordon Art, Baarn, Holland.) *(c)* View of an apatite crystal along its c-axis. This drawing shows a large basal face (with dot marked in its center) and an array of modifying faces around it. The overall symmetry of this crystal requires a 6-fold rotation axis (6) at the location of the dot. There are no mirror planes visible in this orientation of an apatite crystal.

cannot be superimposed on each other). This is the same as the relationship between your right and your left hand.

An *inversion* (*i*) produces an inverted object through an *inversion center*. An inversion involves drawing imaginary lines from every point on the object through the inversion center and out an equal distance on the other side of the inversion center. The inverted object is then "recreated" by connecting the points (Fig. 2.6c). Inversion, like reflection, produces an *enantiomorphic pair*.

Rotation with Inversion

In addition to the symmetrical order generated by operations of rotation axes, there are 1-, 2-, 3-, 4-, and 6-fold rotations that can be combined with inversion and are known as *rotoinversion operations* (see Fig. 2.6*d*). The rotation axes have been illustrated by mo-

tifs that lie in the same plane, as in Fig. 2.10. In combining rotation with inversion it is best to observe the order of a pattern in three dimensions.

Figure 2.11 illustrates the combination of symmetry operations in a 1-fold rotoinversion. This is known as a 1-fold rotoinversion axis, and is symbolized as $\bar{1}$ (read: bar one). The original motif is rotated 360°, so that it returns to its original position and is then inverted through a center. This combination of operations produces the same result as does the presence of a *center of symmetry*. The $\bar{1}$ operation is therefore also referred to as a *center of symmetry*, or *i* (for inversion). The right-hand portion of Fig. 2.11 illustrates how the three-dimensional arrangement of the motif commas appears when projected on the equatorial plane of the globe in Fig. 2.11a. The rotoinversion operations for $\bar{2}$, $\bar{3}$, $\bar{4}$, and $\bar{6}$ (read: bar two, bar three, etc.) are shown in Fig. 2.12. The $\bar{2}$ operation is equivalent to the operation of a mirror plane coinci-

FIG. 2.9. *(a)* Arrangement of pentagons, which individually have 5-fold symmetry axes perpendicular to the page, leads to gaps in the pattern. *(b)* Arrangement of hexagons with 6-fold axes perpendicular to the page, as in honeycombs.

(a)   (b)

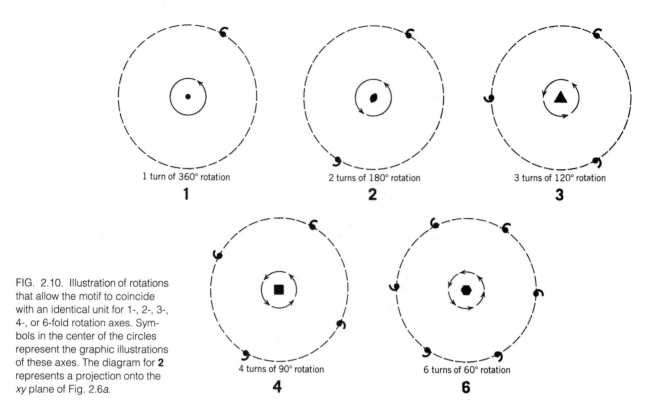

FIG. 2.10. Illustration of rotations that allow the motif to coincide with an identical unit for 1-, 2-, 3-, 4-, or 6-fold rotation axes. Symbols in the center of the circles represent the graphic illustrations of these axes. The diagram for **2** represents a projection onto the *xy* plane of Fig. 2.6*a*.

dent with the equatorial plane of the globe (Fig. 2.12*a*). The $\overline{3}$ operation is the equivalent of a 3-fold rotation axis and inversion (*i*), which is the same as a 3-fold rotation and a center of symmetry. The $\overline{4}$ operation is not resolvable into other operations and as such is unique. The $\overline{6}$ operation is equivalent to a 3-fold axis of rotation with a mirror plane perpendicular to the rotation axis.

It should be noted that the original motif unit (denoted *A* in all illustrations of Figs. 2.11 and 2.12) has

an *enantiomorphic*[1] relationship with the second motif unit (denoted *B*), because of the inversion. The third motif unit (denoted *C*), however, is congruent (similar) with the original motif unit (*A*). All of these symmetry operations (rotation, reflection, and rotoinversion) generate only a finite number of motifs. On the other hand, translation, and translational symmetry opera-

[1]Two enantiomorphic motifs are related by mirror reflection or inversion.

FIG. 2.11. *(a)* Illustration of an operation of rotoinversion, consisting of 360° rotation and subsequent inversion through the center of the globe. *(b)* Projection of the two motif units (*A* and *B*) from the outer skin of the globe onto the equatorial plane. *(c)* Location of the projected motifs on the equatorial plane (see also Fig. 2.12).

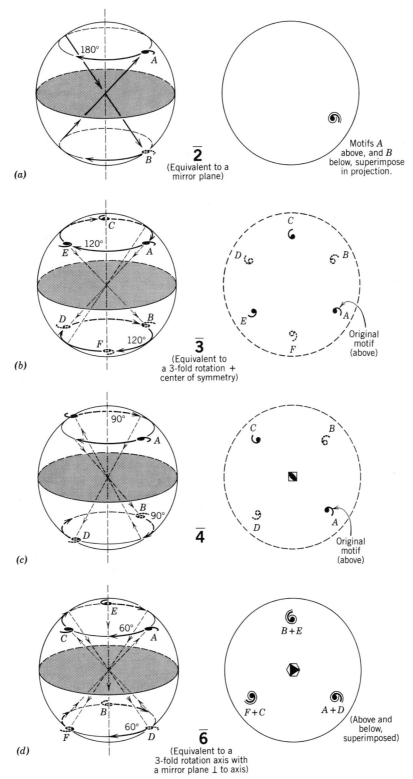

FIG. 2.12. Illustration of operations of rotoinversion on motif units. To go from unit *A* to *B* (to *C*, etc.) involves rotation through an angle α (360°, 180°, 120°, 90°, or 60°) as well as inversion through the center (see Fig. 2.11 for illustration of projection scheme).

tions (such as glide and screw operations; see Chapter 3) will repeat a motif indefinitely. As such, rotation, reflection, and rotoinversion operations are classified as "without translation."

It should be noted that the presence of any (or all) of the above-mentioned symmetry elements (as deduced from the external geometry of crystals) is the fundamental result of the packing of atoms (or ions) and the bonding between them in crystal structures. Figure 2.13 is an illustration of the highly regular (cubic) packing of iron atoms in the crystal structure of iron metal. This internal structure contains all of the elements of symmetry seen in a geometric cube, and additional translational elements as well.

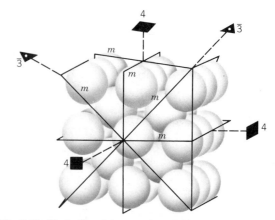

FIG. 2.13. Illustration of the atomic structure of metallic iron. The iron atoms are arranged in what is known as body-centered cubic packing (see Chapter 4). It is obvious from this representation that this highly regular atomic packing contains many of the symmetry elements that have been discussed on the basis of the external shape of crystals. Only a very few of the many symmetry elements present in this crystal structure are shown.

Combinations of Rotations

Until now we have considered patterns generated by a *single* axis of rotation or rotoinversion. We can, however, combine various axes of rotation and generate regular three-dimensional patterns. Symmetry axes must be put together in symmetrically consistent ways, such that an infinite set of axes is not generated. For instance, if one 4-fold axis, *A*, is put at an acute angle to another 4-fold axis, *B*, each will operate on the other, generating an infinite set of axes. To avoid this, axes must be put together at 90°, or at 54°44′, as in the special case of cubic symmetry. Furthermore, all symmetry operators must intersect at a single point.

For example, we might combine a 4-fold rotation axis (4) perpendicular to the plane of the page with a 2-fold rotation axis (2) in the plane of the page. Another example would be a combination of a 6-fold rotation axis (6) perpendicular to the plane of the page with a 2-fold axis (2) in the plane of the page. Both examples are illustrated in Fig. 2.14. The 4- and 6-fold axes in both symmetry combinations are at point *A*, the center of the circle, perpendicular to the page. The 2-fold axis is to the right of *A*, along the east-west direction. The presence of the 4- and 6-fold axes will generate three and five more 2-fold axes, respectively. These generated 2-fold axes are *dashed* (ignore, for the moment, the dotted axes). Although we have generated three and five 2-fold axis extensions, they constitute only two 2-fold axis directions at 90° to each other in Fig. 2.14*b* and three 2-fold axis directions at 120° to each other in Fig. 2.14*c*. Let us now enter a comma (marked by *B* in the drawings) above the page in a position slightly north of the original 2-fold axis. This comma is marked (+), indicating that it lies above the page, in the positive direction of the *z* axis (see Fig. 2.14*a*). The original 2-fold axis (in the east-west direction) will generate another comma from the one given at *B*, namely, on the south side of the 2-fold

axis, and below the page. This generated comma is accompanied by a minus (−) sign, indicating its position below the page. The 4- and 6-fold axes will generate three and five additional motif pairs, respectively, as shown by dashed commas in Figs. 2.14*b* and *c*. If now we carefully observe the arrangement of all of the commas, it becomes clear that we have generated yet another set of 2-fold axes. These axes are dotted and at 45° to the original 2-fold axes in Fig. 2.14*b* and at 30° to the original 2-fold axes in Fig. 2.14*c*. The total symmetry in Figs. 2.14*a* and *b*, therefore, consists of a 4-fold rotation axis perpendicular to the page and two sets of 2-fold axes, the original set in the E–W, N–S directions, and the second at 45° thereto. The total symmetry in Fig. 2.14*c* consists of a 6-fold rotation axis perpendicular to the page and two sets of three 2-fold axes in the plane of the page. The two sets of 2-fold axes are at 30° to each other. In each case, each *set* of 2-fold axes is symmetrically equivalent to the other set. That is, if one 2-fold axis is given, the other symmetry elements generate the other 2-folds; in the example given, as two independent sets of 2-folds. These types of combinations of axes can be represented by a sequence of digits for the types of rotation axes involved. In such symbols, each symmetrically equivalent set of symmetry elements is listed. For the examples in Fig. 2.14 this would result in 422 and 622, respectively. Three-dimensional representations of the locations of the rotation axes in the 422 and 622 combinations are given in Figs. 2.14*d* and *e*.

Other possible combinations of rotational symmetry elements are 222, 32, 23, and 432. Note that

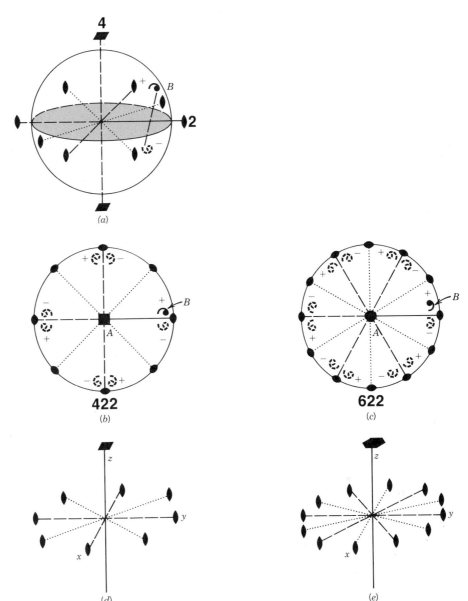

FIG. 2.14. Combination of rotation axes and the generation of ordered patterns from such combinations. *(a)* Perspective view of a vertical 4-fold rotation axis combined with a 2-fold rotation axis E–W in the equatorial plane. *B* is the original motif above the equatorial plane to the back of the E–W 2-fold rotation axis. One motif generated by this E–W axis is also shown. *(b)* and *(c)* Plan views of the location of the symmetry axes and the motif units. *(d)* and *(e)* Three-dimensional sketches of the distribution of symmetry axes. See text for the development of these figures.

in 32 there are only two rotational symmetries indicated rather than three as in the other combinations. Figure 2.15 shows that combining a 3-fold axis with a 2-fold axis in a plane perpendicular to it generates no symmetry axes in addition to three 2-folds. The 432 symmetry axis combination is one of high symmetry with the locations of the axes in specialized positions. Figure 2.16 shows the location of such axes with reference to a cubelike outline. The 4-fold axes are perpendicular to the cube faces, the 3-fold axes are at the corners of the cube, and the 2-fold axes are located at the centers of the edges of the cube. 23 is also cubic symmetry, but now the 2-folds are perpendicular to the cube faces. All cubic or isometric symmetry combinations have a set of four 3-fold axes (along the body diagonals of the cube). These 3-folds meet the face

FIG. 2.15. Combination of a 3-fold rotation axis perpendicular to the page and a 2-fold rotation axis lying (E–W) in the page. The original motif units are on the right-hand side. Generated motif units and symmetry axes are indicated by dashes and dots. The resultant symmetry sequence is 32, not 322. Compare with Fig. 2.14.

32

FIG. 2.16. The location of symmetry axes in 432 with respect to a cubelike outline.

normal axes (4 or 2) at 54°44'. (For a rigorous derivation of the limits on combinations of rotational operators, see Boisen and Gibbs, 1990; the complete reference is given at the end of this chapter.)

Combinations of Rotation Axes and Mirrors

In the previous section we discussed the possibility of combining rotational symmetries as in the sequences 622, 422, 32, etc. Let us consider some examples of combinations of rotation axes and mirror planes. As a general rule, mirror planes within crystals are either perpendicular to or parallel to any rotation axes that are present. In Fig. 2.17a a 4-fold rotation axis is combined with a mirror plane perpendicular to the axis and in Fig. 2.17b a 6-fold axis is combined with a mirror perpendicular to it. In Fig. 2.17a the arrangement of the motif is compatible with the 4-fold rotation axis and is shown by commas above the mirror plane. These are reflected by the mirror plane giving rise to another set of four commas below as shown by the dashed commas. Normally the symmetry elements and motif units are shown in a two-dimensional projection as on the right side of Fig. 2.17a. The motif units above the mirror as well as those below the mirror are projected onto the mirror itself. This causes the commas from above and below to coincide. In order to distinguish motif units that lie above the plane of projection (the mirror plane, in this case) from those below the plane, motif units above the plane are generally shown as solid dots and those below the plane as small open circles. When this convention is used

FIG. 2.17. (a) Combination of a 4-fold symmetry axis and a mirror plane perpendicular to it. The motif units that can be represented by commas are more conventionally shown by solid dots and small open circles in order to differentiate motif units above and below the mirror plane, respectively. (b) Combination of a 6-fold rotation axis and a mirror plane perpendicular to it.

Perspective

Symbol

$$\frac{4}{m}$$

(a)

Projection

m (Solid circle)

$$\frac{6}{m}$$

(b)

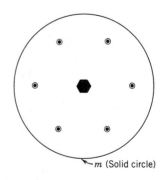

m (Solid circle)

in the projection of Fig. 2.17a, it results in a 4-fold rotation axis surrounded by four motif units (dots) above, and four identical motif units (circles) below the mirror plane. Note that the mirror plane is conventionally shown by a solid circle. This type of combination of symmetry elements is represented by $4/m$ (read: four over m). The symmetry combination in Fig. 2.17b is represented by $6/m$. Other similar combinations are $2/m$ and $3/m$.

In a previous section we derived several combinations of rotation axes such as 622, 422, 222. If we add mirror planes perpendicular to each of the rotation axes, the following symmetry combinations result: $6/m2/m2/m$, $4/m2/m2/m$, and $2/m2/m2/m$. In Figs. 2.18a, b, and c illustrations are given for the combinations 422, $4/m2/m2/m$, and $4mm$ (compare with Fig. 2.14). The motif units are shown as solid dots and equivalent open circles indicating their position above and below the plane of projection, respectively. In 422, we have four motif units above the plane and four below, making a total of eight symmetrically related motifs. In $4/m2/m2/m$ we have a mirror plane perpendicular to the 4-fold axis (shown in Fig. 2.18b as a solid circle), and we have mirror planes perpendicular to each of the four 2-fold rotation axes that lie in the equatorial plane. The trace of the mirror plane perpendicular to the E–W axis coincides with the N–S axis, and so forth. The $4/m2/m2/m$ symmetry combination, therefore, contains four vertical mirror planes in addition to the one horizontal mirror perpendicular to the 4-fold axis. In Figs. 2.18b and c the traces of the mirrors are shown, by convention, as solid lines. In Fig. 2.18b, we generate eight motif units above the equatorial mirror and eight below, making a total of 16 symmetrically related units. Yet another possible combination of 4-fold rotation and mirror planes is expressed by $4mm$. This symbolism denotes one 4-fold rotation axis and four mirror planes that are parallel to (and intersect in) the 4-fold axis and whose traces in the equatorial plane lie N–S, E–W, and 45° to these directions. These mirror planes are called *vertical* mirrors because they are perpendicular to the equatorial plane. In $4/m$, for instance, the mirror is a horizontal mirror *in* the equatorial plane and perpendicular to the 4-fold axis. The total number of motif units related by the combined symmetry elements in $4mm$ is eight, all of which lie on one side of the projection (above the page, as shown in Fig. 2.18c). Other similar combinations are $6mm$, $3m$, and $2mm$.

At this stage it is instructive to evaluate the interdependence of mirrors that intersect each other, and the rotational symmetry (along the line of intersection) that results therefrom. Several examples of such intersections are given in Fig. 2.19. When two vertical mirrors intersect each other at 90°, the vertical line of intersection is equivalent to a 2-fold axis of rotation (Fig. 2.19a). When three vertical mirrors intersect each other at 60°, the vertical line of intersection is one of 3-fold rotation. When four vertical mirrors intersect each other at 45°, the line of intersection is one of 4-fold rotation (Fig. 2.19b). And when six vertical mirrors intersect each other at 30°, the vertical intersection line is one of 6-fold rotation. In Fig. 2.19c an additional horizontal mirror has been added to the configuration in Fig. 2.19a. This horizontal mirror re-

FIG. 2.18. (a) Combination of a 4-fold rotation axis and two sets of 2-fold rotation axes (see also Fig. 2.14). Motif units above the page are shown as solid dots, those below the page as open circles. (b) Combination of a 4-fold rotation axis, four 2-fold rotation axes, and mirror planes perpendicular to each of the axes. (c) Combination of a 4-fold rotation axis and two sets of mirror planes parallel to the 4-fold axis.

422

(a)

$$\frac{4}{m}\frac{2}{m}\frac{2}{m}$$

(b)

4mm

(c)

FIG. 2.19. Illustrations of intersecting mirrors and the resultant lines of intersection, equivalent to rotation axes. *(a)* and *(b)* Perspective and plan views of 2*mm* and 4*mm*. In *(a)* the vertical mirrors are at 90° to each other; in *(b)* the vertical mirrors are at 45° to each other. In *(c)* and *(d)* horizontal mirrors are added to the drawings in *(a)* and *(b)* respectively. The horizontal intersection lines become 2-fold rotations in both illustrations. Compare Fig. 2.19*d* with the plan view in Fig. 2.18*b*.

flects the motifs in a downward direction, and the horizontal intersection lines for all of the mirrors become 2-fold rotations. In Fig. 2.19*d* an additional horizontal mirror has been added to the configuration in Fig. 2.19*b*. Now the motifs at the top are reflected vertically downward and all the horizontal intersections between mirrors are axes of 2-fold rotation.

Before proceeding with further aspects of the morphology of crystals, it should be noted that all of the nontransitional symmetry elements that have been introduced are the easily observed expression (in well-formed crystals or wooden models) of the internal atomic arrangement of the structure. This was

noted in Fig. 2.13 and is shown again in Fig. 2.20. This illustration depicts the regular packing of Na^+ and Cl^- ions in the structure of NaCl, halite. This specific ionic packing scheme is a function of relative ionic sizes (for Na^+ and Cl^-), their electrical charges, and the types of bonds between the ions (these subjects are discussed in detail in Chapter 4). Although the NaCl structure (as shown in Fig. 2.20) also contains translational symmetry elements (as discussed in Chapter 3), it should be obvious that all of the geometric principles discussed thus far are basic to an understanding of the internal structures (not just the morphology) of crystalline materials.

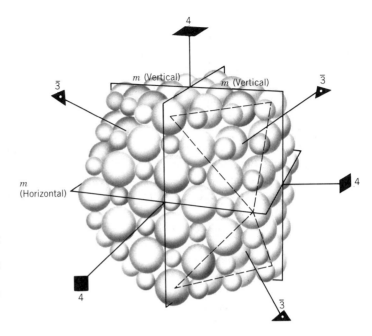

FIG. 2.20. A structure model of the packing of Na⁺ (small spheres) and Cl⁻ (large spheres) in NaCl (halite). The model has a cubo-octahedral outline showing square cube faces and triangular octahedral faces at the corners of the cube. This structure contains all of the symmetry elements that are also present in the morphology of a cube, that is, three 4-fold axes (all are shown), four 3-fold rotoinversion axes (three are shown at corners of the cube), six 2-fold axes (at the cube edge; none are shown), and nine mirror planes in various orientations (only three are shown that are perpendicular to each other).

Résumé of Symmetry Operations Without Translation

In the preceding discussion we introduced several symmetry elements: rotation axes (1, 2, 3, 4, and 6), rotoinversion axes ($\bar{1}$, $\bar{2}$, $\bar{3}$, $\bar{4}$, and $\bar{6}$), a center of symmetry (*i*), and mirror planes (*m*). We have also discussed some of the combinations of rotation axes such as 622, 422, 222, and of rotation axes and mirror planes such as 6/m2/m2/m, 4/m2/m2/m, and 4mm. The number of possible symmetry combinations is not unlimited; indeed the total number of nonidentical symmetry elements and combinations of symmetry elements is only 32. In Table 2.1 they are arranged in a sequence from the lowest rotational symmetry (1) to the highest rotational symmetry (6). In

this table, as in the previous discussion, we have used symbols for the symmetry elements and the combinations of symmetry elements that are referred to as the *Hermann-Mauguin notation* after their inventors. Because of their universal acceptance, they are also called the *international symbols*. The 32 possible elements and combinations of elements are identical to the *32* possible *crystal classes* to which crystals can be assigned on the basis of their morphology (see Table 2.4) or their internal atomic arrangement.

We have illustrated a number of combinations of symmetry elements, but we have not rigorously derived the possible 32 nonidentical symmetry elements or combinations of symmetry elements, which are also known as the *32 point groups*. The word *point*

Table 2.1 **THIRTY-TWO POSSIBLE SYMMETRY ELEMENTS AND COMBINATIONS OF SYMMETRY ELEMENTS**	Increasing Rotational Symmetry ⟶				
Rotation axis only	1	2	3	4	6
Rotoinversion axis only	$\bar{1}(=i)$	$\bar{2}(=m)$	$\bar{3}$	$\bar{4}$	$\bar{6}(=3/m)$
Combination of rotation axes		222	32	422	622
One rotation with perpendicular mirror		2/m	$3/m(=\bar{6})$	4/m	6/m
One rotation with parallel mirrors		2mm	3m	4mm	6mm
Rotoinversion with rotation and mirror			$\bar{3}2/m$	$\bar{4}2m$	$\bar{6}2m$
Three rotation axes and perpendicular mirrors		2/m2/m2/m		4/m2/m2/m	6/m2/m2/m
Additional symmetry combinations present in isometric patterns		23 2/m$\bar{3}$		432 (see Fig. 2.16) $\bar{4}3m$	4/m$\bar{3}$2/m

indicates that the symmetry operations leave one particular point, at least, of the pattern unmoved. The word *group* relates to the mathematical theory of groups which allows for a systematic derivation of all the possible and nonidentical symmetry combinations (see, for example, D. E. Sands, 1975, or Boisen and Gibbs, 1990; for complete references see the end of this chapter).

A crystal, under favorable circumstances of growth, will develop smooth planes or "faces" that may assume regular geometric forms which are the expression of its internal, regular, atomic arrangement. In crystals with well-developed faces, one can recognize the elements of symmetry such as rotation axes, rotoinversion axes, a center of symmetry, and mirror planes. A systematic study of the external forms of crystals leads to 32 possible symmetries or symmetry combinations, which are the same 32 as the point groups noted above. Certain of the 32 *crystal classes* have symmetry characteristics in common with others, permitting them to be grouped together in one of *six crystal systems*. Table 2.2 shows the conventional arrangement of crystal systems and classes. Compare this with Table 2.1 and note that the symmetry elements and combinations are the same in both tables, but that their groupings are somewhat different.

In the subsequent discussion of crystal morphology we will systematically treat the 32 crystal classes (point groups) as arranged within the six crystal systems. That treatment is based on the study and measurement of the external form of crystals. A graphic illustration of these 32 symmetries is given in Fig. 2.21. This figure precedes the systematic treatment of crystal form because it is based on the distribution of motif units, without reference to crystals or crystallographic axes. In Fig. 2.21 the presence or absence of a center of symmetry ($=$ inversion, i) is not indicated by a specific symbol. Of the 32 crystal classes there

are 21 without a center of symmetry and 11 with a center. Table 2.3 distinguishes the 32 crystal classes according to the absence or presence of a symmetry center. Visual inspection of the circles with the distribution of motif units in Fig. 2.21 allows one to determine whether a center of symmetry is present or absent. In such an evaluation one must keep in mind that a center of symmetry inverts the motif unit through the center of the circle; for example, a motif unit at the upper right (above the page) is balanced by a motif unit at the lower left (below the page).

CRYSTAL MORPHOLOGY

As crystals are formed by the repetition in three dimensions of a unit of structure, the limiting surfaces, which are known as the faces of a crystal, depend in part on the shape of the unit. They also depend on the conditions in which the crystal grows. These conditions include all the external influences such as temperature, pressure, nature of solution, direction of flow of the solution, and availability of open space for free growth. The angular relationships, size, and shape of faces on a crystal are aspects of *crystal morphology*.

If a cubic unit cell is repeated in three dimensions to build up a crystal with n units along each edge, a larger cube will result containing n^3 units. With a similar orderly repeat mechanism, different shapes may result, as shown in Fig. 2.3 for distorted cubes, octahedron, and dodecahedron. The octahedral and dodecahedral forms are common in many crystals, but because the unit cell dimensions are on the angstrom level the steps are invisible to the eye and the resulting faces appear as smooth, plane surfaces.

With a given internal structure, a limited number of planes bound a crystal, and only a comparatively few are common. In determining the types of crystal faces that may develop on a crystal we must also consider the internal lattice. Faces are most likely to form on crystals parallel to lattice planes that have a high density of *lattice points* (or *nodes*). The frequency with which a given face is observed is roughly proportional to the number of nodes it intersects in the lattice: the larger the number of nodes, the more common the face, as is illustrated in Fig. 2.22. This rule, known as the *law of Bravais*, is generally confirmed by observations. Although there are exceptions to the law, as pointed out by Donnay and Harker in 1937, it is usually possible to choose the lattice in such a way that the rule holds true.

Table 2.2
THIRTY-TWO CRYSTAL CLASSES
(see also Table 2.4)

Crystal System	Symmetry of Crystal Classes
Triclinic	1 and $\bar{1}$
Monoclinic	2, m, and $2/m$
Orthorhombic	222, 2mm, and $2/m2/m2/m$
Tetragonal	4, $\bar{4}$, $4/m$, 422, 4mm, $\bar{4}$ 2m, and $4/m2/m2/m$
Hexagonal	3, $\bar{3}$, 32, 3m, and $\bar{3}$ 2/m
	6, $\bar{6}$, $6/m$, 622, 6mm, $\bar{6}m2$, and $6/m2/m2/m$
Isometric	23, $2/m\bar{3}$, 432, $\bar{4}3/m$, and $4/m\bar{3}2/m$

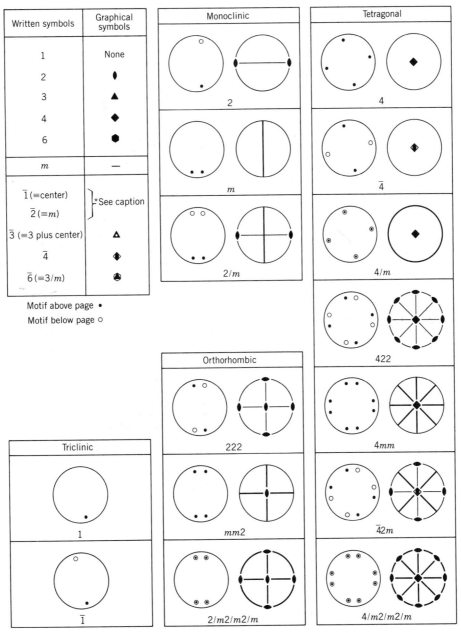

FIG. 2.21. Graphic representation of the distribution of motif units compatible with the symmetry elements of each of the 32 crystal classes (point groups). For all crystal classes, excepting triclinic, there are two circular diagrams, with the left-hand diagram showing the distribution of motif units and the right-hand diagram illustrating the symmetry elements consistent with these motif units. The motif units above the page are equivalent to those below the page, but they are differentiated by dots (above the page) and circles (below the page). The symbols for the symmetry elements are given at the top left corner of the diagram. The presence of a center of symmetry is not shown by any symbol; its presence can be deduced from the arrangement of motif units. Instead of $\bar{2}$ the symbol for a mirror (m) is used. The diagrams for the monoclinic system are shown in what crystallographers refer to as the "second setting," with m vertical (perpendicular to the page) and the 2-fold axis in an east-west orientation. Monoclinic symmetry can also be shown by setting the 2-fold rotation axis perpendicular to the page, and orienting the mirror parallel to the page; this is referred to as the "first setting."

(continued)

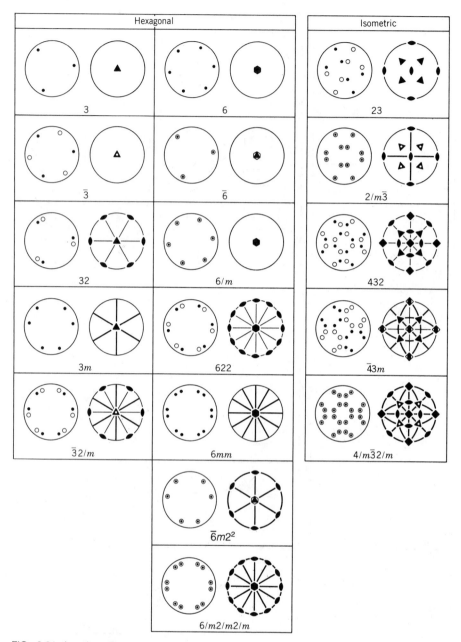

FIG. 2.21. *(continued)*

[2]Because $\bar{6}$ is equivalent to $3/m$, the intersections of the horizontal mirror (perpendicular to the vertical axis) with the vertical mirrors result in 2-fold axes lying in the planes of the vertical mirrors. As such, the traces of the vertical mirrors and the horizontal 2-fold axes are coincident.

Table 2.3
**THE 32 CRYSTAL CLASSES GROUPED
ACCORDING TO THE PRESENCE OR ABSENCE
OF A CENTER OF SYMMETRY**

Crystal System	No Center	Center Present
Triclinic	1	$\bar{1}$
Monoclinic	2, $\bar{2}(=m)$	2/m
Orthorhombic	222, mm2	2/m2/m2/m
Tetragonal	4, $\bar{4}$, 422	4/m, 4/m2/m2/m
	4mm, $\bar{4}$2m	
Hexagonal	3, 32, 3m	$\bar{3}$, $\bar{3}$2/m
	6, $\bar{6}$, 622	6/m, 6/m2/m2/m
	6mm, $\bar{6}$m2	
Isometric	23, 432, $\bar{4}$3m	2/m$\bar{3}$, 4/m$\bar{3}$2/m

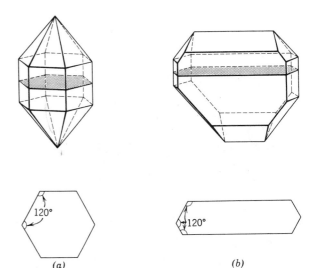

FIG. 2.23. Constancy of interfacial angles as shown in the comparison of a well-formed and highly symmetric quartz crystal (a) and a distorted quartz crystal (b). The sections across the direction of elongation show identical interfacial angles of 120° in both, irrespective of the asymmetric habit in (b).

Because crystal faces have a direct relationship to the internal structure, it follows that the faces have a definite relationship to each other. This fact was observed in 1669 by Nicolaus Steno, who pointed out that the angles between corresponding faces on crystals of quartz are always the same. This observation is generalized today as Steno's law of the constancy of interfacial angles, which states: *The angles between equivalent faces of crystals of the same substance, measured at the same temperature, are constant* (see Fig. 1.3). For this reason crystal morphology is frequently a valuable tool in mineral identification. A mineral may be found in crystals of widely varying shapes and sizes, but the angles between pairs of corresponding faces are always the same. An illustration of such constancy of interfacial angles, in horizontal sections of two quite different-looking quartz crystals, is given in Fig. 2.23.

Although crystals possess a regular, orderly internal structure, different planes and directions within

them have differences in atomic environments. Consider Fig. 2.24 which illustrates the packing of ions in sodium chloride (NaCl), the mineral halite. Any plane parallel to the front face of the cube is composed of half Na^+ and half Cl^- ions. On the other hand, cutting the corner of the cube shows planes containing only Na^+ ions alternating with planes containing only Cl^- ions.

These different atomic arrangements along different crystal planes or directions give rise to *vectorial properties*. Because the magnitude of the property is dependent on direction, it differs for different crystallographic directions. Some of the vectorial properties of crystals are hardness, conductivity for heat and electricity, thermal expansion, speed of light, growth rate, solution rate, and diffraction of X-rays.

Of these properties, some vary continuously with direction within the crystal. Hardness, electrical and heat conductivity, thermal expansion, and the speed of light in crystals are all examples of such *continuous vectorial properties*.

The *hardness* of some crystals varies so greatly with crystallographic direction that the difference may be detected by simple scratch tests. Thus, kyanite, a mineral that characteristically forms elongate bladed crystals, may be scratched with an ordinary pocketknife in a direction parallel to the elongation, but it cannot be scratched by the knife perpendicular to the elongation. Some directions in a diamond crystal are much harder than others. When diamond dust is used for cutting or grinding, a certain percentage of the grains always presents the hardest surface and, hence,

FIG. 2.22. This figure represents one layer of *lattice points* in a cubic lattice. Several lines are possible through the network that include a greater or lesser number of *lattice points* (or nodes). These lines represent the traces of possible crystal planes. It is found that the planes with the highest density of lattice points are the most common, such as AB and AC.

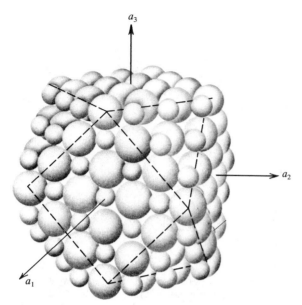

FIG. 2.24. Packing model of halite, NaCl, with cubo-octahedral outline, Na$^+$ small, Cl$^-$ large. Note that the cube faces have sheets of equal numbers of Na$^+$ and Cl$^-$ ions, whereas the octahedral planes (at the corners of the cube) consist of alternating sheets of Na$^+$ and sheets of Cl$^-$ ions. Na$^+$ and Cl$^-$ ions are both surrounded by six closest neighboring ions (six coordination) in a face-centered lattice. This type of structure is also found in PbS, galena, MgO, and many other *AX* compounds.

the dust is capable of cutting along planes in the crystal of lesser hardness. If a perfect sphere cut from a crystal is placed in a cylinder with abrasive and tumbled for a long time, the softer portions of the crystal wear away most rapidly. The nonspherical solid resulting serves as a hardness model for the substance being tested.

The directional character of *electrical conductivity* is of great importance in the manufacture of silicon and germanium diodes, tiny bits of silicon and germanium crystals used to rectify alternating current. In order to obtain the optimum rectifying effect, the small bit of semimetal must be oriented crystallographically, as the conduction of electricity through such crystals varies greatly with orientation.

Ball bearings of synthetic ruby sound very attractive, because the great hardness of ruby would cut down wear and give long life to the bearing. However, when heated, ruby expands differently along different crystallographic directions, and ruby ball bearings would rapidly become nonspherical with the rise of temperature from friction during operation. Because the *thermal expansion* figure of ruby is an ellipsoid of revolution with a circular cross section, however, cylindrical roller bearings (cut parallel to the crystal's 3-fold axis) are practical. Most minerals have unequal coefficients of thermal expansion in different direc-

tions, leading to poor resistance to thermal shock and easy cracking with heating or cooling. SiO$_2$ glass, which has an irregular internal structure as compared with quartz crystal, is more resistant to thermal shock than the mineral.

The *velocity of light* in all transparent crystals, except those that are isotropic (see Chapter 8), varies continuously with crystallographic direction. Of all the vectorial properties of crystals, the optical parameters are most easily determined quantitatively and are expressed as the index of refraction, the reciprocal of the velocity of light in the crystal relative to the velocity of light in air or vacuum.

Discontinuous vectorial properties, on the other hand, pertain only to certain definite planes or directions within the crystal. There are no intermediate values of such properties connected with intermediate crystallographic directions. An example of such a property is *rate of growth*. The rate of growth of a plane in a crystal is intimately connected with the density of lattice points in the plane. We saw that a plane such as *AB* in Fig. 2.22 has a much greater density of nodes than plane *AD*, *AE*, or *AF*. Calculations of the energy involved indicate that the energy of particles in a plane such as *AB*, in which there is a high density of nodes, is less than the energy of particles in less densely populated planes such as *AF*. Hence, the plane *AB* will be the most stable, because in the process of crystallization the configuration of lowest energy is that of maximum stability. Planes *AF*, *AD*, *AE*, and so on, will however grow faster than *AB* because fewer particles need to be added per unit area. In the growth of a crystal from a nucleus, the early forms that appear will be those of relatively high energy and rapid growth. Continued addition of material to these planes will build them out while the less rapidly growing planes lag behind (the rate of growth of a face is inversely proportional to the node density, because fewer nodes require less material to be added for growth, resulting in more rapid growth). Figure 2.25 shows how a crystal, during its development from a nucleus to a larger crystal, may change its external form in various stages. Thus, the edges and corners of a cube may be built out by addition of material to the planes cutting off the corners and edges, whereas little material is added to the cube faces. As growth progresses, the rapidly growing faces disappear, literally growing themselves out of existence, building the slower-growing, more stable forms in the process. After this stage is complete, growth is much slower, as addition is now entirely to the slower-growing, lowest-energy form. Thus, crystals themselves, if taken at various stages of their development, serve as models of the rate of growth for the compound being studied.

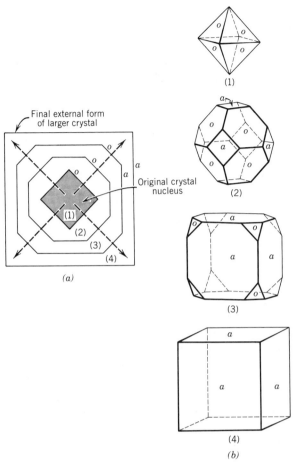

FIG. 2.25. (a) Schematic cross section of a crystal that grew from stage (1), a nucleus (with only *o* faces), via stages (2) and (3) to the final form of (4), with only *a* faces. The arrows are growth vectors representing the direction of fastest crystal growth. Note that the faces perpendicular to these growth vectors (*o* faces) are finally eliminated in stage (4). (b) Illustration of the complete crystals (at the various stages 1 to 4), for which (a) provides the schematic cross section. The form consisting of *o* faces only is an octahedron, the one with *a* faces only is a cube, and the two intermediate forms are combinations of the octahedron and the cube in different stages of development.

The *rate of solution* of a crystal in a chemical solvent is similarly a discontinuous vectorial process, and solution of a crystal or of any fragment of a single crystal may yield a more or less definite solution polyhedron. A clearer illustration of the vectorial nature of the rate of solution is afforded by etch pits. If a crystal is briefly treated with a chemical solvent that attacks it, the faces are etched or pitted. The shape of these pits is regular and depends on the structure of the crystal, the face being attacked, the presence of chemical impurities and inclusions, and the nature of the solvent. Valuable information about the internal geometry of arrangement of crystals may be obtained from a study of such etch pits.

Cleavage may be thought of as a discontinuous vectorial property and, like crystal form, reflects the internal structure, as cleavage always takes place along those planes across which there exist the weakest bonding forces. Those planes are generally the most widely spaced and the most densely populated.

Crystal Symmetry

The external shape of a well-formed crystal reflects the presence or absence of the previously discussed translation-free symmetry elements. These are rotation axes, rotoinversion axes, center of symmetry, and mirror planes. The presence of these symmetry elements can be detected, in a well-formed crystal, by the angular arrangement of the bounding faces and sometimes by their size and shape. In poorly developed or distorted crystals the symmetry is generally not obvious, but can be derived from careful measurement of the angular relation of the bounding faces (see page 53). The recognition of symmetry elements in crystals, or wooden models that display the morphology of perfect crystals, is very similar to the recognition of symmetry in ordered patterns (see pages 109–123). It may be useful to review some aspects of symmetry elements in relation to crystal morphology.

An axis of rotation is an imaginary line through a crystal about which the crystal may be rotated and repeat itself in appearance 1, 2, 3, 4, or 6 times during a complete rotation (see Fig. 2.10). Figure 2.26*a* depicts a 6-fold rotation axis. When rotated about this axis, the crystal repeats itself each 60° or six times in a 360° rotation.

An axis of rotoinversion is an imaginary line that relates rotation about an axis with inversion (see Figs. 2.11 and 2.12). As discussed earlier, $\bar{1}$ is equivalent to a center of symmetry (or inversion, *i*), $\bar{2}$ is equivalent to a mirror plane, $\bar{3}$ is equivalent to a 3-fold rotation plus a center of symmetry, $\bar{4}$ is unique, and $\bar{6}$ is equivalent to a 3-fold rotation axis with a mirror plane perpendicular to the axis. Figure 2.26*b* illustrates the morphological expression of a 4-fold rotoinversion axis.

A center of symmetry is present in a crystal if an imaginary line can be passed from any point on its surface through its center and a similar point is found on the line at an equal distance beyond the center (see Fig. 2.12*a*). This is equivalent to $\bar{1}$, or inversion. A center is represented by *i*, for inversion. Fig. 2.26*c* illustrates a center of symmetry in a crystal.

A mirror plane is an imaginary plane that divides a crystal into halves, each of which, in a perfectly developed crystal, is the mirror image of the other.

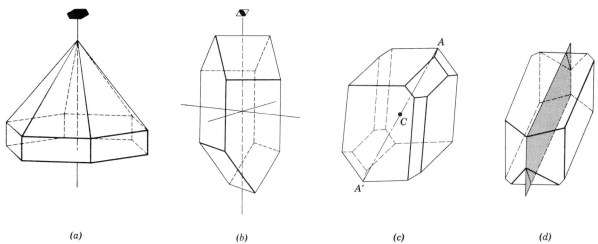

(a)	*(b)*	*(c)*	*(d)*

FIG. 2.26. Translation-free symmetry elements as expressed by the morphology of crystals. *(a)* 6-fold axis of rotation (6); *(b)* 4-fold axis of rotoinversion ($\bar{4}$); *(c)* center of symmetry (*i*); *(d)* mirror plane (*m*).

Figure 2.26*d* illustrates the nature and position of a single mirror in a crystal, also called a *symmetry plane*.

In our earlier discussion of symmetry elements in patterns we noted the existence of 32 point groups or crystal classes (see Tables 2.1 and 2.2). These point groups, uniquely defined by their symmetry content, are listed in Table 2.4. Each of these point groups has, in the past, been given a name, in accordance with the name of the general form in each class. These names are not given in Table 2.4 but are given in subsequent discussion of each of the point groups (or crystal classes). The reason for omitting the generally accepted names of the point groups in Table 2.4 is that *each point group is uniquely defined by its Hermann-Mauguin notation* but not by its name.

As can be seen in Tables 2.2 and 2.4, certain groups of crystal classes have common symmetry characteristics. These groups of crystal classes are known as *crystal systems*. There are six such systems, with the hexagonal system having two subdivisions (hexagonal and rhombohedral) on the basis of the lattice symmetry being hexagonal or rhombohedral (see footnote to Table 2.4).

Crystallographic Axes

In the description of crystals it is convenient to refer the external forms or internal symmetry to a set of three (or four) reference axes. These imaginary reference lines are known as the *crystallographic axes* and are generally taken parallel to the intersection edges of major crystal faces. Such axes are in most instances fixed by the symmetry and coincide with symmetry axes or with normals to symmetry planes. For some crystals there may be more than one choice of crystallographic axes when the selection is made by morphology alone. Ideally the axes should be parallel to and their lengths proportional to the edges of the unit cell.

All crystals, with the exception of those belonging to the hexagonal system (see page 40), are referred to three crystallographic axes designated as *a*, *b*, and *c*; see Fig. 2.27. In the general case (triclinic system) all the axes are of different lengths and at oblique angles to each other. The ends of each axis are designated plus or minus; the front end of *a*, the right-hand end of *b*, and the upper end of *c* are positive; the opposite ends are negative. The angles between the positive ends of the axes are conventionally designated by the Greek letters α, β, and γ. The α angle is enclosed between axial directions *b* and *c*, the β angle between *a* and *c*, and the γ angle between *a* and *b*. In summary the six crystal systems are referred to the following axial directions and axial angles (see also Fig. 2.27):

Triclinic. Three unequal axes all intersecting at oblique angles.

Monoclinic. Three unequal axes, two of which are inclined to each other at an oblique angle and the third perpendicular to the plane of the other two.

Orthorhombic. Three mutually perpendicular axes all of different lengths.

Tetragonal. Three mutually perpendicular axes, two of which (the horizontal axes) are of equal length (a_1 and a_2), but the vertical axis is shorter or longer than the other two.

Hexagonal. Referred to four crystallographic

Table 2.4

**THE THIRTY-TWO CRYSTAL CLASSES
AND THEIR SYMMETRY**

Crystal System	Crystal Class	Symmetry Content	Crystal System	Crystal Class	Symmetry Content
Triclinic	1	none	Hexagonal*	3	$1A_3$
	$\overline{1}$	i		$\overline{3}$	$1\overline{A}_3(= i + 1A_3)$
Monoclinic	2	$1A_2$		32	$1A_3, 3A_2$
	m	$1m$		$3m$	$1A_3, 3m$
	$2/m$	$i, 1A_2, 1m$		$\overline{3}\,2/m$	$1\overline{A}_3, 3A_2, 3m$
Orthorhombic	222	$3A_2$			$(1\overline{A}_3 = i + 1A_3)$
	$mm2$	$1A_2, 2m$		6	$1A_6$
	$2/m2/m2/m$	$i, 3A_2, 3m$		$\overline{6}$	$1\overline{A}_6(= 1A_3 + m)$
Tetragonal	4	$1A_4$		$6/m$	$i, 1A_6, 1m$
	$\overline{4}$	$1\overline{A}_4$		622	$1A_6, 6A_2$
	$4/m$	$i, 1A_4, m$		$6mm$	$1A_6, 6m$
	422	$1A_4\,4A_2$		$\overline{6}m2$	$1\overline{A}_6, 3A_2, 3m$
	$4mm$	$1A_4, 4m$			$(1\overline{A}_6 = 1A_3 + m)$
	$\overline{4}\,2m$	$1\overline{A}_4, 2A_2, 2m$		$6/m2/m2/m$	$i, 1A_6, 6A_2, 7m$
	$4/m2/m2/m$	$i, 1A_4, 4A_2, 5m$	Isometric	23	$3A_2, 4A_3$
				$2/m\overline{3}$	$3A_2, 3m, 4\overline{A}_3$
					$(1\overline{A}_3 = 1A_3 + i)$
				432	$3A_4, 4A_3, 6A_2$
				$\overline{4}\,3m$	$3\overline{A}_4, 4A_3, 6m$
				$4/m\overline{3}2/m$	$3A_4, 4\overline{A}_3, 6A_2, 9m$
					$(1\overline{A}_3 = 1A_3 + i)$

*In this table all crystal classes (point groups) beginning with 6, $\overline{6}$, 3, and $\overline{3}$ are grouped in the hexgonal system. In earlier editions of the *Manual of Mineralogy* the hexagonal system was divided into the hexagonal and rhombohedral divisions. The use of these two subdivisions, as based on the presence of 6 or $\overline{6}$ versus 3 or $\overline{3}$ axes in the morphological symmetry of a crystal, results in confusion when subsequent X-ray investigations show a specific crystal with, for example, 32 symmetry to be based on a hexagonal lattice. This is the case in low quartz, which shows morphological symmetry 32 but is based on a primitive hexagonal lattice, resulting in space group $P3_12$ (or $P3_22$).

The hexagonal system can, however, be divided according to whether the lattice symmetry is hexagonal ($6/m2/m2/m$) or rhombohedral ($\overline{3}2/m$). This results in the following groupings:

Hexagonal (hexagonal lattice division)		Rhombohedral (rhombohedral lattice division)
$6/m2/m2/m$		$\overline{3}\,2/m$
$\overline{6}m2$	$\overline{3}2/m$	$3m$
$6mm$	$3m$	$3\,2$
$6\,2\,2$ and	$3\,2$	$\overline{3}$
$6/m$	$\overline{3}$	3
$\overline{6}$	3	
6		

axes; three equal horizontal axes (a_1, a_2, and a_3) intersect at angles of 120°, the fourth (vertical) is of different length and perpendicular to the plane of the other three.

Isometric. Three mutually perpendicular axes of equal lengths (a_1, a_2, and a_3).

Axial Ratios

In all the crystal systems, with the exception of the isometric, there are crystallographic axes differing in length. If it were possible to isolate a unit cell and measure the dimensions along the edges, which are parallel to the crystallographic axes, we would be able to write ratios between the edge lengths. X-ray crystallographers cannot isolate the cell, but they can measure accurately the cell dimensions in angstrom units (1 Å = 10^{-8} cm). Thus for the orthorhombic mineral sulfur the unit cell dimensions are given as:

$a = 10.47$ Å, $b = 12.87$ Å, and $c = 24.49$ Å (see Fig. 2.28). Using the length of the b axis as a unit of measure one can determine the lengths of a and b relative to this. Thus $a/b : b/b : c/b = X : 1 : Y$. In the case of sulfur we can write the ratios $a : b : c = 0.813 : 1 : 1.903$. The ratios thus express the relative, not the absolute, lengths of the cell edges that correspond to the crystallographic axes.

We saw earlier (page 32) that the number of crystal planes appearing as faces is limited and that the angular relations between them is dependent on the unit cell, that is, the length of its edges and the angles between the edges. Long before X-ray investigations made it possible to determine the absolute dimensions of the unit cell, this relation of crystal morphology to internal structure was recognized, and axial ratios were calculated. By measuring the interfacial angles on a crystal and making appropriate calculations, it is possible to arrive at axial ratios that express the rela-

FIG. 2.27. Illustrations of the conventional system of crystallographic axes adopted for each of the six crystal systems. The three directions are labeled a, b, and c, unless symmetry makes them equivalent. The vertical is labeled c, except in the isometric system where all axes are equal. The two horizonal axes are labeled a and b. If they are equal, they are labeled with the same letter, such as a_1 and a_2. In the hexagonal system there are three equal horizontal axes labeled a_1, a_2 and a_3. See Table 2.9 for the relationship of symmetry notation (as in the Hermann-Mauguin system) and crystallographic axes.

FIG. 2.28. The orthorhombic unit cell of sulfur.

tive lengths of the crystallographic axes. It is interesting to see how closely the axial ratios calculated from unit cell dimensions compare with the older ratios derived from morphological measurements. For example, from morphological measurements and calculations the ratios for sulfur were reported in 1869 as: $a : b : c = 0.8131 : 1 : 1.9034$; from unit cell measurements by X-ray techniques in 1960 they are: $a : b : c = 0.8135 : 1 : 1.9029$.

Face Intercepts

Crystal faces are defined by indicating their intercepts on the crystallographic axes. Thus, in describing a crystal face it is necessary to determine whether it is parallel to two axes and intersects the third, is parallel to one axis and intersects the other two, or intersects all three. In addition, one must determine at what *relative* distance the face intersects the different axes. In

defining the position of a crystal face one must remember that crystal faces are parallel to a family of possible lattice planes. In Fig. 2.29a we have an a–b plane in an orthorhombic net based on the unit cell dimensions of the mineral sulfur.

The lattice plane *AA* is parallel to the b an c axes and intersects the a axis at one length (taken as unit length) along the a axis. The intercepts for this plane would be: $1a, \infty b, \infty c$. Similarly the plane *A'A'*, which is parallel to *AA* but intersects the a axis at two unit lengths, would have intercepts: $2a, \infty b, \infty c$. The plane *BB*, which is parallel to the a and c axes and intersects the b axis at unit distance, has intercepts: $\infty a, 1b, \infty c$. The plane *AB* intersects both horizontal axes (a and b) at unit distance but is parallel to c, which leads to the intercepts: $1a, 1b, \infty c$. A plane that intersects all three axes at unit distances would have intercepts $1a, 1b$, and $1c$. Figure 2.29b shows the development of crystal faces some of which are parallel to the lattice planes shown in Fig. 2.29a. It must be remembered that the face intercepts shown on the faces are strictly relative values and do not indicate any actual cutting lengths.

When intercepts are assigned to the faces of a crystal, without any knowledge of its unit cell dimensions, one face that cuts all three axes is arbitrarily assigned the units 1a, 1b, and 1c. This face, which is referred to as the *unit face*, is generally the largest in case there are several faces that cut all three axes. Figure 2.30 shows an orthorhombic crystal that consists of faces, all of which cut all three crystallographic axes. The largest face (shaded) that intersects all three crystallographic axes at their positive ends is taken as the unit face. Its intercepts are 1a, 1b, and 1c as

FIG. 2.29. (a) Intercepts for some planes in an orthorhombic lattice. (b) Intercepts for some crystal faces on the upper half of an orthorhombic crystal.

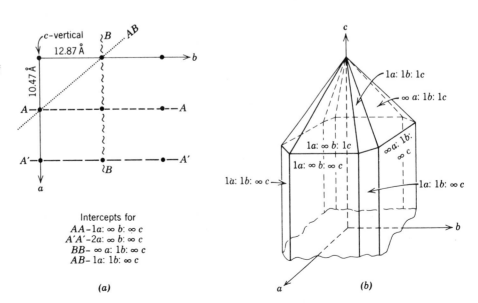

Intercepts for
AA–$1a: \infty b: \infty c$
A'A'–$2a: \infty b: \infty c$
BB–$\infty a: 1b: \infty c$
AB–$1a: 1b: \infty c$

(a)

(b)

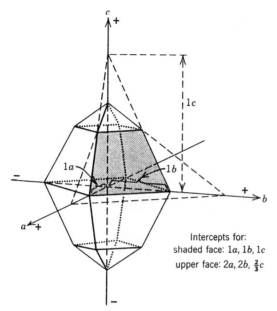

Intercepts for:
shaded face: 1a, 1b, 1c
upper face: 2a, 2b, $\frac{2}{3}$c

FIG. 2.30. Relative intersections of faces on an orthorhombic crystal, all of which cut three crystallographic axes.

shown. The intercepts of the smaller face above it can now be estimated by extending the edges of this face in the directions of the a and b axes. The intercepts for the upper face become 2a, 2b, $\frac{2}{3}$c, with respect to the unit face. These intercepts can be divided by the common factor 2, resulting in 1a, 1b, $\frac{1}{3}$c. This example illustrates that the units 1a and 1b do not represent actual cutting distances but express only relative values. The intercepts of a face have no relation to its size, for a face may be moved parallel to itself for any distance without changing the relative values of its intersections with the crystallographic axes.

Miller Indices

Various methods of notation have been devised to express the intercepts of crystal faces upon the crystal axes. The most universally employed is the system of indices proposed by W. H. Miller, which has many advantages over the system of intercepts discussed above.

The Miller indices of a face consist of a series of whole numbers that have been derived from the intercepts by their inversions and, if necessary, the subsequent clearing of fractions. The indices of a face are always given so that the three numbers (four in the hexagonal system) refer to the a, b, and c axes, respectively, and therefore the letters that indicate the different axes are omitted. Like the intercepts, the Miller indices express a ratio, but for the sake of brevity the ratio sign is also omitted. For the two upper faces in Fig. 2.30 which cut the positive segments of the crystallographic axes, the intercepts are 1a, 1b, 1c, and 2a, 2b, $\frac{2}{3}$c, respectively. Inverting these intercepts leads to: $\frac{1}{1}\frac{1}{1}\frac{1}{1}$ and $\frac{1}{2}\frac{1}{2}\frac{3}{2}$ respectively. For the unit-face this gives a Miller index of (111), and clearing the fractions for $\frac{1}{2}\frac{1}{2}\frac{3}{2}$ by multiplying all by 2 leads to a Miller index of (113) for the other face. The Miller index (111) is read as "one-one-one." Further examples of conversions of intercepts to Miller indices for an isometric lattice and crystal forms are given in Fig. 2.31. Commas are used in Miller indices only when two-digit numbers appear, as in (1, 14, 3). For faces that intersect negative ends of crystallographic axes, a line is placed over the appropriate number, as shown in Fig. 2.32. For example, (1 $\bar{1}$ 1) reads "one, minus one, one" or "one, bar one, one." It should be noted that indices given above for specific faces are

FIG. 2.31. *(a)* Intercepts and Miller indices for some planes in an isometric lattice. *(b)* Intercepts and Miller indices of some faces modifying the corner of a cube.

Intercepts for:
AA$-1:\infty:\infty$
A'A'$-2:\infty:\infty$
BB$-\infty:1:\infty$
AB$-1:1:\infty$

Miller indices for:
AA$-\frac{1}{1}\frac{1}{\infty}\frac{1}{\infty} = 100$
A'A'$-\left(\frac{1}{2}\frac{1}{\infty}\frac{1}{\infty}\right) \times 2 = 100$
BB$-\frac{1}{\infty}\frac{1}{1}\frac{1}{\infty} = 010$
AB$-\frac{1}{1}\frac{1}{1}\frac{1}{\infty} = 110$

(a)

(b)

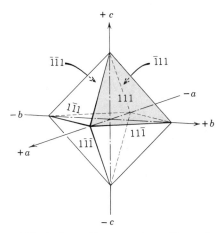

FIG. 2.32. Miller indices with respect to positive and negative ends of crystallographic axes.

the last letter to c. In this notation $h + k + i = 0$ holds invariably. For example, with respect to the three faces indexed in Fig. 2.33:

$$10\overline{1}0, \ 1 + 0 + \overline{1} = 0$$
$$11\overline{2}0, \ 1 + 1 + \overline{2} = 0$$
$$11\overline{2}1, \ 1 + 1 + \overline{2} = 0$$

Form

In its most familiar meaning the term *form* is used to indicate general outward appearance. In crystallography, external shape is denoted by the word *habit*, whereas *form* is used in a special and restricted sense. Thus *a form consists of a group of crystal faces, all of which have the same relation to the elements of symmetry* and display the same chemical and physical properties because all are underlain by like atoms in the same geometrical arrangement. The relationship between form and the elements of symmetry of a crystal is an important one to understand. For example, as in Fig. 2.34 where we have symmetry elements that belong to two crystal classes, $\overline{1}$ (in the triclinic system) and $4/m\,\overline{3}\,2/m$ (in the isometric system) we may wish to develop the full form for the unit face (111). In the case of the $\overline{1}$ symmetry (which is equivalent to a center of symmetry) we develop an additional face by inverting through the origin of the three crystallographic axes. This additional face will have indices $(\overline{1}\,\overline{1}\,\overline{1})$. The form in the case of $\overline{1}$ consists, therefore, of two parallel faces only, and is known as a pinacoid or parallelohedron. In the case of $4/m\,\overline{3}\,2/m$ the symmetry elements will generate seven additional faces for (111) with indices $(\overline{1}11)$, $(1\overline{1}1)$, $(11\overline{1})$, $(\overline{1}\,\overline{1}1)$, $(1\overline{1}\,\overline{1})$, $(\overline{1}11\overline{1})$, and $(\overline{1}\,\overline{1}\,\overline{1})$. This form is known as an octahedron. It should be clear, therefore, that *the number of faces that belongs to a form is determined by the symmetry of the crystal class.*

Although faces of a form may be of different sizes and shapes because of distortion of the crystal, the similarity is frequently evidenced by natural striations, etchings, or growths, as shown in Fig. 2.35. On some crystals the differences between faces of different forms can be seen only after etchings with acid. In Figs. 2.35a and b there are three forms, each of which has a different physical appearance from the others, and in Fig. 2.35c there are two forms.

In our discussion of Miller indices we said that a crystal face may be designated by a symbol enclosed in parentheses as (hkl), (010), or (111). Miller indices also may be used as form symbols and are then enclosed in braces as {hkl}, {010}. Thus in Fig. 2.34 (111)

placed in parentheses. This is to distinguish them from similar symbols used to designate crystal forms (see below), zones, and axial directions (page 48). Furthermore, the $\overline{1}$ ("bar one") that is part of a Miller index representation is *not* the same as $\overline{1}$, a rotoinversion axis.

It is sometimes convenient when the exact intercepts are unknown to use a general symbol (hkl) for the Miller indices; here h, k, and l are, respectively, the reciprocals of rational but undefined intercepts along the a, b, and c axes. The symbol (hkl) would indicate that a face cuts all three crystallographic axes without implying relative units along the axes. If a face is parallel to one of the crystallographic axes and intersects the other two, the general symbols would be written as (0kl), (h0l), and (hk0). A face parallel to two of the axes may be considered to intersect the third at unity, and the indices would, therefore, be: (100), (010), and (001), as well as the negative equivalents such as $(\overline{1}00)$, $(0\overline{1}0)$, and $(00\overline{1})$.

Crystals belonging to the hexagonal system are distinct from other systems because they possess one unique axis that is either 6- or 3-fold in symmetry. This unique axis is perpendicular to the plane which contains three identical axes, labeled a_1, a_2, and a_3. Because of the presence of four crystallographic axes, instead of three, as in the other crystal systems, a four-number system for indexing crystal faces was developed. This is known as the Bravais-Miller system of indexing. The indices are derived from the intercepts on the axes in the same way as the three-number Miller indices. Figure 2.33 illustrates the conversion of intercepts on the four-axis system to Bravais-Miller indices for three differently oriented crystal faces. The general symbol in this index system is $(hk\overline{i}l)$, in which the first three letters refer to the a_1, a_2, and a_3 axes and

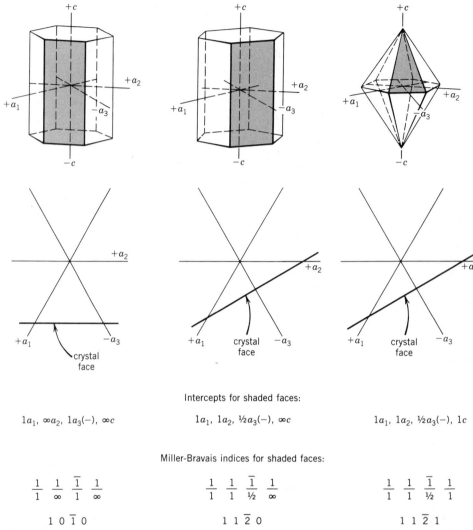

Intercepts for shaded faces:

$1a_1, \infty a_2, 1a_3(-), \infty c$ $1a_1, 1a_2, \frac{1}{2}a_3(-), \infty c$ $1a_1, 1a_2, \frac{1}{2}a_3(-), 1c$

Miller-Bravais indices for shaded faces:

$\dfrac{1}{1} \quad \dfrac{1}{\infty} \quad \dfrac{\bar{1}}{1} \quad \dfrac{1}{\infty}$ $\dfrac{1}{1} \quad \dfrac{1}{1} \quad \dfrac{\bar{1}}{\frac{1}{2}} \quad \dfrac{1}{\infty}$ $\dfrac{1}{1} \quad \dfrac{1}{1} \quad \dfrac{\bar{1}}{\frac{1}{2}} \quad \dfrac{1}{1}$

$1\ 0\ \bar{1}\ 0$ $1\ 1\ \bar{2}\ 0$ $1\ 1\ \bar{2}\ 1$

FIG. 2.33. Derivation of the four-digit Bravais-Miller index from the intercepts of three different crystal faces in the hexagonal system.

refers to a specific face, whereas {111} embraces all faces of the form. In choosing a form symbol it is desirable to select, if possible, the face symbol with positive digits: {111} rather than {1$\bar{1}$1}, {010} rather than {0$\bar{1}$0}.

In each crystal class there is a form, the faces of which intersect all of the crystallographic axes at different lengths; this is the general form, {hkl}. All other forms that may be present are *special forms*. In the orthorhombic, monoclinic, and triclinic crystal systems, {111} is a general form, for the unit length along each of the axes is different. In the crystal systems of higher symmetry in which the unit distances along two or more of the crystallographic axes are the same, a general form must intersect the like axes at different multiples of the unit length. Thus {121} is a general

form in the tetragonal system but a special form in the isometric system and {123} is a general form in the isometric system. The concept of a *general form* can also be related to the symmetry elements of a specific crystal class. *An (hkl) face will not be parallel or perpendicular to a symmetry axis or plane, regardless of the crystal class.* A *special form*, however, consists of faces that are parallel or perpendicular to any of the symmetry elements in the crystal class. For most crystal classes the general form contains a larger number of faces than any of the special forms of that same class.

In Fig. 2.34 we developed a two-faced form and an eight-faced form. The two-faced form {111} for crystal class $\bar{1}$ is referred to as an *open form*, because it consists of two parallel faces that do not enclose

$\bar{1}$

$\frac{4}{m}\bar{3}\frac{2}{m}$

Pinacoid {111}

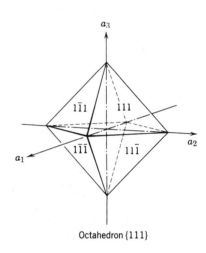

Octahedron {111}

FIG. 2.34. Development of a form from one face with Miller index (111) in the $\bar{1}$ and the $4/m\,\bar{3}\,2/m$ crystal classes. In the case of $\bar{1}$ symmetry only a two-faced form, a pinacoid, develops. In the illustration of the isometric class the pole to the (111) face coincides with the position of the 3-fold rotoinversion axis. Faces at the top of the crystal are generally shown by dots, those at the bottom by open circles. Because of the complexity of the isometric figure only the positions of the four top faces of the octahedron are shown by Miller indices; there exists a similar set of four faces underneath.

space (see also Fig. 2.36). The eight-faced form {111} in crystal class $4/m\,\bar{3}\,2/m$ is a *closed form* as the eight faces together enclose space (see Figs. 2.32 and 2.36).

A crystal will often display several forms in combination with one another but may have only one, provided that it is a closed form. Because any combination of forms must enclose space, a minimum of two open forms is necessary. The two may exist together or be in combination with other closed or open forms.

Names of Forms
In this, as in previous editions of this textbook, we will follow a scheme of form nomenclature originally proposed by Groth in 1895 and modified by A. F. Rogers

FIG. 2.35. Different appearances of different forms. *(a)* Quartz—hexagonal. *(b)* Apophyllite—tetragonal. *(c)* Pyrite—isometric.

(a)   *(b)*   *(c)*

in 1935 (see References). This scheme recognizes 48 different types of crystal forms as distinguished by the angular relations of the crystal faces. Of these 48 forms, 32 are the general forms found in the 32 crystal classes (or point groups); 10 are special, closed forms of the isometric system; and 6 are special open forms (prisms) of the hexagonal and tetragonal systems. In this scheme of nomenclature the name of each of the 32 general forms, {hkl} (or {hkīl} in the hexagonal system), becomes the descriptive name of each of the 32 crystal classes. For example, in $\bar{1}$ the general form {hkl} is a 2-faced form known as a pinacoid, and the descriptive name for $\bar{1}$ is pinacoidal class. Similarly, the name of the general form {hkl} in $4/m2/m2/m$ (of the tetragonal system) is ditetragonal-dipyramid; accordingly, the class $4/m2/m2/m$ is commonly referred to as the ditetragonal-dipyramidal class. This nomenclature for general forms and crystal classes has been widely used in the English-speaking world.

A somewhat different scheme, in which only 47 (instead of 48 as above) different form types are distinguished, has been recommended by crystallographers of the Fedorov Institute of Leningrad in 1925. The reason for the reduction in this scheme, from 48 to 47 different forms and form names, is that the *dome* ({hkl} form in class *m*) and the sphenoid ({hkl} form in class 2) are not distinguished as different forms; each one is a *dihedron* (i.e., a form composed of two nonparallel faces) as recognized in the scheme of the Fedorov Institute (see Table 2.5). This nomenclature has recently received more international acceptance than that of Groth and Rogers, because the naming of the forms is more logical and somewhat easier to learn for the beginning student. In the internationally recommended scheme the name of the form is determined by the number and shape of its faces, and is based on Greek words, as explained in the footnote to Table 2.5. Although the names defined by the Fedorov system are more logical, some also tend to become longer.

In Tables 2.5 and 2.6 we list names of the 48 forms in the Groth-Rogers nomenclature as well as the equivalent names for the 47 forms as defined by crystallographers of the Fedorov Institute. In the subsequent systematic discussion of forms for each of the 32 crystal classes (or point groups) we will first list the form name for the Groth-Rogers scheme, and add, between parentheses, the internationally recommended form name. That is, if the two naming schemes are different (many of the names are identical in the two schemes; see Tables 2.5 and 2.6), we will give dual nomenclature in the text as well as the captions of many of the figures. We will not logically carry this duality through all of the text discussion, however, because of the difficulty of reading a name that is consistently followed by a synonym in parentheses. Note that, although form names are often useful, such names are not absolutely necessary because a form is uniquely defined by a combination of its Miller indices and the Hermann-Mauguin notation of its point group symmetry.

Illustration and Description of Forms

The forms listed in Tables 2.5 and 2.6 are illustrated in Fig. 2.36. In the case of prisms, pyramids, and dipyramids (nos. 5 through 25 in Fig. 2.36) the three-dimensional representations cannot properly convey the shapes of the cross sections. For this reason the various prisms numbered 5 through 11 are accompanied by a cross section perpendicular to the long axis of the prism; the shapes of these cross sections represent sections perpendicular to the long axes of the pyramids (nos. 12 through 18) and dipyramids (nos. 19 through 25) as well. The total symmetry content of each of these forms is discussed in detail in a subsequent section entitled "The Thirty-two Crystal Classes." Here follow only brief descriptions of the forms shown in Fig. 2.36 (the numbers in parentheses in the following descriptions all relate to Fig. 2.36).

Pedion (*monohedron*). A single face comprising a form (1).

Pinacoid (*parallelohedron*). An open form made up of two parallel faces (2).

Dome (*dihedron*). An open form consisting of two nonparallel faces symmetrical with respect to a mirror plane (*m*) (3).

Sphenoid (*dihedron*). Two nonparallel faces symmetrical with respect to a 2-fold rotation axis (4). In the internationally recommended system of nomenclature both the *dome* and the *sphenoid* are referred to as a *dihedron*. This is because both forms consist of two intersecting faces (dihedron) and because the nomenclature does not reflect the symmetry element responsible for the form.

Prism. An open form composed of 3, 4, 6, 8, or 12 faces, all of which are parallel to the same axis. Except for certain prisms in the monoclinic system, the axis is one of the principal crystallographic axes (5–11).

Pyramid. An open form composed of 3, 4, 6, 8, or 12 nonparallel faces that meet at a point (12–18).

Table 2.5

THE NAMES OF THE 33 (OR 32) DIFFERENT TYPES OF NONISOMETRIC CRYSTAL FORMS*

Name According to Groth-Rogers[†]	No. of Faces	Internationally Recommended Name (after Fedorov Institute)
1. *Pedion*	1	*Monohedron*
2. *Pinacoid*	2	*Parallelohedron*
3. *Dome*	2 ⎫	*Dihedron*
4. *Sphenoid*	2 ⎭	
5. Rhombic prism	4	Rhombic prism
6. Trigonal prism	3	Trigonal prism
7. Ditrigonal prism	6	Ditrigonal prism
8. Tetragonal prism	4	Tetragonal prism
9. Ditetragonal prism	8	Ditetragonal prism
10. Hexagonal prism	6	Hexagonal prism
11. Dihexagonal prism	12	Dihexagonal prism
12. Rhombic pyramid	4	Rhombic pyramid
13. Trigonal pyramid	3	Trigonal pyramid
14. Ditrigonal pyramid	6	Ditrigonal pyramid
15. Tetragonal pyramid	4	Tetragonal pyramid
16. Ditetragonal pyramid	8	Ditetragonal pyramid
17. Hexagonal pyramid	6	Hexagonal pyramid
18. Dihexagonal pyramid	12	Dihexagonal pyramid
19. Rhombic dipyramid	8	Rhombic dipyramid
20. Trigonal dipyramid	6	Trigonal dipyramid
21. Ditrigonal dipyramid	12	Ditrigonal dipyramid
22. Tetragonal dipyramid	8	Tetragonal dipyramid
23. Ditetragonal dipyramid	16	Ditetragonal dipyramid
24. Hexagonal dipyramid	12	Hexagonal dipyramid
25. Dihexagonal dipyramid	24	Dihexagonal dipyramid
26. Trigonal trapezohedron	6	Trigonal trapezohedron
27. Tetragonal trapezohedron	8	Tetragonal trapezohedron
28. Hexagonal trapezohedron	12	Hexagonal trapezohedron
29. *Tetragonal* scalenohedron	8	*Rombic* scalenohedron
30. *Hexagonal* scalenohedron	12	*Ditrigonal* scalenohedron
31. Rhombohedron	6	Rhombohedron
32. Rhombic *disphenoid*	4	Rhombic *tetrahedron*
33. Tetragonal *disphenoid*	4	Tetragonal *tetrahedron*
Total no. of forms = 33		Total no. of forms = 32

*The number of faces of each form is given. The numbers on the left refer to Fig. 2.36. In this listing the eight synonymous terms are in italics.

[†]The terminology in the first column is a compromise of generally accepted names used in the English-speaking world. The internationally recommended system of nomenclature is based on the generally accepted terms for geometric shapes, with adjectives derived from Greek words. The adjectives give the shape of the cross section of the ideal geometric form, either at midheight or near an apex. In the rhombic prism, for example, the cross section of the prism is rhomb-shaped. The Greek prefixes for 1, 2, 3, 4, 5, 6, 8, and 12, are *mono, di, tri, tetra, penta, hexa, octa,* and *dodeca;* the word *gonia* means angle, as in hexagon; *hedron* means face, as in polyhedron; *rhomb,* a planar figure with four equal sides and opposite angles equal (two acute, two obtuse); *prism,* three or more faces intersecting in parallel edges; *pyramid,* three or more faces whose edges intersect in a point; *trapezohedron,* a form with faces that are trapezoids (four-sided faces in which no two sides are parallel but two adjacent sides are equal); and *scalenohedron,* in which faces are scalene triangles (triangles in which no two sides are equal).

Dipyramid. A closed form having 6, 8, 12, 16, or 24 faces (19–25). A dipyramid can be considered as formed from two pyramids by reflection of one of them into the other across a horizontal mirror plane.

Trapezohedron. A closed form that has 6, 8, or 12 faces in all, with 3, 4, or 6 upper faces offset from 3, 4, or 6 lower faces (26–28). These forms are the result of a 3-, 4-, or 6-fold axis combined with perpendicular 2-fold axes. In addition there

Table 2.6

THE NAMES OF THE 15 DIFFERENT TYPES OF FORMS IN THE ISOMETRIC SYSTEM, ACCORDING TO TWO SYSTEMS OF NOMENCLATURE*

Name According to System of Groth-Rogers	No. of Faces	Internationally Recommended Name (after Fedorov Institute)
34. Cube	6	Hexahedron
35. Octahedron	8	Octahedron
36. Dodecahedron (rhombic)	12	Rhomb-dodecahedron
37. Tetrahexahedron	24	Tetrahexahedron
38. Trapezohedron	24	Tetragon-trioctahedron
39. Trisoctahedron	24	Trigon-trioctahedron
40. Hexoctahedron	48	Hexaoctahedron
41. Tetrahedron	4	Tetrahedron
42. Tristetrahedron	12	Trigon-tritetrahedron
43. Deltoid dodecahedron	12	Tetragon-tritetrahedron
44. Hextetrahedron	24	Hexatetrahedron
45. Gyroid	24	Pentagon-trioctahedron
46. Pyritohedron	12	Dihexahedron
47. Diploid	24	Didodecahedron
48. Tetartoid	12	Pentagon-tritetrahedron

In the internationally recommended scheme a prefix describes the shape (as in rhomb) or the number of edges (as in trigon).

*The number of faces of each form is given. The number on the left refers to Fig. 2.36.[†]

[†]See footnote to Table 2.5 for the derivation of the terms.

is an isometric *trapezohedron* (*tetragon-triocta-hedron*; no. 38) a 24-face form. In a well-developed single trapezohedron each face is a trapezium.

Scalenohedron. A closed form with 8 or 12 faces (29 and 30), grouped in symmetrical pairs. In the *tetragonal scalenohedron* (*rhombic scalenohedron*), pairs of upper faces are related by an axis of 4-fold rotoinversion to pairs of lower faces. The 12 faces of the *hexagonal scalenohedron* (*ditrigonal scalenohedron*) display three pairs of upper faces and three pairs of lower faces in alternating positions. The pairs are related by the center of symmetry, which coexists with a 3-fold axis in rotoinversion axis $\bar{3}$. In an ideally developed scalenohedron, each face is a scalene triangle.

Rhombohedron. A closed form composed of six faces of which three faces at the top alternate with three faces at the bottom, the two sets of faces being offset by 60° (no. 31). Rhombohedrons are found only in point groups $\bar{3}\ 2/m$, 32, and $\bar{3}$.

Disphenoid (*rhombic* or *tetragonal tetrahedron*). A closed form consisting of two upper faces that alternate with two lower faces, offset by 90° (32 and 33).

The specialized forms in the isometric system are numbered 34 through 48. Every one of these forms contains four 3-fold rotation axes or $\bar{3}$ rotoinversion axes. One such direction (in the upper right, positive quadrant) is indicated in the illustrations. These forms are discussed more fully on pages 63 through 100.

In crystal drawings it is convenient to indicate the faces of a form by the same letter. The choice of which letter shall be assigned to a given form rests largely with the person who first describes the crystal. However, there are certain simple forms that, by convention, usually receive the same letter. Thus the three pinacoids that cut the a, b, c axes are lettered respectively a, b, and c, the letter m is usually given to {110} and p to {111} (Fig. 2.37).

Zones

One of the conspicuous features on many crystals is the arrangement of a group of faces with parallel intersection edges. Considered collectively, these faces form a *zone*. A line through the center of the crystal that is parallel to the lines of face intersections is called the *zone axis*. In Fig. 2.38 the faces m', a, m, and b are in one zone, and b, r, c, and r' are in another. The lines, designated [001] and [100], are the zone axes.

A zone is indicated by a symbol similar to that for Miller indices of faces, the generalized expression of which is [uvw]. Any two nonparallel faces determine a zone. Assume one wishes to determine the zone axis of two such faces, (hkl) and (pqr). The method usually used is to write twice the symbol of one face and directly beneath, twice the symbol of the other face. The first and last digits of each line are

Non-isometric forms

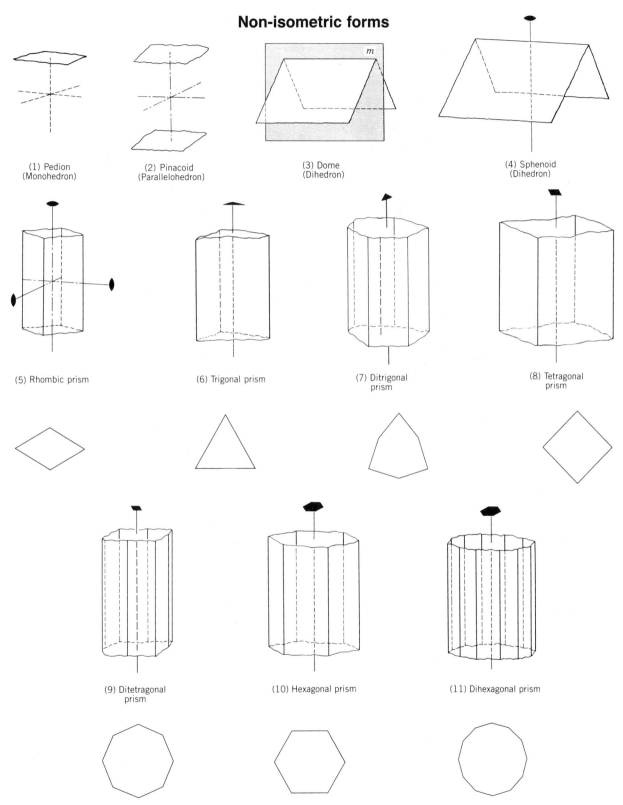

(1) Pedion
(Monohedron)

(2) Pinacoid
(Parallelohedron)

(3) Dome
(Dihedron)

(4) Sphenoid
(Dihedron)

(5) Rhombic prism

(6) Trigonal prism

(7) Ditrigonal
prism

(8) Tetragonal
prism

(9) Ditetragonal
prism

(10) Hexagonal prism

(11) Dihexagonal prism

FIG. 2.36. The 48 (or 47) different crystal forms and some of their symmetry elements.

(continued)

Non-isometric forms (cont'd)

(12) Rhombic pyramid   (13) Trigonal pyramid   (14) Ditrigonal pyramid   (15) Tetragonal pyramid   (16) Ditetragonal dipyramid

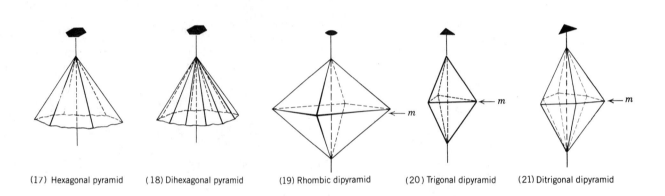

(17) Hexagonal pyramid   (18) Dihexagonal pyramid   (19) Rhombic dipyramid   (20) Trigonal dipyramid   (21) Ditrigonal dipyramid

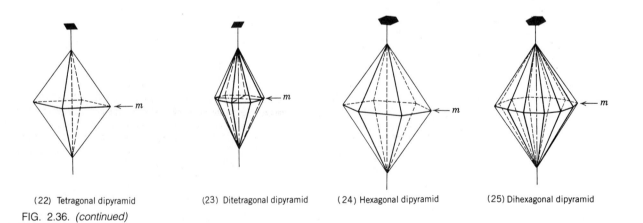

(22) Tetragonal dipyramid   (23) Ditetragonal dipyramid   (24) Hexagonal dipyramid   (25) Dihexagonal dipyramid

FIG. 2.36. *(continued)*

Non-isometric forms (cont'd)

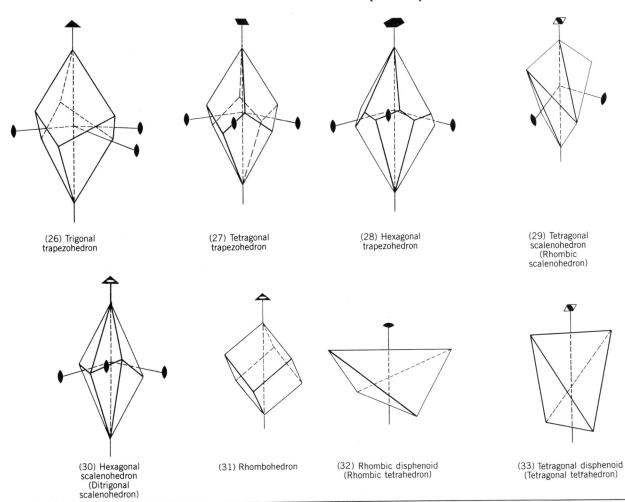

(26) Trigonal trapezohedron

(27) Tetragonal trapezohedron

(28) Hexagonal trapezohedron

(29) Tetragonal scalenohedron (Rhombic scalenohedron)

(30) Hexagonal scalenohedron (Ditrigonal scalenohedron)

(31) Rhombohedron

(32) Rhombic disphenoid (Rhombic tetrahedron)

(33) Tetragonal disphenoid (Tetragonal tetrahedron)

Isometric forms

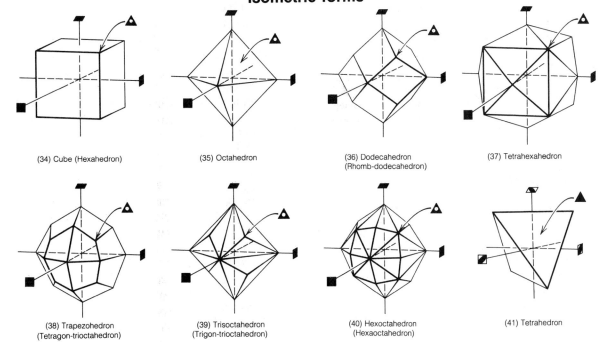

(34) Cube (Hexahedron)

(35) Octahedron

(36) Dodecahedron (Rhomb-dodecahedron)

(37) Tetrahexahedron

(38) Trapezohedron (Tetragon-trioctahedron)

(39) Trisoctahedron (Trigon-trioctahedron)

(40) Hexoctahedron (Hexaoctahedron)

(41) Tetrahedron

FIG. 2.36. *(continued)*

(continued)

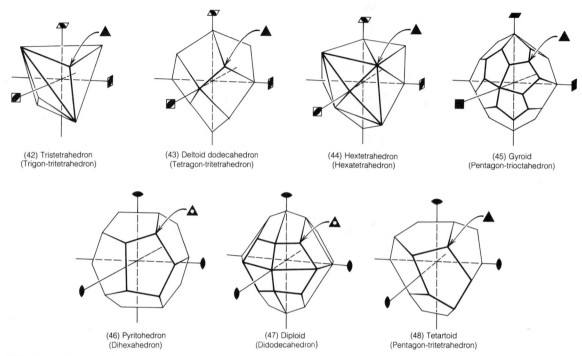

(42) Tristetrahedron
(Trigon-tritetrahedron)

(43) Deltoid dodecahedron
(Tetragon-tritetrahedron)

(44) Hextetrahedron
(Hexatetrahedron)

(45) Gyroid
(Pentagon-trioctahedron)

(46) Pyritohedron
(Dihexahedron)

(47) Diploid
(Didodecahedron)

(48) Tetartoid
(Pentagon-tritetrahedron)

FIG. 2.36. *(continued)*

disregarded and the remaining numbers, joined by sloping arrows, multiplied. In each set the product of 2 is subtracted from the product of 1 as:

$$
\begin{array}{c|ccccc}
h & k & l & h & k & l \\
  & 1 & 1 & 1 & & \\
  & 2 & 2 & 2 & & \\
p & q & r & p & q & r \\
\hline
\end{array}
$$

$$kr - lq, \ lp - hr, \ hq - kp$$

$$
\begin{array}{c|ccccc}
1 & 1 & 0 & 1 & 1 & 0 \\
  & 1 & 1 & 1 & & \\
  & 2 & 2 & 2 & & \\
0 & 1 & 0 & 0 & 1 & 0 \\
\hline
\end{array}
$$

$$0 - 0, \ 0 - 0, \ 1 - 0 \text{ or } [001]$$

As a specific example assume *m*, Fig. 2.37 is (*hk*0) with index (110) and *b* is (010). The zone axis is thus

FIG. 2.37. Conventional lettering of forms on crystal drawings.

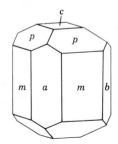

[001] as shown above. It should be noted that the zone symbols are enclosed in square brackets, as [*uvw*], to distinguish them from face and form symbols.

Crystal Habit

The term *habit* is used to denote the general shapes of crystals as cubic, octahedral, prismatic. Because habit is controlled by the environment in which crystals grow, it may vary with locality. At one place it may be equant and at others tabular or fibrous. Only rarely do crystals present an ideal geometrical shape. But even in asymmetric crystals evidence of the symmetry is present in the physical appearance of the faces and in the constancy of the interfacial angles.

FIG. 2.38. Crystal zones and zone axes.

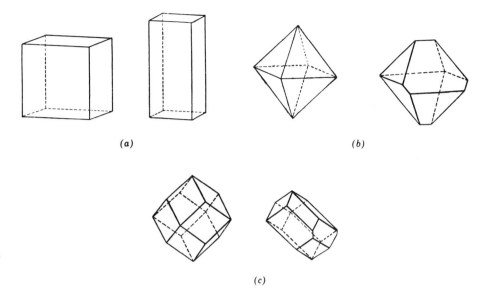

FIG. 2.39. Examples of some crystal habits in the isometric system. *(a)* Cube and distorted cube. *(b)* Octahedron and asymmetric octahedron. *(c)* Dodecahedron and distorted dodecahedron.

Given in Fig. 2.39 are various crystal forms, first ideally developed and then distorted.

MEASUREMENT OF CRYSTAL ANGLES

As stated by Steno's Law, the angles between equivalent faces of the same substance are constant, regardless of whether the crystal is misshapen or ideal. The angles that are measured between the normals to the crystal faces characterize a crystal and should be measured carefully. A graphic representation of the distribution of the angles, and the normals to the crystal faces, may define the symmetry of the crystal and therefore its crystal class. Angles are measured with *goniometers*, and for accurate work a *reflecting goniometer* is used. A crystal mounted on this instrument can be rotated about a zone axis and will reflect a collimated beam of light from its faces through a telescope to the eye. The angle through which a crystal must be turned in order to throw successive beams of light from two adjacent faces into the telescope determines the angle between the faces. Figure 2.40 illustrates a modern, two-circle reflecting goniometer.

A simpler instrument used for approximate work and with larger crystals is known as a *contact goniometer* (Fig. 2.41a). In using a contact goniometer it is imperative that the plane determined by the two arms of the goniometer be exactly at right angles to the edge between the measured faces. It becomes clear from the construction in Fig. 2.41b that the angle between the poles to the crystal faces, the *internal angle*, is measured. Thus in Fig. 2.41a the angle should read as 40°, not as 140°.

CRYSTAL PROJECTIONS

A crystal projection is a means of representing the three-dimensional crystal on a two-dimensional plane surface. Different projections are used for different purposes, but each is made according to some definite rule so that the projection bears a known and reproducible relationship to the crystal. The crystal drawings in this book are known as *clinographic projections* and are a type of perspective drawing which yields a portrait-like picture of the crystal in two dimensions.

FIG. 2.40. Two-circle reflecting goniometer with crystal mounted at the right, a telescope at the upper left, and a light source at the lower left. (Courtesy of Huber Diffraktionstechnik, X-ray Crystallography Equipment, Rimstig, Germany.) Compare with Fig. 1.5d.

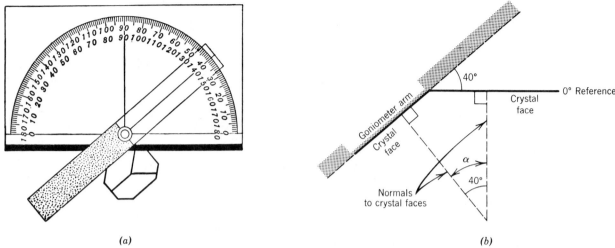

FIG. 2.41. Contact goniometer. (b) Schematic enlargement of (a) showing the measurement of the internal angle α.

Spherical Projection

This is the best means of conveying the appearance of a crystal and generally serves much better for the purpose than a photograph.

Spherical Projection

Because the actual size and shape of the different faces on a crystal are chiefly the result of accidents of growth, we wish to minimize this aspect of crystals in projecting, but at the same time emphasize the angular relation of the faces. We can do this using the *spherical projection*. We can envisage the construction of such a projection in the following way (see Fig. 2.42). Imagine a hollow model of a crystal containing a bright point source of light. Now let us place this model within a large hollow sphere of translucent material in such a way that the light source is at the center of the sphere. If we now make a pinhole in each of the faces so that the ray of light emerging from the hole is perpendicular to the face, these rays of light will fall on the inner surface of the sphere and make bright spots. The whole situation resembles a planetarium in which the crystal model with its internal light source and pinholes is the projector and the translucent sphere is the dome. If we now mark on the sphere the position of each spot of light, we may remove the model and have a permanent record of its faces. Each of the crystal faces is represented on the sphere by a point called the face *pole*. This is the spherical projection.

The position of each pole and thus its angular relationship to other poles can be described using angular coordinates on the sphere. This is done in a manner similar to the location of points on the earth's surface by means of longitude and latitude. For example, the angular coordinates 74°00′ west longitude, 40°45′ north latitude, locate a point in New York City. This means that the angle, measured at the center of the earth, between the plane of the equator and a line from the center of the earth through that point in New York is exactly 40°45′; and the angle between the Greenwich meridian and the meridian passing through the point in New York, measured west in the plane of the equator, is exactly 74°00′. These relations are shown in Fig. 2.43.

A similar system may be used to locate the poles of faces on the spherical projection of a crystal. There is one major difference between locating points on a spherical projection and locating points on the earth's surface. On the earth, the latitude is measured in degrees north or south from the equator, whereas the angle used in the spherical projection is the colatitude, or *polar angle*, which is measured in degrees from the north pole. The north pole of a crystal projection is thus at a colatitude of 0°, the equator at 90°. The colatitude of New York City is 49°15′. This angle is designated in crystallography by the Greek letter ρ (rho).

The "crystal longitude" of the pole of a face on the spherical projection is measured, just as are longitudes on earth, in degrees up to 180°, clockwise and counterclockwise from a starting meridian analogous to the Greenwich meridian of geography. To locate this starting meridian, the crystal is oriented in conventional manner with the (010) face to the right of the crystal. The meridian passing through the pole of this face is taken as zero. Thus, to determine the crystal longitude of any crystal face, a meridian is passed

FIG. 2.42. Spherical projection of some isometric forms.

FIG. 2.43. Latitude and longitude of New York City.

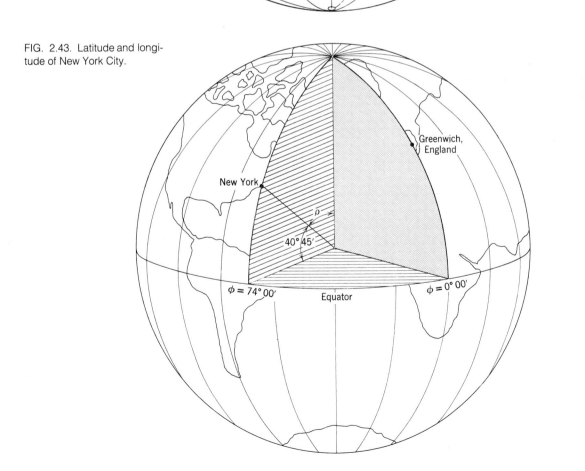

through the pole of that face, and the angle between it and the zero meridian is measured in the plane of the equator. This angle is designated by the Greek letter ϕ (phi).

If any plane is passed through a sphere, it will intersect the surface of the sphere in a circle. The circles of maximum diameter are those formed by planes passing through the center and having a diameter equal to the diameter of the sphere. These are called *great circles*. All other circles formed by passing planes through the sphere are *small circles*. The meridians on the earth are great circles, as is the equator, whereas the parallels of latitude are small circles.

The spherical projection of a crystal brings out interesting zonal relationships, for the poles of all the faces in a zone lie along a great circle of the projection. In Fig. 2.42 (spherical projection) faces (001), (101), (100), (10$\bar{1}$), and (00$\bar{1}$) lie in a zone with the zone axis [010]. Because the great circle along which the poles of these faces lie passes through the north and south poles of the projection, it is called a *vertical great circle*. The zone axis is always perpendicular to the plane containing the face poles; thus all vertical circles have horizontal zone axes.

Stereographic Projection

In reducing the spherical projection of a crystal to a plane surface, it is important that the angular relations of the faces be preserved in such a way as to reveal the true symmetry. This can best be done with the *stereographic projection*.

The stereographic projection is a representation in a plane of half of the spherical projection, usually the northern hemisphere. The plane of the projection is the equatorial plane of the sphere, and the *primitive* circle (the circle outlining the projection) is the equator itself. If one were to view the poles of crystal faces located in the northern hemisphere of the spherical projection with the eye at the south pole, the intersection of the lines of sight with the equatorial plane would be corresponding points representing the poles on the stereographic projection. We can thus construct a stereographic projection by drawing lines from the south pole to the face poles in the northern hemisphere. The points corresponding to these poles on the stereographic projection are located where these lines intersect the equatorial plane. The relationship of the two projections is illustrated in Fig. 2.44.

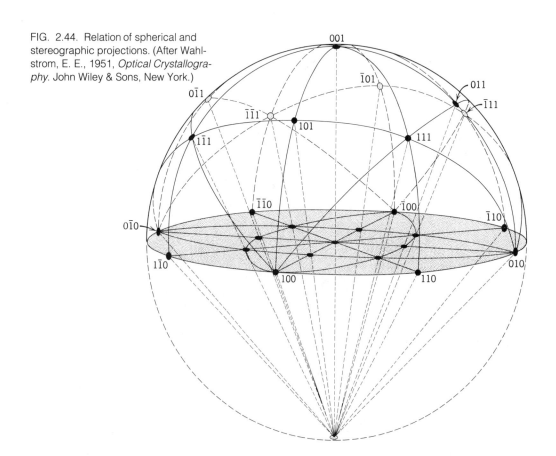

FIG. 2.44. Relation of spherical and stereographic projections. (After Wahlstrom, E. E., 1951, *Optical Crystallography.* John Wiley & Sons, New York.)

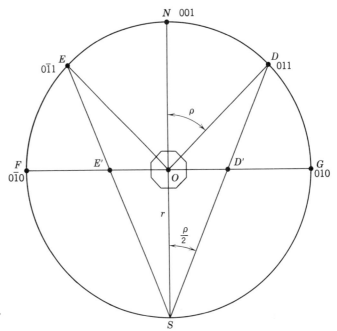

FIG. 2.45. Section through sphere of projection showing relation of spherical to stereographic poles.

In practice one plots poles directly on the stereographic projection. It is, therefore, necessary to determine stereographic distances in relation to angles of the spherical projection. Figure 2.45 shows a vertical section through the spherical projection of a crystal in the plane of the "zero meridian," that is, the plane containing the pole of (010). N and S are, respectively, the north and south poles of the sphere of projection, O is the center of the projected crystal. Consider the face (011). OD is the perpendicular to the face (011), and D is the pole of this face on the spherical projection. The line from the south pole, SD, intersects the trace of the plane of the equator, FG, in the point D', the stereographic pole of (011). The angle NOD will be recognized as the angle ρ (rho). In order to plot D' directly on the stereographic projection, it is necessary to determine the distance OD' in terms of angle ρ. Because ΔSOD is isosceles, $\angle ODS = \angle OSD$. $\angle ODS + \angle OSD = \angle NOD = \rho$. Therefore, $\angle OSD = \rho/2$. $OS = r$, the radius of the primitive of the projection.

$$\tan \rho/2 = OD'/r \quad \text{or} \quad OD' = r \tan \rho/2$$

To sum up, in order to find the stereographically projected distance from the center of the projection of the pole of any face, find the natural tangent of one-half of ρ of that face and multiply by the radius of the projection. The distance so obtained will be in whatever units are used to measure the radius of the primitive circle of the projection.

In addition to determining the distance a pole should lie from the center of the projection, it is also necessary to determine its "longitude" or ϕ (phi) angle. Because the angle is measured in the plane of the equator, which is also the plane of the stereographic projection, it may be laid off directly on the primitive circle by means of a circular protractor. It is first necessary to fix the "zero meridian" by marking a point on the primitive circle to represent the pole of (010). A straight line drawn through this point and the center of the projection is the zero meridian. With the protractor edge along this line and the center point at the center of the projection, the ϕ angle can be marked off. On a construction line from the center of the projection through this point lie all possible face poles having the specified ϕ angle. Positive ϕ angles are laid off clockwise from (010); negative ϕ angles are laid off counterclockwise, as shown in Fig. 2.46.

In order to plot the pole of the face having this given ϕ value, it is necessary to find the natural tangent of one-half of ρ, to multiply by the radius of the projection, and to lay off the resulting distance along the ϕ line. Although any projection radius may be chosen, one of 10 cm is usually used. This is large enough to give accuracy but not be unwieldy, and at the same time simplifies the calculation. With a 10 cm radius, it is only necessary to look up the natural tangent, move the decimal point one place to the right, and plot the result as centimeters from the center of the projection.

When the poles of crystal faces are plotted stereographically as explained above, their symmetry of arrangement should be apparent (see Fig. 2.47). We have seen (page 56) that a great circle in the spherical

FIG. 2.46. Stereographic projection of crystal faces.

projection is the locus of poles of faces lying in a crystal zone. When projected stereographically, vertical great circles become diameters of the projection; all other great circles project as circular arcs that subtend a diameter. The limiting case of such great circles is the primitive of the projection, which is a great circle common to both the spherical and stereographic projections. The poles of vertical crystal faces lie on the primitive and thus are projected without angular distortion.

FIG. 2.47. Stereographic projection of some isometric crystal faces.

Stereographic Net

Both the measurement and plotting of angles on the stereographic projection are greatly facilitated by means of a stereographic net. A net with radius of 5 cm is shown in Fig. 2.48 and a net with radius of 10 cm is given on the inside of the back cover. This type of net is also called the Wulff net, named after G. V. Wulff, Russian crystallographer (1863–1925). Both great and small circles are drawn on the net at intervals of 1° or 2°. In Fig. 2.48 the intervals are 2°.

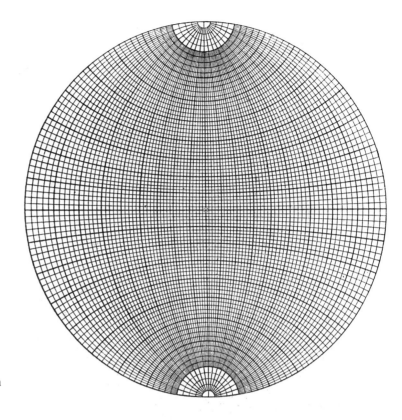

FIG. 2.48. Stereographic (Wulff) net. Radius equals 5 centimeters. See inside back cover for a stereographic net of 10 centimeter radius.

In practice the use of a net with a 10 cm radius, mounted on a stiff backing, is most convenient. If such a net is not available, it can be constructed by photocopying the net on the back cover and mounting it on thin poster board. If a 5 cm net is considered adequate, Fig. 2.48 can be photocopied and backed by thin mounting board. The exact center of the net can now be pierced by a thumbtack from the back. The sharp point of the thumbtack, at the center of the stereographic net, will function as the pivot about which a sheet of tracing paper (or an $8\frac{1}{2} \times 11$ in. sheet of onionskin paper) can be rotated. Figure 2.49 shows this superposition of a semitransparent sheet on top of the stereographic net. The user should trace the outer (primitive) circle onto the overlay and should also mark the E–W and N–S directions. The eastern end of the E–W line should be marked $\varphi = 0°$, the southern end of the N–S line as $\varphi = 90°$, and the northern end of this same line as $\varphi = -90°$. Now the φ angles can be plotted directly along the primitive circle in clockwise (+) or counterclockwise (−) directions. The angle ρ can be located directly along the two vertical great circles (the N–S and E–W directions that intersect in the center of the projection); only along these two directions are graduations available for the direct plotting of angles.

The ρ angle of any face that projects at the center of the projection (at the location of the thumbtack) is equal to 0°. Any face that lies on the outside perimeter of the primitive circle has a ρ angle $= 90°$. Therefore any ρ angle between 0° and 90° is plotted outward from the center of the projection (away from the location of the thumbtack) along the E–W or N–S directions. If a combination of φ and ρ angles is such that the φ angle is neither 0° nor 90°, and the ρ is also neither 0° nor 90°, the transparent overlay must be rotated about the center until the φ direction of the specific plane coincides with either the N–S or E–W direction. Only then can the ρ angle of that plane be plotted. For example (see Fig. 2.49), if a plane has a $\varphi = 30°$ and a $\rho = 60°$, the location of the φ angle of the pole to the plane is marked by a short line at $\varphi = 30°$ on the circumference; subsequently this $\varphi = 30°$ location is rotated to be coincident with the E–W or N–S lines, and the $\rho = 60°$ is plotted directly using the graduations available on the underlying stereographic net. This is the location of the face pole, of a face with $\varphi = 30°$, $\rho = 60°$.

Instead of measuring φ and ρ angles the beginning student usually measures interfacial angles. Such interfacial angles can often be expressed as angles with respect to (010) or (001) of the crystal (if indeed the crystal shows the presence of these two forms). The measured interfacial angles, or the φ and ρ angles available in reference works in the literature, can be plotted easily with the aid of a stereographic net. We

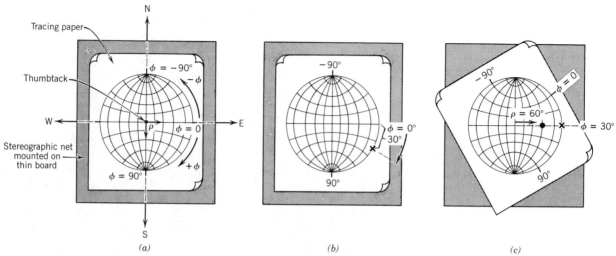

FIG. 2.49. *(a)* Illustration of the use of a stereographic net (mounted on thin board), pierced in its center by a thumbtack from the back, and overlain by a somewhat transparent paper (tracing paper or onionskin paper). The primitive circle, as well as the locations of $\varphi = 0°$, $\varphi = 90°$, and $\varphi = -90°$, must always be marked on the transparent overlay before any angles are plotted. *(b)* To project the pole of a plane with $\varphi = 30°$ and $\rho = 60°$, the angle $\varphi = 30°$ is plotted (x) on the primitive in a clockwise direction from $\varphi = 0°$. *(c)* The direction of $\varphi = 30°$ has been rotated to coincide with the E–W direction and the angle $\varphi = 60°$ can be measured directly along the vertical great circle. The black dot is the pole of the crystal face with $\varphi = 30°$, $\rho = 60°$.

will illustrate the plotting of the face poles for two types of crystals: (1) for an orthorhombic crystal in which the three axial directions (*a, b,* and *c*) are at 90° to each other, and (2) for a monoclinic crystal where two of the three crystallographic directions are non-orthogonal (not 90°).

Projection of an Orthorhombic Crystal

Figure 2.50*a* is a crystal drawing of the orthorhombic mineral anglesite, $PbSO_4$, and Fig. 2.50*b* lists the φ and ρ angles of the crystal faces that lie in the positive quadrant. In Fig. 2.50*c* we illustrate the process of locating the poles of these faces on the stereographic projection. The starting point, as in all projections, is (010), face *b*. The pole of this face should be placed on the primitive at 0°. The interfacial angles $b \wedge n = 32\frac{1}{2}°$ and $b \wedge m = 52°$, can be measured and plotted as angles on the primitive. Face *c* is (001); it makes an angle of 90° with *b* and its pole should be placed at the center of the projection. Face *o* is in a zone with faces *c* and *b* and thus has a φ angle of 0°. Its ρ angle, $c \wedge o = 52°$, can be measured directly and plotted along the vertical great circle. Face *d* lies in a vertical zone at 90° to the zone *c, o, b*. It thus has $\varphi = 90°$ and $c \wedge d = 39\frac{1}{4}°$ can be plotted along the vertical circle of the net. The pole of face *y* cannot be plotted directly, but the angles $b \wedge y = 45°$ and $c \wedge y = 57°$ can be measured. To locate this pole the projection is rotated 90° (to the N–S direction) so that *b* lies along

the radii of the small circles of the net, and a tracing of the 45° circle is made. This small circle is the locus of all poles 45° from *b*. The locus of $c \wedge y = 57°$ is a circle centered on the pole of 001, which is the center of the stereographic projection where the thumbtack pierces the paper. This circle is one of colatitude (as explained earlier under spherical projection) and can be most easily drawn with a compass, after the spread of the compass has been measured along, for example, the E–W direction of the stereographic net at $\rho = 57°$. Where this colatitude circle intersects the small circle of $b \wedge y = 45°$ is the location of the pole of *y*.

To check this position, measure the angle $m \wedge y = 38°$. This is most easily done by locating the pole to *m* at the $\varphi = 90°$ position (southern end) of the stereographic net and reading the angular difference between *m* and *y* along one of the inclined great circles. Indeed one can trace a small circle with *m* at $\varphi = 90°$, which will intersect the pole of *y*. Now that the pole *y* is located its φ and ρ angles can be read directly from the stereographic net. The φ value will read $32\frac{1}{2}°$ and the ρ value will be 57°.

In Fig. 2.50*d* we have plotted *all* the faces of the crystal of anglesite shown in Fig. 2.50*b*, not just those in the upper, positive quadrant of the stereographic projection as in Fig. 2.50*c*. The distribution of these face poles is consistent with the combination of symmetry elements $2/m2/m2/m$. There are three 2-fold axes at 90° to each other as well as three mirror planes

	φ	ρ
c(001)	— — —	0°
b(010)	0°	90°
n(110)	32½°	90°
m(210)	52°	90°
o(011)	0°	52°
d(101)	90°	39½°
y(111)		57°
y(111)	angle b∧y = 45°	

(a)

(b)

2/m 2/m 2/m

(c)

(d)

FIG. 2.50. *(a)* A crystal of the orthorhombic mineral anglesite with symmetry 2/m2/m2/m. *(b)* Listing of φ and ρ angles, as reported in the literature, and originally measured by reflecting goniometer for the faces of the crystal of anglesite shown in *(a)*. *(c)* A stereographic projection of all of the faces, in the positive quadrant, of the mineral anglesite. *(d)* The location of all of the face poles, with their Miller indices, as well as the symmetry elements consistent with the location of these face-poles. A horizontal cross section of the crystal in *(a)* is shown as well.

that intersect at the center of the stereographic projection. Two of these mirrors stand vertically as shown along the N–S and E–W directions and the third mirror is parallel to the page, in the plane of the primitive circle.

Projection of a Monoclinic Crystal

The location of the crystallographic axes in a monoclinic crystal is somewhat less straightforward than in the orthorhombic case, described above. In this text we locate the 2-fold rotation axis in monoclinic crystals in a horizontal, east-west direction. This orientation is referred to as the "second setting" (in the "first setting" the 2-fold axis is vertical). This means that a possible mirror plane (as in 2/m), which is perpendicular to the 2-fold axis, stands vertically. If the mon-

oclinic crystal shows an elongated habit, the direction of elongation is commonly chosen as parallel to the c axis. Further, if there is a prominent sloping plane, such as c in Fig. 2.51a, the a axis is taken as parallel to it and at right angles to b.

The monoclinic crystal shown in Fig. 2.51a is that of diopside, $CaMgSi_2O_6$. It is oriented as described above with the mirror plane vertical, the 2-fold axis parallel to the b axis, and horizontal (in an E–W direction). As shown in a cross section in Fig. 2.51c, the c axis is parallel to the elongation of the crystal, and the a axis is parallel to the sloping c faces (001) and (00$\bar{1}$). In Fig. 2.51b there is a listing of the pertinent interfacial angles measured on the crystal shown in Fig. 2.51a. The interfacial angle c ∧ a ((001) ∧ (100)) = 74° is shown in the cross section of the crystal, as

Interfacial Angles

ca, $(001) \wedge (100) = 74°$
mm, $(110) \wedge (1\bar{1}0) = 93°$
cu, $(001) \wedge (111) = 34°$
uu, $(111) \wedge (1\bar{1}1) = 48°$
cm, $(001) \wedge (110) = 79°$

(a)

(b)

(c)

(d)

(e)

(f)

(g)

(h)

2/m

FIG. 2.51. Plotting of face poles for a monoclinic crystal. *(a)* A crystal of the monoclinic mineral diopside. *(b)* Interfacial angles, as reported in the literature and originally measured by reflecting goniometer, for the faces of the monoclinic crystal of diopside, shown in *(a)*. *(c)* A vertical cross section through the diopside crystal showing the location of the *c* axis (parallel to the elongation of the crystal) and of the *a* axis (parallel to the (001) face). The β angle of 106° and the interfacial angle $c \wedge a = 74°$ (between (001) and (100)) are shown. *(d)* The upper half of the diopside crystal centered within the sphere of the spherical projection. Note the location of the piercing points of the +*c* and +*a* axes as well as the poles of *c* (001), *a* (100) and the two sets of faces marked *m* ((110) and (1$\bar{1}$0)) and *u* ((111) and 1$\bar{1}$1)). *(e)* A cross section through the sphere in *(d)*;

this section contains the *a* and *c* axes and the poles of (100) and (001). It should be noted that the (+) end of the *a* axis lies below the equatorial plane of the stereographic projection (the primitive circle) and that the (−) end of *a* axis lies above it. Therefore in projecting the piercing points of −*a* and +*a* from the sphere onto the stereographic projection, the "end points" of *a* move inward along the N–S direction (great circle) by 16° each. The −*a* end projects from above (as a solid dot) and the +*a* end projects from below (as an open circle). *(f)* Plotting of the poles of *b* ((010) at φ = 0°), *a* at φ = 90°, the two faces *m* with an interfacial angle = 93° (symmetrically on either side of *a*), and *c* (001) with an interfacial angle with *a* = 74°. The *c* axis is at the center of the projection and the −end of the *a* axis lies 16° inward from the pole to (100).

62

is the one nonorthogonal angle (β) between the three crystallographic axes *a*, *b*, and *c*. In descriptions of monoclinic minerals in this book, the angle between the *a* and *c* crystallographic axes, known as the β angle, is taken as the angle that is not equal to 90°. In this crystal the β angle is 106°. Because of the presence of this nonorthogonal angle between the two axial directions of *a* and *c*, the location of one of these two axes will be different from what we have seen in the stereographic projection of, for example, the orthorhombic crystal in Fig. 2.50. In order to clarify this, we have drawn the upper part of the diopside crystal as surrounded by the sphere of the spherical projection in Fig. 2.51*d*. This figure shows that the pole to (001) is not coincident with the piercing point of the (+) end of the *c* axis; similarly the pole of (100) is not coincident with the piercing point of the (+) end of the *a* axis.

A cross section parallel to the *a–c* plane of the crystal in the spherical projection is shown in Fig. 2.51*e*. The angular relations shown here are identical to those shown in Fig. 2.51*c*. In this cross section it is important to note the inclined position of the *a* axis. Only the (−) end of the *a* axis is above the trace of the stereographic projection (horizontal diameter of figure); however, the (+) end of the *a* axis lies below it. When the ends of the *a* axis are projected onto the stereographic projection, the (−) end will project as indicated in Fig. 2.51*e*. Its piercing point will be located 16° in from the primitive circle; its projected location is shown as an opaque dot in Fig. 2.51*f*. The (+) end of the *a* axis lies below the trace of the stereographic projection and must be projected upward. The point where it pierces the spherical projection is indicated as an open circle 16° in from the primitive. This is also shown in Fig. 2.51*f*. Once the *a* and *c* axes are located, the various face poles can be plotted. All those in the vertical zone (// *c* axis) can be located on the perimeter of the stereographic net; these are *b*, *m* and *a*.

The interfacial angle *c* ∧ *a* ((001) ∧ (100) = 74°) implies that the pole of the (001) face is 16° away from the piercing point of the *c* axis, in a southward direction along the N–S vertical great circle. This is also shown in Fig. 2.51*f*. The only poles still to be located are for the two inclined faces *u* which are symmetrically oriented with respect to the *a–c* plane (the mirror plane). The interfacial angle *c* ∧ *u* = 34°. This angle can be plotted with respect to *c* (001) after the *c* pole has been moved to the E–W vertical great circle. The 34° angle between *c* and *u* can now be plotted because these two faces lie in the same zone; that is, they both lie on the same great circle. In this case the great circle passing through *c* and *u* is inclined by 16° from the vertical. Two such inclined great circles are shown with solid lines in Fig. 2.51*f*. Once both poles of the two *u* faces in the upper hemisphere are plotted the *u* ∧ *u* interfacial angle (as given in Fig. 2.51*a*) can be verified. The faces *b*, *u*, *u*, *b* lie in the same zone. As such the poles to these four faces lie on the same (inclined) great circle. On this circle (solid in Fig. 2.51*g*) the interfacial angle of 48° can be measured. The interfacial angle *c* ∧ *m* = 79° can be verified in a similar manner: the faces *m*, *c*, *m* lie in the same zone; as such their poles lie on the same inclined great circle.

In Fig. 2.51*h* we have plotted *all* of the face poles of the diopside crystal, and have assigned a Miller index to each pole. Furthermore, the symmetry elements compatible with the distribution of the face poles are also shown; this turns out to be 2/*m* with the 2-fold axis horizontal in an E–W direction and the mirror plane in a vertical position perpendicular to the 2-fold axis.

THE THIRTY-TWO CRYSTAL CLASSES

In the following section the 32 classes listed in Table 2.4 are described under the crystal systems in which they are grouped. The crystal systems will be treated in order of increasing symmetry. Within each system, however, the classes will be discussed in order of decreasing symmetry. *The symmetry of each class is given by the Hermann-Mauguin notation but it is also shown by means of stereograms, giving projections of all the faces of the general forms. These are the forms from which the classes derive their names.* In the stereograms it is necessary to show faces in the southern hemisphere as well as in the northern hemisphere in order to give completely the symmetry of the class. This is done by superimposing stereographic projections of the two hemispheres, with the poles in the

---

The interfacial angle *c* ∧ *u* ((001) ∧ (111)) is listed as 34° in *(b)*. With *c* and *u* on the same great circle this 34° angle can be measured, symmetrically on either side of *c* (001). The projection shows that *m*, *u*, *c* lie on the same great circle and therefore in the same zone. *(g)* Here all the face poles of the crystal in *(a)* are plotted. Those marked *u* and shown as black dots are for face poles in the upper part of the sphere (now projected onto the primitive circle), and those shown as open circles are for poles projected upward from the lower part of the sphere. The interfacial angles *u* ∧ *u* ((111) ∧ (11$\bar{1}$)), and *c* ∧ *m* ((001) ∧ (110)) are shown along great circles as 48° and 79°, respectively (see listing in *b*). *(h)* The location of all of the face poles and their Miller indices and the symmetry consistent with the distribution of the face poles.

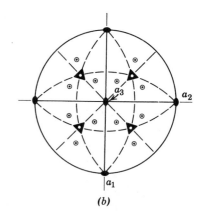

FIG. 2.52. Crystal with symmetry $2/m\bar{3}$. The stereo-gram shows a mirror plane at right angles to each of the 2-fold rotation axes, and four 3-fold axes of rotoinversion. The solid primitive circle denotes a horizontal mirror plane; faces at the bottom of the crystal lie directly below those at the top.

(a)

(b)

northern hemisphere represented by solid points and those in the southern hemisphere by open circles. Thus, if two poles lie directly one above the other on the sphere, they will be represented by a solid point surrounded by a circle. A vertical face is represented by a single point on the primitive, for, although such a pole would appear on projections of both top and bottom of the crystal, it represents but one face.

Figure 2.52a is a drawing of a crystal with a hor-izontal symmetry plane. The stereogram of this crys-tal, Fig. 2.52b, consequently, has for all faces solid points circled to indicate corresponding faces at the top and bottom of the crystal. Figure 2.53a is a draw-ing of a crystal lacking a horizontal mirror plane. Its stereogram, Fig. 2.53b, has 12 solid points as poles of faces in the northern hemisphere and an additional 12, independent circles as poles of faces in the south-ern hemisphere.

The number of minerals (and synthetic crystalline compounds) falling within each of the six crystal sys-tems is highly variable. A listing of 3837 crystalline compounds and their distribution among the 32 crys-tal classes is given in Bloss (1971, p. 28). For these compounds the distribution is as follows:

Triclinic	2%
Monoclinic	21
Orthorhombic	20
Tetragonal	12
Hexagonal	19
Isometric	26

Within these systems the largest number of mineral species is concentrated in the crystal class with the highest symmetry of each crystal system. Such classes, with the highest symmetry in a crystal system, are called the *holohedral* classes, from the Greek *holos* meaning "whole" or "complete." These are: $\bar{1}$ in tri-clinic, $2/m$ in monoclinic, $2/m2/m2/m$ in ortho-rhombic, $4/m2/m2/m$ in tetragonal, $\bar{3}2/m$ in the hexagonal system (when based on a rhombohedral lattice), $6/m2/m2/m$ in the hexagonal system (when based on a hexagonal lattice), and $4/m\bar{3}2/m$ in the isometric system.

In the following systematic treatment of the 32 crystal classes, stereographic illustrations of the var-ious crystal forms and their accompanying symmetry elements are extensively used. In these the same graphic symbols for symmetry elements and crystal-

FIG. 2.53. Crystal with symmetry $\bar{4}3m$. Stereogram shows three 4-fold axes of rotoinversion, six mirror planes, and four 3-fold rotation axes. The broken primitive circle indi-cates lack of a horizontal mirror plane and faces at the top of the crystal do not lie above those at the bottom.

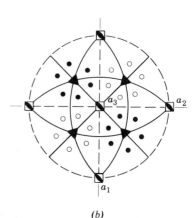

(a)

(b)

lographic directions are used throughout. A listing of these graphic symbols is given in Table 2.7. These are the standard symbols used in the *International Tables for X-ray Crystallography* (1969).

Throughout our discussion of the 32 crystal classes, as well as in subsequent chapters of this book, we will use the unabbreviated notation in describing their symmetry content. For example, we will use the symmetry representation $2/m2/m2/m$ for the orthorhombic dipyramidal class. Crystallographers commonly use abbreviated symbols, and would represent $2/m2/m2/m$ by simply *mmm*. Similarly abbreviated

Hermann-Mauguin notations are given in Table 2.8 for the six crystal classes in which it is possible to choose between a lengthier (and more descriptive) and an abbreviated notation. The reason for the use of the short forms is one of brevity. To the crystallographer *mmm* means three mutually perpendicular mirror planes that are perpendicular to the *a*, *b*, and *c* axes of the orthorhombic system, respectively. The lines of intersection between the three sets of intersecting mirrors turn out to be axes of 2-fold rotation (see Figs. 2.19 and 2.65); in other words, *mmm* implies the same symmetry content as $2/m2/m2/m$.

Table 2.7 GRAPHIC SYMBOLS USED IN STEREOGRAPHIC ILLUSTRATIONS

Axes of rotation	Written symbol	Graphic symbol
1-fold	1	none
2-fold	2	◗
3-fold	3	▲
4-fold	4	■
6-fold	6	⬢
Axes of rotoinversion		
1-fold	$\bar{1}$	(equivalent to a center of symmetry or *i*, inversion)*
2-fold	$\bar{2}$	(equivalent to a mirror, *m*)
3-fold	$\bar{3}$	◮
4-fold	$\bar{4}$	◈
6-fold	$\bar{6}$	⬡
Center of symmetry	*i*	none (see footnote)*
Mirror planes	*m* (horizontal, parallel to the plane of the page)	⬤ *m*
		solid line along primitive circle.
	m (vertical, perpendicular to the plane of the page)	*m* *m* —Both solid lines
	m absent, in both horizontal and vertical positions	
Crystallographic directions	*a, b, c*	- - - - *a, b,* or *c* broken lines labeled with the appropriate letter
		If *m* includes a crystallographic direction, a solid line is used—with *a, b,* or *c*

*A center of symmetry is shown in writing by the letter i for inversion, which is equivalent to $\bar{1}$. If a symmetry center occurs at the center of the sphere of projection, its presence is not shown by a symbol on the stereogram, but it can be detected from the arrangement of poles of equivalent faces (see also Fig. 2.21).

Table 2.8
LISTING OF THE EQUIVALENT LONG AND ABBREVIATED FORMS OF SIX CRYSTAL CLASSES

System	Long Form	Short Form
Orthorhombic	$2/m2/m2/m$	mmm
Tetragonal	$4/m2/m2/m$	$4/mmm$
Hexagonal	$\bar{3}2/m$	$\bar{3}m$
	$6/m2/m2/m$	$6/mmm$
Isometric	$2/m\bar{3}$	$m3$
	$4/m\bar{3}2/m$	$m3m$

Similar reasoning holds for the elimination of the 2-fold rotational symbols in the short forms $4/mmm$ and $6/mmm$.

In the case of $\bar{3}2/m$ the combination of the 3-fold rotoinversion axis and three vertical mirror planes (see Fig. 2.91) generates three 2-fold axes that bisect the angles between the mirrors. As such the 2-fold axes in $\bar{3}2/m$ are the inherent result of the symmetry implied in the short form $\bar{3}m$. The short forms for the isometric symbols are less obvious. The short symbol $m3m$ implies (1) a mirror plane parallel to the cube face (for the first m in $m3m$), (2) a 3-fold axis that runs from corner to corner in the cube (see Fig. 2.101), and (3) a symmetry plane that runs diagonally across a cube, from edge to edge (referring to the last m in $m3m$). The interaction of these three symmetry elements will produce all of the symmetry content of the more descriptive form $4/m\bar{3}2/m$. If the diagonal mirror plane were absent from $m3m$, the overall symmetry would read only $m3$. These two symmetry elements combine to generate the symmetry of the nonabbreviated form $2/m\bar{3}$.

In the following systematic treatment of the six crystal systems and the 32 crystal classes, the introductory paragraphs to each crystal system will address (among other matters) the relationship of the symmetry elements (as expressed by the Hermann-Mauguin notation) and the orientation of the crystallographic axes. Table 2.9 will provide the student with a synopsis of the conventions in such orientations (see also Fig. 2.27).

In the description of forms on the following pages the geometrically perfect model of the form is considered in each case. It should be kept in mind that in nature this ideal is rarely obtained, and that crystals not only are frequently malformed but also are usually bounded by a combination of forms. In the laboratory study of crystal forms wooden models may be used that similarly represent highly idealized crystals: their symmetry-related faces have been made equal in size and shape by the manufacturer of the models. Such perfection aids the student in the study of symmetry content and the recognition of forms.

Throughout this book the drawings of the external forms of crystals are in a standardized orientation with the c axis vertical, the b axis east-west, and the a axis toward the observer. Because of this standardized orientation the position of the axes has been omitted in most crystal drawings.

Triclinic System

Crystallographic Axes. In the triclinic system the crystal forms are referred to three crystallographic axes of unequal lengths that make oblique angles with each other (Fig. 2.54). The three rules to follow in orienting a triclinic crystal and thus in determining the position of the crystallographic axes are: (1) The most pronounced zone should be vertical. The axis of this zone then becomes c; (2) {001} should slope forward and to the right; (3) two forms in the vertical zone should be selected: one as {100} the other as {010}. The directions of the a and b axes are determined by the intersections of {010} and {100}, respectively, with {001}. The b axis should be longer than the a axis. In reporting on the crystallography of a new triclinic mineral or one that has not been recorded in the literature, the convention should be followed that $c < a < b$. The relative lengths of the three axes and the angles between them can be established only by X-ray diffraction techniques. The angles between the positive ends of b and c, c and a, and a and b are designated respectively as α, β, and γ (see Fig. 2.54).

FIG. 2.54. Triclinic crystal axes.

$$\bar{1}$$

Symmetry—i. The symmetry consists of a 1-fold axis of rotoinversion, which is equivalent to the center of symmetry, or inversion (i). Figure 2.55 illustrates a triclinic pinacoid (or parallelohedron) and its stereogram. This class is referred to as the *pinacoidal class* after its general {hkl} form.

Forms. All the forms are *pinacoids* (*parallelohedrons*) and thus consist of two identical and parallel faces. Once a crystal is oriented, the Miller indices of the pinacoid establish its position.

Table 2.9

CHARACTERISTIC SYMMETRY, AND RELATIONSHIPS BETWEEN CRYSTAL AXES AND SYMMETRY NOTATION OF CRYSTAL SYSTEMS

Crystal Class	System	Characteristic Symmetry	Hermann-Mauguin Notation
$1, \bar{1}$	Triclinic	1-fold (inversion or identity) symmetry only	Because of low symmetry, no crystallographic constraints.
$2, m, 2/m$	Monoclinic	one 2-fold rotation axis and/or one mirror	The 2-fold axis is taken as the b axis, and the mirror (the a–c plane) is vertical (second setting).
222, $mm2$ $2/m2/m2/m$	Orthorhombic	three mutually perpendicular directions about which there is binary symmetry (2 or m)	The symbols refer to the symmetry elements in the order a, b, c; 2-fold axes coincide with the crystallographic axes.
$4, \bar{4}, 4/m$ 422, $4mm$, $\bar{4}2m$, $4/m2/m2/m$	Tetragonal	one 4-fold axis	The 4-fold axis refers to the c axis; the second symbol (if present) refers to the axial directions (a_1 and a_2); the third symbol (if present) to directions at 45° to a_1 and a_2.
$6, \bar{6}, 6/m$ $622, 6mm$ $\bar{6}m2, 6/m2/m2/m$ $3, \bar{3}, 32,$ $3m, \bar{3}2/m$	Hexagonal*	one 6-fold axis one 3-fold axis	The first number refers to the c axis; the second and third symbols (if present) refer respectively to symmetry elements parallel to and perpendicular to the crystallographic axes a_1, a_2, and a_3.
$23, 2/m\bar{3},$ $432, \bar{4}3m,$ $4/m\bar{3}2/m$	Isometric	four 3-fold axes each inclined at 54°44′ to the crystallographic axes (see Figs. 2.16 and 2.101)	The first number refers to the three crystallographic axes a_1, a_2, and a_3; the second number refers to four diagonal directions of 3-fold symmetry (between corners of a cube); the third number or symbol (if present) refers to six directions between the edges of a cube (see Fig. 2.101).

*The accepted orientation of the symmetry elements in two crystal classes of the hexagonal system is not straightforward. These are $\bar{6}m2$ and $3m$. The location of the 6- or 3-fold axis is unambiguous. However, the location of the next symmetry element is not obvious. In $\bar{6}m2$, the third entry (2-fold rotation axes) coincides with the perpendiculars to a_1, a_2, and a_3; the m's are coincident with these same directions. In $3m$ the m's are located in directions perpendicular to a_1, a_2, and a_3.

$\bar{1}$

 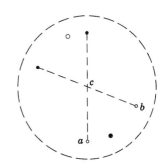

FIG. 2.55. Triclinic pinacoid (or parallelohedron) and stereogram.

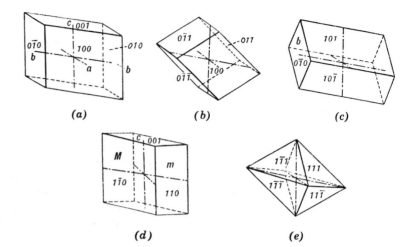

FIG. 2.56. Triclinic pinacoids (or parallelo-hedrons). *(a)* Front {100}, side {010}, and base {001}. *(b)* {011} positive, {0$\bar{1}$1} negative. *(c)* {101} positive, {10$\bar{1}$} negative. *(d)* {110} positive, {1$\bar{1}$0} negative. *(e)* Four different forms.

1. {100}, {010}, and {001} *Pinacoids.* Each of these pinacoids intersects one crystallographic axis and is parallel to the other two. The front or *a* pinacoid, {100}, intersects the *a* axis and is parallel to the other two; the side or *b* pinacoid, {010}, intersects only the *b* axis; the basal or *c* pinacoid, {001}, intersects only the *c* axis.

2. {0*kl*}, {*h0l*}, and {*hk*0} *Pinacoids.* The {0*kl*} form is parallel to *a* and can be positive {0*kl*}, or negative {0$\bar{k}l$}; the {*h0l*} form is parallel to *b*, {*h0l*} positive, and {$\bar{h}0l$} negative; the {*hk*0} form is parallel to *c*, {*hk*0} positive, and {*h$\bar{k}$*0} negative.

3. {*hkl*} *Pinacoids.* {*hkl*} positive right, {*h$\bar{k}$l*} positive left, {*hk$\bar{l}$*} negative right, {$\bar{h}kl$} negative left. Each of these two-faced forms can exist independently of the others.

Various combinations of the above pinacoids are illustrated in Fig. 2.56.

Among the minerals that crystallize in $\bar{1}$ are:

amblygonite	polyhalite
chalcanthite	rhodonite
microcline	turquoise
pectolite	ulexite
plagioclase	wollastonite
feldspars	

Of the above-mentioned minerals microcline, rhodonite, and chalcanthite are fairly common as well-formed crystals (see Fig. 2.57).

1

Symmetry. There is merely a 1-fold rotation axis, which is equivalent to no symmetry. Figure 2.58 illustrates a triclinic *pedion* (or *monohedron*) and its

stereogram. This class is referred to as the *pedial class* after its {*hkl*} form.

Forms. The general form {*hkl*} as well as all other forms are pedions (or monohedrons), and thus each face stands by itself.

A few very rare minerals belong to this class. Axinite was originally classified as 1 but has been found to have symmetry $\bar{1}$.

See Table 2.10 for a listing of forms in the triclinic system.

FIG. 2.57. Triclinic crystals. *(a)* Rhodonite. *(b)* Chalcanthite.

FIG. 2.58. Triclinic pedion (or monohedron) and stereogram.

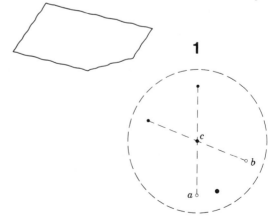

Table 2.10
DISTRIBUTION OF FORMS IN THE TRICLINIC, MONOCLINIC, AND ORTHORHOMBIC SYSTEMS*

No. of Faces	Name of Form	Class								Unique Form for
		1	$\bar{1}$	2	m	2/m	222	2mm	2/m2/m2/m	
1	Pedion	+		+	+			+		
2	Pinacoid		+	+	+	+	+	+	+	
2	Dome ⎫			+	+			+		
2	Sphenoid ⎭									
4	Prism					+	+	+	+	
4	Disphenoid						+			222
4	Pyramid							+		2mm
8	Dipyramid								+	2/m2/m2/m

*From Buerger, M. J., 1956, *Elementary Crystallography.* John Wiley & Sons, New York, 528 pp.

Monoclinic System

Crystallographic Axes. Monoclinic crystals are referred to three axes of unequal lengths. The only restrictions in the angular relations are that $a \wedge b$ (γ) and $c \wedge b$ (α) = 90°. For most crystals the angle between $+a$ and $+c$ is greater than 90°, but in rare instances it too may equal 90° (e.g., staurolite). In such cases the monoclinic symmetry is not apparent from the morphology and the crystal is said to be pseudo-orthorhombic. The 2-fold rotation axis or the direction perpendicular to the mirror plane is usually taken as the *b* axis; the *a* axis is inclined downward toward the front; and *c* is vertical. This orientation, known as the "second setting," is traditional for mineralogists.[3]

Figure 2.59 represents the crystallographic axes of the monoclinic mineral orthoclase, with β = 116°01'. Although the direction of the *b* axis is fixed by symmetry, the directions that serve as the *a* and *c* axes are matters of choice and depend on crystal habit and cleavage. If crystals show an elongated development (prismatic habit) parallel to a direction in the *a* −*c* plane, that direction often serves as the *c* axis. Further, if there is a prominent sloping plane (or planes), such as planes *c* or *r* in the drawings of Fig. 2.62 the *a* axis may be taken as parallel to these. It is quite

[3]Some crystallographers orient monoclinic crystals according to the "first setting" in which the 2-fold axis or the normal to the mirror plane is chosen as the *c* axis instead of the *b* axis. Here, all monoclinic crystals are referred to the "second setting."

FIG. 2.59. Monoclinic crystal axes.

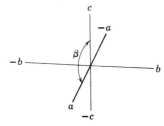

possible that there may be two, or even more, choices that are equally good, but in the description of a new mineral it is conventional to orient the crystals such that $c < a$.

Cleavage is also an important factor in orienting a monoclinic crystal. If there is good pinacoidal cleavage parallel to the *b* axis, as in orthoclase, it is usually taken as the basal cleavage. If there are two equivalent cleavage directions, as in the amphiboles and pyroxenes, they are usually taken to be vertical prismatic cleavages.

2/m

Symmetry—*i*, 1A₂, 1*m*. The axis of 2-fold rotation is chosen as the *b* axis, and the *a* and *c* axes lie in the mirror plane, which is perpendicular to the *b* axis (Fig. 2.60a). The stereogram in Fig. 2.60*b* shows the symmetry of an {*hkl*} prism, the general form. Because the *a* axis slopes down and to the front, it does not lie in the equatorial plane; the positive end intersects the sphere of projection in the southern hemisphere (for further discussion see "Projection of a Monoclinic Crystal," page 61). This class is referred to as the *prismatic class*, because the general form {*hkl*} is a *prism*.

Forms. There are only two types of forms in this monoclinic class: *pinacoids* (*parallelohedrons*) and *prisms*.

1. *Pinacoids.* (See Fig. 2.61.) Front or *a* pinacoid, {100}, side or *b* pinacoid, {010} and basal or *c* pinacoid, {001}. There are also {*h0l*} and {$\bar{h}0l$} pinacoids; these two pinacoids are independent of each other and the presence of one does not necessitate the presence of the other (see Figs. 2.61 and 2.62).

2. *Prisms.* The four-faced prism {*hkl*} is the general form, but {*0kl*} and {*hk0*} are prisms as well. The {*0kl*} prism intersects the *b* and *c* axes and is parallel to the *a* axis. The general form can occur as two inde-

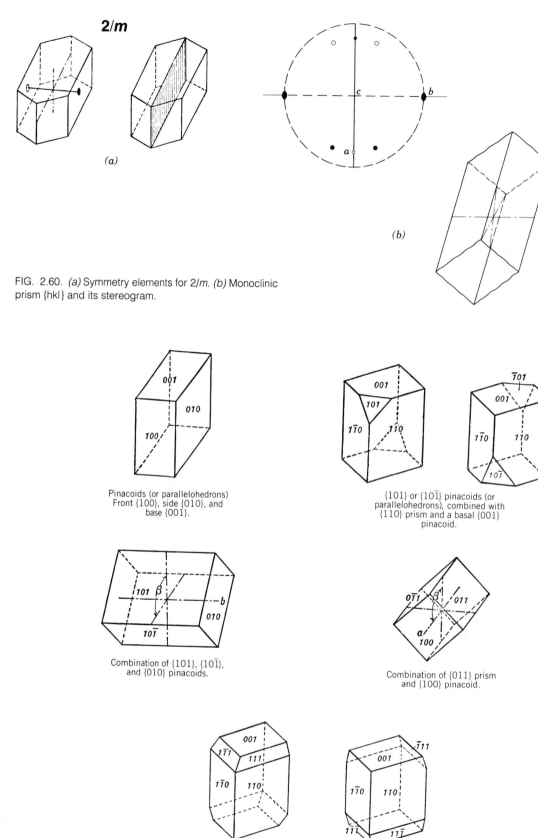

2/m

(a)

(b)

FIG. 2.60. (a) Symmetry elements for 2/m. (b) Monoclinic prism {hkl} and its stereogram.

Pinacoids (or parallelohedrons)
Front {100}, side {010}, and base {001}.

{101} or {10$\bar{1}$} pinacoids (or parallelohedrons), combined with {110} prism and a basal {001} pinacoid.

Combination of {101}, {10$\bar{1}$}, and {010} pinacoids.

Combination of {011} prism and {100} pinacoid.

Combinations of prism {111} (or {$\bar{1}$11}) and prism {110} and basal pinacoid {001}.

FIG. 2.61. Commonly developed forms and form combinations in 2/m.

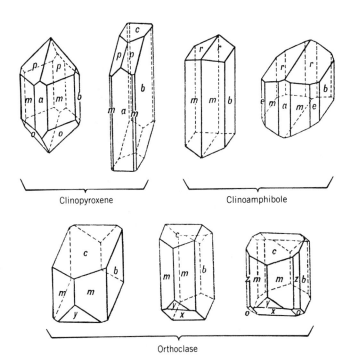

FIG. 2.62. Monoclinic crystals with 2/m symmetry. Forms: a {100}, b {010}, c {001}, m {110}, p {111}, o {$\bar{2}$21}, r {011}, e {120}, x {$\bar{1}$01}, y {$\bar{2}$01}, z {130}.

pendent prisms {hkl} and {$\bar{h}$kl}. Prisms are illustrated in Figs. 2.61 and 2.62.

The only form in this crystal class that is fixed by the choice of the b axis as the 2-fold rotation axis is the {010} pinacoid. The other forms may vary with the choice of the a and c axes. For instance, the {100} pinacoid, {001} pinacoid, and {h0l} pinacoids may be converted into each other by a rotation about the b axis. In the same manner the prisms can be changed from one position to another.

Many minerals crystallize in the monoclinic, prismatic class. Some of the most common are:

azurite	kaolinite
borax	malachite
chlorite	mica (group)
clinoamphibole (group)	orpiment
clinopyroxene (group)	orthoclase
datolite	realgar
epidote	spodumene
gypsum	talc
heulandite	titanite

m

Symmetry—1m. There is one vertical mirror plane (010) that includes the a and c crystallographic axes. The general form {hkl} in this class is a *dome* (*dihedron*). Figure 2.63 shows this form and its stereogram. This class is referred to as the *domatic class.*

Forms. The *dome* is a two-faced form symmetrical across a major plane, in contrast to a sphenoid (Fig. 2.64), which is symmetrical about an axis of 2-fold rotation. There are two possible independent orientations of the dome, {hkl} and {$\bar{h}$kl}. The form {010} is a pinacoid, but all faces lying across the mirror plane such as {100}, {$\bar{1}$00}, {00$\bar{1}$}, and {h0l}, {$\bar{h}$0l} are pedions.

The rare minerals hilgardite, $Ca_2ClB_5O_8(OH)_2$ (Fig. 2.63b) and clinohedrite, $CaZnSiO_3(OH)_2$, crystallize in this class.

2

Symmetry—1A$_2$. The b crystallographic axis is an axis of 2-fold rotation. Figure 2.64 shows the general {hkl} form, a *sphenoid* (*dihedron*), and its stereogram. This class is known as the *sphenoidal class.*

Forms. With the absence of an a–c symmetry plane, the b axis is polar, and different forms are present at opposite ends. The {010} pinacoid of class 2/m becomes two pedions, {010} and {0$\bar{1}$0}. Similarly the {0kl}, {hk0} and {hkl} prisms in 2/m degenerate into pairs of enantiomorphic sphenoids. A *sphenoid* is a two-faced form symmetrical about a 2-fold rotation axis, b, in contrast to a *dome* with symmetry across a mirror plane. The general form, the sphenoid, is enantiomorphic with Miller indices {hkl} and {h$\bar{k}$l}.

Mineral representatives in the sphenoidal class are rare, but chief among them are the members of the halotrichite isostructural group, of which picker-

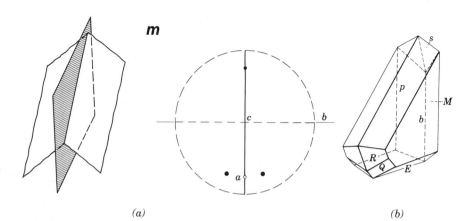

m

FIG. 2.63. *(a)* Monoclinic dome (or dihedron) {*hkl*} and its stereogram. *(b)* The mineral hilgardite with symmetry *m*.

(*a*)

(*b*)

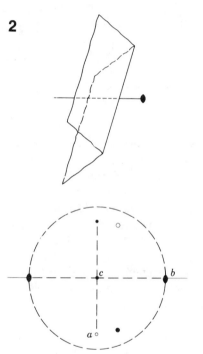

2

FIG. 2.64. Monoclinic sphenoid (or dihedron) {*hkl*} and its stereogram.

ingite, $MgAl_2(SO_4)_4 \cdot 22H_2O$, is the most common member.

See Table 2.10 for a listing of forms in the monoclinic system.

Orthorhombic System

Crystallographic Axes. The crystal forms in the orthorhombic system are referred to three crystallographic axes of unequal length that make angles of 90° with each other (see Fig. 2.65a). The relative lengths of the axes, or the axial ratios, must be determined for each orthorhombic mineral. In orienting an orthorhombic crystal, the convention is to set the crys-

tal such that $c < a < b$. In the past, however, this convention has not necessarily been observed, and it is customary to confirm to the orientation given in the literature. One finds, therefore, that any one of the three axes may have been chosen as *c*. The longer of the other two is then taken as *b* and the shorter as *a*.

The decision as to which of the three axes should be chosen as the vertical axis rests largely on the crystal habit of the mineral. If its crystals commonly show an elongation in one direction, this direction is usually chosen as the *c* axis (see topaz crystals in Fig. 2.66). If, on the other hand, the crystals show a prominent pinacoid and therefore are tabular, this pinacoid is usually taken as {001} with *c* normal to it (see barite and celestite crystals in Fig. 2.66). Cleavage also aids in orienting orthorhombic crystals. If, as in topaz, there is one pinacoidal cleavage, it is taken as {001}. If, as in barite, there are two equivalent cleavage directions, they are set vertical and their intersection edges determine *c*. After the orientation has been determined, the length of the axis chosen as *b* is taken as unity, and the relative lengths of *a* and *c* are given in terms of it. Figure 2.65a represents the crystallographic axes for the orthorhombic system.

In the Hermann-Mauguin notation for the orthorhombic system, the symbols refer to the symmetry elements in the order *a, b, c*. For example, in the class *mm*2, the *a* and *b* axes lie in vertical mirror planes and *c* is an axis of 2-fold rotation.

2/m2/m2/m

Symmetry—*i*, $3A_2$, 3*m*. The three axes of 2-fold rotation coincide with the three crystallographic axes; perpendicular to each of the axes is a mirror plane (Fig. 2.65b). The general form, rhombic dipyramid {*hkl*}, and its stereogram are shown in Fig. 2.65c. This class is named the *rhombic-dipyramidal class*.

2/m 2/m 2/m

FIG. 2.65. *(a)* Orthorhombic crystal axes. *(b)* Rotational axes and mirror planes in 2/m2/m2/m. *(c)* The rhombic dipyramid {*hkl*} and its stereogram. *(d)* A crystal of hypersthene (a member of the orthopyroxene series) showing the rhombic dipyramid, *o*.

Forms. There are three types of forms in this class: pinacoids, prisms, and dipyramids.

1. *Pinacoid (parallelohedron).* The pinacoid, consisting of two parallel faces, can occur in three different crystallographic orientations. These are: {100}, front or *a* pinacoid, which intersects *a* and is parallel to *b* and *c*; {010}, side or *b* pinacoid, which intersects *b* and is parallel to *a* and *c*; and {001}, basal or *c* pinacoid, which intersects *c* and is parallel to *a* and *b* (see Fig. 2.66).

2. *Rhombic Prisms.* Rhombic prisms consist of four faces that are parallel to one axis and intersect the other two. In the {0*kl*} prism the faces are parallel to *a* but intersect *b* and *c*; in the {*h0l*} prism the faces are parallel to *b* but intersect *a* and *c*; and in the {*hk*0} prism, the faces are parallel to *c* but intersect *a* and *b*. Examples of {011}, {101}, and {110} prisms are given in Fig. 2.66. Because all prisms intersect two axes and parallel the third, one prism will be transformed into another by a different choice of axes.

3. *Rhombic Dipyramid* {*hkl*}. A rhombic dipyramid has eight triangular faces, each of which intersects all

three crystallographic axes. In Fig. 2.66 is an illustration of the unit dipyramid {111}.

Combinations. Practically all orthorhombic crystals consist of combinations of two or more forms. Characteristic combinations for various minerals are given in Fig. 2.66.

There are many mineral representatives in this class. Among the more common are the following:

andalusite	goethite
anthophyllite (and	marcasite
other	olivine
orthorhombic	sillimanite
amphiboles)	stibnite
aragonite (group)	sulfur
barite (group)	topaz
brookite	
chrysoberyl	
columbite	
cordierite	
enstatite (and other	
orthopyroxenes)	

Pinacoids.

Rhombic prism {011} and pinacoid {100}.

Prism {101} and pinacoid {010}.

Prism {110} and pinacoid {001}.

Rhombic dipyramid.

Sulfur crystals. Forms: p {111}, s (113), both dipyramids. n {011}, prism, c {001}, basal pinacoid.

Celestite crystal. Forms: c {001}, o {011}, d {101}, m {210}, l {104}.

Topaz crystals. Forms: c {001}, b {010}, pinacoids. f {011}, y {021}, d {101}, m {110}, l {120}, prisms. o {111}, dipyramid.

Barite crystals.

FIG. 2.66. Commonly developed forms and form combinations in 2/m2/m2/m.

mm2

Symmetry—2m, 1A₂. The 2-fold rotation axis coincides with the c crystallographic axis. Two mirror planes at right angles to each other intersect in this axis. A rhombic pyramid {hkl} and its stereogram are shown in Fig. 2.67. This class is referred to as the *rhombic-pyramidal class.*

Forms. Because of the absence of a horizontal mirror, the forms at the top of the crystal are different from those at the bottom. The rhombic dipyramid thus

FIG. 2.67. (a) Rhombic pyramid {hkl} and its stereogram. (b) A hemimorphite crystal showing a rhombic pyramid {hkl}, v, at the lower end.

mm2

Hemimorphite

(a)

(b)

becomes two rhombic pyramids, {hkl} at the top, and {hkl̄} at the bottom. Similarly, {0kl} and {h0l} prisms do not exist, but each occurs as two domes. These domes have indices {0kl} and {0kl̄} as well as {h0l} and {h0l̄}. There are also pedions, {001} and {00ī}, and {hk0} prisms.

Only a few minerals crystallize in this class; the most common representatives are hemimorphite, $Zn_4Si_2O_7(OH)_2 \cdot H_2O$ (Fig. 2.67) and bertrandite, $Be_4Si_2O_7(OH)_2$.

222

Symmetry—$3A_2$. The three axes of 2-fold rotation coincide with the crystallographic axes. There is no other symmetry. Figure 2.68 illustrates the general form {hkl}, the rhombic disphenoid (rhombic tetrahedron), and gives a stereogram of the right disphenoid. This class is known as the *rhombic-disphenoidal class*.

Forms. The *rhombic disphenoid (rhombic tetrahedron)* is a closed form consisting of two upper faces that alternate with two lower faces, offset by 90°. Each of its faces is a scalene triangle. It resembles a tetragonal disphenoid in which each face is an isosceles triangle. There are two disphenoids; the right {hkl} and left {hk̄l} are enantiomorphic (see Fig. 2.68). Pinacoids and prisms may also be present in this class.

Although there are several representative minerals crystallizing in this class, they are all comparatively rare. The most common is epsomite, $MgSO_4 \cdot 7H_2O$.

See Table 2.10 for a listing of forms in the orthorhombic system.

Orthorhombic Axial Ratios

To express the relative lengths of the axes in the orthorhombic system there are two ratios, $a : b$ and $b : c$, where the length of b is taken as unity. They are given as $a : b : c = - : 1 : -$, and can be calculated using ϕ and ρ angles.

As an example, consider the face of a general form ABC, Fig. 2.69. Assume this to be face (132) of aragonite with $\phi = 28°11'$, $\rho = 50°48'$. OP is the face normal, and OD is normal to AB. Therefore, angle BOD is ϕ and angle COP is ρ. Reducing the indices (132) to intercepts we find $AO = 6a$, $OC = 3c$, and $OB = 2b = 2$, because $b = 1$. Hence, we may find a (see Fig. 2.69b) by

$$\cot \phi = \frac{6a}{2b} \quad \text{or} \quad a = \frac{b \cot \phi}{3}$$

$$b = 1 \quad a = \frac{1 \times \cot 28°11'}{3} \quad a = 0.6221$$

In triangle COD (Fig. 2.69c), $\tan \rho = 3c/OD$. In triangle BOD, $\cos \phi = OD/2$. Setting both of these expressions equal to OD,

$$OD = \frac{3c}{\tan \rho} = 2 \cos \phi \quad \text{or}$$

$$c = \frac{2 \tan \rho \cos \phi}{3}$$

Substituting ϕ and ρ values of face (132)

$$c = \frac{2 \tan 50°48' \cos 28°11'}{3} = 0.7205$$

Using Miller indices h, k, and l rather than intercepts, a and c relative to b are obtained using the formulas: $a = h/k \cot \phi$ and $c = l/k \tan \rho \cos \phi$. Crystallographic calculations generally involve the variables: (1) axial ratios, (2) indices, and (3) angles ϕ and ρ. When two of these variables are known, the third may be calculated using the above formulas.

FIG. 2.68. Enantiomorphic forms of the rhombic disphenoid (or rhombic tetrahedron) {hkl}, left and right, and a stereogram of the right form.

222

Left   Right

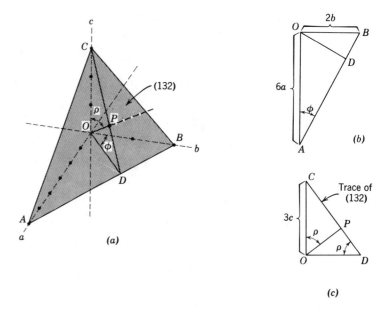

FIG. 2.69. Illustrations of angular measurements in an axial ratio calculation using the (132) face in the orthorhombic system.

The ϕ and ρ angles may be obtained from interfacial angles. For example (Fig. 2.70a), the angle between b (010) and another face, such as m in the [001] zone is the ϕ angle of m. Other faces, such as p and o, in the same horizontal zone with m, also have $b \wedge m$ as their ϕ angle. When c (001) is present (Fig. 2.70a), the ρ angles of f and o are respectively, $c \wedge f$ and $c \wedge o$. When (010) and (001) are not present (Fig. 2.70b), ϕ and ρ angles must be calculated from interfacial angles. For example, $\phi m = 90° - (m \wedge m')/2$; and $\rho f = (f \wedge f')/2$.

Tetragonal System

Crystallographic Axes. The forms of the tetragonal system are referred to three crystallographic axes that make right angles with each other. The two horizontal axes, a, are equal in length and interchangeable, but the vertical axis, c, is of a different length. Figure 2.71a represents the crystallographic axes for the tetragonal mineral zircon with c less than a. Figure 2.71b represents the crystallographic axes of the mineral octahedrite with c greater than a. Figure 2.71 also shows the proper orientation of the crystallographic axes and the method of their notation. When the general form symbols are used, $h < k$.

In the Hermann-Mauguin notation of the symmetry elements in the tetragonal system, the first part of the symbol (consisting of 4 or $\bar{4}$) refers to the c axis, whereas the second or third parts refer to the axial (a_1 and a_2) and diagonal symmetry elements, respectively.

4/m2/m2/m

Symmetry—i, $1A_4$, $4A_2$, $5m$. The vertical crystallographic axis is one of 4-fold rotation. There are four horizontal axes of 2-fold symmetry, two of which coincide with the crystallographic axes (a_1 and a_2) and the others at 45° to them. There are five mirror

FIG. 2.71. Tetragonal crystal axes.

FIG. 2.70. Topaz crystals.

4/m 2/m 2/m

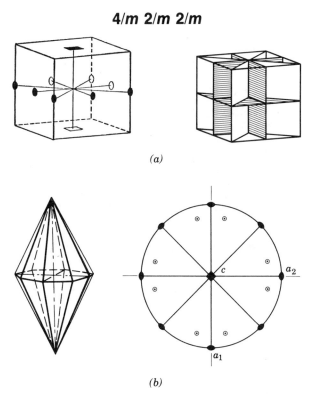

(a)

(b)

FIG. 2.72. (a) Symmetry axes and planes for 4/m2/m2/m. (b) The ditetragonal dipyramid {hkl} and its stereogram.

planes perpendicular to the symmetry axes. One of the horizontal symmetry axes lies in each of the vertical mirror planes. The positon of the axes and mirrors is shown in Fig. 2.72. The general form {hkl}, the ditetragonal dipyramid, is illustrated in Fig. 2.72 with its stereogram. This class is known as the *ditetragonal-dipyramidal class*.

Forms

1. *Basal Pinacoid* {001}. A form composed of two parallel faces perpendicular to the 4-fold axis and thus parallel with the horizontal *m*. It is shown in combination with various prisms in Fig. 2.73.

2. *Tetragonal Prisms* {010} and {110}. The {010} prism consists of four faces that are perpendicular to the 2-fold axes of the first kind and are therefore parallel with the mirrors in the first 2/m in the symbol. The {110} prism has its faces perpendicular to the 2-fold axes of the second kind and thus is parallel with the mirrors in the second 2/m in the symbol.

3. *Ditetragonal Prism* {hk0}. Consists of eight rectangular vertical faces, each of which intersects the two horizontal crystallographic axes unequally. There are various ditetragonal prisms, depending on their differing relations to the horizontal axes. A common form represented in Fig. 2.73 has indices {120}.

4. *Tetragonal Dipyramids* {hhl} and {0kl}. The {hhl} dipyramid has eight isosceles triangular faces, each of which intersects all three crystallographic axes, with equal intercepts upon the two horizontal axes. There are various such dipyramids, depending on the inclination of their faces to *c*. The unit dipyramid {111} (Fig. 2.73), which intersects all the axes at their unit lengths, is most common. Indices of other similar dipyramids are {221}, {331}, {112}, {113}, and so on, or, in general, {hhl}. The {0kl} dipyramid is composed of eight isosceles triangular faces, each of which intersects one horizontal axis and the vertical axis and is parallel to the second horizontal axis. There are various such dipyramids with different intersections upon the vertical axis. The most common is the unit dipyramid {011} (Fig. 2.73). Miller indices for other similar dipyramids are {021}, {031}, {012}, {013}, or, in general {0kl}.

5. *Ditetragonal Dipyramid* {hkl}. Composed of 16 triangular faces, each of which intersects all three of the crystallographic axes, cutting the two horizontal axes at different lengths. There are various ditetragonal dipyramids, depending on the different intersections on the crystallographic axes. One of the most common is the dipyramid {131}, shown in Fig. 2.73.

Several common minerals crystallize in 4/m2/m2/m. Major representatives are rutile (TiO$_2$), anatase (TiO$_2$), cassiterite (SnO$_2$), apophyllite (KCa$_4$Si$_8$O$_{20}$(OH,F)·8H$_2$O), zircon (ZrSiO$_4$), and vesuvianite (Ca$_{10}$Mg$_2$Al$_4$(SiO$_4$)$_5$(Si$_2$O$_7$)$_2$(OH$_4$)).

Tetragonal Combinations. Characteristic combinations of forms in this class, as found on crystals of different minerals, are shown in Fig. 2.73.

$\overline{4}2m$

Symmetry—1$\overline{A}_4$, 2A$_2$, 2*m*. The 4-fold rotoinversion axis is chosen as the *c* axis and the axes of 2-fold rotation as the two *a* axes. At 45° to the *a* axes are two vertical mirror planes intersecting in the vertical axis (see Fig. 2.74a). Figure 2.74b illustrates a tetragonal scalenohedron (rhombic scalenohedron) {hkl} and its stereogram. This class is known as the *tetragonal-scalenohedral class*.

Forms

1. *Tetragonal Disphenoids* (tetragonal tetrahedrons) {hhl} *positive*, {h$\overline{h}$l} *negative* are the only important forms in this class. They consist of four isosceles triangular faces that intersect all three of the crystallographic axes, with equal intercepts on the two horizontal axes. There may be different disphenoids, depending on their varying intersections with the vertical axis. Two different disphenoids and a combina-

FIG. 2.73. Commonly developed forms and form combinations in 4/m2/m2/m.

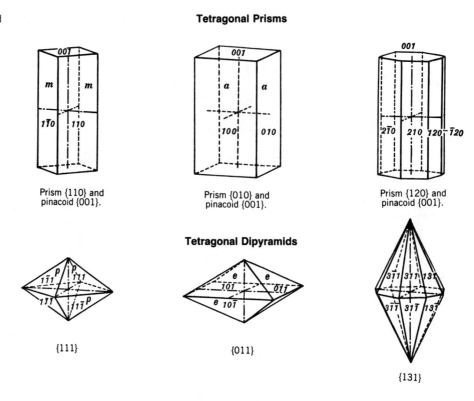

Tetragonal Prisms

Prism {110} and pinacoid {001}.

Prism {010} and pinacoid {001}.

Prism {120} and pinacoid {001}.

Tetragonal Dipyramids

{111}

{011}

{131}

Tetragonal Crystals
Forms: e {011}, u {021}, c {001}, a {010}, m {110}, x {211}.
In this illustration the Miller indices for the forms are based on the knowledge of the orientation of the unit cell. If the forms were indexed on the basis of morphology e would be {111}, a {110}, and u {221}.

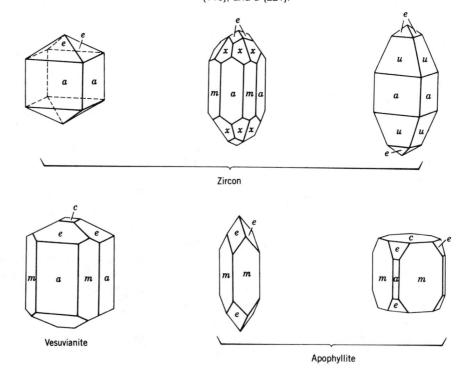

Zircon

Vesuvianite

Apophyllite

$\overline{4}2m$

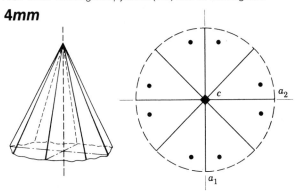

(a)

(b)

(c)

FIG. 2.74. *(a)* Symmetry axes and planes for $\overline{4}2m$. *(b)* The tetragonal scalenohedron (or rhombic scalenohedron) {*hkl*} and its stereogram. *(c)* Tetragonal disphenoids {*hhl*} and {$h\overline{h}l$} (also known as tetragonal tetrahedron) and combination of the two types.

tion of a positive and a negative disphenoid are shown in Fig. 2.74c.

The tetragonal disphenoid differs from the tetrahedron in the isometric system in that its vertical crystallographic axis is not of the same length as the horizontal axes. The only common mineral in this class is chalcopyrite, crystals of which ordinarily show only the disphenoid {112}. This disphenoid closely resembles a tetrahedron, and it requires accurate measurements to prove its tetragonal character.

2. *Tetragonal Scalenohedron (rhombic scalenohedron)* {*hkl*}. This form, Fig. 2.74*b*, if it were to occur by itself, is bounded by eight similar scalene triangles. It

is a rare form and observed only in combination with others. Other forms that may be present are: pinacoid, tetragonal prisms, ditetragonal prisms, and tetragonal dipyramids.

Chalcopyrite ($CuFeS_2$) and stannite (Cu_2FeSnS_4) are the only common minerals that crystallize in this class.

4mm

Symmetry—$1A_4$, $4m$. The vertical axis is a 4-fold rotation axis, and four mirror planes intersect in this axis. Figure 2.75 illustrates the general form {*hkl*}, a *ditetragonal pyramid*, and its stereogram. This class is known as the *ditetragonal-pyramidal class*.

Forms. The lack of a horizontal symmetry plane gives rise to different forms at the top and bottom of crystals of this class. There are pedions {001} and {00$\overline{1}$}. The {*hhl*} and {*h0l*} tetragonal pyramids have corresponding lower forms, {$hh\overline{l}$} and {$h0\overline{l}$}. The ditetragonal pyramid {*hkl*} is an upper form, whereas {$hk\overline{l}$} is the lower form. Tetragonal as well as ditetragonal prisms may be present.

Diaboleite, $Pb_2CuCl_2(OH)_4$, a rather rare mineral, is the only mineral representative in this crystal class.

422

Symmetry—$1A_4$, $4A_2$. The vertical axis is one of 4-fold rotation and there are four 2-fold axes at right angles to it. In other words, the five symmetry directions are occupied by the rotation axes. Mirror planes and the center of symmetry are lacking. (The symmetry axes are the same as those of the class $4/m2/m2/m$.) The general {*hkl*} form is the tetragonal trapezohedron, and the class is referred to as the *tetragonal-trapezohedral class.*

FIG. 2.75. Ditetragonal pyramid {*hkl*} and its stereogram.

4mm

422

Left

Right

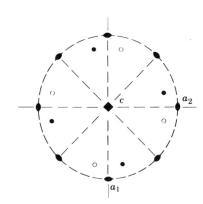

FIG. 2.76. Enantiomorphic forms of the tetragonal trapezohedron, left $\{h\bar{k}l\}$ and right $\{hkl\}$, and a stereogram of the right form.

Forms. The tetragonal trapezohedron has eight faces, corresponding to half the faces of the ditetragonal dipyramid. The enantiomorphic forms are right $\{hkl\}$ and left $\{h\bar{k}l\}$ (Fig. 2.76). The other forms that may be present are the same as in $4/m2/m2/m$.

Phosgenite, $Pb_2CO_3Cl_2$, is the only mineral representative in this class.

4/m

Symmetry—i, $1A_4$, $1m$. There is only the vertical 4-fold rotation axis with a symmetry plane perpendicular to it. Figure 2.77 illustrates a tetragonal dipyramid and its stereogram. This class is known as the *tetragonal-dipyramidal class* after the general form $\{hkl\}$.

Forms. The tetragonal dipyramid, $\{hkl\}$, is an eight-faced form having four upper faces directly above four lower faces. This form by itself appears to have higher symmetry, and it must be in combination with other forms to reveal the absence of vertical symmetry planes. The pinacoid $\{001\}$ and tetragonal prisms $\{hk0\}$ may be present. The tetragonal prism $\{hk0\}$ is equivalent to the four alternate faces of the ditetragonal prism and is present in those classes of

the tetragonal system that have no vertical mirror planes or horizontal 2-fold rotation axes.

Mineral representatives in this class are: scheelite ($CaWO_4$), powellite ($CaMoO_4$), fergusonite ($YNbO_4$), and members of the scapolite series ($Na_4Al_3Si_9O_{24}Cl$ to $Ca_4Al_6Si_6O_{24}CO_3$). Figure 2.77 illustrates a crystal of fergusonite in which the tetragonal dipyramid z reveals the true symmetry of this class.

$\bar{4}$

Symmetry—$1\bar{A}_4$. The vertical axis is a 4-fold axis of rotoinversion. There is no other symmetry. Figure 2.78 illustrates a tetragonal disphenoid (tetragonal tetrahedron) and its stereogram. This class is referred to as the *tetragonal-disphenoidal class*.

Forms. The *tetragonal disphenoid* (*tetragonal tetrahedron*) $\{hkl\}$ is a closed form composed of four isosceles triangles. In the absence of other modifying faces, the form appears to have two vertical symmetry planes, giving it the symmetry $\bar{4}2m$. The true symmetry is shown only in combination with other forms. The pinacoid and tetragonal prisms may be present. Other tetragonal disphenoids are $\{hhl\}$ and $\{0kl\}$.

FIG. 2.77. Tetragonal dipyramid $\{hkl\}$ and its stereogram. By itself this form appears to have higher symmetry. In the crystal of fergusonite the presence of this form (z) reveals the true symmetry, $4/m$.

4/m

Fergusonite

$\overline{4}$

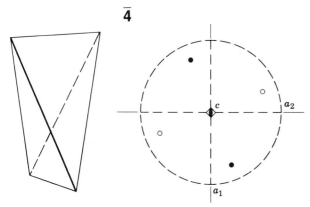

FIG. 2.78. Tetragonal disphenoid (or tetragonal tetrahedron) {hkl} and its stereogram.

The only mineral representative in this class is the rare mineral cahnite, $Ca_2B(AsO_4)(OH)_4$.

4

Symmetry—$1A_4$. The vertical axis is one of 4-fold rotation. There is no other symmetry. The general form {hkl}, a tetragonal pyramid, and its stereogram are shown in Fig. 2.79. This class is known as the *tetragonal-pyramidal class* after the general form {hkl}.

Forms. The *tetragonal pyramid* is a four-faced form. The upper form {hkl} is different from the lower form {hk$\bar{l}$}, and each has a right- and left-hand variation. There are thus two enantiomorphic pairs of tetragonal pyramids. Other tetragonal pyramids are {hhl} and {0kl}. Pedions and tetragonal prisms may also be present.

As in some other classes, the true symmetry is not shown morphologically unless the general form occurs in combination with other forms. Figure 2.79

shows a crystal of wulfenite, $PbMoO_4$. Other mineral representatives are unknown.

See Table 2.11 for a listing of forms in the tetragonal system.

Tetragonal Axial Ratios

The axial ratio of a tetragonal crystal is expressed as $a:c$, with the length of the two equal a axes taken as unity. It is calculated from ϕ and ρ angles derived from interfacial angles by the general formula

$$c = \frac{l}{k} \tan \rho \cos \phi$$

where k and l are Miller indices.

Tetragonal crystals are so oriented that (010), perpendicular to a_2, has $\phi = 0°$. Thus, for {0kl} forms, $\cos \phi = 1$ and the formula becomes: $c = (l/k) \tan \rho$. For the trigonometry involved, consider the calculation of c from angular measurements of face (021) (Fig. 2.80a,b). $\tan \rho_{021} = CO/OA$. $CO = 2c$, $AO = a = 1$. Thus, $c = \tan \rho_{021}/2$. For the unit form {011}, $c = \tan \rho$.

For {hh0} and {hhl} tetragonal forms, $\phi = 45°$, $\cos \phi = 0.7071$ and

$$c = \frac{l}{k} \tan \rho \cdot 0.7071$$

Consider the calculation of c using the angular measurements of face (221) (Fig. 2.80a,c,d). In triangle AOB, $OB = \cos 45°$. In triangle COB, $OC = 2c$, $\tan \rho_{221} = OC/OB$. Thus $c = (\tan \rho \cos 45°)/2$.

When (001), face c in Fig. 2.81a, is present, ρ of face p can be measured directly as $c \wedge p$. If (001) is not present (Fig. 2.81b), ρ of face e can be determined as $90° - a$ (010) $\wedge e(011)$ and ρ of face $s = 90° - m$ (110) $\wedge s$ (111). If the pyramidal form and a prism

FIG. 2.79. Tetragonal pyramid {hkl} and its stereogram. Crystals of wulfenite show this form (n) consistent with only 4.

4

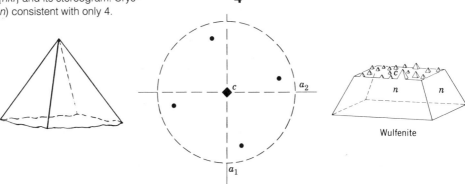

Wulfenite

Table 2.11

DISTRIBUTION OF FORMS IN THE TETRAGONAL SYSTEM*

No. of Faces	Name of Form	Class 4	$\bar{4}$	4/m	422	4mm	$\bar{4}2m$	4/m2/m2/m	Unique Form for
1	Pedion	+				+			
2	Pinacoid		+	+	+		+	+	
4	Tetragonal prism	+	+	+	+	+	+	+	
4	Tetragonal pyramid	+				+			
4	Tetragonal disphenoid		+				+		
8	Ditetragonal prism				+	+	+	+	
8	Tetragonal dipyramid			+	+		+	+	
8	Tetragonal trapezohedron				+				422
8	Tetragonal scalenohedron						+		$\bar{4}2m$
8	Ditetragonal pyramid					+			4mm
16	Ditetragonal dipyramid							+	4/m2/m2/m

*From Buerger, M. J., 1956, *Elementary Crystallography.* John Wiley & Sons, New York, 528 pp.

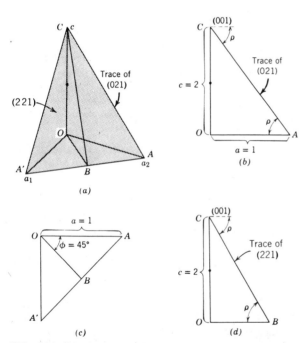

FIG. 2.80. Tetragonal crystal and angular relationships for the calculation of axial ratios using the faces (021) and (221).

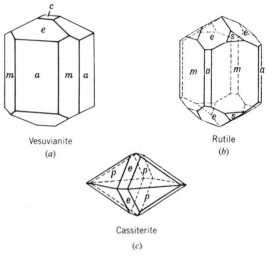

FIG. 2.81. Tetragonal crystals. Forms: a {010}, c {001}, m {110}, e {011}, p {111}, s {221}.

do not lie in a horizontal zone (Fig. 2.81c), measure the interfacial angle over the top of the crystal as $p \wedge p'$ (where p and p' are faces of the same form differing in ϕ by 180°). ρ of p is one-half this interfacial angle.

Hexagonal System

All of the crystal classes in the hexagonal system can be based on a hexagonal lattice, whether their symbol begins with 6, $\bar{6}$, 3, or $\bar{3}$. The five beginning with 3 or $\bar{3}$ can also be based on a rhombohedral lattice.

Crystallographic Axes. The forms of the hexagonal system are referred to four crystallographic axes as proposed by Bravais. Three of these, designated a_1, a_2, and a_3, lie in the horizontal plane and are of equal length with angles of 120° between the positive ends; the fourth axis, c, is vertical. When properly oriented, one horizontal crystallographic axis, a_2, is left to right, and the other two make 120° angles on either side of it (Fig. 2.82a). The positive end of a_1 is to the front and left, the positive end of a_2 is to the right, and the positive end of a_3 is to the back and left. Figure 2.82b shows the four axes in clinographic projection. In stating the indices for any face of a hexagonal crystal, four numbers (the Bravais-Miller symbol) must be given. The numbers expressing the reciprocals of the intercepts of a face on the

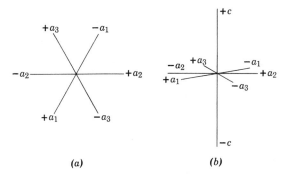

FIG. 2.82. Hexagonal crystal axes.

axes are given in the order a_1, a_2, a_3, c. Therefore, $(11\bar{2}1)$, which represents the intercepts $3a_1$, $3a_2$, $-3/2a_3$, $3c$, refers to a face that cuts the positive ends of the a_1 and a_2 axes at twice the distance it cuts the negative end of the a_3 axis; it cuts the c axis at the same relative number of units (3) as it cuts the a_1 and a_2 axes. The general Bravais-Miller form symbol is $\{hki\bar{l}\}$ with $h > k$. The third digit of the index is the sum of the first two times -1; or, stated another way, $h + k + i = 0$ (see also Fig. 2.33 and page 43).

In the Hermann-Mauguin notation the first number refers to the principal axis of symmetry coincident with c. The second and third symbols, if present, refer respectively to symmetry elements parallel with and perpendicular to the crystallographic axes a_1, a_2, and a_3.

6/m2/m2/m

Symmetry—*i*, 1A_6, 6A_2, 7*m*. The vertical axis is an axis of 6-fold rotation. There are six horizontal axes of 2-fold rotation, three of which coincide with the crystallographic axes (a_1, a_2, and a_3); the other three lie midway between them. There are seven mirror planes, each perpendicular to one of the symmetry axes. See Fig. 2.83 for the location of the symmetry elements and a drawing of the general form, the dihexagonal dipyramid, as well as its stereogram. This class is known as the *dihexagonal-dipyramidal class*, after the general form.

Forms

1. *Pinacoid* {0001}. Composed of two parallel faces, perpendicular to the 6-fold axis, and thus parallel with the horizontal *m*. It is commonly referred to as a basal pinacoid. It is shown in combination with various prisms in Fig. 2.84.

2. *Hexagonal Prisms* {10$\bar{1}$0} and {11$\bar{2}$0}. The {10$\bar{1}$0} prism consists of six vertical faces, each of which intersects two of the horizontal crystallographic

axes equally and is parallel to the third. The faces of this prism are parallel to the 2-fold axes of the first kind. The {11$\bar{2}$0} prism also has six vertical faces, but each of these intersects two of the horizontal axes equally and the intermediate horizontal axis at one-half this distance. These two types of hexagonal prisms are geometrically identical forms; the distinction between them is only in orientation. See Fig. 2.84 for illustrations.

3. *Dihexagonal Prism* {hki0}. This form consists of 12 vertical faces, each of which intersects all three of the horizontal crystallographic axes at different lengths. There are various dihexagonal prisms, depending on their intercepts with the horizontal axes. A common dihexagonal prism with indices {21$\bar{3}$0} is shown in Fig. 2.84.

4. *Hexagonal Dipyramids* {h0$\bar{h}$l} and {hh$\overline{2h}$l}. The {h0$\bar{h}$l} hexagonal dipyramid consists of 12 isosceles triangular faces, each of which intersects two horizontal crystallographic axes equally, is parallel to the third, and intersects the vertical axis. Various hexagonal dipyramids are possible, depending on the inclination of the faces to the c axis. The unit form has the indices {10$\bar{1}$1} (see Fig. 2.84). The {hh$\overline{2h}$l} dipyramid

FIG. 2.83. *(a)* Symmetry axes and planes for 6/*m*2/*m*2/*m*. *(b)* The dihexagonal dipyramid {hki$\bar{l}$} and its stereogram.

6/m 2/m 2/m

(a)

(b)

Hexagonal Prisms

$\{10\bar{1}0\}$ $\{11\bar{2}0\}$ $\{21\bar{3}0\}$

Hexagonal Dipyramids

$\{10\bar{1}1\}$ $\{11\bar{2}2\}$ $\{21\bar{3}1\}$

Beryl.

FIG. 2.84. Commonly developed forms and form combinations in $6/m2/m2/m$.

is also composed of 12 isosceles triangular faces. Each face intersects two of the horizontal axes equally and the third (the intermediate horizontal axis) at one-half this distance; each face also intersects the vertical axis. Various such dipyramids are possible, depending on the inclination of the faces to c. A common form has the indices $\{11\bar{2}2\}$ (see Fig. 2.84).

5. *Dihexagonal Dipyramid* $\{hk\bar{i}l\}$. This is composed of 24 triangular faces. Each face is a scalene triangle that intersects all three of the horizontal axes differently and also intersects the vertical axis. A common form, $\{21\bar{3}1\}$, as well as various combinations of forms in this class are shown in Fig. 2.84.

Beryl, $Be_3Al_2Si_6O_{18}$, affords the best example of a mineral representative in this class. Other minerals are molybdenite, MoS_2, pyrrhotite, $Fe_{1-x}S$, and nickeline (synonymous with niccolite), $NiAs$.

$\bar{6}m2$

Symmetry—$1\bar{A}_6$, $3A_2$, $4m$. The 6-fold rotoinversion axis is chosen as the c axis. Note that $\bar{6}$ is equivalent to a 3-fold axis of rotation with a mirror perpendicular to it (see page 34). Three mirror planes intersect in the vertical axis and are perpendicular to the three horizontal crystallographic axes a_1, a_2, and a_3. The three horizontal 2-fold axes lie in the vertical mirror planes. A ditrigonal dipyramid, $\{hk\bar{i}l\}$, and its stereogram are shown in Fig. 2.85. This class is known as the *ditrigonal-dipyramidal class*, after the general form.

Forms. The *ditrigonal dipyramid* $\{hk\bar{i}l\}$ is a 12-faced form with six faces at the top of the crystal and six at the bottom. Additional forms that may be present are: pinacoid, trigonal prisms, hexagonal

$\bar{6}m2$

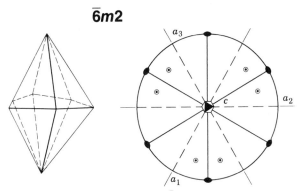

FIG. 2.85. Ditrigonal dipyramid $\{hk\bar{i}l\}$ and its stereogram. In the conventional orientation of the symmetry elements of this crystal class the 2-fold rotation axes are located in directions perpendicular to a_1, a_2, and a_3 and lie in the vertical mirror planes (see footnote to Table 2.9).

prism, ditrigonal prisms, trigonal dipyramids, and hexagonal dipyramids.

Benitoite, $BaTiSi_3O_9$, is the only mineral that has been described as crystallizing in this class.

6mm

Symmetry—$1A_6$*, 6m.* The 6-fold rotation axis is chosen as the *c* axis, and six vertical mirror planes intersect in this axis. Figure 2.86 shows a dihexagonal pyramid and its stereogram. This class is known as the *dihexagonal-pyramidal class* after the $\{hk\bar{i}l\}$ general form.

Forms. The forms of this class are similar to those of class $6/m2/m2/m$, but inasmuch as a horizontal mirror plane is lacking, different forms appear at the top and bottom of the crystal. The dihexagonal pyramid is thus two forms: $\{hk\bar{i}l\}$ upper and $\{hk\bar{i}l\}$ lower. The hexagonal-pyramidal forms are: $\{h0\bar{h}l\}$ up-

per and $\{h0\bar{h}l\}$ lower; and $\{hh\overline{2h}l\}$ upper and $\{hh\overline{2h}l\}$ lower. The pinacoid cannot exist here, but instead there are two pedions $\{0001\}$ and $\{000\bar{1}\}$. Hexagonal prisms and the dihexagonal prism may be present.

Wurtzite, ZnS, greenockite, CdS, and zincite, ZnO, are the most common mineral representatives in this class. Figure 2.86 shows a zincite crystal with a hexagonal prism terminated above by a hexagonal pyramid and below by a pedion.

622

Symmetry—$1A_6$*, 6A*$_2$. The symmetry axes are the same as those in class $6/m2/m2/m$ (see Fig. 2.83a), but mirror planes and the center of symmetry are lacking. This class is referred to as the *hexagonal-trapezohedral class*, after its general form $\{hk\bar{i}l\}$.

Forms. The *hexagonal trapezohedrons* $\{hk\bar{i}l\}$ *right* and $\{ih\bar{k}l\}$ *left* are enantiomorphic forms, each with 12 trapezium-shaped faces (see Fig. 2.87). Other forms that may be present are the pinacoid, hexagonal prisms, dipyramids, and dihexagonal prisms.

High quartz, SiO_2, and kalsilite, $KAlSiO_4$, are the only mineral representative in this class.

6/m

Symmetry—i*, *$1A_6$*, 1m.* There is only the vertical 6-fold rotation axis with a symmetry plane perpendicular to it. Figure 2.88 illustrates a hexagonal dipyramid $\{hk\bar{i}l\}$, and its stereogram. This class is known as the *hexagonal-dipyramidal class*.

Forms. The general forms of this class are the *hexagonal dipyramids*, $\{hk\bar{i}l\}$ positive, $\{kh\bar{i}l\}$ negative. These forms consist of 12 faces, six above and six below, which correspond in position to one-half the

FIG. 2.86. Dihexagonal pyramid $\{hk\bar{i}l\}$ and its stereogram. The same pyramid (*p*) as shown in zincite crystals.

6mm

Zincite.

622

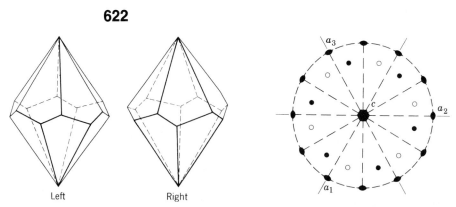

FIG. 2.87. Enantiomorphic, left- and right-handed hexagonal trapezohedrons, {$ih\bar{k}l$} and {$hk\bar{i}l$}, respectively. A stereogram of the left-handed form.

6/m

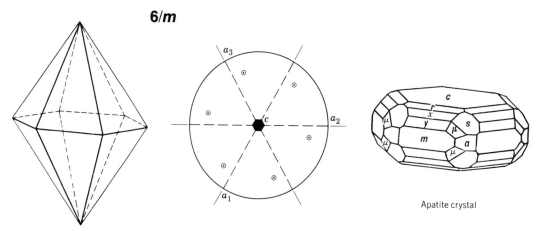

Apatite crystal

FIG. 2.88. Hexagonal dipyramid {$hk\bar{i}l$} and its stereogram. This form by itself appears to have higher symmetry, but in combination with other forms it reveals its low symmetry content. The form (μ) is the hexagonal dipyramid in the apatite crystal.

faces of a dihexagonal dipyramid. Other forms that may be present are pinacoid and prisms.

This class has as its chief mineral representatives the minerals of the apatite group, $Ca_5(PO_4)_3(OH, F, Cl)$. The dipyramid revealing the symmetry of the class is rarely seen but is illustrated as face μ in Fig. 2.88.

$\bar{6}$

Symmetry—$1\bar{A}_6$ (= $1A_3$ + $1m$). The vertical axis is a 6-fold axis of rotoinversion ($\bar{6}$) which is equivalent to a 3-fold axis of rotation with a symmetry plane perpendicular to it ($3/m$). Figure 2.89 shows the general form {$hkil$}, a trigonal dipyramid, and its stereogram. This class is named the *trigonal-dipyramidal class.*

Forms. There are four *trigonal dipyramids* of the general form with six faces each, three above and

three below, corresponding to six faces of the dihexagonal dipyramid. The symmetry does not permit hexagonal forms. Thus, instead of hexagonal prisms there are two trigonal prisms. Similarly, in place of hexag-

FIG. 2.89. Trigonal dipyramid {$hk\bar{i}l$} and its stereogram.

$\bar{6}$

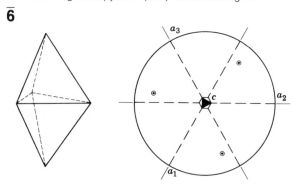

onal dipyramids there are two trigonal dipyramids. The pinacoid {0001} may be present.

There is no example of an authenticated mineral or other crystalline substance belonging to this crystal class.

6

Symmetry—1A$_6$.   The vertical axis is one of 6-fold rotation. There is no other symmetry. A hexagonal pyramid and its stereogram are shown in Fig. 2.90. This class is referred to as the *hexagonal-pyramidal class* after the general form {hkīl}.

Forms.   The forms of this class are similar to those of class 6/m, but because a horizontal mirror plane is lacking, different forms are present at the top and bottom of the crystal. The hexagonal pyramid is thus four 6-faced forms: upper positive and negative, and lower positive and negative. Pedions, hexagonal pyramids, and prisms may be present.

The form development of crystals is rarely sufficient to enable one without other evidence to place unequivocally a crystal in this class. The mineral nepheline, (Na,K)AlSiO$_4$, is the chief mineral representative.

6

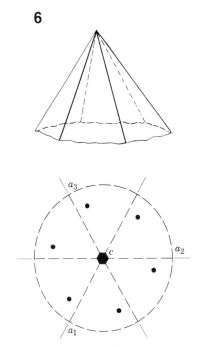

FIG. 2.90. Hexagonal pyramid {hkīl} and its stereogram.

FIG. 2.91. *(a)* Symmetry axes and planes for $\bar{3}2/m$. *(b)* Relationship between the rhombohedron {h0h̄l} and a hexagonal dipyramid {hkīl}. *(c)* Hexagonal scalenohedron (or ditrigonal scalenohedron) {hkīl} and its stereogram. *(d)* Relationship between the scalenohedron and a dihexagonal dipyramid.

$\bar{3}2/m$

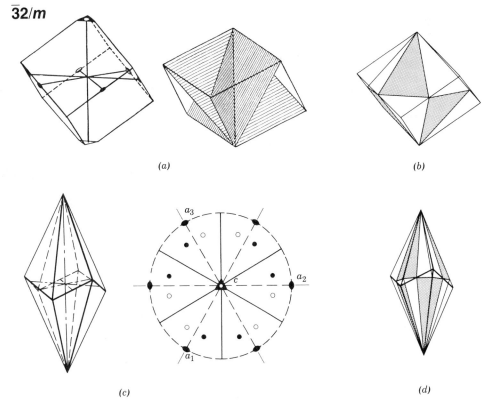

(a)   *(b)*

(c)   *(d)*

$\bar{3}\,2/m$

Symmetry—$1\bar{A}_3$, $3A_2$, $3m$. The 3-fold rotoinversion axis is the vertical axis and the three 2-fold rotation axes are the three horizontal crystallographic axes (a_1, a_2, and a_3). There are three vertical mirror planes bisecting the angles between the horizontal axes (see Fig. 2.91a). Figure 2.91c illustrates the general form $\{hk\bar{i}l\}$, a hexagonal scalenohedron (ditrigonal scalenohedron) and its stereogram. This class is known as the *hexagonal-scalenohedral class.*

Forms

1. *Rhombohedron $\{h0\bar{h}l\}$ positive, $\{0h\bar{h}l\}$ negative.* The rhombohedron is a form consisting of six rhomb-

shaped faces, which correspond in their positions to the alternate faces of a hexagonal dipyramid $\{h0\bar{h}l\}$. The relation of these two forms to each other is shown in Fig. 2.91b. The rhombohedron may also be thought of as a cube deformed in the direction of one of the axes of 3-fold rotoinversion. The deformation may appear either as an elongation along the rotoinversion axis, producing an acute solid angle, or compression along the rotoinversion axis, producing an obtuse solid angle. Depending on the angle, the rhombohedron is known as acute or obtuse.

Depending on the orientation the rhombohedron may be positive or negative (Fig. 2.92). There are var-

FIG. 2.92. Commonly developed forms and form combinations in $\bar{3}2/m$.

Rhombohedron $\{10\bar{1}1\}$, positive

Rhombohedron $\{01\bar{1}1\}$, negative

Scalenohedron $\{21\bar{3}1\}$, positive

Scalenohedron $\{12\bar{3}1\}$, negative

Calcite crystals. Forms: e $\{01\bar{1}2\}$ and f $\{02\bar{2}1\}$, negative rhombohedrons. r $\{10\bar{1}1\}$ and M $\{40\bar{4}1\}$ positive rhombohedrons. m $\{10\bar{1}0\}$, prism. c $\{0001\}$, basal pinacoid. v $\{21\bar{3}1\}$, scalenohedron.

Chabazite.

Corundum.

ious rhombohedrons that differ from each other in the inclination of their faces to the *c* axis. The index symbol of the unit positive rhombohedron is $\{10\bar{1}1\}$ and of the unit negative rhombohedron $\{01\bar{1}1\}$.

2. *Scalenohedron* (*ditrigonal scalenohedron*) $\{hk\bar{i}l\}$ *positive*, $\{kh\bar{i}l\}$ *negative*. This form consists of 12 scalene triangular faces corresponding in position to alternate *pairs* of faces of a dihexagonal dipyramid (Fig. 2.91*d*). The scalenohedron is differentiated from the dipyramid by the zigzag appearance of the middle edges.

There are many different scalenohedrons; most frequently seen is $\{21\bar{3}1\}$, a common form on calcite; indicated by *v* in crystals in Fig. 2.92.

The rhombohedron and scalenohedron of this class may combine with forms found in classes of higher hexagonal symmetry. Thus they may be in combination with hexagonal prisms, dihexagonal prisms, hexagonal dipyramid, and pinacoid (see calcite, chabazite, and corundum crystals in Fig. 2.92).

Several common minerals crystallize in this class. Chief among them is calcite ($CaCO_3$) and the other members of the calcite group. Other minerals are corundum (Al_2O_3), hematite (Fe_2O_3), brucite ($Mg(OH)_2$), nitratite ($NaNO_3$, synonymous with soda niter), arsenic (As), millerite (NiS), antimony (Sb), and bismuth (Bi).

3*m*

Symmetry—$1A_3$, 3*m*. The vertical axis is a 3-fold rotation axis, and three mirror planes intersect in this axis. In the Hermann-Mauguin notation of this class the 3 refers to the vertical *c* axis and the *m* refers to three planes perpendicular to the three horizontal axes a_1, a_2, and a_3. These three mirror planes intersect in the vertical 3-fold axis. A ditrigonal pyramid $\{hk\bar{i}l\}$ and its stereogram are shown in Fig. 2.93. This class is known as the *ditrigonal pyramidal class*, after the general form.

Forms. The forms are similar to those of the class $\bar{3}2/m$ but with only half the number of faces. Because of the lack of 2-fold rotation axes, the faces at the top of the crystals belong to different forms from those at the bottom. There are four possible *ditrigonal pyramids*, with indices $\{hk\bar{i}l\}$, $\{kh\bar{i}l\}$, $\{hk\bar{i}\bar{l}\}$, and $\{kh\bar{i}\bar{l}\}$. Other forms that may be present are pedions, hexagonal prisms and pyramids, trigonal pyramids, trigonal prisms, and ditrigonal prisms. There are four possible trigonal pyramids with indices $\{h0\bar{h}l\}$, $\{0h\bar{h}l\}$, $\{0h\bar{h}\bar{l}\}$, and $\{h0\bar{h}\bar{l}\}$.

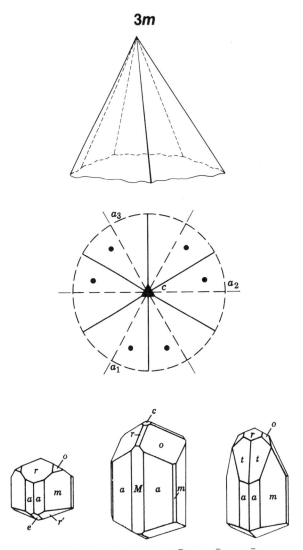

Tourmaline crystals: Forms: $r = \{10\bar{1}1\}$, $o\{02\bar{2}1\}$, $t\{21\bar{3}1\}$, $c\{0001\}$, $a\{11\bar{2}0\}$, $M\{10\bar{1}0\}$, $m\{01\bar{1}0\}$, $e\{10\bar{1}2\}$, $r'\{01\bar{1}1\}$.

FIG. 2.93. (*a*) Ditrigonal pyramid $\{hk\bar{i}l\}$ and its stereogram. In the conventional orientation of the symmetry elements of this crystal class the mirror planes are perpendicular to a_1, a_2, and a_3 (see footnote to Table 2.9). (*b*) Tourmaline crystals showing 3*m* symmetry.

Tourmaline (Fig. 2.93) is the most common mineral crystallizing in this class, but in addition there are members of the proustite (Ag_3AsS_3)-pyrargyrite (Ag_3SbS_3) series and alunite ($KAl_3(SO_4)_2(OH)_6$).

32

Symmetry—$1A_3$, $3A_2$. The four axial directions are occupied by the rotation axes. The vertical

32

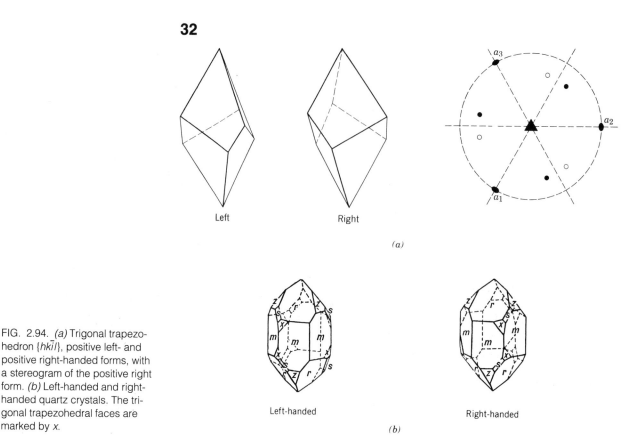

FIG. 2.94. *(a)* Trigonal trapezo-hedron {$hk\bar{i}l$}, positive left- and positive right-handed forms, with a stereogram of the positive right form. *(b)* Left-handed and right-handed quartz crystals. The trigonal trapezohedral faces are marked by *x*.

crystallographic axis is an axis of 3-fold rotation and the three horizontal crystallographic axes are axes of 2-fold rotation. The symmetry axes are similar to those in class $\bar{3}2/m$, but planes of symmetry are lacking. Figure 2.94a shows a positive left and a positive right trigonal trapezohedron and a stereogram for the positive right form. The class is referred to as the *trigonal-trapezohedral class*, after the general {$hk\bar{i}l$} form.

Forms. There are four trigonal trapezohedrons, each made up of six trapezium-shaped faces. Their Miller indices are: {$hk\bar{i}l$}, {$i\bar{k}hl$}, {$kh\bar{i}l$}, and {$ki\bar{h}l$}. These forms can be grouped into two enantiomorphic pairs, each with a right and left form (see Fig. 2.94a). Other forms that may be present are: pinacoid, trigonal prisms, hexagonal prism, ditrigonal prisms, and rhombohedrons.

Low-temperature quartz is the most common mineral crystallizing in this class, but only rarely can faces of the trigonal trapezohedron be observed. When this form is present, the crystals can be distinguished as right-handed or left-handed (Fig. 2.94b), depending on whether, with a prism face fronting the observer, the trigonal trapezohedral faces, *x*, truncate the edges between prism and the top rhombohedron

faces at the right or at the left. The faces marked *s* are trigonal dipyramids.

Cinnabar, HgS, and the rare mineral berlinite, AlPO$_4$, also crystallize in this class.

$\overline{3}$

Symmetry—1$\bar{A}_3$. The vertical axis is a 3-fold axis of rotoinversion. This is equivalent to a 3-fold rotation axis and a center of symmetry. Figure 2.95 illustrates a rhombohedron and its stereogram. This class is known as the *rhombohedral class* after the general {$hk\bar{i}l$} form.

Forms. As general forms of this class there are four different *rhombohedrons*, each corresponding to six faces of the dihexagonal-dipyramid. If one of these appeared alone on a crystal, it would have the morphological symmetry of class $\bar{3}2/m$. It is only in combination with other forms that its true symmetry becomes apparent. The {0001} pinacoid and hexagonal prisms may be present.

Dolomite, CaMg(CO$_3$)$_2$, is the most common mineral crystallizing in this class; other representa-

$\overline{3}$

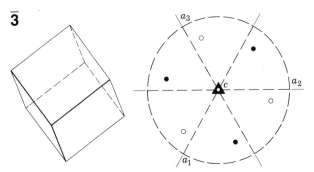

FIG. 2.95. Rhombohedron $\{hk\overline{i}l\}$ and its stereogram.

3

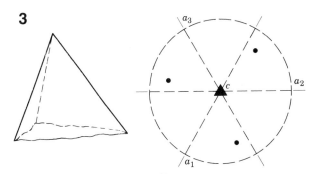

FIG. 2.96. Trigonal pyramid $\{hk\overline{i}l\}$ and its stereogram.

tives are ilmenite, $FeTiO_3$, willemite, Zn_2SiO_4, and phenakite, Be_2SiO_4.

3

Symmetry—$1A_3$. One 3-fold rotation axis is the only symmetry. Figure 2.96 shows a trigonal pyramid and its stereogram. This class is known as the *trigonal-pyramidal class* after the general $\{hk\overline{i}l\}$ form.

Forms. There are eight *trigonal pyramids* of the general form, $\{hk\overline{i}l\}$, four above and four below, each corresponding to three faces of the dihexagonal dipyramid. In addition there are trigonal pyramids above and equivalent, but independent pyramids below. Pedions and several different trigonal prisms may be present. In combination with a pedion a trigonal

pyramid appears to have the symmetry of $3m$ with three vertical symmetry planes (Fig. 2.93). It is only when several trigonal pyramids are in combination with one another that the true symmetry is revealed.

Possibly the mineral gratonite, $Pb_9As_4S_{15}$, belongs in this class; there are no other mineral representatives.

See Table 2.12 for a listing of forms in the hexagonal system.

Axial Ratios in the Hexagonal System

With the exception of the hexagonal system, crystals are oriented so that (010) is to the right when $\phi = 0°00'$. In the hexagonal system the negative end of the a_3 axis is taken as $\phi = 0°$. According to this conven-

Table 2.12
DISTRIBUTION OF FORMS IN THE HEXAGONAL SYSTEM*

No. of Faces	Name of Form	3	$\overline{3}$	32	3m	$\overline{3}2/m$	6	$\overline{6} = 3/m$	6/m	622	6mm	$\overline{6}m2 = 3/mm2$	6/m2/m2/m	Unique Form for
1	Pedion	+			+		+				+			
2	Pinacoid		+	+		+		+	+	+		+	+	
3	Trigonal prism	+		+	+			+				+		
3	Trigonal pyramid	+			+									
6	Ditrigonal prism			+	+							+		
6	Hexagonal prism		+	+	+	+	+		+	+	+	+	+	
6	Trigonal dipyramid		+					+				+		
6	Rhombohedron		+	+		+								
6	Trigonal trapezohedron			+										32
6	Ditrigonal pyramid				+									3m
6	Hexagonal pyramid				+		+				+			
12	Hexagonal dipyramid				+			+	+			+	+	
12	Hexagonal scalenohedron				+									$\overline{3}2/m$
12	Dihexagonal prism				+					+	+		+	
12	Ditrigonal dipyramid											+		$\overline{6}m2$
12	Hexagonal trapezohedron									+				$\overline{6}22$
12	Dihexagonal pyramid										+			6mm
24	Dihexagonal dipyramid												+	6/m2/m2/m

*From Buerger, M. J., 1956, *Elementary Crystallography*. John Wiley & Sons, New York, 528 pp.

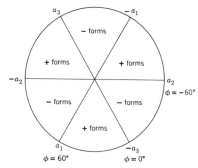

FIG. 2.97. Distribution of rhombohedral forms.

FIG. 2.98. Beryl crystal with forms $m\{10\bar{1}0\}$, $p\{10\bar{1}1\}$, $s\{22\bar{4}1\}$, $0\{11\bar{2}2\}$, and $c\{0001\}$.

FIG. 2.99. Illustrations of the measurements necessary to calculate axial ratios using hexagonal forms $\{10\bar{1}1\}$ and $\{11\bar{2}2\}$.

tion ϕ of $\{hh\overline{2h}l\}$ forms is 0°, whereas ϕ of $\{h0\bar{h}l\}$ forms is 30°. This apparent inconsistency has definite advantages in working with rhombohedral crystals; it gives positive forms $+\phi$ values and negative forms $-\phi$ values (see Fig. 2.97).

To determine an axial ratio of a crystal, the forms must first be indexed and their ϕ and ρ angles determined. For many crystals ϕ and ρ can be measured directly as interfacial angles. For other crystals it may be necessary to project the measurable interfacial angles and determine ϕ and ρ from the projection (see page 60).

Consider the drawing of the beryl crystal, Fig. 2.98. It is oriented so that m and p are $\{h0\bar{h}0\}$ and $\{h0\bar{h}l\}$, respectively, and therefore $\phi = 30°$. o and s are $\{hh\overline{2h}l\}$ forms with $\phi = 0°$. The angles $c \wedge p$, $c \wedge o$, $c \wedge s$ are, respectively, the ρ angles of p, o, and s. If c were not present, the ρ angle of p could be determined as the complement of $m \wedge p$. If only faces of form p were present, an interfacial angle could be measured over the top of the crystal between p and p' (p' a face at the back of the crystal 180° removed from p). ρp equals half of this measured angle.

An axial ratio in the hexagonal system expresses the length of c in terms of a as unity or $a:c = 1:?$. For ease of calculations the $-a_3$ axis is taken as unity because it is at the position of $\phi = 0°$. (See distance OA in Fig. 2.99.) The reciprocal of the intercept on this axis is $i = -(h + k)$.

The formula for determining c from the angles of the general form is:

$$c = \frac{l \tan \rho_{hkil} \cos \phi}{h + k}$$

If a $\{hh\overline{2h}l\}$ form is used, $\phi = 0°$, and $\cos \phi = 1$; hence $\cos \phi$ disappears from the equation, leaving:

$$c = \frac{l \tan \rho_{hh\overline{2h}l}}{h + h}$$

A face of the form $\{11\bar{2}2\}$ intersects $-a_3$ and c at unit distances but a_1 and a_2 at twice unity (Figs. 2.99a and c). Thus for this form $l = 2$, $h + k = 2$, and $c = \tan \rho$. For the form $\{11\bar{2}1\}$, $c = \tan \rho/2$.

For the $\{10\bar{1}1\}$ form: $\phi = 30°$, $\cos \phi = 0.8660$, and the equation becomes:

$$c = \frac{l \tan \rho_{ho\bar{h}l} \times 0.8660}{h + 0}$$

Isometric System

Crystallographic Axes. The crystal forms of classes of the isometric system are referred to three axes of equal length that make right angles with each other. Because the axes are identical, they are interchangeable, and all are designated by the letter a.

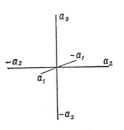

FIG. 2.100. Isometric crystal axes.

When properly oriented, one axis, a_1, is horizontal and oriented front to back, a_2 is horizontal and right to left, and a_3 is vertical (see Fig. 2.100).

In the Hermann-Mauguin notation the first number (4, $\bar{4}$, or 2) refers to the three crystallographic axes a_1, a_2, and a_3. If the number is 4 or $\bar{4}$, it means that there are three 4-fold axes of rotation or inversion

FIG. 2.101. (a) Symmetry axes and (b) planes for $4/m\bar{3}2/m$, and (c) the hexoctahedron $\{hkl\}$ and its stereogram. The location of one $\bar{3}$ is noted in the model.

$4/m\bar{3}2/m$

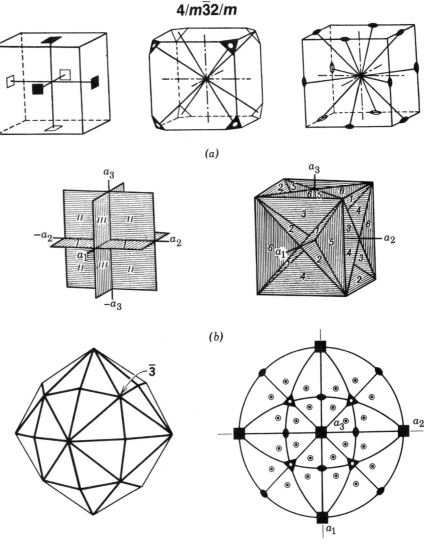

coincident with the three crystallographic axes. If it is 2, there are three 2-fold axes, coincident with the three crystallographic axes. The second number ($\bar{3}$ or 3) refers to the four diagonal directions of 3-fold symmetry, between the corners of a cube (see Fig. 2.101a). The third number or symbol (if present) refers to symmetry elements between the six pairs of opposing cube edges (see Fig. 2.101a). If it is 2 (as in 432), there are six 2-fold axes perpendicular to the edges; if it is m (as in $\bar{4}3m$), there are six mirror planes; if it is $2/m$ (as in $4/m\bar{3}2/m$), there are six 2-fold axes with mirrors perpendicular to them.

Form Symbols. Although the symbol of any face of a crystal form might be used as the form symbol, it is conventional, when possible, to use one in which h, k, and l are all positive. In forms that have two or more faces with h, k, and l positive, the rule followed is to take the form symbol with $h < k < l$. For example, the form with a face symbol (123) also has faces with symbols (132), (213), (231), (312), and (321). Following the rule, {123} would be taken as the form symbol, for there $h < k < l$.

In giving the φ and ρ angles of a form, it is customary to give those for only one face; the others can be determined by knowing the symmetry. The face for which these coordinates are given is the one with the smallest φ and ρ values. This is the face of the form in which $h < k < l$.

$4/m\bar{3}2/m$

Symmetry—$3A_4$, $4\bar{A}_3$, $6A_2$, $9m$. The three crystallographic axes are axes of 4-fold rotation. There are also four diagonal axes of 3-fold rotoinversion; these axes emerge in the middle of each of the octants formed by the intersection of the crystallographic axes. Further, there are six diagonal directions of 2-fold rotation, each of which bisects one of the angles between two of the crystallographic axes. There is also a center of symmetry because $\bar{3}$ is equivalent to $3 + i$. These symmetry elements are shown in Fig. 2.101a.

This class has nine mirror planes. Three of them are known as the axial planes, because each includes two crystallographic axes, and six are called diagonal planes, because each bisects the angle between two of the axial planes (see Fig. 2.101b). This combination of symmetry elements defines the highest symmetry possible in crystals. Every crystal form and every combination of forms that belongs to this class must show its complete symmetry. It is important to remember that in this class the three crystallographic axes are

axes of 4-fold rotation. Thus one can easily locate the crystallographic axes and properly orient the crystal.

The hexoctahedron, the general form from which this *hexoctahedral class* derives its name, is shown in Fig. 2.101c with a stereogram.

Forms. Illustrations of the most common forms and combinations of forms in this class are given in Figs. 2.102 and 2.103. Fig. 2.36 illustrates the 15 isometric forms.

1. *Cube (Hexahedron)* {001}. The cube is composed of six square faces that make 90° angles with each other. Each face intersects one of the crystallographic axes and is parallel to the other two.

2. *Octahedron* {111}. The octahedron is composed of eight equilateral triangular faces, each of which intersects all three of the crystallographic axes equally. When in combination with a cube, the octahedron can be recognized by its eight similar faces, each of which is equally inclined to the three crystallographic axes. It should be noted that the faces of an octahedron truncate symmetrically the corners of a cube.

3. *Dodecahedron (Rhomb-Dodecahedron)* {011}. The dodecahedron is composed of 12 rhomb-shaped faces. Each face intersects two of the crystallographic axes equally and is parallel to the third. Figure 2.102 shows a simple dodecahedron as well as combinations of a dodecahedron and a cube, of a dodecahedron and an octahedron, and of a cube, octahedron, and dodecahedron. Note that the faces of a dodecahedron truncate the edges of both the cube and the octahedron.

4. *Tetrahexahedron* {0kl}. The tetrahexahedron is composed of 24 isosceles triangular faces, each of which intersects one axis at unity and the second at some multiple and is parallel to the third. There are a number of tetrahexahedrons that differ from each other with respect to the inclination of their faces. The most common is {012}. The indices of other forms are {013}, {014}, {023}, and so on, or in general, {0kl}. It is helpful to note that the tetrahexahedron, as its name indicates, resembles a cube each of whose faces have been raised to accommodate four others. Figure 2.102 shows a simple tetrahexahedron as well as a cube with its edges beveled by the faces of a tetrahexahedron.

5. *Trapezohedron (Tetragon-Trioctahedron)* {hhl}. The trapezohedron is composed of 24 trapezium-shaped faces, each of which intersects one of the crystallographic axes at unity and the other two at equal multiples. There are various trapezohedrons with their faces having different angles of inclination, but the most common is {112} (Fig. 2.103). This form is also called a tetragon-trioctahedron to indicate that each

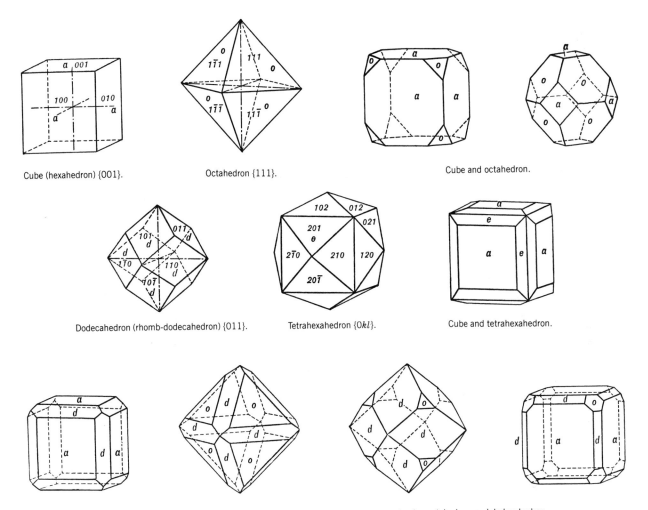

Cube (hexahedron) {001}.

Octahedron {111}.

Cube and octahedron.

Dodecahedron (rhomb-dodecahedron) {011}.

Tetrahexahedron {0kl}.

Cube and tetrahexahedron.

Combinations of cube and dodecahedron, octahedron and dodecahedron, and cube, octahedron, and dodecahedron.

FIG. 2.102. Some of the commonly developed forms and form combinations in 4/m$\bar{3}$2/m (see also Fig. 2.103).

of its faces has four edges and to distinguish it from another 24-faced form, the trisoctahedron (or trigon-trioctahedron).

Figure 2.103 shows the common trapezohedron *n* {112} truncating the edges of the dodecahedron. Both forms by themselves and in combination are common on the mineral garnet.

6. *Trisoctahedron (Trigon Trioctahedron)* {*hll*}. The trisoctahedron is composed of 24 isosceles triangular faces, each of which intersects two of the crystallographic axes at unity and the third axis at some multiple. There are various trisoctahedrons, the faces of which have different inclinations, but most common is {122} (Fig. 2.103). The trisoctahedron, like the trapezohedron, is a form that may be conceived as an octahedron, each face of which has been raised to accommodate three others. It is frequently spoken of

as the trigon-trioctahedron, indicating that its faces have three edges and thus differ from those of the trapezohedron (or tetragon-trioctahedron). Figure 2.103 shows a combination of an octahedron and trisoctahedron.

7. *Hexoctahedron (Hexaoctahedron)* {*hkl*}. The hexoctahedron is composed of 48 triangular faces, each of which intersects all three crystallographic axes at different lengths. There are several hexoctahedrons that have varying ratios of axial intercepts. A common hexoctahedron has indices {123}. Other hexoctahedrons have indices {124}, {135}, and so on, or, in general, {*hkl*}. Figure 2.103 shows a simple hexoctahedron as well as combinations with other isometric forms.

Determination of Indices of Forms. In determining the forms present on any crystal in this class it

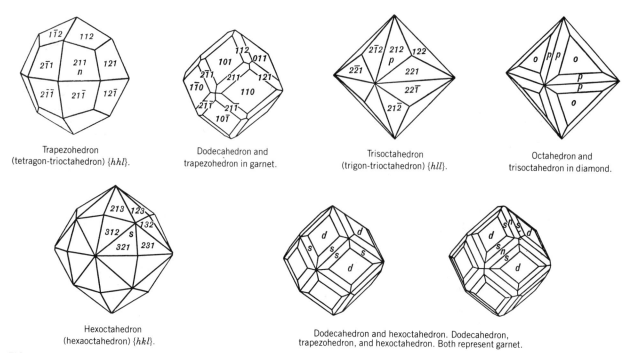

Trapezohedron
(tetragon-trioctahedron) {hhl}.

Dodecahedron and
trapezohedron in garnet.

Trisoctahedron
(trigon-trioctahedron) {hll}.

Octahedron and
trisoctahedron in diamond.

Hexoctahedron
(hexaoctahedron) {hkl}.

Dodecahedron and hexoctahedron. Dodecahedron,
trapezohedron, and hexoctahedron. Both represent garnet.

FIG. 2.103. Additional, commonly developed forms and form combinations in 4/m$\overline{3}$2/m (see also Fig. 2.102).

is first necessary to locate the crystallographic axes (axes of 4-fold symmetry). Once the crystal has been oriented by these axes, the faces of the cube, dodecahedron, and octahedron are easily recognized, because they intersect respectively one, two, and three axes at unit distances. The indices can be quickly obtained for faces of other forms that truncate symmetrically the edges between known faces. The algebraic sums of the h, k, and l indices of two faces give the indices of the face symmetrically truncating the edge between them. Thus, in Fig. 2.103 the algebraic sum of the two dodecahedron faces (101) and (011) is (112), or the indices of a face of a trapezohedron.

Occurrence of Isometric Forms in Class 4/m$\overline{3}$2/m. The cube, octahedron, and dodecahedron are the most common isometric forms. The trapezohedron is also frequently observed as the only form on a few minerals. The other forms, the tetrahexahedron, trisoctahedron, and hexoctahedron, are rare and are ordinarily observed only as small truncations in combinations.

A large group of minerals crystallize in this class. Some of the most common are:

analcime	galena	silver
copper	garnet	spinel group
cuprite	gold	sylvite
diamond	halite	uraninite
fluorite	lazurite	

$\overline{4}3m$

Symmetry—$3\overline{A}_4$, $4A_3$, $6m$. The three crystallographic axes are axes of 4-fold rotoinversion. The four diagonal axes are axes of 3-fold rotation, and there are six diagonal mirror planes, the same planes shown in Fig. 2.101b for class 4/m$\overline{3}$2/m. The location of all of these symmetry elements is shown in Fig. 2.104a. The general form, the hextetrahedron, and its stereogram are illustrated in Fig. 2.104c. This class is known as the *hextetrahedral class*.

Forms

1. *Tetrahedron* {111} *positive*, {1$\overline{1}$1} *negative*. The tetrahedron is composed of four equilateral triangular faces, each of which intersects all the crystallographic axes at equal lengths. The tetrahedral form can be considered as derived from the octahedron in class 4/m$\overline{3}$2/m by the omission of the alternate faces and the extension of the others, as shown in Fig. 2.104b. The positive tetrahedron {111} is shown in Fig. 2.104a. If the other four faces of the octahedron in Fig. 2.104b had been extended, the tetrahedron resulting would have had a different orientation, as shown in Fig. 2.105. This is the negative tetrahedron, {1$\overline{1}$1}. The positive and negative tetrahedrons are geometrically identical. The existence of both must be recognized for they may occur together as shown in Fig. 2.105. If the positive and negative tetrahedron are equally developed on the same crystal, the combi-

$\overline{4}3m$

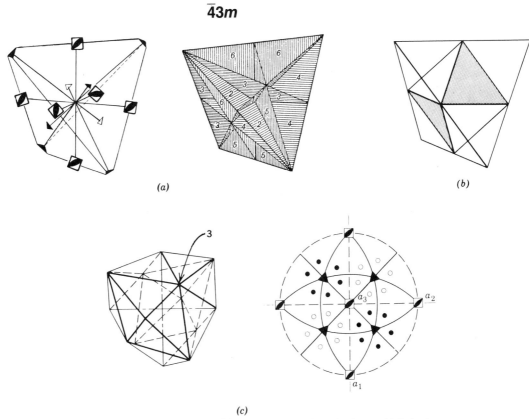

(a)

(b)

(c)

FIG. 2.104. *(a)* Symmetry axes and planes for $\overline{4}3m$. *(b)* Relation of octahedron and tetrahedron. *(c)* Hextetrahedron {*hkl*} and its stereogram. The location of one 3 is noted in the model.

nation could not be distinguished from an octahedron unless, as often happens, the faces of the two forms showed different lusters, etchings, or striations that would serve to differentiate them.

Figure 2.105 illustrates the forms and many of the form combinations that are common in this class.

2. *Tristetrahedron (Trigon-Tritetrahedron)* {*hhl*} *positive,* {*hh̄l*} *negative.* These forms with 12 faces can be conceived as a tetrahedron, each face of which has been raised to accommodate three others. The positive form may be made negative by a rotation of 90° about the vertical axis.

3. *Deltoid Dodecahedron (Tetragon-Tritetrahedron)* {*hll*} *positive,* {*h̄ll*} *negative.* This is a 12-faced form in which three four-sided faces occur in place of one face of the tetrahedron.

4. *Hextetrahedron (Hexatetrahedron)* {*hkl*} *positive,* {*hk̄l*} *negative.* The hextetrahedron has 24 faces that can be viewed as a tetrahedron, each face of which has been raised to accommodate six others.

Members of the tetrahedrite-tennantite series, $(Cu, Fe, Zn, Ag)_{12}Sb_4S_{13}$ to $(Cu, Fe, Zn, Ag)_{12}As_4S_{13}$, are the only common minerals that ordinarily show

distinct hextetrahedral forms. Sphalerite, ZnS, occasionally exhibits them, but commonly its crystals are complex and distorted.

432

Symmetry—$3A_4$, $4A_3$, $6A_2$. This class has all the symmetry axes of the $4/m\overline{3}2/m$ class but lacks mirror planes and a center of symmetry (Fig. 2.106). This class is known as the *gyroidal class* after the general form.

Forms. The *gyroid (pentagon-trioctahedron)* {*hkl*} *right,* {*khl*} *left.* These two forms each have 24 faces and are enantiomorphic, with a left-handed and a right-handed form (see Fig. 2.106). All the forms of class $4/m\overline{3}2/m$, with the exception of the hexoctahedron, can be present in the gyroidal class.

For many years cuprite was considered gyroidal, but more recent studies have shown that it is probably hexoctahedral. With the elimination of cuprite, no known mineral crystallizes in this class.

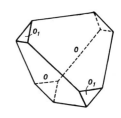

Tetrahedrons
Positive {111} and negative {1$\bar{1}$1}.

Combination (+) and (−).

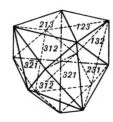

Tristetrahedron.
(trigon-tritetrahedron) {hhl}.

Deltoid dodecahedron
(tetragon-tritetrahedron) {hll}.

Positive hextetrahedron
(hexatetrahedron) {hkl}.

Combinations of cube and tetrahedron.

Tetrahedron and dodecahedron.

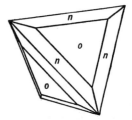

FIG. 2.105. Commonly developed forms and form combinations in $\bar{4}3m$.

Dodecahedron, cube, and tetrahedron.

Tetrahedron and tristetrahedron.

FIG. 2.106. Enantiomorphic forms of the gyroid {hkl}, with left and right forms, and a stereogram of the left-handed form. The location of one 3 is noted on the right-handed model.

432

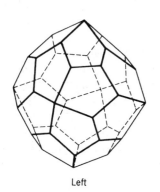

Left

Right

Gyroid (pentagon–trioctahedron)

98

$2/m\overline{3}$

Symmetry—$3A_2$, $4\overline{A}_3$, $3m$. The three crystallographic axes are axes of 2-fold rotation; the four diagonal axes, each of which emerges in the middle of an octant, are axes of 3-fold rotoinversion; the three axial planes are mirror planes. This class has a center of symmetry because $\overline{3}$ is equivalent to $3 + i$. The combination of symmetry elements and the positive diploid and its stereogram are shown in Fig. 2.107. This class is known as the *diploidal class* after the general form.

Forms

1. *Pyritohedron* (*Pentagonal Dodecahedron* or *Dihexahedron*) {h0l} *positive*, {0kl} *negative*. This form consists of 12 pentagonal faces, each of which intersects one crystallographic axis at unity, intersects the second axis at some multiple of unity, and is parallel to the third. A rotation of 90° about a crystallographic axis brings the positive pyritohedron into the negative position. There are a number of pyritohedrons that differ from each other with respect to the inclination of their faces. The most common positive pyritohedron has indices {102} (Fig. 2.108). This figure also shows the corresponding negative pyritohedron.

2. *Diploid* (*Didodecahedron*) {hkl} *positive*, {khl} *negative*. The diploid is a rare form composed of 24 faces that correspond to one-half of the faces of a hexoctahedron. The diploid may be pictured as having two faces built up on each face of the pyritohedron. As in the case of the pyritohedron, a rotation of 90°

$2/m\overline{3}$

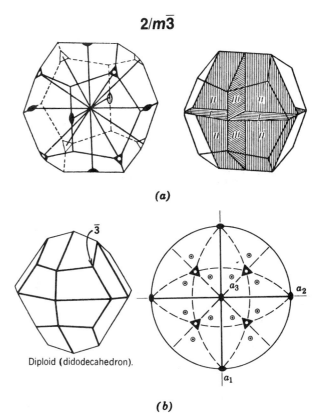

(a)

Diploid (didodecahedron).

(b)

FIG. 2.107. *(a)* Symmetry axes and planes for $2/m\overline{3}$, and *(b)* the diploid {hkl} and its stereogram. The location of one $\overline{3}$ is noted on the model.

FIG. 2.108. Commonly developed forms and form combinations in $2/m\overline{3}$.

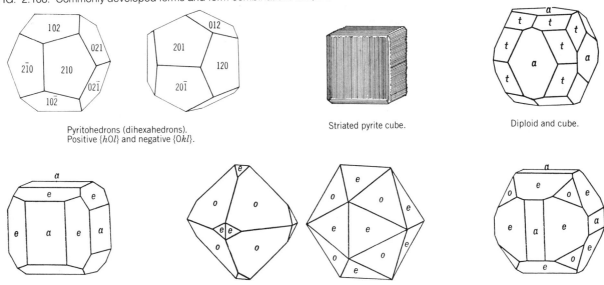

Pyritohedrons (dihexahedrons).
Positive {h0l} and negative {0kl}.

Striated pyrite cube.

Diploid and cube.

Cube and pyritohedron.

Octahedron and pyritohedron.

Pyritohedron, cube, and octahedron.

about one of the crystallographic axes brings the positive diploid into the negative position.

In addition to the pyritohedron and the diploid, the cube, dodecahedron, octahedron, trapezohedron, and trisoctahedron may be present. On some crystals these forms may appear alone and so perfectly developed that they cannot be distinguished from the forms of class $4/m\overline{3}2/m$. This is often true of octahedrons and cubes of pyrite. Usually, however, they will show by the presence of striation lines or etch figures that they conform to the symmetry of class $2/m\overline{3}$. This is shown in Fig. 2.108 by a cube of pyrite with characteristic striations showing the lower symmetry. Figure 2.108 also shows combinations of the pyritohedron with forms of the hexoctahedral class, as well as a combination of cube and diploid {124}.

The chief mineral crystallizing in this class is pyrite (FeS_2); other rarer minerals are members of the skutterudite-nickel skutterudite series ($CoAs_{2-3}$ to $NiAs_{2-3}$), gersdorffite (NiAsS), and sperrylite ($PtAS_2$).

23

Symmetry—$3A_2$, $4A_3$. The three crystallographic axes are axes of 2-fold rotation, and the four diagonal directions are axes of 3-fold symmetry. Figure 2.109 shows drawings of the positive left and positive right tetartoids and a stereogram of a positive right form. This class is known as the *tetartoidal class*.

Forms. There are four separate forms of the *tetartoid* (*pentagon-tritetrahedron*). These are: positive right {hkl}, positive left {khl}, negative right {k̄h̄l}, and negative left {hk̄l}. They comprise two enantiomorphic pairs, positive right and left, and negative right and left. Other forms that may be present are the cube, dodecahedron, pyritohedron, tetrahedron and

deltoid dodecahedron. Cobaltite, (Co,Fe)AsS, is the most common mineral representative crystallizing in this class.

See Table 2.13 for a listing of forms in the isometric system.

Characteristics of Isometric Crystals

Four 3-fold symmetry axes are common to all isometric crystals. Symmetrically developed crystals are equidimensional in the three directions of the crystallographic axes. The crystals commonly show faces that are squares, equilateral triangles, or these shapes with truncated corners. All forms are *closed forms*. Thus, crystals are characterized by a large number of similar faces; the smallest number of any form of the hexoctahedral class is six.

Some important *interfacial angles* of the isometric system that may aid in the recognition of the most common forms are as follows:

Cube (100) $\wedge$ cube (010) = 90°00′

Octahedron (111) $\wedge$ octahedron ($\overline{1}11$) = 70°32′

Dodecahedron (011) $\wedge$ dodecahedron (101) = 60°00′

Cube (100) $\wedge$ octahedron (111) = 54°44′

Cube (100) $\wedge$ dodecahedron (110) = 45°00′

Octahedron (111) $\wedge$ dodecahedron (110) = 35°16′

INTERGROWTHS OF CRYSTALS

In the previous section of this chapter we have described the morphology (external form) and inherent symmetry of, at times, highly idealized single crystals.

FIG. 2.109. Enantiomorphic forms of the tetartoid. Positive left and positive right and a stereogram of the positive right {hkl} form. The location of one 3 is noted on the positive right model.

23

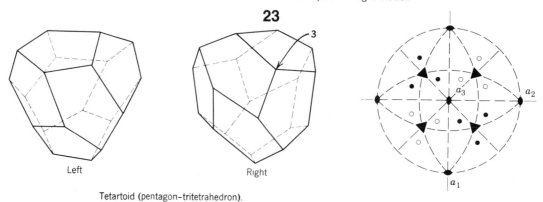

Left   Right

Tetartoid (pentagon–tritetrahedron).

Table 2.13

DISTRIBUTION OF FORMS IN THE ISOMETRIC SYSTEM*

No. of Faces	Name of Form	Class					Unique Form for
		23	432	$2/m\bar{3}$	$\bar{4}3m$	$4/m\bar{3}2/m$	
4	Tetrahedron	+			+		
6	Cube	+	+	+	+	+	
8	Octahedron		+	+		+	
12	Dodecahedron	+	+	+	+	+	
	Pyritohedron	+		+			
	Tristetrahedron	+			+		
	Deltoid dodecahedron	+			+		
	Tetartoid	+					23
24	Tetrahexahedron		+		+	+	
	Trapezohedron		+	+		+	
	Trisoctahedron		+	+		+	
	Hextetrahedron				+		$\bar{4}3m$
	Diploid			+			$2/m\bar{3}$
	Gyroid		+				432
48	Hexoctahedron					+	$4/m\bar{3}2/m$

*From Buerger, M. J., 1956, *Elementary Crystallography*. John Wiley & Sons, New York, 528 pp.

Such well-formed (euhedral), relatively large, single crystals are rare and much sought after by mineral collectors and museum curators. But most minerals occur as random aggregates of grains in the rocks of the earth. These grains are generally anhedral (lacking external faces), but, being crystalline, they possess an internal order evidenced by cleavage, optical properties, and X-ray diffraction.

There are, however, some relatively common intergrowth patterns of well-formed crystals (as well as anhedral grains) that are *not* random in nature. Such are *parallel growths* of the same crystalline substance and crystallographically oriented over-growths of one crystalline substance on another of different composition. The latter is known as *epitaxis*. Yet another type of crystallographically controlled (nonrandom) intergrowth occurs when two crystals of the same substance are related by a symmetry element that is not normally present in either individual crystal. Such crystallographically controlled intergrowths are called *twins* or *twinned crystals*. We will briefly discuss some of the nonrandom overgrowths (parallel growth and epitaxis), and we will subsequently discuss at greater length the morphological aspects of twinning because of its relatively common occurrence.

Parallel Growth and Epitaxis

An aggregate of identical crystals with their crystallographic axes and faces parallel is called a *parallel growth*. Such aggregates, although they may at first appear to represent several crystals, are a single crystal because the internal (atomic) structure remains unchanged in orientation throughout the specimen. Figure 2.110 illustrates some types of parallel growth as found in quartz and barite. Such overgrowths are likely to occur because they represent, on an atomic scale (at the interface of the two or more individuals), a lower potential energy for the crystals than if they were intergrown randomly.

When two compositionally different crystalline substances show a nonrandom overgrowth, it is known as *epitaxis*. Although the two intergrown crystals will tend to have different structures (and unit cell sizes) because they are compositionally distinct, there will be planes in their internal structures where there is a good fit (or the least amount of misfit) between the two individuals. As shown in Fig. 2.111a, staurolite (monoclinic) may occur in parallel growth with kyanite (triclinic) with the (010) plane of staurolite parallel with the (100) plane of kyanite. In this occurrence the (010) in staurolite is similar in atomic spacings to the (100) in kyanite. Figure 2.111b illustrates the epitaxial overgrowth of a plagioclase feldspar (oligoclase) on a crystal of microcline. The common direction between the two different structures is [100] with the planes of attachment being (001) in plagioclase and (001) and (010) in microcline. These three planes have a fairly good (but not quite perfect) fit between them in terms of the internal structure and atomic spacing for the two minerals involved.

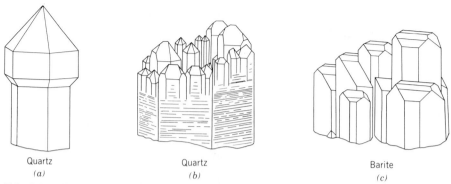

Quartz
(a)

Quartz
(b)

Barite
(c)

FIG. 2.110. Examples of parallel growth. *(a)* Overgrowth of a larger crystal of quartz on a smaller one, forming the shape of a scepter. *(b)* The termination of a large quartz cyrstal in a collection of smaller crystals, all in parallel orientation. *(c)* Parallel intergrowth of barite crystals.

Twinning

A twin is a symmetrical intergrowth of two (or more) crystals of the same substance. Such crystallographically controlled intergrowths are called *twinned crystals.* The two or more individuals of the twinned aggregate are related by a symmetry element that is absent in a single (untwinned) crystal. The new symmetry element (*twin element*) brings one individual (crystal) into coincidence with another individual (crystal) in a twinned position. It is generally necessary to do careful morphological measurements (by reflecting goniometer) as well as X-ray diffraction studies (mainly by X-ray precession methods) to distinguish a twin from a random intergrowth of crystals.

The operations (*twin elements*) that may relate a crystal to its twinned counterpart are: (1) reflection by a mirror plane (*twin plane*); (2) rotation about a crystal direction common to both (*twin axis*) with the angular rotation normally 180°; and (3) inversion about a

point (*twin center*). Twinning is defined by a *twin law,* which states whether there is a center, an axis, or a plane of twinning, and gives the crystallographic orientation for the axis or plane. A twin plane is identified by its Miller index (e.g., (010)) and a twin axis direction is identified by a zone symbol (e.g., [001]).

The surface on which two individuals are united is known as the *composition surface.* If this surface is a plane, it is called the *composition plane.* The composition plane is commonly, but not invariably, the twin plane. If the twin law can be defined only by a twin plane, however, the twin plane is always parallel to a possible crystal face but never parallel to a plane of symmetry. The twin axis is a zone axis or a direction perpendicular to a rational lattice plane; but it can never be an axis of even rotation (2-, 4-, 6-fold) if the twin rotation involved is 180°. In some crystals a 90° rotation about a 2-fold axis can be considered a twin operation.

Twinned crystals are usually designated as either *contact twins* or *penetration twins.* Contact twins have a definite composition surface separating the two individuals, and the twin is defined by a twin plane such as ($\bar{1}\,\bar{1}1$) in Fig. 2.112a. This ($\bar{1}\,\bar{1}1$) plane is one of four possible and crystallographically equivalent directions in the octahedron {111} of the isometric system. Therefore, if one wishes to describe all possible octahedral twin planes, one uses the {111} form symbol instead of the notation (111) for a specific plane. Penetration twins are made up of interpenetrating individuals having an irregular composition surface, and the twin law is usually defined by a twin axis direction (e.g., [111] or [001]; see Figs. 2.112d to f).

Repeated or *multiple twins* are made up of three or more parts twinned according to the same law. If all the successive composition surfaces are parallel, the resulting group is a *polysynthetic twin* (Figs. 2.113a, b, and c). If successive composition planes are

FIG. 2.111. Examples of epitaxis. *(a)* Parallel growth of staurolite ($Fe_2Al_9O_6(SiO_4)_4(O,OH)_2$) and kyanite ($Al_2SiO_5$). *(b)* Oligoclase ($NaAlSi_3O_8$ with about 13% substitution by $CaAl_2Si_2O_8$) overgrowths on microcline ($KAlSi_3O_8$). (This figure from Kern, R. and Gindt, R., 1958, *Bulletin Societé Française Min. Cryst.*, vol. 81, p. 264.)

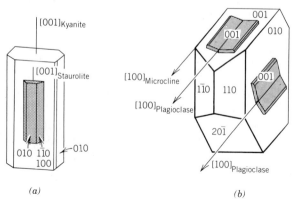

(a)

(b)

Contact Twins

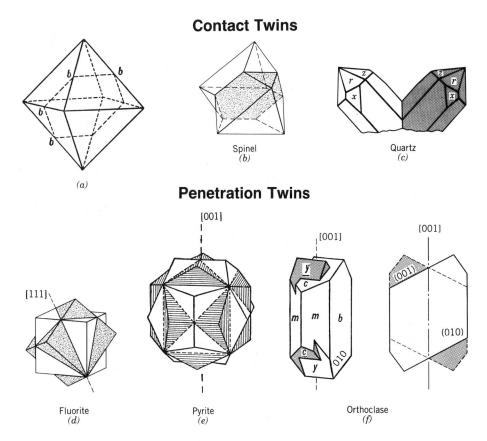

FIG. 2.112. *(a)* Octahedron with possible twin plane *b*-*b*($\overline{1}11$). This is one of four octahedral directions in the form {111}. *(b)* Octahedral twinning {111} as shown by spinel. *(c)* Right- and left-handed quartz crystals twinned along (11$\overline{2}$2), the *Japan twin* law. *(d)* Two interpenetrating cubes of fluorite twinned on [111] as the twin axis. *(e)* Two pyritohedral crystals (of pyrite) forming an *Iron Cross*, with twin axis [001]. *(f)* Orthoclase exhibiting the *Carlsbad twin* law in which two interpenetrating crystals are twinned by 180° rotation about the *c* axis, [001] direction. The schematic cross section, parallel to (010), reveals the presence of the 2-fold twin axis along [001].

Spinel *(b)*

Quartz *(c)*

Penetration Twins

Fluorite *(d)*

Pyrite *(e)*

Orthoclase *(f)*

Polysynthetic Twins

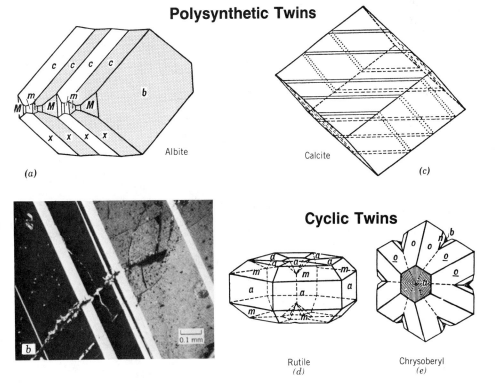

FIG. 2.113. *(a)* Albite polysynthetically twinned on {010}. *(b)* The same polysynthetic twinning as in *a* but as seen in a polarizing microscope. The dark and light lamellae in albite are related by reflection across (010). *(c)* Polysynthetic twinning in calcite on ($\overline{1}$012) which is one of the three directions of the negative rhombohedron. *(d)* Cyclic twin in rutile with the twin planes parallel to faces of the form {011}. *(e)* Cyclic twin in chrysoberyl with the twin planes parallel to faces of the form {031}.

Albite *(a)*

Calcite *(c)*

Cyclic Twins

Rutile *(d)*

Chrysoberyl *(e)*

not parallel, a *cyclic twin* results (Figs. 2.113*d* and *e*). When a large number of individuals in a polysynthetic twin are closely spaced, crystal faces or cleavages crossing the composition planes show striations, owing to the reversed positions of adjacent individuals.

Twinning in the lower symmetry groups generally produces a resulting aggregate symmetry higher than that of each individual because the twin planes, or twin axes, are added symmetry elements.

The origin of twinning will be discussed in Chapter 3 (pages 164–168) after we have introduced the concepts of order and lattice. Here we will concern ourselves only with the morphological expression of twinning.

Common Twin Laws

Triclinic System

The feldspars best illustrate the twinning in the triclinic system. They are almost universally twinned according to the *albite law*, with the {010} twin plane, as shown in Figs. 2.113*a* and *b*. Another important type of twinning in triclinic feldspar is according to the *pericline law*, with [010] the twin axis. When, as frequently occurs in microcline, albite and pericline twins are closely interwoven, a typical cross-hatched or "tartan" pattern can be seen under a polarizing microscope (Fig. 2.114). In addition, triclinic feldspars twin according to the same laws as monoclinic feldspars (see below).

Monoclinic System

In the monoclinic system twinning on {100} and {001} is most common. Figure 2.115 illustrates gypsum with

FIG. 2.114. Photomicrograph of transformation twinning (see p. 156) in microcline. The specimen is viewed under a microscope with crossed polarizers. The section of the photograph is approximately parallel to (001). The twin laws represented are albite with twin and composition plane (010), and pericline with twin axis direction [010].

{100} the twin plane (*swallow-tail twin*). This same figure also shows three twin laws that occur in the mineral orthoclase. Two of these are contact twins: a *Manebach twin* with {001} as the twin plane, and a *Baveno twin* with {021} as the twin plane. The most common twin in orthoclase is the *Carlsbad twin*, an interpenetration twin in which the *c* axis, [001], is the twin element. In this case the two individuals are united on an irregular surface roughly parallel to (010).

Orthorhombic System

In the orthorhombic system the twin plane is most commonly parallel to a prism face. The contact twin of aragonite and the cyclic twins of aragonite and cerussite are all twinned on {110} (see Figs. 2.116*a* and

FIG. 2.115. Examples of common twin laws in monoclinic crystals.

Monoclinic Twins

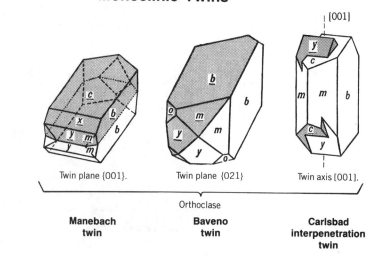

Gypsum.
Twin plane {100}.

Twin plane {001}.

Twin plane {021}

Twin axis [001].

Orthoclase

Swallow-tail twin

Manebach twin

Baveno twin

Carlsbad interpenetration twin

Orthorhombic Twins

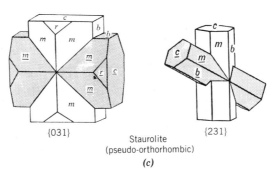

FIG. 2.116. Examples of common twins in orthorhombic crystals. *(a)* Contact and cyclic twinning on {110} in aragonite. *(b)* A cyclic twin on {110} in cerussite. *(c)* Staurolite twinned on {031} and {231}. The staurolite structure is actually monoclinic with β = 90°; it therefore appears pseudo-orthorhombic. It is illustrated here because of its orthorhombic-looking morphology.

b). The pseudohexagonal appearance of the cyclically twinned aragonite results from the fact that (110) ∧ (1$\bar{1}$0) is nearly 60°.

The mineral staurolite, which is monoclinic with a β angle of 90°, is pseudo-orthorhombic and morphologically appears orthorhombic. It is commonly found in two types of penetration twins. In one, with {031} as twin plane, a right angle cross results; in the other, with twin plane {231}, a 60° cross is formed (Fig. 2.116*c*).

Tetragonal System
The most common type of twin in the tetragonal system has {011} as the twin plane. Crystals of cassiterite and rutile, twinned according to this law, are shown in Fig. 2.117.

Hexagonal System
In the hexagonal system carbonates, especially calcite, serve as excellent illustrations of three twin laws. The twin plane may be {0001}, with c the twin axis (Fig. 2.118*a*), or it may be the positive rhombohedron {10$\bar{1}$1}. But twinning on the negative rhombohedron {01$\bar{1}$2} is most common and may yield contact twins or polysynthetic twins as the result of pressure (Fig. 2.118*b*). The ease of twinning according to this law

can be demonstrated by the artificial twinning of a cleavage fragment of Iceland spar by the pressure of a knife blade.

In the class 622, quartz shows several types of twinning. Figure 2.118*c* illustrates the *Brazil law* with the twin plane parallel to {11$\bar{2}$0}. Here, right- and left-

FIG. 2.117. Examples of common twin laws in tetragonal crystals.

Tetragonal Twins

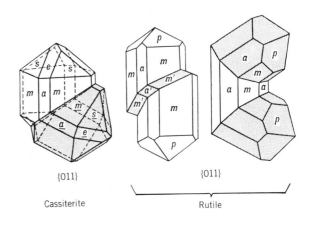

Cassiterite   Rutile

Hexagonal Twins

FIG. 2.118. Examples of twins in the hexagonal system. *(a)* and *(b)* Various twins in calcite. The calcite twin on the right is artificial and can be produced by pressure with a knife edge. *(c)* A *Brazil twin* in quartz. *(d)* A *Dauphiné twin* in quartz formed by rotation of 180° about the c-axis is, [0001]; see also Fig. 3.55. *(e)* A *Japan twin* in quartz.

handed individuals have formed a penetration twin. Figure 2.118d shows a *Dauphiné twin*, a penetration twin with c the twin axis. Such twins are composed either of two right-hand or two left-hand individuals (see also page 168). Figure 2.118e illustrates the *Japan law* with the twin plane {11$\bar{2}$2}. The reentrant angles usually present on twinned crystals do not show on either Brazil or Dauphiné twins.

Isometric System

In the holohedral class of the isometric system (4/m$\bar{3}$2/m) the twin axis, with a few rare exceptions, is a 3-fold symmetry axis, and the twin plane is thus parallel to a face of the octahedron. Figures 2.112a and b show an octahedron with plane bb a possible twin plane, as well as an octahedron twinned according to this law, forming a contact twin. This type of twin is especially common in gem spinel and hence is called a *spinel twin*. Figure 2.112d shows two cubes

forming a penetration twin with the 3-fold rotoinversion axis [111] the twin axis.

In the class 2/m$\bar{3}$, two pyritohedrons may form a penetration twin (Fig. 2.112e) with a 90° rotation about the twin axis [001]. This twin is known as the *iron cross*.

REFERENCES AND SUGGESTED READING

Bloss, F. D., 1971, *Crystallography and Crystal Chemistry.* Holt, Rinehart and Winston, New York, 545 pp.

Boisen, M. B., Jr. and Gibbs, G. V., 1990, *Mathematical Crystallography*, rev. ed. *Reviews in Mineralogy*, v. 15, Mineralogical Society of America, 406 pp.

Boldyrev, A. K., 1936, Are there 47 or 48 simple forms possible in crystals? *American Mineralogist*, v. 21, pp. 731–734.

Buerger, M. J., 1978, *Elementary Crystallography: An Introduction to the Fundamental Geometric Features of Crystals*, rev. ed. MIT Press, Cambridge, Mass., 528 pp.

————, 1971, *Introduction to Crystal Geometry*. McGraw-Hill Book Co., New York, 204 pp.

Donnay, J. D. H. and Curien, H., 1958, Nomenclature des 47 formes cristallines. *Bulletin Societé Française Minéralogie et Cristallographie*, v. 81, pp. XLIV–XLVII.

Goldschmidt, V., 1913–1923, *Atlas der Kristallformen* [9 volumes and 9 atlases]. Heidelberg, Universitätsbuchhandlung.

International Tables for X-ray Crystallography, 1969, v. 1, N. F. M. Henry and K. Lonsdale, eds. Symmetry Groups, International Union of Crystallography, Kynoch Press, Birmingham, England, 558 pp.

Klein, C., 1989, *Minerals and Rocks: Exercises in Crystallography, Mineralogy, and Hand Specimen Petrology*. John Wiley & Sons, New York, 402 pp.

Phillips, F. C., 1971, *An Introduction to Crystallography*. John Wiley & Sons, New York, 351 pp.

Rogers, A. F., 1935, A tabulation of crystal forms and discussion of form-names. *American Mineralogist*, v. 20, pp. 838–851.

Sands, D. E., 1975, *Introduction to Crystallography*. W. A. Benjamin, Inc., New York, 166 pp.

CHAPTER 3

CRYSTALLOGRAPHY: INTERNAL ORDER AND SYMMETRY

In Chapter 1 we gave a definition of a mineral as follows: A mineral is a naturally occurring homogeneous solid with a definite (but generally not fixed) chemical composition and a highly ordered atomic arrangement. It is usually formed by inorganic processes. In this definition for a mineral, or any crystalline substance, the fundamental clause is an *ordered atomic arrangement*. This is equivalent to saying that crystalline materials are *periodic* in their atomic build-up; that is, they internally exhibit *periodic translations*, along a set of chosen coordinate axes. Such atomic order (or periodicity) distinguishes the crystalline state from liquids and gases, as well as glasses.

In Chapter 2 we discussed the external form and inherent symmetries of minerals. This led us to the formulation of the 32 crystal classes, or point groups. Before the discovery in 1912 of X-ray diffraction by minerals, it had long been suspected that the regular (and symmetrical) external form of euhedral crystals was a reflection of some type of internal order. Indeed, René J. Haüy had suggested as early as 1784 that crystals were built by stacking together tiny identical building blocks, which he referred to as "integral molecules" (see Fig. 1.4). The concept of "integral molecules" is essentially that of unit cells in modern X-ray crystallography.

Now that we are about to address the *three-dimensional periodicity* of mineral atomic structures, we should look once more at the statement that the internal structure of minerals is based on an *ordered atomic arrangement*. This statement implies that a certain atom (or ion) is present in exactly the same structural (atomic) site throughout an essentially infinite atomic array. An atom in the same atomic site means that it is surrounded by an identical arrangement of neighboring atoms, throughout the structure which consists of millions of unit cells with dimensions on the order of 5 to 20 angstroms (Å, equivalent to 0.5 to 2 nanometers, nm). Such complete order is present in "ideal" crystals, of which examples of the internal structure are shown in Figs. 1.12 and 3.19. However, high-resolution transmission electron microscopy (HRTEM), with magnification levels on the order of $1,000,000\times$, has shown that atomic arrangements at and below the level of the unit cell may deviate considerably from perfect crystalline periodicity. Such enormous magnifications, which allow for the resolution of patterns and atoms on an angstrom level, are extremely far removed from those of visual observation of minerals (at $1\times$ magnification) and those of optical microscopy (at magnification levels of about $1000\times$). This means that a mineral that appears to be homogeneous during optical microscopic observation may be found to be inhomogeneous when studied by X-ray diffraction and, even more strikingly, with HRTEM techniques. Such inhomogeneities may be due to structural misfits (defects) or to chemical zonation. Furthermore, all structures show increased thermal vibrations of the atoms as a function of increased temperature. This may result in a random distribution of atoms (or ions) that at lower temperatures would be located in one or more specific structural (atomic) sites. Such randomization of atoms in a structure as a function of increased temperature is referred to as "disorder." The imperfections in crystals and the disorder of atoms in an otherwise periodic structure are considered as local phenomena in an overall well-crystallized structure with three-dimensional periodicities.

In this chapter we will first discuss ordered arrangements, beginning with examples of one- and two-dimensional order before we evaluate three-dimensional order. We will subsequently introduce the concepts of structural disorder and imperfections (defects), the metamict and amorphous state, and structural aspects of twinning.

A crystal structure may be thought of as a repetition of a motif or a group of atoms on a lattice, or periodic array of points. The ordered patterns that characterize crystalline materials represent a lower energy state than random patterns. Intuitively, a brick wall constructed with carefully positioned bricks in an ordered arrangement provides a more stable (and less energetic configuration) than a brick wall made with a random arrangement of identical bricks (see Fig. 3.1a). The brick in these walls can be considered as the motif and can be replaced by a comma, as in Fig. 3.1b. This analogy shows how an ordered pattern is generated by a motif repeated in a regular sequence of new locations. Any motion that brings the original motif into coincidence with the same motif elsewhere in the pattern is referred to as an *operation*. Many wallpaper patterns, for example, are based on a two-dimensional pattern in which the motif (sprays of flowers, dots, figurines) is arranged in a regular geometrical pattern. More abstract and less symmetrical arrangements, as are found in many contemporary wallpapers, may have a much less clearly defined pattern.

TRANSLATION DIRECTIONS AND DISTANCES

As stated earlier, a crystal is a homogeneous solid possessing, long range, *three-dimensional internal order*. Such order is the result of the repeat of motif units (these are chemical units—for example, copper atoms) by regular translations in three dimensions. The three-dimensional pattern is said to be *homogeneous* if the angles and distances from one motif to surrounding motifs in one location of the pattern are the same in all parts of the pattern (see Fig. 3.2).

FIG. 3.1 *(a)* Two-dimensional ordered and more random arrangement of bricks in a brick wall. *(b)* The brick patterns are represented by commas as motifs.

FIG. 3.2 *(a)* A two-dimensional pattern with translation components t_1 and t_2 at 90° to each other. *(b)* A two-dimensional pattern with translation components t_1 and t_2 at <90° to each other. *(c)* A three-dimensional pattern with translation components t_1, t_2, and t_3. None of the translation components make 90° with each other.

(a)

(b)

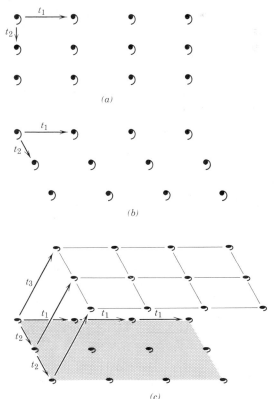

(c)

Figure 3.2a illustrates a two-dimensional array of motifs (commas). The order in such an array can be expressed in terms of two translations (t_1 and t_2) at 90° to each other. A somewhat less symmetrical pattern of motifs is shown in Fig. 3.2b, where the t_1 direction is the same as in Fig. 3.2a but the t_2 translation is at an angle <90° to the t_1 translation. These two illustrations can be thought of as infinite strings of units along the t_1 direction that have been repeated by parallel and identical infinite strings along a translation direction t_2. The translations, as marked by t_1 and t_2, are vectors.

A three-dimensional ordered pattern can be obtained by adding yet another translation component (t_3) that does not lie in the plane of t_1 and t_2 (see Fig. 3.2c). This results in a pattern that is infinite in three dimensions. In crystals, such a pattern is not exactly infinite, although it is generally considered so. The magnitude of the translations in inorganic crystals is on the order of 1 to 10 angstroms[1] (Å, where Å = 10^{-8} cm), because that is the scale of ionic radii in crystals. This means that a dimension of 1 cm in a crystal would contain approximately 100 million translations; indeed this can be considered infinite!

It is often convenient to ignore the actual shape of the motif units in a pattern and to concentrate only on the geometry of the repetitions in space. If the motifs (commas in Fig. 3.2) are replaced by points, we have a regular pattern of points that is referred to as a lattice. *A lattice is, therefore, an imaginary pattern of points (or nodes) in which every point (node) has an environment that is identical to that of any other point (node) in the pattern. A lattice has no specific origin, as it can be shifted parallel to itself.*

A three-dimensional crystal structure can be viewed as the result of three-dimensional translations acting upon motif units (the chemical units of the structure). The translations inside such a crystal structure are extremely small (on an atomic scale) and cannot be seen by the naked eye. The sizes of these translations are expressed in nanometers (10^{-7} cm) or angstroms. The only way in which such extremely small distances can be imaged is by transmission electron microscopy (see Fig. 1.12). It is very important, therefore, to realize that the external form of a crystal, although an expression of its internal structure, is translation-free. The symmetry elements, observable in the external form development of crystals, are therefore also translation-free.

Let us now develop lattices systematically, beginning with one-dimensional rows.

[1] Named after the Swedish physicist Anders Jonas Ångström, 1814–1874.

ONE-DIMENSIONAL ORDER (ROWS)

A sequence of equally spaced equivalent points (or motifs) along a line represents order in one dimension, or a *row* (see Fig. 3.3). In such a row the magnitude of the *unit translation* (in this case b) determines the spacing. The motif, the unit of pattern or the atom printed at each lattice point, determines the ultimate pattern. Figure 3.3 illustrates several rows of objects with different spacings, b, along a direction defined as y, and with different motifs. Such rows can be found as borders along illustrations, in wallpaper, along friezes, and in the structures of crystalline materials.

TWO-DIMENSIONAL ORDER (PLANE LATTICES)

Two-dimensional order is the result of regular translations in two different directions, designated x and y. Figure 3.4a shows a two-dimensional, ordered array of motifs (commas, in this instance) on which various choices of the x and y coordinate axes (the axes along which the translations take place) have been superimposed. The units of the translation distance are marked as a (or a_1, a_2, etc.) and b (or b_1, b_2, etc.).

Such translational patterns can be described by translation vectors, where the vectors are noted as **a** and **b**, the magnitudes of the translations as a and b, and the coordinate axes along which the translations are repeated as x and y. The angle between the x and y axes is denoted by γ.

Figure 3.4b shows how a regular two-dimensional pattern is produced with two different spacings (the unit of translation along the row = b; the unit of translation between the rows = a). This arrangement is an oblique array of commas because we chose an angle γ between the x and y directions that is neither 90° nor 60° nor 120°. In order to visualize the array of motifs (commas) without reference to the shape of the motif, it is standard practice to replace each motif with a point (thus eliminating any sense of shape or symmetry of the motif) and to connect such points (or nodes) by lines. This creates a two-dimensional *net* or *plane lattice*, as shown in Fig. 3.4c. The regularly spaced points (or nodes) represent the locations of motifs, which in chemical structures may be atoms, ions, molecules, or ionic complexes. The smallest building unit in the two-dimensional pattern of Fig. 3.4c is that of the shaded parallelogram; this is known as the unit cell. If this *unit cell* is repeated indefinitely by translations a and b along directions x and y, the array shown in Fig. 3.4b and the lattice shown in Fig. 3.4c will result.

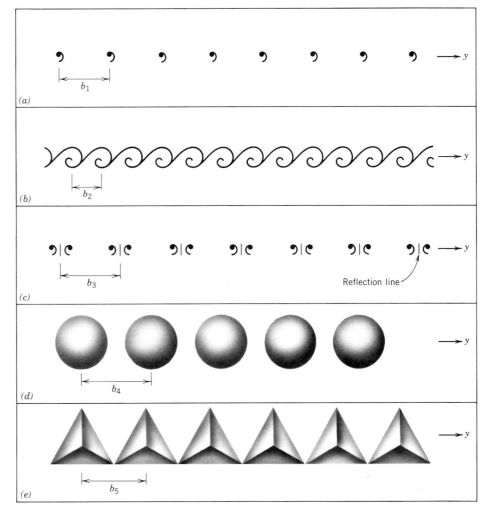

FIG. 3.3 Rows of objects at various spacings b along a translation direction y.
(a) Regularly spaced asymmetric motifs, commas (asymmetric means "without any kind of symmetry"). (b) A scroll design of asymmetric motifs. (c) A row of equally spaced motifs in which one part of the motif is related to the other by a line of reflection (marked by a short line between pairs of motifs).
(d) A row of spheres that may represent atoms in a structure. (e) A row of tetrahedra that may represent anionic complexes such as tetrahedral SiS_4 or GeO_4 groups.

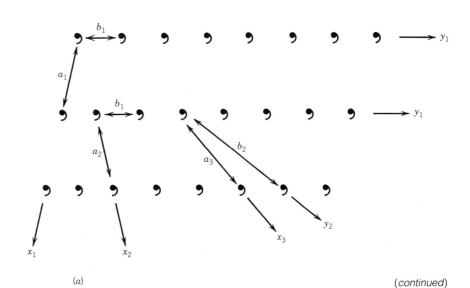

FIG. 3.4 (a) A two-dimensional, ordered array of motifs showing three different ways to generate the array by combining translation directions (x_1, x_2, x_3; y_1, y_2) and distances (a_1, a_2, a_3; b_1, b_2). These three different choices all generate the same pattern.

(a)

(continued)

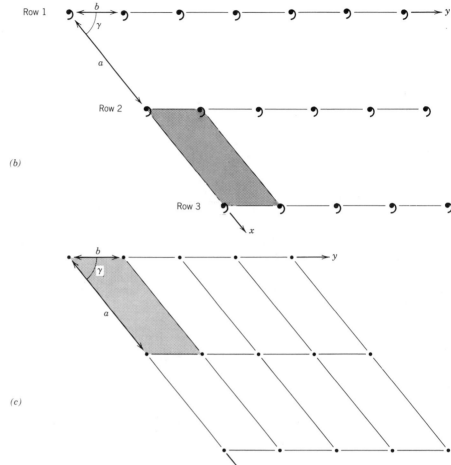

(b)

(c)

FIG. 3.4 (continued)
(b) Yet another choice of translation directions and translation distances. The angle between the two directions of *x* and *y* is γ. This arrangement represents an oblique (γ ≠ 90°), two-dimensional (planar), and ordered array of motifs. *(c)* A planar lattice based on the array of motifs in *(b)*. A lattice is by definition *infinitely extending*, but only finite portions can be shown in illustrations. The shaded parallelogram in *(b)* represents the smallest unit of pattern and a unit cell in the lattice of *(c)*.

There are only five possible and distinct plane lattices (also known as *nets*). The five choices are the result of repeating a row (with translation distance *b* along direction *y*) along direction *x* with repeat distance *a*. The five plane lattice types that result depend on the choice of the angle γ (between directions *x* and *y*; that is, whether γ = 90°, 60°, or some other angle) and on the size of *a* relative to *b* (that is, whether *a* = *b*, or not). These five plane lattices (nets), illustrated in Fig. 3.5, represent the only possible ways to arrange points *periodically* in two dimensions.

In Fig. 3.5*a*, row no. 1 is indefinitely repeated by translations along direction *x*, with translation distance *a*. Here *a* is unequal to *b*; the γ angle is not 90°, and an *oblique lattice* (or *clinonet*) results.

In Fig. 3.5*b*, row no. 1 is indefinitely repeated by translations along direction *x*, with translation distance *a*. Here *a* is unequal to *b*; the angle γ = 90°, and a *primitive rectangular lattice* (or *orthonet*) results (*primitive* implies nodes occur at the corners of the chosen unit cell only).

In Fig. 3.5*c*, row no. 1 is indefinitely repeated by translations along direction *x*, with translation distance *a*, and with angle γ such that cos γ = *a*/2*b*. The resulting lattice (or net) is conventionally described in terms of two orthogonal directions (*x* and *y'*), resulting in centering of the net inside the rectangular unit cell choice. This is known as a *centered rectangular lattice* (or *centered orthonet*). The same array of nodes can be described by two vectors (**a'** and **b'**, where **a'** = **b'** and γ' ≠ 90°, 60°, or 120°), resulting in a *primitive lattice* with a diamond shape. This alternate, primitive lattice choice is referred to as a *diamond lattice*. Either of these two unit cell choices (primitive or nonprimitive [= centered]) will, when repeated indefinitely along two directions, produce the pattern of nodes in Fig. 3.5*c*.

In Fig. 3.5*d*, row no. 1 is indefinitely repeated by translations in direction *x*, with a translation distance of *a*, such that *a* = *b* (or $a_1 = a_2$) and γ = 60°. This results in a *hexagonal lattice* (or *hexanet*).

In Fig. 3.5*e*, row no. 1 is indefinitely repeated by

FIG. 3.5 Development of the five distinct plane lattices (or nets) by the indefinite repeat of a row (along direction y, with specified translation distance b), along direction x with repeat distance a; γ is the angle between x and y. The total symmetry content of each of the unit cell choices is given in the right-hand column. Rotational axes are shown by standard symbols, mirrors by broad lines, and glide lines by dashed lines.

Oblique net
$a \neq b$
$\gamma \neq 90°$

p2

Rectangular net
$a \neq b$
$\gamma = 90°$

p2mm

Centered rectangular net
$\cos \gamma = \dfrac{a}{2b}$
$a \neq b$

Diamond net
$a' = b'$
$\gamma' \neq 90°, 60°,$ or $120°$

c2mm

Hexagonal net
$a = b$ (or $a_1 = a_2$)
$\gamma = 60°$

p6mm

Square net
$a = b$ (or $a_1 = a_2$)
$\gamma = 90°$

p4mm

translations in direction x, with a translation distance of a, such that $a = b$ (or $a_1 = a_2$) and $\gamma = 90°$. This results in a *square lattice* (or *tetranet*).

The smallest units of repeat in these lattices outline the unit cells (shaded in Fig. 3.5). They range from a parallelogram (Fig. 3.5a), to *two types of rectangle* (Figs. 3.5b and c), to a *diamond shape* (Fig. 3.5c), to a *rhombus* (Fig. 3.5d), and a *square* (Fig. 3.5e). In Fig. 3.5c there is a choice of two differently shaped and sized unit cells. The preferable choice is the rectangular (centered, and larger) unit cell because of the orthogonality of its shape (orthogonal means "involving right angles or perpendiculars"), and because the mirror lines inherent in the symmetry of the pattern are parallel to the axial directions of the unit cell. Of the various choices of unit cells in Fig. 3.5, only one contains a central node, namely, the *centered rectangular lattice*. All other choices contain only corner nodes and are referred to as *primitive lattices* (e.g., primitive oblique lattice, etc.).

The five different plane lattices, or nets, in Fig. 3.5 should have been drawn as indefinitely extending. However, limitations on illustration size restrict the number of points (or nodes) that can be shown. The shaded unit cell in the vertical column of nets, on the left side of the figure, is enlarged in the column on the right to illustrate the distribution of symmetry elements in one unit cell of each of the patterns. The locations of the various rotational symmetry axes (2, 3, 4, and 6), perpendicular to the page, are shown by the standard symbols introduced in Chapter 2. The locations of mirrors perpendicular to the page are shown by broad black lines. But remember, there are no mirrors or symmetry axes in the plane of the page (that is, parallel to the page). In the lattices in Figs. 3.5d and e, the dashed lines represent *glide lines*, which are a combination of translation and reflection.

The listing of the symmetry content for each lattice type is similar to that introduced in Chapter 2. However, each symmetry listing is preceded by a small letter *p* (for primitive) or *c* (for centered) to indicate the lattice choice.

Various types of two-dimensional ordered patterns can be seen in our daily surroundings. For example, brick or tiled walls can be regarded as the result of repeating a motif (e.g., a brick or a ceramic tile) along two different translation directions (t_1 and t_2) parallel to the wall. Figure 3.6 shows some examples

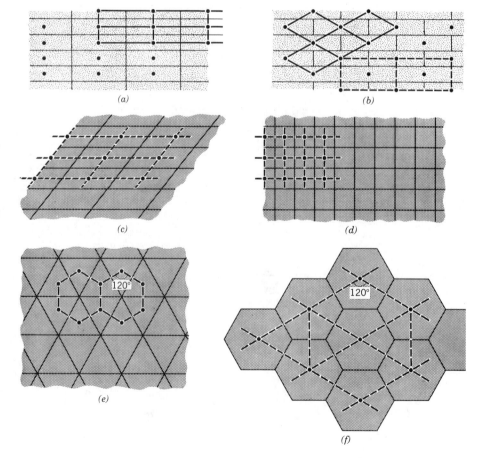

FIG. 3.6 (a) and (b) represent brick walls; (c) through (f) are possible arrangements of tiles. In each case the original motifs (bricks or tiles) are in part replaced by lattice points (nodes). When these are connected, the shape of the lattice becomes obvious. These shapes are: (a) a rectangular lattice; (b) a centered rectangular lattice, or a diamond lattice; (c) an oblique lattice; (d) a square lattice; (e) a hexagonal lattice; and (f) a hexagonal lattice (with a centering node) or a noncentered rhombus lattice.

as well as their lattice types and unit cell choices. Such two-dimensional coverings of a wall by tiles are known as *tesselations*.

It is useful at this time to illustrate that an observer can have a number of choices of unit cell, once the lattice array of motifs has been established. Figure 3.7 shows a regular array of nodes that was originally generated by the two vectorial directions **a** and **b** and angle γ, as shown in the primitive, oblique unit cell marked *A*. Another unit cell choice, which through infinite repeat along two vectors in the plane of the page would have generated the same pattern, is unit cell *B*. The spacings and angle are the same as in *A*. It is an equivalent primitive, oblique unit cell. A third choice, designated as *C*, is also primitive and oblique but with $a_1 = a_2$ and with a different subtended angle. All the other choices, *D*, *E*, and *F*, are *nonprimitive* or *multiple* unit cells, because each contains in addition to parts of each of the four corner nodes (a total of one node), one more node per unit cell. In *D* the additional node is centered (a *centered unit cell*), and in *E* and *F* it is made up of two half nodes on two of the sides. Generally, the smallest unit cell (e.g., *A*, *B*, or *C*) or an orthogonal and centered unit cell (e.g., *D*) would be the most appropriate choice. Figure 3.8 illustrates how a wallpaper designer has used the concepts of symmetry and translation in repetition of a

FIG. 3.8 Choice of alternate unit cells in a wallpaper design. The motif of the wallpaper contains symmetry *2mm*.

motif. Each motif in this design contains a 2-fold rotation axis perpendicular to the page and two reflection lines (*m*) perpendicular to each other (planar point group symmetry 2mm). The pattern of the motifs can be described by a primitive, oblique unit cell with edges a_1 and *b*, but the most appropriate choice, in view of the 2mm symmetry content of the motifs, would be a rectangular centered unit cell with edges a_2 and *b*. For three-dimensional patterns several rules have been made to restrict the possible choices of unit cells (see page 124).

Rotation Angle Restrictions

Now that we have introduced the concepts of two-dimensional ordered arrays and plane lattices, we can evaluate geometrically why certain rotational axes are possible and others (e.g., 5-fold rotation) are not. Figure 3.9 illustrates the geometric restrictions on rotation axes in ordered arrangements that also contain translation. If the motif units, represented by large nodes in Fig. 3.9, are part of an ordered arrangement, then the distances *AB* and *BC* must be equal. If the motif at *B* contains a rotation axis with the axis $\perp$ to the plane of the figure, then the translations require similar axes at *A* and *C*. Furthermore, if points *D*, *E*, *F*, and *G* are related to *B* by a rotation, then $BC = BD = BE = BF = BG = t$. This also means that the dis-

FIG. 3.7 A regular array of nodes that can be generated by the infinite repeat of various unit cell choices (*A*, *B*, *C*, *D*, *E*, or *F*) along two vectors **a** and **b**.

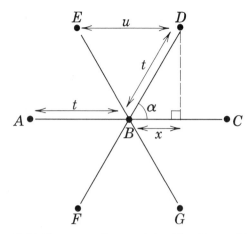

FIG. 3.9 Motifs separated by translation (t) and a possible axis of rotation, perpendicular to the page, at each of the motif units. One axis of rotation at motif B is shown.

tance ED, which lies on a line parallel to AC, must be equal to AB or a multiple thereof. In other words, $ED = u = mt$ where m = integer. If the rotation by which A, F, G, C, D, and E are related is through an angle α, the following geometric relations hold:

$$\cos \alpha = x/t \quad \text{and also} \quad x = \tfrac{1}{2}ED = \tfrac{1}{2}u$$
$$\therefore \cos \alpha = \tfrac{1}{2}u/t = u/2t$$
$$\therefore 2t \cos \alpha = u$$

Combining $u = mt$ and $u = 2t \cos \alpha$ gives

$$mt = 2t \cos \alpha \quad \text{or} \quad \cos \alpha = m/2$$

where m is an integer. This leads to restrictions on the solutions possible for the angle of rotation α. For $m = 2$

$$m/2 = 1, \alpha = 0° \text{ or } 360°$$

For $m = 1$,

$$m/2 = 1/2, \alpha = 60°$$

For $m = 0$,

$$m/2 = 0, \alpha = 90°$$

For $m = -1$,

$$m/2 = -1/2, \alpha = 120°$$

For $m = -2$,

$$m/2 = -1, \alpha = 180°$$

Any other integral values of m produce values of cos α greater or less than ± 1, which is possible but mathematically meaningless. Other rotation angles produce noninteger values of m. For example, a 5-fold rotation axis would require an angle of rotation of 72°. This leads to a value for cos 72° = 0.30902. Such a number cannot equal $m/2$ in which m must be an integer. Therefore, a 5-fold rotation axis is not possible in an ordered, crystalline structure. Five-fold symmetry is, however, not uncommon in objects in the biological world (e.g., in the distribution of petals in a geranium flower).

Symmetry Content of Planar Motifs

Two-dimensional motifs, as are often seen in wallpaper designs, in printed cloth, in ceramic tiles, and elsewhere, can display highly variable symmetry contents. However, because such motifs are printed on one side of a paper (with the other side generally blank), there are no symmetry elements that lie in the plane of the paper. That is, there is no mirror plane parallel to the paper, nor are there axes of rotational symmetry parallel to the paper. However, there may be a number of symmetry elements perpendicular to the plane of the drawing. These are mirror lines (m) (in three-dimensional patterns, m's are referred to as mirror planes; in two-dimensional patterns they are known as mirror lines; their reflection operations are equivalent) and rotation axes (1, 2, 3, 4, and 6). Although one can create an independent motif unit with 5-, or 11-, or more-fold rotational axes, the symmetry content of motif units that are part of a repetitive and ordered (crystalline) array can contain only one of five rotational axes (1, 2, 3, 4, and 6). There are only ten possible symmetry contents for two-dimensional motifs that, through regular translation, can become part of two-dimensional ordered patterns. These are shown in Fig. 3.10. The letters in this figure refer to the rotational symmetry (perpendicular to the page) inherent in the motif, and the m's note the location of mirror lines in various directions; the graphic symbols are the same as those used in Chapter 2. These ten different symmetry contents represent the symmetry about a central (stationary) point and are referred to as the *ten planar point groups*. There are ten because each of the six symmetry elements, 1, 2, 3, 4, 6, and m, can occur individually, and the other four consist of possible combinations of rotational symmetry and mirrors, as in 2mm, 3m, 4mm, and 6mm. The significance of these symbols or groups of symbols is the same as discussed in Chapter 2 for the Hermann-Mauguin (international) notation of point groups (or crystal classes). The numerals refer to rotations about a point. The m's refer to reflection lines. The m's in 2mm and

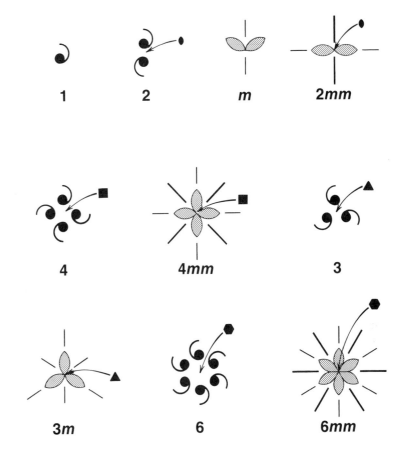

FIG. 3.10 The symmetry content of two-dimensional motifs. Locations of mirror lines (m) are shown by solid lines, and rotational axes by the standard symbols.

3m and the first m in 4mm and 6mm refer to reflection lines that are coincident with the axial directions. These lines are at right angles in orthogonal arrays but at 120° to each other in 6-fold and 3-fold patterns. Thus, in 4mm the first m refers to two reflection lines at right angles to each other (see Fig. 3.10). The second m in 4mm and 6mm refers to intermediate reflection lines, which in 4mm are at 45° to the first set of lines and in 6mm are at 30°. The ten planar point groups shown in Fig. 3.10 are the two-dimensional analogues to the 32 three-dimensional crystal classes (point groups) discussed in Chapter 2.

Symmetry Content of Plane Lattices

The arrangements of nodes (or lattice points) in the five planar lattices reflect various inherent symmetry elements as shown in Fig. 3.5. Illustrations of the complete symmetry of some of the lattice types are given in Fig. 3.11. This illustration shows the motif distributions and symmetry elements compatible with selected lattice types. In Fig. 3.11a we have a primitive oblique unit cell based on a regular distribution of commas. No rotation axis except 1 (which is equivalent to 0 or 360° rotation) is compatible with this pattern; mirror lines are also absent. Because the lattice is primitive (p) and contains only 1-fold rotational symmetry (1), it is referred to as p1. Figure 3.11b shows the same lattice as in a, but now it represents the repeat of two commas related by 2-fold rotation. It is still a primitive oblique unit cell, and it lacks mirror lines. It is described as p2. All of the possible locations of 2-fold rotations are shown in Fig. 3.11c, in an enlarged unit cell drawing of Fig. 3.11b. The 2-fold rotations at the corner nodes are probably obvious. However, there are others, halfway along each of the cell edges, as well as at a location in the very center of the cell. Figure 3.11c shows, by dashed lines, how some of the commas are related by such 2-fold rotations. It is standard procedure (e.g., in *International Tables for X-ray Crystallography*, vol. 1) to represent the total symmetry content of a lattice such as p2 without reference to motif units. This is shown in Fig. 3.11d. A square lattice choice, as in Fig. 3.11e, contains points of 4-fold rotation not only at the corner nodes, but there is an additional 4-fold rotation point at the center if the distribution of commas (as motifs) is carefully evaluated. Furthermore, there are 2-fold rotations at the centers of the edges of the square. The total symmetry content of this primitive square lattice, referred to as p4, is shown in Fig. 3.11f.

It was noted in Fig. 3.3 that reflection lines (mirror

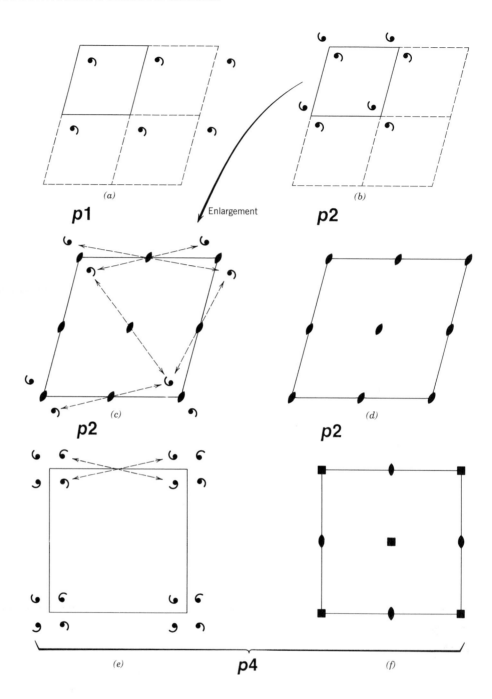

FIG. 3.11 Examples of rotational symmetry elements compatible with a primitive oblique lattice (*a* through *d*) and a primitive square lattice (*e* and *f*). See text for explanation.

symmetry, *m*) are compatible with four of the five planar lattice types (no reflections can be present in the oblique lattice). Mirror symmetry has been discussed in Chapter 2, page 22. Because we are dealing here with the symmetry content of two-dimensional lattices that contain translational elements, we must also concern ourselves with the possible combination of reflection and translation. Such a combined operation (*m* + translation) is referred to as a *glide* operation, a *glide line*, or a *glide reflection*. A glide reflection causes a motif to be reflected across a reflection line and to be translated parallel to the reflection line. Figure 3.12 shows how motif units are related by a glide line (or glide plane) that has a translation component of *t*/2, where *t* is the shortest translation paralleling the glide plane.

Consideration of the ten two-dimensional (planar) point groups (Fig. 3.10) in conjunction with the five plane lattices (Fig. 3.5), and the possibility of glide reflections (*g*) in addition to (or in place of) possible mirror reflections (*m*) leads to the so-called two-dimensional *plane groups*.

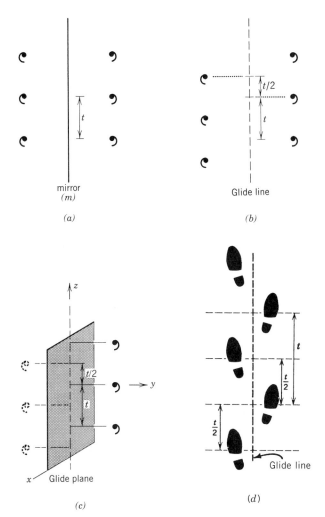

FIG. 3.12 Mirror symmetry and glide planes, *(a)* A two-dimensional array of an asymmetric motif with a spacing of *t* related by a reflection line, or a mirror perpendicular to the page. *(b)* A two-dimensional view of motifs that are related, across a glide line, with a glide component of *t*/2. *(c)* A three-dimensional illustration of a glide plane with glide component *t*/2. (In graphic illustrations, reflection lines and the traces of mirrors perpendicular to the page are shown by solid lines; glide lines and the traces of glide planes perpendicular to the page are shown by dashed lines.) *(d)* Human tracks showing relationship of motifs (footprints) by a glide line. Glide component *t*/2.

Two-Dimensional Plane Groups

Two-dimensional plane groups represent the infinite repetition of motifs on a plane. In the evaluation of such two-dimensional groups we must concern ourselves with the shapes (oblique, rectangular, hexagonal, or square) and the possible multiplicity (*p* for primitive and *c* for centered) of the plane lattice types. We must also take into account the translation-free point groups (2, 3, 4, 6, *m*, 2*mm*, 3*m*, 4*mm*, and 6*mm*) and their compatibility with the lattice types (see Fig. 3.11) as well as the possible presence of glide

reflections (*g*) in place of or in addition to possible mirror reflections (*m*).

Because we have already given examples of combinations of lattice type with planar point group (e.g., *p*1, *p*2, and *p*4 in Fig. 3.11) we will now concern ourselves with combinations of lattice types (*p* or *c*) and *m* or *g*. Both of these symmetry elements are compatible with four of the five plane lattices; an oblique lattice is the only one that *cannot* accommodate these operations. Figure 3.13 shows examples of a rectangular lattice (primitive as well as centered) containing *m* or *g* operations, or both, and an example of a square lattice containing both *m* and *g* operations.

In order to derive all possible two-dimensional plane groups one must consider systematically all possible combinations of lattice types with permissible symmetry elements (or groupings of elements). As we have already seen (Figs. 3.11*a* and *b*), for an oblique cell only *p*1 and *p*2 are appropriate. For a rectangular cell we have other choices of symmetry elements (*m* and *g*) as well as the choice of a primitive (*p*) or a centered (*c*) cell. For example, for point group *m* we might expect *pm*, *pg*, *cm*, and *cg* as possible two-dimensional plane groups. Similarly, for 2*mm* we might consider *pmm*, *cmm*, *pmg*, *pgg*, *cmg*, and *cgg* as permissible two-dimensional plane group choices. When we look at a tabulation of all possible two-dimensional plane groups compatible with point group symmetries *m* and 2*mm* (see Table 3.1), we find that only seven of the above ten possibilities actually occur. Table 3.1 lists the 17 possible two-dimensional plane groups. The reason for this relatively small number is that not all combinations lead to new or different plane groups. Furthermore, the interaction of the symmetry of the motif (planar point group) with the symmetry of the various plane lattices affects the overall resultant symmetry content of the planar pattern. The final pattern displays the symmetry of the lattice when the symmetry elements of the motif are aligned with the corresponding symmetry elements of the lattice. If the motif has less symmetry than the lattice, the pattern will express the motif's lesser degree of symmetry, with the symmetry elements of the motif aligned with the corresponding symmetry elements of the lattice. Figure 3.14 illustrates the 17 possible plane patterns (known as plane groups). More detailed illustrations of each of these two-dimensional plane groups are given in *International Tables for X-ray Crystallography*, vol. 1, pp. 57–72.

It has been mentioned (see also Fig. 3.6) that tessellations (two-dimensional coverings of a wall by tiles) can represent various patterns of the 17 two-dimensional plane groups. Artistic and often complicated drawings by the Dutch graphic artist M. C.

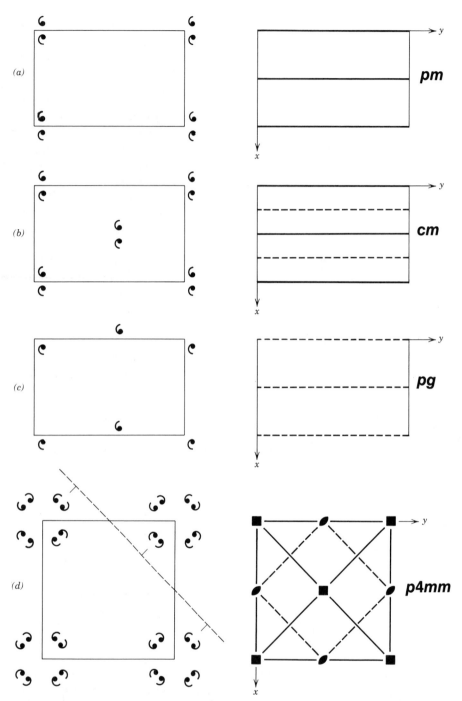

FIG. 3.13 Examples of mirror and glide symmetry compatible with a rectangular lattice type (*a* through *c*) and a square lattice (*d*). In the left-hand column are given the shape of the unit cell and the distribution of the motifs with respect to the outline of the unit cell. In the right-hand column only the distribution of the symmetry elements with respect to the unit cell outline is given. Crystallographic directions are indicated by axes *x* and *y*. Mirror lines (*m*) are shown by solid heavy lines and glide lines (*g*) by dashed lines. Rotation axes are shown by the standard symbols (see Tables 2.7 and 3.3).

(*a*) A primitive rectangular cell with mirror lines parallel to the *y* axis.

(*b*) A centered rectangular cell. The combination of centering and mirror lines produces glide lines parallel to and interleaved with the mirrors.

(*c*) A primitive rectangular cell with glides parallel to the *y* axis.

(*d*) In the left-hand illustration the location of one of four possible diagonal glide lines is shown on the basis of the distribution of motif clusters. In the right-hand column the total symmetry of this square planar array consists of 4-fold rotations at corners and the center of the cell; 2-fold rotations at the centers of edges; mirror lines parallel to the two axes *x* and *y*; mirror lines in two diagonal positions; and glide lines interleaved with the diagonal mirror lines.

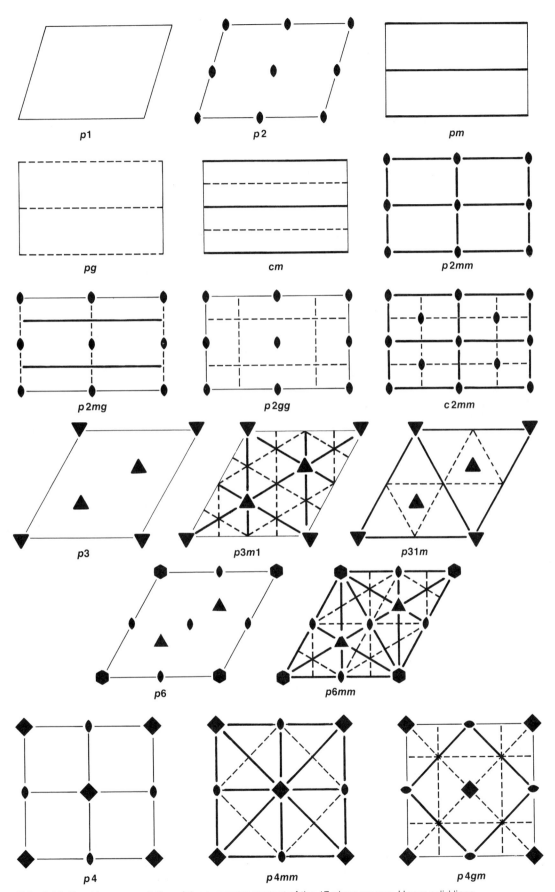

FIG. 3.14 Graphic representation of the symmetry content of the 17 plane groups. Heavy solid lines and dashed lines represent mirrors and glide lines, respectively, perpendicular to the page.

Table 3.1

TWO-DIMENSIONAL POINT GROUPS AND SPACE GROUPS*

Lattice	Point Group	Plane Group
Oblique *p*	1	*p*1
	2	*p*2
Rectangular *p* and *c*	*m*	*p m*
		p g
		c m
	2*m m*	*p* 2 *mm*
		p 2 *m g*
		p 2 *g g*
		c 2 *m m*
Square *p*	4	*p* 4
	4*m m*	*p* 4 *m m*
		p 4 *g m*
Hexagonal *p*	3	*p*3
	3 *m*	*p*3*m*1[†]
		*p*31*m*[†]
	6	*p*6
	6 *m m*	*p* 6 *m m*

*From *International Tables for X-ray Crystallography*, 1969, v. 1, N. F. M. Henry and K. Lonsdale, eds.: Symmetry Groups. International Union of Crystallography, Kynoch Press, Birmingham, England.

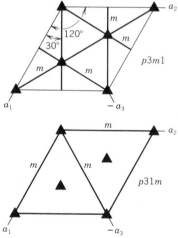

[†]There are two distinct groups for 3*m*—*p*3*m*1 and *p*31*m*. They have the same total symmetry content and shape. However, the conventional location of cell edges (as defined by three axes a_1, a_2, and a_3) differs by 30° in the two groups. In *p*3*m*1 the mirror lines bisect the 60° angle between cell edges; in *p*31*m* the reflection lines coincide with the cell edges.

Escher are commonly filled with designs of fish, horsemen, and birds. These drawings have been compiled by Caroline H. MacGillavry in a book entitled *Fantasy and Symmetry: The Periodic Drawings of M. C. Escher* (Harry N. Abrams, New York, 1976). Two of the Escher drawings from this book have been redrawn in Fig. 3.15. Following is a brief passage from the introduction to this book (p. IX):

It occurred to several scientists attending this meeting [Fifth International Congress of the International Union of Crystallography, held in Cambridge, England, in 1960] that Escher's periodic drawings [in an exhibition arranged for that same meeting by crystallographers J. D. H. and Gabrielle Donnay] would make excellent material for teaching the principles of symmetry. These patterns are complicated enough to illustrate clearly the basic concepts of translation and other symmetry, which are so often obscured in the clumsy arrays of little circles, pretending to be atoms, drawn on blackboard by teachers of crystallography classes. On the other hand, most of the designs do not present too great difficulties for the beginner in the field.

The student who is interested in the scientific as well as esthetic aspects of two-dimensional periodic patterns should consult the book by MacGillavry (and other books on M. C. Escher listed in the reference section to this chapter). Any of these books are probably available in an art or architecture library. Figure 3.15 also contains redrawn illustrations of two of the 13 designs by François Brisse (1981), which he prepared especially for the 12th Congress of the International Union of Crystallography, held in Ottawa, Canada, in 1981.

The arrangements of motifs in the four illustrations of Fig. 3.15 represent four of the 17 possible two-dimensional plane groups listed in Table 3.1 and illustrated in Fig. 3.14. The best way for the student to evaluate the shape and size of the unit cell, as well as the symmetry content of an infinitely extending periodic drawing, is to place a sheet of transparent paper over it. On this transparent overlay one can substitute opaque circles (nodes) for the smallest motif, or a part of the motif (a motif unit). If symmetry is present, it is best to locate the nodes on the location of such symmetry elements (rotations, *m* or *g*). Once the nodes have been located, the lattice of the design can be chosen by drawing lines between the nodes. It is especially instructive to use the standard symbols for rotations perpendicular to the page (⬧ ▲ ■ ⬢) and for mirrors (solid lines) and glides (dashed lines). This will reveal the similarity of the two-dimensional plane group (as deduced from these four artistic illustrations) with the representations in Fig. 3.14. (see also Klein, 1989; complete reference at end of chapter).

THREE-DIMENSIONAL ORDER

Until now we have discussed aspects of order in one direction and in two-dimensional patterns. The con-

(a) Fish and boats

(b) Birds and fish

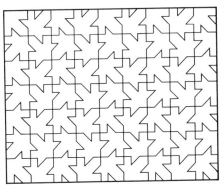

(c) The sailing ship *Bluenose*

(d) Canadian maple leaf

FIG. 3.15 Illustration of two-dimensional plane group symmetry in periodic drawings. *(a)* and *(b)* are redrawn from illustrations (plates 1a and 2, respectively) by M. C. Escher, as published by Caroline H. MacGillavry (see reference list); copyright © 1962 and 1963, respectively, M. C. Escher/Cordon Art, Baarn, Holland. *(c)* and *(d)* are redrawn from François Brisse in his publication entitled "La Symétrie bidimensionelle et le Canada" (see reference list at the end of this chapter). The two-dimensional plane groups represented by the illustrations are: *(a)* p1, *(b)* p2, *(c)* pg, and *(d)* p4gm.

cepts presented thus far are basic to an understanding of the regularity expressed by three-dimensional objects such as crystalline matter. In an assessment of three-dimensional order, a third direction (vector) will be needed to describe the distribution of nodes in a three-dimensional (space) lattice. Many of the symmetry elements introduced earlier will also be found in three-dimensional periodic structures. Rotational symmetry about a point will become rotational symmetry about a line (or axis); reflection (*m*) or glide reflection (*g*) across a line becomes a mirror plane or a glide plane, respectively. However, we will also need to introduce some new symmetry operations in three-dimensional patterns that are not present in planar periodicity. These new operations combine rotation and translation and are known as *screws* (see Figs. 3.21 and 3.22); the direction along which a screw operation takes place is known as a *screw axis*.

In order to fully describe the unit cell shape of the lattice (of a three-dimensionally ordered structure) and the total symmetry content compatible with this lattice, we will need to consider (as we did in planar patterns) combinations of lattice type with translation-free symmetry elements (e.g., *m*, rotation axes, and inversion) and translational symmetry elements (screw axes and glide planes). Such combinations will lead to 230 possible three-dimensional space groups, as compared with 17 two-dimensional plane groups. Each of these space groups belongs to one of the 32 point groups (see Chapter 2). Each space group will also be built upon a specific lattice type. In three-dimensional arrays there will be a choice of 14 different lattice types (these 14 include primitive and multiple lattices) as compared with only five lattice types for the two-dimensional patterns.

Three-Dimensional Lattices

Three-dimensional lattices can be constructed by adding one additional translation direction (vector) to the plane lattices of Fig. 3.5; this third vector must not lie in the plane of the two-dimensional nets. Vector space is referred to three non-coplanar axes, *x*, *y*, and *z*, which intersect at the origin. The unit cell vectors are denoted as **a**, **b**, and **c**, and the unit cell translations along *x*, *y*, and *z*, respectively, are noted as *a*, *b*, and *c* (see Fig. 3.16). The *x*, *y*, and *z* coordinate axes are commonly referred to as the *a*, *b*, and *c* axes (see, for example, such usage in Chapter 2). The unit cell dimensions are expressed in angstrom or nanometer units.

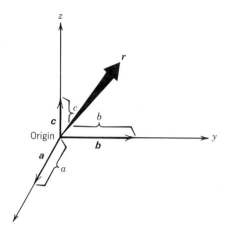

FIG. 3.16 *(a)* Representation of vectors **a**, **b**, and **c**, coordinate axis directions x, y, and z, and unit cell translations a, b, c, along directions x, y, and z, respectively. A general vector, **r**, in this three-dimensional space can be expressed as a linear combination of **a**, **b**, **c**, such that $\mathbf{r} = x\,\mathbf{a} + y\,\mathbf{b} + z\,\mathbf{c}$, where x, y, and z are real numbers. The x, y, and z coordinate axes (associated with cell edges a, b, and c, respectively) are commonly referred to as the a, b, and c axes (see, e.g., Chapter 2).

Figure 3.17 illustrates the various three-dimensional (space) lattice types that result when the five planar nets (shown in Fig. 3.5) are stacked in various ways along a third direction (z). The space lattices that result may be primitive or nonprimitive. *A primitive space lattice is a parallelepiped with lattice points only at its corners.* If the unit cell is nonprimitive, the centering may occur on a pair of opposite faces of the unit cell and is called *A-, B-,* or *C-centered*, depending on whether the centering takes place along the direction of the x, y, or z axis. The centering may also occur on all faces of the unit cell and is then referred to as *F* (for face-centered), or it may be present in the center of the unit cell and is referred to as *I*, body-centered (*I* from the German word *innenzentrierte*). These various types of unit cells are shown in Fig. 3.17, as well as the two choices of unit cells in space lattices that are derived from stacking a hexagonal net (see Fig. 3.5*d*). Space lattice no. 10, in Fig. 3.17, is based on a net with two equal translations (a_1 and a_2) that make an angle of 120° with each other. The unit cell in lattice no. 11, in Fig. 3.17, is known as a rhombohedral (*R*) unit cell in which the translation directions are a_R and the angles between the three equivalent edges of the unit cell are α_R.

In our prior discussion of unit cell choices in planar patterns, we noted that for a specific array of nodes (e.g., see Fig. 3.7), a considerable number of unit cell choices is possible. In order to reduce the number of choices (in three-dimensional arrays), crystallographers have drawn up the following restrictions as to unit cell choice:

1. The edges of the unit cell should coincide, if possible, with symmetry axes of the lattice.

2. The edges should be related to each other by the symmetry of the lattice.

3. The smallest possible cell should be chosen in accordance with (1) and (2).

Clearly, any regular three-dimensional array of nodes can be outlined by a primitive lattice. However, it is frequently desirable and appropriate to choose a nonprimitive unit cell. In Table 2.4 we outlined the 32 nonidentical symmetry elements and combinations of symmetry elements in terms of the crystal classes and crystal systems. *The types of space lattices compatible with these 32 point groups are known as the 14 Bravais lattices* and are shown in Fig. 3.17. These lattice types are unique, as was shown by Auguste Bravais (1811–1863), after whom they are named. That is, *they represent the only possible ways in which points can be arranged periodically in three dimensions.*

In Fig. 3.18 the 14 lattice types are arranged by crystal system. The names of the crystal systems reflect the characteristic symmetries of the lattice types. For example, in the triclinic system, which includes the symmetries 1 and $\bar{1}$, the unit cell compatible with these symmetries has no constraints, and thus its shape is one of low symmetry. The isometric system, however, contains very high symmetry ($4/m\bar{3}2/m$, $\bar{4}32$, $43m$, $2/m\bar{3}$, and 23), which is reflected in the unit cell having the highest symmetry constraints. There is a primitive lattice for each of the six crystal systems, and centered lattices occur in five of them. It should also be noted that only one face-centered lattice (namely, *C*) is shown in Fig. 3.18. If the lattice had been chosen in such a way as to be *A*-centered or *B*-centered rather than *C*-centered, this would not introduce a new category of lattice type. The *A*-, *B*-, and *C*-centered lattices are symmetrically identical and can be converted into each other by an appropriate exchange of the crystallographic axes. Table 3.2 provides a synopsis of the above discussion. The right-hand column in Table 3.2 has the heading "Multiplicity of Cell." This allows for a numerical distinction between *primitive* and *nonprimitive* lattice choices. A *primitive* lattice has nodes only at the corners. Each corner node is shared between eight adjoining cells. For such a primitive lattice, there are eight corner nodes, of which $\frac{1}{8}$th of each node contributes to the cell. That is, it has a multiplicity of $8 \times \frac{1}{8} = 1$. In a face-centered cell, each node on a face is shared between two adjoining cells. Therefore,

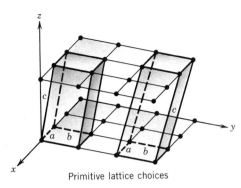

Primitive lattice choices

(1)   Stacking of an oblique net (or plane lattice) at an arbitrary angle results in *primitive triclinic lattices*.

Primitive

(2)   Stacking of a primitive rectangular net in a vertical direction (z), with $x \wedge z$ angle (β) $\neq$ 90°, leads to a *primitive monoclinic lattice*.

Centered on (001)

(3)   Stacking of a centered rectangular net in a vertical direction (z), with $x \wedge z$ angle (β) $\neq$ 90°, results in a *centered monoclinic lattice*.

Primitive

(4)   Stacking of a primitive rectangular net in a vertical direction (z), with the $x \wedge z$ angle = 90°, leads to a *primitive orthorhombic lattice*.

Centered on (001)

(5)   Stacking of a centered rectangular net in a vertical direction (z), with the $x \wedge z$ angle = 90°, leads to a *centered orthorhombic lattice*.

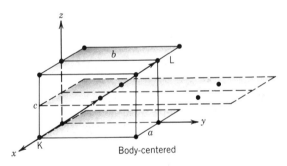

Body-centered

(6)   Stacking of a primitive rectangular net along the direction between nodes K and L results in an orthorhombic lattice with a central node. This is an *orthorhombic body-centered lattice*.

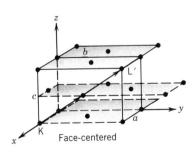

Face-centered

(7)   Stacking of a centered rectangular net along the direction between nodes K and L' (on the front face) leads to centering on all faces of the three-dimensional lattice. This is a *face-centered orthorhombic lattice*.

Primitive

(8)   Stacking of a square net along the z direction, with angle $x \wedge z =$ 90°, and with the c translation $\neq a_1$ or a_2, results in a *primitive tetragonal lattice*.

FIG. 3.17 Stacking of the five nets (plane lattices; see Fig. 3.5) in various ways (as specifically noted in this figure) leads to the 14 possible space lattices. These 14 lattice types are also known as the 14 Bravais lattices (see also Fig. 3.18).

(*continued*)

Body-centered

(9)   Stacking of the same net as in (8) but now in a direction between nodes K and L results in a *body-centered tetragonal lattice.*

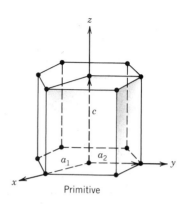

Primitive

(10)   Stacking of a hexagonal net in a *z* direction such that angle $x \wedge z = 90°$ leads to a primitive hexagonal lattice. If this lattice choice is rotated 3 times about *z*, it results in a *c-centered hexagonal lattice.*

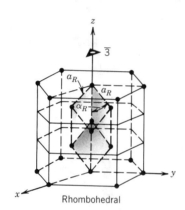

Rhombohedral

(11)   A hexagonal net can also be stacked along the edge directions of a rhombohedron (a_R). This results in a *rhombohedral lattice*, the edge directions of which are symmetrical with respect to the $\overline{3}$ axis along the *z* direction.

Primitive

(12)   Stacking of a square net along the *z* direction, with $x \wedge z$ angle = 90° and with *c* translation = a_1, and a_2, results in a *primitive isometric lattice.*

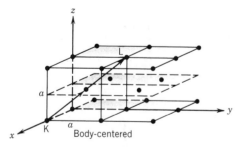

Body-centered

(13)   Stacking of a square net along the direction between nodes K and L (a body diagonal) results in a *body-centered isometric lattice.*

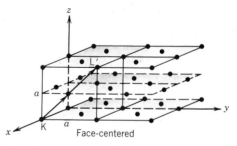

Face-centered

(14)   Stacking of a square net along the direction between nodes K and L' (along the front face), results in a *face-centered isometric lattice.*

FIG. 3.17 (*continued*)

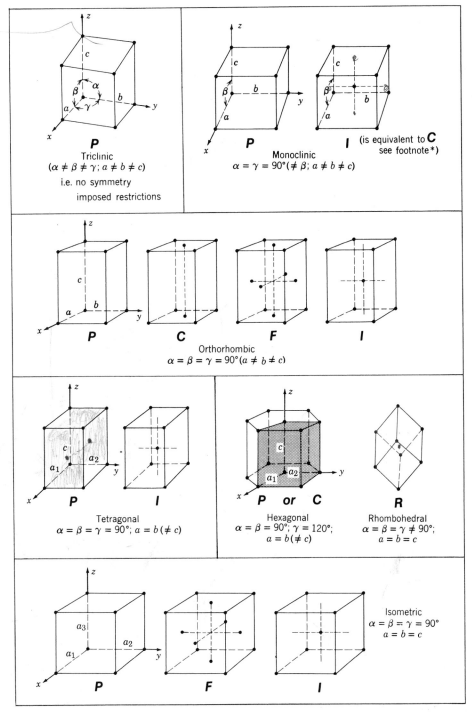

FIG. 3.18 The 14 unique types of space lattices, known as the Bravais lattices. Axial lengths are indicated by a, b, and c and axial angles by α, β, and γ. Each lattice type has its own symmetry constraints on lengths of edges a, b, and c and angles between edges, α, β, and γ. In the notations the nonequivalence of angles or edges that usually exist but are not mandatory are set off by parentheses.

*In the monoclinic system the unit cell can be described by a body-centered or a C-face-centered cell by a change in choice of the length of the a axis and the angle β. Vectorially these relations are: $a_I = c_c + a_c$; $b_I = b_c$; $c_I = -c_c$; and $a_I \sin \beta_I = a_c \sin \beta_c$. Subscripts I and c refer to the unit cell types.

Table 3.2

DESCRIPTION OF SPACE LATTICE TYPES AND DISTRIBUTION OF THE 14 BRAVAIS LATTICES AMONG THE SIX CRYSTAL SYSTEMS

Name and Symbol	Location of Nonorigin Nodes	Multiplicity of Cell
Primitive (P)	. . . .	1
Side-centered (A)	Centered on A face (100)	2
(B)	Centered on B face (010)	2
(C)	Centered on C face (001)	2
Face-centered (F)	Centered on all faces	4
Body-centered (I)	An extra lattice point at center of cell	2
Rhombohedral (R)	A primitive rhombohedral cell	1

Primitive (P) in each of the 6 crystal systems	=	6
Body-centered (I) in monoclinic, orthorhombic, tetragonal, and isometric	=	4
Side-centered (A = B = C) in orthorhombic	=	1
Face-centered (F) in orthorhombic and isometric	=	2
Rhombohedral (R) in hexagonal	=	1
	Total =	14

the total node content of a side-centered cell is: $8 \times \frac{1}{8} = 1$ (for corner nodes) $+ 2 \times \frac{1}{2} = 1$ (for face-centered nodes), resulting in a multiplicity of $1 + 1 = 2$. Any node in the interior of a cell, as in a body-centered choice, belongs only to its own cell.

The shape and size of unit cells of minerals are most commonly determined by X-ray diffraction techniques (see Chapter 7). During the last decade, how-ever, high-resolution transmission electron microscopy (HRTEM) has allowed the direct observation of projected images of crystal structures on photographic plates. Such a structure image is shown in Fig. 3.19 for the mineral cordierite. The dark parts of the photograph outline the projected image of the structure and the superimposed lines outline a rectangular unit cell for cordierite.

FIG. 3.19 High magnification structure image of an a–b section through the mineral cordierite. An orthorhombic unit cell is outlined and distances are given in angstroms. The insert shows the idealized structure of cordierite, as determined by X-ray diffraction techniques. The scales of the idealized structure and the electron transmission image are identical. (From Buseck and Iijima, *American Mineralogist*, v. 59, pp. 1–22, 1974.)

FIG. 3.20 A drawing of the structure of low (α) quartz (SiO_2), with the normally vertical z axis tilted at a small angle to better show the repeat distance c (of the unit cell) along the z direction. The primitive hexagonal space lattice, outlined by the various parallelepipeds, shows that each unit cell (with edges a_1, a_2, and c) contains a complete and representative unit of the repeating pattern of the structure. (From Boisen, M.G. and Gibbs, G.V., 1990, *Mathematical Crystallography*, rev. ed. *Reviews in Mineralogy*, v. 15. Mineralogical Society of America, Washington, D.C.)

○ Silicon
◯ Oxygen

The internal (atomic) structure of crystalline materials is generally determined by a combination of X-ray, neutron, and electron diffraction techniques, and may be supplemented by a combination of spectroscopic methods. These methods, used singly or in combination, provide a quantitative three-dimensional reconstruction of the location of the atoms (or ions), the chemical bond types and their orientations, and the overall internal symmetry of the structure. From such structural information, the appropriate three-dimensional (space) lattice is derived. Figure 3.20 illustrates the derivation of the appropriate lattice type and unit cell from the structure of low (α) quartz (SiO_2). This illustration also shows that the equivalent points (or "identipoints") in a lattice are generally not atomic positions. Rather, as in Fig. 3.20, they are geometric points in the structure that have the same angle and distance relationships to the Si and O atoms (and hence to the 3-fold and 2-fold axes of the structure).

Screw Axes and Glide Planes

In Chapter 2 we discussed the translation-free symmetry operations (rotation, rotoinversion, mirror, and center of symmetry), and in this chapter we have introduced pure translation (as in lattices), as well as the concept of a glide line in two-dimensional patterns. In three-dimensional periodic arrays we find two symmetry elements that combine a symmetry operation (rotation or mirror reflection) with a translation component. A rotational operation with translation (t) parallel to the axis of rotation is known as a *screw operation*, and a mirror reflection with a translation component ($t/2$ or $t/4$) parallel to the mirror is known as a *glide operation*.

The 2-, 3-, 4-, and 6-fold rotational operations can all be combined with a translation. (A 1-fold rotation axis combined with a translation is equivalent to a translation only.) Figure 3.21 compares the differences in motif distribution for a 4-fold rotation and a 4-fold screw operation. Rotation alone, through an angle (α), about an imaginary axis, generates a sequence of the motif along a circle. For a 90° rotation angle a pattern with four motifs is generated. In the 4-fold screw operation the four motifs are generated from the original unit in a three-dimensional, helical path (as in a screw motion; see Fig. 3.21b). Screw axes are said to be *isogonal* (from the Greek meaning "same angle") with the equivalent rotational axes. This means that 4-fold screw operations rotate the motif through 90° angles while translating the motif parallel to the rotation axis. All possible screw axes, isogonal with rotational axes, are illustrated in Fig. 3.22.

As with any screw motion, *screw axes are right- or left-handed*. A right-handed screw may be defined as one that advances away from the observer when rotated clockwise. The screw axis symbols consist of the symbols for rotation axes (2, 3, 4, and 6) followed by a subscript that represents the fraction of the translation (t) inherent in the operation. For example, 2_1 means that $\frac{1}{2}t$ (obtained by placing the subscript over the main axis symbol, as in a fraction) is the translation involved. For a 3-fold rotation there are two possible screw axes, namely, 3_1 and 3_2. The translation component in both screw axes is $\frac{1}{3}t$, but a convention allows for the distinction between the directions of the screw. When the ratio of the subscript to the number of the rotation axis (as $\frac{1}{3}$ for 3_1) is less than $\frac{1}{2}$, the screw is right-handed. When this ratio is more than $\frac{1}{2}$, it is left-handed (as in 3_2), and when the ratio is $\frac{1}{2}$, the

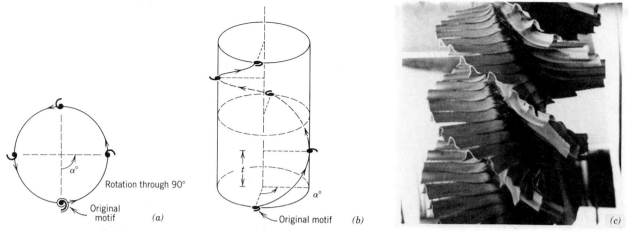

FIG. 3.21 Generation of patterns by 4-fold rotation (a) and a combination of translation and rotation (b), which results in a screw motion (4_1). (c) Example of a many-fold screw axis in the vertical direction of a palm. The screw operation is shown by the trimmed remains of the leaves. The rotational symmetry is approximately 30. Photographed in Darwin, Northern Territory, Australia. As discussed in Chapter 2 (page 22), independent objects can have unlimited rotational symmetry, whereas such symmetry is limited to 1, 2, 3, 4, and 6 in ordered arrays.

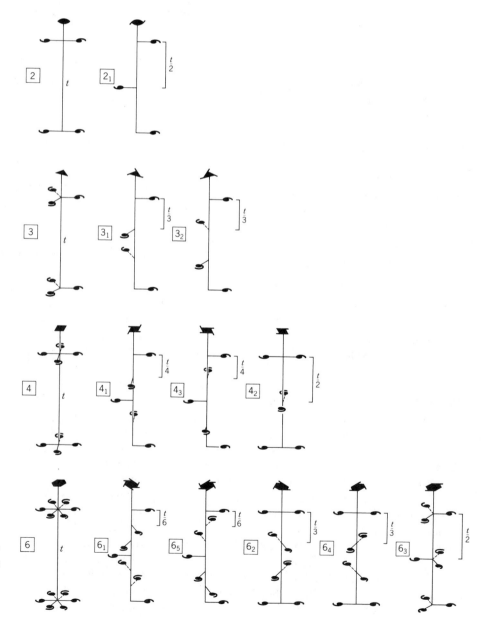

FIG. 3.22 Repetition of motif units by screw axes. The left column represents rotation axes and the columns to the right represent isogonal screw axes. The symbols at the top of the rotation and screw axes are internationally accepted. For projections of these screw axes, see Fig. 3.23.

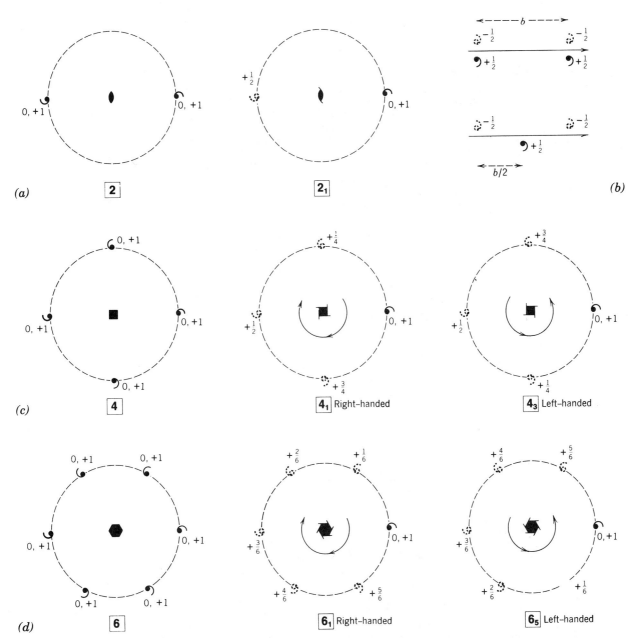

FIG. 3.23 Examples of several axes of rotation and some of their isogonal screw axes. The effect of these symmetry operations on motifs is shown as well. All diagrams are set such that the a and b axes (x and y directions) lie in the plane of the page. The height above the plane of the page, along the $+c$ axis (in the $+z$ direction), is indicated by a ($+$) sign. For example, 0 and $+1$ mean that the motif lies in the plane of the page (at zero height along the c axis) and is also repeated along the c axis by a unit lattice repeat to $+1c$. Fractions (e.g., $+\frac{1}{2}$, $+\frac{1}{3}$, $+\frac{1}{4}$) refer to heights above the page along the c axis (z direction). Motif units that do not lie in the plane of the page (but have been projected from above) are shown by a dashed comma design. (a) Two-fold rotation axis and isogonal screw axis. The circles show the effect of these axes on motifs when the axes are oriented perpendicular to the page. (b) These same axes when oriented parallel to the plane of the page (or, when they lie within the plane of the page). (c) Four-fold rotation axis and two enantiomorphic 4-fold screw axes. The rotational directions of the screws ("handedness") are shown by the arrows. (d) Six-fold rotation axis and two enantiomorphic 6-fold screw axes.

screw is considered neutral in direction (see Fig. 3.22). In other words, 3_1 and 3_2 are an *enantiomorphous pair of screw axes*, with 3_1 right-handed and 3_2 left-handed. Similarly, the following pairs are enantiomorphous: 4_1 and 4_3, 6_1 and 6_5, and 6_2 and 6_4. An-

other way of representing the operation of symmetry axes (rotation and screw axes) is shown in Fig. 3.23. Here the motif units are projected onto the plane of the page (from above the page), and many of the resulting drawings are similar to the illustrations in Figs.

Table 3.3

SYMBOLS FOR SYMMETRY AXES
(ALL GRAPHIC SYMBOLS ARE FOR AXIS NORMAL TO THE PAGE, UNLESS OTHERWISE NOTED)

Symbol	Symmetry Axis	Graphic Symbol	Type of Translation (if Present)	Symbol	Symmetry Axis	Graphic Symbol	Type of Translation (if Present)
1	1-fold rotation	None	None	4	4-fold rotation	■	None
$\bar{1}$	1-fold rotoinversion	○	None	4_1	4-fold screw (right-handed)		$\frac{1}{4}c$
2	2-fold rotation	⬤	None	4_2	4-fold screw (neutral)		$\frac{2}{4}c = \frac{1}{2}$
		(parallel to paper)		4_3	4-fold screw (left-handed)		$\frac{3}{4}c$
2_1	2-fold screw	⬤	$\frac{1}{2}c$	$\bar{4}$	4-fold rotoinversion	◩	None
		(parallel to paper)	$\frac{1}{2}a$ or $\frac{1}{2}b$	6	6-fold rotation	⬢	None
3	3-fold rotation	▲	None	6_1	6-fold screw (right-handed)		$\frac{1}{6}c$
3_1	3-fold screw (right-handed)	▲	$\frac{1}{3}c$	6_2	6-fold screw (right-handed)		$\frac{2}{6}c$
3_2	3-fold screw (left-handed)	▲	$\frac{2}{3}c$	6_3	6-fold screw (neutral)		$\frac{3}{6}c = \frac{1}{2}$
$\bar{3}$	3-fold rotoinversion	▲	None	6_4	6-fold screw (left-handed)		$\frac{4}{6}c$
				6_5	6-fold screw (left-handed)		$\frac{5}{6}c$
				$\bar{6}$	6-fold rotoinversion	⬣	None

2.10 to 2.12 for rotation and rotoinversion axes, respectively. The fractions next to the motifs in Fig. 3.23 represent the distance the units lie above the surface of the page. The fractions that represent t/n are preceded by a plus sign (+) to indicate that the motif units lie above the page (in the positive direction of z in an x, y, z coordinate system). Table 3.3 lists the conventional symbols for all types of rotational symmetry.

In addition to the repetition of a motif by a mirror reflection, a regular pattern can be generated by a combination of a mirror reflection and a translation. This operation is referred to as a *glide plane* or *glide reflection*. Figure 3.12c shows how motifs are related to a glide plane that has a translation component of $t/2$. In our prior discussion of planar patterns (see Figs. 3.13b and c) we noted how a glide line repeats motifs on either side of the line with a periodicity of half the lattice translation. In three-dimensional patterns a wider variety of glide movements can occur.

Specific glide directions can be identified in two- and three-dimensional patterns and expressed in terms of a set of axes such as $x, y,$ and z. However, many crystallographers refer the internal order as well

as external morphology of crystals to three axes, $a, b,$ and c. The c axis is vertical and the a and b axes lie in a plane that does not contain c. If the glide component $(t/2)$ in a three-dimensional ordered arrangement is parallel to the a axis, it is referred to as an a *glide* and is represented by the symbol a. Similarly, if the glide component $(t/2)$ is parallel to the b or c axes, the glide is referred to as a b or c *glide*, respectively. If the glide component can be represented by $a/2 + b/2, a/2 + c/2, b/2 + c/2,$ or $a/2 + b/2 + c/2,$ it is referred to as a *diagonal glide* and is represented by the symbol n. If the glide component can be represented by $a/4 + b/4, b/4 + c/4, a/4 + c/4,$ or $a/4 + b/4 + c/4,$ it is known as a *diamond glide* and is symbolized by the letter d. It should be obvious that in the *diamond (d) glide* the simultaneous translations are one-fourth of the cell edges, whereas for a *diagonal (n) glide* the translation components are equal to one-half of the cell edges. Table 3.4 gives the standard symbols used in graphic illustrations of glide planes and mirrors. Figure 3.24 illustrates the results of mirror and glide operations in a three-dimensional sketch. Figure 3.25 shows the results of these same operations in two-dimensional motif distributions.

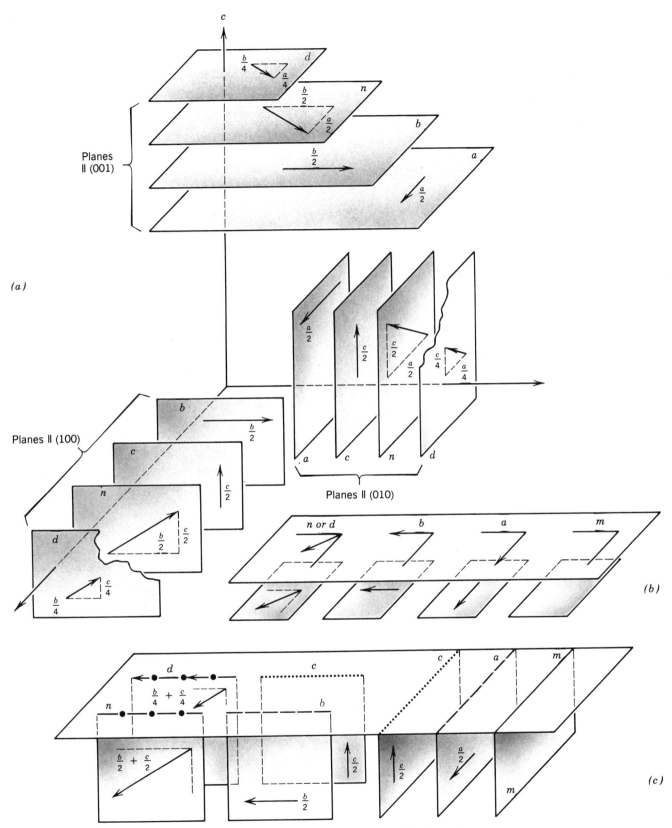

FIG. 3.24 (a) Sketch of the various glide planes and their translation components with reference to orthorhombic coordinate axes. (b) Symbols for glide and mirror planes when such planes are parallel to the plane of projection (001) or (0001). (c) Symbols for glide and mirror planes when such planes are perpendicular to the standard plane of projection. (Adapted from Bloss, F. D., 1971, *Crystallography and Crystal Chemistry: An Introduction*, Figs. 7.9 and 7.11, copyright © 1971 by Holt, Rinehart, and Winston, Inc., New York. Reprinted by permission of F. D. Bloss.)

Table 3.4

SYMBOLS FOR MIRROR AND GLIDE PLANES*

Symbol	Symmetry Plane	Graphic Symbol		Nature of Glide Translation
		Normal to Plane of Projection	Parallel to Plane of Projection[†]	
m	Mirror	————	⌐ 120°	None
a, b		– – – – –	↑←	$a/2$ along [100] or $b/2$ along [010]
c	Axial glide plane	··············	None	$c/2$ along the c axis
n	Diagonal glide plane	—·—·—·—	↗	$a/2 + b/2$; $a/2 + c/2$; $b/2 + c/2$; or $a/2 + b/2 + c/2$ (tetragonal and isometric)
d	Diamond glide plane	—·— ←·—·— / —·—· →·—	↗	$a/4 + b/4$; $b/4 + c/4$; $a/4 + c/4$; or $a/4 + b/4 + c/4$ (tetragonal and isometric)

*From *International Tables for X-ray Crystallography*, 1969, v. 1, N. F. M. Henry and K. Lonsdale, eds.: Symmetry Groups. International Union of Crystallography, Kynoch Press, Birmingham, England.
[†]When planes are parallel to the paper, heights other than zero are indicated by writing the z coordinate next to the symbol (e.g., $\frac{1}{4}$ or $\frac{3}{8}$). The arrows indicate the direction of the glide component.

SPACE GROUPS

When we combine the 14 possible space lattice types (Bravais lattices) with the symmetry inherent in the 32 crystal classes (the translation-free point group symmetries), as well as the two symmetry operations that involve translation (screws and glides), we arrive at the concept of *space groups*. *Space groups*, therefore, *represent the various ways in which motifs* (such as atoms in crystals) *can be arranged in space in a homogeneous array* (homogeneous meaning that each motif is equivalent to every other motif in the pattern).

In the last decade of the nineteenth century three men of different nationalities and interests derived independently how many unique patterns could occur in three-dimensional periodic arrays. These were E. von Federov, a Russian crystallographer; Artur Schoenflies, a German mathematician; and William Barlow, a British amateur. Their unanimous conclusion was that there are 230 such unique patterns; they are known as the 230 space groups.

Because the derivation of all 230 is a lengthy and complex assignment, we will, as examples, present a small number of lower symmetry space groups to illustrate some of the concepts involved. It should be remembered that the presence of periods of translation (as in lattices or glide planes or screw axes) cannot be detected morphologically because the translations involved are on the order of 1 to 10 Å. The translation-free symmetry combinations are point groups, whereas space groups define the symmetry

and translations in space. If we were to ignore the translation components in the 230 space groups, we would arrive at the 32 point groups.

Space groups have the following characteristics. (1) They are based on one of the 14 Bravais lattices that is compatible with a specific point group, and (2) they are isogonal with one of the 32 point groups. Isogonal implies that rotation and screw axes with the same rotational repeat have the same rotational angle (e.g., 60° in a 6-fold rotation or 6-fold screw axis). This means that the screw axes 6_1, 6_2, 6_3, 6_4, and 6_5 are isogonal with the rotation axis 6. In other words, *the point group is the translation-free residue of a family of possible isogonal space groups.*

A specific point group designation consists of a series of symmetry elements, as in $2/m2/m2/m$. For each of the specific point group symmetry elements there is a possible space group element. That is, instead of a mirror plane perpendicular to the first 2-fold rotation axis and a mirror plane perpendicular to the second 2-fold rotation axis, there may be glide planes in their places that would be symbolized as $2/b2/a2/m$. The space group symbol is additionally preceded by a symbol that designates the general lattice type (P, A, B, C, I, F, or R). The full symbol for a space group that is isogonal with $2/m2/m2/m$ might be $I2/b2/a2/m$. Another example of a space group isogonal with $2/m2/m2/m$ would be $P2_1/b2_1/c2_1/a$, and so on.

Let us now illustrate some aspects of space groups in the triclinic and monoclinic systems. In the

FIG. 3.25 Examples of symmetry planes and the distribution of motifs about them. In all drawings the plane of the paper is considered the *ab* plane. The location of a motif above the plane (in the (+) direction of *c*) is marked by +, or +$\frac{1}{4}$, +$\frac{1}{2}$, and so on. The (−) *c* direction is shown by (−) signs next to the motif. In the right-hand figure of *(a)* the motif units superimpose exactly; the upper motif is shown by •, the lower equivalent by a small circle about it. Note conventional symbols; see also Table 3.4. *(a)* A mirror plane. *(b)* A glide plane with a glide component parallel to the *b* axis (*b*/2). *(c)* A glide plane with a glide component parallel to the *c* axis (*c*/2). *(d)* A diagonal glide plane (*n*) with simultaneous glide components parallel to the *a* and *b* axes (*a*/2 + *b*/2). *(e)* A diamond glide plane (*d*) with simultaneous glide components parallel to the *a* and *b* axes (*a*/4 + *b*/4).

triclinic system only two possible space groups can occur, namely, *P*1 and *P*$\bar{1}$. Here we have combined the two possible point groups in the triclinic system, 1 and $\bar{1}$ (see Table 2.4), and the only possible lattice type, *P* (see Fig. 3.18). In the monoclinic system we have three possible point groups (2, *m*, and 2/*m*) and two possible lattice types (*P* and *I*) to consider. Let us restrict ourselves to a consideration of one of the point groups (2) and the two possible lattice types (*P* and *I*; *I* in the monoclinic system can be transferred into *A*, *B*, or *C* by a different choice of coordinate axes; see footnote to Fig. 3.18). The four possible space group notations are *P*2, *P*2$_1$, *I*2, and *I*2$_1$. Figure 3.26 illustrates the arrangement of motifs (commas) about the 2-fold rotation and 2-fold screw axes in relation to a monoclinic unit cell. Of the four possible space groups only three turn out to be unique because *I*2 and *I*2$_1$ are equivalent in their arrangement of symmetry elements, except for a change in the location of the origin chosen for the lattice. The three unique space groups for the monoclinic point group 2 are therefore *P*2, *P*2$_1$, and *I*2 (which is equivalent to *C*2). A more complete discussion of some additional space groups can be found in Buerger (1978).

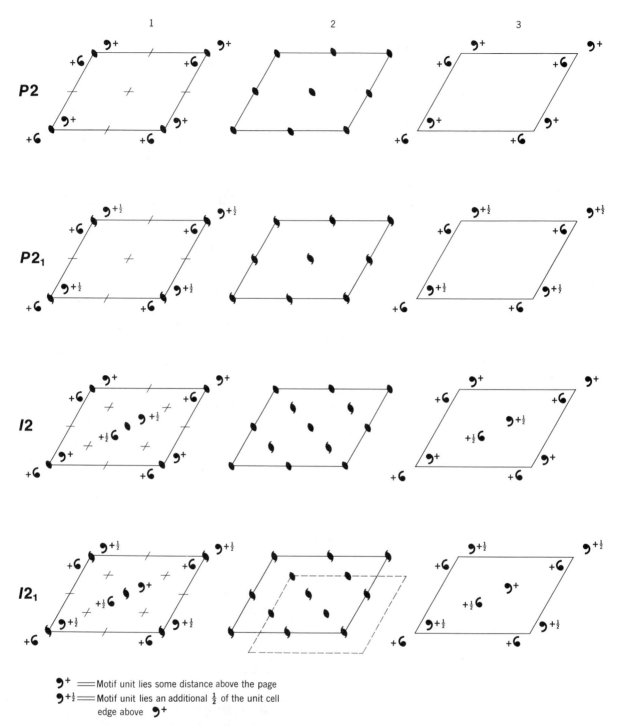

FIG. 3.26 Derivation of the monoclinic space groups $P2$, $P2_1$, $I2$, and $I2_1$. Column 1 depicts only the lattice type, the motif units and the symmetry element specified for each lattice point (corners and centers). In the illustrations in column 1 it should become clear that additional symmetry elements are inherent in the unit cell at the points indicated by $+$. Column 2 depicts all the symmetry elements inherent in the pattern. Column 3 shows the arrangement of motif units only. For space group $I2_1$ it becomes clear that the symmetry contained in the unit cell is equivalent to the symmetry in the unit cell for $I2$. Only a shift in origin of the unit cell is needed to obtain a symmetry pattern identical to that for $I2$ (column 2). The dashed unit cell indicates the actual shift necessary. In other words $I2$ and $I2_1$ are identical and therefore only one of the two arrangements is unique. In Table 3.5 the space group $I2$ is listed as $C2$ because these two notations are equivalent (see footnote to Fig. 3.18).

Table 3.5 gives a listing of the 32 crystal classes (point groups) and the 230 space groups compatible therewith. When one compares the listing in the column entitled "Crystal Class" with the entries in the column "Space Groups," the isogonality of the relationships of these two groupings becomes obvious. The reason for including Table 3.5 is *not* for the beginning student to memorize all or many of the unique (230) space groups. It is included to illustrate the ease with which a space group symbol can be reduced to its isogonal point group (crystal class). All of the entries in the "Space Group" column contain as the first entry a symbol for the lattice type. Furthermore, many space group representations contain notations for

Table 3.5

THE 230 SPACE GROUPS, AND THE ISOGONAL 32 CRYSTAL CLASSES (POINT GROUPS). THE SPACE GROUP SYMBOLS ARE, IN GENERAL, UNABBREVIATED*

Crystal Class	Space Group
1	$P1$
$\bar{1}$	$P\bar{1}$
2	$P2, P2_1, C2$
m	Pm, Pc, Cm, Cc
$2/m$	$P2/m, P2_1/m, C2/m, P2/c, P2_1/c, C2/c$
222	$P222, P222_1, P2_12_1 2, P2_12_12_1, C222_1, C222, F222, I222, I2_12_12_1$
$mm2$	$Pmm2, Pmc2_1, Pcc2, Pma2, Pca2_1, Pnc2, Pmn2_1, Pba2, Pna2_1, Pnn2, Cmm2, Cmc2_1, Ccc2, Amm2, Abm2, Ama2, Aba2, Fmm2, Fdd2, Imm2, Iba2, Ima2$
$2/m2/m2/m$	$P2/m2/m2/m, P2/n2/n2/n, P2/c2/c2/m, P2/b2/a2/n, P2_1/m2/m2/a,$ $P2/n2_1/n2/a, P2/m2/n2_1/a, P2_1/c2/c2/a, P2_1/b2_1/a2/m, P2_1/c2_1/c2_1/n,$ $P2/b2_1/c2_1/m, P2_1/n2_1/n2/m, P2_1/m2_1/m2/n, P2_1/b2/c2_1/n,$ $P2_1/b2_1/c2_1/a, P2_1/n2_1/m2_1/a, C2/m2/c2/m, C2/m2/c2_1/a,$ $C2/m2/m2/m, C2/c2/c2/m, C2/m2/m2/a, C2/c2/c2/a, F2/m2/m2/m,$ $F2/d2/d2/d, I2/m2/m2/m, I2/b2/a2/m, I2/b2/c2/a, I2/m2/m2/a$
4	$P4, P4_1, P4_2, P4_3, I4, I4_1$
$\bar{4}$	$P\bar{4}, I\bar{4}$
$4/m$	$P4/m, P4_2/m, P4/n, P4_2/n, I4/m, I4_1/a$
422	$P422, P42_12, P4_122, P4_12_12, P4_222, P4_22_12, P4_322, P4_32_12, I422, I4_122$
$4mm$	$P4mm, P4bm, P4_2cm, P4_2nm, P4cc, P4nc, P4_2mc, P4_2bc, I4mm, I4cm, I4_1md, I4_1cd$
$\bar{4}2m$	$P\bar{4}2m, P\bar{4}2c, P\bar{4}2_1m, P\bar{4}2_1c, P\bar{4}m2, P\bar{4}c2, P\bar{4}b2, P\bar{4}n2, I\bar{4}m2, I\bar{4}c2, I\bar{4}2m, I\bar{4}2d$
$4/m2/m2/m$	$P4/m2/m2/m, P4/m2/c2/c, P4/n2/b2/m, P4/n2/n2/c, P4/m2_1/b2/m,$ $P4/m2_1/n2/c, P4/n2_1/m2/m, P4/n2_1/c2/c, P4_1/m2/m2/c, P4_2/m2/c2/m,$ $P4_2/n2/b2/c, P4_2/n2/n2/m, P4_2/m2_1/b2/c, P4_2/m2_1/n2/m,$ $P4_1/n2_1/m2/c, P4_2/n2_1/c2/m, I4/m2/m2/m, I4/m2/c2/m, I4_1/a2/m2/d,$ $I4_1/a2/c2/d$
3	$P3, P3_1, P3_2, R3$
$\bar{3}$	$P\bar{3}, R\bar{3}$
32	$P312, P321, P3_112, P3_121, P3_212, P3_221, R32$
$3m$	$P3m1, P31m, P3c1, P31c, R3m, R3c$
$\bar{3}2/m$	$P\bar{3}1m, P\bar{3}1c, P\bar{3}m1, P\bar{3}c1, R\bar{3}m, R\bar{3}c$
6	$P6, P6_1, P6_5, P6_2, P6_4, P6_3$
$\bar{6}$	$P\bar{6}$
$6/m$	$P6/m, P6_3/m$
622	$P622, P6_122, P6_522, P6_222, P6_422, P6_322$
$6mm$	$P6mm, P6cc, P6_3cm, P6_3mc$
$\bar{6}m2$	$P\bar{6}m2, P\bar{6}c2, P\bar{6}2m, P\bar{6}2c$
$6/m2/m2/m$	$P6/m2/m2/m, P6/m2/c2/c, P6_3/m2/c2/m, P6_2/m2/m2/c$
23	$P23, F23, I23, P2_13, I2_13$
$2/m\bar{3}$	$P2/m\bar{3}, P2/n\bar{3}, F2/m\bar{3}, F2/d\bar{3}, I2/m\bar{3}, P2_1/a\bar{3}, I2_1/a\bar{3}$
432	$P432, P4_232, F432, F4_132, I432, P4_332, P4_132, I4_132$
$\bar{4}3m$	$P\bar{4}3m, F\bar{4}3m, I\bar{4}3m, P\bar{4}3n, F\bar{4}3c, I\bar{4}3d$
$4/m\bar{3}2/m$	$P4/m\bar{3}2/m, P4/n\bar{3}2/n, P4_2/m\bar{3}2/n, P4_2/n\bar{3}2/m, F4/m\bar{3}2/m, F4/m\bar{3}2/c,$ $F4_1/d\bar{3}2/m, F4_1/d\bar{3}2/c, I4/m\bar{3}2/m, I4_1/a\bar{3}2/d$

*From *International Tables for X-ray Crystallography*, 1969, v. 1, N. F. M. Henry and K. Lonsdale, eds.: Symmetry Groups. International Union of Crystallography, Kynoch Press, Birmingham, England.

screw axes and glide planes. Lattice type, screw, and glide operations all contain translational elements. If all of these translational elements are removed and the translation-free equivalent operations are substituted for them, the point group, which is isogonal with the space group, will result. For example, the last space group in Table 3.5 is listed as $I4_1/a\bar{3}2/d$. Because lattice types cannot be reflected in translation-free symbols for crystal classes (point groups), the lattice type must be dropped altogether. The remaining symbols must become translation-free as well. This means that a and d glides must be replaced by the translation-free mirror symbol m. This leads to the point group notation $4/m\bar{3}2/m$ in the isometric system.

If screw axes are present, a similar process allows easy conversion from space group to point group. For example, in space group $P6_122$, the screw axis is replaced by the isogonal 6-fold rotation and the lattice type (P) is dropped. This results in the isogonal point group 622 in the hexagonal system. The ease in relating point and space group symmetries is one of the powerful aspects of the Hermann-Mauguin (international) notation. It allows the reader to "translate" a space group notation for a crystalline material into the isogonal and simpler point group notation.

One aspect of the Hermann-Mauguin notation that may need some clarification is the use by crystallographers of what is known as "abbreviated symbols." In our discussion of point groups we have always used the complete symbol such as $2/m2/m2/m$ and $4/m2/m2/m$. The abbreviated symbolism for these two point groups would be mmm and $4/mmm$, respectively (see also Table 2.8). The reason for the abbreviation for $2/m2/m2/m$ is the understanding that the three mutually perpendicular mirror planes intersect each other along 2-fold axes. Similar reasoning explains the abbreviation $4/mmm$. In much of this text, for reasons of clarity, we will use only complete point group symbols. Abbreviated symbols are generally used in the literature for space group notation, however. For example, a mineral with point group $4/m\bar{3}2/m$ may have its space group reported as $Fm3m$, because the presence of 4-fold and 2-fold rotation axes is implied. Such abbreviated symbols for space group notation will be used in the four chapters on systematic mineralogy.

Another system of space group notation may be encountered in the literature, especially in older texts. This is known as the *Schoenflies notation*. Because this notation does not logically follow from the Hermann-Mauguin point group symbols, it is not used here. References at the end of this chapter will enable you to find the Schoenflies system.

Illustrations of Space Groups

In Chapter 2 we discussed and illustrated the symmetry content of each of the 32 crystal classes (point groups). In our present discussion of three-dimensional periodic symmetry we would have to discuss and illustrate each of the 230 space groups if our aim were analogous, in-depth coverage of both subjects. This would be a formidable task which would require a great deal of text and many illustrations. We judge this to be inappropriate in a textbook for beginning students in mineralogy and/or geology. Readers interested in systematic coverage of space groups should consult some of the standard references on the subject, three of which are profusely illustrated. These are: *International Tables for X-ray Crystallography*, vols. 1 and A; *Elementary Crystallography*, by M. J. Buerger; and *Mathematical Crystallography*, by M. B. Boisen, Jr., and G. V. Gibbs (complete references for these volumes are given at the end of this chapter). Although systematic, in-depth coverage of space groups is outside the realm of this text, we do wish to introduce the reader to some of the general aspects of space groups (and their illustrations) now that we have defined all of the necessary "operators" of space group symmetry. In the subsequent pages we will illustrate and discuss a few representative space groups that are found in common (rock-forming) minerals. After that we will discuss and illustrate the relationship of external morphology (point group symmetry) to internal structure and the space group link between these two concepts. We will do this for three well-known minerals.

In the illustrations of Fig. 3.27 the conventions of the *International Tables for X-ray Crystallography* (vol. 1) are used. The coordinate axes are oriented as follows. The a axis is toward the reader, the b to the right (in an E–W position), and the c axis is perpendicular to the paper. The origins of the drawings are in the upper left corner. The motifs are represented by small circles instead of commas (as used extensively until this part of the book). The heights of motif locations are indicated by $+$ or $-$ next to them: $+$ means a distance upward (along c) from the page, and $-$ means an equivalent distance in a downward direction. Motifs and symmetry operators can be accompanied by fractions (e.g., $\frac{1}{4}, \frac{1}{2}, \frac{1}{3}$) indicating a fractional distance (upward or downward) within the unit cell. The open circle ($\bigcirc$) is considered a right-hand motif, whereas an open circle with a small comma inside ($\odot$) is the left-hand equivalent. These two symbols are thus enantiomorphic and can be related by a mirror, glide, or inversion operation. The two types of circles are analogous to the enantiomorphic commas used in prior illustrations. The symbols for the various sym-

FIG. 3.27 Examples of illustrations of space groups. (Reproduced from *The International Tables for X-ray Crystallography*, vol. 1; see text for discussion.) (*continued*)

metry operators are given in Tables 3.3 and 3.4. An inversion point (center of symmetry) is noted by a very small circle. If its height is not given, it is assumed to be zero (in the plane of the page). If a mirror exists parallel to the plane of projection (the plane of the page), the superimposed motifs are indicated by the small circle used for motifs, but now divided vertically down the middle (⬀). This symbol is accompanied by + and − signs at its side to indicate that one motif (above, +) superimposes with another below (−). The enantiomorphic relation in such an occurrence is shown by the small comma inside one half of the circle (e.g., ⬀).

The left-hand diagrams in Fig. 3.27 show the distribution of motifs in the unit cell of the three-dimensional periodic array, and the right-hand diagrams locate the various symmetry elements with respect to this unit cell.

Figure 3.27a shows the monoclinic space group C2/m, which is common in rock-forming minerals such as clinoamphiboles and sanidine. The space group illustrations are oriented such that the y axis (with the b cell edge) is the 2-fold axis in an east-west

position. This means that we have adopted the second setting for this space group (in the first setting the c axis would be the unique 2-fold axis).[1]

This monoclinic space group, C2/m, illustrates an important aspect of space group notation. The notation shows only 2-fold rotation axes, but the right-hand side of Fig. 3.27a also shows 2-fold screw axes interleaved with the 2-fold rotation axes. Similarly, only m's are given in the space group symbol, but glide planes are interleaved with them. This is related to the choice of a c-centered lattice (C). In space groups with nonprimitive lattices, screw axes and glide planes are introduced because of centering. These new elements, however, are not noted in the symbols. Therefore, particular attention should be paid to such occurrences in nonprimitive lattice types.

[1]Because the monoclinic system is referred to three axes a, b, and c and a β angle—which is not 90°—the projection of lattices in this system, with b to the right—in the plane of the page—and c perpendicular to the page, will not allow the a axis to lie in the plane of the page. Instead, an inclined a axis is projected onto the page and the north-south direction is one of $a\sin\beta$ instead of a, as in orthogonal systems.

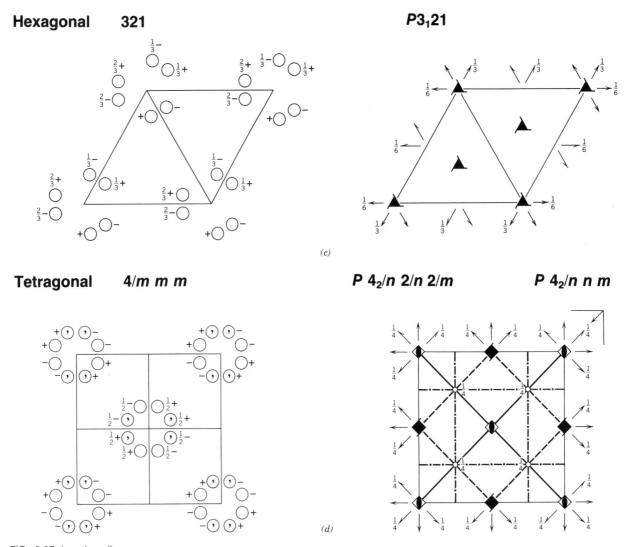

FIG. 3.27 *(continued)*

The *C2/m* diagrams also show the presence of *m*'s and 2-fold rotation axes at distances of one-half the appropriate cell lengths. These extra symmetry elements are oriented in accordance with the symmetry elements of the space group; such extra elements are present in all space groups. Centers of symmetry (inversion) are indicated by small open circles.

Figure 3.27*b* illustrates the orthorhombic space group $P2_1/n2_1/m2_1/a$, which is isogonal with point group $2/m2/m2/m$. The abbreviated (short) symbol for this point group is *mmm*, and the analogous short symbol for the space group is *Pnma*. This space group notation is one of six possible choices; these reflect different ways of orienting an orthorhombic cell with respect to the *a*, *b*, and *c* axes. The six equivalent permutations are: *Pnma*, *Pbnm*, *Pmcn*, *Pnam*, *Pmnb*, and *Pcmn*. The structure of the common rock-forming mineral olivine is based on this space group. The left-hand side of Fig. 3.27*b* shows the locations of motifs,

and the right-hand side illustrates the lattice type and symmetry elements compatible therewith. All 2-fold rotation axes in the point group are represented by 2-fold screw axes in the space group (each axial direction, *a*, *b*, and *c*, is coincident with 2_1 symmetry). The diagonal (*n*) glide perpendicular to the *a* axis is shown by the E–W dot-dash lines; the *m*'s perpendicular to the *b* axis by the solid lines, and the *a* glide by the graphic symbol for a plane parallel to the plane of the page (but at one-fourth of the unit cell above the plane of projection).

The space group of the common form of quartz (low quartz) is illustrated in Fig. 3.27*c*. This is $P3_121$, which is one of a pair of enantiomorphic space groups, $P3_121$ and $P3_221$. These are isogonal with point group 321 (also referred to as 32). The unit cell has a primitive hexagonal outline. The location of motifs in the left-hand diagram immediately reveals the 3-fold screw axes at the four corners of the unit cell;

two additional locations are present in the centers of the triangular halves of the unit cell. The locations of the 2-fold rotation axes are coincident with the a_1, a_2, and a_3 axis directions. The last numeral, 1, refers to directions at 30° to the axial directions of the unit cell (the diagonal directions) and indicates that there is no symmetry in these directions.[2] This space group also contains 2-fold screw axes that occur halfway along the cell translations and are parallel to the 2-fold rotation axes. These are the result of combining a rotational axis with a nonparallel axial translation (e.g., the 2-fold rotation axes that are not at 90° to the edges of the unit cell).

The tetragonal space group $P4_2/nnm$ (that of zircon) is illustrated in Fig. 3.27d. This space group is isogonal with point group $4/mmm$. The full (unabbreviated) symbols for space and point groups are $P4_2/n2/n2/m$ and $4/m2/m2/m$, respectively. The square lattice and inherently high symmetry content are obvious in the illustrations. The 4-fold rotation axes of the point group appear as alternating and parallel 4-fold rotoinversion and 4-fold screw axes (neutral as a result of 4_2). The n glide (in $4_2/n$) is parallel to the plane of the page at location $+\frac{1}{4}c$. The axial directions (a_1 and a_2) contain 2-fold rotation axes and have vertical n glides parallel and perpendicular to the a_1 and a_2 axes (these locations are shown by the "dot-dash" lines). The diagonal directions (at 45° to the a_1 and a_2 axes) contain mirror planes as well as 2-fold rotation axes with interleaved 2-fold screw axes. Centers of symmetry are located at $+\frac{1}{4}c$.

Space Groups as an Expression of Morphology and Structure

In Chapter 2 we derived the inherent symmetries of the 32 crystal classes (point groups) as based on the morphology of crystals. In this chapter we have evaluated order in one, two, and three dimensions. The three-dimensional order as expressed by space groups represents the 230 possible ways in which average three-dimensional periodic arrays can be arranged.

Our approach to the discussion of space groups, based on periodic patterns of nodes in space, has been more abstract than the discussion of point groups (in Chapter 2) as based on observed crystal symmetry.

In spite of the abstract approach, space groups are an elegant and powerful "shorthand" for the characterization of the internal structure of crystalline material. We know that a point group is the translation-free residue of an isogonal space group. In other words, an atomic pattern that might contain screw axes and/or glide planes will exhibit an external form which reflects only rotation axes and/or mirrors. The angular relations between the faces of a crystal will not be affected by the specific nature of the internally present axes and planes. Screw axes and glide planes cause displacements in the internal structure because of their translational components, but such displacements are so small they cannot be observed morphologically and are therefore absent in our evaluation of point group symmetry. Indeed, such translations are so small (2 to 5 Å) that only X-ray and electron diffraction techniques can resolve the distances.

A clear understanding of what space groups mean can be obtained from correlations between point group symmetry (as based on morphology), internal structure, and the resultant space group symbolism. The most informative way to evaluate an internal structure is by the visual study and inspection of three-dimensional models of the structure (see the section "Crystal Structure" in the succeeding pages). However, in our illustrations in Figs. 3.28, 3.29, and 3.30 (as in all other structure illustrations in this book) we are restricted to two-dimensional representations of three-dimensional arrangements. In Fig. 3.28 we will correlate the symmetry deduced from the morphology of diopside, $CaMgSi_2O_6$ (one of the members of the clinopyroxene series), with the internal structure and the space group derived therefrom. The diopside crystal shown in Fig. 3.28a has a point group symmetry $2/m$, as shown in Fig. 3.28b. In our orientation of the crystal we have chosen the b axis as the unique, 2-fold axis (this is referred to as the second setting). Figure 3.28c is a sketch of the simplest possible lattice (primitive) compatible with point group symmetry $2/m$. In this sketch the locations of all 2-fold rotation axes and mirror planes are shown; locations of centers of symmetry (inversion) are not indicated. In order to arrive at the correct (not just the simplest) choice of lattice type (and unit cell), one must refer to the internal structure of diopside (and similar clinopyroxenes). Figures 3.28d and e show two sections through this type of structure. Figure 3.28d is a plan view (down the c axis), whereas Fig. 3.28e is a view of part of the b–c plane in the structure. The shaded parts of Fig.

---

[2]The numeral 1 is commonly used in space group notation in order to distinguish different orientations of the same overall symmetry content. For example, point group 32 has, among others, isogonal space groups $P312$ and $P321$. The symmetry content is the same in both representations. In $P312$, however, there is no symmetry along the a_1, a_2, and a_3 axes, and the 2-fold rotation axes lie in directions perpendicular to a_1, a_2, and a_3. In $P321$ the a_1, a_2, and a_3 axes are the directions of 2-fold rotation and the directions perpendicular to a_1, a_2, and a_3 lack any symmetry. Other examples are: $P3m1$ and $P31m$; $P32/m1$ and $P312/m$ (see footnote to Table 3.1). Examples in the monoclinic system are: $P121$, $P12_11$, and $C121$ for space groups compatible with crystal class 2 (second setting). Here the b axis is one of 2-fold rotation symmetry, and the a and c axial directions lack symmetry (as noted by the two numerals 1 in the a and c locations of the symbol).

Monoclinic   2/m   2nd Setting

(a)

(b)

(c)

(d)

(e)

(f)

C2/c

(g)

FIG. 3.28 Relationship of point group symmetry to space group representation for diopside, a member of the clinopyroxene group. *(a)* Crystal of diopside. *(b)* Stereogram of 2/m symmetry. *(c)* Sketch of a possible monoclinic, primitive lattice compatible with 2/m. Locations of 2-fold axes and mirrors are noted. *(d)* A view of the clinopyroxene structure along the *c* axis. (Redrawn from Cameron, M. and Papike, J. J., 1980, Crystal chemistry of pyroxenes, in *Reviews in Mineralogy*, v. 7.) Centering of the cell is outlined by the distribution of shaded units in the structure. Rotation and screw axes parallel to *b* are noted, as are glide planes (marked *g*). *(e)* The location of vertical glide planes, with a *c*/2 component, in an idealized tetrahedral pyroxene chain. *(f)* A sketch of the centered, monoclinic lattice, as based on the structural information of parts *(d)* and *(e)*. For simplicity only some of the rotation axes, screw axes and glide planes are shown. *(g)* An illustration of the *C2/c* space group of diopside.

3.28*d* identify repeats of some of the motifs in the structure; these are the cross sections of tetrahedral-octahedral-tetrahedral (*t–o–t*) chains that are parallel to the *c* axis (see the section on the Pyroxene Group in Chapter 13 for further discussion). These structural units are repeated in a centered pattern in the plan view; in three dimensions this is compatible with a *C*-centered monoclinic lattice (*C*). The locations of 2-fold rotation axes (parallel to the *b* axis) and interleaved 2-fold screw axes (the result of the centering) are shown. The vertical section in Fig. 3.28*e* shows the presence of a glide component of *c*/2 in the tetrahedral chain. This is expressed by the *c* in the space group symbol *C*2/*c*, in place of *m* in the point group notation 2/*m*. The locations of these glide planes are shown in the plan of the pyroxene structure in Fig. 3.28*d* and are marked *g*. Interleaved, halfway between these glides, will be additional glide planes that result from the centering of the unit cell choice. In Fig. 3.28*f* a sketch is shown of the three-dimensional lattice that is appropriate to the structure of diopside; it is monoclinic and *C*-centered. For purposes of clarity, the drawing shows only some of the symmetry elements present; additional glides, 2-fold rotation, and screw axes would be present in locations equivalent to those shown in Fig. 3.28*c*. In Fig. 3.28*g* the conventional space group representation is given for *C*2/*c*. In this discussion of the pyroxene structure we did not locate the various positions of inversion symmetry (centers of symmetry). It is conventional to choose the origin of a space group at the location of such a center of symmetry. It is thus that the 2-fold rotation and screw axes appear at $+\frac{1}{4}$ in Fig. 3.28*g*; the centers, at the positions noted, lie $\frac{1}{4}c$ below the symmetry axes.

Figure 3.29 illustrates the correlation between point group symmetry, structure, and resultant space group for the mineral beryl, $Be_3Al_2Si_6O_{18}$. Figure 3.29*a* shows the hexagonal crystal form of beryl, and the resultant point group symmetry is shown in Fig. 3.29*b*. The choice of lattice type for this mineral must obviously be hexagonal. Two sections through the crystal structure of beryl are shown in Figs. 3.29*c* and *d*. The horizontal projection onto (0001) (Fig. 3.29*c*) shows the unique 6-fold Si_6O_{18} rings. If the centers of four of these rings are chosen as the positions for possible lattice nodes, the rhombus shape of the lattice becomes obvious (a primitive hexagonal lattice choice, as in *P* of the space group symbol). The 6-fold symmetry about the centers of the rings is also clear. In a vertical cross section of the structure (Fig. 3.29*d*) the location of horizontal mirrors through the centers of the rings is shown (the hexagonal outline of and horizontal mirrors through these rings account for

6/*m* in the space group symbol). Be and Al provide cross-links between the Si_6O_{18} rings; Be is in tetrahedral (4-fold) coordination and Al in octahedral (6-fold) coordination. The Be ions occupy positions where three directions of 2-fold rotation axes intersect (222) and the Al ions occupy locations where vertical 3-fold axes intersect with horizontal 2-fold axes (32). These symmetry locations are superimposed on the horizontal section through the beryl structure in Fig. 3.29*e*. For reasons of clarity not all possible symmetry elements have been shown in this representation. The location of the horizontal mirror (perpendicular to the *c* axis) is indicated by the 120° bracket at the bottom right side of the figure. The many horizontal 2-fold rotation axes are located at $\frac{1}{4}c$ above the plane of the diagram. Figure 3.29*d* shows that the structure contains 6-fold rings that are not equivalent in orientation to the rings above and below. These rotations between interleaved rings (along the *c* direction) are the result of reflections by vertical glide planes (marked as *c*'s in the space group symbol). Figure 3.29*f* shows the conventional representation of the space group P6/*mcc*, which reflects the beryl structure.

The last illustration of correlating point group, space group, and internal structure is that of the mineral diamond, C, in Fig. 3.30. Figures 3.30*a* and *b* depict a possible morphology of diamond and the point group symmetry derived therefrom. Figure 3.30*c* shows a unit cell of diamond based on a cubic space lattice. The structure of diamond consists of carbon atoms in tetrahedral coordination with four nearest carbon neighbors. The lines joining carbon atoms in the upper left part of the unit cell of Fig. 3.30*c* show this coordination by the dotted outline of a tetrahedron. Upon careful inspection of the distribution of carbon atoms in the unit cell, it becomes clear that each of the faces of the cube contains (in addition to carbon atoms at all of the corners) one carbon atom at its center. This "all-face-centered" distribution of nodes in the lattice is reflected in *F* of the space group notation. When this same unit cell is projected onto a horizontal cube face at the bottom of the unit cell, the drawing in Fig. 3.30*d* results. This projection reveals the presence, at positions halfway between the carbon atoms (or projected carbon atoms), of two types of 4-fold screw axes. These two enantiomorphic screw axis types (4_1 and 4_3) are located in alternate rows and are parallel to the 4-fold rotation axes of the point group notation. The helical paths for two of the screw rotations are shown. The symmetry about each of the carbon atoms is $\overline{4}3m$, because of the tetrahedral coordination. A projection of the symmetry elements in the space group of diamond is shown in Fig. 3.30*e*. In addition to the 4_1 and 4_3 axes there are 4-fold ro-

Hexagonal   6/m 2/m 2/m

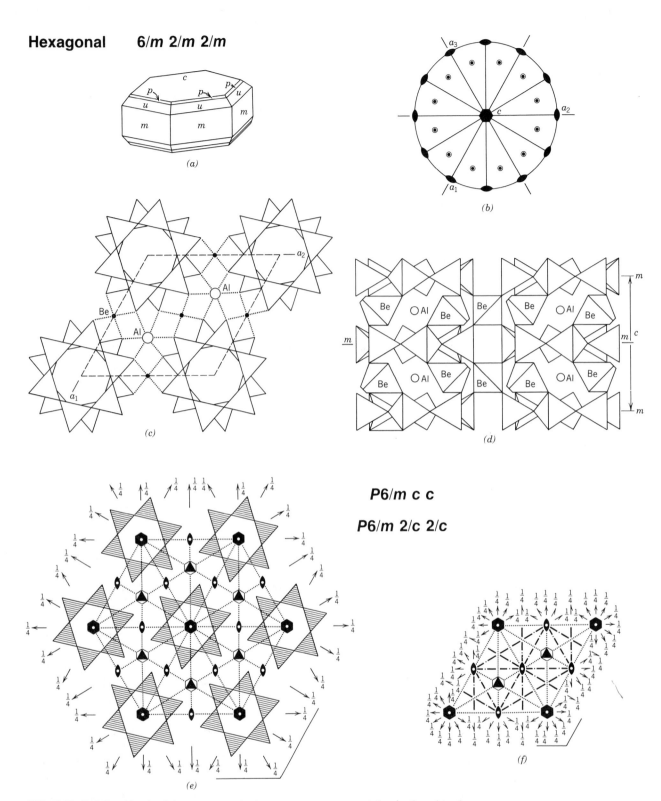

FIG. 3.29. Relationship of point group symmetry to space group representation for the mineral beryl. *(a)* A beryl crystal flattened on (0001), typical of cesium-rich beryl. *(b)* A stereogram of the point group symmetry 6/m2/m2/m as reflected by the morphology. *(c)* The structure of beryl as projected onto (0001). A unit cell is shown by dashed lines. *(d)* A vertical section through the beryl structure. *(e)* A projection of some of the structural elements of beryl and their relationship to each other as shown by the presence of some of the symmetry elements. (From Shubnikov, A. V. and Koptsik, V. A., 1974, *Symmetry in Science and Art*. Plenum Press, New York.) *(f)* The conventional representation of the space group *P*6/*m*2/*c*2/*c* which is compatible with the beryl structure.

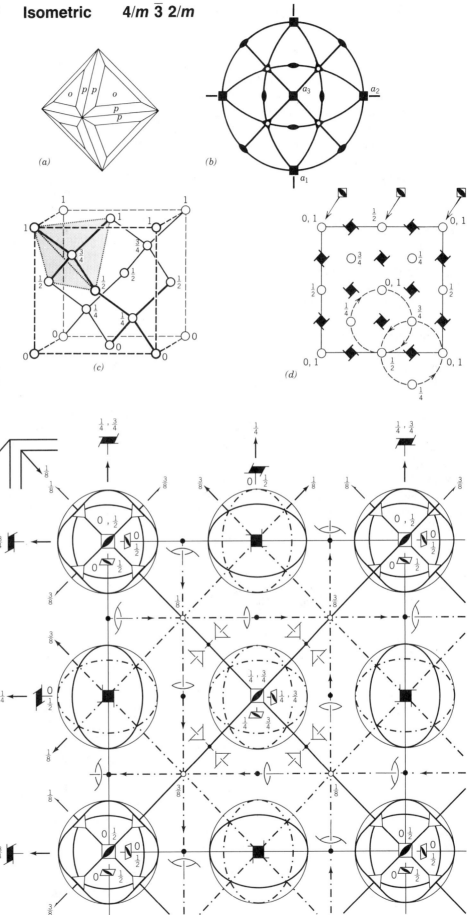

FIG. 3.30. Relationship of point group symmetry to space group representation for diamond. *(a)* A diamond crystal showing an octahedron *(o)* modified by a trisoctahedron *(p)*. *(b)* A stereogram of the point group symmetry $4/m\overline{3}2/m$ compatible with the morphology shown in *(a)*. *(c)* A sketch of a unit cell of the diamond structure. *(d)* A projection of the diamond structure in *(c)* onto a horizontal cube face. The carbon atoms are related to each other by 4-fold screw axes of opposite hand (4_1 and 4_3). Four-fold rotoinversion axes are coincident with the positions of each of the carbon atoms. Only three are shown by the graphical symbol (◨) so as not to clutter the diagram. *(e)* The conventional representation of the space group $F4_1/d\overline{3}2/m$, which is compatible with the diamond structure. (From *International Tables for Crystallography*, 1983, vol. A; see reference list for complete citation)

Isometric $4/m\,\overline{3}\,2/m$

(a)

(b)

(c)

(d)

$Fd\,3m$

$F4_1/d\,\overline{3}\,2/m$

(e)

toinversion axes (◤) interspaced with the screw ro-tations. Centers of symmetry (small circles with frac-tions such as $\frac{1}{8}$ and $\frac{3}{8}$) are noted. Diamond glide planes (*d*) (shown by dot-dash lines with arrows) are parallel to the cube faces of the unit cell. These planes are inclined to the plane of the figure because they have translation components of the type *b*/4 + *c*/4, *a*/4 + *c*/4. Vertical mirror planes (and interleaved glide planes) occur in diagonal directions.

CRYSTAL STRUCTURE

A knowledge of the structure of crystalline materials is basic to the understanding of properties such as cleavage, hardness, density, melting point, refractive index, X-ray diffraction pattern, solid solution, and exsolution. Furthermore, our knowledge of atomic and ionic sizes, and of the nature of the chemical bonds in crystals, derives from the precise determi-nations of crystal structures.

A structural crystallographer determines the structure of a crystalline material through the interpre-tation of the interrelationship of various properties such as external symmetry, X-ray diffraction effects, electron diffraction patterns, density, and chemical composition. The resulting crystal structure provides us with a knowledge of the geometric arrangement of all of the atoms (or ions) in the unit cell, and the bond-ing and coordination between them. The symmetry of

the internal, atomic arrangement of a crystal is ex-pressed by its space group, whereas the morphology of the crystal is reflected in the point group symmetry.

Let us illustrate briefly the chemical and physical interrelationships in the derivation of the simple struc-ture of sodium chloride, NaCl. The external mor-phology of halite is consistent with $4/m\overline{3}2/m$ sym-metry. X-ray diffraction data indicate that the unit cell has an *a* axis dimension of 5.64 Å and that the unit cell is not primitive but face-centered (*F*). The space group of halite is $F4/m\overline{3}2/m$ (*Fm3m* in abbreviated form). A chemical analysis yields 39.4 weight percent Na and 60.6 weight percent Cl. Division of these weight percentages figures by the appropriate atomic weights yields an atomic ratio for Na:Cl of 1:1, which is expressed as NaCl. The density of halite is 2.165 g/cm³. If the density (*D*) and the volume (*V*) of the unit cell (a^3) are known, then the number of for-mula units per unit cell (*Z*) can be calculated from $D = (Z \times M)/(N \times V)$, where *M* is molecular weight and *N* is Avogadro's number (see page 259). This number for halite is 4, which means that the unit cell contains 4 NaCl units, or $4Na^+$ and $4Cl^-$ ions. It is now possible to derive the structural arrangement of the ions in NaCl. The Na^+ (or Cl^-) ions must be lo-cated in a face-centered cubic array. The radius ratio of Na^+ to Cl^- would predict a coordination of 6 about each of the ions (see Chapter 4). A reasonable and indeed correct interpretation of the crystal struc-ture is given in Fig. 3.31.

FIG. 3.31. Sodium chloride (NaCl) structure based on face-centered cubic lattice type. Note that each ion is surrounded by six neighboring ions of opposite charge outlining octahedral polyhedra (see also Fig. 3.36).

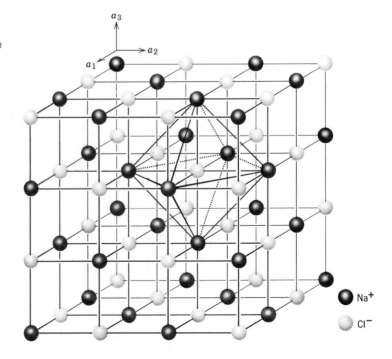

Na⁺

Cl⁻

The Determination of Crystal Structures

The orderly arrangement of atoms in a crystal is known as the *crystal structure*. It provides information on the location of all the atoms, bond positions and bond types, space group symmetry, and the chemical content of the unit cell.

The first steps in determining the atomic structure of a crystal are the measurement of its unit cell size and the evaluation of its space group. The systematic measurement of the geometric distribution of diffracted X-ray beams on single crystal X-ray photographs in one or more crystallographic orientations (precession photographs are most convenient; see Chapter 7) yields information about the geometry of that crystal's unit cell, specifically the lengths of the unit cell edges and the angles between them. Information about the space group and the crystal structure (i.e., the symmetry of the atomic arrangement and the coordinates of atoms within the unit cell) is contained in the intensities of the diffracted beams, which are measured either from X-ray single crystal photographs, or more commonly with quantum counting detectors on single-crystal diffractometers. The diffracted beams are identified by Bragg indices *hkl* associated with the lattice planes of the same indices.

The relationship between the intensity (*I*) of the diffracted beam associated with lattice planes having Bragg diffraction indices *hkl* and coordinates $x_jy_jz_j$ of atom *j* in the unit cell is

$$I = k\left[\sum_j f_j e^{i2\pi(hx_j + ky_j + lz_j)}\right]^2 = kF_{hkl}^2$$

where k is a term containing several physical constants and experimental factors, including a scale factor; f_j is the scattering factor of atom *j* (which depends on atomic number and scattering angle and includes corrections for atomic thermal motion); and the summation is over all atoms in the unit cell. The summation is termed the *structure factor*, F_{hkl}, and its value clearly depends on the kinds of atoms in the cell and their positions.

Because the intensities of diffracted beams are related to F_{hkl}^2 rather than F_{hkl}, the atomic coordinates cannot be extracted straightforwardly from measured intensities. Over the past 60 years a great deal of effort has been devoted to methods by which these intensities can be made to yield atomic positions, with remarkable success. Modern methods are based either on analysis using Fourier techniques, which are very powerful, particularly when the positions of a small number of heavy atoms are known, or on so-called direct methods, which are essentially statistical in

nature and have become extremely successful as computing power has increased. An illustration of a now-classical electron density map (and the structure derived therefrom) by W. L. Bragg in 1929 is given in Fig. 3.32. Once the atom positions are known crudely, they can be refined to high precision, using least-squares analysis on many measured F_{hkl}^2, a procedure that can yield amplitudes of atomic thermal vibration as well as site occupancies of substituting atomic species in minerals that are members of solid solution series. Accurate interatomic distances (bond lengths) can then be calculated from refined atomic coordinates. In recent years highly accurate electron density maps, calculated as a Fourier series in which the structure factors are the Fourier coefficients, have yielded information about bonding and the spatial distribution of valence electrons. Examples of such detailed electron density maps are given in "Molecules as models for bonding in silicates," by G. V. Gibbs (1982); complete reference is given at the end of this chapter.

Supplementing structure analyses based on diffraction phenomena (these are X-ray, neutron, and electron diffraction techniques), a variety of spectroscopic methods, including infrared, optical, Mössbauer, and resonance techniques such as nuclear magnetic resonance (NMR) yield information about the local environments or ionization states of certain atomic species. Thus a complete structural picture of a mineral at the atomic level usually requires information obtained by both diffraction and spectroscopy.

Illustration of Crystal Structures

As in the case of illustrations of the external morphology, where clinographic and stereographic projections are used to illustrate crystal forms, the three-dimensional arrangement of crystal structures must be represented on a two-dimensional page. The three-dimensional structure is commonly projected onto the page by the use of *fractional cell coordinates, x, y,* and *z*. For example, as in Fig. 3.33a the location of an atom within a cell with edges a, b, and c can be described in terms of distances from the origin whose components are parallel with the edges. These components are *x', y',* and *z',* where $x = x'/a$, $y = y'/b$, and $z = z'/c$; *x, y,* and *z* range from 0 to 1. A listing of the fractional coordinates of all atoms (and/or ions) in a unit cell defines the crystal structure and a two-dimensional projection can be constructed. Usually such projections are made onto one of the cell faces or onto planes perpendicular to a specific direction in the crystal (e.g., the crystallographic axes). The frac-

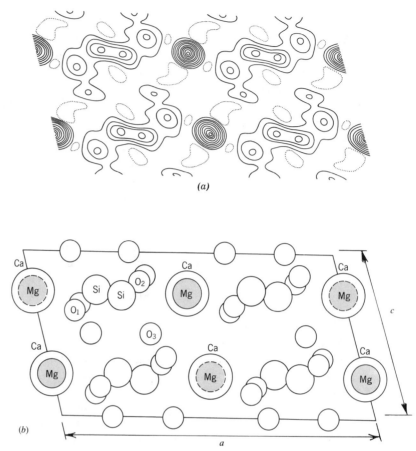

FIG. 3.32. *(a)* The summation of Fourier series for diopside, $CaMgSi_2O_6$, projected on (010). The distribution of scattering matter is indicated by contour lines drawn through points of equal density in the projection. *(b)* The atomic positions of diopside (with space group *C* 2/*c*) projected on (010), as derived from the distribution in *(a)*. (Redrawn after Bragg, W. L., 1929, *Zeitschrift für Kristallographie*, v. 70, p. 488.)

tional heights of atoms above the plane of projection are noted adjacent to them, although values of 0 to 1 corresponding to positions on the bottom and top faces of the unit cell are often omitted. The coordinates 0, 0, 0 specify an atom at the origin of the unit cell.

As an example of a structure projection, a projection on the (0001) plane of high tridymite, SiO_2, is shown (Fig. 3.33*b*). Two unit cells are shown. The *x* and *y* coordinates of the oxygen and silicon ions in the structure are shown by their locations in the a_1–a_2 plane. The *z* coordinates of the ions are shown by fractions indicating their location above the page of the projection. Instead of using fractional coordinates, structure projections may be represented by atomic positions accompanied by numbers ranging from 0 to 100. An atom in the lower face of the unit cell is indicated by 0, one on the top face by 100, and inter-

mediate heights are given accordingly. An atom marked 75, for example, is located three-quarters of the way up the unit cell. Two examples of such illustrations are given in Fig. 3.34.

For many purposes it is useful to represent a crystal structure in terms of coordination polyhedra instead of the locations of the atoms or ions. Figure 3.35 illustrates the structure of high tridymite in terms of the linkage of SiO_4 tetrahedra (compare with Fig. 3.33*b*). To aid in the visualization of complex crystal structures, crystal structure models can be built or obtained commercially. Such models reproduce the internal atomic arrangement on an enormously magnified scale (e.g., 1 Å might be represented by 1 cm in a model). Several types of models are illustrated in Figs. 3.36 and 3.37. A *close-packed model* is most realistic in terms of the representation of the filling of space on an atomic scale (see Fig. 3.36*a*). It is difficult

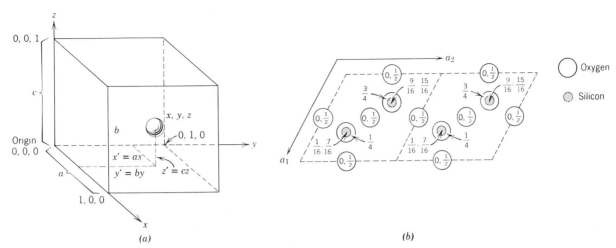

(a)   (b)

FIG. 3.33. *(a)* Fractional cell coordinates (x, y, z) of an atom in a unit cell as outlined. *(b)* Atomic arrangement in high tridymite polymorph of SiO_2, projected on (0001). Oxygens occur in the base of the unit cell (0), one quarter ($\frac{1}{4}$), halfway ($\frac{1}{2}$), and three-quarters of the way up ($\frac{3}{4}$). In this projection some oxygen positions superimpose, as shown by the z coordinates 0, $\frac{1}{2}$. Silicon positions, above and below oxygen locations, occur at $\frac{1}{16}$, $\frac{7}{16}$, $\frac{9}{16}$, and $\frac{15}{16}$ of the z axis.

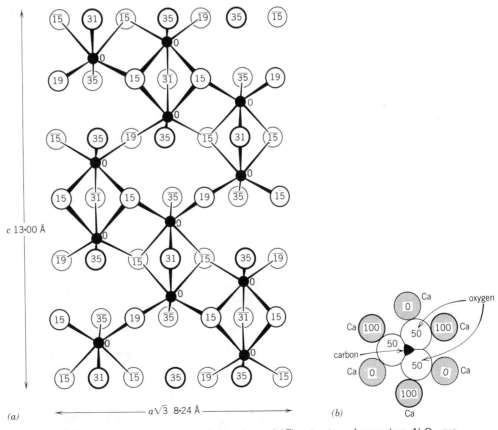

(a)   (b)

FIG. 3.34. Examples of projections of atomic structures. *(a)* The structure of corundum, Al_2O_3, projected on ($2\bar{1}\bar{1}0$). The heights of oxygens relate to the plane of aluminum ions at zero. Compare this illustration with the polyhedral representation in Fig. 11.2b. *(b)* Projection of some of the ion positions in calcite, $CaCO_3$, onto (0001). The upper layer of Ca^{2+} ions is shown by the number 100; the Ca^{2+} ions in the lower layer by 0. Note the triangular outline of the CO_3 group.

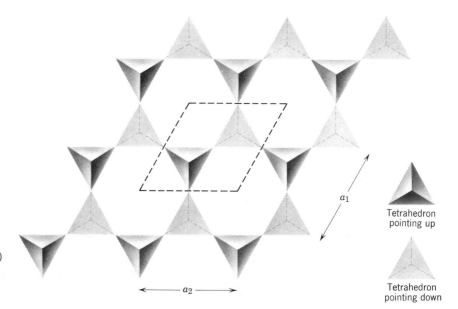

FIG. 3.35. A single sheet of the high tridymite structure (polymorph of SiO_2) projected onto the (0001) plane. Unit cell is outlined. Compare with Fig. 3.33b.

Tetrahedron pointing up

Tetrahedron pointing down

to visualize the interior of such a model, however. A *polyhedral model* (Figs. 3.36a and 3.36b)[3] represents the coordination polyhedra instead of specific atomic locations. Note the different appearances of two models in Fig. 3.36, both of which represent the NaCl structure. A polyhedral model of a complex structure, such as in Fig. 3.37a, is most useful in visualizing the overall symmetry and the regular repeat units of the structure. This often-idealized regularity of the polyhedral arrangement also allows relatively easy identification of the space group elements in the structure. An *open "ball and stick" model* (Fig. 3.37b) allows a

view of the coordination inside the model and portrays the irregularly coordinated environment of ions in a complex structure.

ISOSTRUCTURALISM

Although it might seem that uraninite, UO_2, and fluorite, CaF_2, would have little in common, their X-ray powder diffraction patterns show analogous lines, although different in spacing and intensity. Structure analysis reveals that there are four U^{4+} ions in uraninite around each oxygen, whereas eight O^{2-} are grouped about each uranium. In fluorite, four Ca^{2+} are grouped about each fluorine, and eight F^{1-} are packed about each calcium. Uraninite and fluorite

[3]Polyhedral models are available from Polyhedral Models, P.O. Box 1121, Minneapolis, Minnesota 55440. Ball and stick models are available from Crystal Structures Limited, Bottisham, Cambridge, England.

FIG. 3.36. (a) Close-packed model of NaCl structure. Na^+ ions light; Cl^- ions dark. (b) Polyhedral representation of octahedral co-ordination of Cl^- about Na^+ (as well as Na^+ about Cl^-) in NaCl.

(a)   (b)

FIG. 3.37. *(a)* Polyhedral structure model representing a monoclinic amphibole. Tetrahedral linking of SiO_4 (Si_4O_{11} chains parallel to *c* axis) and octahedral coordination of cations between chains. Dark octahedra represent M_4 positions in structure. *(b)* Expanded "ball and stick" model of the structure of biotite $K(Mg,Fe)_3 (AlSi_3O_{10})$-$(OH)_2$. Top and bottom layers represent K^+ ions in 12 coordination with oxygen. Central layer represents Mg^{2+} and Fe^{2+} in 6-coordination with O^{2-} and $(OH)^-$.

have structures that are analogous in every respect, although their unit cell dimensions are different and other properties are, of course, totally dissimilar. Their space groups are identical, *Fm3m* (equivalent to $F4/m\bar{3}2/m$). These two substances are said to be *isostructural* or *isotypous*, and belong to the *same structure type*. Occasionally the term *isomorphism* is used instead of *isostructuralism*. Crystals in which the centers of the constituent atoms occupy geometrically similar positions, regardless of the size of the atoms or the absolute dimensions of the structure, are said

to belong to the same structure type. For example, all *Fm3m* crystals in which there are equal numbers of cations and anions in 6-coordination belong to the NaCl (halite) structure type. A few of the large number of minerals of diverse composition belonging to this structure type are KCl, sylvite; MgO, periclase; NiO, bunsenite; PbS, galena; MnS, alabandite; AgCl, chlorargyrite; and TiN, osbornite. All of these substances display the same space group as that of halite, *Fm3m*. The relative sizes of cations and surrounding anions are of major importance in the type of packing of atoms, and the subsequent resulting structure. This is covered extensively in Chapter 4 as part of the "co-ordination principle."

The two minerals stishovite, SiO_2, and rutile, TiO_2, are also isostructural. Their structures conform to the tetragonal space group $P4_2/mnm$ (or $P4_2/m2/n2/m$) and in both the cation (Si^{4+} or Ti^{4+}) is surrounded by six oxygen neighbors, in octahedral packing about the cation. In all other known forms of SiO_2 (including quartz), as well as all other silicates found in the Earth's crust, silicon is surrounded by four neighboring oxygens, in tetrahedral packing. Stishovite is a very high pressure polymorphic form (see below) of quartz (see also Fig. 3.38a).

Of great importance in mineralogy is the concept of the *isostructural group*: a group of minerals related

FIG. 3.38. (a) Stability relations of the SiO_2 polymorphs. Pressure is expressed in kilobars = 1000 bars, where 1 bar = 0.987 atmosphere. (b) P-T diagram for H_2O. Six polymorphic types are indicated by I, II, III, V, VI, and VII. (After Bridgman, P. W., 1935, *Jour. Chem. Phys.*, v. 5, p. 965, and *Phase Diagrams Ceramists*, copyright 1964 by the American Ceramic Society.)

(a)

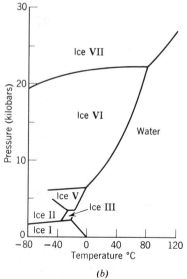

(b)

Table 3.6
ARAGONITE GROUP OF ORTHORHOMBIC ISOSTRUCTURAL CARBONATES (SPACE GROUP *Pmcn*)

Mineral	Chemical Composition	Cation Size* (Å)	Unit Cell Dimensions (Å)			Volume (Å³)	Specific Gravity (G)	Cleavage Angle 110 ∧ 1̄10
			a	*b*	*c*			
Aragonite	$CaCO_3$	Ca^{2+} 1.18	4.96	7.97	5.74	226.91	2.94	63°48'
Strontianite	$SrCO_3$	Sr^{2+} 1.45	5.11	8.41	6.03	259.14	3.78	62°41'
Cerussite	$PbCO_3$	Pb^{2+} 1.49	5.19	8.44	6.15	269.39	6.58	62°46'
Witherite	$BaCO_3$	Ba^{2+} 1.47	5.31	8.90	6.43	303.87	4.31	62°12'

*Because the metal ions are coordinated to 9 oxygens, the ionic radii (see page 188) are given for 9-fold coordination.

to each other by analogous structure, generally having a common anion and frequently displaying extensive ionic substitution. Many groups of minerals are isostructural, of which the barite group of sulfates, the calcite group of carbonates, and the aragonite group of carbonates are perhaps the best examples. The extremely close relationship that exists among the members of many groups is illustrated by the aragonite group listed in Table 3.6.

POLYMORPHISM

The ability of a specific chemical substance to crystallize with more than one type of structure (as a function of changes in temperature, pressure, or both) is known as *polymorphism* (from the Greek meaning "many forms"). The various structures of such a chemical element or compound are known as *polymorphic forms*, or *polymorphs*. Examples of some polymorphous minerals are given in Table 3.7; Fig. 3.38 shows the stability regions of various polymorphs, in terms of temperature (*T*) and pressure (*P*), for the chemical systems SiO_2 and H_2O.

Three major types of mechanisms are recognized

by which one polymorphic form of a substance can change to another. These are: *reconstructive, displacive*, and *order-disorder* polymorphism.

The reason why a constant chemical composition may have different structural arrangements is that some structural configurations represent greater (or lesser) internal (structural) energies (*E*) than others. The relative internal energy of a specific polymorph may be a function of temperature, pressure, or both. A higher internal energy, as a function of increasing temperature, is caused by higher frequencies of vibrations of the atoms. Figure 3.39*a* shows the abrupt differences in the relative internal energy levels as a function of temperature for three polymorphs that are related by reconstructive transformations; these three structures are completely different from each other. Figure 3.39*b* shows the continuous increase in internal energy level in one polymorph (form 1) as a function of increasing temperature, until at a specific temperature the transformation takes place. This is what occurs in a displacive transformation, as a function of increasing temperature. Pressure can also be a major driving force in polymorphic transformations. Increasing pressure favors the development of structural arrangements that result in an increase in the density of atomic packing (as reflected in increased density (*D*),

Table 3.7
EXAMPLES OF POLYMORPHOUS MINERALS

Composition	Mineral Name	Crystal System and Space Group	Hardness	Specific Gravity
C	Diamond	Isometric—$Fd3m$	10	3.52
	Graphite	Hexagonal—$P6_3/mmc$	1	2.23
FeS_2	Pyrite	Isometric—$Pa3$	6	5.02
	Marcasite	Orthorhombic—$Pnnm$	6	4.89
$CaCO_3$	Calcite	Rhombohedral—$R\bar{3}c$	3	2.71
	Aragonite	Orthorhombic—$Pnam$	$3\frac{1}{2}$	2.94
SiO_2	Low quartz	Hexagonal—$P3_121$	7	2.65
	High quartz	Hexagonal—$P6_222$		2.53
	High tridymite	Hexagonal—$P6_3/mmc$	7	2.20
	High cristobalite	Isometric—$P2_13(?)$	$6\frac{1}{2}$	2.20
	Coesite	Monoclinic—$C2/c$	$7\frac{1}{2}$	3.01
	Stishovite	Tetragonal—$P4/mnm$		4.30

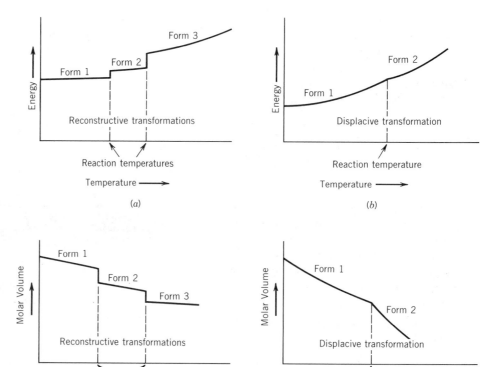

(a)

(b)

FIG. 3.39. Variation in internal energy (E) as a function of temperature for (a) three reconstructive polymorphs and (b) two displacive polymorphic forms. (Adapted from Buerger, M. J., 1961, Polymorphism and phase transformations. *Fortschr. Miner.*, v. 39, pp. 9–14.) Variation in molar volume as a function of pressure for (c) three reconstructive polymorphs and (d) two displacive polymorphic forms.

(c)

(d)

or specific gravity (G); see Chapter 6) and a decrease in molar volume (molar volume is the volume occupied by one mole of substance). Figure 3.39c shows abrupt discontinuities in the molar volume values, whereas Fig. 3.39d shows no discontinuities; here, however, the slopes of the molar volumes are discontinuous.

In a *reconstructive* polymorphic reaction the internal rearrangement in going from one form to an-

other is extensive. It involves the breaking of atomic bonds and a reassembly of the structural units in a different arrangement (see Fig. 3.40). This type of transformation requires a large amount of energy, is not readily reversed, and is sluggish. An example of such a polymorphic reaction is the change from tridymite or cristobalite to low quartz. All three are polymorphs of SiO_2 (see Fig. 3.35 for the tridymite structure and Fig. 3.42 for the quartz structure). Cris-

FIG. 3.40. Schematic representation of reconstructive polymorphism. The transformation of a hypothetical structure (a) made of octahedral coordination polyhedra to (b) requires the breaking of bonds and complete rearrangement of the octahedral units. (Adapted from Buerger, M. J., 1961, Polymorphism and phase transformations. *Fortschr. Miner.*, v. 39, pp. 9–14.)

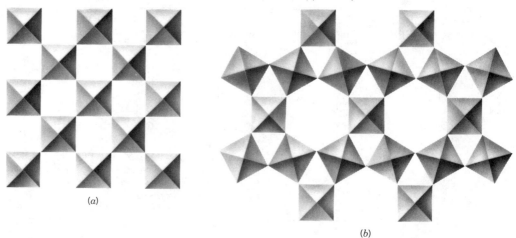

(a)

(b)

tobalite and tridymite are formed at high temperatures and relatively low pressures (see Fig. 3.38a), such as in SiO_2-rich lava flows. Table 3.7 lists the specific gravity values as 2.20 for both forms; these are the lowest values for any of the SiO_2 polymorphs listed. A high activation energy is needed to change the cristobalite (or tridymite) SiO_2 network into the arrangement of the low quartz structure. Cristobalite and tridymite are metastable in terms of atmospheric conditions; however, both minerals are abundantly present in very old terrestrial volcanic flows as well as in Precambrian lunar lavas. Such persistence of metastable minerals testifies to the fact that high energy is required to activate a reconstructive polymorphic transformation. Coesite and stishovite are forms of SiO_2 that are stable in the high to very high pressure part of the stability diagram in Fig. 3.38a. Coesite and stishovite both occur in meteorite craters, as a result of meteorite impacts on Earth. Coesite is also found in kimberlites, which are high pressure host rocks for diamond and which originate in the upper part of the Earth's mantle. In Table 3.7 note the high specific gravity for coesite and the very high value for stishovite, as compared to those of the other SiO_2 polymorphs. Stishovite is unusually densely packed, on an atomic level, for a silicate structure. In its structure each Si is surrounded by six neighboring oxygens (in octahedral coordination), whereas all other SiO_2

polymorphs contain Si with four oxygen neighbors in tetrahedral coordination. The importance of variable pressure regimes in the formation of some polymorphs is also seen in Fig. 3.38b, where the stability fields of the various polymorphs of ice are strongly dependent on the pressure applied to the system H_2O. The differences between the polymorphs for C, for FeS_2, and for $CaCO_3$ (Table 3.7) are such that a major reworking and rearrangement of the structure is needed to go from one structure type to the other; each of these examples represents reconstructive polymorphism (see also graphite-diamond relations and the discussion of calcite-aragonite on pages 315 and 317, respectively).

In a *displacive* polymorphic reaction the internal adjustment in going from one form to another is very small and requires little energy. The structure is generally left completely intact, and no bonds between ions are broken; only a slight displacement of atoms (or ions) and readjustment of bond angles ("kinking") between ions is needed. This type of transformation occurs instantaneously and is easily reversible. Figure 3.41a is a schematic representation of a possible tetrahedral structure (with relatively high symmetry) "kinking" to produce two structural arrangements with lesser symmetry. These may be related by a symmetry element such as a twin plane as shown in Fig. 3.41b. A similar displacive transformation occurs

FIG. 3.41. Schematic representation of displacive polymorphism. *(a)* A hypothetical structure consisting of an infinite network of tetrahedra. It represents a fairly open atomic arrangement with relatively high symmetry. This would be the high-temperature polymorphic form. *(b)* The same infinite network of tetrahedra as shown in *(a)* but in a less symmetrical and somewhat collapsed ("kinked") arrangement. This represents the lower temperature polymorph. The two oppositely "kinked" structures are mirror images of each other. Similar symmetrical relationships exist between domains of structure that show transformation twinning (see Fig. 3.55b).

(a)

(b)

m

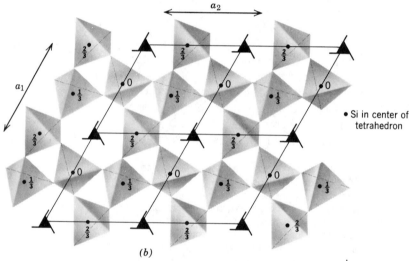

FIG. 3.42. *(a)* Projection of the tetrahedral SiO_2 framework of high quartz onto (0001). Four unit cells and 6_2 axes are shown. Fractional heights represent locations of centers (Si) of tetrahedra. *(b)* Projection of the tetrahedral SiO_2 framework of low quartz onto (0001). Four unit cells and some 3_2 axes are shown. Fractional heights represent locations of centers (Si) of tetrahedra. These two figures illustrate a displacive transformation.

when the high quartz form of SiO_2 is cooled to below 537°C (at atmospheric pressure; see also Fig. 3.38*a*) and rearranges its structure to that of low quartz. The difference between the two forms of quartz is expressed by their space groups (high quartz, $P6_222$; low quartz, $P3_221$) and can be portrayed in a basal projection of the SiO_2 framework for both forms (see Fig. 3.42). The structural arrangement in the low-temperature form is slightly less symmetric and somewhat more dense than in the high-temperature form. The transition from high to low quartz can be viewed as the result of "kinking" of atomic bonds in the original high quartz structure. Because the high-temperature form of quartz (high quartz) shows a higher symmetry than the low-temperature form, twinning ("transformation twinning") may result in the transition from high to low quartz. Such twins, known as *Dauphiné twins* (see Fig. 3.55), represent a megas-

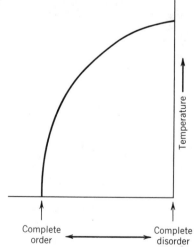

FIG. 3.43. The relationship of structural order and temperature. The higher the temperature the more disordered the distribution of atoms in specific structural sites. This would apply, for example, to the distribution of Al^{3+} versus Si^{4+} among tetrahedrally coordinated sites in the feldspar structure (see Fig. 3.45).

copic expression of the presence of obversed and reversed units of structure in a low quartz crystal.

Yet another type of polymorphism is referred to as an *order-disorder* transformation. This is commonly observed in metallic alloys but also occurs in minerals. To appreciate this transformation, it should be noted that perfect order occurs only at absolute zero (0 kelvin, which is equivalent to $-273.15°C$). An increase in temperature disturbs the perfect order of a structure, until at some high temperature a totally disordered (totally random) state is obtained (see Fig. 3.43). As such, there is no definite transition point between perfect order and complete disorder. In a state of perfect order, the atoms in a structure are arranged in specific crystallographic site locations. At high temperatures, close to but below the melting point of a substance, atoms (or ions) tend to become completely disordered and are ready to break away from the structure. Slow cooling of a mineral structure will permit the original randomized ions (at high temperatures) to select specific sites in the structure and become more ordered. Figure 3.44a is a qualitative illustration of various degrees of disorder.

A more quantitative statement of order-disorder is shown in Fig. 3.44b. This shows an alloy of composition AB with 50% of A and 50% of B existing in various states of disorder, of which a totally disordered and a perfectly ordered state are two extreme condi-

FIG. 3.44. *(a)* Distribution patterns of black-and-white squares. I: statistically random; II: somewhat less random, toward a chessboardlike pattern; III: even less random with segregations of larger black-and-white domains. (From Laves, F., reproduced in Correns, C. W., 1967, *Introduction to Mineralogy.* Springer-Verlag, New York, p. 92. Original reference to Laves could not be located.) *(b)* Schematic illustration of order-disorder polymorphism in alloy *AB* (see text for explanation).

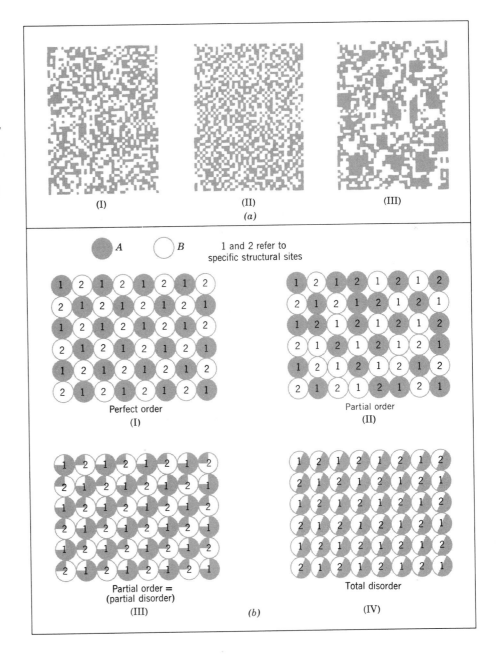

tions. In a perfectly ordered state atoms of A are arranged in perfect and regular repeat with respect to B atoms (see Fig. 3.44b-I). Atom A is always in structural site 1 and atom B in structural site 2. Two examples of somewhat disordered states for the atomic arrangement of such an alloy are shown in Figs. 3.44b-II and III. Here the distributions are not perfectly ordered and yet also are not random. In Fig. 3.44b-II the ratio $A:B$ is still 1:1, but the atoms A and B are in a partially disordered array. In Fig. 3.44b-III, out of every four no. 1 sites three are occupied by A (on the average) and one is occupied by B (on the average). The opposite holds for site no. 2. In other words, in Fig. 3.44b-III the probability ratio for site 1 being occupied by atoms of A rather than B is 3:1. A state of total disorder for the alloy AB is shown schematically in Fig. 3.44b-IV. Total disorder, on an atomic scale, implies equal probability of finding A or B in a specific site in the structure. In other words, the probability of a given atomic site being occupied by one type of atom instead of another equals 1. This means that for a graphic representation, as in Fig. 3.44b-IV, each atomic position is indicated to represent A as well as B occupancy (in a statistical sense).

An example of order-disorder polymorphism in a mineral is shown by potassium feldspar ($KAlSi_3O_8$), in which Al occupies a structural position identical with and replacing Si in the mineral. The high-temperature form, sanidine, with space group $C2/m$, shows a disordered distribution of Al in the aluminosilicate framework. The low-temperature K-feldspar, low microcline with space group $C\bar{1}$, however, shows an ordered distribution of Al in the aluminosilicate framework. States of intermediate order (which are equivalent to intermediate disorder) are present between that of high-temperature sanidine and low-temperature microcline. Figure 3.45 is a schematic illus-

tration of Al-Si order-disorder in a tetrahedral ring of a structure; this is similar but not identical to the order-disorder in $KAlSi_3O_8$.

POLYTYPISM

A special kind of polymorphism, known as *polytypism*, occurs when two polymorphs differ only in the stacking of identical, two-dimensional sheets or layers. As a consequence, the unit cell dimensions parallel to the sheets will be identical in the two polytypes. However, the atomic spacings between the sheets (or layers) will be related to each other as multiples or submultiples. Polytypism is a well-known feature of SiC, ZnS, the micas, and other layer silicates. The only difference between sphalerite (ZnS) and its polymorph wurtzite (ZnS) is that the S atoms in sphalerite are in cubic closest packing, whereas in wurtzite they are in hexagonal closest packed arrays. However, wurtzite shows extensive polytypism as reflected in c dimensions of the unit cell. Examples of wurtzite polytypes and their c dimensions are: $4H$, 12.46 Å; $6H$, 18.73 Å; $8H$, 24.96 Å; and $10H$, 31.20 Å (H refers to hexagonal cell). The "thickness" of the basic ZnS layer in these polytypes is 3.12 Å. The examples show multiples of this basic unit in their c dimension.

Micas and other layer silicates consist of infinitely extending tetrahedral silicate sheets (of composition $Si_2O_5(OH)$) that are stacked in various ways along the c axis. The inherent symmetry of such sheets is monoclinic. Because of the hexagonal symmetry about the (OH) group in the plan view of the $Si_2O_5(OH)$ tetrahedral sheets, there are six alternate directions (see Fig. 3.46a) in which $Si_2O_5(OH)$ sheets can be stacked. These six directions can be represented by three

FIG. 3.45. Schematic representation of order-disorder of Al^{3+} and Si^{4+} in four linked tetrahedra that may be part of a tetrahedral framework structure (the composition of the ring is $AlSi_3O_8$.) *(a)* The probability of finding Al^{3+} is equal for all four tetrahedra, that is, 1/4 Al is statistically distributed over the four tetrahedral sites (*total disorder*). This is compatible with a 2-fold rotation axis as shown. *(b)* The Al-Si distribution is such that each of two tetrahedra (on average) contains 1/2 Al, whereas the other two tetrahedra contain Si only. This *partial order* is still compatible with 2-fold rotation symmetry. *(c)* All of the Al (one Al^{3+} per ring) is concentrated in one tetrahedron. This is a state of *complete order*. The ring now has lost the 2-fold rotation symmetry.

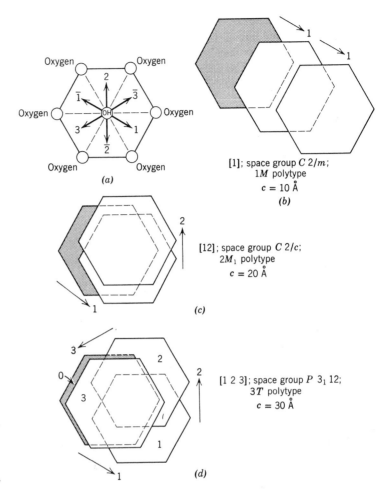

FIG. 3.46. Schematic illustration of some possible stacking polymorphs (*polytypes*) in mica. *(a)* Three vectorial directions for possible location of the (OH) group in an Si_2O_5(OH) sheet which is stacked above or below the hexagonal ring shown. *(b)* Stacking of Si_2O_5(OH) sheets in the same direction. *(c)* Stacking in two directions at 120° to each other. *(d)* Stacking in three directions at 120° to each other. Here, sheet 3 should lie directly above sheet 0 but is offset slightly for illustrative purposes.

vectors (1, 2, and 3, and negative directions $\bar{1}$, $\bar{2}$, and $\bar{3}$) at 120° to each other. If the stacking of Si_2O_5(OH) sheets is always in the same direction (see Fig. 3.46*b*), the resulting structure will have monoclinic symmetry, with space group $C2/m$; this is referred to as the 1*M* (*M* = monoclinic) polytype with stacking sequence [1]. If the stacking sequence of Si_2O_5(OH) sheets consists of alternating 1 and $\bar{1}$ directions (in opposing directions along the same vector), a structure results that is best described as orthorhombic with space group $Ccm2_1$. This stacking sequence can be expressed as [1 $\bar{1}$] and the polytype is known as 2*O*(*O* = orthorhombic). If the stacking sequence consists of two vectors at 120° to each other, that is, [1 2], another monoclinic polytype results; this is known as 2*M*$_1$ with space group $C2/c$ (see Fig. 3.46*c*). When all three vectors come into play, such as in [1 2 3], a trigonal polytype, 3*T* (*T* = trigonal), results that is compatible with either of two enantiomorphic space groups, $P3_1 12$ or $P3_2 12$. Additional polytype stacking sequences are possible, but most micas that commonly show polytypism belong to 1*M* ($C2/m$), 2*M*$_1$ ($C2/c$) or 3*T* ($P3_1 12$), polytypes.

METAMICT MINERALS

Minerals described as metamict originally formed as crystalline solids, but their crystal structure has been destroyed, to various degrees, by radiation from radioactive elements present in the original structure (see "Radioactivity, pages 218–220).

All metamict minerals are radioactive, and it is believed that the structural breakdown results mainly from the bombardment of alpha particles emitted from radioactive uranium or thorium contained in these minerals. The decay of a radioactive element can result in the complete destruction of the periodic structure around it. Usually this is caused by daughter elements of different size and charge from the parent element. However, the process is not well understood, for some minerals containing less than 0.5% of these elements may be metamict, whereas other minerals containing high percentages of them may be crystalline.

The various stages of metamictization (the various degrees to which the original crystalline structure has been destroyed) are determined by a combination

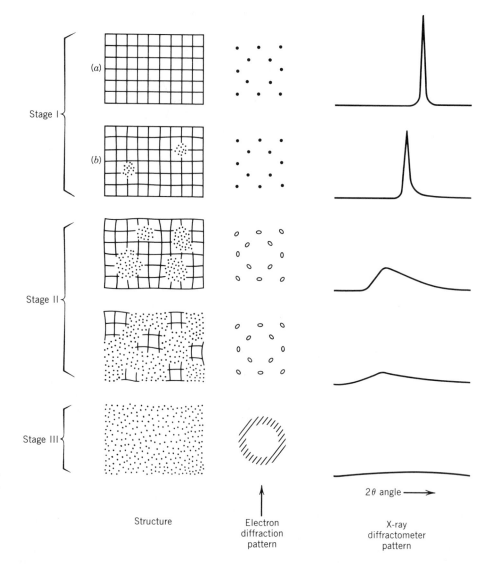

FIG. 3.47. Schematic representation of the progressive damage done to the structure of zircon as a result of radioactive decay (see text). Electron diffraction patterns and/or X-ray diffraction patterns can be used, in addition to high-resolution electron microscopy (HRTEM), to establish the amount of structural damage. (After Murakimi, T., Chakoumakos, B. C., Ewing, R. C., Lumpkin, G. R., and Weber, W. J., 1991, Alpha-decay event damage in zircon. *American Mineralogist*, v. 76, pp. 1510–1532.)

of X-ray diffraction and high resolution transmission electron microscopy (HRTEM) techniques. A well-ordered structure gives well-defined X-ray diffraction patterns, shows optical microscopic interference effects, and has ordered electron beam images (electron diffraction and structural projections). Different stages of destruction of the original structure are reflected in changes in the diffraction and structural patterns. Figure 3.47 shows several such stages of destruction of the original crystal structure of zircon ($ZrSiO_4$) owing to the presence of radioactive uranium and thorium. Stage I(a) shows a well-ordered structure with well-established periodicities throughout; this is the unaltered structure. Stage I(b) shows some destruction of the structure and the appearance of some aperiodic domains (amorphous regions). Stage II shows large increases in the volume percentage of aperiodic (amorphous) domains as a result of further destruction of the

structure. In the final stage, III, all periodicities of the original structure are lacking and the material is totally amorphous. The changes in the electron diffraction and X-ray diffraction patterns are the evidence on which the structural pictures are based. Figure 3.48 is an HRTEM illustration of metamict domains in an otherwise crystalline matrix of olivine ((Mg, Fe)$_2$ SiO_4). In this instance the damage was artificially induced by high-voltage ion beam irradiation of the originally completely crystalline olivine.

The original nonmetamict mineral may well have shown good cleavage, in addition to the well-defined electron and X-ray diffraction patterns in Fig. 3.47. The metamict product will not show cleavage but instead will show conchoidal fracture. Many metamict minerals are bounded by crystal faces and are thus amorphous pseudomorphs after an earlier crystalline mineral. When a metamict mineral is heated, its crys-

FIG. 3.48. High-resolution transmission electron microscopic (HRTEM) structural image of radiation-induced metamict domains inside a well-crystallized host of olivine ((Mg, Fe)$_2$ SiO_4). The damage was artificially produced by subjecting the original olivine crystal to a high-kilovoltage ion beam. Scale bar is in nanometers (nm): 1 nm = 10 angstroms. (From Wang, L. M., Miller, M. L., and Ewing, R. C., 1991, High-resolution TEM observation of displacement cascades in krypton-irradiated silicate minerals, in *Proceedings of the 49th Annual Meeting of the Electron Microscopy Society of America*, EMSA, pp. 910–911).

tal structure may be reconstituted and its density increased.

MINERALOIDS (NONCRYSTALLINE MINERALS)

The definition of a mineral states that "a mineral . . . has a highly ordered atomic arrangement." There are, however, a number of noncrystalline, natural solids classed as amorphous. Figure 3.49 is a schematic illustration, on an atomic scale, of the amorphous state.

Amorphous minerals include gel minerals and glasses. Gel minerals (mineraloids) are usually formed under conditions of low temperature and low pressure and commonly originate during weathering processes. They characteristically occur in mammillary, botryoidal, and stalactitic masses. The ability of amorphous materials to absorb all kinds of ionic species accounts for their often wide variations in chemical composition. Commonly amorphous minerals are limonite, $FeO \cdot OH \cdot nH_2O$, and allophane, a hydrous aluminum silicate.

The structure of a silica-rich glass, such as volcanic glass, is said to have *short-range order* but lacks *long-range order*. The Si^{4+} and Al^{3+} ions in such a glass occur in tetrahedral coordination as they do in crystalline compounds. In glasses, however, such tetrahedral coordination polyhedra do not repeat their pattern over more than a few angstrom units. In other words, *long-range order* is absent although *short-range order* (as shown by the presence of coordination tetrahedra) is present.

A well-known example of a partly amorphous material is opal. Its chemical composition can be represented as $SiO_2 \cdot nH_2O$ with an average range of H_2O content from 4 to 9 weight percent. Opal was originally considered to be completely without internal structure; however, careful electron beam studies show that it contains an ordered arrangement of very small SiO_2 spheres (see Fig. 6.21).

PSEUDOMORPHISM

The existence of a mineral with the outward crystal form of another mineral species is known as *pseudomorphism*. If a crystal of a mineral is altered so that

FIG. 3.49. Schematic illustration of the atomic arrangement in an amorphous alloy *(a)* and a crystalline alloy *(b)*.

(a)

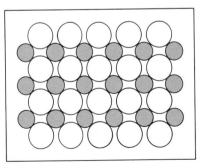

(b)

the internal structure or chemical composition is changed but the external form is preserved, it is called a *pseudomorph* (from the Greek meaning "false form"). The chemical composition and structure of a pseudomorph belong to one mineral species, whereas the crystal form corresponds to another. For example, pyrite, FeS_2, may change to limonite, $FeO \cdot OH \cdot nH_2O$, but it will preserve all the external features of the pyrite. Such a crystal is described as a pseudomorph of *limonite after pyrite*. Pseudomorphs usually are further defined according to the manner in which they were formed, as by:

1. *Substitution.* In this type of pseudomorph there is a gradual removal of the original material and a corresponding and simultaneous replacement of it by another with no chemical reaction between the two. A common example of this is the substitution of silica for wood fiber to form petrified wood. Another example is quartz, SiO_2, after fluorite, CaF_2.

2. *Encrustation.* In the formation of this type of pseudomorph a crust of one mineral is deposited over crystals of another. A common example is quartz encrusting cubes of fluorite. The fluorite may later be carried away entirely by solution, but its former presence is indicated by the casts left in the quartz.

3. *Alteration.* In this type of pseudomorph there has been only a partial addition of new material, or a partial removal of the original materials. The change of anhydrite, $CaSO_4$, to gypsum, $CaSO_4 \cdot 2H_2O$, and the change of galena, PbS, to anglesite, $PbSO_4$, are examples of alteration pseudomorphs. A core of the unaltered mineral may be found in such pseudomorphs.

STRUCTURAL COMPLEXITIES AND DEFECTS

In our prior discussion we assumed that atomic structures, as expressed by space group notation and morphology, are perfect. That is, we assumed that a crystalline material consists of an ordered, three-dimensional, repetitive array of atoms and/or ions. In X-ray diffraction studies, on which almost all structure analyses of crystalline material are based, it is assumed that such repetitive, periodic order exists. However, specialized X-ray studies and applications of the transmission electron microscope (TEM) and the more recently developed high resolution transmission electron microscope technique (HRTEM) have shown that *structural defects* (or *imperfections*) on an atomic scale are common in three-dimensional structures. Such *imperfections* affect basic properties of crystalline materials such as strength, conductivity, mechanical deformation, and color. Before describing the var-

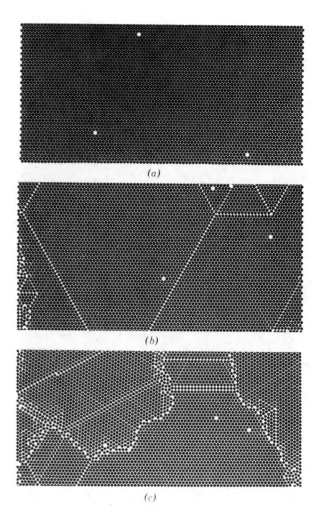

FIG. 3.50. A two-dimensional array of small spheres as a model for a crystal structure (6000 high precision stainless steel balls between two acrylic covers). *(a)* Regular, close packing of atoms with only three point defects in the pattern. *(b)* Point and line defects in the pattern. *(c)* A mosaic of domains separated by defective boundaries. (Photographs courtesy of Artorium, Inc., Montreal, Canada.)

ious types of imperfections let us qualitatively illustrate some aspects of their occurrence. In Fig. 3.50 we have photographed a two-dimensional array of very small spheres.[4] This represents a two-dimensional model of a crystal. In Fig. 3.50a we see an almost perfectly ordered array of spheres. In Fig. 3.50b we see structurally perfect domains with enclosed point defects ("holes") and traversed by linear defects. In Fig. 3.50c we see separate, well-ordered blocks of structure separated by highly defective boundaries. These regions would represent *domains* (of slightly differing atomic orientation) within a single crystal; as

[4]This device known as *Atomix* is available from François Dallégret Artorium, Inc., 353 Prince Albert, Montreal, Quebec, H3Z 2N9, Canada.

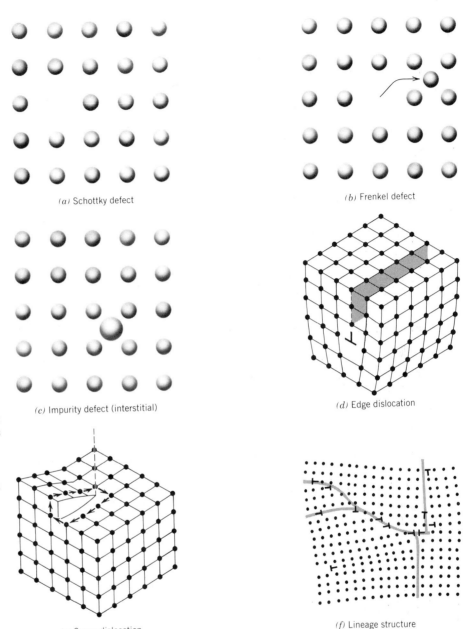

(a) Schottky defect

(b) Frenkel defect

(c) Impurity defect (interstitial)

(d) Edge dislocation

(e) Screw dislocation

(f) Lineage structure

FIG. 3.51. Schematic representations of defects in crystal structures. *(a)* An ion (or atom) is missing from the structure. *(b)* An ion (or atom) is displaced from its normal site. *(c)* An interstitial impurity is randomly lodged in an otherwise regular structure. *(d)* A plane of atoms that stops along a dislocation line (the edge dislocation is shown by the usual graphic symbol—an upside-down T). *(e)* A screw dislocation line about which atomic planes wind in a helical form. *(f)* A crystal made of a mosaic of domains that differ only slightly in orientation. The irregular zones (which are planes in three dimensions) of defects are *lineage structures*.

such the single crystal is said to consist of a mosaic of slightly misoriented blocks (domains). Imperfections in crystal structures are generally classified according to their geometry such as *point defects, line defects,* or *plane defects*. Several of these are illustrated schematically in Fig. 3.51.

Examples of point defects are *Schottky* and *Frenkel defects*. In a *Schottky defect* some cations (or anions) are absent from their normal sites in the crystal structure. If a crystal with such defects is to retain its electrical neutrality, the total charge on the cation vacancies must equal that of the anion vacancies. A *Frenkel defect* represents the absence of a cation (or

anion) from its proper structural site location and a mislocation of this same cation (or anion) in an interstitial site. This is more common for cations than anions because anions tend to be larger. Yet another type of point defect is known as an *impurity defect*. Such defective points are the result of the addition of a foreign ion (1) to an interstitial position in the crystal structure or (2) in place of one or several of the normally present ions. Impurity defects will change the chemical composition of the original "perfect" crystal, but the amounts of impurities are generally so small (in the range of parts per million or parts per billion) that they cannot be chemically detected.

However, some property such as color may be strongly affected by such trace impurities.

Line defects involve, as the term implies, concentrations of defects along linear features in a crystal structure. Such line defects are commonly known as *dislocations* because they create an offset in the crystal structure. There are two types: *edge dislocations* and *screw dislocations*. When a plane of atoms or ions in a crystal structure terminates in a line instead of continuing as would be required in a "perfect" crystal, it is said to contain an *edge dislocation* (see Fig. 3.51*d*). The presence of such line defects allows a crystal to deform under stress, by the atomic slippage of these linear defects throughout the structure. Such *slip planes* in the crystal structure represent planes of atomic misfit and therefore of lesser coherence. *Screw dislocations* are structural defects arranged along a screw axis that normally is not present in the structure. Figure 3.51*e* shows that the upper atomic surface of the schematic structure resembles a spiral ramp, centered about a vertical dislocation line, which is equivalent to a screw axis direction. Such spiral steps are of great importance during crystal growth because new ions or atoms, which are added to the outside surface of a growing crystal, are best housed along the ledge (see also Fig. 2.2).

Plane defects represent two-dimensional zones along which slightly misoriented blocks within a single crystal are joined. In ideally perfect crystals the whole internal structure is regarded as a rigorously continuous and symmetric repeat of unit cells. This implies short-range as well as long-range order. In less ideal (more real) crystals, blocks of structure may be slightly misoriented, as in a mosaic pattern. Each block in the mosaic has short-range order, but the whole crystal lacks long-range order. Figure 3.51*f* shows irregular zones (lines) of edge dislocations spaced at irregular intervals. In a three-dimensional structure these zones (or lines) will be irregular planar features along which ions (or atoms) have an irregular structural environment. These zones of irregularity are also known as *lineage structures*. The structures on either side of such lineages are slightly misoriented with respect to each other. When lineages are present, a continuous crystal structure must be viewed as made up of volumes of nearly perfect structure that occur in a mosaic of slight misorientation with each other.

Another planar defect is a *stacking fault* in which a regular sequence of layers (e.g., along the *c* axis of a structure) is interrupted by an improperly positioned layer, or when one of the ideally required layers is missing. Examples of this are sequences of ions (or atoms) in hexagonal closest packing (*AB, AB, AB,* . . .) interrupted by a layer from a cubic closest packing sequence (*ABC, ABC, ABC,* . . .) (see Chapter 4 for a discussion on closest packing).

Two other types of structural defects are exemplified by *omission solid solution* and *color centers*. Omission solid solution is discussed under "Compositional Variation in Minerals" (Chapter 5) and color centers under "Color" (Chapter 6).

Figure 3.52 shows several structural complexities and defects as resolved by high resolution transmission electron microscope (HRTEM) studies. Figure 3.52*a* is a structure image of crocidolite, a fibrous (asbestiform) variety of the amphibole riebeckite. It consists of a mosaic of domains that show slightly different crystallographic orientations, separated by grain boundaries (lineage structure). Figure 3.52*b* shows a structure image of grossular garnet in which a very small region (within the otherwise highly periodic and ordered garnet matrix) is dislocated, and separated from the matrix by stacking faults. Figure 3.52*c* shows a deformed, curled region (*C*) in an otherwise well-layered matrix of lizardite (Lz), a variety of serpentine. Figure 3.52*d* depicts modulations in the structure of antigorite, a variety of serpentine. Modulations in the antigorite structure (see Fig. 13.87*a*) are wave-like features that result from a small dimensional misfit that occurs between the tetrahedral and octahedral layers. This misfit is accommodated by a curling of the structural layers, much like what happens when a telephone book is bent. A continuous curl results in the chrysotile structure (see Figs. 13.87 and 13.89).

Figure 3.53*a* shows a variety of local defects and complexities in an amphibole structure image, and Fig. 3.53*b* is a schematic interpretation of these same features.

TWINNING

In the last part of Chapter 2 we discussed the morphological expression of twinned crystals. Now that we have a better understanding of crystal structures and lattices on which they are based, we will evaluate some of the internal aspects of twinning.

The more nearly perfect the periodic atomic arrangement of a crystal is, the lower its internal energy will be. Twinning is a deviation from perfection, and therefore the internal energy of a twin should be slightly greater than that of a "perfect" untwinned individual.

Under certain conditions of growth, two or more crystals may form a rational, symmetrical intergrowth.

FIG. 3.52. Illustrations of structural complexities and defects. *(a)* HRTEM structure image of croci-dolite, a fibrous variety of riebeckite, viewed down the *c* axis. The locations of the *a** and *b* direc-tions are shown (*a** = *a* sin β). Very slightly rotated domains (= subgrains) are separated from each other along grain boundaries. The straight line features inside several of the domains are errors in the widths of the structural chains of amphiboles. (From Ahn, J. H. and Buseck, P. R., 1991, Microstructures and fiber-formation mechanisms of crocidolite asbestos. *American Mineralogist*, v. 76, pp. 1467–1478; see also Fig. 3.50.) *(b)* HRTEM structure image of a small dislocated region in an otherwise highly regular (ordered) matrix of grossular garnet. The central dislocated region is separated from the surrounding matrix by stacking faults. Crystallographic directions are shown as [011] and [100]. [1$\bar{1}$1]tr and [$\bar{1}$01]tr represent the traces of the [1$\bar{1}$1] and [$\bar{1}$01] directions, respec-tively. Both traces represent stacking faults. Scale is in nanometers. (From Allen, F. M., Smith, B. K., and Buseck, P. R., 1987, Direct observation of dissociated dislocations in garnet. *Science*, v. 238, pp. 1695–1697; photo by J. Berry.) *(c)* HRTEM structure image of locally deformed (curled) serpen-tine (C) inside a matrix of well-layered lizardite (Lz), also a variety of serpentine. The curled region occurs along a cleavage fracture. Phl = phlogopite mica. Scale is in nanometers. (From Sharp, T. G., Otten, M. T., and Buseck, P. R., 1990, Serpentinization of phologopite phenocrysts from a mi-caceous kimberlite. *Contributions to Mineralogy and Petrology*, v. 104, pp. 530–539.) *(d)* HRTEM im-age of modulations in antigorite, a variety of serpentine. The modulations are about 50 Å in length. The *c** direction = *c* sin β. The large white arrows show defective boundaries, which may be out-of-phase or twin boundaries. (From Buseck, P. R. and Cowley, J. M., 1983, Modulated and intergrowth structures in minerals and electron microscope methods for their study. *American Mineralogist*, v. 68, pp. 18–40; photo by G. Spinnler.)

FIG. 3.53. *(a)* HRTEM structure image of crocidolite, a fibrous amphibole, viewed down the *c* axis of the structure. Note the scale bar of 90 Å (1 angstrom = 10^{-8} cm = 0.1 nanometer). Most of the field is composed of a regular pattern, which is the structure image of adjoining double (2) chains in the amphibole structure (see also page 488–494). However, the image also shows quintuple (5) and single (1) chain-width errors (these are marked by the appropriate numbers on the photograph). These defects appear as lines in the projected image, but represent continuous planar features in the three-dimensional structure. The a^* ($a^* = a \sin \beta$) and *b* directions are shown. *(b)* A schematic representation using I-beams to illustrate the complexities and defects observed in the structure image in *(a)*. The arrow at the left top of the figure points toward a disrupted region caused by structural mismatch. Similar mismatch occurs just "north" and "south" of the quintuple (5) chain. The mismatch to the "south" of the quintuple chain has been accommodated by an orderly arrangement of two double (2) chain widths and one single (1) chain width. (*a* and *b* from Ahn, J. H. and Buseck, P. R., 1991, Microstructures and fiber-formation mechanisms of crocidolite asbestos. *American Mineralogist*, v. 76, pp. 1467–1478.)

Such a *crystallographically controlled intergrowth* is called a *twin*. A twin thus comprises two (or more) "twinned crystals." The lattice directions of one crystal in a twin bear a definite crystallographic relation to the lattice directions of the other cyrstal. Figure 3.54 illustrates the twin relationship of two different orientations of the same orthorhombic lattice. As noted earlier (Chapter 2), the operations that relate a crystal to its twinned counterpart can be: (1) reflection by a mirror (*twin plane*); (2) rotation about a line (*twin axis*) (although there are exceptions, the rotation is usually through an angle of 180°); and (3) inversion about a point (*twin center*). The twin operation is known as the *twin law*, in which the *twin elements* (twin mirror, twin axis) are parallel with lattice elements (namely, net, row).

The description of a *twin law* consists of stating the crystallographic orientation of the twin element, axis, or plane. The two twinned orientations of an orthorhombic lattice (shown in Fig. 3.54) are related by

a twin reflection in ($\bar{1}$10); this is also the composition plane, of which *XX'* is the trace. Along this twin plane there is a perfect coincidence of lattice nodes of crystals I and II, because the nodes belong to both twinned orientations, a fact that explains why the contact between the two crystals coincides with the twin plane. The dashed lattice is called the *twin lattice*, and it pervades the whole twin. Its cell is a triple cell in the individual crystals but is a primitive cell in the twin lattice (note the two inside nodes, which are not restored by the twinning—they are reflected, not translated, from crystal I to crystal II). The extension of this twin lattice across the twin plane may be exact ($\omega = 0$) or it may show a slight deviation at the composition plane ($\omega > 0$). This possible deviation is expressed by ω, the *twin obliquity*, which is the angle between the normal to the twin plane and the lattice row that is almost normal to it. This angle ranges from 0 to a maximum of about 6° in observed twins. The twin lattice is either the same as the crystal lattice ($n = 1$) or it is

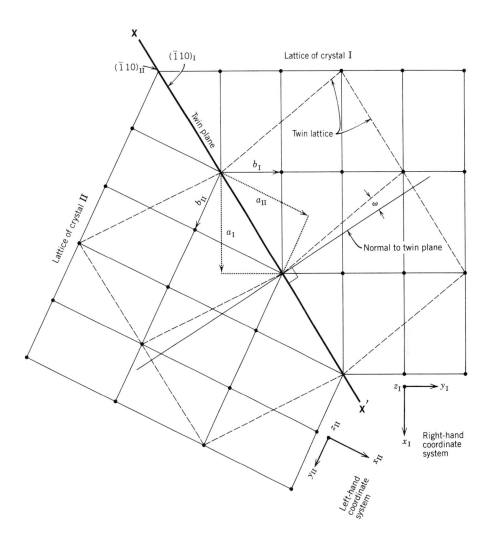

FIG. 3.54. Cross section through the orthorhombic lattices of two crystals twinned on XX' ($\bar{1}$10). See text for discussion.

a superlattice thereof, as in Fig. 3.54, where $n = 3$. (In a superlattice, one or more translation periodicities are multiples of those in the related lattice.) The *twin index, n,* is defined as (volume per node in twin lattice)/(volume per node in crystal lattice). The extension of the twin lattice through crystals I and II is a fundamental condition for the stability of the twin: the smaller the obliquity and the index, the more frequent the twin will be.

Origin of Twinning

The various mechanisms by which twins are formed have been discussed by Buerger (1945) in terms of *growth twins, transformation twins,* and *gliding* (or *deformation*) twins.

Growth twins are the result of an emplacement of atoms, or ions (or groups of atoms or ions) on the outside of a growing crystal in such a way that the regular arrangement of the original crystal structure (and, therefore, its lattice) is interrupted. For example,

in Fig. 3.54 the line *XX'* may be considered the trace of an external face of a growing crystal. An ion or atom (or group of ions) would have the choice of attaching itself at structural sites that represent a continuation of the nodes in the lattice of crystal I or in nodes compatible with the lattice of crystal II. In the former case the original structure is continued without interruption, but in the latter a twinned relationship results. *Growth twinning* therefore reflects "accidents" during free growth ("nucleation errors") and can be considered as *primary twinning*. Growth twins can be recognized in mineral specimens by the interpenetration of two twinned crystals, or the presence of a single boundary between them. Most of the crystal drawings of twins in Chapter 2 represent growth twins.

Transformation twins occur in preexisting crystals and represent *secondary twinning*. Transformation twinning may result when a crystal that formed at high temperature is cooled and subsequently rearranges its structure to one whose symmetry is different from that of the high-temperature form. For example high

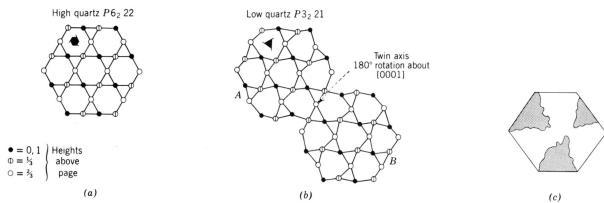

FIG. 3.55. *(a)* The distribution of silicon atoms in high quartz. The atomic positions are given in terms of thirds above the page. *(b)* A part of the low quartz structure (*A*) twinned with respect to another part (*B*) by a rotation axis of 180° perpendicular to the page. Note the presence of 3-fold screw axes instead of 6-fold screw axes as in high quartz. *(c)* The 3-fold distribution of sectors of low quartz twinned according to the Dauphiné Law. The composition surfaces are generally irregular. This type of twin is visible only on sawn and etched basal sections.

quartz, with space group $P6_222$, if cooled below 573°C, transforms to low quartz, with space group $P3_221$. Figure 3.55*a* illustrates the high quartz structure, whereas Fig. 3.55*b* shows the structure of low quartz in which part *A* is related to part *B* by a 180° rotation (see also Figs. 3.41 and 3.42). In the transition to the low quartz structure the original high quartz structure has a choice of two orientations, related by 180° rotation, for the trigonal structural arrangement of low quartz (see Fig. 3.55*b*). The relationship of these two orientations is known as a *Dauphiné twin* and is expressed as a 180° rotation about [0001]. If such a twin of low quartz is heated above 573°C, the Dauphiné twinning will disappear and untwinned high quartz will form spontaneously. Figure 3.55*c* shows a basal section of low quartz with the 3-fold distribution of the twinned orientations of low quartz.

Another example of transformation twinning is provided by $KAlSi_3O_8$, which occurs in three different structural forms: high-temperature sanidine ($C2/m$),

lower temperature orthoclase ($C2/m$), and lowest temperature microcline ($C\bar{1}$); see also "Polymorphism," (page 158). Microcline invariably shows the microscopic twin pattern shown in Fig. 2.114, which is known as *microcline* or *"tartan" twinning.* The cross-hatched pattern, which is usually visible only between crossed polarizers, consists of two types of twins, twinned according to the *albite* and *pericline* laws. This combination of twinned relations is a key feature in the symmetry change from monoclinic (as in sanidine) to triclinic (as in microcline). In this transition from higher temperature, the mirror plane and the 2-fold rotation axis are lost, leading to nucleation of triclinic "domains" that are related by twinning.

Gliding (or *deformation*) *twins* result when a crystal is deformed by the application of a mechanical stress. Deformation twinning is another type of *secondary twinning*. If the stress produces slippage of the atoms on a small scale, a twin may result (Fig. 3.56). If the movement of the atoms is large, slip or gliding may occur without twinning, which may finally lead to the rupture of the crystal. Deformation twinning is common in metals and is frequently present in metamorphosed limestones, as shown by the presence of polysynthetically twinned calcite (see Fig. 2.113*c*). Similarly, plagioclase feldspars from metamorphic terrains may show deformation twinning.

FIG. 3.56. Deformation twinning in an oblique lattice due to the application of mechanical stress as indicated by the arrows. Note that the amount of movement of the first layer above and parallel with the twin plane in *(b)* is less than that of the successive layers further removed from the twin plane.

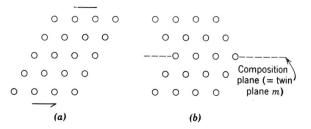

REFERENCES AND SUGGESTED READING

Bloss, F. D., 1971, *Crystallography and Crystal Chemistry.* Holt, Rinehart and Winston, New York, 545 pp.
Boisen, M. B., Jr., and Gibbs, G. V., 1990, *Mathematical Crystallography*, rev. ed. *Reviews in Mineralogy*, v. 15.

Mineralogical Society of America, Washington, D.C., 460 pp.

Bragg, W. L. and Claringbull, G. F., 1965, *Crystal Structures of Minerals.* Cornell University Press, Ithaca, 409 pp.

Brisse, F., 1981, La Symétrie bidimensionelle et le Canada. *Canadian Mineralogist*, v. 19, pp. 217–224 [all illustrations are in color].

Buerger, M. J., 1978, *Elementary Crystallography: An Introduction to the Fundamental Geometric Features of Crystals*, rev. ed. MIT Press, Cambridge, Mass., 528 pp.

———, 1971, *Introduction to Crystal Geometry.* McGraw-Hill Book Co., New York, 204 pp.

———, 1945, The genesis of twin crystals. *American Mineralogist*, v. 30, pp. 469–482.

Donnay, G. and Donnay, J. D. H., 1974, Classification of tri-periodic twins. *Canadian Mineralogist*, v. 12, pp. 422–425.

Donnay, J. D. H., 1955, *Twinning.* Encyclopaedia Britannica, pp. 827–828L.

Escher, M. C., 1968, *The Graphic Works of M. C. Escher.* Hawthorn Books, New York, 98 pp.

Gibbs, G. V., 1982, Molecules as models for bonding in silicates. *American Mineralogist*, v. 67, pp. 421–451.

International Tables for Crystallography, 1983, v. A, Space Group Symmetry, T. Hahn, ed. International Union of Crystallography, D. Reidel Publishing Co., Boston, 854 pp.

International Tables for X-ray Crystallography, 1969, v. 1, N. F. M. Henry and K. Lonsdale, eds.: Symmetry Groups. International Union of Crystallography, Kynoch Press, Birmingham, England, 558 pp.

Klein, C., 1989, *Minerals and Rocks: Exercises in Crystallography, Mineralogy, and Hand Specimen Petrology.* John Wiley & Sons, Inc., New York, 402 pp.

MacGillavry, C. H., 1976, *Fantasy and Symmetry: The Periodic Drawings of M. C. Escher.* Harry N. Abrams, New York, 84 pp.

Schattschneider, D., 1990, *Visions of Symmetry: Notebooks, Periodic Drawings, and Related Work of M. C. Escher.* W. H. Freeman and Co., New York, 354 pp.

Shubnikov, A. V. and Koptsik, V. A., *Symmetry in Science and Art.* Plenum Press, New York, 420 pp.

Stevens, P. S., 1984, *Handbook of Regular Patterns: An Introduction to Symmetry in Two Dimensions.* MIT Press, Cambridge, Mass., 400 pp.

CHAPTER 4

CRYSTAL CHEMISTRY

The chemical composition of a mineral is of fundamental importance, for many of its properties are in a great measure dependent on it. However, such properties depend not only on the chemistry but also on the geometrical arrangement of the constituent atoms or ions and the nature of the electrical forces that bind them together. Thus, for an understanding of minerals one must consider their chemistry, their bonding, and their structure. In Chapter 3 we discussed how motifs can be arranged in three-dimensional, ordered, and repetitive patterns. Such motifs are the atoms, ions, ionic groups, and occasional molecules in mineral structures. In this chapter we will evaluate various chemical parameters of minerals, with special emphasis on atomic structure and the types of bonds that provide for the coherence of the "chemical motifs" in the crystal structure.

As a result of common interests among crystallographers and chemists an interdisciplinary science, *crystal chemistry*, came into being. The goal of this science is the elucidation of the relationships among the chemical composition, internal structure, and physical properties of crystalline matter. In mineralogy it serves as a unifying thread upon which often apparently unrelated facts of descriptive mineralogy may be strung. As a basis for discussion of the crystal chemical and structural aspects of minerals, it is necessary to consider first some of the elementary concepts of atoms and ions and their bonding forces in crystalline materials.

THE ATOM

An atom is the smallest subdivision of matter that retains the characteristics of the element. It consists of a very small, massive nucleus composed of *protons* and *neutrons* surrounded by a much larger region thinly populated by *electrons* (see Table 4.1). Although atoms are so small that their images can barely be resolved even with the highest magnification of the transmission electron microscope, their sizes can be deduced from measurements of interatomic distances. The atomic radii are expressed in nanometer or angstrom units (1 nanometer = 10 angstroms). For ex-

Table 4.1
SOME ATOMIC PARTICLES*

Particle	Symbol	Atomic Mass Units	Relative Charge
Electron	e	0.0005486	-1
Proton	p	1.007276	$+1$
Neutron	n	1.008665	0

*Consideration of other particles discovered in studies of high-energy physics is unnecessary in this context.

FIG. 4.1 Schematic representation of the hydrogen atom as based on the model of Niels Bohr. A single electron (e^-) moves in an orbit around the nucleus, as in a planetary system.

ample, the smallest atom, hydrogen, has a radius of only 0.46 Å, whereas the largest, cesium, has a radius of 2.72 Å (see Table 4.7). Each proton carries a unit positive charge; the neutron, as the name implies, is electrically neutral; and each electron carries a unit negative charge (see Table 4.1). Because the atom as a whole is electrically neutral, there must be as many electrons as protons. The weight of the atom is concentrated in the nucleus, because the mass of an electron is only 1/1837 that of the proton. Although the electrons and nuclei are both extremely small, the electrons move so rapidly about the nuclei that they give to the atoms relatively large effective diameters, up to 100,000 times the diameter of the nucleus.

The simplest atom is that of hydrogen, which consists of one proton and one electron (see Fig. 4.1). Atoms of the other natural elements have from two electrons (helium) to 92 electrons (uranium).

The fundamental difference between atoms of the different elements lies in the electrical charge of the nucleus. This positive charge is the same as the number of protons, and this number, equal to the

number of electrons in an uncharged atom, is called the *atomic number* (= Z). The sum of the protons and neutrons determines the *characteristic mass*, or mass number of an element. Atoms of the same element but with differing numbers of neutrons are called *isotopes*. For example, oxygen (Z = 8) has three isotopes, the most common of which has a nucleus with eight protons and eight neutrons (= 16); this is known as ^{16}O. Rarer and heavier isotopes of oxygen carry eight protons and nine or ten neutrons; these are ^{17}O and ^{18}O, respectively. Similarly, hydrogen, H, can exist in several isotopic forms. The element H (Z = 1) consists of one proton and one electron; the H isotope with one neutron in the nucleus is ^2H, known as deuterium (D), and the isotope with two neutrons in the nucleus is ^3H, tritium (T). A schematic illustration of some isotopes is given in Fig. 4.2.

Chemical Elements and the Periodic Table

All the chemical elements are listed in alphabetical order in Table 4.2. In this table their chemical symbols, atomic number (Z), and atomic weights are also given. Although such a table is useful as a reference for the information it contains, it provides no insight into the well-known regularity of the atomic build-up. Instead the *periodic table* (see Fig. 4.3) is the widely used chemical listing of all the elements. In this table the elements are arranged in order of increasing atomic number (Z) because this order exhibits best the periodic repetition of chemical and physical properties of the elements. In other words, the charge on the

FIG. 4.2 Schematic illustration of the content of protons and neutrons in the nucleus of several elements. The isotopes in the first column have equal numbers of protons and neutrons; those in the second column have an extra neutron; those in the third column have two extra neutrons.

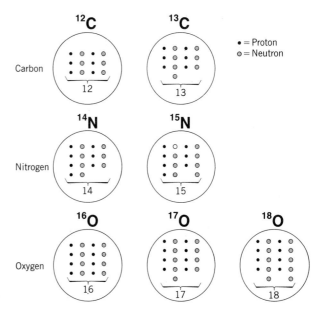

Table 4.2
ALPHABETICAL LISTING OF THE ELEMENTS, THEIR SYMBOLS, ATOMIC NUMBER, AND ATOMIC WEIGHTS

Name	Symbol	Atomic Number, Z	Atomic Weight*	Name	Symbol	Atomic Number, Z	Atomic Weight*
Actinium	Ac	89	227.0278	Mercury	Hg	80	200.59
Aluminum	Al	13	26.98154	Molybdenum	Mo	42	95.94
Americium	Am	95	(243)	Neodymium	Nd	60	144.24
Antimony	Sb	51	121.75	Neon	Ne	10	20.179
Argon	Ar	18	39.948	Neptunium	Np	93	237.0482
Arsenic	As	33	74.9216	Nickel	Ni	28	58.70
Astatine	At	85	~210	Niobium	Nb	41	92.9064
Barium	Ba	56	137.33	Nitrogen	N	7	14.0067
Berkelium	Bk	97	(247)	Nobelium	No	102	(259)
Beryllium	Be	4	9.01218	Osmium	Os	76	190.2
Bismuth	Bi	83	208.9804	Oxygen	O	8	15.9994
Boron	B	5	10.81	Palladium	Pd	46	106.4
Bromine	Br	35	79.904	Phosphorus	P	15	30.97376
Cadmium	Cd	48	112.41	Platinum	Pt	78	195.09
Calcium	Ca	20	40.08	Plutonium	Pu	94	(244)
Californium	Cf	98	(251)	Polonium	Po	84	(209)
Carbon	C	6	12.011	Potassium	K	19	39.0983
Cerium	Ce	58	140.12	Praeseodymium	Pr	59	140.9077
Cesium	Cs	55	132.9054	Promethium	Pm	61	(145)
Chlorine	Cl	17	35.453	Protoactinium	Pa	91	231.0359
Chromium	Cr	24	51.996	Radium	Ra	88	226.0254
Cobalt	Co	27	58.9332	Radon	Rn	86	(222)
Copper	Cu	29	63.546	Rhenium	Re	75	186.207
Curium	Cm	96	(247)	Rhodium	Rh	45	102.9055
Dysprosium	Dy	66	162.50	Rubidium	Rb	37	85.4678
Einsteinium	Es	99	(252)	Ruthenium	Ru	44	101.07
Erbium	Er	68	167.26	Samarium	Sm	62	150.4
Europium	Eu	63	151.96	Scandium	Sc	21	44.9559
Fermium	Fm	100	(257)	Selenium	Se	34	78.96
Fluorine	F	9	18.998403	Silicon	Si	14	28.0855
Francium	Fr	87	(223)	Silver	Ag	47	107.868
Gadolinium	Gd	64	157.25	Sodium	Na	11	22.98977
Gallium	Ga	31	69.72	Strontium	Sr	38	87.62
Germanium	Ge	32	72.59	Sulfur	S	16	32.064
Gold	Au	79	196.9665	Tantalum	Ta	73	180.9479
Hafnium	Hf	72	178.49	Technetium	Tc	43	98.906
Helium	He	2	4.00260	Tellurium	Te	52	127.60
Holmium	Ho	67	164.9304	Terbium	Tb	65	158.9254
Hydrogen	H	1	1.0079	Thallium	Tl	81	204.37
Indium	In	49	114.82	Thorium	Th	90	232.0381
Iodine	I	53	126.9045	Thulium	Tm	69	168.9342
Iridium	Ir	77	192.22	Tin	Sn	50	118.69
Iron	Fe	26	55.847	Titanium	Ti	22	47.90
Krypton	Kr	36	83.80	Tungsten	W	74	183.85
Lanthanum	La	57	138.9055	Uranium	U	92	238.029
Lawrencium	Lr	103	(260)	Vanadium	V	23	50.9415
Lead	Pb	82	207.2	Xenon	Xe	54	131.30
Lithium	Li	3	6.941	Ytterbium	Yb	70	173.04
Lutetium	Lu	71	174.967	Yttrium	Y	39	88.9059
Magnesium	Mg	12	24.305	Zinc	Zn	30	65.38
Manganese	Mn	25	54.9380	Zirconium	Zr	40	91.22
Mendelevium	Md	101	(258)				

*A value given in parentheses denotes the mass of the most stable known isotope.

SOLID BLACK

173

nucleus (which is reflected in the atomic number) and the number of electrons in the neutral atom are what determines the order in which the elements occur. The table is organized into *vertical columns*, also called *groups*, with Roman numerals (I, II, . . . VIII); these numbers are equal to the numbers of electrons contained in their outermost shell (this is discussed in more detail on page 180); the *horizontal rows*, also known as *periods*, are designated by Arabic numerals (1, 2, . . . , 7).

In the periodic table (Fig. 4.3), most of the elements are *metals*, but those that are shaded, in the right-hand part, are *nonmetals*. Metals have distinct properties such as high electrical conductivity, metallic luster, generally high melting points, ductility (can be drawn into wires), and malleability (can be hammered into sheets). Nonmetals, in contrast, show poor electrical conductivity, do not have the characteristic luster of metals, and, as solids, are brittle. The far right-hand column (VIII) contains the inert, noble gases.

The two long rows below the main part of the table are known as the *lanthanide series*, or *rare earths* (Z 58 through 71), and the *actinide series* (Z 90 through 103). These elements belong in the body of the table but are placed below simply to conserve space. Most of the elements listed in the table occur naturally; however, elements with atomic numbers 43, 61, and 93 through 103 are known only from synthesis.

The Bohr Model of the Atom

The first widely accepted picture of the atom was that developed in 1913 by the Danish physicist Niels Bohr. This model is based on the fact that when an electrical charge is passed through a tube containing hydrogen, light is emitted, and the light's spectrum is found to consist of several sharp lines with specific wavelength (λ) values (see Fig. 4.4). Four of these lines, in the visible part of the spectrum, can be seen with the naked eye, and the others, in the ultraviolet region, can be recorded on photographic film. Because of the specific and characteristic wavelength (λ) values of the light lines emitted by all chemical elements heated to a high temperature, Niels Bohr concluded that electrons of the elements occur in specific energy levels at various distances away from the nu-

FIG. 4.4 The line spectrum of hydrogen obtained by refraction through a glass prism. The wavelengths of the lines in the visible spectrum are given in nanometers (nm). The lines to the left of these are in the ultraviolet region of the spectrum. The distinct line separation is due to the fact that the light energy released has only distinct quantities of energy, and none in between. (From Brady, J. E. and Humiston, G. E., 1982, *General Chemistry: Principles and Structure*, 3rd ed. John Wiley & Sons, New York; copyright © John Wiley & Sons; reprinted by permission.)

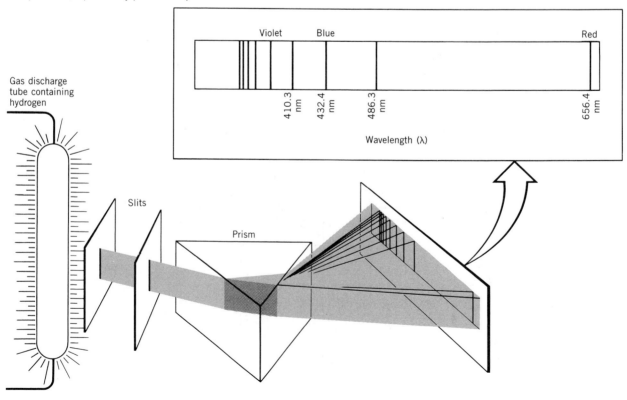

cleus. He postulated that when an electron absorbs energy it jumps to a higher energy level, and when it loses energy it drops to a lower energy level (see Fig. 4.6). This led Niels Bohr to conclude that electrons occur in discrete, or *quantized*, energy levels. The energy of the emitted radiation is proportional to the wavelength of the radiation, in accordance with the Einstein equation

$$E = \frac{h\,c}{\lambda}$$

where E is the energy, c is the velocity of light, λ is the wavelength of radiation emitted, and h is a proportionality constant known as Planck's constant. This is a universal constant of nature which relates the energy of a quantum of radiation to the frequency of the oscillator that emitted it. Its numerical value is 6.62517×10^{-27} erg sec.

In the Bohr model of the atom the electrons are visualized as circling the nucleus in "orbits" (as in a planetary system), or energy levels, at distances from the nucleus depending on their energies. Figure 4.1 is a schematic illustration of the hydrogen atom as based on the Bohr theory. At the center is the *nucleus*, which, except in the hydrogen atom, is made up of *protons* and *neutrons* (the *hydrogen* nucleus is made up of a single proton).

In chemical elements with more than one electron, the electrons are distributed in shells designated by n (e.g., $n = 1, n = 2, n = 3$, etc.) and known as the *principal quantum numbers*. The corresponding electron shells (or orbits) are designated K, L, M, N, etc. (see Fig. 4.5). For this model, Bohr derived an equation for the energy of the electron, as follows:

$$E = -A\,\frac{1}{n^2}$$

where A is a constant involving knowledge of the mass and charge of the electron and Planck's constant, and n is the principal quantum number, which identifies the electron orbit. From this equation it becomes clear that the energy of an electron in a particular orbit depends on the value of n. The greatest absolute value of energy is represented by $n = 1$, because this yields the largest value for the fraction $1/n^2$. Because of the negative sign in the equation this is the lowest (most negative) energy, E (see Fig. 4.6). In Fig. 4.6 the energy (E) scale is expressed in negative units, in accordance with the above equation. Although this may appear unusual, the important fact to note is that the electron orbits represent energy levels with various distinct energy differences between them.

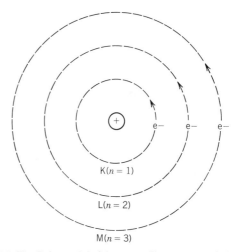

FIG. 4.5 The Bohr model of the atom. Electrons travel along specific orbits of fixed energy levels (known as K, L, M, N, . . . shells, with principal quantum number $n = 1, 2, 3, 4, . . . , \infty$).

The Schrödinger Model of the Atom

Although the concept of electrons circling the nucleus in well-defined orbits (as in the Bohr model) gained wide acceptance, it did not satisfactorily explain a number of important observations. It was not applicable to the line spectra of complex atoms (with atomic numbers higher than those of hydrogen). Additional spectral lines were interpreted as the result of the axial spin (clockwise or counterclockwise) of the

FIG. 4.6 Schematic representation of the energy levels of electrons (for $n = 1, 2, 3, 4, . . . , \infty$) in the Bohr atomic model. The two vertical "up" arrows show two possibilities for the increase in energy level of an electron as a result of the absorption of energy; the three "down" arrows show three possibilities for how electron energy is lost as a result of emission of energy.

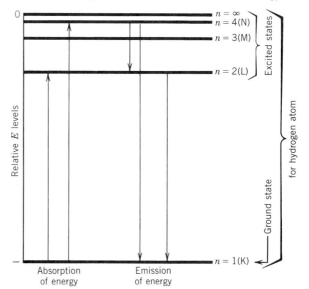

electrons, which was not accounted for in the Bohr model. A single quantum number, *n*, furthermore, did not account for possible elliptical electron orbits. It also gave no basic understanding for the quantization of energy between various orbital levels, and it failed to explain why an orbiting electron did not radiate energy.

In 1923 the French physicist Louis-Victor de Broglie demonstrated that electrons, instead of behaving solely as particles whose positions can be determined in space (as in the Bohr model), had properties identical to those of waves. The wavelength (λ) of a particle with mass *m* and velocity *v* is expressed as

$$\lambda = \frac{h}{mv}$$

where *h* is Planck's constant.

With electrons having wave-like properties, it becomes impossible to visualize them as being in a specific place at a particular time. This notion, expressed as the *uncertainty principle*, was introduced by the German scientist Werner Heisenberg. It implies that the motion of an electron around the nucleus of an atom cannot be satisfactorily described in terms of orbits, be they circular or elliptical.

In 1926 all of the above developments were incorporated into a new atomic model by Erwin Schrödinger, and were expressed as a wave equation. In this equation the electrons are described by wave functions, and the theoretical model is founded on the quantum properties of energy, that is, the theory of *quantum mechanics*. The Schrödinger equation relates the probability of finding an electron at a given time, in a specific place, to the mass and potential energy of the particle at that time and place.

One form of the Schrödinger equation is as follows:

$$\frac{\partial^2\psi}{\partial x^2} + \frac{\partial^2\psi}{\partial y^2} + \frac{\partial^2\psi}{\partial z^2} + \frac{8\pi^2 m}{h^2}(E - V)\psi = 0$$

where ψ is a wave function in terms of three coordinate axial directions (*x*, *y*, *z*), *m* is the mass of the electron, *E* is the total energy of the electron, *V* is the potential energy of the electron at some specified point, and *h* is Planck's constant.

The Schrödinger equation is also stated as follows:

$$\nabla^2\psi + \frac{8\pi^2 m}{h^2}(E - V)\psi = 0$$

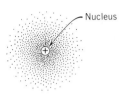

FIG. 4.7 Schematic illustration of the electron probability density distribution (ψ²) of an electron around a nucleus (an electron cloud). The dots do not show the location of the electron but only the probability of finding the electron in that location. The greater the density of dots, the greater the probability of finding the electron in that region.

where the ∇^2 symbol is a differential operator with

$$\nabla^2\psi = \frac{\partial^2\psi}{\partial x^2} + \frac{\partial^2\psi}{\partial y^2} + \frac{\partial^2\psi}{\partial z^2}$$

The unique contribution of the Schrödinger equation is its representation of physical observations in a way that had not been possible in earlier developed models. The function ψ(*x*, *y*, *z*), which occurs in the wave equation, describes the behavior of an electron. The square of this function, ψ²(*x*, *y*, *z*), defines a region in space (*x*, *y*, *z*) where the electron may be found with a certain probability. This is shown schematically in Fig. 4.7. This illustration might be considered the final print of millions of superimposed photographs of the position of an electron in three-dimensional space, defined by *x*, *y*, and *z* axial directions. With the electron in rapid motion, the final composite photograph would produce an array of dots (like a cloud) with dense and more openly spaced regions. The dense regions would be those of a high probability of finding an electron, and those of low density would represent a lower probability of locating an electron. The wave theory, therefore, portrays the motion of electrons only in terms of the probability of finding a certain electron within a small volume unit. It does not describe the movement of electrons in well-defined, simple orbits.

The electron behavior summarized by the Schrödinger wave equation can be compared to the vibrations of a string stretched between two fixed points. Figure 4.8 shows some of the ways in which a string can vibrate. In Fig. 4.8*a* the string vibrates in the simplest way, that is, between two fixed end points. In Fig. 4.8*b* the string distorts like a standing sinusoidal wave, with a central point where there is no displacement. The points of no displacement are known as *nodes*. Figure 4.8*c* shows a string vibrating between two fixed end nodes, and with two nodes at one-third and two-thirds of the length of the string. Each of these different vibration patterns can be defined by a fundamental audio frequency, *ν* audio (where *ν* audio =

Number of
half wavelengths

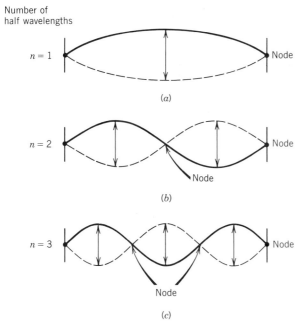

$n = 1$ Node

(a)

$n = 2$ Node

Node

(b)

$n = 3$ Node

Node

(c)

FIG. 4.8 Relationship between principal quantum n and the number of half wavelengths in a standing wave. In solutions to the wave equation, nodes represent regions with no electron density (see Fig. 4.10).

v/λ; v = velocity of sound and λ = wavelength of the string) and a principal quantum number, n. In the one-dimensional example of a string, the quantum number n gives the number of half-wavelengths in the vibrations, and $(n + 1)$ is the number of nodes (including the nodes at the ends of the string). In the case of a three-dimensional electron wave, nodes may form along each of the three axial directions.

In order to completely specify the position of an electron in three-dimensional space, *three quantum numbers* are needed. These are the *principal quantum number, n*; the *azimuthal quantum number* (or *orbital shape quantum number*), *l*; and the *magnetic quantum number, m*. These three quantum numbers follow from the solution of the Schrödinger wave equation and represent specified parameters in the mathematical formulation of ψ.

In three dimensions, the *principal quantum number, n*, is a function of the distance r of the electron from the nucleus. The probability of finding the electron at a distance r from the nucleus is given by $4\pi r^2\psi^2$. For the s-orbital[1] in the hydrogen atom, this is plotted in Fig. 4.9. This curve shows that the electron occupies all of the small specified volume around the nucleus, but that it is most frequently found at distance r_o, the radius predicted in the Bohr model of

[1]The wave functions that describe the motions of an electron are known as "orbitals," to distinguish them from the "orbits" in the Bohr atomic model.

the atom. The *principal quantum number, n*, reflects the effective volume (or mean radius) of an electron orbital and can have any positive integral value from 1 to infinity, 1, 2, 3, . . . , ∞. It also reflects the energy levels, or shells, in an atom. The larger the value of n, the greater is the average energy of the levels belonging to the shell. As in the Bohr theory, $n = 1$ defines the K shell, $n = 2$ defines the L shell, $n = 3$ defines the M shell, and so on. n determines the position of the horizontal rows of the periodic table (see Fig. 4.3).

The *azimuthal quantum number* (or *orbital shape quantum number*), *l*, determines the general shape of the region in which an electron moves (that is, the shape of the orbital), and to some degree its energy. For a given shell, *l* may have values of 0, 1, 2, 3, . . . , to a maximum of $(n - 1)$ for that shell. This means that for the K shell, with $n = 1$, the only value of *l* that is possible is $l = 0$. When $n = 2$, two values of *l* are possible, 0 and 1, resulting in two subshells for the L shell. The values of *l* that are possible for each value of n are given in Table 4.3. It follows from this table that the number of subshells in any given shell is equal to its value of n.

The various states of *l* (0, 1, 2, 3, . . . , $n - 1$) have been given letter designations as follows:

value of *l*	0 1 2 3 4 5 6 . . .
subshell designation	*s p d f g h i* . . .

The letters *s, p, d,* and *f* are abbreviations of the spectroscopic terms *sharp, principal, diffuse,* and *fundamental*, respectively. Our discussion will

FIG. 4.9 The radial distribution function ($4\pi r^2\psi^2$) for the s-orbital plotted against increasing distance (r) from the nucleus. The vertical axis is a measure of the probability of finding the electron at a specific distance from the nucleus. The maximum in this function coincides with r_0, the radius of the smallest orbit in the Bohr atomic model.

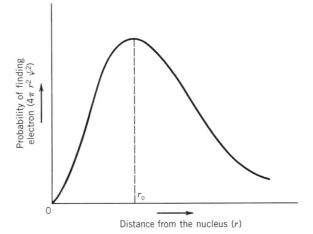

Probability of finding electron ($4\pi r^2\psi^2$)

r_0

0   Distance from the nucleus (r)

Table 4.3

SUMMARY OF THE THREE QUANTUM NUMBERS

Principal Quantum Number, n (Shell)	Azimuthal Quantum Number, l (Subshell)	Subshell Designation	Magnetic Quantum Number, m (Orbital)	Number of Orbitals in Subshell	Maximum Number of Electrons	
1 (K)	0	1s	0	1	2	2
2 (L)	0	2s	0	1	2	8
	1	2p	-1, 0, +1	3	6	
3 (M)	0	3s	0	1	2	18
	1	3p	-1, 0, +1	3	6	
	2	3d	-2, -1, 0, +1, +2	5	10	
4 (N)	0	4s	0	1	2	32
	1	4p	-1, 0, +1	3	6	
	2	4d	-2, -1, 0, +1, +2	5	10	
	3	4f	-3, -2, -1, 0, +1, +2, +3	7	14	

FIG. 4.10 Surfaces showing the angular dependence of the function ψ^2 for s, p, and d-orbitals of the hydrogen atom. These angular wave functions can be regarded as probability distributions of electrons.

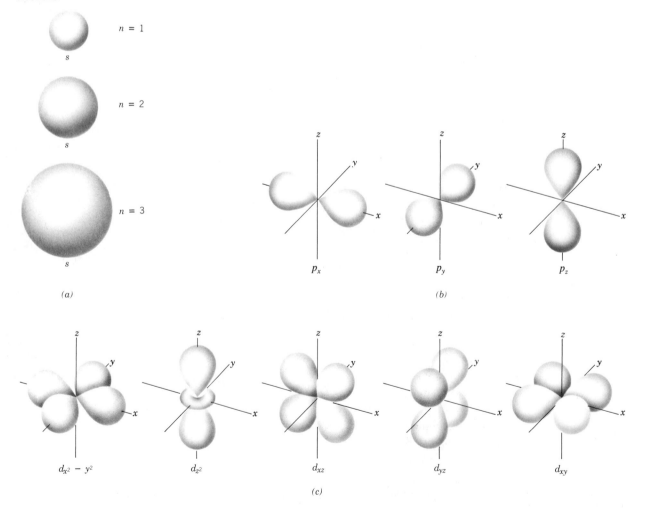

be limited to the *s*, *p*, *d*, and *f* subshells, because they are the ones populated by electrons in atoms in their lowest energy state, or ground state. The *s* orbital is spherical in shape, the *p* orbital is quasi-dumbbell-shaped and axially directed, and the *d* orbital has various shapes (see Fig. 4.10).

To designate a subshell within a given shell, the value of *n* (for the shell) is followed by the letter designation of the subshell. That is, the 2*s* subshell is a subshell of the second shell ($n = 2$), with $l = 0$. The 3*d* subshell is a subshell of the third shell ($n = 3$), with $l = 2$.

The *magnetic quantum number, m*, restricts the orientation and shape of each type of orbital. It has integer values that range from $-l$ to $+l$. When $l = 0$, only one value of *m* is permitted, $m = 0$. This means that the *s* subshell has only one orbital (the *s* orbital).

Table 4.3 gives a summary of the interrelationship of the *n*, *l*, and *m* quantum numbers and the number of subshell orbitals.

In addition to the three quantum numbers, *n*, *l*, and *m*, which follow from the solution of the wave equation, there is a fourth quantum number, the *spin quantum number, s*, which defines the direction of spin of the electron in space. Because an electron can spin only in one of two directions, it has only two values, namely $+\frac{1}{2}$ and $-\frac{1}{2}$. A spinning electron behaves as a small magnet and will produce a magnetic field while moving around its orbit, both from its orbital motion and from its spin. An orbiting electron produces a magnetic field, just like the magnetic field produced by an electric current moving through a coiled wire. In addition, the electron's axial spin can produce a magnetic field. In Fig. 4.11, which illustrates an electron orbital and the spinning motion of an electron, the overall magnetic field is represented by *H*. Because the spinning electron behaves as a small magnet, there will be an interaction between *H* (the magnetic field strength) and the field produced by the axial spin of the electron. The axial spin of the electron will either reinforce or oppose the field strength (*H*), depending on whether the spin is clockwise or anticlockwise. Note that in this picture of the electron's spin, the electron is depicted as a charged particle, not a wave function.

It is useful, as in the discussion of magnetism (Chapter 6), to keep account of the electron spin directions in an atom. This is commonly done by representing an electron with its associated spin (in one direction) by an arrow pointing up, ↑, and an electron with a reverse spin (in the opposite direction) by an arrow pointing down, ↓. To indicate the distribution of electrons among orbitals, the arrows are placed over bars that symbolize orbitals, such as for

H $\dfrac{\uparrow}{1s}$ He $\dfrac{\uparrow\downarrow}{1s}$

Li $\dfrac{\uparrow\downarrow}{1s}\dfrac{\uparrow}{2s}$ Be $\dfrac{\uparrow\downarrow}{1s}\dfrac{\uparrow\downarrow}{2s}$

and

Ne $\dfrac{\uparrow\downarrow}{1s}\dfrac{\uparrow\downarrow}{2s}\underbrace{\dfrac{\uparrow\downarrow}{}\dfrac{\uparrow\downarrow}{}\dfrac{\uparrow\downarrow}{}}_{2p}$

Two electrons paired in the same orbital have spin directions opposing each other (see discussion of the Pauli exclusion principle, below). The magnetic moments of such paired electrons nullify each other (see Table 6.5), whereas an effective magnetic moment is the result of unpaired electrons in outer orbitals. The spin of unpaired electrons is a major factor in the magnetic properties of atoms. This will be further discussed under "Magnetic Properties" in Chapter 6.

In conclusion, each electron in an atom can be assigned values for four quantum numbers, *n*, *l*, *m*, and *s*, which determine the orbital in which the electron occurs and the direction in which the electron spins.

There is, however, a restriction on the values these quantum numbers may have. This is known as the *Pauli exclusion principle*, which states that no two electrons in any one atom may have all four quantum numbers the same. This has the effect of limiting the number of electrons in any given orbital to two, and it also requires that the spins of the two electrons be in opposite directions (with *s* values of $+\frac{1}{2}$ and $-\frac{1}{2}$).

FIG. 4.11 The orbital and spin motions of an electron. The magnetic field *H* is the result of the electron moving along the orbital. The spin of the electron can reinforce or oppose the magnetic field *H*, depending on the rotation direction of the spin (clockwise or anticlockwise).

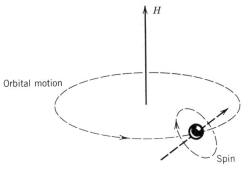

Table 4.4

QUANTUM NOTATION AND ELECTRON DISTRIBUTION

Shell and Main Energy Levels (n)	Energy Sublevels		Number of Orbitals		Maximum Number of Electrons	
K ($n = 1$)	$1s$	($l = 0$)	1	1	2	2
L ($n = 2$)	$2s$	($l = 0$)	1 }		2 }	
	$2p$	($l = 1$)	3 }	4	6 }	8
M ($n = 3$)	$3s$	($l = 0$)	1 }		2 }	
	$3p$	($l = 1$)	3 }	9	6 }	18
	$3d$	($l = 2$)	5 }		10 }	
N ($n = 4$)	$4s$	($l = 0$)	1 }		2 }	
	$4p$	($l = 1$)	3 }		6 }	
	$4d$	($l = 2$)	5 }	16	10 }	32
	$4f$	($l = 3$)	7 }		14 }	
O ($n = 5$)	$5s$	($l = 0$)	1 }		2 }	
	$5p$	($l = 1$)	3 }		6 }	
	$5d$	($l = 2$)	5 }	16	10 }	50*
	$5f$	($l = 3$)	7 }		14 }	
P ($n = 6$)	$6s$	($l = 0$)	1		2 }	
	$6p$	($l = 1$)	3		6 }	72*
	$6d$	($l = 2$)	5		10 }	
Q ($n = 7$)	$7s$	($l = 0$)	1		2	98*

*This number is not reached in naturally occurring atoms.

Because of the restriction of only two electrons per orbital, the maximum number of electrons that can occur in the various (s, p, d, and f) subshells is also limited. For example, the s subshell has only one orbital and therefore can house only a maximum of two electrons; the p subshell has three orbitals and therefore can house a maximum of six electrons. This is illustrated in Table 4.4.

The relative energies of shells, subshells, and orbitals in atoms with more than one electron are illustrated in Fig. 4.12. This shows that the energy of a shell increases with increasing value of the principal quantum number, n. It also shows that, as the value of n increases, there is an overlap in the energy levels of subshells for $n = 3$ and higher. That is, the $4s$ subshell has a lower relative energy level than the $3d$ subshell. Such overlap becomes even more common in the higher shells. This sequence of energy levels is critical in the determination of the arrangement of electrons in the atom. A complete table of electron configurations of the elements is given in Table 4.5.

Electron Configuration and the Periodic Table

In the earlier, introductory discussion to the periodic table (Fig. 4.3), we noted that the table is arranged by increasing atomic number (Z). Now that we have discussed the orbital structure of the atom, we can address this periodicity in greater detail. In general, elec-

trons tend to occupy higher energy orbitals only after lower energy levels have been filled (see Table 4.5).

The periodic table in Fig. 4.3 is organized such that the vertical columns (identified by I, II, III, IV, . . . , VIII) list atoms whose outer shells contain the number of electrons equal to the Roman numeral at the top of the column. For example, all elements in

FIG. 4.12 Relative energies of the orbitals in neutral, many-electron isolated atoms.

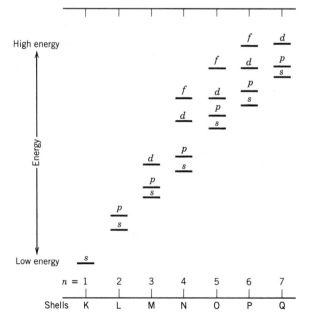

Table 4.5
ELECTRON CONFIGURATIONS OF THE ATOMS

Shell / Element	K 1s	L 2s	2p	M 3s	3p	3d	N 4s	4p	4d	4f	O 5s	5p	5d	5f	5g	P 6s	6p	6d	Q 7s
1. H	1																		
2. He	2																		
3. Li	2	1																	
4. Be	2	2																	
5. B	2	2	1																
6. C	2	2	2																
7. N	2	2	3																
8. O	2	2	4																
9. F	2	2	5																
10. Ne	2	2	6																
11. Na	2	2	6	1															
12. Mg	2	2	6	2															
13. Al	2	2	6	2	1														
14. Si	2	2	6	2	2														
15. P	2	2	6	2	3														
16. S	2	2	6	2	4														
17. Cl	2	2	6	2	5														
18. Ar	2	2	6	2	6														
19. K	2	2	6	2	6		1												
20. Ca	2	2	6	2	6		2												
21. Sc	2	2	6	2	6	1	2												
22. Ti	2	2	6	2	6	2	2												
23. V	2	2	6	2	6	3	2												
24. Cr	2	2	6	2	6	5	1												
25. Mn	2	2	6	2	6	5	2												
26. Fe	2	2	6	2	6	6	2												
27. Co	2	2	6	2	6	7	2												
28. Ni	2	2	6	2	6	8	2												
29. Cu	2	2	6	2	6	10	1												
30. Zn	2	2	6	2	6	10	2												
31. Ga	2	2	6	2	6	10	2	1											
32. Ge	2	2	6	2	6	10	2	2											
33. As	2	2	6	2	6	10	2	3											
34. Se	2	2	6	2	6	10	2	4											
35. Br	2	2	6	2	6	10	2	5											
36. Kr	2	2	6	2	6	10	2	6											
37. Rb	2	2	6	2	6	10	2	6			1								
38. Sr	2	2	6	2	6	10	2	6			2								
39. Y	2	2	6	2	6	10	2	6	1		2								
40. Zr	2	2	6	2	6	10	2	6	2		2								
41. Nb	2	2	6	2	6	10	2	6	4		1								
42. Mo	2	2	6	2	6	10	2	6	5		1								
43. Tc	2	2	6	2	6	10	2	6	5		2								
44. Ru	2	2	6	2	6	10	2	6	7		1								
45. Rh	2	2	6	2	6	10	2	6	8		1								
46. Pd	2	2	6	2	6	10	2	6	10										
47. Ag	2	2	6	2	6	10	2	6	10		1								
48. Cd	2	2	6	2	6	10	2	6	10		2								
49. In	2	2	6	2	6	10	2	6	10		2	1							
50. Sn	2	2	6	2	6	10	2	6	10		2	2							
51. Sb	2	2	6	2	6	10	2	6	10		2	3							
52. Te	2	2	6	2	6	10	2	6	10		2	4							
53. I	2	2	6	2	6	10	2	6	10		2	5							
54. Xe	2	2	6	2	6	10	2	6	10		2	6							
55. Cs	2	2	6	2	6	10	2	6	10		2	6				1			
56. Ba	2	2	6	2	6	10	2	6	10		2	6				2			
57. La	2	2	6	2	6	10	2	6	10		2	6	1			2			

(continued)

Table 4.5
(*continued*)

Element	K	L		M			N				O					P			Q
	1s	2s	2p	3s	3p	3d	4s	4p	4d	4f	5s	5p	5d	5f	5g	6s	6p	6d	7s
*58. Ce	2	2	6	2	6	10	2	6	10	1	2	6	1			2			
*59. Pr	2	2	6	2	6	10	2	6	10	2	2	6	1			2			
*60. Nd	2	2	6	2	6	10	2	6	10	3	2	6	1			2			
*61. Pm	2	2	6	2	6	10	2	6	10	4	2	6	1			2			
*62. Sm	2	2	6	2	6	10	2	6	10	5	2	6	1			2			
*63. Eu	2	2	6	2	6	10	2	6	10	6	2	6	1			2			
*64. Gd	2	2	6	2	6	10	2	6	10	7	2	6	1			2			
*65. Tb	2	2	6	2	6	10	2	6	10	8	2	6	1			2			
*66. Dy	2	2	6	2	6	10	2	6	10	9	2	6	1			2			
*67. Ho	2	2	6	2	6	10	2	6	10	10	2	6	1			2			
*68. Er	2	2	6	2	6	10	2	6	10	11	2	6	1			2			
*69. Tm	2	2	6	2	6	10	2	6	10	12	2	6	1			2			
*70. Yb	2	2	6	2	6	10	2	6	10	13	2	6	1			2			
*71. Lu	2	2	6	2	6	10	2	6	10	14	2	6	1			2			
72. Hf	2	2	6	2	6	10	2	6	10	14	2	6	2			2			
73. Ta	2	2	6	2	6	10	2	6	10	14	2	6	3			2			
74. W	2	2	6	2	6	10	2	6	10	14	2	6	4			2			
75. Re	2	2	6	2	6	10	2	6	10	14	2	6	5			2			
76. Os	2	2	6	2	6	10	2	6	10	14	2	6	6			2			
77. Ir	2	2	6	2	6	10	2	6	10	14	2	6	7			2			
78. Pt	2	2	6	2	6	10	2	6	10	14	2	6	9			1			
79. Au	2	2	6	2	6	10	2	6	10	14	2	6	10			1			
80. Hg	2	2	6	2	6	10	2	6	10	14	2	6	10			2			
81. Tl	2	2	6	2	6	10	2	6	10	14	2	6	10			2	1		
82. Pb	2	2	6	2	6	10	2	6	10	14	2	6	10			2	2		
83. Bi	2	2	6	2	6	10	2	6	10	14	2	6	10			2	3		
84. Po	2	2	6	2	6	10	2	6	10	14	2	6	10			2	4		
85. At	2	2	6	2	6	10	2	6	10	14	2	6	10			2	5		
86. Rn	2	2	6	2	6	10	2	6	10	14	2	6	10			2	6		
87. Fr	2	2	6	2	6	10	2	6	10	14	2	6	10			2	6		1
88. Ra	2	2	6	2	6	10	2	6	10	14	2	6	10			2	6		2
89. Ac	2	2	6	2	6	10	2	6	10	14	2	6	10			2	6	1	2
*90. Th	2	2	6	2	6	10	2	6	10	14	2	6	10			2	6	2	2
*91. Pa	2	2	6	2	6	10	2	6	10	14	2	6	10	2		2	6	1	2
*92. U	2	2	6	2	6	10	2	6	10	14	2	6	10	3		2	6	1	2
*93. Np	2	2	6	2	6	10	2	6	10	14	2	6	10	4		2	6	1	2
*94. Pu	2	2	6	2	6	10	2	6	10	14	2	6	10	6		2	6		2
*95. Am	2	2	6	2	6	10	2	6	10	14	2	6	10	7		2	6		2
*96. Cm	2	2	6	2	6	10	2	6	10	14	2	6	10	7		2	6	1	2
*97. Bk	2	2	6	2	6	10	2	6	10	14	2	6	10	8		2	6	1	2
*98. Cf	2	2	6	2	6	10	2	6	10	14	2	6	10	10		2	6		2
*99. Es	2	2	6	2	6	10	2	6	10	14	2	6	10	11		2	6		2
*100. Fm	2	2	6	2	6	10	2	6	10	14	2	6	10	12		2	6		2
*101. Mv	2	2	6	2	6	10	2	6	10	14	2	6	10	13		2	6		2
*102. No	2	2	6	2	6	10	2	6	10	14	2	6	10	14		2	6		2

*Lanthanide and actinide elements; some configurations uncertain.

column I (Z = 1, 3, 11, 19, 37, 55, and 87) have only one electron in the *s* orbital of the outer shell (see also Table 4.5). All elements in column II have two *s* orbital electrons in the outer shell. Elements in column III contain three electrons (two *s* and one *p*). Those in column IV have four electrons (two *s* and two *p*), and so forth, until column VIII, which contains all those elements (except He) with eight outer shell electrons (two *s* and six *p*). Helium has two electrons in the *s* orbital, completely filling the K shell.

The horizontal *rows* in Fig. 4.3 are numbered 1 through 7; these are equivalent to the K, L, M, N, . . . shells. In a left-to-right sequence in any given row, the outer shell of an atom is progressively filled, beginning with the *s* orbital and ending with the *p* orbitals. Atoms with Z = 21 to 30 (in row 4), with Z = 39 to

48 (in row 5), and with $Z = 57$ to 80 (in row 6) are called the *transition elements* because the orbital electrons, in excess of those present in calcium ($Z = 20$), fill in inner shells. For example, in row 4, elements with $Z = 21$ to 30 fill the $3d$ orbital of the more interior M shell; in row 6, elements with $Z = 57$ to 80 first fill $4f$ and subsequently $5d$ orbitals on the interior side of the P shell.

The symbolism for summarizing an atom's electron configuration is as follows: the symbol of each orbital is followed by an exponent indicating the number of electrons present in the orbital. The symbol for atomic silicon ($Z = 14$) is $1s^2 2s^2 2p^6 3s^2 3p^2$. This signifies two electrons in the $1s$ and $2s$ orbitals, six in the $2p$ orbital, and two in the $3s$ and $3p$ orbitals. A listing of the symbolic notation of the electronic structure for elements $Z = 1$ to 37 is given in Table 4.6; this should be compared with Table 4.5.

The preceding discussion has shown that the periodic table of elements is not just a tabulation, but rather the ordered result of basic chemical properties that depend on the nature of the outer electrons, the *valence electrons*. These are the electrons available for chemical bonding. As a result of the similarity of the

Table 4.6

FIRST IONIZATION POTENTIALS, ELECTRONEGATIVITY VALUES, AND ELECTRONIC STRUCTURE OF THE ELEMENTS THROUGH ATOMIC NUMBER 37

Z	Element	First Ionization Potential, in Electron Volts (e.v.)*,†	Electronegativity‡	Electronic Structure
1	H	13.598	2.1	$1s^1$
2	He	24.587	0	$1s^2$
3	Li	5.392	1.0	$1s^2 2s^1$
4	Be	9.322	1.5	$1s^2 2s^2$
5	B	8.298	2.0	$1s^2 2s^2 2p^1$
6	C	11.260	2.5	$1s^2 2s^2 2p^2$
7	N	14.534	3.0	$1s^2 2s^2 2p^3$
8	O	13.618	3.5	$1s^2 2s^2 2p^4$
9	F	17.422	4.0	$1s^2 2s^2 2p^5$
10	Ne	21.564	0	$1s^2 2s^2 2p^6$
11	Na	5.139	0.9	$[Ne]3s^1$
12	Mg	7.646	1.2	$[Ne]3s^2$
13	Al	5.986	1.5	$[Ne]3s^2 3p^1$
14	Si	8.151	1.8	$[Ne]3s^2 3p^2$
15	P	10.486	2.1	$[Ne]3s^2 3p^3$
16	S	10.360	2.5	$[Ne]3s^2 3p^4$
17	Cl	12.967	3.0	$[Ne]3s^2 3p^5$
18	Ar	15.759	0	$[Ne]3s^2 3p^6$
19	K	4.341	0.8	$[Ar]4s^1$
20	Ca	6.113	1.0	$[Ar]4s^2$
21	Sc	6.54	1.3	$[Ar]3d^1 4s^2$
22	Ti	6.82	1.5	$[Ar]3d^2 4s^2$
23	V	6.74	1.6	$[Ar]3d^3 4s^2$
24	Cr	6.766	1.6	$[Ar]3d^5 4s^1$
25	Mn	7.435	1.5	$[Ar]3d^5 4s^2$
26	Fe	7.870	1.8	$[Ar]3d^6 4s^2$
27	Co	7.86	1.8	$[Ar]3d^7 4s^2$
28	Ni	7.635	1.8	$[Ar]3d^8 4s^2$
29	Cu	7.726	1.9	$[Ar]3d^{10} 4s^1$
30	Zn	9.394	1.6	$[Ar]3d^{10} 4s^2$
31	Ga	5.999	1.6	$[Ar]3d^{10} 4s^2 4p^1$
32	Ge	7.899	1.8	$[Ar]3d^{10} 4s^2 4p^2$
33	As	9.81	2.0	$[Ar]3d^{10} 4s^2 4p^3$
34	Se	9.752	2.4	$[Ar]3d^{10} 4s^2 4p^4$
35	Br	11.814	2.8	$[Ar]3d^{10} 4s^2 4p^5$
36	Kr	13.999		$[Ar]3d^{10} 4s^2 4p^6$
37	Rb	4.177	0.8	$[Kr]5s^1$

*e.v. = electron volt = 23 kilocalories/mole.

†From Lide, D. R., ed., 1991, *CRC Handbook of Chemistry and Physics*, 72nd ed. CRC Press, Boca Raton, Fla.

‡From Pauling, L., 1960, *Nature of the Chemical Bond*, 3rd ed. Cornell University Press, Ithaca, N.Y., 644 pp.

...iemical character of certain elements (because of their similarity in outer electron configurations), such elements often have similar chemical behavior and can be found in similar crystallographic sites within minerals.

THE ION

Until this point, we have concerned ourselves with atoms in which the number of protons and electrons is equal. However, elements in the periodic table can be divided into two groups: those that have a tendency to give up electrons, and those that are capable of acquiring electrons. Those that are electron donors are the *metals* (in the left-hand part of the periodic table, Fig. 4.3) and those that are electron acceptors are the *nonmetals* (in the right-hand part of the periodic table). When one or more electrons are lost from the electron configuration of an atom, a *cation* is formed, and when electrons are added, an *anion* results. This can be expressed as:

$$X_{atom} - e^- \rightarrow X^+_{cation}$$

and

$$X_{atom} + e^- \rightarrow X^-_{anion}$$

In either process, energy is involved. The energy required to remove the most weakly held electron from a neutral atom (to infinity) is known as the *first ionization potential*. This value (listed for a representative number of elements in Table 4.6) expresses how strongly the nucleus of a neutral atom attracts an electron in a partially filled orbital. Figure 4.13a illustrates that the first ionization potential values increase with increasing atomic number, within each period. This increase coincides with the progressive filling of electron orbitals and expresses the reluctance of atoms to lose electrons from orbitals that are nearly completely filled. It further shows that inert gases (He, Ne, Ar, Kr) have maximal values and alkali metals (Li, Na, K, Rb) have minimal values. This means that the electron configurations of the inert gases are more stable than those of the alkali metals. As such, a relatively small amount of energy is needed to remove an electron from an alkali metal, thus producing a stable monovalent (1 +) ion. Figure 4.13a also shows that it takes more energy to remove an electron from beryllium than it does to remove an electron from lithium or boron. An unpaired electron, whether it is in a 2s orbital (as in lithium) or in a 2p orbital (as in boron; see Table 4.6), is evidently less well bonded to the atom

than a paired electron, as in the filled 2s orbital of beryllium. A similar effect is seen in period 3 in the first ionization potential of magnesium versus that of sodium and aluminum. The ionization potentials of elements in group 4 with partially filled *d* orbitals (see Table 4.6) vary little with atomic number. The general trend, within each period, is one of low ionization potentials (elements acting as electron donors, and as such metallic in their character) to high ionization potentials (elements acting as electron acceptors, and as such nonmetallic in character).

Ionization potentials that express the energy needed to remove additional electrons (i.e., more than one electron) are very much larger than those of the first ionization reported in Table 4.6. Such higher values reflect the much greater energy needed to remove an electron from an atom that has already acquired a positive charge, as well as the higher energy needed to remove an additional electron from remaining electrons that fill the orbital in which they occur. Because of these energy barriers, elements involved in chemical reactions tend to lose only their *valence* electrons, which are those that reside outside filled orbitals.

Elements in column I in Fig. 4.3, which include Li, Na, K, Rb, easily lose the one and only outer electron, as seen in Fig. 4.13. This results in the formation of monovalent cations such as Li^+, Na^+, K^+, and Rb^+. For elements in column II (Be, Mg, Ca, Sr, Ba), there are similarly low ionization potential values (see Fig. 4.13 and Table 4.6), suggesting that relatively little energy is needed to make these atoms into divalent cations such as Be^{2+}, Mg^{2+}, Ca^{2+}, and so on. Similar considerations hold for the formation of trivalent (e.g., Al^{3+}) and tetravalent (e.g., Si^{4+}) cations.

Several elements are found in more than one *valence* or *oxidation state*. For example, iron (Fe) can occur in a divalent state (*ferrous* iron, Fe^{2+}) or in an even more oxidized trivalent (*ferric* iron, Fe^{3+}) state. The electronic configuration of atomic Fe is $1s^2 2s^2 2p^6 3s^2 3p^6 3d^6 4s^2$ (Table 4.6). In Fe^{2+}, the atom loses the two 4s electrons, but it can lose one 3d electron as well, making Fe^{3+}, which then has a half-filled subshell. Because in a crystalline substance *d* orbitals have energy differences that are similar to wavelengths in visible light, transition elements, such as iron, tend to play a major role in the coloration of minerals (see Chapter 6).

Although ionization potentials are useful predictors of some chemical properties, Linus Pauling developed an additional concept, known as *electronegativity*. It is a measure of the ability of an atom *in a crystal structure or molecule* to attract electrons to its outer shell. It is represented as a dimensionless number (see Tables 4.6 and 4.10) which is calculated

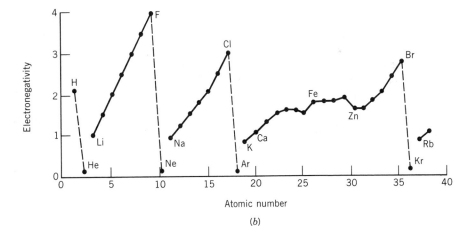

FIG. 4.13 (a) Variation of the first ionization potential as a function of increasing atomic number (Z) for the first 37 elements. (b) Variation of the electronegativity for the same elements as shown in (a).

from the known bond strengths between atoms in molecules. Elements with low electronegativity are electron donors and those with high values are electron acceptors. Ranges of electronegativity values are illustrated in Fig. 4.13b. Here it can be seen that, within a specified period, electronegativity values rise as a function of increasing atomic number. It is also clear from Fig. 4.13b that the electronegativity values of elements within column I (H, Li, Na, K, Rb, and so on; see Table 4.6, and Fig. 4.3) or within column II (Be, Mg, Ca, Sr, and so on) decrease with increasing

atomic number. The same holds for their values of ionization potential. This leads to the conclusion that the bond strength (or binding energy) between the nucleus and the first valence electron of an element (in a specified group) decreases as the volume of the atom in the group increases. This implies that the large atoms hold their outer valence electrons more loosely than do smaller atoms. The electronegativity values of the noble gases are set at zero because these atoms do not attract electrons. The concept of electronegativity is especially useful in assessing the type of bond formed between different atoms. It is well known that elements with very different electronegativity values will tend to form bonds that are *ionic* in character, whereas those with similar electronegativities form *covalent* bonds (for further discussion of bond type, see page 201). As such, the *difference* in electronegativity values is a very useful parameter in the estimation of the chemical bonding mechanism.

Atomic and Ionic Radii

The sizes of atoms or ions are difficult to define but even more difficult to measure experimentally. The radius of an atom is defined by the radius of the maximum radial charge density of the outermost shells of the atom, but the effective radius of an atom (or ion) is also dependent on the type and number of neighboring atoms and/or ions, and on the charge of the ion. In a crystal of a pure metal, where identical atoms are bonded to each other, the radius of the individual atom is assumed to be one-half the bond length. Such measurements provide sizes for atomic radii (see Fig. 4.14a and Table 4.7). In ionic crystals, where two oppositely charged ions are bonded together, however, the distance between a positive and a negative ion is the sum of two different radii (see Fig. 4.14b). This distance is determined by electrostatic forces. There is, between any pair of oppositely charged ions, an attractive electrostatic force that is directly proportional to the product of their charges and inversely proportional to the square of the distance between their centers. This is known as Coulomb's law, formulated in 1787 by the French physicist Charles Coulomb. It is stated as follows:

$$F = k \frac{(q^+)(q^-)}{d^2}$$

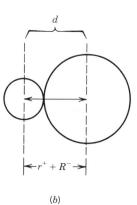

FIG. 4.14 (a) Metallic radii are half the distance between the centers of two adjoining metal atoms. (b) Ionic radii are defined as the distance between centers of the cation and anion, with the radius of e.g., the anion well known ($r\,O^{2-}$ in 3-coordination = 1.36 Å; $r\,F^-$ in 3-coordination = 1.30 Å). Therefore, the radius of Mg^{2+} is obtained by subtracting 1.36 Å from the internuclear distance between Mg^{2+} and O^{2-} in MgO.

where F is the force of attraction between two oppositely charged ions, q^+ and q^- are the charges on the ions, d is the distance between them, and k is a proportionality constant. A force operating over a certain distance can also be expressed as energy (E). In Fig. 4.15 the curve representing the attractive force is the

Table 4.7													
METALLIC RADII (IN Å)*													
Li	Be												
1.57	1.12												
Na	Mg	Al											
1.91	1.60	1.43											
K	Ca	Sc	Ti	V	Cr	Mn	Fe	Co	Ni	Cu	Zn	Ga	Ge
2.35	1.97	1.64	1.47	1.35	1.29	1.37	1.26	1.25	1.25	1.28	1.37	1.53	1.39
Rb	Sr	Y	Zr	Nb	Mo	Tc	Ru	Rh	Pd	Ag	Cd	In	Sn	Sb
2.50	2.15	1.82	1.60	1.47	1.40	1.35	1.34	1.34	1.37	1.44	1.52	1.67	1.58	1.61
Cs	Ba	La	Hf	Ta	W	Re	Os	Ir	Pt	Au	Hg	Tl	Pb	Bi
2.72	2.24	1.88	1.59	1.47	1.41	1.37	1.35	1.36	1.39	1.44	1.55	1.71	1.75	1.82

*The values refer to 12-coordination. From Wells, A. F., 1991, *Structural Inorganic Chemistry*, 5th ed. Clarendon Press, Oxford, England, 1382 pp.

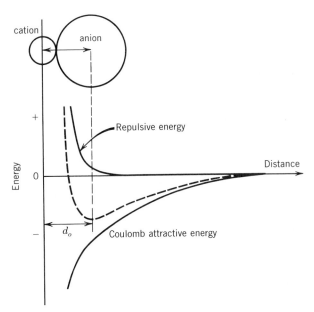

FIG. 4.15 Electrostatic interaction between a cation and anion. The attractive force acts over longer distances than the repulsion. The attractive and repulsive forces add to produce a resultant (dashed curve) in which the minimum value corresponds to the equilibrium distance (d_0) between the centers of the cation and anion.

lower one, with (−) energy values. When ions approach each other under the influence of these forces, repulsive forces are also set up. These repulsive forces arise from the interaction of the negatively charged electron clouds and from the opposition of the positively charged nuclei; they increase rapidly with diminishing internuclear distance. In Fig. 4.15 this is represented by the upper curve with (+) energy values. The distance at which these repulsive forces balance the attractive forces is the characteristic interionic spacing (bond length) for a pair of ions. This is shown in Fig. 4.15 by the minimum value in the curve that is the resultant of the attractive and repulsive forces.

In the simplest case, when both cations and anions are fairly large and of low charge and both have numerous symmetrically disposed neighbors of opposite sign, ions may be regarded as spheres in contact. Sodium chloride, in which both cation and anion are monovalent, fairly large, and surrounded by six neighbors of opposite polarity, is a good example. In such crystals the interionic distance may be regarded as the sum of the radii of the two ions in contact.

If one of the ionic radii in an internuclear distance is well known from prior experimental measurements, the radius of the other ion can be obtained (see Fig. 4.14b). For example, once Linus Pauling (in 1927) determined the radius of 6-coordinated O^{2-} to be 1.40 Å, the radii of many cations that are bonded ionically to oxygen were obtained by subtracting the value of 1.40 Å from the measured bond length between cation-oxygen pairs. Such measurements provide ionic radii.[2] However, the ionic radius of an ion may not be constant from one crystal structure to another. This is due to possible changes in bond type and coordination number (= the number of closest neighbors around a specific atom or ion). Because of such influences, the values of ionic radii given in tabulations generally represent averaged values.

Shannon and Prewitt (1969; revised by Shannon, 1976) evaluated in great detail the variation of ionic size as a function of coordination number. For example, they found considerable variation in the ionic size of O^{2-} as a function of coordination number, ranging from 1.35 Å in 2-fold coordination to 1.42 Å in 8-fold coordination (see Table 4.8). The ionic radius values reported by Shannon and Prewitt (1969) and Shannon (1976) (many of which are listed in Table 4.8; for a complete listing see Shannon's Table 1) are referred to as *effective* ionic radii. The term effective is used because they were determined empirically from highly accurate data for a very large number of oxide structures. According to Shannon and Prewitt (1969) and Shannon (1976), these radii may well represent the best possible fit because they reproduce interatomic distances in a wide variety of crystalline solids. It is clear from Table 4.8 that many of the radii (be they for anions or cations) vary as a function of coordination number (**C.N.** = the number of atoms that surround a particular atom or ion in a structure; coordination number is discussed in detail in a subsequent section). For examples of anion radius changes see O^{2-} and F^- in Table 4.8. Radius changes in cations as a function of coordination number are clear for many of those listed in Table 4.8. For example, the K^+ radius = 1.38 Å for 6-coordination, 1.51 Å for 8-coordination, and 1.59 Å for 10-coordination. The increase in the cation radius reflects the expansion of the cation into the space (or void) provided by the surrounding anions. Figure 4.16 graphically illustrates the expansion of the radius of some selected cations as a function of coordination number.

[2]In practice, the value of one anion, usually oxygen, is assumed. Cation radii are then calculated from interatomic distances in oxides. These cation radii are subsequently used to calculate other anion radii (from interatomic measurements of, for example, chlorides or sulfides). These are then used to calculate other cation radii, which in turn are used to calculate oxygen radii. The process is repeated iteratively so as to arrive at a self-consistent set of ionic radii.

Table 4.8
EFFECTIVE IONIC RADII (IN Å) FOR IONS COMMONLY FOUND IN MINERALS*

Main group elements

Row	I	II	III	IV	V	VI	VII
2	Li^+ 0.59(4), 0.74(6), 0.92(8)	Be^{2+} 0.16(3), 0.27(4), 0.45(6)	B^{3+} 0.11(4), 0.27(6)	C^{4+} −0.08(3), 0.15(4), 0.16(6)	N^{5+} −0.10(3), 0.13(6)	O^{2-} 1.36(3), 1.38(4), 1.40(6), 1.42(8)	F^- 1.31(4), 1.33(6)
3	Na^+ 0.99(4), 1.02(6), 1.18(8), 1.24(9), 1.39(12)	Mg^{2+} 0.57(4), 0.72(6), 0.89(8)	Al^{3+} 0.39(4), 0.48(5), 0.54(6)	Si^{4+} 0.26(4), 0.40(6)	P^{5+} 0.17(4), 0.29(5), 0.38(6)	S^{2-} 1.84(4); S^{6+} 0.12(4), 0.29(6)	Cl^- 1.81(6)
4	K^+ 1.38(6), 1.51(8), 1.55(9), 1.59(10), 1.64(12)	Ca^{2+} 1.00(6), 1.12(8), 1.18(9), 1.23(10), 1.34(12)	Ga^{3+} 0.47(4), 0.55(5), 0.62(6)	Ge^{4+} 0.39(4), 0.53(6)	As^{3+} 0.58(6); As^{5+} 0.34(4), 0.46(6)	Se^{2-} 1.98(6)	Br^- 1.96(6)
5	Rb^+ 1.52(6), 1.61(8), 1.66(10), 1.72(12)	Sr^{2+} 1.18(6), 1.26(8), 1.36(10), 1.44(12)	In^{3+} 0.62(4), 0.80(6), 0.92(8)	Sn^{4+} 0.69(6), 0.81(8)	Sb^{3+} 0.76(6); Sb^{5+} 0.60(6)	Te^{2-} 2.21(6)	I^- 2.20(6)
6	Cs^+ 1.67(6), 1.74(8), 1.81(10), 1.85(11), 1.88(12)	Ba^{2+} 1.35(6), 1.42(8), 1.47(9), 1.52(10), 1.61(12)	Bi^{3+} 0.96(5), 1.03(6), 1.17(8)	Pb^{2+} 1.19(6), 1.29(8), 1.35(9), 1.40(10)			

Transition elements

Row 4	Row 5	Row 6
Sc^{3+} 0.75(6), 0.87(8)	Y^{3+} 0.90(6), 1.02(8)	La^{3+} 1.03(6), 1.16(8), 1.22(9), 1.27(10)
Ti^{4+} 0.42(4), 0.61(6), 0.74(8)	Zr^{4+} 0.72(6), 0.78(7), 0.84(8), 0.89(9)	Hf^{4+} 0.71(6), 0.76(7), 0.83(8)
V^{5+} 0.36(4), 0.46(5), 0.54(6)	Nb^{5+} 0.64(6), 0.69(7), 0.74(8)	Ta^{5+} 0.64(6), 0.69(7), 0.74(8)
Cr^{3+} 0.62(6); Cr^{4+} 0.41(4), 0.55(6); Cr^{6+} 0.26(4)	Mo^{4+} 0.65(6); Mo^{6+} 0.41(4), 0.59(6)	W^{6+} 0.42(4), 0.51(5), 0.60(6)
Mn^{2+} 0.83(6), 0.96(8); Mn^{3+} 0.65(6); Mn^{4+} 0.53(6)		Re^{4+} 0.63(6); Re^{7+} 0.38(4), 0.53(6)
Fe^{2+} 0.63(4), 0.78(6), 0.92(8); Fe^{3+} 0.65(6), 0.78(8)		
Co^{2+} 0.74(6), 0.90(8)	Rh^{4+} 0.60(6)	
Ni^{2+} 0.55(4), 0.69(6)	Pd^{2+} 0.64(4), 0.86(6)	Pt^{2+} 0.80(6)
Cu^+ 0.46(2), 0.77(6); Cu^{2+} 0.57(4), 0.65(5), 0.73(6)	Ag^+ 1.15(6), 1.28(8)	
Zn^{2+} 0.60(4), 0.74(6), 0.90(8)	Cd^{2+} 0.58(4), 0.74(6), 0.90(8)	Hg^{2+} 0.96(4), 1.02(6), 1.14(8)

Th^{4+} 0.94(6), 1.05(8), 1.09(9), 1.13(10)	U^{4+} 0.89(6), 1.00(8); U^{6+} 0.52(4), 0.73(6)

*Numbers in parentheses are the coordination numbers of the ions. Radii in upright digits are from Shannon (1976). Radii in italics are from Pauling (1960), revised and supplemented by Ahrens (1952). For complete references see reference list at end of chapter.

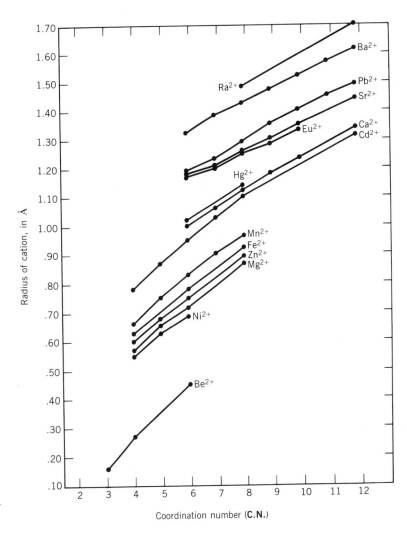

FIG. 4.16 The change in ionic radius as a function of coordination number (**C.N.**) for some selected cations (From Shannon, 1976; for complete reference see list at end of chapter.)

It is instructive to compare the metallic radii, given in Table 4.7, with the ionic radii for the same elements, given in Table 4.8. In all cases the ionic radius of cations is considerably smaller than the metallic radius, for the same element. This is the result of the loss of one or more outer electrons, and the resultant reduction in the overall size of the electron cloud. Anions, on the other hand, because they gain electrons, are larger than the corresponding neutral atom. In a specific crystal structure, the measured radius of a given element may be somewhere between the element's atomic and ionic radii because the bond type in that specific structure is a mixture of several bond types, such as ionic, covalent, and/or metallic (to be discussed in a subsequent section).

A regular change in ionic size is reflected by the arrangement of the elements in the periodic table. For elements of the same column, the ionic radii increase as the atomic number increases. For example, in Table 4.8, column II, the smallest ion is Be^{2+}, with a

radius of 0.16 to 0.45 Å. The last ion listed in column II is Ba^{2+}, with a radius ranging from 1.35 to 1.61 Å. While ionic radii generally increase with increasing atomic number, the trivalent ions of the lanthanide elements decrease in radius with increasing atomic number, from La^{3+} ($Z = 57$), with a radius of 1.16 Å (8-fold coordination), to Lu^{3+} ($Z = 71$), with a radius of 0.98 Å (8-fold coordination). This feature, known as the *lanthanide contraction*, is the result of building up inner electron orbitals before adding to a new outer orbital (see Table 4.5). As a result of the increasing nuclear charge, and the fairly weak "shielding" of this positive charge by inner-shell $4f$ electrons, an increased attraction is exerted on the outer-shell $5s$, $5p$, and $5d$ electrons, which draws them in more tightly, causing a decrease in ionic radius.

For positive ions with the same electronic structure, the radii decrease with increasing charge. For example, the ionic radii of the metallic elements in the third horizontal row, all of which have two electrons in the first shell and eight in the second shell, decrease (for 6-coordination) from Na^+, with a radius of approximately 1.02 Å, to P^{5+}, with a radius of 0.38 Å. The size of these ions with identical electron configurations decreases because the increased nuclear charge exerts a greater pull on the electrons, thus decreasing the effective radius of the ion.

For an element that can exist in several valence states (ions of the same element with different charges), the higher the charge on the positive ion, the smaller is its radius. For example, Mn^{2+} = 0.83 Å, Mn^{3+} = 0.65 Å, and Mn^{4+} = 0.53 Å. This decrease in size is due to the greater pull exerted by the nucleus on a reduced electron cloud.

In addition to variations in size as a function of coordination number and bond type, there can also be considerable change in the *shape* of some atoms and ions. Atoms and ions are not rigid bodies but respond to external electrical forces by dilation and deformation. A larger number of neighboring ions tends to distend the central ion as a function of increasing coordination number; a smaller number allows it to collapse a little. Some distortion of shape may accompany the distention of ions. These effects are collectively called *polarization* and are of considerable importance in crystal structures. If the apparent shape and size are strongly affected by its structural environment, the ion is said to have a high polarizability; if, on the other hand, it behaves essentially as a rigid sphere in all environments, it is said to have a low polarizability. Generally, large monovalent anions with a noble gas electronic structure are most easily polarized. The greater the polarization between two neighboring ions, the more the electron density is localized between the two nuclei and the more covalent is the nature of the bond between them. Figure 4.17 is a schematic illustration of polarization of a large monovalent anion by a highly charged smaller cation.

COORDINATION OF IONS

When oppositely charged ions unite to form a crystal structure in which the binding forces are dominantly electrostatic, each ion tends to gather to itself, or to *coordinate*, as many ions of opposite sign as size per-

mits. When the atoms are linked by simple electrostatic bonds, they may be regarded as approximately spherical in shape, and the resulting geometry is simple. The coordinated ions always cluster about the central coordinating ion in such a way that their centers lie at the apices of a polyhedron. Thus, in a stable crystal structure, each cation lies at the approximate center of a *coordination polyhedron* of anions. The number of anions in the polyhedron is the *coordination number* (**C.N.**) of the cation with respect to the given anion, and is determined by their relative sizes. Thus, in NaCl each Na^+ has six closest Cl^- neighbors and is said to be in 6-coordination with Cl (**C.N.** 6). In fluorite, CaF_2, each calcium ion is at the center of a coordination polyhedron consisting of eight fluorine ions and hence is in 8-coordination with respect to fluorine (**C.N.** 8). See Fig. 4.18 for illustrations of both structures.

Anions may also be regarded as occupying the centers of coordination polyhedra formed of cations. In NaCl each chloride ion has six sodium neighbors and hence is in 6-coordination with respect to sodium. Because both sodium and chlorine are in 6-coordination, there must be equal numbers of both, in agreement with the formula, NaCl. On the other hand, examination of the fluorite structure reveals that each fluorine ion has four closest calcium neighbors and hence is in 4-coordination with respect to calcium (**C.N.** 4). Although these four calcium ions do not touch each other, they form a definite coordination polyhedron about the central fluorine ion in such a way that the calcium ions lie at the apices of a regular tetrahedron (see Fig. 4.18*b*). Because each calcium ion has eight fluorine neighbors (**C.N.** 8), whereas each fluorine ion has only four calcium neighbors, it is obvious that there are twice as many fluorine as calcium ions in the structure. This accords with the formula CaF_2, and with the valences for calcium and fluorine.

It is easily seen that the relative sizes of the calcium and fluorine ions would permit a structure containing equal numbers of each with both ions in 8-coordination. The fact that in fluorite only half the possible calcium sites are filled calls attention to an important restriction of crystal structures; namely, *the total numbers of ions of all kinds in any stable ionic crystal structure must be such that the crystal as a whole is electrically neutral*. That is, the total number of positive charges must equal the total number of negative charges; hence, in fluorite there can be only half as many divalent positive calcium ions as there are monovalent fluorine ions.

Figure 4.18*c* illustrates the sphalerite, ZnS, structure in which Zn^{2+} is in tetrahedral coordination with S^{2-}.

FIG. 4.17 Polarization effect of a small, highly charged cation on a large anion.

Halite structure

(a)

Fluorite structure

(b)

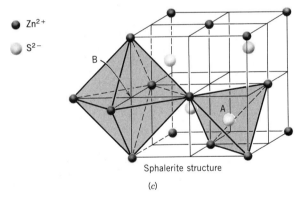

Sphalerite structure

(c)

FIG. 4.18 Visualization of coordination polyhedra in various structure types. *(a)* The halite, NaCl, structure ($F4/m\overline{3}2/m$ = $Fm3m$) with the ions arranged on a face-centered cubic lattice. Both Na^+ and Cl^- are in 6-coordination (**C.N.** 6) with each other. A coordination polyhedron about Na^+ is shown; it is an octahedron. *(b)* The fluorite, CaF_2, structure ($Fm3m$) with the Ca^{2+} ions arranged on a face-centered cubic lattice. F^- is coordinated to four Ca^{2+} (**C.N.** 4); this is tetrahedral in shape. Each Ca^{2+}, however, is coordinated to eight neighboring F^- (**C.N.** 8); this is cubic in shape. *(c)* The sphalerite, ZnS, structure ($F\overline{4}3m$), with Zn^{2+} ions on a face-centered cubic lattice. Each S^{2-} (in position marked *A*) is surrounded by four Zn^{2+} ions, in tetrahedral coordination (**C.N.** 4). The position marked *B* is empty and has octahedral surroundings.

Radius Ratio

Although each ion in a crystal affects every other ion to some extent, the strongest forces exist between ions that are nearest neighbors. These are said to constitute the *first coordination shell*. The geometric arrangement of this shell, and hence the coordination number, is a function of the relative sizes of the coordinating ions. However, we have already seen that the effective size of an ion is not constant but depends to varying degrees on the total number of closest neighboring ions, its coordination number (**C.N.**), its polarizability, and the type of bonding involved. *Because of these influences, it is unrealistic to regard ions and atoms as rigid spheres with well-established constant radii.* Nonetheless, the average sizes of ions are useful as a first approximation in the prediction of approximate interatomic distances in unknown structures. Furthermore, the ratio of the radius of a cation to that of an adjoining anion can be used reasonably successfully in predicting the number of closest neighbors.

The relative size of ions is generally expressed as a radius ratio: radius ratio = $R_A : R_X$, where R_A is the radius of the cation and R_X the radius of the anion, in angstrom units. The radius ratio of sodium and chlorine in halite, NaCl, is therefore:

$$R_{Na^+} = 1.02 \text{ Å } (\textbf{C.N. } 6) \qquad R_{Cl^-} = 1.81 \text{ Å } (\textbf{C.N. } 6)$$

$$R_{Na^+} : R_{Cl^-} = 1.02/1.81 = 0.56$$

The radius ratio of calcium and fluorine in fluorite, CaF_2, is

$$R_{Ca^{2+}} = 1.12 \text{ Å } (\textbf{C.N. } 8) \qquad R_{F^-} = 1.31 \text{ Å } (\textbf{C.N. } 4)$$

$$R_{Ca^{2+}} : R_{F^-} = 1.12/1.31 = 0.86$$

When two or more cations are present in a structure, coordinated with the same anion, separate radius ratios must be computed for each.

When coordinating and coordinated ions (or atoms) are the same size, the radius ratio is 1. Trial with a tray of identical spheres, such as Ping-Pong balls, reveals that spherical units may be arranged in three dimensions in either of two ways, called *hexagonal closest packing* (HCP) and *cubic closest packing* (CCP). In either arrangement, each sphere is in contact with 12 closest neighbors (**C.N.** 12). With reference to Fig. 4.19*a*, it is clear that a sphere can be surrounded by six spheres of equal size, all of which lie in the same plane and touch each other. This is referred to as a *hexagonal closest packed layer* because of the hexagonal arrangement of spheres (marked *A* in Fig. 4.19*a*). Between the spheres one can distinguish two

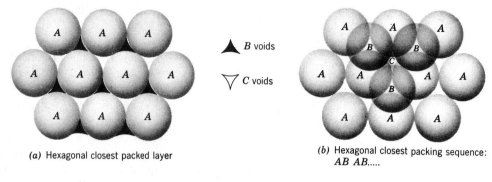

(a) Hexagonal closest packed layer

▲ B voids

▽ C voids

(b) Hexagonal closest packing sequence: *AB AB.....*

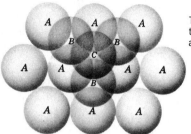

(c) Cubic closest packing sequence: *ABC ABC.....*

The plane of this drawing turns out to be perpendicular to the 3-fold axis at the corner of a cube

FIG. 4.19 Closest packing of like spheres.

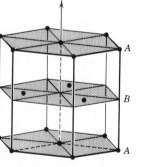

FIG. 4.20 (a) Two representations of a lattice based on hexagonal closest packing (HCP). (b) Two representations of a lattice based on cubic closest packing (CCP). Upon inspection of the location of nodes in this lattice it is found to be all-face-centered (F; or FCP).

[0001]

Hexagonal close-packed (HCP)

(a)

[111]

A C B
(111)

A

Cubic close-packed (CCP)

(b)

types of voids, B and C, on the basis of the orientation of their triangular shapes. In Fig. 4.19b a part of a second hexagonal closest packed layer is superimposed on top of the B voids. This sequence of stacking can be represented by the combination of letters AB. If a third layer is now put on top of the second layer but with the spheres resting in the dimples of the second layer above the A spheres in the first layer, we obtain a sequence ABA, which can be extended upward by another layer of spheres on top of the B voids, giving rise to an indefinite stacking sequence, ABABAB . . . , which is known as *hexagonal closest packing* (HCP). If, however, we choose to stack the third layer on top of the second layer of the AB sequence in the dimples that directly overlie the C voids in the first

layer, we form a three-layer sequence ABC, as in Fig. 4.19c. We can continue this type of stacking sequence indefinitely, resulting in an ABCABCABC . . . sequence, or *cubic closest packing* (CCP). Figure 4.20 illustrates the hexagonal lattice that underlies the HCP arrangement and the cubic lattice that underlies the CCP arrangement. The space group based on the HCP stacking would be $P6_3/m2/m2/c$, and the space group compatible with CCP stacking is $F4/m\overline{3}2/m$ (the faces of the cubic lattice are populated by spheres, as expressed by F). Twelve-fold coordination is rare in minerals, with the exception of the native metals; they, and many alloys, have structures based on HCP or CCP. Some two-dimensional examples of hexagonal closest packing are given in Fig. 4.21.

FIG. 4.21 Examples of hexagonal closest packing (HCP). *(a)* The hexagonal closest packing of C_{60} molecules ("buckminsterfullerene") as revealed by high-resolution transmission electron microscopy (HRTEM). These complex carbon molecules are of synthetic origin and have not been found in nature. Scale is in nanometers (nm): 1 nm = 10 Å. (For further discussion, see Curl, R. F. and Smalley, R. E., 1991, Fullerenes. *Scientific American*, v. 265, pp. 54–63.) *(b)* A computer calculated image that replicates the observed structure image in *(a)*. (*a* and *b* from Wang, S. and Buseck, P. R., 1991, Packing of C_{60} molecules and related fullerenes in crystals: A direct view. *Chemical Physics Letters*, v. 182, pp. 1–4. *(c)* Illustration of the structure of C_{60}, "buckminsterfullerene," named after the American engineer and philosopher R. Buckminster Fuller because of his invention of the geodesic dome which underlies the structure of C_{60}. *(d)* Hexagonal closest packing on a very large scale, as seen in everyday life when steel, concrete, or plastic pipes are stacked.

When the coordinating cation is somewhat smaller than the anions, 8-coordination may result. This is also called *cubic coordination* because the centers of the anions lie at the eight corners of a cube (Fig. 4.22a). If we consider a cubic coordination polyhedron in which the anions (considered as rigid spheres) touch each other as well as the central cation (also considered a rigid sphere), we may compute the limiting value of radius ratio for **C.N.** = 8. Allowing the radius of the anion to equal unity, the radius of the

FIG. 4.22 *(a)* Cubic or 8-coordination of X ions about an A ion. $R_A : R_X > 0.732$. *(b)* Limiting condition for cubic coordination. *(c)* Octahedral or 6-coordination of X ions about an A ion. $R_A : R_X = 0.732-0.414$. *(d)* Limiting condition for octahedral coordination. *(e)* Tetrahedral or 4-coordination of X ions about an A ion. $R_A : R_X = 0.414-0.255$. *(f)* Limiting condition for tetrahedral coordination. *(g)* Triangular or 3-coordination of X ions about an A ion. $R_A : R_X = 0.225-0.155$. *(h)* Limiting condition for triangular coordination. *(i)* Linear or 2-coordination of X ions about an A ion. $R_A : R_X < 0.155$.

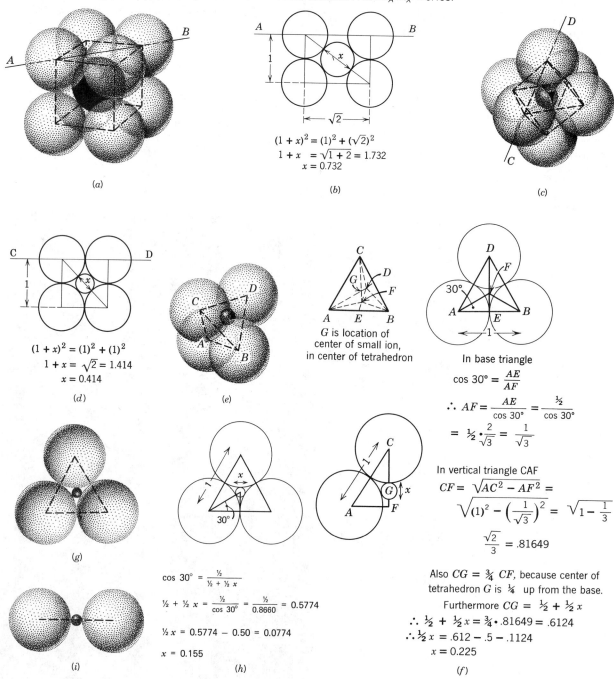

$$(1 + x)^2 = (1)^2 + (\sqrt{2})^2$$
$$1 + x = \sqrt{1 + 2} = 1.732$$
$$x = 0.732$$

(a) *(b)* *(c)*

$$(1 + x)^2 = (1)^2 + (1)^2$$
$$1 + x = \sqrt{2} = 1.414$$
$$x = 0.414$$

(d) *(e)*

G is location of center of small ion, in center of tetrahedron

In base triangle
$$\cos 30° = \frac{AE}{AF}$$
$$\therefore AF = \frac{AE}{\cos 30°} = \frac{½}{\cos 30°}$$
$$= ½ \cdot \frac{2}{\sqrt{3}} = \frac{1}{\sqrt{3}}$$

In vertical triangle CAF
$$CF = \sqrt{AC^2 - AF^2} =$$
$$\sqrt{(1)^2 - \left(\frac{1}{\sqrt{3}}\right)^2} = \sqrt{1 - \frac{1}{3}}$$
$$\frac{\sqrt{2}}{3} = .81649$$

Also $CG = \frac{3}{4} CF$, because center of tetrahedron G is ¼ up from the base.
Furthermore $CG = ½ + ½ x$
$$\therefore ½ + ½ x = \frac{3}{4} \cdot .81649 = .6124$$
$$\therefore ½ x = .612 - .5 - .1124$$
$$x = 0.225$$

(g)

$$\cos 30° = \frac{½}{½ + ½ x}$$
$$½ + ½ x = \frac{½}{\cos 30°} = \frac{½}{0.8660} = 0.5774$$
$$½ x = 0.5774 - 0.50 = 0.0774$$
$$x = 0.155$$

(i) *(h)* *(f)*

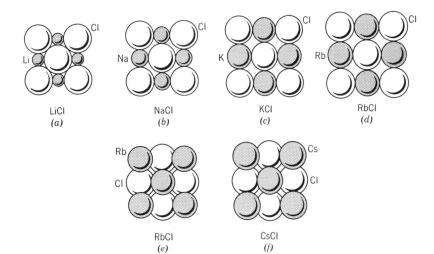

FIG. 4.23 Change in coordination from 6 to 8 in alkali chlorides. *(a), (b), (c),* and *(d)* have sodium chloride type structures with 6-coordination; *(e)* and *(f)* have cesium chloride type structures with 8-coordination.

cation for this limiting condition must be 0.732 (Fig. 4.22*b*). Hence, cubic coordination has a calculated maximum stability for radius ratios between 0.732 and 1.000.

For values of radius ratio less than 0.732, 8-coordination is not as stable as 6-coordination, in which the centers of the coordinated ions lie at the apices of a regular octahedron. Six-fold coordination is accordingly called *octahedral coordination* (Fig. 4.22*c*). We may, as before, calculate the limiting value of radius ratio for the condition in which the six coordinated anions touch each other and the central cation. The lower limit of radius ratio for stable 6-coordination is found to be 0.414 (Fig. 4.22*d*). Hence we may expect 6 to be the common coordination number when the radius ratio lies between 0.732 and 0.414. Na and Cl in halite, Ca and oxygen in calcite, the *B*-type cations in spinel, and many cations in silicates are examples of 6-coordination.

Figure 4.23 illustrates the change from 6- to 8-coordination in the alkali chlorides with increasing ionic radius of the cation. It is interesting to note that rubidium may be in either 6- or 8-coordination, and thus rubidium chloride is polymorphous. Li^+, Na^+, and K^+ have radius ratios with Cl^- between 0.414 and 0.732, and therefore go into 6-coordination with Cl^-. Cs^+ has a radius ratio with Cl^- between 0.732 and 1.00, and thus goes into 8-coordination with Cl^-. Rb^+ has a radius ratio with Cl^- close to 0.732, and therefore can have either 6- or 8-coordination with Cl^-; this is the origin of its polymorphism.

For values of radius ratio less than 0.414, 6-coordination is not as stable as 4-coordination, in which the centers of the coordinated ions lie at the apices of a regular tetrahedron. Four-fold coordination is therefore called *tetrahedral coordination* (Fig. 4.22*e*). Calculating the limiting value of radius ratio

for the condition in which four coordinated anions touch each other and the central ion, we find it to be 0.225 (Fig. 4.22*f*). Therefore, we may expect 4 to be the common coordination number when the radius ratio lies between 0.414 and 0.225. Tetrahedral coordination is typified by the SiO_4 group in silicates, by the *A*-type ion in spinel, and by the sphalerite (ZnS) and diamond structures.

Triangular or 3-coordination is stable between limits of 0.225 and 0.155 (see Figs. 4.22*g* and *h*) and is common in nature in CO_3, NO_3, and BO_3 groups.

Linear or 2-coordination (see Fig. 4.22*i*) is very rare in ionic bonded crystals. Examples are the uranyl group $(UO_2)^{2+}$, the nitrite group $(NO_2)^{-2}$, and copper with respect to oxygen in cuprite, Cu_2O. Figure 4.24 illustrates the various regular coordination polyhedra as a function of radius ratio and coordination number.

Examples of 5-, 7-, 9-, and 10-coordination are also known. Such coordination numbers are possible only in complex structures in which the anions are not closely packed.

The coordination polyhedra in experimentally determined structures are almost always distorted. The smaller and more strongly polarizing the coordinating cation, or the larger and more polarizable the anion, the greater is the distortion and the wider the departure from the theoretical radius ratio limits. Also, *if the bonding mechanism is not dominantly ionic, radius ratio considerations may not be safely used to determine the coordination number.*

Obviously, every ion in a crystal structure has some effect on every other ion—it is attractive if the charges are opposite, repulsive if the same. Hence, ions tend to group themselves in crystal structures in such a way that cations are as far apart as possible yet consistent with the coordination of the anions that will

Minimum Radius Ratio $R_A : R_X$	Coordination Number C. N.	Packing Geometry		
< 0.155	2	Linear		
0.155	3	Corners of an equilateral triangle (triangular coordination)		
0.255	4	Corners of a tetrahedron (tetrahedral coordination)		
0.414	6	Corners of an octahedron (octahedral coordination)		
0.732	8	Corners of a cube (cubic coordination)		
1.0	12	Corners of a cuboctahedron (close packing)		

FIG. 4.24 Atomic packing schemes.

result in electrical neutrality. Thus, when cations share anions between them, they do so in such a way as to place themselves as far apart as possible. Hence, the coordination polyhedra formed around each are linked commonly through corners and edges but generally not through faces (Fig. 4.25). Cations tend to share as small a number of anions as possible, and

sharing of as many as three or four anions is relatively rare.

At this stage, it may be instructive to list the most common ionic charges, the most common coordination polyhedra, and ionic sizes of the 11 most common elements in the Earth's crust. This information is given in Table 4.9. Knowledge of this listing will be

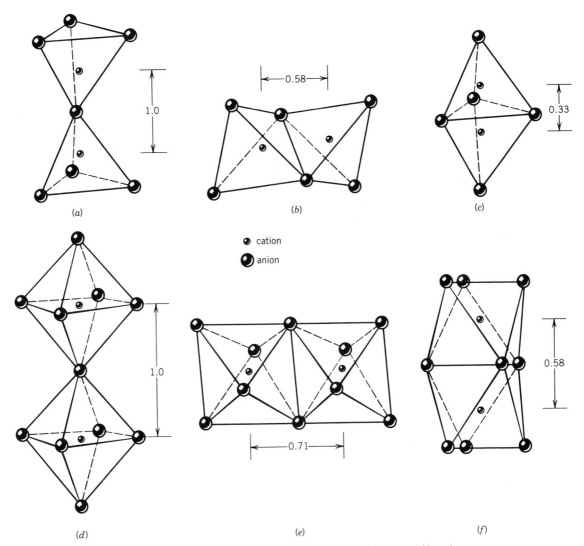

FIG. 4.25 *(a)* Tetrahedrons sharing corners, as is commonly found in crystal structures. Here the cation-cation distance is given as 1.0. *(b)* Tetrahedrons sharing edges; very uncommon. The cation-cation distance is reduced from 1.0 to 0.58. In this case cation-cation repulsion would occur, causing severe polyhedral distortion. *(c)* Tetrahedrons sharing faces; this is never found when both tetrahedra are occupied by a cation. The cation-cation distance is reduced to 0.33 from 1.0 (in *a*) and 0.58 (in *b*). *(d)* Octahedrons sharing corners, as is common in crystal structures. The cation-cation distance is given as 1.0. *(e)* Octahedrons sharing edges is also common. The cation-cation distance is reduced to 0.71 (but the cations are still considerably farther apart than in the case of tetrahedrons sharing edges; see *b*). *(f)* Octahedrons sharing faces; this is not uncommon in crystal structures. The cation-cation distance is reduced to 0.58 from 1.0 (in *d*) and 0.71 (in *f*). Face-sharing of octahedrons is possible because the cation-cation distance is larger for octahedrons than tetrahedrons. Furthermore, cations in octahedral coordination tend to have lower charges (e.g., Mg^{2+}, Fe^{2+}) than cations in tetrahedral coordination (e.g., Si^{4+}, Al^{3+}); such, the repulsive force between cations inside octahedrons is generally less than between cations inside tetrahedrons.

most helpful in the subsequent discussion of silicate and oxide structures, as well as in understanding aspects of solid solution (Chapter 5).

Pauling's Rules

Every stable crystal structure bears witness to the operation of some broad generalizations that determine

the structure of solid matter. These principles were enunciated in 1929 by Linus Pauling in the form of the following five rules:

Rule 1. A coordination polyhedron of anions is formed about each cation, the cation-anion distance being determined by the radius sum and the coordination number (i.e., number of nearest neighbors) of

Table 4.9

SOME COMMON ELEMENTS (EXCLUSIVE OF HYDROGEN) THAT OCCUR IN ROCK-FORMING MINERALS, ARRANGED IN DECREASING IONIC SIZE

Ion	Coordination Number with Oxygen		Ionic Radius Å
O^{2-}			1.36 (3)
K^+	8–12		1.51 (8)–1.64 (12)
Na^+	8–6	cubic to	1.18 (8)–1.02 (6)
Ca^{2+}	8–6	octahedral	1.12 (8)–1.00 (6)
Mn^{2+}	6		0.83 (6)
Fe^{2+}	6		0.78 (6)
Mg^{2+}	6	octahedral	0.72 (6)
Fe^{3+}	6		0.65 (6)
Ti^{4+}	6		0.61 (6)
Al^{3+}	6		0.54 (6)
Al^{3+}	4		0.39 (4)
Si^{4+}	4	tetrahedral	0.26 (4)
P^{5+}	4		0.17 (4)
S^{6+}	4		0.12 (4)
C^{4+}	3	triangular	−0.08 (3)

*The first column lists the most common ionic (valency) states of the elements. The second column lists their most common coordination with respect to oxygen, and the third column lists ionic sizes for specific coordinations (the number in parentheses is **C.N.**). A complete listing of elemental abundances is given in Table 5.1.

the cation by the radius ratio. This was discussed in the previous section.

Rule 2. The electrostatic valency principle. In a stable crystal structure the total strength of the valency bonds that reach an anion from all the neighboring cations is equal to the charge of the anion. This needs some further examination. The strength of an electrostatic bond (e.v.) may be defined as an ion's valence charge (z) divided by its coordination number (n): e.v. = z/n, expressed in absolute values. The resulting number, called the *electrostatic valency* (e.v.), is a measure of the strength of any of the bonds reaching the coordinating ion from its nearest neighbors. For instance, in NaCl, the Cl^- is surrounded by six Na^+ neighbors in octahedral coordination, and each of the bonds reaching Na^+ has a strength (e.v.) of $\frac{1}{6}$. This means that six bonds between a central Na^+ and six octahedrally coordinated closest neighbors of Cl^- completely and exactly neutralize the charge on the central Na^+ (see Fig. 4.26a). The Cl^- also has six neighbors (Na^+), so that the e.v. for each of the bonds reaching Cl^- is $\frac{1}{6}$. As such, the charge on the Cl^- is neutralized by six bonds of $\frac{1}{6}$ reaching this central ion from six (octahedrally coordinated) Na^+ ions. Figures 4.26b and c give examples of electrostatic valencies in additional coordination polyhedra. The electrostatic valency rule is very helpful in the evaluation of

the polyhedral nature of crystal structures. In a stable structure the sum of the electrostatic valencies from cations in coordination polyhedra exactly balances the charge on the anion that is shared among these coordination polyhedra. For example, in the structure of grossular garnet ($Ca_3Al_2Si_3O_{12}$), Ca^{2+} is in cubic coordination (**C.N.** = 8), Al^{3+} is in octahedral coordination (**C.N.** = 6), and Si^{4+} is in tetrahedral coordination (**C.N.** = 4). Their bonds, therefore, have

FIG. 4.26 Illustration of the neutralization of a central ion by bonds from the nearest neighbors. Each of these bonds has an electrostatic valency (e.v.). The total of all bonds with specific e.v.'s neutralizes the central ion. *(a)* Octahedral coordination in the halite structure (see Fig. 4.18a). *(b)* Cubic coordination of F^{-1} around Ca^{2+} in the fluorite structure (see Fig. 4.18b). *(c)* Tetrahedral coordination of Ca^{2+} around F^- in the fluorite structure (see Fig. 4.18b).

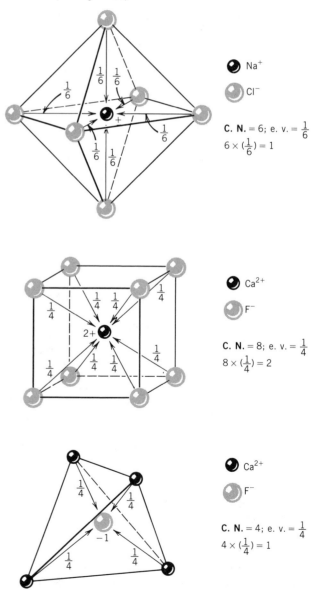

electrostatic valencies of $\frac{2}{8} = \frac{1}{4}$, $\frac{3}{6} = \frac{1}{2}$, and $\frac{4}{4} = 1$, respectively. In order to satisfy (neutralize) the 2^- charge on a shared oxygen, that oxygen ion must belong to *two* cubic Ca^{2+} polyhedra, *one* Al^{3+} octahedron, and *one* Si^{4+} tetrahedron ($2 \times \frac{1}{4} + \frac{1}{2} + 1 = 2$). This is the fundamental linkage of coordination polyhedra in grossular garnet.

Crystals in which all bonds are of equal strength are called *isodesmic*. This generalization is so simple that it seems trivial, but in some cases unexpected results emerge from calculation of electrostatic valencies. For example, minerals of the spinel group have formulas of the type AB_2O_4, where A is a divalent cation such as Mg^{2+} or Fe^{2+} and B is a trivalent cation such as Al^{3+} or Fe^{3+}. Such compounds have been called aluminates and ferrates, by analogy with such compounds as borates and oxalates. This nomenclature suggests that ionic clusters or radicals are present in the structure. X-ray data reveal that the A ions are in 4-coordination, whereas the B ions are in 6-coordination. Hence, for the A ions, e.v. $= \frac{2}{4} = \frac{1}{2}$, and for B ions, e.v. $= \frac{3}{6} = \frac{1}{2}$. All bonds are the same strength, and such crystals are isodesmic multiple oxides.

When small, highly charged cations coordinate larger and less strongly charged anions, compact, firmly bonded groups result, as in the carbonates and nitrates. If the strength of the bonds within such groupings is computed, the numerical value of the electrostatic valency is always greater than one-half the total charge on the anion. This means that in such groups, the anions are more strongly bonded to the central coordinating cation than they can possibly be bonded to any other ion. For example, in the carbonate group, C^{4+} is in 3-coordination with O^{2-}, and hence we may compute the e.v. $= \frac{4}{3} = 1\frac{1}{3}$. This is greater than one-half the charge on the oxygen ion, and hence a functional group, or radical, exists. This is the carbonate triangle, the basic structural unit of carbonate minerals (see Fig. 4.27a). Another example is the sulfate group. O^{2-} is in 4-coordination with S^{6+}; hence the e.v. $= \frac{6}{4} = 1\frac{1}{2}$. Because this is greater than one-half the charge on the oxygen ion, the sulfate radical forms a tightly knit group, and oxygen is more strongly bonded to sulfur than it can be bound to any other ion in the structure. This is the tetrahedral unit (see Fig. 4.27b) that is the fundamental basis of the structure of all sulfates. Compounds such as sulfates and carbonates are said to be *anisodesmic*.

Of course, it must be understood that if the tightly bonded groups are regarded as single structural units, then in a compound such as $CaCO_3$, calcite, all Ca to $(CO_3)^{2-}$ group bonds are of equal strength and resemble those of an isodesmic crystal. Similarly, in simple sulfates like barite, all Ba to $(SO_4)^{2-}$ group bonds

Triangular $(CO_3)^{2-}$ group
C. N. = 3; e. v. $= \frac{4}{3} = 1\frac{1}{3}$

e. v. $= 1\frac{1}{3}$

Charge remaining on each oxygen $= -\frac{2}{3}$

(a)

Tetrahedral $(SO_4)^{2-}$ group
C. N. = 4; e. v. $= \frac{6}{4} = 1\frac{1}{2}$

e. v. $= 1\frac{1}{2}$

Charge remaining on each oxygen $= -\frac{1}{2}$

(b)

FIG. 4.27 Examples of polyhedral anionic groupings about a central cation, in which the electrostatic valence that the central cation contributes to each coordinating anion is greater than half the anion's charge. Therefore, the anion is more tightly bonded to the central cation than it can be to any other cation in the structure. These discrete bonding units (or complex ions) occur in what are known as anisodesmic structures.

are equal in strength. The crystals are, however, called anisodesmic because of the presence of the strong C–O and S–O bonds in addition to the weaker Ca to $(CO_3)^{2-}$ group and Ba to $(SO_4)^{2-}$ group bonds.

Logic requires yet another case: that in which the strength of bonds joining the central coordinating cation to its coordinated anions equals exactly half the charge of the anion. In this case each anion may be bonded to some other unit in the structure just as strongly as it is to the coordinating cation. The other unit may be an identical cation, and the anion shared between two cations may enter into the coordination polyhedra of both. Let us consider the case of silicon-oxygen groupings, in which the oxygens are in tetrahedral coordination about the central Si^{4+}. The e.v. of the bonds between oxygen and Si^{4+} is $\frac{4}{4} = 1$. This is half the bonding strength of the oxygen ion. Consequently, an SiO_4 tetrahedron may link to some other ion just as strongly as to the central Si^{4+} ion. If this ion is another Si^{4+}, two tetrahedra may combine, linked through a common oxygen to form a single $(Si_2O_7)^{6-}$ group. In similar fashion, SiO_4 tetrahedra may join, or *polymerize*, to form chains, sheets, or networks by sharing oxygen ions. Such crystals are called *mesodesmic*. The most important example is the silicates.

All ionic crystals may be classified on the basis of the relative strengths of their bonds into isodesmic and anisodesmic crystals.

Rule 3. The existence of edges, and particularly of faces, common to two anion polyhedra in a coordinated structure decreases its stability (see Fig. 4.25). This effect is large for cations with high valency and small coordination number and is especially large when the radius ratio approaches the lower limit of stability of the polyhedron.

Rule 4. In a crystal containing different cations, those of high valency and small coordination number tend *not* to share polyhedral elements with each other, and when they do, the shared edges contract (to put more negative charge between the cations), and the cations are displaced from their polyhedral centers, away from the shared edge or face (to minimize cation-cation repulsion).

Rules 1 through 4 are all designed to maximize cation-anion attractions and to minimize anion-anion and cation-cation repulsion.

Rule 5. The principle of parsimony. The number of essentially different kinds of constituents in a crystal tends to be small, because, characterisically, there are only a few types of contrasting cation and anion sites. Thus, in structures with complex compositions, a number of different ions may occupy the same structural position (site). These ions must be considered as a single "constituent."

This can be illustrated with respect to the polyhedral representation of the amphibole structure (see Fig. 4.28), where the number of different crystallographic sites is limited but the range of chemical constituents in the amphibole composition is large. Figure 4.28 shows the presence of tetrahedral *T* sites (labeled *T1* and *T2*), octahedral *M* sites (labeled *M1*, *M2*, and *M3*, because of slightly different octahedral size and

FIG. 4.28 Polyhedral representation of the structure of monoclinic amphibole (*C2/m*) projected down the *a* axis. Polyhedral sites are marked as follows: *T* = tetrahedral; *M1*, *M2*, and *M3* = regular octahedral; *M4* = distorted cubic; *A* is irregular coordination to 10 to 12 neighboring oxygens and (OH); (OH) notes the specific location of hydroxyl in the formula. (After Papike, J. J. et al., 1969, *Mineralogical Society of America, Special Paper*, no. 2, p. 120.)

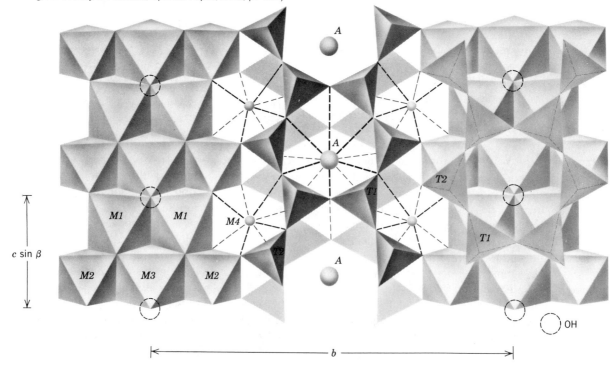

shape), an *M4* site (with irregular 8-coordination), a very large and irregularly coordinated *A* site, and the location of $(OH)^-$ groups. The tetrahedral sites will accommodate Si^{4+} and Al^{3+}. The *M1*, *M2*, and *M3* sites house Mg^{2+}, Fe^{2+}, Mn^{2+}, Al^{3+}, Fe^{3+}, and Ti^{4+}. The *M4* sites can be the locations for Mg^{2+}, Mn^{2+}, Fe^{2+}, Ca^{2+}, and Na^{2+}. The *A* site may be occupied by Na^+ or K^+, and the $(OH)^-$ location also houses Cl^- and F^-. As such, as many as 13 different ions are distributed among five distinctly different crystallographic sites (tetrahedral, regular octahedral, distorted cubic, the very large *A* site, and the (OH) location).

BONDING FORCES IN CRYSTALS

The forces that bind together the atoms (or ions, or ionic groups) of crystalline solids are electrical in nature. Their type and intensity are largely responsible for the physical and chemical properties of minerals. Hardness, cleavage, fusibility, electrical and thermal conductivity, compressibility, and the coefficient of thermal expansion are directly related to binding forces. In general, the stronger the average bond, the harder is the crystal, the higher its melting point, and the smaller its coefficient of thermal expansion. The great hardness of a diamond is attributed to the very strong electrical forces linking its constituent carbon atoms. The minerals periclase, MgO, and halite, NaCl, are isostructural, yet periclase melts at 2820°C, whereas halite melts at 801°C. The greater amount of heat energy required to separate the atoms in periclase indicates that it has a stronger electrical bond than halite.

These electrical forces are chemical bonds and can be described as belonging to one of five principal bond types: ionic, covalent, metallic, van der Waals, and hydrogen bonding. It should be understood that this classification is one of expediency and that transitions, or hybrids, may exist between all types. The electrical interaction of the ions or atoms constituting the structural units determines the properties of the resulting crystal. It is the similarity in properties among crystals having similar types of electrical interaction that justifies the use of the classification of bonding mechanisms.

The bonding forces linking the atoms of silicon and oxygen in quartz display in almost equal amount the characteristics of the ionic and the covalent bond. The strong bonding between Si^{4+} and O^{2-} ions in SiO_2 causes the electron densities of the two ions to be localized between the nuclei of Si and O. Such distortions of the electron clouds from generally

FIG. 4.29 A map of the electron density distribution of an Si–O–Si bond in coesite, one of the high-pressure polymorphs of SiO_2. This map displays the difference between the total electron density and the electron density prior to formation of the bond. The solid contours represent positive electron density, the dotted contours negative density, and the long dashed contour is the zero density contour. The contour interval is 0.1 electrons per $Å^3$. The distance between O and Si is 1.61 Å. (From Geisinger, K. L. and Gibbs, G. V., 1983. An X-ray diffraction study of the electron distribution in coesite. *Geological Society of America, Abstracts with Programs*, v. 15, p. 580.)

spherical to more ellipsoidal shapes have been determined directly from X-ray diffraction intensity measurements. Figure 4.29 shows a considerable build-up of charge density in the Si–O bond of coesite (SiO_2), with the strongest charge concentration in the vicinity of the more electronegative oxygen atoms. Another example of a crystal displaying two bond types is galena, PbS. This mineral exhibits characteristics related to the metallic bond (Pb–Pb), as expressed by good electrical conductivity and bright metallic luster, and of the ionic bond (Pb–S), as shown by excellent cleavage and brittleness. Furthermore, many minerals (such as mica) contain two or more bond types of different character and strength.

Ionic Bond

An assessment of the chemical activity of elements in terms of the occupancy of their outer orbitals with valence electrons leads to the conclusion that all atoms have a strong tendency to achieve an inert gas configuration with a completely filled valence shell. The noble gases—helium, neon, argon, krypton, and xenon—have completely filled valence shells and are almost completely inert (see Table 4.5). An ionic bond is achieved when one or more electrons in the valence shell of an atom are transferred to the valence shell of another, so that they both achieve an inert gas configuration. Sodium, for example, has a single valence electron in its outer orbital which it loses readily, leaving the atom with an unbalanced positive charge and the noble gas configuration of neon. Atomic chlorine, on the other hand, needs to acquire one electron to achieve the noble gas structure of argon (see Tables 4.5 and 4.6).

The above electron losses and gains can be stated as

$$Na(1s^2 2s^2 2p^6 3s^1) \rightarrow Na^+(1s^2 2s^2 2p^6) + e^-$$

$$Cl(1s^2 2s^2 2p^6 3s^2 3p^5) + e^- \rightarrow Cl^-(1s^2 2s^2 2p^6 3s^2 3p^6)$$

The electron lost by the sodium is picked up by the chlorine. Once formed, the Na^+ and Cl^- attract each other because of their opposite charges. *The attraction between oppositely charged ions constitutes the ionic (or electrostatic) bond* (see Fig. 4.30). The formation of this bond is the result of *the exchange of electron(s) of the metal atom (forming a cation) to the nonmetal atom (forming an anion)*.

In a crystal of sodium chloride characteristic properties may be recognized: cubic crystal habit, cleavage, specific gravity, index of refraction, and so forth. These properties in no way resemble those of the shining metal (Na) or the greenish, acrid gas (Cl_2), which are the elemental constituents of the substance. Touching the crystal to the tongue yields the taste of the solution. In other words, the properties conveyed into the crystal by its constituent elements are the properties of the *ions*, not the elements.

Physically, ionically bonded crystals are generally of moderate hardness and specific gravity, have fairly high melting points, and are poor conductors of electricity and heat. The lack of electrical conductivity in the ionic bonding of crystals is due to the stability of the ions, which neither gain nor lose electrons easily. Because the electrostatic charge constituting the ionic bond is evenly spread over the ion, a cation will tend to surround itself with as many anions as can be fitted around it. This means that the ionic bond is non-directional, and the symmetry of the resultant crystals is generally high (see Fig. 4.30 and Table 4.11 on page 208).

The strength of the ionic bond, u, depends on two factors: (1) the center-to-center spacing between the ions (r), and (2) the product of their charges (q):

$$u = (A q_1 q_2)/r$$

where A is a numerical quantity, the *Madelung constant*. This expression is similar to that for the Coulombic force (F) operating over a distance d, where F is defined by the equation on page 186 (see also Fig. 4.14b).

The effect of increased interionic distance on the strength of the ionic bond is readily seen when we compare the halides of sodium. Figure 4.31a shows the melting points and interionic distances for these compounds. It is apparent that the strength of the bond, as measured by the melting temperature, is inversely proportional to the bond length. The melting temperatures of the fluorides of the alkali metals illustrate that it does not matter whether the size of the anion or cation is varied: the shorter the bond length, the stronger is the bond (see Fig. 4.31c). LiF is an exception to this generalization, which is explained by anion-anion (F–F) repulsion in a structure having a very small cation.

The charge of the coordinated ions has an even more powerful effect on the strength of the bond. A comparison of the absolute values of the melting temperatures for the alkali-earth oxides (Fig. 4.31b), which are divalent compounds ($q = 2$), with the absolute values for the monovalent alkali fluorides ($q = 1$), in which the interionic spacings are closely comparable (Fig. 4.31c), reveals the magnitude of the effect. Although the interionic distance is almost the same for corresponding oxides and fluorides, the bonds uniting the more highly charged ions are much stronger. Figure 4.32 illustrates the effect of interionic spacing and charge on hardness. All substances cited as examples in these two figures have the same structure and may be regarded as ionically bonded.

Covalent Bond

We have seen that ions of chlorine may enter into ionically bonded crystals as stable units of structure because, by taking an electron from a metal such as

FIG. 4.30 An idealized ionic structure in two dimensions.

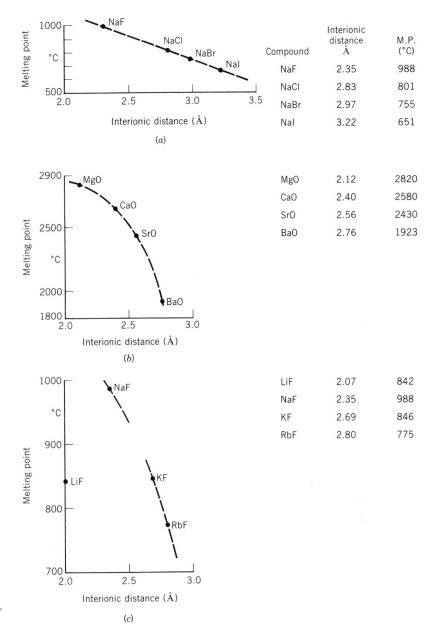

Compound	Interionic distance Å	M.P. (°C)
NaF	2.35	988
NaCl	2.83	801
NaBr	2.97	755
NaI	3.22	651

(a)

MgO	2.12	2820
CaO	2.40	2580
SrO	2.56	2430
BaO	2.76	1923

(b)

LiF	2.07	842
NaF	2.35	988
KF	2.69	846
RbF	2.80	775

(c)

FIG. 4.31 Melting point versus interionic distance in ionic-bonded compounds. (Data from *Handbook of Chemistry and Physics*, 72nd ed. CRC Press, Boca Raton, Fla., 1991.)

Na, they achieve a filled outer orbital. A single atom of chlorine with an incomplete valence orbital is in a highly reactive condition. It seizes upon and combines with almost anything in its neighborhood. Generally, its nearest neighbor is another chlorine atom, and the two unite in such a way that two electrons, one from each chlorine atom, do double duty in the outer orbitals of both atoms, and both thus achieve the stable inert gas configuration. As a result of this sharing of an electron, the two atoms of chlorine are strongly bound together.

The bonding mechanism between two chlorine atoms can be symbolically stated as follows:

$$\cdot \ddot{\underset{\cdot\cdot}{Cl}} \colon + \cdot \ddot{\underset{\cdot\cdot}{Cl}} \colon \rightarrow \colon \ddot{\underset{\cdot\cdot}{Cl}} \colon \ddot{\underset{\cdot\cdot}{Cl}} \colon$$

Here the small dots represent the outer-shell (valence) electrons in the s and p orbitals of the M shell in chlorine (this notation is known as Lewis symbols, after the American chemist Gilbert N. Lewis, 1875–1946). The valency shell is completed by the sharing of electrons between the two Cl on the right side of the equation. A schematic representation of this type of covalent bonding is shown in Fig. 4.33.

The number of covalent bonds that an atom may form can commonly be predicted by counting the number of electrons required to achieve a stable elec-

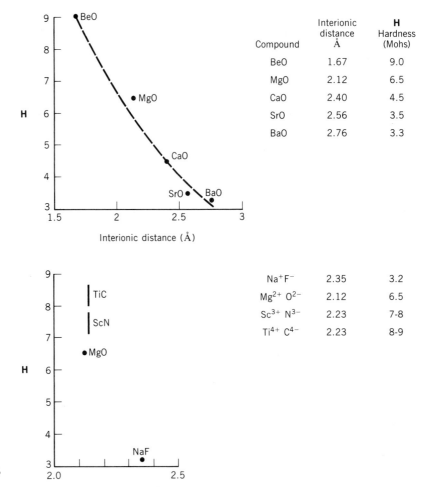

Compound	Interionic distance Å	**H** Hardness (Mohs)
BeO	1.67	9.0
MgO	2.12	6.5
CaO	2.40	4.5
SrO	2.56	3.5
BaO	2.76	3.3

Na^+F^-	2.35	3.2
$Mg^{2+}O^{2-}$	2.12	6.5
$Sc^{3+}N^{3-}$	2.23	7-8
$Ti^{4+}C^{4-}$	2.23	8-9

FIG. 4.32 Hardness versus interionic distance and charge in ionic-bonded compounds. (Data from Evans, R. C., 1952, *Crystal Chemistry.* Cambridge University Press, London.)

tron configuration (such as that of a noble gas). Carbon, for example, has four electrons in its valence shell; through sharing four additional electrons, it achieves the noble gas configuration of neon.

This *electron-sharing* or *covalent* bond is the strongest of the chemical bonds. Minerals so bonded are characterized by general insolubility, great stability, and very high melting points. They do not yield ions in the dissolved state and are nonconductors of electricity both in the solid state and in solution. Because the electrical forces constituting the bond are sharply localized in the vicinity of the shared electron, the bond is highly directional, and the symmetry of the resulting crystals is likely to be lower than where ionic bonding occurs (see Table 4.11). In chlorine, the bonding energy of the atom is entirely consumed in linking to one neighbor, and stable Cl_2 molecules result that show little tendency to link one to another. Certain other elements—in general, those near the middle of the periodic table, such as carbon, silicon,

aluminum, and sulfur—have two, three, and four vacancies in their outer orbitals. They therefore can form up to four covalent bonds with neighboring atoms. In some cases, these are multiple shared pairs of electrons between two atomic centers. This may result in very stable groups, which in turn may link together to form larger aggregates or groups.

Carbon is an outstanding example of such an atom. The four valence electrons in each carbon are sufficient to fill the bonding orbitals by electron sharing with four other carbon atoms, forming a very stable, firmly bonded configuration having the shape of

FIG. 4.33 Schematic representation of the electron distribution between two covalently bonded atoms.

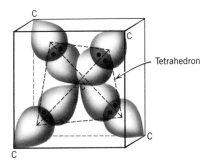

FIG. 4.34 Schematic representation of the overlap of orbitals in a C atom (at the center of the cube) with similar orbitals from four carbon atoms at the corners. This represents the covalent bonding in diamond.

a tetrahedron with a central carbon atom bonded to four others at the apices (see Fig. 4.34). Every carbon atom is linked in this way to four others, forming a continuous network. The energy of the bonds is strongly localized in the vicinity of the shared electrons, producing a very rigid structure—that of diamond, the hardest natural substance. Because all the valence electrons in diamond are used for bonding, none can move freely to conduct electricity.

Covalent Atomic Radii. In covalently bonded structures, the interatomic distance is generally equal to the arithmetic mean of the interatomic distances in crystals of the elemental substances. Thus, in diamond, the C–C spacing is 1.54 Å; in metallic silicon the Si–Si distance is 2.34 Å. We may therefore suppose that if these atoms unite to form a compound, SiC, the silicon-carbon distance will be near 1.94 Å, the arithmetic mean of the elemental spacings. X-ray measurement determines this spacing in the familiar synthetic abrasive, silicon carbide, as 1.93 Å.

Estimation of the Character of the Bonding Mechanism

It is now generally recognized that there is some electron sharing in most ionically bonded crystals, whereas atoms in covalently bonded substances often have some electrostatic charges. Assessment of the relative proportions of ionic to covalent character is based in part on the polarizing power and polarizability of the ions involved. Compounds of a highly polarizing cation with an easily polarized anion, such as AgI, may show strongly covalent character. In contrast, AgF, because of the lower polarizability of the smaller fluorine ion, is a dominantly ionically bonded compound.

Bonds between elements of the first and seventh columns of the periodic table and between the second and the sixth columns are dominantly ionic. Examples

are the alkali halides and the alkali-earth oxides. Bonds between like atoms or atoms close together in the periodic table will be covalent.

Linus Pauling, in 1939, provided a method by which the percentage of ionic character of a chemical bond can be estimated. The basis for this method is his scale of electronegativities of the elements (see Tables 4.6 and 4.10 for a partial listing of these values; see also Fig. 4.13b). *Electronegativity* is a measure of the ability of an atom to attract electrons to itself, and is expressed in a set of dimensionless numbers. Elements with low electronegativity are electron donors, and those with high values act as electron acceptors. The *differences* in the electronegativity values of the elements are an expression of the ionic character of the bond formed by their atoms. This difference is expressed as $X_A - X_B$, where X_A is the electronegativity of element A bonded to element B, with its own electronegativity value X_B. Linus Pauling used this difference in electronegativity values in an equation to estimate the ionic character of a single bond. This equation is:

$$\text{amount of ionic character} = 1 - e^{-1/4(x_A - x_B)}$$

This function is shown as a curve in Fig. 4.35. The usefulness of this function (and curve) can be illustrated by plotting, as an example, the differences in electronegativity values for the elements of the four compounds listed in the lower half of Fig. 4.32. These are NaF, MgO, ScN, and TiC. The electronegativity values of the elements are given in Table 4.11; the values for $(X_A - X_B)$ for these compounds are as follows:

| NaF | 3.1 | ScN | 1.7 |
| MgO | 2.3 | TiC | 1.0 |

Table 4.10
ELECTRONEGATIVITY OF ELEMENTS*

Li	Be	B		C	N	O	F
1.0	1.5	2.0		2.5	3.0	3.5	4.0
Na	Mg	Al		Si	P	S	Cl
0.9	1.2	1.5		1.8	2.1	2.5	3.0
K	Ca	Sc	Ti	Ge	As	Se	Br
0.8	1.0	1.3	1.5	1.8	2.0	2.4	2.8
Rb	Sr	Y	Zr	Sn	Sb	Te	I
0.8	1.0	1.2	1.4	1.8	1.9	2.1	2.5
Cs	Ba						
0.7	0.9						

*After Pauling, L., 1960, *The Nature of the Chemical Bond.* Cornell University Press, Ithaca, N.Y.

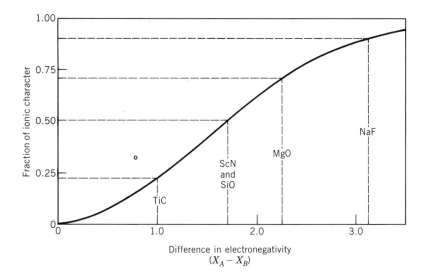

FIG. 4.35 Curve relating the amount of ionic character of a bond *A–B* to the difference in electronegativity $(X_A - X_B)$ of the atoms. Several compounds listed in Fig. 4.32 are plotted as well (see text). (After Pauling, L., 1960, *The Nature of the Chemical Bond.* Cornell University Press, Ithaca, N.Y., 1960.)

The plot of these values on Fig. 4.35 shows that NaF is essentially ionic in its bond type (~92%) and that TiC is somewhat less than 25% ionic in its bond type; the two other compounds are intermediate between these values. Similarly, one can evaluate the ionic character of bond types between silicon-oxygen, aluminum-oxygen, and boron-oxygen. The values for $(X_A - X_B)$ for these turn out to be 1.7 (for Si–O), 2.0 (for Al–O), and 1.5 (for B–O) (see Table 4.11). The 1.7 value for the Si–O bond is plotted in Fig. 4.35 and shows the bond type to be 50% ionic in character. The Al–O bond is 63% ionic, and the B–O bond only 44% ionic.

The above shows that electronegativity, along with ionic size and valence, is helpful in predicting the chemical behavior of elements. It follows from Fig. 4.35 that compounds made of elements with very different values of electronegativity are more ionic than compounds made up of elements close to each other in electronegativity.

Metallic Bond

Metallic sodium is soft, lustrous, opaque, and sectile, and conducts heat and electrical current well. X-ray diffraction analysis reveals that it has the regular repetitive pattern of a true crystalline solid. The properties of the metal differ so from those of its salts that we are led to suspect a different mechanism of bonding. Because sodium, like all true metals, conducts electricity, electrons are free to move readily through the structure. So prodigal with their electrons are sodium and its close relatives, cesium, rubidium, and potassium, that the impact of the radiant energy of light knocks a considerable number entirely free of the structure. This photoelectric effect, on which such in-

struments as exposure meters depend, shows that the electrons are very weakly tied into the metal structure. We may thus postulate that the structural units of true metals are actually the atomic nuclei plus nonvalence electron orbitals bound together by the aggregate electrical charge of a cloud of valence electrons that surrounds the nuclei. Many of the electrons owe no affinity to any particular nucleus and are free to drift through the structure or even out of it entirely, without disrupting the bonding mechanism. This is schematically illustrated in Fig. 4.36, which shows positively charged ions (from which the valence electrons have been removed) in a dense cloud of valence electrons. The attractive force between the nuclei with their filled electron orbitals (but lacking valence electrons) and the cloud of negative electrons holds such structures together. This type of bond is fittingly called the *metallic bond*. To it metals owe their high plasticity, tenacity, ductility, and conductivity, as well as their

FIG. 4.36 A schematic cross section through the structure of a metal. Each circle with a positive charge represents a nucleus with filled, nonvalence electron orbitals of the metal atoms. The mobile electrons are represented by the cloud around the atoms (light gray shading). A possible electron path between the nuclei is shown.

generally low hardness. Among minerals, only the native metals display pure metallic bonding. Table 4.11 gives a brief listing of some of the properties related to metallic bond type in crystalline materials.

van der Waals Bond

Molecules such as N_2, O_2, and Cl_2 form molecular solids despite the fact that all the valence orbitals are occupied either by nonbonding electrons, or by electrons used in covalent bonding to form dimers (a dimer is a molecule created from two identical simpler units). For example, if we take energy away from a Cl_2 gas by cooling it, the molecules ultimately will collapse into the close-packed, chaotic liquid state. If still more heat energy is removed, the amplitude of vibrations of the Cl_2 molecules is further reduced, and ultimately the minute, stray electrical fields existing about the essentially electrically neutral atoms will serve to lock the sluggishly moving molecules into the orderly structure of the solid state. This phenomenon of the solidification of chlorine takes place at very low temperatures, and warming above $-102°C$ will permit the molecules to break the very weak bonds and return to the disordered state of a liquid.

Neutral molecules such as Cl_2 may develop a small concentration of positive charge at one end, with a corresponding dearth of positive charge (which results in a small negative charge) at another end. This is the result of electrons in the occupied orbitals of the interacting atoms synchronizing their motions to avoid each other as much as possible. Figure 4.37 is a schematic illustration of electrons in orbitals having synchronized their motions such that an instantaneous and weak dipole attraction is produced between the two atoms. This weak dipole can induce a similar effect in neighboring atoms, which will cause the whole molecular structure to be bonded together by this weak dipole effect. Such bonding is especially effective over large distances in molecular structures. In the formation of crystals these molecules are aligned with negative poles against positive poles of neighboring molecules. Such is the mechanism that bonds the Cl_2 molecules in solid Cl_2. This weak bond, which ties neutral molecules and essentially uncharged structure units into a cohesive structure by virtue of small residual charges on their surfaces, is called the *van der Waals* (or *residual*) bond and is the weakest of the chemical bonds. Common in organic compounds and solidified gases, it is not often encountered in minerals, but when it is, it generally defines a zone of ready cleavage and low hardness. An example is the mineral graphite, which consist of covalently bonded sheets of carbon atoms linked only

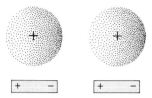

FIG. 4.37 Polarization of an atom, because of an increase in the concentration of electrons on one side of the atom, causes a dipole effect. The weak dipole attraction is that of the van der Waals bond.

by van der Waals bonds (see Fig. 4.38 and Table 4.11).

The most common form of crystalline sulfur is made up of discrete S_8 molecules with a cyclic structure (see Figs. 4.39 and 10.7). Within the ring there is pure covalent bonding, but adjacent rings are held together by van der Waals forces, which account for the low hardness (H $= 1\frac{1}{2}$ to $2\frac{1}{2}$) and melting point (at 112.8°C) of sulfur.

Hydrogen Bond

Polar molecules can form crystalline structures by the attraction between the oppositely charged ends of molecules (see Fig. 4.40a). The *hydrogen bond* is an electrostatic bond between a positively charged hydrogen ion and a negatively charged ion such as O^{2-} and N^{3-}. Because hydrogen has only one electron, when it transfers this electron to another more elec-

FIG. 4.38 Perspective sketch of the graphite structure with covalent bonding between carbon atoms within layers and residual (van der Waals) bonding between layers. Note large separation (3.35 Å) between layers.

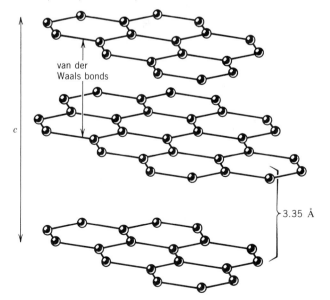

Table 4.11
EXAMPLES OF PROPERTIES CONFERRED BY THE PRINCIPAL TYPES OF CHEMICAL BOND

Property	Bond Type			
	Ionic (Electrostatic)	Covalent (Electron-shared)	Metallic	van der Waals (Residual)
Bond strength	Strong	Very strong	Variable strength, generally moderate	Weak
Mechanical	Hardness moderate to high, depending on interionic distance and charge; brittle	Hardness great Brittle	Hardness low to moderate; gliding common; high plasticity; sectile, ductile, malleable	Crystals soft and somewhat plastic
Electrical	Poor conductors in the solid state; melts and solutions conduct by ion transport	Insulators in solid state and melt	Good conductors; conduction by electron transport	Insulators in both solid and liquid state
Thermal (melting point = m.p.; coefficient of thermal expansion = coef.)	m.p. moderate to high depending on interionic distance and charge; low coef.	m.p. high; low coef.; atoms and molecules in melt	Variable m.p. and coef.; atoms in melt	Low m.p.; high coef.; liquid crystal molecules in melt
Solubility	Soluble in polar solvents to yield solutions containing ions	Very low solubilities	Insoluble, except in acids or alkalis by chemical reaction	Soluble in organic solvents to yield solutions
Structure	Nondirected; gives structures of high coordination and symmetry	Highly directional; gives structures of lower coordination and symmetry	Nondirected; gives structures of very high coordination and symmetry	Nondirected; symmetry low because of shape of molecules
Examples	Halite, NaCl; Fluorite, CaF_2; most minerals	Diamond, C; Sphalerite, ZnS; molecules of O_2; organic molecules; graphite (strong bond)	Copper, Cu; Silver, Ag; Gold, Au; Electrum, (Au,Ag); most metals	Sulfur (weak bond); organic compounds; graphite (weak bond)

FIG. 4.39 S_8 rings occur in the crystal structure of sulfur. These rings are linked to each other by van der Waals bonds. Compare with Fig. 10.7.

tronegative ion in ionic bonding, the remaining proton of the hydrogen nucleus becomes unshielded. This positive ion has the ability to form weak hydrogen bonds with other negative ions or with the negative ends of polar molecules, such as H_2O. The closeness of approach allows the formation of a dipole-dipole bond which is relatively weak when compared with an ionic or covalent bonding mechanism. However, it is considerably stronger than the van der Waals bond.

Ice is an especially good example of bonding in an intermolecular structure. The shape of an H_2O molecule is polar (see Fig. 4.40c), and because of it the two hydrogen atoms in the H_2O molecule provide the bonding to two other neighboring H_2O molecules. Two additional neighboring H_2O molecules in turn provide H atoms to provide two more hydrogen bonds. Therefore each oxygen atom is bound to four neighboring oxygen atoms, in a tetrahedral arrangement, by intervening hydrogen bonds. As in the case

of the van der Waals bonds, the hydrogen bonds are weak, but there are many of them per unit volume of structure. Figure 4.40d is an illustration of the tetrahedrally bonded structure of one of the polymorphs of ice. It is a well-known fact that ice is less dense than water at the melting temperature. At the onset of melting this relatively open network structure collapses and in the resulting liquid the H_2O molecules are more densely packed than in the solid.

Hydrogen bonding is common in hydroxides in which the $(OH)^-$ group does not behave strictly as a spherical anionic group but is more realistically represented by an asymmetrical coordination, which produces a dipole effect. Hydrogen bonding is also present in many of the layer silicates such as micas and clay minerals, which contain hydroxyl groups.

Crystals with More than One Bond Type

Among naturally occurring substances, with their tremendous diversity and complexity, the presence of only one type of bonding is rare, and two or more bond types coexist in most minerals. Where this is so, the crystal shares in the properties of the different bond types represented, and often strongly directional properties result. Thus, in the mineral graphite, the cohesion of the thin sheets that make up the mineral structure is the result of the strong covalent bonding

FIG. 4.40 (a) Schematic representation of the packing of polar molecules in a crystalline solid. Charges of opposite sign are arranged as closest neighbors. (b) Model of a hydrogen bond. (c) A water molecule and the bond angle between H–O–H. (d) Hydrogen bonding as shown by one of the polymorphs of ice. The coordination is tetrahedral and similar to that in diamond.

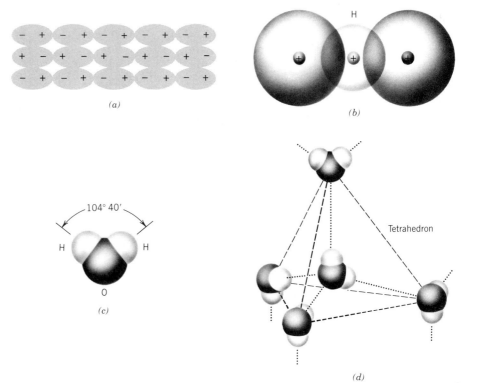

in the plane of the sheets, whereas the excellent cleavage between the sheets reflects the weak van der Waals bonds joining the sheets together (see Fig. 4.38). The layer silicates, which consist of sheets of strongly bonded silica tetrahedra with a relatively weak ionic and/or hydrogen bond joining the sheets together through the cations, similarly reflect, in their remarkable basal cleavage between the sheets, the difference in strength of the two bond types (see Fig. 13.85). As we shall see, the prismatic habit and cleavage of the pyroxenes and amphiboles, and the chunky, blocky habit and cleavage of the feldspars, may similarly be traced to the influence of relatively weak bonds joining together more strongly bonded structure units having a chain, band, or block shape (see also Chapter 6).

EXAMPLES OF SOME COMMON STRUCTURE TYPES

In the visualization of inorganic crystal structures it is often useful to view the larger anions as a closest packed array with the smaller cations housed in the interstitial positions (interstices). It was shown in Fig. 4.19 that close-packed layers of spheres can be stacked into a hexagonal closest packed sequence (HCP: *ABAB* . . .) or a cubic closest packed sequence (CCP: *ABCABC* . . .). In Fig. 4.41a the interstices that arise in such close-packed sequences are shown as tetrahedral and octahedral in coordination. These tetrahedral and octahedral interstitial sites may or may not be fully occupied by cations, as we will see in various of the following structures. In structures in which the cations are larger than can be housed in octahedral coordination ($R_A : R_X$ values between 0.73 and 1.0; see Fig. 4.24) they commonly occur in *simple cubic packing* (SCP), with the cations at the center of

a cube and the anions at the eight corners (see Fig. 4.41b).

The important structure types that will be reviewed here are those of NaCl (halite), CsCl (cesium chloride), ZnS (sphalerite), CaF_2 (fluorite), TiO_2 (rutile), $CaTiO_3$ (perovskite), $MgAl_2O_4$ (spinel), and the silicates.

NaCl Structure

In this *AX* structure type the anions are in CCP, with the cations filling all the octahedral sites. All tetrahedral interstices are empty. Values of $R_A : R_X$ in the range of 0.73 to 0.41 favor octahedral coordination (see Fig. 4.24). The cations and anions in this structure occur in edge-sharing octahedra (see Fig. 4.42), with each of the 12 edges of an octahedron shared with a neighboring octahedron. This structure type is adopted by a large number of *AX* compounds in the appropriate range of radius ratios. Examples are:

Halides: LiF, LiCl, LiBr, LiI
NaF, NaCl, NaBr, NaI
KF, KCl, KBr, KI
RbF, RbCl, RbBr, RbI
Oxides: MgO, CaO, SrO, BaO, NiO
Sulfides: MgS, CaS, MnS, PbS.

The structure of pyrite (FeS_2) can be considered as a derivative of the NaCl structure (see Fig. 10.17) in which Fe^{2+} is in the Na position and covalently bonded S_2 pairs in the Cl position.

CsCl Structure

The cesium chloride structure instead of the NaCl structure is adopted by *AX* compounds when the ra-

FIG. 4.41 (a) Hexagonal closest packing sequence (HCP) with tetrahedral interstices (or voids, marked *B*) and octahedral interstices (marked *C*). (b) Simple cubic packing (SCP) results when $R_A : R_X$ lies between 0.73 and 1.0. The relatively large cation is surrounded by eight nearest neighbors in cubic coordination.

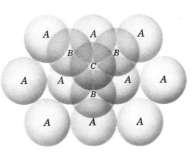

B voids are tetrahedral
C voids are octahedral

(a)

(b)

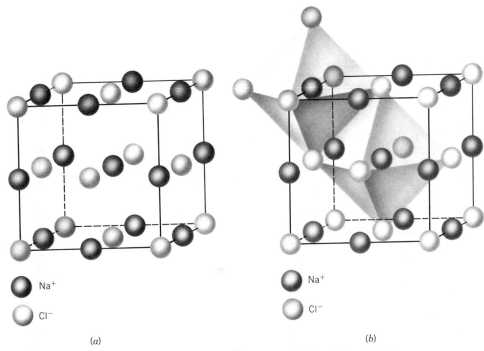

FIG. 4.42 *(a)* The structure of NaCl, halite ($F4/m\bar{3}2/m = Fm3m$). The Na^+ and Cl^- ions are arranged in a face-centered cubic lattice. *(b)* The same structure showing the edge-sharing octahedrons about the Na^+. Similar edge-sharing octahedrons could be drawn about Cl^-.

dius ratio $R_A : R_X$ is greater than 0.73 (see Fig. 4.24). Therefore, the anions (X) are in simple cubic packing (SCP) and the cations fill the large interstices between them. The overall structure (see Fig. 4.43) is made up of centered cubes that share faces with six other neighboring cubes. This feature makes this structure an unattractive choice for highly charged cations (see Pauling's Rule 3).

Examples of *AX* compounds that exhibit this structure are:

$$CsCl, \; CsBr, \; CsI, \; (NH_4)Cl, \; (NH_4)Br$$

FIG. 4.43 The structure of CsCl ($P4/m\bar{3}2/m = Pm3m$). The ions are distributed in a primitive cubic lattice. Each cation is surrounded by eight neighbors, and so is each anion.

Sphalerite (ZnS) Structure

The radius ratio, $R_{Zn} : R_s = 0.32$ (radius of $Zn^{2+} = 0.60$ Å in 4-coordination; $S^{2-} = 1.84$ Å in 4-coordination; see Table 4.8) predicts that Zn^{2+} is in tetrahedral coordination with the neighboring S^{2-} (see Fig. 4.44*a*). The sphalerite structure may be considered as a derivative of the diamond, C, structure (see Fig. 4.44*b*) in which half the carbon atoms in the diamond structure are replaced by Zn and the other half by S. In comparison with the sphalerite structure, diamond may be viewed as having C atoms in the cation positions of the *AX* compound, as well as C atoms in the anion positions. The tetrahedral arrangement of C about C is a result of the covalent bonding with four valence electrons of neighboring carbons (see Fig. 4.34). Derivatives of the sphalerite structure, such as chalcopyrite, $CuFeS_2$, and tetrahedrite, $Cu_{12}Sb_4S_{13}$, are illustrated in Fig. 10.15. SiC (silicon carbide; carborundum) is isostructural with sphalerite.

CaF₂ Structure

For AX_2 compounds in which the radius ratio ($R_A : R_X$) exceeds 0.73 (see Fig. 4.24), the fluorite structure may be adopted. In this structure the Ca^{2+} ions are arranged at the corners and face centers of a cubic unit cell and F^- are at the centers of the eight equal cube-

FIG. 4.44 *(a)* The structure of sphalerite, ZnS ($F\bar{4}3m$). Both Zn^{2+} and S^{2-} are in a face-centered cubic array. Zn^{2+} is in tetrahedral coordination with four S^{2-} neighbors. *(b)* The structure of diamond, C ($F4_1/d\bar{3}2/m$ = $Fd3m$). The carbons are arranged in a face-centered cubic lattice. The coordination of C by four nearest carbon neighbors is tetrahedral.

● Zn^{2+}

○ S^{2-}

○ C^{4+}

(a)

(b)

lets into which the cell may be mentally divided (see Fig. 4.45). Each Ca^{2+} is surrounded by eight F^- in cubic coordination, and each F^- is surrounded by four Ca^{2+} at the corners of a tetrahedron. The fluorite structure may be derived from the CsCl structure by replacing Cl^- with F^- and *every other* Cs^+ with Ca^{2+}. This leaves alternate cubic interstices vacant and results in the octahedral cleavage of fluorite and isostructural minerals. The cubic coordination of F^- about a central Ca^{2+} is the result of the radius ratio of $R_{Ca} : R_F \sim .75$. These cubic coordination polyhedra share edges only; similarly, the tetrahedral coordination polyhedra share only edges. This arrangement allows for the maximum separation of the Ca^{2+} cations from each other (see Fig. 4.45). Because the *A* cation has double the charge of the *X* anion, the number of anions in the structure must be double that of the cations in order to achieve electrostatic neutrality. As such, the resulting general formula is AX_2. This structure type is adopted by large number of AX_2 halides and oxides. Examples are:

Halides: CaF_2, SrF_2, BaF_2, CuF_2, CdF_2, $SrCl_2$, $BaCl_2$

Oxides: ZrO_2, HfO_2, CeO_2, UO_2

Rutile (TiO_2) Structure

The rutile structure is based on HCP packing, with Ti filling half the octahedral interstitial positions. AX_2 compounds that exhibit $R_A : R_X$ ratios between about 0.73 and 0.41 (see Fig. 4.24) may adopt the rutile structure in which the *A* cation is octahedrally coordinated (**C.N.** = 6) to the *X* anion (see Fig. 4.46). The oxygen anions are coordinated by three cations in a triangular array (**C.N.** = 3). The structure consists of octahedrons that are linked along horizontal edges. This linking pattern forms strips of octahedrons parallel to the *c* axis, and these bands are cross-linked to each other by corner-sharing of neighboring octahedrons. This structure results in the {110} prismatic cleavage of rutile, parallel to *c*. A large number of inorganic compounds assume the rutile structure. Examples are:

MgF_2, NiF_2, CoF_2, FeF_2, MnF_2, ZnF_2

TiO_2, MnO_2, SnO_2, WO_2, PbO_2, SiO_2 (stishovite, a high-pressure polymorph of quartz; see Fig. 3.38a)

Perovskite (ABO_3) Structure

This structure type is based on CCP of oxygen, with one-quarter of the oxygens replaced by a large *A* ca-

FIG. 4.45 The structure of fluorite, CaF_2 ($F4/m\bar{3}2/m$ = $Fm3m$). Ca^{2+} ions are arranged in a face-centered cubic lattice. The F^- ions are in simple cubic packing (SCP) with Ca^{2+} occupying the void at the centers of *alternating* cubic interstices.

● Ca^{2+}

○ F^-

Fluorite structure

Ti^{4+}

O^{2-}

FIG. 4.46 Two views of the structure of rutile, TiO$_2$ ($P4_2/m2_1/n2/m$ = $P4_2/mnm$). *(a)* Standard orientation of two unit cells of rutile stacked in the c direction. The octahedrons share two horizontal edges with adjacent octahedrons, forming bands parallel to the vertical *(c)* axis. *(b)* The array of bands of edge-sharing octahedrons (parallel to c) running into the page. Chains of edge-sharing octahedrons running parallel to c are clearly seen, cross-linked by corner-sharing octahedrons. The HCP stackings are shown by *ABAB.* . . . (From Waychunas, G. A., 1991, Crystal chemistry of oxides and hydroxides, in *Oxide Minerals. Reviews in Mineralogy*, v. 25, pp. 11–68.)

tion. This large cation position is in 10-coordination with the surrounding oxygens. The B cations occur in octahedrons that share only apices (see Fig. 4.47). The valence of the A and B ions is not specified; however, the total valence of both ions ($A + B$) must be equal to 6 (to balance the O$_3^{2-}$ in the formula ABO_3). The perovskite structure type is adopted by many compounds. Examples arranged in columns for the charge of the cations in the A and B sites are:

$A^{1+}B^{5+}$	$A^{2+}B^{4+}$	$A^{3+}B^{3+}$
NaNbO$_3$	CaTiO$_3$	LaCrO$_3$
KNbO$_3$	SrTiO$_3$	YAlO$_3$
KTaO$_3$	BaZrO$_3$	LaAlO$_3$
	BaTiO$_3$	

The perovskite structure type is of especial interest because it is considered to be a common structure type under very high-pressure conditions, as in the central and deep mantle (see page 216).

Spinel (AB_2O_4) Structure

The spinel structure type consists of a CCP array of oxygens in which one-eighth of the tetrahedral interstices (A) and half of the octahedral interstices (B) are occupied by cations. All spinels contain two differing cations, or at least two different valences of the same

FIG. 4.47 The structure of perovskite, CaTiO$_3$ ($P2_1/c2_1/m2_1/n$ = $Pcmn$) in a perspective projection looking down the c axis. (Modified after Smyth, J. R. and Bish, D. L., 1988, *Crystal Structures and Cation Sites of the Rock Forming Minerals*. Allen and Unwin, Boston.) Layers of anion-sharing octahedra (containing Ti^{4+}) are oriented perpendicular to the c axis. Ca^{2+} is in 10-coordination.

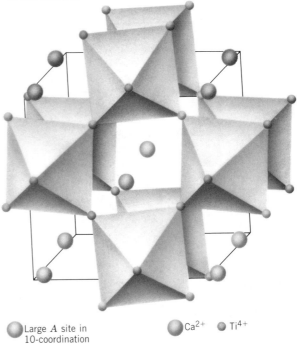

Large A site in 10-coordination

Ca^{2+} Ti^{4+}

cation in the ratio of 2:1. Spinels are classified as *normal* or *inverse* spinels, depending on where the more abundant of the cations is housed. If it occurs in the octahedral site it is classified as *normal*. If it is equally split between the octahedral and tetrahedral sites, it is *inverse*.

The CCP layers of oxygen are stacked parallel to {111}, resulting in alternating layers of octahedral sites and tetrahedral sites (see Fig. 4.48). Occupied octa-

FIG. 4.48 The spinel (AB_2O_4) structure ($F4_1/d\bar{3}2/m = Fd3m$). (a) Alternating layers parallel to {111} of octahedral and octahedral-tetrahedral polyhedra, as based upon approximate cubic closest packing. (b) A close-packed layer of oxygen in the spinel structure, projected onto the {111} plane. The large spheres are oxygen, and the cation layers on either side of the oxygen layer are shown as well. (a and b redrawn after Waychunas, G. A., 1991, Crystal chemistry of oxides and hydroxides, in *Oxide Minerals. Reviews in Mineralogy*, v. 25.)

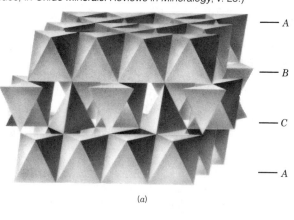

— A

— B

— C

— A

(a)

● Octahedral

○ Tetrahedral

1Å

(b)

hedra are joined along edges to form rows and planes parallel to {111} of the structure, and tetrahedra provide cross-links between layers of octahedra (Fig. 4.48a). A plan view of an oxygen layer parallel to {111} and its coordination with cations is given in Fig. 4.48b.

In the general formula of spinel (ABO_4), the smaller tetrahedral A site is commonly occupied by Mg^{2+}, Fe^{2+}, Mn^{2+}, Zn^{2+} and the larger octahedral B site by Al^{3+}, Cr^{3+}, or Fe^{3+}. The coordination polyhedra about the various cations in spinel are not what might be predicted on the basis of the ionic sizes of the cations. Because Mg^{2+} is larger than Al^{3+}, one would expect Mg to occur in the octahedral B site and Al in the tetrahedral A site. In the normal spinel structure (e.g., $MgAl_2O_4$), however, the general concepts of radius ratio do not apply; indeed, the larger cation is in the smaller polyhedron, and vice versa. Only when crystal field stabilization energies are considered instead of geometric aspects of the ions does it become clear why the larger cation may occupy tetrahedral sites.

The spinel structure has a coordination scheme similar to that of the silicates of the olivine series, Mg_2SiO_4 to Fe_2SiO_4. This compositional series can be represented as $X_2^{2+}Y^{4+}O_4$. Although this is not the same as $A^{2+}B_2^{3+}O_4$, as in spinel, in both cases the overall cation charge is identical. If one compares the structure of an Mg-Fe olivine with that of a possible Mg-Fe spinel, one finds that the spinel structure is about 12% denser than the olivine structure of the same composition. This leads to the conclusion that the spinel form of olivine must be abundant in the mantle as a result of very high confining pressures (see Fig. 4.51). Some examples of spinel compositions are:

Spinel	$MgAl_2^{3+}O_4$
Hercynite	$FeAl_2^{3+}O_4$
Gahnite	$ZnAl_2^{3+}O_4$
Chromite	$Fe^{2+}Cr_2^{3+}O_4$
Magnesiochromite	$Mg^{2+}Cr_2^{3+}O_4$

For further discussion of spinel, see pages 375–376.

Silicate Structures

Silicates, compounds consisting of abundant oxygen and silicon, are the major mineral components of the Earth's crust. The oxygen in these structures is in close packing with the cations in various coordination polyhedra (see Table 4.9) between the oxygens. In *all* crustal silicates, silicon is in tetrahedral coordination with oxygen. Only in high-pressure phases, such as stishovite (a polymorph of SiO_2), is Si found to occur

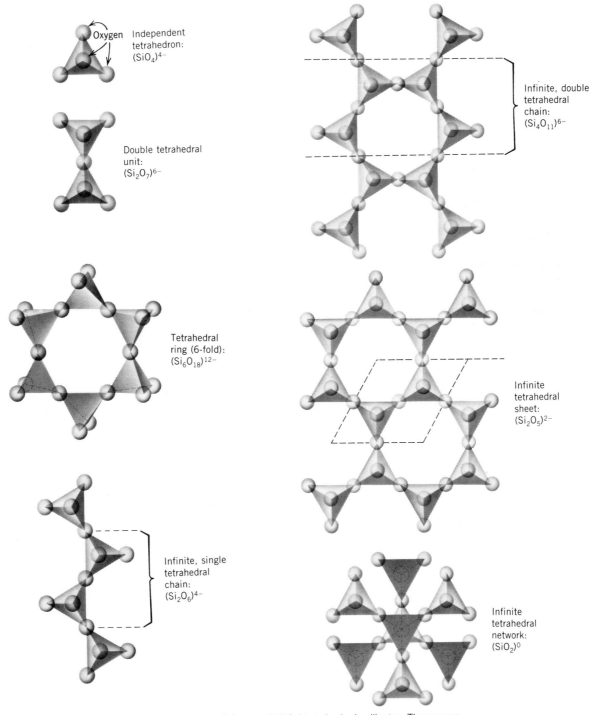

FIG. 4.49 Examples of some of the common linkages of (SiO_4) tetrahedra in silicates. The oxygen that links two tetrahedra is known as the "bridging" oxygen.

in 6-coordination with oxygen. As such, the tetrahedral $(SiO_4)^{4-}$ is a fundamental "packing unit" of silicates. The various ways in which this $(SiO_4)^{4-}$ group can link itself to other $(SiO_4)^{4-}$ groups by sharing one, two, three, or all four corner oxygens of the tetrahedron are fundamental to any classification of silicates.

An overview of these tetrahedral linking (or polymerization) schemes is given in Fig. 4.49. The basic underlying reason for the ability of the $(SiO_4)^{4-}$ tetrahedron to link itself in so many ways to other $(SiO_4)^{4-}$ tetrahedra is the fact that the electrostatic valency (e.v.) of the bonds between Si^{4+} and oxygen is $\frac{4}{4} =$

1. This is exactly half the bonding strength of the oxygen ion. Consequently, the SiO_4 tetrahedron may link itself through a "bridging" oxygen to another SiO_4 tetrahedron. Individual silicate structures are discussed in detail in Chapter 13.

In addition to the classification of silicates used in this text, there are other classifications proposed by Liebau (1985) and Zoltai (1960) (see references at the end of this chapter).

CHANGES IN STRUCTURE TYPE AS A FUNCTION OF HIGH PRESSURES

Until now we have discussed various aspects of coordination polyhedra and structure types in minerals that are common on the Earth's crust. A very interesting question that can be asked, and which has been very actively researched over the last twenty years, is: What type of structural arrangements are found in minerals that occur below the Earth's crust, in various parts of the mantle?

Figure 4.50 is a cross section of the Earth, giving the depth locations of the upper and lower mantle and the transition zone between them. On the left-hand side, this cross section gives depth in kilometers, and on the right the pressures associated with these depths. Clearly, the overall density of materials in the

deeper parts of the Earth must be considerably greater than on the Earth's crust. As such, the atomic packing schemes of common rock-forming minerals, deeper down, must be different from those observed in minerals in crustal rocks.

The average mean density (see Chapter 6 for discussion of density and specific gravity) of rocks in the crust is only about 2.8 gm/cm^3, whereas the mean density of the Earth as a whole is 5.52 gm/cm^3. This means that rocks below the crust (i.e., below the Mohorovičić discontinuity) must increase their density as a function of changing composition and/or atomic packing of the elements that constitute the minerals. The Earth's core is generally accepted to consist mainly of Fe and Ni, with minor amounts of S, C, and Si. Such a composition satisfies the geophysical parameters of density and magnetism of the core. The mantle beween the Mohorovičić discontinuity (at 36 km below continents) and the core (at 2900 km; see Fig. 4.50) is thought to consist of mixtures of silicates, oxides, minor sulfides, and lesser metal. A density profile for the upper mantle as a function of distance below the Earth's surface is given in Fig. 4.51.

The makeup of the upper part of the mantle (between 36 and 220 km) has been evaluated by (1) direct mineralogic and petrologic study of rock types that, through explosive pipes, have come to the sur-

FIG. 4.50 Major subdivisions of the Earth's interior. The pressure is expressed in kilobars, where 1 kilobar = 1000 bars, and 1 bar = 0.987 atmosphere. (From Liu, L. and Bassett, W. A., 1986, *Elements, Oxides, Silicates: High-Pressure Phases with Implications for the Earth's Interior.* Oxford University Press, New York.)

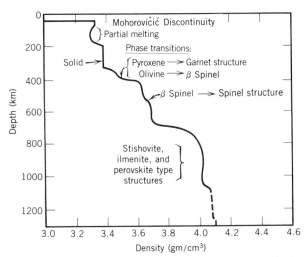

FIG. 4.51 Possible changes in mineral assemblages and in the structural packing of minerals as a function of the changing density in the mantle. Below 670 km, it is likely that all silicon is in 6-coordination with oxygen. (Adapted from Ringwood, A. E., 1975, *Composition and Petrology of the Earth's Mantle.* McGraw-Hill, New York.)

face of the Earth from upper mantle regions, and (2) high-pressure, high-temperature experiments in the laboratory of silicate-rich compositions. (An illustration of a high-density mineral assemblage from such a deep-rooted pipe, a kimberlite pipe, is given in Fig. 4.52). These evaluations suggest that the overall chemical composition of the upper part of the mantle is made of Fe-Mg–rich silicates, which are common in olivine-rich rocks (peridotite) and basalt. If a specific mixture of these two rock types is assumed (commonly referred to as ultramafic), experimentalists can subject such a composition to high pressures (and temperatures) in order to achieve changes in mineral compositions and structures (changes in atomic and ionic packing) as a function of increasing pressure or depth in the upper mantle. Some of these experimentally based findings are shown in Fig. 4.51.

In order to achieve a generally increasing density, as shown by the density curve in Fig. 4.51, without changing the overall chemical composition of the mantle material, structural changes must occur in the constituent minerals that allow for denser packing of the ions and atoms in these minerals as a function of increasing depth. In other words, polymorphic changes would be expected to take place as a function of increasing pressure. Figure 4.51 suggests that, between 300 and 400 km depth, the pyroxene structure adopts a denser, garnet-type structure and that olivine converts to a denser packing than is present in the structure of olivine as found in the Earth's crust

(this denser polymorph is referred to as β spinel in Fig. 4.51). At somewhat greater depths, this β polymorph of olivine would give way to a silicate structure, based on the relatively denser packing of spinel.

At depths below 670 km it is likely that *all silicon is in 6-coordination* instead of 4-coordination as in minerals in crustal rocks. In the transition zone, between about 250 and 670 km, silicates will adopt structure types in which silicon is both in 4- and 6-coordination. Examples of expected reactions in the transition zone of the mantle are:

$$Mg^{[6]}Si^{[4]}O_3 \rightarrow Mg^{[6]}_2Si^{[4]}O_4 + Si^{[6]}O_2$$

enstatite	olivine	stishovite
structure	structure	rutile structure

$$Mg^{[6]}_2Si^{[4]}O_4 \rightarrow Mg^{[6]}_2Si^{[4]}O_4$$

olivine	spinel
structure	structure

Superscripts in square brackets, e.g., [4] and [6], signify **C.N.** for the cations.

The most abundant types of densely packed silicate structures to be found below the 670 km zone (in the lower mantle) will be the rutile and perovskite types. SiO_2, for example, would be expected to occur as stishovite, with Si in 6-coordination with oxygen.

FIG. 4.52 Photomicrograph of an association of coesite (c), kyanite (k), omphacite (o), and garnet (g). Coesite is a high-pressure polymorph of SiO_2; kyanite is the high-pressure polymorph of Al_2SiO_5; and omphacite is a high-pressure form of clinopyroxene. This association is part of a nodule collected in the Roberts Victor kimberlite pipe in South Africa. It is concluded to have originated in the upper parts of the Earth's mantle. (From Smyth, J. R. and Hatton, C. J., 1977, A coesite-sanidine grospydite from the Roberts Victor kimberlite. *Earth and Planetary Science Letters*, v. 34, pp. 284–290.)

Common mineral compositions rich in Mg and Ca, such as olivines and pyroxenes, would be expected to occur in a perovskite structure type. In the upper part of the lower mantle the following transformations can be expected (from L. Liu and W. A. Bassett (1986); for complete reference see listing at end of this chapter):

$$Mg^{[6]}Si^{[4]}O_3 \rightarrow \tfrac{1}{2}Mg^{[6]}_2Si^{[4]}O_4 + \tfrac{1}{2}Si^{[6]}O_2 \rightarrow$$

enstatite	modified	stishovite
structure	spinel	rutile
	structure	structure

$$\rightarrow \tfrac{1}{2}Mg^{[6]}_2Si^{[4]}O_4 + \tfrac{1}{2}Si^{[6]}O_2 \rightarrow$$

| spinel | stishovite |
| structure | rutile structure |

$$\rightarrow Mg^{[6]}Si^{[6]}O_3 \rightarrow Mg^{[8]}Si^{[6]}O_3$$

| ilmenite* | perovskite |
| structure | structure |

*For a discussion of the ilmenite, $FeTiO_3$, structure, see page 374.

Examples of the greatly increased densities, reflecting much denser packing of elements on an atomic scale, of a few common compositions are given in Table 4.12.

Even though the silicate and oxide crystal structures to be expected in the deep mantle region will be very difficult, if not impossible, to reproduce experimentally, it is to be expected that they will be based on dense, close-packed oxygens in which the interstitial polyhedra will show much more edge and even face-sharing than is expected at lower pressure such as on the Earth's crust. Also, higher pressure favors a higher **C.N.** for a given cation, such as **C.N.** = 6 for Si^{4+}, instead of **C.N.** = 4.

Radioactivity

The nuclei of some elements are unstable and can change (decay) spontaneously to different kinds of nuclei, with the release of radioactive energy in the process. As radioactivity is a statistically random process, the probability that a nucleus will decay in a given time interval is expressed in terms of a decay constant, λ, the fraction of the radioactive nuclei present that will decay in a unit of time. The equation that expresses the decay is:

$$\frac{dP}{dt} = -\lambda P$$

where P is the number of parent atoms at time t and λ is the *disintegration* or *decay constant*, whose units are reciprocal time. The decay constants are now well known for many unstable elements, from laboratory measurements. It has been found that these decay constants are truly constant in any terrestrial environment. This observation is basic to any radioactive "time clock," which provides radiometric ages of minerals and rocks.

Examples of geologically important unstable nuclei are the following isotopes: ^{40}K, ^{87}Rb, ^{232}Th, ^{238}U, and ^{235}U (see also Table 4.13). In the decay process of such unstable nuclei, nuclear particles (including alpha or beta particles) are emitted, as are gamma rays. An *alpha particle* consists of two protons and two neutrons that are tightly bonded together. The alpha particle, which is identical with the nucleus of ^4He, is ejected from the unstable nucleus during alpha decay. In such decay the atomic number Z of the nucleus decreases by two because of the removal of two protons, and the mass number decreases by four. An example of such decay is:

$$^{238}_{92}U \rightarrow ^{234}_{90}Th + \alpha + \gamma + energy$$

A *beta particle* is a negatively charged particle with the mass of an electron that is emitted from the unstable parent nucleus. Nuclei do not contain electrons, so we must regard the origin of this electron as the decay of a neutron into a proton, with the emission of the beta particle ($n = p + \beta^-$). By this process the charge of the nucleus is increased by 1, and the parent nucleus becomes the nucleus of the next higher element in the periodic table:

$$^{87}_{37}Rb \rightarrow ^{87}_{38}Sr + \beta^- + energy$$

Table 4.12
COMPOSITIONS AND MEASURED DENSITIES OF THREE COMMON CRUSTAL MINERAL TYPES AS COMPARED WITH THEIR CALCULATED DENSITIES IN HIGH-PRESSURE POLYMORPHS WITH SiO_6 OCTAHEDRA*

Mineral	Density	C.N. (Si)
SiO_2 (quartz)	2.65	4
SiO_2 (stishovite)	4.29*	6 (rutile structure type)
$CaSiO_3$ (wollastonite)	2.91	4
$CaSiO_3$	4.25*	6 (perovskite structure type)
$MgSiO_3$ (enstatite)	3.21	4
$MgSiO_3$	4.10*	6 (perovskite structure type)

*High-pressure polymorphic data from Hazen, R. M. and Finger, L. W., 1991, Predicted High-Pressure Mineral Structures with Octahedral Silicon, *Geophysical Laboratory Annual Report*, Carnegie Institution of Washington, no. 2250, pp. 101–107.

Table 4.13
RADIOACTIVE ISOTOPES USED IN RADIOMETRIC DATING, THEIR HALF-LIVES, AND DAUGHTER ELEMENTS

Element	Isotope	Half-life* (in years)	Ultimate Daughter Elements†
Potassium	^{40}K	1.28×10^9 yr	^{40}Ca and ^{40}Ar
Rubidium	^{87}Rb	5×10^{11} yr	^{87}Sr
Thorium	^{232}Th	1.41×10^{10} yr	^{208}Pb and ^4He
Uranium	^{238}U	$.4.51 \times 10^9$ yr	^{206}Pb and ^4He
	^{235}U	7.1×10^8 yr	^{207}Pb and ^4He

*Half-life is the time required for one half of the original number of radioactive atoms to decay.
†The daughter elements are the new atoms formed at the expense of the disintegrated ones, the parent elements.

In the decay series of $^{238}U \rightarrow {}^{206}Pb$, $^{235}U \rightarrow {}^{207}Pb$, and $^{232}Th \rightarrow {}^{208}Pb$, which form the basis for three independent methods of age determination, several alpha and beta particles are produced in a sequence of intermediate radioactive products.

The process of disintegration can also occur by *electron capture*, in which an orbital electron is captured by the nucleus; this converts a proton to a neutron ($p + e^- = n$). The nuclear charge decreases by one without any significant change in mass. An example of this reaction is:

$$^{40}_{19}K + \text{orbital } e^- \rightarrow {}^{40}_{18}Ar + \gamma + \text{energy}$$

In the above equations γ stands for high-energy electromagnetic radiation emitted by an excited nucleus as it drops into a less-excited state. Such emitted radiation is usually a by-product of alpha and beta decay and electron capture. Gamma radiation is on the short wavelength side of X-radiation (see Fig. 7.1).

In minerals with considerable amounts of U and Th, this radiation effect is readily measured and is therefore a diagnostic property in the characterization of radioactive minerals. Examples of radioactive minerals are uraninite (UO_2), thorianite (ThO_2), and autunite ($Ca(UO_2)_2(PO_4)_2 \cdot 10{-}12H_2O$). The radiation is most easily measured, in the laboratory and in the field, by a Geiger counter or scintillation counter. The presence of radiation can also be shown by placing unexposed film in an opaque wrapper against a radioactive specimen. The radiation will expose the film; this is known as *autoradiography* (see Fig. 4.53).

As a result of the production of energetic alpha particles (and the associated recoil energy these particles impart to the daughter nucleus), beta particles, and gamma radiation, and the changes in ionic size in going from parent to daughter elements, the crystal structure in which these processes occur is generally profoundly affected. For this reason, most U- and Th-rich minerals have undergone partial or complete destruction of their structures, leading to various stages of *metamictization* (see "Metamict Minerals," page 159).

The production of gamma radiation, as a result of the various decay processes, is of major importance

FIG. 4.53 Autoradiograph of a dendritic aggregate of uraninite, UO_2. Locality: Ruggles Pegmatite, Grafton Center, New Hampshire (Harvard Mineralogical Collection). (From Frondel, C. 1958, *Systematic Mineralogy of Uranium and Thorium.* U.S. Geological Survey Bulletin no. 1064, 400 pp.)

in the exploration for radioactive minerals. Not only can Geiger counters be used in field exploration programs, but airborne gamma-ray spectrometers can detect and distinguish the radiation emanating from ^{238}U, ^{232}Th, and ^{40}K decay. Such airborne surveys detect mainly radioactive materials that occur on the uppermost surface of the Earth, because even a thin cover will effectively block the radiation.

REFERENCES AND SUGGESTED READING

Ahrens, L. H., 1952, The use of ionization potentials. *Geochemica et Cosmochimica Acta*, v. 2, pp. 155–169.

Bloss, F. D., 1971, *Crystallography and Crystal Chemistry: An Introduction*. Holt, Rinehart and Winston, New York, 545 pp.

Brady, J. E., 1990, *General Chemistry*, 5th ed. John Wiley & Sons, New York, 944 pp.

Bragg, W. L. and Claringbull, G. F., 1965, *Crystal Structure of Minerals*. Cornell University Press, Ithaca, N.Y., 409 pp.

Evans, R. C., 1966, *An Introduction to Crystal Chemistry*, 2nd ed. University Press, Cambridge, England, 410 pp.

Faure, G., 1986, *Principles of Isotope Geology*, 2nd ed. John Wiley & Sons, New York, 589 pp.

Frondel, C., 1958, *Systematic Mineralogy of Uranium and Thorium*. U.S. Geological Survey Bulletin no. 1064, 400 pp.

Gray, H. B., 1973, *Chemical Bonds: An Introduction to Atomic and Molecular Structure*. Benjamin/Cummings, Menlo Park, Calif., 232 pp.

Liebau, F., 1985, *Structural Chemistry of Silicates, Structure, Bonding, and Classification*. Springer-Verlag, New York, 347 pp.

Liu, L. and Bassett, W. A., 1986, *Elements, Oxides, Silicates: High-Pressure Phases with Implications for the Earth's Interior*. Oxford University Press, New York, 250 pp.

Mason, B. and Moore, C. B., 1982, *Principles of Geochemistry*, 4th ed. John Wiley & Sons, New York, 344 pp.

Pauling, L., 1960, *The Nature of the Chemical Bond*, 3rd ed. Cornell University Press, Ithaca, N.Y., 644 pp.

Pimentel, G. C. and Spratley, R. D., 1969, *Chemical Bonding Clarified Through Quantum Mechanics*. Holden-Day, San Francisco, 344 pp.

Shannon, R. D. and Prewitt, C. T., 1969, Effective ionic radii in oxides and fluorides. *Acta Crystallographica*, v. 25, pp. 925–946.

Shannon, R. D., 1976, Revised effective ionic radii and systematic studies of interatomic distances in halides and chalcogenides. *Acta Crystallographica*, v. A32, pp. 751–767.

Shriver, D. F., Atkins, P. W. and Langford, C. H., 1990, *Inorganic Chemistry*. W. H. Freeman, New York, 706 pp.

Smyth, J. R. and Bish, D. L., 1988, *Crystal Structures and Cation Sites of the Rock-Forming Minerals*. Allen and Unwin, Boston, 332 pp.

Wells, A. F., 1991, *Structural Inorganic Chemistry*, 5th ed. Clarendon Press, Oxford, England, 1382 pp.

Zoltai, T., 1960, Classification of silicates and other minerals, with tetrahedral structures. *American Mineralogist*, v. 45, pp. 960–973.

CHAPTER 5

MINERAL CHEMISTRY

The chemical composition of the majority of known minerals shows a considerable range in elemental constituents. This is to a large degree the result of two or more chemical species substituting for each other in a specific atomic site of the mineral structure. Because of common substitutions of, for example, Mg^{2+} for Fe^{2+} in octahedral sites, Ca^{2+} for Na^+ in 8- to 10-fold sites, and Al^{3+} for Si^{4+} in tetrahedral sites in structures, the chemical composition of a specific mineral can be highly variable. Therefore, the often complex chemical analytical results of minerals (reported as chemical analyses) are the basic expression of their great variability in chemistry.

In this chapter we will give a brief overview of some of the most commonly used analytical techniques; we will assess the processes that underlie chemical substitution and exsolution ("unmixing"); and we will evaluate mineral analyses, their recalculation schemes, and their graphic representation.

But before we discuss these aspects of mineral chemistry, it is instructive to look briefly at some aspects of the chemistry of the Earth's crust. Patterns of mineral distribution and assessments of our mineral wealth are directly related to the abundance and distribution of elements in the Earth's crust. Although abundant chemical and mineralogical data are available for the Earth's crust, important chemical and mineralogical information has also been obtained

from extraterrestrial samples as provided by meteorites and lunar materials.

CHEMICAL COMPOSITION OF THE EARTH'S CRUST

Geophysical investigations indicate a division of the Earth into a crust, mantle, and core (see Fig. 4.50). The crust is approximately 36 km thick under the continents and from 10 to 13 km thick under the oceans. The boundary between the crust and the underlying upper mantle is referred to as the Mohorovičić discontinuity. The upper part of the crust, which consists of the materials immediately underfoot, is composed of a relatively large percentage of sedimentary rocks and unconsolidated materials. However, this sedimentary cover forms but a thin veneer on an underlying basement of igneous and metamorphic rocks. Clarke and Washington (1924) estimated that the upper 10 miles of the crust consists of 95% igneous rocks (or their metamorphic equivalents), 4% shale, 0.75% sandstone, and 0.25% limestone. The average composition of igneous rocks, therefore, would closely approximate the average crustal composition. To this effect, Clarke and Washington compiled 5159 "superior" analyses of igneous rocks, the average of which (Table 5.1) represents the average composition

Table 5.1

AVERAGE AMOUNTS OF THE ELEMENTS IN CRUSTAL ROCKS, IN WEIGHT PERCENT FOR THE COMMON ELEMENTS (AS INDICATED BY %) AND IN PARTS PER MILLION FOR THE LESS ABUNDANT ELEMENTS*

Atomic Number	Element	Crustal Average	Granite (G-1)	Diabase (W-1)
1	H	0.14%	0.04%	0.06%
3	Li	20	22	15
4	Be	2.8	3	0.8
5	B	10	1.7	15
6	C	200	200	100
7	N	20	59	52
8	O	46.60%	48.50%	44.90%
9	F	625	700	250
11	Na	2.83%	2.46%	1.60%
12	Mg	2.09%	0.24%	3.99%
13	Al	8.13%	7.43%	7.94%
14	Si	27.72%	33.96%	24.61%
15	P	0.10%	0.04%	0.06%
16	S	260	58	123
17	Cl	130	70	200
19	K	2.59%	4.51%	0.53%
20	Ca	3.63%	0.99%	7.83%
21	Sc	22	2.9	35
22	Ti	0.44%	0.15%	0.64%
23	V	135	17	264
24	Cr	100	20	114
25	Mn	0.09%	0.02%	0.13%
26	Fe	5.00%	1.37%	7.76%
27	Co	25	2.4	47
28	Ni	75	1	76
29	Cu	55	13	110
30	Zn	70	45	86
31	Ga	15	20	16
32	Ge	1.5	1.1	1.4
33	As	1.8	0.5	1.9
34	Se	0.05	0.007	0.3
35	Br	2.5	0.4	0.4
37	Rb	90	220	21
38	Sr	375	250	190
39	Y	33	13	25
40	Zr	165	210	105
41	Nb	20	24	9.5
42	Mo	1.5	6.5	0.57
44	Ru	0.01		
45	Rh	0.005		<0.001
46	Pd	0.01	0.002	0.025
47	Ag	0.07	0.05	0.08
48	Cd	0.2	0.03	0.15
49	In	0.1	0.02	0.07
50	Sn	2	3.5	3.2
51	Sb	0.2	0.31	1.01
52	Te	0.01	<1	<1
53	I	0.5	<0.03	<0.03
55	Cs	3	1.5	0.9
56	Ba	0.04%	0.12%	0.02%
57	La	30	101	9.8
58	Ce	60	170	23
59	Pr	8.2	19	3.4
60	Nd	28	55	15
62	Sm	6.0	8.3	3.6
63	Eu	1.2	1.3	1.1
64	Gd	5.4	5	4
65	Tb	0.9	0.54	0.65
66	Dy	3.0	2.4	4
67	Ho	1.2	0.35	0.69
68	Er	2.8	1.2	2.4

(continued)

Table 5.1
(*continued*)

Atomic Number	Element	Crustal Average	Granite (G-1)	Diabase (W-1)
69	Tm	0.5	0.15	0.30
70	Yb	3.4	1.1	2.1
71	Lu	0.5	0.19	0.35
72	Hf	3	5.2	2.7
73	Ta	2	1.5	0.50
74	W	1.5	0.4	0.5
75	Re	0.001	<0.002	<0.002
76	Os	0.005	0.00007	0.0003
77	Ir	0.001	0.00001	0.003
78	Pt	0.01	0.0019	0.0012
79	Au	0.004	0.004	0.004
80	Hg	0.08	0.1	0.2
81	Tl	0.5	1.2	0.11
82	Pb	13	48	7.8
83	Bi	0.2	0.07	0.05
90	Th	7.2	50	2.4
92	U	1.8	3.4	0.58

*From Mason, B. and Moore, C. B., 1982, *Principles of Geochemistry*. Copyright © 1982 by John Wiley & Sons, Inc., New York.

of the continental crust. It turns out that this average composition is intermediate between that of granite and basalt (or its coarser grained equivalents, diabase and gabbro), which are the two most common igneous rock types (see Table 5.1). If the sampling by Clarke and Washington had included an appropriate number of basalts from the deep ocean basins, their average would have been more representative of the average crust than of just the average continental crust. Consideration of the extensive basalts on the ocean floor would lower the average Si, K, and Na values but would increase the proportion of Fe, Mg, and Ca.

It should be noted that eight elements make up approximately 99 weight percent of the Earth's crust; of these, oxygen is by far the most abundant. This predominance is even more apparent when the figures are recalculated to atom percent and volume percent (Fig. 5.1). As such, the Earth consists almost entirely of oxygen compounds, especially silicates but also oxides and carbonates. Thus the minerals referred to as the "rock-forming" minerals are, with few exceptions, members of these groups. In terms of number of atoms, oxygen exceeds 60%. If the volumes of the most common ions are considered, oxygen is found to constitute about 94% of the total volume of the crust. In other words, the Earth's crust, on an atomic scale, consists essentially of a close packing of oxygen anions with interstitial metal ions, chiefly Si.

It is noteworthy that many elements important to our economy have very low values for their average abundances in the crust (see Table 5.1). For example, Cu (atomic number, $Z = 29$) = 55 parts per million (ppm), Pb ($Z = 82$) = 13 ppm, and Hg ($Z = 80$) = 0.08 ppm. On the other hand, the less commonly used element Zr ($Z = 40$) is more abundant (165 ppm) than

FIG. 5.1 The eight most common elements in the Earth's crust. (*From Mason, B. and Moore, C. B., 1982, *Principles of Geochemistry*, 4th ed. John Wiley & Sons, Inc., New York. †Ionic radii taken from Table 4.8. Numbers in parentheses refer to coordination number.)

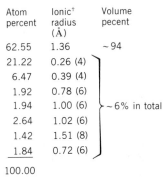

	Weight percent*		Atom percent	Ionic† radius (Å)	Volume percent
O	46.60		62.55	1.36	~94
Si	27.72		21.22	0.26 (4)	
Al	8.13		6.47	0.39 (4)	
Fe	5.00		1.92	0.78 (6)	
Ca	3.63		1.94	1.00 (6)	~6% in total
Na	2.83		2.64	1.02 (6)	
K	2.59		1.42	1.51 (8)	
Mg	2.09		1.84	0.72 (6)	
	98.59		100.00		

Weight percent

Cu. Similarly, Ga (Z = 31) is more abundant (15 ppm) than Hg. Clearly, in order to produce metals needed for our economy, one must locate areas of high concentrations in order to make the mining profitable. Copper is not extracted from rocks of average composition, but from ore deposits in which copper has been concentrated and is present in specific copper-bearing minerals (ore minerals).

Some elements, for example Rb (Z = 37), are dispersed throughout common minerals and are never concentrated. Rb does not form specific Rb compounds, but is housed in K-rich minerals, and therefore is an example of what is called a *dispersed element*. Other elements are strongly concentrated in specific minerals. For example, Zr is concentrated in zircon ($ZrSiO_4$), and Ti in rutile (TiO_2) and ilmenite ($FeTiO_3$).

Although we cannot directly determine the average composition of the Earth as a whole, assumptions and calculations can be made on the basis of meteorite samples. These are generally concluded to represent materials analogous to materials within the Earth. The composition of iron meteorites (mainly FeNi alloy) is believed to be very similar to the composition of the Earth's core. Meteorites with about 50% metal and 50% silicate are probably representative of the lower mantle, and stony (silicate) meteorites with little metal are similar to materials in the crust. On the basis of the known average compositions of these types of meteorites and the known volumes of the core, mantle, and crust, one can arrive at an average estimate for the composition of the Earth as a whole. Mason and Moore (1982) made such a calculation, with the following results, in weight percent: Fe, 34.63%; O, 29.53%; Si, 15.20%; Mg, 12.70%; Ni, 2.39%; S, 1.93%; Ca, 1.13%; Al 1.09%; and seven other elements (Na, K, Cr, Co, Mn, P, and Ti) each in amounts from 0.1 to 1%. Although all of the above elements in the calculated average Earth abundance (except for Ni and S) are also major elements in the average crustal abundance listing (Table 5.1 and Fig. 5.1), they appear in a different order because of the consideration of materials from the Fe-Ni-S–rich and O- and H_2O-poor mantle and core.

CHEMICAL ANALYTICAL TECHNIQUES

Chemical analyses of minerals (and rocks) are obtained by a variety of analytical techniques. Prior to about 1947, quantitative mineral analyses were obtained mainly by "wet" analytical techniques, in which the mineral is dissolved by some appropriate means. Determination of the element(s) in solution was then usually made by one or more of the following techniques: (1) *colorimetry*, which involves color-forming reactions in solution and subsequent comparison of intensities of beams of visible light transmitted through the analysis solution and a set of color-graded series of standard solutions; (2) *volumetric (or titrimetic) analysis*, which entails determining the volume of a solution of known concentration that is needed to react quantitatively with a solution of a weighed or volumetrically measured quantity of material. The weight of the element being determined is calculated from the volume of reagent used; and (3) *gravimetric analysis*, which involves the precipitation of elements in solution by formation of insoluble compounds that are subsequently dried or ignited, and weighed.

Since 1960 the majority of analyses have been made by instrumental techniques such as atomic absorption spectroscopy, X-ray fluorescence analysis, electron microprobe analysis, and optical emission spectroscopy. Each of these techniques has its own specific sample preparation requirements and fairly well-established detection limits and error ranges. The results of any analysis are generally presented in a table of weight percentages of the elements or oxide components in the mineral analyzed. "Wet" analytical techniques allow the quantitative determination of various oxidation states of cations (such as Fe^{2+} vs. Fe^{3+}) and also the determination of the H_2O content of hydrous minerals. Instrumental methods generally do not provide information as to the oxidation state of elements or the presence of H_2O.

Minerals submitted for chemical analysis ideally consist of one mineral species only (the one being analyzed) and must be free of weathering or other alteration products and inclusions. Because it is often very difficult, if not impossible, to separate as much as $\frac{1}{10}$ gram to 1 gram of clean material for analysis, analysis results not infrequently must be accompanied by a statement concerning the amount of impurities. An instrumental method, such as electron probe microanalysis, allows for quantitative analysis, *in situ* (in a polished section), of mineral grains as small as 1 micron in diameter (1 μm = 10^{-3} mm). All of the above-mentioned techniques require special analytical skills and often very expensive equipment (see below).

It may be helpful at this point to distinguish a *qualitative* from a *quantitative* chemical analysis. A *qualitative* analysis involves the detection and identification of all the constituents of a compound ("What is present?"). A *quantitative analysis* involves the determination of the weight percentages (or parts per million composition) of elements in a compound

("How much of each is present?"). A preliminary qualitative analysis is commonly very helpful in deciding on the methods to be followed in a quantitative analysis. In present-day practice, a quantitative analysis is commonly done through a combination of methods. For example, a gravimetric analysis procedure is often combined with atomic absorption spectrometry; an electron microprobe analysis may be supplemented by data obtained gravimetrically.

Wet Chemical Analysis

After the sample that is to be analyzed has been ground to a fine powder, the first choice to be made in a wet procedure is that of the best method of decomposition of the sample. Normally this is one of several acids, such as hydrochloric acid (HCl), sulfuric acid (H_2SO_4), or hydrofluoric acid (HF), or a mixture of acids. Commonly, fusion (heating at high temperature with a solid flux) is necessary to render the sample soluble. It is this dissolution of the sample that gives the overall technique the name of "wet" analysis. Once the sample is completely in solution, the next steps involve the appropriate colorimetric, volumetric, or gravimetric procedures to determine the desired element(s). The selection of the appropriate technique is dictated by the concentration of the element(s) in the sample and the ease with which an element can be separated from interfering elements. Concentration ranges of elements are related to techniques, as follows:

Method	Concentration of Element in Sample
Gravimetric	low % to 100%
Volumetric	low % to 100%
Colorimetric	parts per million to low %

Gravimetric Analysis

The term *gravimetric analysis* implies that the amount of a product produced in a chemical reaction is obtained by weighing. For example, in the reaction

$$mC + nR \rightarrow C_mR_n \downarrow$$

C is the constituent, R is the reagent, and C_mR_n is the precipitated product, the amount of which is determined by weighing. The precipitation processes, and the selection of the appropriate reagents to cause precipitation of a specific element from the original solution, must also be a separation process. That is, only one specific element is to be precipitated at one time.

The final procedure in quantitative gravimetric analysis is the computation of the weight percent values of the elements (or oxide components) according to the formula:

$$P = \frac{w}{W} \times 100$$

where P is the percentage of the constituent sought, w is the weight of the constituent, and W is the weight of the sample.

The college student is introduced to wet chemical analytical techniques in chemistry courses in qualitative and/or quantitative chemical analysis. However, the training involved in becoming a professional chemical analyst is lengthy and demanding of skill and patience. Many of the best wet analytical procedures are time-consuming. For these reasons, newer instrumental analytical techniques have replaced many of the wet analytical procedures. Instrumental techniques, if the instrumentation is kept in good running order, are generally more efficient in the production of chemical analytical data, especially when large numbers of samples are routinely analyzed for a specific set of elements. Some instrumental methods may also be uniquely sensitive to the detection of elements present in only trace amounts, such as at the parts per million or parts per billion level. Such detection sensitivity is generally outside the realm of most gravimetric and volumetric procedures.

Atomic Absorption Spectroscopy

This analytical technique is commonly considered another "wet" analytical procedure because the original sample must be completely dissolved in a solution before it can be analyzed. It is impossible to adequately describe the impact atomic absorption spectroscopy (AAS) has had on the chemical determination of elements since the technique was first introduced by Alan Walsh in 1955. A classical wet chemist of that time instantly appreciated the speed, accuracy, and elimination of the need for most chemical separations that came with AAS. While it is true that much AAS work is now being supplemented by inductively coupled plasma (ICP) and ICP–mass spectrometric (ICP-MS) methods, the low cost, accuracy, and ease of operation of AAS will ensure it a prominent place in the analytical laboratory for years to come.

The basic concepts of AAS can be explained with reference to Fig. 5.2. The energy source in this technique is a light source (a hollow cathode lamp), the energy of which ranges from the visible to the ultra-

FIG. 5.2 Schematic representation of the major components in an atomic absorption spectrometer.

violet portion of the electromagnetic spectrum, with wavelengths down to about 190 nm (see Fig. 5.3). The energy of a quantum (see also Chapter 4) is proportional to the wavelength of the radiation, in accordance with the Einstein equation

$$E = \frac{hc}{\lambda}$$

where E is the energy of the quantum, c is the speed of light, λ is the wavelength of the radiation, and h is a proportionality constant known as Planck's constant. The cloud in Fig. 5.2 ideally represents atoms that are free from any molecular bonding forces. When the light energy is equivalent to the energy required to raise the atom from its low energy levels to higher energy levels, it is absorbed and causes excitation of the atom. In Fig. 5.2 the light beam enters the sample cloud and the absorption of the light beam in the cloud is monitored by a spectrometer, across

the atomic cloud. In order to determine element concentrations by this analytical technique, the atoms must be completely free of any of the bonding that exists in the solid or liquid state, because the electrons will not be free to absorb specific wavelengths from the lamp if they are bound to surrounding atoms. To achieve this, the sample solution is aspirated as a fine mist into a flame, where it is converted to an atomic vapor. The amount of light radiation that is measured by the spectrometer (after absorption in the atomic cloud) is expressed as

$$A = \log I_0/I$$

where A is the *absorbance*, I_0 is the incident light intensity, and I the transmitted light intensity.

The incident light intensity is supplied by a hollow cathode lamp. This lamp, at high lamp current, causes emission of a line spectrum that is characteristic of the element being excited in the cathode tube.

FIG. 5.3 The electromagnetic spectrum with the visible region expanded. Wavelength (λ) values are expressed in nm (nanometers), where 1 nm = 10 angstroms and 1 nm = 10^{-9} meters.

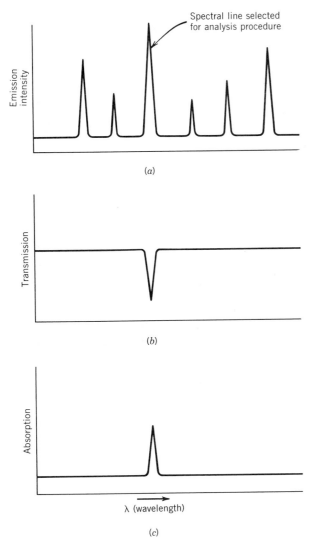

FIG. 5.4 *(a)* Line spectrum emitted by the hollow-cathode lamp. This is the same spectrum as that of the element to be analyzed. *(b)* The element in the atomic vapor (above the flame of the burner) absorbs the energy of the specific spectral line selected for the analysis. *(c)* The photomultiplier tube records the absorption of the specific spectral line.

From this light spectrum a single high-intensity line of specific wavelength (λ) is selected as the light source (Fig. 5.4a). λ is selected so that the spectral line emitted by the cathode lamp is the same as that of the element being analyzed in the atomic vapor.

Prior to the vaporization and analysis procedure, a spectrometer (consisting of a photomultiplier tube, a monochromator, and associated electronics) is preset on the wavelength (λ) of the emission spectral line (of the hollow cathode tube). This is achieved with the aid of the monochromator, which allows the isolation of a specific spectral line. The sample being vaporized absorbs energy at the λ value of the cath-

ode tube (Fig. 5.4b), and the final reduction in intensity (due to absorption) is measured by the photomultiplier tube (Fig. 5.4c). The detectibility for some elements, such as magnesium and sodium, is extremely high, with quantitative determinations possible in the ppm range. Figure 5.5 is a photograph of a commercially available atomic absorption spectrometer.

X-Ray Fluorescence Analysis

This analytical technique, also known as *X-ray emission spectrography*, is used in most research laboratories that study the chemistry of inorganic substances, but it is also routinely used in a wide range of industrial applications. Examples of such applications are in the mining industry (for quality control of the product shipped to the consumer), in the glass and ceramics industry, in the manufacture of metals and alloys, and in environmental protection and pollution control applications.

The analysis sample in this technique is ground to a fine powder and subsequently compressed into a circular pellet, or into a disc with the admixture of a binder. This sample preparation is distinctly different from the procedures used in the "wet" techniques discussed earlier. The pellet or disc of sample is irradiated (for a short period of time) with X-rays generated in a high-intensity X-ray tube (see Fig. 5.3 for the λ range of X-rays). These incident X-rays from the X-ray tube are, to a considerable extent, absorbed according to Beer's Law,

$$\log \frac{I_0}{I} = K_d \, \Delta d$$

where I_0 is the incident X-ray intensity, I is the intensity of the X-ray beam that was not absorbed in the sample, K_d is a proportionality constant, and Δd is the thickness of the sample. The X-ray energy that is absorbed in the sample results in the generation of an *X-ray emission spectrum that is characteristic* for each element in the sample. In the process of the absorption of X-ray energy in the sample, electrons are dislodged from the innermost shells (known as K, L, M; see Chapter 4). An expelled electron (from, e.g., the K shell) must be replaced, and it is highly probable that the vacancy will be filled from the next outer shell (the L shell) rather than from a more remote shell. This creates a new vacancy, which is filled from the next shell, and so on. Electrons that "fall into" inner electron shells move from higher to lower energy levels and as a result emit energy in the form of characteristic

FIG. 5.5 Automated atomic absorption spectrometer. This specific instrument is the Smith-Hieftje 8000 Automatic AA Spectrophotometer. The hollow cathode lamp is at the left, the flame atomizer is in the center, and the spectrometer and automatic readout system are on the right. In the front of the instrument is an automated sample system by which 150 sample solutions can be analyzed automatically. (Courtesy of Thermo Jarrell Ash Corp., Franklin, Mass.)

X-radiation (see Fig. 5.6). The groupings of spectral lines are classified as K-, L-, or M-spectra in terms of the patterns in which the outer electrons fall into the lower energy states. These generated characteristic X-rays are known as *secondary X-rays* and the emission phenomenon is called *X-ray fluorescence*. Each element has characteristic spectral lines with specific wavelengths superimposed on a low-intensity continuous background spectrum. Examples of two characteristic K-spectra for two different elements are given in Fig. 5.7.

The generated (fluoresced or secondary) X-ray spectrum may consist of a very large number of spectral lines in a sample that consists of more than one or two elements. Such a spectrum must be resolved into its spectral line components, such that the lines can be identified by wavelengths (in nm or Å) that are

specific to the elements responsible for their production. This is achieved with an X-ray spectrometer consisting of a diffracting crystal (with known spacing between adjoining atomic planes) and an X-ray detector. The crystal will diffract the various λ values of the impinging X-rays according to the Bragg equation

$$n \lambda = 2d \sin \theta$$

where n is generally a small number, ranging from 1 to 2 or 3 (known as the "order of diffraction"), λ is the wavelength of a specific spectral line, d is the distance between a specific set of atomic planes in the diffracting crystal, and θ is the angle over which the X-ray is reflected by the crystal (see also Chapter 7).

The intensity and position of each spectral line diffracted by the crystal in the spectrometer is re-

FIG. 5.6 Schematic illustration of the production if K and L characteristic spectra as a result of electrons cascading from upper to lower energy levels in the atomic structure. X-ray emission lines are labeled with a capital letter representing the shell whose vacancy is being filled. The Greek letter is α if the electron that fills the vacancy originates in the next highest shell, β if the electron comes from two shells up, and so on. The Arabic numeral subscripts indicate specific subshells (1 for s, 2 for p) for the origin of the electron that fills the vacancy.

corded by an electronic X-ray counting device (generally a scintillation counter or flow proportional counter). This is traced on a recorder or presented on a high-resolution visual display terminal. A schematic diagram of the major components that constitute an X-ray fluorescence analyzer is given in Fig. 5.8.

Qualitative X-ray fluorescence analysis involves identification of the various spectral lines with the elements responsible for them (see Fig. 5.9). Quantitative analysis is more involved because each X-ray intensity must be quantitatively compared with that of a standard (of known composition) of the same elemental makeup. Both peak and background intensities near the peak are counted to permit estimation of peak heights. On-line computers handle quantitative correction programs in extremely short time.

X-ray fluorescence analysis can be used for the determination of major elements (those in the one to many tens of percent range). However, it is also very sensitive to accurate determinations of some trace element components (e.g., Y, Zr, Sr, Rb, in the ppm range) because of near-zero background. A commercially available automated instrument is shown in Fig. 5.10.

Electron Probe Microanalysis

Electron probe microanalysis methods are based on the same principles as those outlined for X-ray fluorescence, except that the original energy source is not an X-ray tube but instead is a finely focused beam of electrons. Electrons are charged particles that can be focused using electrical fields. (X-rays, by contrast, cannot be focused, they can only be collimated.) As the name implies, this is a *microanalytical* technique, because it allows the qualitative and/or quantitative analysis of a minute volume of material.

The energy source used for generating electrons that will dislodge inner-shell electrons in the sample to be analyzed is a tungsten filament that at very high temperatures and high voltage becomes a source of free electrons (inside the high vacuum of the electron gun and column of the instrument; see Fig. 1.9). These electrons can be focused into a very fine beam through a set of electromagnetic lenses between the electron source (the W filament) and the sample to be analyzed (see Fig. 1.9). Under fairly optimal operating conditions, this electron beam can be focused to a smaller than 1 μm diameter. Because electrons in the beam impinge upon the sample at high velocity, they will penetrate a distance that may be about three times larger than the diameter of the beam (see Fig. 5.11). The minimal analysis volume that can be obtained will range from about 10 to 20 μm^3, which in weight is approximately 10^{-11} grams (for a silicate material).

In this minute analysis volume, the incident higher energy electrons displace inner-shell electrons of the constituent atoms of the sample. Outer-shell

FIG. 5.7 Characteristic K-spectra for the elements Mo and Cu, superimposed on a continuous spectrum. This continuous spectrum is considered part of the background X-ray intensity in analytical techniques.

FIG. 5.8 Schematic illustration of the major components in an X-ray fluorescence analyzer.

electrons fill these inner-shell vacancies, losing energy by the emission of characteristic X-rays (just as in X-ray fluorescence, discussed above; see Figs. 5.6 and 5.7). These characteristic X-ray spectra are analyzed by a crystal spectrometer (as in the case of X-ray fluorescence) or by an energy dispersive X-ray analysis system. A qualitative electron microprobe analysis is relatively quick and easy to perform, once the sample is in the proper condition for analysis (see below). A quantitative analysis may involve some complex correction procedures that address various interactions between the generated X-rays and their travel path through the sample before detection by the spectrometer. Such corrections, known as matrix corrections, are generally performed by a high-speed computer that is on-line with the instrument. See Fig. 1.8

for a photograph of a computer-automated electron microprobe.

Because the analytical technique involves a finely focused electron beam that can cause considerable heating of the sample analysis area, and because a minute analysis volume is involved, the sample preparation for electron microprobe techniques is very different from that for X-ray fluorescence. Most commonly, the analysis sample is a polished section (or polished thin section) of a mineral, or rock, or other solid material to be analyzed. The polished surface can be accurately located under the finely focused electron beam by a precise and generally computer-controlled stage mechanism. This allows for the location of grains, or analysis area, as small as 1 μm in diameter, in the field of a high-powered opti-

FIG. 5.9 Chart recording of the X-ray fluorescence spectrum obtained of elements present in a genuine bank note. The elements responsible for specific peaks are identified. The W peaks are due to the X-ray tube used with a W target; they are not part of the chemical composition of the bank note. The horizontal scale is in angles of 2θ, expressed as degrees. (From Liebhafsky, H. A., Pfeiffer, H. G., Winslow, E. H. and Zemany, P. D., 1960, *X-ray Absorption and Emission in Analytical Chemistry*. John Wiley & Sons, New York, 357 pp.)

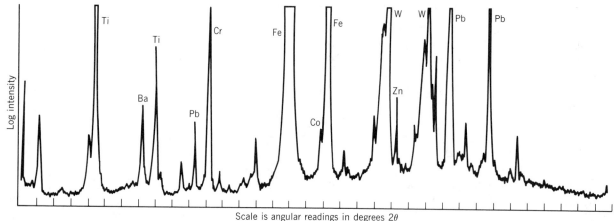

Scale is angular readings in degrees 2θ

FIG. 5.10 The Siemens SRS 303 X-ray fluorescence spectrometer. The left-hand cabinet contains the X-ray source, X-ray spectrometer, and an automatic sample changer. The desk area, on the right, is the control and data handling center of the equipment. (Courtesy of Siemens Industrial Automation, Inc., Madison, Wis.)

cal (transmitted and reflected light) microscope, which is built into the electron beam column.

The absolute detection limit for most elements analyzed by this technique is not as good as that for

FIG. 5.11 Schematic representation of the very small volume irradiated by electrons incident upon the polished surface of a sample. The characteristic X-rays generated are dispersed and analyzed by a spectrometer (with a diffracting crystal, as in X-ray fluorescence) or an energy dispersive analysis system (see Fig. 1.8 for a photograph of an electron microprobe and Fig. 1.9 for a schematic cross section of the instrument.)

Finely focused
electron beam

X-rays to be analyzed
by spectrometer

Polished surface
of analysis sample

1 μm

Analysis volume
~ 10 to 20 μm³

X-ray fluorescence because of the presence of a continuum (background) spectrum. However, the ability to obtain a quantitative chemical analysis on a minute volume of material (or a specific mineral grain) is the main reason for the great popularity of this technique in studies of minerals, rocks, ceramics, alloys, and so on.

Optical Spectrographic Analysis

Optical emission spectrography is based on the fact that atoms of elements may be excited (energized), in a flame, arc, spark, or discharge tube, to emit radiant energy. When this energy is dispersed, either by a prism or by a grating, it can be photographed or recorded as a spectrum. The number of lines and the intensity of the lines in this spectrum are determined primarily by the electronic configuration of the atom. Because atoms have different electronic structures, each element has a characteristic spectrum that can be used to identify it as present in the analysis sample. Quantitative analysis by this technique entails measuring the brightness of the spectral lines as recorded on photographic plates, or as measured by photomultiplier tubes.

FIG. 5.12 Schematic illustration of the main components in one type of optical emission spectrograph.

Figure 5.12 shows a schematic layout of the major components of an optical emission spectrograph. The analysis sample is ground to a fine powder and loaded into the bottom electrode, where a high-temperature spark, or arc, between the electrodes vaporizes it. With the atoms in this excited (vapor) state, they emit light of definite wavelengths (a characteristic spectrum) that results from the falling of electrons from outer electron shells into vacancies in inner shells caused by the original excitation of the atoms (see Fig. 5.6). The radiation emitted during this return to the "ground state" is the result of electrons falling from a higher level (E_2) to a lower energy level (E_1), which can be expressed as $E_2 - E_1 = hc/\lambda$.

In optical emission spectrography, complex spectral patterns result from the large numbers of different electronic transitions. In order to disperse the spectrum, a prism or grating is used (a grating is made by ruling a very large number of very finely spaced parallel lines on a mirror). The resulting spectrum is recorded on a large photographic plate, or recorded electronically by photomultiplier tubes. The resultant spectrum is commonly analyzed by computer methods.

Standard optical spectrographs used for research and quality control are costly and can be operated only by a skilled technician. However, several small, relatively low-priced spectroscopes and spectrographs are also available. Because of their simplicity of operation, they may be used as a qualitative determinative tool in the elementary mineralogy laboratory. One such instrument, the Vreeland spectroscope, is illustrated in Fig. 5.13. Figure 5.14 shows

schematically how a spectrum produced in an arc is analyzed in this spectroscope. Part of the light source is dispersed via an optical prism and subsequently by a diffraction grating. This dispersed spectrum is viewed with an eyepiece (spectroscope) or recorded on film (spectrograph). In the Vreeland instrument, while the spectrum from the vaporizing mineral is being observed, standard spectra of many different elements can be brought into the field of view for comparison, and the elements composing the mineral can be identified.

FIG. 5.13 Vreeland spectroscope. (Courtesy of Spectrex Co., Redwood City, Calif.)

FIG. 5.14 Sketch of optical path in a Vreeland spectroscope.

COMPOSITIONAL VARIATION IN MINERALS

As already noted, it is the exceptional mineral that is a *pure substance*, and most minerals display extensive variation in chemical composition. Compositional variation is the result of substitution, in a given structure, of an ion or ionic group for another ion or ionic group. This type of process, referred to as *ionic substitution*, or *solid solution*, occurs among minerals that are isostructural (see Chapter 3). A concise definition of solid solution is as follows: *a solid solution is a mineral structure in which specific atomic site(s) are occupied in variable proportions by two or more different chemical elements (or groups)*. The main factors that determine the amount of solid solution taking place in a crystal structure are:

1. *The comparative sizes of the ions, atoms, or ionic groups* that are substituting for each other. Generally a wide range of substitution is possible if the *size difference* between the ions (or atoms) is *less than about 15%*. If the radii of the two elements differ by 15 to 30%, substitution is limited or rare, and if the radii differ by more than 30%, little substitution is likely.

2. *The charges of the ions involved in the substitution.* If the charges are the same, as in Mg^{2+} and Fe^{2+}, the structure in which the ionic replacement occurs will remain electrically neutral. If the charges are not the same, as in the case of Al^{3+} substituting for Si^{4+}, additional ionic substitutions elsewhere in the structure must take place in order to maintain overall electrostatic neutrality.

3. *The temperature at which the substitution takes place.* There is, in general, a greater tolerance toward atomic substitution at higher temperatures when thermal vibrations (of the overall structure) are greater and the sizes of available atomic sites are larger. There-

fore, in a given structure, one expects a greater variability in its composition at higher temperature than at lower temperature.

The types of solid solution can be discussed in terms of *substitutional, interstitial,* and *omission* solid solution mechanisms.

Substitutional Solid Solution

The simplest types of ionic substitutions are *simple cationic* or *anionic* substitutions. In a compound of the type A^+X^-, A^+ may be partly or wholly replaced by B^+. In this instance there is no valence change. Such a substitution is illustrated by the substitution of Rb^+ in the K^+ position of KCl or biotite. A simple anionic substitution can be represented in an A^+X^- compound in which part or all of X^- can be replaced by Y^-. An example is the incorporation of Br^- in the structure of KCl in place of Cl^-. An example of a *complete binary solid solution series* (meaning substitution of one element by another over the total possible compositional range, as defined by two end member compositions) is provided by olivine $(Mg,Fe)_2SiO_4$. Mg^{2+} can be replaced in part or completely by Fe^{2+}; the *end members* of the olivine series between which there is complete solid solution are Mg_2SiO_4 (forsterite) and Fe_2SiO_4 (fayalite). Another example of a complete solid solution series is given by $(Mn,Fe)CO_3$, which occurs as a series from $MnCO_3$ (rhodochrosite) to $FeCO_3$ (siderite). An example of a complete anionic series between two compounds is given by KCl and KBr. The size of the two anions is within 10% of each other, allowing for complete substitution of Cl^- and Br^-, and vice versa (see Fig. 5.15). Table 5.3 is an example of extensive, but not complete, simple cation solid solution in the mineral sphalerite, in which Fe^{2+} is the main replacement for Zn^{2+} in the structure.

If in a general composition of $A^{2+}X^{2-}$ a cation B^{3+} substitutes for some of A^{2+}, electrical neutrality can be maintained if an identical amount of A^{2+} is at the same time replaced by cation C^+. This can be represented by

$$2A^{2+} \rightleftarrows 1B^{3+} + 1C^+$$

with identical total electrical charges on both sides of the equation. This type of substitution is known as *coupled substitution*. The substitution of Fe^{2+} and Ti^{4+} for $2Al^{3+}$ in the corundum (Al_2O_3) structure of the gem sapphire is an example of such coupled substitution. The plagioclase feldspar series can be represented in terms of two end members, $NaAlSi_3O_8$ (albite) and $CaAl_2Si_2O_8$ (anorthite). The complete solid

Simple cationic:

$Mg^{2+} \rightleftarrows Fe^{2+}$ (complete) Mg^{2+} (6) = 0.72Å
$Fe^{2+} \rightleftarrows Mn^{2+}$ (complete) Fe^{2+} (6) = 0.78Å
$Na^{+} \rightleftarrows K^{+}$ (partial) Mn^{2+} (6) = 0.83Å
$Mg^{2+} \rightleftarrows Mn^{2+}$ (partial) Na^{+} (8) = 1.18Å
                                                                                   K^{+} (8) = 1.51Å

Simple anionic:

$Br^{-} \rightleftarrows Cl^{-}$ (complete) Br^{-} (6) = 1.96Å
                                                                        Cl^{-} (6) = 1.81Å
$I^{-} \rightleftarrows Cl^{-}$ (partial) I^{-} (6) = 2.20Å

Coupled cationic: *Extent of solid solution:*

$Na^{+}Si^{4+} \rightleftarrows Ca^{2+}Al^{3+}$ complete at high
2 cations: 2 cations: temperature in
total charge 5+   total charge 5+   plagioclase

$Ca^{2+}Mg^{2+} \rightleftarrows Na^{+}Al^{3+}$ limited, as in omphacite,
2 cations: 2 cations: a member of the
total charge 4+   total charge 4+   pyroxene group

$Mg^{2+}Al_2^{3+} \rightleftarrows Fe^{2+}{}_2Ti^{4+}$ extensive, as in the
3 cations: 3 cations: spinel group
total charge 8+   total charge 8+

$\square° = Si^{4+} \rightleftarrows Na^{+}Al^{3+}$ extensive, as in
2 sites: 2 cations in 2   arfvedsonite, a sodium
total charge 4+   sites: amphibole
$\square°$ = vacancy   total charge 4+

FIG. 5.15 Examples of various types of substitutional solid solution. Ionic radii are taken from Table 4.8. Parentheses after the ions enclose the coordination number.

solution between these two end member compositions illustrates coupled substitution:

$$Na^{1+}Si^{4+} \rightleftarrows Ca^{2+}Al^{3+}$$

This means that for each Ca^{2+} that replaces Na^{1+} in the feldspar structure, one Si^{4+} is replaced by Al^{3+} in the Si-O framework. The equation shows that the total electrical charges on both sides of the equation are identical; as such the structure remains neutral. An example of only limited coupled solid solution is provided by two pyroxenes, diopside ($CaMgSi_2O_6$) and jadeite ($NaAlSi_2O_6$). The coupled replacement can be represented as: $Ca^{2+}Mg^{2+} \rightleftarrows Na^{+}Al^{3+}$. Although the pyroxene structure is neutral with either type of cation pair, the atomic sites and coordination polyhedra do not allow for a complete solid solution of this type (see Fig. 5.15).

Atomic sites that in some structure are *unfilled (vacant)* may become partly or wholly occupied due to a substitutional scheme in which a vacancy is involved. For example, a partial coupled substitution in the amphibole tremolite, $\square Ca_2Mg_5Si_8O_{22}(OH)_2$, where $\square$ is the normally vacant A site in the structure, may result from

$$\square \, Si^{4+} \leftrightarrows Na^{1+}Al^{3+}$$

In this coupled substitutional scheme Al^{3+} replaces Si^{4+} in the tetrahedral position, and the additional Na^{1+} is housed in a site that is normally vacant (shown as $\square$) (see Fig. 5.15).

Interstitial Solid Solution

Between atoms, or ions, or ionic groups of a crystal structure interstices exist, which normally are considered voids. When ions or atoms are located in these structural voids, we speak of *interstitial substitution* or *interstitial solid solution.* In some crystal structures these voids may be channel-like cavities, as in beryl, $Be_3Al_2Si_6O_{18}$. In this ring silicate, large ions as well as molecules can occupy the tubular cavities of the superimposed rings (see Fig. 5.16). Considerable amounts of K^+, Rb^+, Cs^+, and H_2O as well as CO_2 are reported in beryl analyses, housed interstitially in the hexagonal channels. H_2O and CO_2 molecules are weakly bonded to the internal oxygens of the 6-fold Si_6O_{18} rings. The large monovalent alkalis K^+, Rb^+, and Cs^+ are also located inside these rings, but they are much more strongly bonded, because of the following coupled substitutional mechanisms:

$$Si^{4+} \rightleftarrows Be^{2+} + 2R^+$$

and

$$Si^{4+} \rightleftarrows Al^{3+} + R^+$$

where R represents K^+, Rb^+, or Cs^+. In the first substitution two monovalent alkali cations are housed inside the interstice of the hexagonal Si_6O_{18} chain; in the second, only one cation is positioned there.

Another common example of interstitial solid solution is found in the zeolite group of silicates (see Chapter 13). The zeolites constitute a group of tectosilicates in which (SiO_4) and (AlO_4) tetrahedra are linked together in a very open framework. Within this skeletal framework there may be large holes and continuous channels, with openings ranging from 2 to 7 Å in diameter. These channels provide easy access to the interior of the crystals, and they also accommodate water molecules. Because the water molecules are weakly tied, by hydrogen bonding, to the surrounding Al-Si framework, mild heating will remove them from these interstitial positions without collapse of the framework. Upon removal of the heat and in the presence of liquid water or steam, zeolites rehydrate by re-incorporation of H_2O molecules in the interstices from which the molecules were originally removed.

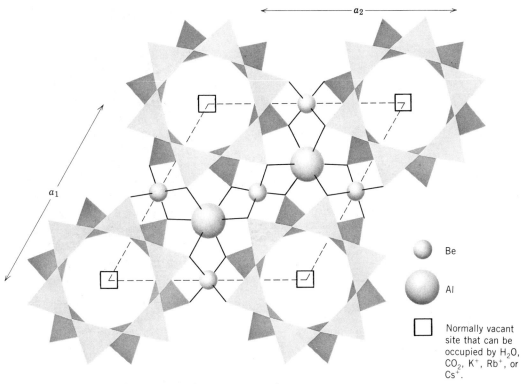

FIG. 5.16 Plan of the hexagonal structure of beryl, $Be_3Al_2Si_6O_{18}$ projected onto the basal plane (0001). The Si_6O_{18} rings at two different heights are shown. The hexagonal channels are the locus for large alkali ions and neutral molecules such as H_2O and CO_2. A unit cell is outlined by dashed lines.

Omission Solid Solution

Omission solid solution occurs when a more highly charged cation replaces two or more other cations for charge balance. The substitution can take place only in one atomic site, leaving other sites vacant or omitted. For instance, when Pb^{2+} substitutes for K^+ in the blue-green variety of microcline feldspar ($KAlSi_3O_8$), known as amazonite, it replaces *two* K^+ ions, but occupies only one site:

$$K^+ + K^+ \leftrightarrows Pb^{2+} + \square$$

creating a lattice vacancy $\square$, which can become a color center (see Chapter 6).

The best-known mineralogical example of this type of solid solution is provided by pyrrhotite, $Fe_{(1-x)}S$. In pyrrhotite, sulfur occurs in hexagonal closest packing and iron in 6-coordination with sulfur. If each octahedral site were occupied by Fe^{2+}, the formula of pyrrhotite would be FeS. In pyrrhotites, however, there is a variation in the percentage of *vacancies* in the octahedral sites, causing the composition to range from Fe_6S_7 through $Fe_{11}S_{12}$, close to FeS. The formula is generally expressed as $Fe_{(1-x)}S$, with x ranging from 0 to 0.2. Minerals, such as pyrrhotite, in

which a particular structural site is incompletely filled are known as *defect structures* (see Fig. 5.17 for a structural image of pyrrhotite). When Fe^{2+} is absent from some of the octahedral sites in pyrrhotite, with the sulfur net completely intact, the structure is not electrically neutral. It is very likely that some of the Fe is in the Fe^{3+} state to compensate for the deficiency in Fe^{2+}. If this is so, a neutral pyrrhotite formula can be written as $(Fe^{2+}_{1-3x}Fe^{3+}_{2x})\square_x S$, where $\square$ represents vacancies in the cation position. If x in this example were 0.1, there would be 0.1 vacancies, 0.7 Fe^{2+}, and 0.2 Fe^{3+} in the structure of pyrrhotite. In other words, Fe^{3+} is accommodated in the pyrrhotite structure by the following substitution: $Fe^{2+} + Fe^{2+} + Fe^{2+} \leftrightarrows Fe^{3+} + Fe^{3+} + \square$, which creates a vacancy $\square$.

Another mineral example of omission solid solution (with defects, or cation vacancies) is maghemite, γ-Fe_2O_3, an oxidation product of magnetite, Fe_3O_4 (or $Fe_2^{3+}Fe^{2+}O_4$). Maghemite is an iron-deficient spinel-type structure in which the tetrahedral and octahedral cation sites show considerable vacancies. The maghemite composition may be stated as follows: $Fe^{3+}Fe^{3+}_{1.67}\square_{0.33}O_4$, where $\square$ denotes a cation vacancy.

FIG. 5.17 High-resolution transmission electron microscope (HRTEM) structure image of pyrrhotite in which the white spots correspond to columns of iron atoms; these columns are aligned perpendicular to the plane of the photograph. The less intense spots represent columns from which some of the iron atoms are missing (omission solid solution). The columns are alternately more or less fully occupied by iron atoms. The white square is 3 Å to a side. (From Pierce, L. and Buseck, P. R., 1974, Electron imaging of pyrrhotite superstructures. *Science*, v. 186, pp. 1209–1212; copyright 1974 by the AAAS. See also Buseck, P. R., 1983, Electron microscopy of minerals. *American Scientist*, v. 71, pp. 175–185.)

EXSOLUTION

In our prior discussion of solid solution in minerals, it was pointed out that if two ions (or atoms) have very different sizes, the amount of substitution for each

FIG. 5.18 *(a)* Schematic two-dimensional representation of elements *A* and *B* in a completely disordered array (at high temperature). *(b)* Separation of elements *A* and *B* to *A*-rich and *B*-rich regions = exsolution (at lower temperature).

other would be expected to be limited. However, it was also pointed out that increased temperature would aid the substitution of ions of divergent size. That is, at elevated temperatures, the structure of a mineral expands overall and the amplitudes of vibra-

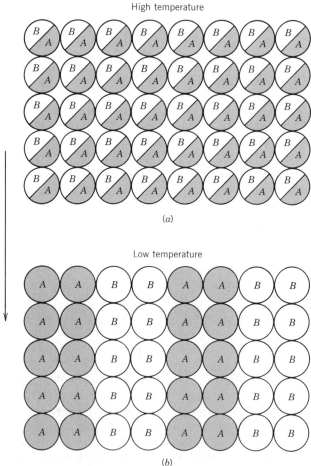

tion of the atoms become larger. As temperature increases, previously distinct structural sites become more similar on average, and finally the sites may become indistinguishable. The interchange of cations between sites leads to chemical disorder (see Chapter 3), where the average chemical content of each site may become the same. In other words, where there might have been only limited solid solution between, for example, Na^+ (radius = 1.18 Å for **C.N.** = 8) and K^+ (radius = 1.51 Å for **C.N.** = 8; from Table 4.8) at low temperature in a silicate (e.g., K-Na feldspar series), it is well known that at high temperature, such as about 1000°C, this same silicate series shows complete miscibility (or solid solution) between the $NaAlSi_3O_8$ and $KAlSi_3O_8$ end members.

Here we will address what happens when an originally homogeneous high-temperature mineral, containing cations of considerably different size, is allowed to cool to low temperature. The term *exsolution* refers to *the process whereby an initially homogeneous solid solution separates into two (or possibly more) distinct crystalline minerals without the addition or removal of material to or from the system. This means that no change in the bulk composition takes place* (see Fig. 5.18). Exsolution or "unmixing" is analogous to a well-known phenomenon that occurs when a mixture of oil and vinegar (vinaigrette) is made. The oil and vinegar are energetically stirred or shaken so as to produce a homogeneous but cloudly liquid suspension of very fine particles of both (oil and vinegar). When this mixture is allowed to stand for a short time, the original components will separate out ("unmix") and produce two clear liquids, oil and vinegar, as the end members.

Exsolution generally, although not necessarily, occurs on cooling. Some relatively coarse exsolution features can be observed under a high-power microscope when the scale of exsolution is about 1 μm (1 μm = 10^{-3} mm). An example of microscope exsolution in amphiboles is shown in a photomicrograph in Fig. 5.19. Exsolution lamellae that separate from the originally homogeneous host mineral are generally crystallographically oriented, as seen in Fig. 5.19. In amphiboles the orientation of exsolved lamellae is commonly parallel to {001} and {100}. In alkali (Na-K) feldspars, exsolution lamellae can often be seen in hand specimens to be approximately parallel to {100}. Such coarse-grained intergrowths in alkali feldspars consist of Na-rich feldspar lamellae exsolved from a K-rich host; they are known as *perthites* (see Fig. 13.129). If exsolution is on such a scale that it can be resolved only microscopically, it is referred to as *microperthite*; and if X-ray diffraction techniques are needed to resolve extremely fine (submicroscopic) la-

mellae, the intergrowth is known as *cryptoperthite*. These types of crystallographically controlled intergrowths resulting from exsolution are common in many mineral systems: alkali feldspars, pyroxenes, amphiboles, and iron oxides, to name a few examples. Examples of extremely fine exsolution lamellae in amphiboles and pyroxenes are given in Fig. 5.20.

The origin of exsolution textures is best illustrated with reference to a schematic temperature-composition (*T-X*) diagram in Fig. 5.21. The horizontal axis represents the compositional variation, in terms of molecular percentages, between two possible silicates that are isostructural. The hypothetical silicates are monoclinic and are represented as ASi_xO_y and BSi_xO_y (for pyroxene and amphibole groups, for example). The ionic sizes of A and B differ by about 25%. If we were to collect from the literature the chemical analyses for all low-temperature geological occurrences of these two minerals, we might conclude that there is very limited solid solution between the two end members, as shown graphically by the low-temperature (T_1) composition bar. These same two hypothetical silicates might also be present in high-temperature basalt flows that cooled very rapidly. At such high temperature (T_2) we may find a complete solid solution series (shown by shading in Fig. 5.21). What happens to the extent of solid solution between

FIG. 5.19 Photomicrograph of relatively coarse exsolution lamellae of hornblende (hbl) in grunerite (gru) and grunerite (gru) lamellae in hornblende (hbl). The lamellae are oriented parallel to (001) and (100), respectively.

.05 mm

(a)   (b)

FIG. 5.20 (a) Bright-field transmission electron microscope (TEM) image of cummingtonite exsolution lamellae in a glaucophane host. The narrow lamellae of ferromagnesian amphibole occur in two symmetrically related orientations, nearly parallel to (2 8 $\bar{1}$) and (2 $\bar{8}$ $\bar{1}$) of the C2/m clinoamphibole structure. These unusual orientations represent planes of best dimensional fit between the glaucophane and cummingtonite structures. This amphibole appeared perfectly homogeneous in the optical microscope and in the electron microprobe. (From Smelik, E. A. and Veblen, D. R., 1991, Exsolution of cummingtonite from glaucophane: A new direction for exsolution lamellae in clinoamphiboles. _American Mineralogist_, v. 76, pp. 971–984.) (b) High-resolution transmission electron microscope (HRTEM) image showing a sodic pyroxene lamella (Napx) parallel to (100) in an augite matrix. (From Otten, M. T. and Buseck, P. R., 1987, TEM study of the transformation of augite to sodic pyroxene in eclogitized ferrogabbro. _Contributions to Mineralogy and Petrology_, v. 96, pp. 529–538).

FIG. 5.21 (a) Schematic T-X (X = composition and is expressed as molecular or atomic %) diagram for hypothetical silicates ASi_xO_y and BSi_xO_y. At temperature T_2 a complete solid solution exists between the two end members. At lower temperatures a miscibility gap occurs. The arrows along the edge of the gap indicate the change in compositions of the coexisting silicates with decreasing T. The process of exsolution is discussed in the text. (b) Cross sections of mineral (with composition X_1 in FIG. 5.21a) illustrating sequence of appearance of possible exsolution lamellae. (1) homogeneous, at T_2. (2) exsolution lamellae parallel to (001) at T_4. (3) additional set of exsolution lamellae parallel to (100) at T_3, and coarsened lamellae parallel to (001).

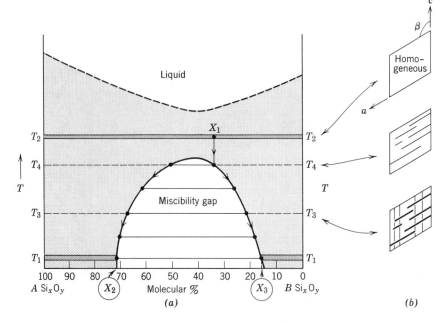

T_2 and T_1 is outlined by the *miscibility gap* in the diagram. This gap represents a temperature-composition field in which solid solution between the end members decreases gradually from higher to lower T. For specific compositions in this region two minerals coexist, rather than one homogeneous mineral, as above the gap.

Let us take a specific composition (X_1) in the region above the miscibility gap at T_2. At this temperature, the structure of the silicate is in a high-energy state and will allow relatively easy accommodation of A and B (although these differ by as much as 25% in ionic radius) in the same atomic sites. In other words, the A and B ions randomly occupy the various cationic sites in the structure (disorder; see Fig. 5.18a). On very slow cooling of this composition a temperature (T_4) is reached at which the structure is unable to accommodate a completely random distribution of A and B ions. Stresses in the structure resulting from the difference in size of the cations cause diffusion of A^+ ions to produce silicate regions with a preponderance of A^+; simultaneous B^+ ion diffusion results in silicate regions rich in B^+. This is the beginning of the process known as *unmixing*, which means that an originally homogeneous mineral segregates into two chemically different minerals (see Fig. 5.18b). The miscibility gap continually widens as the temperature is lowered from T_4. Under equilibrium conditions, the compositions of lamellae of the exsolution intergrowth that form as a function of decreasing temperature are given by points along the edge of the miscibility gap; such compositional points are joined by tielines. At the lowest temperature, T_1, we may find a rather coarse-grained, crystallographically well-oriented intergrowth of (A,B) Si_xO_y lamellae with the specific composition (X_2) in a host of $(B,A)Si_xO_y$ with specific composition (X_3). Although the segregation of ions within the silicate structure, which began at T_4, is complex on an atomic level, it is due mainly to the inability of a single structure to house ions to disparate sizes in a random distribution when decreasing temperature lowers the energy of the crystal. Fig. 5.22 is an illustration of the free energy (G)–composition curve for the exsolution process. It shows the high free energy content of the high-temperature disordered compositional region, and the much lower free energy states of the two exsolved minerals, whose compositions are marked as x_2 and x_3 on the compositional bar. The free energy, also known as the Gibbs free energy (G), is defined as $G = E + PV - TS$, where E = internal energy, PV is a pressure-volume term (which for solids at atmospheric pressure is negligible compared to other thermodynamic quantities; see also Chapter 9), and TS is a temperature-entropy term.

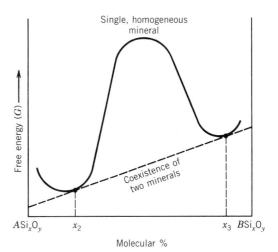

FIG. 5.22 Free energy (G)-composition curve illustrating the free energy changes during exsolution. The dashed line, which is tangential to the two minima in the free energy curve, gives the composition of the coexisting minerals at a specific temperature.

Although exsolution intergrowths are common in many mineral groups, the resulting textures are rarely visible in hand specimens. Perthite in the alkali feldspar series is an example of exsolution on a large scale, visible in hand specimen. The schematic sequence of cross sections in Fig. 5.21b illustrates what may be seen under the microscope as a result of the processes depicted in Fig. 5.21a. An originally homogeneous crystal of composition X_1 develops sets of differently oriented exsolution lamellae between temperatures T_4 and T_3.

These segregation processes all take place within the silicate structure in the solid state. The T-X diagram depicting this sequence of events (Fig. 5.21) is, therefore, in general referred to as a *subsolidus phase diagram*. (Other types of phase diagrams will be discussed in Chapter 9.)

The process of unmixing, on an atomic scale, is one of atomic migration (diffusion). An increase in temperature (and an increase in kinetic energy) will increase the mobility of an atom (or ion) and its chances of breaking away from its neighbors and moving to a new position. This means that diffusion rate is strongly temperature dependent. As the temperature is lowered, as in the unmixing process depicted in Fig. 5.21, the mobility of the atoms may become so limited that the original homogeneous mineral cannot achieve the pure end member compositions shown in Fig. 5.18b. A composition may result that is a compromise between lowering the thermodynamic free energy of the system, and the decrease in the mobility of the atoms. The phase or assemblage of phases with the lowest free energy will be the most stable. In the exsolution process, the exsolved phases (the two minerals) have a much lower free energy than the original homogeneous, high-temperature and disordered composition (see Fig. 5.22).

Exsolution lamellae may range from visible in hand specimens (as in microcline perthite) to so small that they can be resolved only by very high-resolution transmission electron microscopy (see Fig. 5.20). The scale of the exsolution lamellae is a good indicator of the cooling rate of the rock in which this intergrowth texture occurs. In slowly cooled rocks, more time is available for diffusion (and unmixing), and as a result the exsolution texture is coarser. In rocks that are cooled extremely quickly ("quenched"), the high-temperature, homogeneous, disordered mineral may be preserved, although in a metastable state. In a slightly slower but still rapid cooling process, very fine exsolution textures may develop.

RECALCULATION OF CHEMICAL ANALYSES

Elements such as gold, arsenic, and sulfur occur in the native state and their formulas are the chemical symbols of the elements. Most minerals, however, are compounds composed of two or many more elements and their formulas, recalculated from the results of quantitative chemical analyses, indicate the atomic proportions of the elements present. Thus, in galena, PbS, there is one atom of sulfur for each atom of lead, and in $CuFeS_2$ there is one atom of sulfur for each atom of copper and iron. Such specific atomic proportions underlie the definition of a mineral, which states that "a mineral . . . has a definite (but generally not fixed) chemical composition." Very few minerals, however, have a truly constant composition; examples of specific and essentially constant compositions are provided by quartz, SiO_2, and kyanite, Al_2SiO_5. Such minerals are often referred to as *pure substances*. The great majority of minerals show large variations in composition within specific atomic sites of their structures. For example, ZnS, sphalerite, shows a large range of Fe content. The Fe substitutes for Zn in the sphalerite structure, and in general terms an Fe-rich (ferroan) sphalerite composition can be expressed as $(Zn, Fe)S$. Here the total cation-anion ratio is still fixed, 1 : 1, however the Zn and Fe contents are variable.

As noted above, a quantitative chemical analysis provides the basic information for the atomic formula of a mineral. A weight percent analysis in terms of metals, or oxides, gives us a listing of what elements are present and in what quantities, but it gives us no direct information about how the elements (or ions) occur in the structure of the mineral. A quantitative weight percent analysis should add up close to 100%. Slight variations above or below 100% occur because of cumulative small errors that are part of the analytical procedure. As an example, let us consider the analysis of chalcopyrite in Table 5.2. The percentages in column 1, reported by the chemist, represent the percentages by weight of the various elements. Because the elements have different atomic weights, these percentages do not represent the ratios of the different atoms. To arrive at their relative proportions, the weight percentage in each case is divided by the atomic weight of the element. This gives a series of numbers (column 3), the *atomic proportions*, from which the *atomic ratios* can be quickly derived. In the analysis of chalcopyrite, these ratios are Cu : Fe : S = 1 : 1 : 2; thus $CuFeS_2$ is the chemical formula.

The reverse procedure, that of calculating the percentage composition from the formula, is shown in the lower half of Table 5.2. The formula of chalcopyrite is $CuFeS_2$, which makes for a gram-formula weight of 183.53. The calculated Cu, Fe, and S weight percentage values are very similar to the measured percentages reported in column 1 of Table 5.2. The calculated values are slightly different from the determined values, probably because of a slight experimental error in the analysis technique.

Let us now examine the recalculation of some sphalerite analyses, which show a considerable range in Fe, and of troilite, FeS (Table 5.3). The upper half of the table shows the weight percentages for the various elements and the lower part of the table provides atomic proportions obtained by dividing each weight

Table 5.2
CHALCOPYRITE (CuFeS₂) RECALCULATIONS

	1 Weight Percent	2 Atomic Weights	3 Atomic Proportions	4 Atomic Ratios	
Cu	34.30	63.54	0.53982	1	
Fe	30.59	55.85	0.54772	1	approx.
S	34.82	32.07	1.08575	2	
Total	99.71				

	Atomic Weights		
Cu	63.54	Cu(in %) $= \dfrac{63.54}{183.53} \times 100\% = 34.62$
Fe	55.85	Fe(in %) $= \dfrac{55.85}{183.53} \times 100\% = 30.43$
S(2×)	(32.07) × 2	S(in %) $= \dfrac{64.14}{183.53} \times 100\% = 34.94$
Total	183.53	Total 99.99

Table 5.3
RECALCULATION OF SEVERAL SPHALERITE ANALYSES AND OF TROILITE

	Weight Percentage*				
	1	**2**	**3**	**4**	**5**
Fe	0	0.15	7.99	18.25	63.53
Mn	0	0	0	2.66	0
Cd	0	0	1.23	0.28	0
Zn	67.10	66.98	57.38	44.67	0
S	32.90	32.78	32.99	33.57	36.47
Total	100.00	99.91	99.59	99.43	100.00
	Recalculated in Terms of Atomic Proportions				
Fe	0	0.003 ⎤	0.143 ⎤	0.327 ⎤	1.137
Mn	0	0 ⎪	0 ⎪	0.048 ⎪	0
Cd	0	0 ⎬ 1.027	0.011 ⎬ 1.032	0.002 ⎬ 1.060	0
Zn	1.026	1.024 ⎦	0.878 ⎦	0.683 ⎦	0
S	1.026	1.022	1.029	1.047	1.137
(Zn + Fe + Mn + Cd):S	1 : 1	1 : 1	1 : 1	1 : 1	1 : 1
Formulas	ZnS	$(Zn_{99.7}Fe_{0.3})S$	$(Zn_{85.1}Fe_{13.8}Cd_{1.1})S$	$(Zn_{64.4}Fe_{30.8}Mn_{4.5}Cd_{0.2})S$	FeS
Fe : Zn	0 : 100	0.3 : 99.7	14.0 : 86.0	32.4 : 67.6	100 : 0

*Analyses 1–4 from *Dana's System*, v. 1, 1944; analysis 5 for pure troilite FeS.

percentage figure by the appropriate atomic weight. One of the lower lines in Table 5.3 shows that the ratio of (Zn + Fe + Mn + Cd) : S is constant in the sphalerite analyses, namley 1 : 1, and that it is 1 : 1 in the troilite as well. However, the Zn in the sphalerite structure is partially substituted for by variable amounts of Fe, Mn, and Cd. The maximum amount of Fe reported in Table 5.3 is in column 4, which shows 18.25 weight percent Fe. This recalculates to 0.327 atomic proportions out of a total of 1.060 atomic proportions for (Zn + Fe + Mn + Cd). These analyses of naturally occurring sphalerites lead to the conclusion that the representation of sphalerite compositions as ZnS is a great oversimplification. Only analysis 1 (Table 5.3) is pure ZnS and analysis 5 is pure FeS. Analyses 2, 3, and 4 contain variable amounts of Fe, Mn, and Cd. Analysis 4, which contains the largest amounts of elements other than Zn and S, can serve to illustrate some of the useful atomic proportion recalculations. Its total cation content (in atomic percentage) is 1.060. Fe is $0.327/1.060 \times 100\% = 30.8\%$; similarly, Mn, Cd, and Zn values are 4.5, 0.2, and 64.4%, respectively. These cation proportion values can be used as subscripts to represent the specific composition of the sample of sphalerite in column 4; this leads to the formulation $(Zn_{64.4}Fe_{30.8}Mn_{4.5}Cd_{0.2})S$. The other specific formulas in Table 5.3 were obtained in a similar manner. It is obvious from the analyses that Zn and Fe are the main variables. Often it is necessary to obtain the cation distribution of such *major* elements only. For analysis 4 in Table 5.3 we then must eliminate from consideration the quantities of Cd and Mn. The total of (Fe + Zn) in the atomic

proportions column = 1.010, rather than 1.060 for (Fe + Mn + Cd + Zn). If now we wish to know the Fe versus Zn percentage in this analysis, we factor $0.327/1.010 \times 100\% = 32.4\%$. This value, and those similarly obtained for the other analyses, are listed in the last line of Table 5.3. These Fe : Zn ratios show that a selection of sphalerites may show *a compositional range from pure* ZnS to $(Zn_{0.68}Fe_{0.32})S$. The mineral troilite is found only in meteorites and shows no Zn in its chemical composition; it can therefore be considered as a compound of constant composition, FeS. The large variation of Fe in sphalerites is due to the substitution of Fe for Zn in the same structural site in the sphalerite structure; this is also known as *solid solution* of Fe in ZnS (see page 233). The largest amount of Fe solid solution in the analyses of Table 5.3 is shown by analysis 4, with composition $(Zn_{67.6}Fe_{32.4})S$.

The majority of minerals such as silicates, oxides, carbonates, phosphates, sulfates, and so on, are compounds containing large amounts of oxygen. By convention the analyses of such minerals are reported as percentages of oxides, rather than as percentages of metals. Calculations similar to those outlined above are performed with the oxide components to arrive at molecular proportions of the oxides, rather than atomic proportions of the elements. Table 5.4 gives an example of the recalculation of an analysis of gypsum. The analytically determined oxide components in column 1 are divided by the molecular weights of the corresponding oxides (column 2) to arrive at molecular proportions (column 3). From the molecular ratios in column 4 we see that $CaO : SO_3 : H_2O =$

Table 5.4

RECALCULATION OF GYPSUM AND OLIVINE ANALYSES

Gypsum	1 Weight Percent	2 Molecular Weights	3 Molecular Proportions	4 Molecular Ratios		
CaO	32.44	56.08	0.57846	1		
SO$_3$	46.61	80.08	0.58211	1	approx.	
H$_2$O	20.74	18.0	1.15222	2		
Total	99.79					

Olivine	1 Weight Percent	2 Molecular Weights	3 Molecular Proportions	4 Atomic Proportions Cations	Oxygens	6 On Basis of 4 Oxygens	7 Atomic Ratios
SiO$_2$	34.96	60.09	0.58179	Si 0.5818	1.1636	0.989	1
FeO	36.77	71.85	0.51176	Fe^{2+} 0.5118	0.5118	0.870	
MnO	0.52	70.94	0.00733	Mn 0.0073	0.0073	0.012 } 2.022	2
MgO	27.04	40.31	0.67080	Mg 0.6708	0.6708	1.140	
Total	99.29				2.3535		

Olivine in terms of end member compositions: Mg = 1.140, Fe = 0.870; total; 2.010. Percentage Mg = 56.7; percentage Fe = 43.3; percentage Mg$_2$SiO$_4$ = % forsterite = % Fo = 56.7%; percentage Fe$_2$SiO$_4$ = % fayalite = % Fa = 43.3%.

1 : 1 : 2, and the composition can be written as CaO·SO$_3$·2H$_2$O, or as CaSO$_4$·2H$_2$O. The latter formula is preferred because it avoids creating the erroneous impression that the mineral is composed of discrete oxide molecules.

In the lower part of Table 5.4 the chemical analysis of a member of the olivine series is given. The desired result of a recalculation of the analysis again is a statement of the molecular proportions of the oxide components, or of the cation components. The steps in going from columns 1 to 3 are the same as in the gypsum analysis. Column 4 lists the values for the atomic proportions of the various atoms. The atomic proportions of the metal atoms are the same as the corresponding molecular proportions in column 3, because, for example, one "molecule" of SiO$_2$ contributes 1 Si and one "molecule" of FeO contributes 1 Fe. However, the number of oxygens contributed by each molecular proportion is not the same as the numbers listed in column 3. For each of the divalent metals there is one oxygen atom, but for each silicon there are two oxygens. In other words, one molecular proportion of FeO contributes 1 oxygen, but one molecular proportion of SiO$_2$ contributes two oxygens. This is reflected in the numbers of column 5.

The total number of oxygens contributed by the atomic proportions in column 5 is 2.3535. From crystal structure data we know that olivine, with the general formula of (Mg, Fe)$_2$SiO$_4$, has four oxygens per formula unit. In order to arrive at the cation proportions in terms of four oxygens, we multiply each of the cation numbers in column 4 with the 4/2.3535 ratio. This leads to the numbers in columns 6 and 7. The final formulation for this olivine is

(Mg$_{1.12}$Fe$_{.86}$Mn$_{.01}$)SiO$_4$, obtained by assuming 2.0 for total cations, instead of 2.022 as calculated. Such a formula is often expressed in terms of *end member compositions*. In olivine these are Mg$_2$SiO$_4$, forsterite, and Fe$_2$SiO$_4$, fayalite. The amounts of forsterite (Fo) and fayalite (Fa) in the general formula are directly proportional to the atomic proportions of Mg and Fe, or molecular proportions of MgO and FeO in the analysis. For example, in Table 5.4, this leads to Fo$_{57}$ and Fa$_{43}$. If the olivine composition ranges mainly between Mg and Fe compositions, it suffices to state the final recalculation as Fo$_{57}$.

The olivine structure is relatively simple, with Mg and Fe substituting for each other in the same structural position (site). Many silicates, however, have various structural sites among which elements can be distributed. An example would be a member of the diopside-hedenbergite series (CaMgSi$_2$O$_6$ − CaFeSi$_2$O$_6$) in the pyroxenes. Table 5.5 gives a chemical analysis, as well as the various recalculation steps for pyroxene similar to that outlined for olivine. In column 1 of Table 5.5 percentages are reported for FeO as well as Fe$_2$O$_3$. In analyses made by instrumental techniques Fe^{3+} and Fe^{2+} cannot be distinguished, and total Fe is then reported as FeO only, or Fe$_2$O$_3$ only. Column 2 provides molecular proportions, obtained by dividing the weight percent figures by the appropriate molecular weights. Column 3 lists the cation proportions for each oxide "molecule." Note, for example, that Al$_2$O$_3$ contains 2Al, therefore 0.0295 is multiplied by 2. Column 4 lists the number of oxygens contributed by each oxide "molecule." Note that SiO$_2$ contains two oxygens per "molecule"; the molecular proportion figure (0.8384) is therefore

Table 5.5
RECALCULATION OF PYROXENE ANALYSIS

	1 Weight Percent	2 Molecular Proportions	3 Cation Proportions	4 Number of Oxygens	5 Cations on Basis of 6 Oxygens	6 Cation Assignment		7 End Member Recalculation
SiO_2	50.38	0.8384	0.8384	1.6768	1.875	Si	1.875 ⎫	MgO as $MgSiO_3$
Al_2O_3	3.01	0.0295	0.0590	0.0885	0.132 → Al	Al	0.125 ⎭ 2.0	(enstatite = En)
						Al	0.007 ⎫	FeO as $FeSiO_3$ (ferrosilite = Fs)
TiO_2	0.45	0.0056	0.0056	0.0112	0.012	Ti	0.012 ⎪	CaO as $CaSiO_3$
Fe_2O_3	1.95	0.0122	0.0244	0.0366	0.055	Fe^{3+}	0.055 ⎬ 1.033	(wollastonite = Wo)
FeO	4.53	0.0630	0.0630	0.0630	0.141	Fe^{2+}	0.141 ⎪ ≈ 1	Using molecular
MnO	0.09	0.0013	0.0013	0.0013	0.003	Mn	0.003 ⎪	proportions:
MgO	14.69	0.3643	0.3643	0.3643	0.815	Mg	0.815 ⎭	
								MgO = 0.3644
CaO	24.32	0.4321	0.4321	0.4321	0.966	Ca	0.966 ⎫	FeO = 0.0630
Na_2O	0.46	0.0074	0.0148	0.0074	0.033	Na	0.033 ⎬ 1.006	CaO = 0.4321
K_2O	0.15	0.0016	0.0032	0.0016	0.007	K	0.007 ⎭ ≈ 1	Total = 0.8595
Total	99.94			Total O = 2.6828				%/En = 42.39
								%Fs = 7.33
			Oxygen factor $\dfrac{6}{2.6828} = 2.236469$					%Wo = 50.27

multiplied by 2. The total number for all oxygens contributed by all the oxide "molecules" in column 4 is 2.6828. From crystal structure data we know that a pyroxene of this type has a formula such as $Ca(Mg,Fe)(Si,Al)_2O_6$. The formula is therefore recalculated on the basis of six oxygens, and the numbers in column 3 are multiplied by 2.23647 to obtain the figures in column 5. These cations are subsequently assigned to specific atomic sites in the pyroxene structure. If necessary, sufficient Al is added to Si so that the sum Si + Al = 2.0 in column 6; the remainder of the Al is added to the sum of the intermediate size cations. Intermediate size cations (Al, Ti, Fe^{3+}, Fe^{2+}, Mn, Mg) are assigned to the M1 cation position (see Fig. 13.48) in the structure; these cation numbers together amount to 1.033, in agreement with the general formula $Ca_1(Mg, Fe)_1(Si, Al)_2O_6$. The remaining larger cations (Ca, Na, K) are assigned to the M2 position in the pyroxene structure, and their total is 1.006, again approximately 1.

Pyroxene analyses are often recalculated in terms of *end member compositions*. The analysis in Table 5.5 contains major amounts of CaO, MgO, and FeO and can therefore be recalculated in terms of $CaSiO_3$ (wollastonite), $MgSiO_3$ (enstatite), and $FeSiO_3$ (ferrosilite) end members. The calculations are given in column 7, showing that the weight percent analysis can be recast as $Wo_{50.3}En_{42.4}Fs_{7.3}$.

Complex hydrous silicates such as amphiboles are recalculated by the same sequence of steps as illustrated for olivine and pyroxene, but the H_2O content is evaluated as (OH) groups in the amphibole

structure. Table 5.6 gives an example of a Na-containing actinolite analysis. A general actinolite formula is $Ca_2(Mg,Fe)_5Si_8O_{22}(OH)_2$. Column 1 lists weight percentages for $H_2O(+)$ and $H_2O(-)$. The $H_2O(+)$ is considered part of the actinolite structure but the $H_2O(-)$ is not and is therefore neglected in subsequent calculations (see footnote to Table 5.6). Column 2 lists the molecular proportions, column 3 the cation proportions, and column 4 gives the total contribution of (O,OH) for each of the "molecules" in column 2. The sum in column 4 divided into 24 gives the ratio by which the whole analysis must be multiplied in order to put it on a basis of 24(O,OH). Column 3 is now multiplied by the factor 8.4803, giving the results in column 5 on the basis of 24(O,OH). In general it will be found that Si is less than 8.0 and Al is added to bring (Si,Al) to 8.0. The remaining Al, if any, is added to the cation group immediately below consisting of Fe^{3+}, Fe^{2+}, and Mg. The total for this group of intermediate size cations is 5.111, close to 5.0, and the total for the remaining larger size cations is 2.068, close to 2.0. The intermediate size cations enter the M1, M2, and M3 sites in the amphibole structure and the larger cations enter M4 (see Fig. 13.66). Mn could be distributed among M1, M2, and M3 as well as M4 because its ionic size lies between that of Ca^{2+} and Mg^{2+}. The (OH) content turns out to be about 2. The recalculated analysis now has the general form $(Ca, Na, Mn)_2(Fe, Mg)_5(Si, Al)_8O_{22}(OH)_2$. Instrumental analysis results of hydrous minerals such as amphiboles do not provide information on the oxidation state of iron (Fe^{2+} vs. Fe^{3+}), for example, or

Table 5.6

RECALCULATION OF AMPHIBOLE ANALYSIS

	1 Weight Percent	2 Molecular Proportions	3 Number of Cations	4 Total Number of (O, OH)	5 Cations on Basis of 24 (O, OH)		6 Cations on Basis of 23 Oxygens
SiO_2	56.16	0.9346	0.9346	1.8692	7.926 ⎱	7.958	7.931
Al_2O_3	0.20	0.0019	0.0038	0.0057	0.032 ⎰	≈8	0.032
Fe_2O_3	1.81	0.0113	0.0226	0.0339	0.192		0.192
FeO	6.32	0.0880	0.0880	0.0880	0.746 }	5.111 ≈5	0.747
MgO	19.84	0.4921	0.4921	0.4921	4.173		4.176
MnO	2.30	0.0324	0.0324	0.0324	0.275 ⎱		0.275
CaO	9.34	0.1665	0.1665	0.1665	1.412 }	2.068	1.413
Na_2O	1.30	0.0210	0.0420	0.0210	0.356 ⎰	≈2	0.356
K_2O	0.14	0.0015	0.0030	0.0015	0.025		0.025
$H_2O(+)$	2.16	0.1198	0.2396	0.1198	2.032	≈2	
$H_2O(-)$*	0.48	—		—			
Total	100.05			2.8301			

$$\frac{24}{2.8301} = 8.4803; \quad 2.8301 - 0.1198 = 2.7103; \quad \frac{23}{2.7103} = 8.4861$$

*This H_2O is weakly held by the powdered sample and is removed by drying at low temperature. It is assumed to be absorbed from the atmosphere, not structurally bound.

on the presence of structural water. Such anhydrous analytical results are frequently recalculated on an anhydrous basis. For the analysis in Table 5.6 such a basis would be 23 oxygens, replacing $2(OH)^-$ by one O^{2-}. Although this is not correct in terms of the structure of amphiboles, it does allow for comparison among analytical results in which H_2O was not determined. In column 6 the actinolite is recalculated on an anhydrous basis and the cation results are similar to, but not identical with, the results in column 5.

Additional examples of recalculations of mineral analyses followed by recalculation exercises are given in Klein, C., 1989, *Minerals and Rocks: Exercises in Crystallography, Mineralogy, and Hand Specimen Petrology.* John Wiley & Sons, New York, 402 pp.

GRAPHIC REPRESENTATION OF MINERAL COMPOSITION

Because most minerals show solid solution features ranging from partial to complete, it is frequently useful to show such variations in composition in graphic form. A composition graph can be constructed after we have chosen the chemical components pertaining to the mineral compositions to be graphed. The mineral kyanite, Al_2SiO_5, is known to be essentially constant in composition. A chemical analysis of a pure kyanite yields the following data, in terms of weight percent: Al_2O_3, 62.91, and SiO_2, 37.08. Selecting Al_2O_3 and SiO_2 as the two end components of a bar graph, we can position the kyanite composition on it

in terms of weight percent of the oxides (see Fig. 5.23a). In order to plot a mineral composition in terms of the weight percent of its components we need to obtain its chemical analysis or we can recalculate the weight percentages of the components from the formula as shown in Table 5.2. However, formulas that provide direct information about atomic proportions of elements or molecular proportions of components can be used directly for graphing of compositions. The formula of kyanite can be written as $1Al_2O_3 + 1SiO_2 = 1Al_2SiO_5$. In other words, the composition of kyanite can be expressed in terms of the molecular proportions of the oxide components Al_2O_3 and SiO_2. This is shown in Fig. 5.23b.

In Table 5.3 we recalculated several sphalerite analyses. These compositions can be shown graphically (see Fig. 5.23c) in terms of two end members, ZnS and FeS. Intermediate compositions can be obtained directly from the Zn : Fe ratios in the last line of Table 5.3. After these numbers are plotted, and with the addition of further literature analyses as well, the solid solution extent of FeS in sphalerite can be generalized on the basis of a population of analysis points. Although only very few points are shown on the bar graph in Fig. 5.23c, it can be stated that sphalerite shows a partial solid solution series from ZnS to at least $Zn_{0.676}Fe_{0.324}S$. This statement is shown graphically by the shading. Troilite, on the other hand, shows no Zn content and must appear as a single point on the bar graph, indicating a total lack of Zn substitution. The graph shows that between troilite

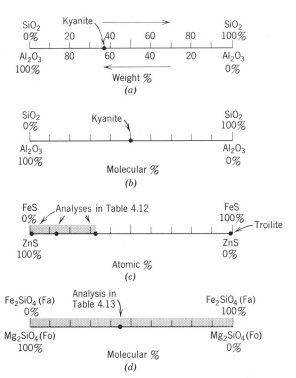

FIG. 5.23 Graphic representations of chemical compositions in terms of bar diagrams. (a) Kyanite (Al_2SiO_5) composition in terms of weight percentage of the oxides. (b) Kyanite in terms of molecular percentage of the oxides. (b) Sphalerite analyses from Table 5.3 in terms of atomic percentages of Zn and Fe. (d) Complete solid solution series of olivine and composition of olivine from Table 5.4 in terms of molecular percentages of forsterite and fayalite.

and the most Fe-rich sphalerite plotted there is a region in which homogeneous compositions of Fe-rich sphalerites are not found. In Table 5.4 we recalculated an olivine analysis in terms of molecular percentages of MgO and FeO; such percentages can also be expressed as Mg_2SiO_4(Fo) and Fe_2SiO_4(Fa), the two end member compositions of the olivine solid solution series. Figure 5.23d shows a bar graph depicting the complete solid solution series between Fo and Fa as well as the specific composition of the olivine of Table 5.4. Bar diagrams of this type form the horizontal axis in *variation diagrams* that relate changes in physical properties to variation in composition (see Figs. 6.8 and 7.17). Composition bars also form the horizontal axis in temperature-composition diagrams (see Fig. 5.21 and Chapter 9).

As we have already seen in Tables 5.4 to 5.6, mineral analyses can be very complex and may show substitution of several elements in a specific atomic site of the mineral structure. In order to portray the variation of at least three components instead of just

two, as in Fig. 5.23 a *triangular diagram* is frequently used. Triangular coordinate graph paper is available commercially[1] with ruling for each 1% division. Let us first illustrate the use of a triangular composition diagram by plotting some fairly simple end member compositions. In Figure 5.24a we list the mineral names, formulas, and compositions in terms of weight percentages of the oxides for two end members of the orthopyroxene series, enstatite and ferrosilite (also known as orthoferrosilite to reflect the orthorhombic nature of the mineral) and the two end members of the olivine series, forsterite and fayalite. On the basis of the data given in Fig. 5.24a we have two choices for a triangular plot, namely, on the basis of weight percentages, or on the basis of the relative cation proportions in the formulas. We will do the first plot on the basis of the weight percentage values.

In Fig. 5.24b we have chosen the corners of the triangle as 100 weight percent SiO_2, 100 weight percent MgO, and 100 weight percent FeO. This allows us to plot all four compositions because they are made up only of these three oxide components. Enstatite and forsterite contain 40.16 and 57.30 weight percent MgO, respectively. The MgO scale extends from 0 to 100% on the left side of the triangle, and all we need to do is locate the MgO values (we could have chosen the SiO_2 values as well, and we would have plotted the identical compositional points). The right-hand side of the triangle extends between 100% SiO_2 and 100% FeO, and the direction of increasing FeO (from 0% to 100%) is along the right side from top to bottom. Here we have located the compositions of the minerals in (a) on the basis of the given FeO weight percentage values. We have joined the end members within each series by a dashed line to show where intermediate compositions would plot. The two lines are not quite parallel because of the slight differences in MgO content (57.30 − 40.16 = 17.14 weight percent) and FeO content (70.51 − 54.46 = 16.05 weight percent) of the Mg and Fe end member compositions. The final plot, in weight percent, is easy enough to achieve *if* indeed we have the weight percentage values at hand. In Fig. 5.24c we have chosen the elements Si, Mg, and Fe for the corners. In order to plot the four minerals in terms of cation proportions we might conclude that each of the mineral analyses would have to be recalculated from weight percentage, via molecular proportions, to cation proportions, and so on, as shown in Tables 5.5 and 5.6 for two different silicates. This is the correct procedure for complex mineral compositons; for simple end mem-

[1]From Keuffel and Esser Co.

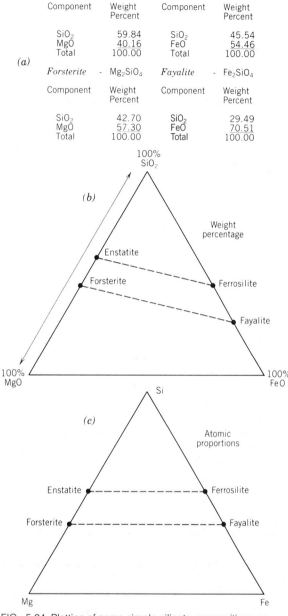

Enstatite	- MgSiO$_3$	*Ferrosilite*	- FeSiO$_3$
Component	Weight Percent	Component	Weight Percent
SiO$_2$	59.84	SiO$_2$	45.54
MgO	40.16	FeO	54.46
Total	100.00	Total	100.00

Forsterite	- Mg$_2$SiO$_4$	*Fayalite*	- Fe$_2$SiO$_4$
Component	Weight Percent	Component	Weight Percent
SiO$_2$	42.70	SiO$_2$	29.49
MgO	57.30	FeO	70.51
Total	100.00	Total	100.00

(a)

(b)

100% SiO$_2$

Weight percentage

Enstatite

Forsterite

Ferrosilite

Fayalite

100% MgO

100% FeO

(c)

Si

Atomic proportions

Enstatite

Ferrosilite

Forsterite

Fayalite

Mg

Fe

FIG. 5.24 Plotting of some simple silicate compositions on triangular coordinate paper.

ber compositions for which we know the exact formulas, however, that is unnecessary. Let us inspect the formulas.

Enstatite contains one Mg and one Si atom (or ion) per formula. The same ratio holds for ferrosilite, with one Fe and one Si. In other words, enstatite contains one Mg out of a total of two cations (Mg + Si = 2); the situation is analogous in ferrosilite with one Si to one Fe. On such a scale our compositions will plot at one-half for Mg, Fe, or Si along the two slanting

sides of the triangle. If the components had been expressed as 100% Si, 100% Mg, and 100% Fe, the points would have been at the 50 : 50% locations. The plotted positions are shown and have been joined by a dashed line that is parallel to the bottom edge of the triangle.

Now let us plot the two compositions in the olivine series. Forsterite contains two Mg and one Si, which is a total of three cations; fayalite contains two Fe and one Si, again a total of three cations. This means that the plotted compositions must lie at two-thirds of Mg (or Fe) from the 1 Mg (or 1 Fe) corners. In terms of percentages, the Mg and Fe values would locate at 66.6%. These two points are shown and have also been connected by a dashed line, which again is parallel to the base, and as such also parallel to the line for intermediate compositions in the orthopyroxene series. A weight percentage diagram, as in Fig. 5.24*b*, is commonly used in the plotting of rock and mineral compositions in igneous rocks, because the investigator is mainly interested in the possible changes in composition of the melt from which an igneous rock formed. However, mineralogists and petrologists who deal with sedimentary and metamorphic rocks much prefer the atomic representation. Such a diagram is especially easy to construct if end member formulas are known or are easily located. The regularity of the diagram (with parallel lines) is also easy to remember.

Now that we have plotted simple compositions let us see what is involved if we wish to plot the composition of a pyroxene as listed in Table 5.5. This analysis has been recalculated in terms of three components, CaSiO$_3$ (wollastonite), MgSiO$_3$ (enstatite), FeSiO$_3$ (ferrosilite). The ratios of these three components are the same as they would be for CaO, MgO, and FeO (these two sets of alternate components are completely interchangeable). In Fig. 5.25 we will use the three mineral components, wollastonite (Wo), enstatite (En), and ferrosilite (Fs), which mark the corners of the triangular graph paper. Any pyroxene composition that involves only two of the three components can be plotted along an edge of the triangle, identical to the procedure in the bar graphs of Fig. 5.23. For example, diopside (CaMgSi$_2$O$_6$) can be plotted along the left edge of the triangle: 1CaSiO$_3$ + 1MgSiO$_3$ = 1CaMgSi$_2$O$_6$. Similarly, hedenbergite (CaFeSi$_2$O$_6$) can be plotted along the right edge, according to 1CaSiO$_3$ + 1FeSiO$_3$ = 1CaFeSi$_2$O$_6$. Compositions along the bottom edge can be expressed in general as (Mg, Fe)SiO$_3$. A specific compositional location can be obtained from a formulation (Fe$_{0.80}$Mg$_{0.20}$)SiO$_3$, which can be restated as 80 molecular percent fer-

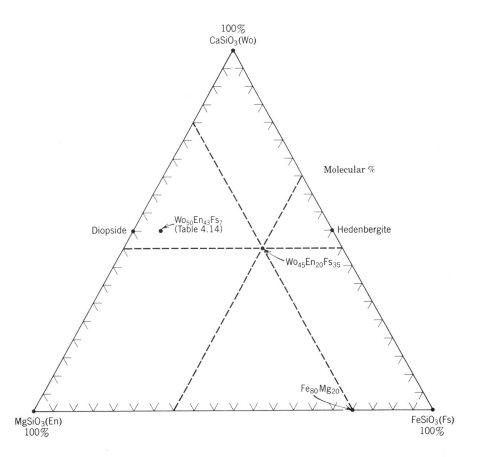

FIG. 5.25 Plotting of chemical compositions on triangular coordinate paper. The specific compositions represented are for the pyroxene family of silicates. See text for explanation of plotting procedure.

rosilite (Fs) and 20 molecular percent enstatite (En). This composition is shown in Fig. 5.25.

A more general pyroxene composition that shows the presence of all three components may be expressed as $Wo_{45}En_{20}Fs_{35}$. This is shorthand for a pyroxene containing 45 molecular percent CaO(Wo), 20 molecular percent MgO(En), and 35 molecular percent FeO(Fs). In order to locate this composition on the triangular graph paper, note that the join En-Wo represents 0% Fs; similarly the En-Fs join represent 0% Wo, and so forth. The distance between the corner and the opposite side of the triangle is graduated in "%" lines from 100% at the corner to 0% along the join. For example, the $CaSiO_3$ corner represents 100% $CaSiO_3$ and horizontal lines between this corner and the base of the triangle mark the change from 100% to 0% $CaSiO_3$. If one wishes to plot $Wo_{45}En_{20}Fs_{35}$, locate first the Wo_{45} line (marked on Fig. 5.25), then locate the En_{20} line (also marked). Where the two lines intersect, the composition is located on the Fs_{35} line. The position of the composition, $Wo_{50}En_{43}Fs_7$, from Table 5.5 is also shown in Fig. 5.25.

Like bar graphs, triangular diagrams are frequently used to depict the extent of solid solution in minerals. Figure 5.26 shows the extent of solid solution among minerals in the chemical system SiO_2-MgO-FeO. In this representation we can show the olivine series $(Mg,Fe)_2SiO_4$ and the orthorpyroxene series $(Mg,Fe)SiO_3$, both of which show an essentially complete solid solution between the end member compositions. SiO_2 (quartz and its various polymorphs) shows no substitution in terms of FeO or MgO and is drawn as a point, indicating its constant composition. Forsterite (Mg_2SiO_4) can be written as $2MgO + 1SiO_2 = 1Mg_2SiO_4$; of the three oxide "molecules," one is SiO_2 and the other two are MgO. On the left side of the triangle, Fo is therefore located at $\frac{1}{3}$ away from the MgO corner. Enstatite $(MgSiO_3)$ can be written as $1MgO + 1SiO_2 = 1MgSiO_3$, which is located halfway between the MgO and SiO_2 corners. The plotting process here is identical to that in Fig. 5.24c.

In addition to depicting graphically the extent of solid solution, triangular diagrams are useful in showing the minerals that make up (coexist in) a specific rock type. In such diagrams, known as *assemblage diagrams*, the minerals that coexist with each other (i.e., touch each other along the borders of their grains) are connected by *tielines*. Such tielines denote

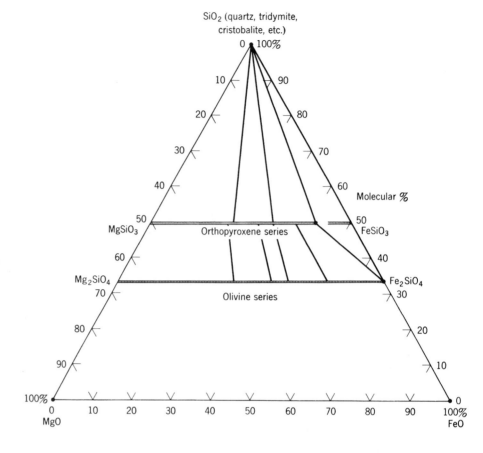

FIG. 5.26 Triangular representation in terms of molecular percent of minerals in the chemical system MgO-FeO-SiO$_2$. The olivine and orthopyroxene series show complete solid solution series between Mg and Fe end members, as shown by the continuous bars. Lines connecting SiO$_2$ to various orthopyroxene compositions, and between orthopyroxene and olivine compositions, are *tielines*; these connect coexisting minerals in rocks. The triangle outlines a critobalite-orthopyroxene (Fs$_{83}$)-fayalite coexistence.

the fact that two, three, or four minerals are found next to each other in a rock. In Fig. 5.26 some tielines have been drawn for possible olivine-orthopyroxene pairs, as well as for an assemblage, in a different rock, of cristobalite, Fe-rich orthopyroxene, and fayalite. This mineral assemblage could be found in a high-temperature, Fe-rich basalt. Such a three-mineral coexistence is denoted by an *assemblage triangle*, of which all of the three sides are tielines.

Triangular diagrams would tend to limit us to representation of only three components, which may be simple elements, compound oxides, or more complex components as expressed by mineral formulas. In order to represent more than three components on one triangle, some components are frequently combined and some may not be considered in the graphic representation. For example, in an attempt to represent carbonate compositions in the system CaO-MgO-FeO-MnO-CO$_2$, we must somehow reduce the possible number of compositional variables. Because all carbonates contain CO$_2$ there will be little gained in using the CO$_2$ component to show the small CO$_2$ variations that do exist. We will therefore ignore CO$_2$ in the graphic representation. We may further decide to combine the FeO and MnO because Fe^{2+} and Mn^{2+}

substitute freely for each other in the carbonate structure and because Mn^{2+} typically is far less abundant in most environments than Fe^{2+}. This leaves three components, CaO, MgO, and (FeO + MnO). Figure 5.27a shows that a complete solid solution series may exist between magnesite (MgCO$_3$) and siderite + rhodochrosite (FeCO$_3$ + MnCO$_3$); the completeness of this series is well documented (see also Fig. 12.7). It also shows an almost complete series between dolomite, CaMg(CO$_3$)$_2$, and ankerite, CaFe(CO$_3$)$_2$ + kutnahorite, CaMn(CO$_3$)$_2$. Calcite (CaCO$_3$) shows a very limited ionic substitution by Mg, Fe, and Mn. Tielines show coexistences of magnesite-dolomite, calcite-dolomite, and calcite-ankerite-siderite in the triangle. The three-mineral assemblage is not uncommon in banded Precambrian iron-formations. In chemically more complex systems, such as are exhibited by most rocks, additional combinations of components are necessary in order to portray the chemistry of the minerals in the rock on a triangular diagram. A frequently used presentation of this type is known as an *ACF diagram*, in which A = Al$_2$O$_3$ + Fe$_2$O$_3$ − (Na$_2$O + K$_2$O); C = CaO; and *F* = MgO + MnO + FeO, all in molecular proportions. Figure 5.27b shows an example of such a diagram for rocks of medium-grade

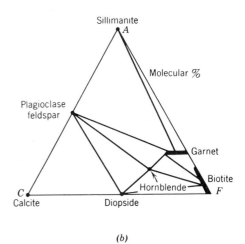

FIG. 5.27 *(a)* Compositions of carbonates in the system CaO-MgO-FeO-MnO-CO$_2$. The CO$_2$ component is not considered in this diagram. (FeO + MnO) is a combined component. A complete solid solution series exists between dolomite and ankerite. A complete series also exists between magnesite and siderite. Tielines connect possible coexisting carbonate members. The three-mineral triangle portrays the coexistence of calcite-ankerite-siderite. *(b)* Mineral compositions in metamorphosed carbonate rocks, pelitic schists, and amphibolites in terms of an *ACF* diagram. See text.

metamorphic origin. Triangles connect compositions of three coexisting minerals in these rocks. Additional discussion of the procedures in plotting mineral compositions on triangular diagrams, and exercises on the graphic representation of mineral compositions, are given in Klein (1989).

REFERENCES AND SUGGESTED READING

Adler, I., 1966, *X-ray Emission Spectrography in Geology*. Elsevier Publishing Co., New York, 258 pp.

Clarke, F. W. and Washington, H. S., 1924, The composition of the Earth's crust. U.S. Geological Survey Professional Paper 127, 117 pp.

Goldstein, J. I., Newbury, D. E., Echlin, P., Joy, D. C., Fiori, C. and Lifshin, E., 1981, *Scanning Electron Microscopy and X-ray Microanalysis*. Plenum Press, New York, 673 pp.

Ingle, J. D. and Crouch, S. R., 1988, *Spectrochemical Analysis*. Prentice Hall, Englewood Cliffs, N.J., 590 pp.

Klein, C., 1989, *Minerals and Rocks: Exercises in Crystallography, Mineralogy, and Hand Specimen Petrology*. John Wiley & Sons, New York, 402 pp.

Laitenen, H. A. and Harris, S., 1975, *Chemical Analysis*, 2nd ed. McGraw-Hill Book Co., New York, 611 pp.

Liebhafsky, H. A., Pfeiffer, H. G., Winslow, E. H. and Zemany, P. D., 1960, *X-ray Absorption and Emission in Analytical Chemistry*. John Wiley & Sons, New York, 357 pp.

Mason, B. and Moore, C. B., 1982, *Principles of Geochemistry*, 4th ed. John Wiley & Sons, New York, 344 pp.

Skoog, D. A. and Leary, J. L., 1992, *Principles of Instrumental Analysis*, 4th ed. Saunders College Publishing, New York, 700 pp.

Strobel, H. A. and Heineman, W. R., 1989, *Chemical Instrumentation: A Systematic Approach*, 3rd ed. John Wiley & Sons, New York, 1210 pp.

Wenk, H. R., 1976, *Electron Microscopy in Mineralogy*. Springer-Verlag, New York, 564 pp.

Willard, H. H., Merritt, L. M., Dean, J. A. and Settle, F. A., 1981, *Instrumental Methods of Analysis*, 6th ed. D. Van Nostrand Co., New York, 1030 pp.

CHAPTER 6

PHYSICAL PROPERTIES OF MINERALS

The physical properties of minerals are a direct result of their chemical and structural characteristics; as such we will draw, in various parts of our discussion, on subjects covered in Chapters 2 through 5. In this chapter we examine mainly those physical properties that can be determined by inspection or by relatively simple tests. Because they are determined in hand specimens, they are important in the rapid recognition of minerals. Other physical properties, such as those determined by X-ray or optical techniques, require special and often sophisticated equipment and may involve elaborate sample preparation. X-ray and optical properties are discussed separately in Chapters 7 and 8. Crystal symmetry, morphology, and structure (also physical properties) are treated in Chapters 2 and 3.

CRYSTAL HABITS AND AGGREGATES

The habit or appearance of single crystals as well as the manner in which crystals grow together in aggregates are of considerable aid in mineral recognition. Terms used to express habit and state of aggregation are given below, and several are illustrated in Fig. 6.1.

1. Minerals in isolated or distinct crystals may be described as:
    (a) *Acicular.* Slender, needlelike crystals.

(b) *Capillary and filiform.* Hairlike or threadlike crystals.
    (c) *Bladed.* Elongated crystals flattened like a knife blade.
2. For groups of distinct crystals the following terms are used:
    (a) *Dendritic.* Arborescent, in slender divergent branches, somewhat plantlike.
    (b) *Reticulated.* Latticelike groups of slender crystals.
    (c) *Divergent or radiated.* Radiating crystal groups.
    (d) *Drusy.* A surface covered with a layer of small crystals.
3. Parallel or radiating groups of individual cyrstals are described as:
    (a) *Columnar.* Stout, columnlike individuals.
    (b) *Bladed.* An aggregate of many flattened blades.
    (c) *Fibrous.* Aggregate of slender fibers, parallel or radiating.
    (d) *Stellated.* Radiating individuals forming starlike or circular groups.
    (e) *Globular.* Radiating individuals forming small spherical or hemispherical groups.
    (f) *Botryoidal.* Globular forms resembling, as the word, derived from the Greek, implies, a "bunch of grapes."

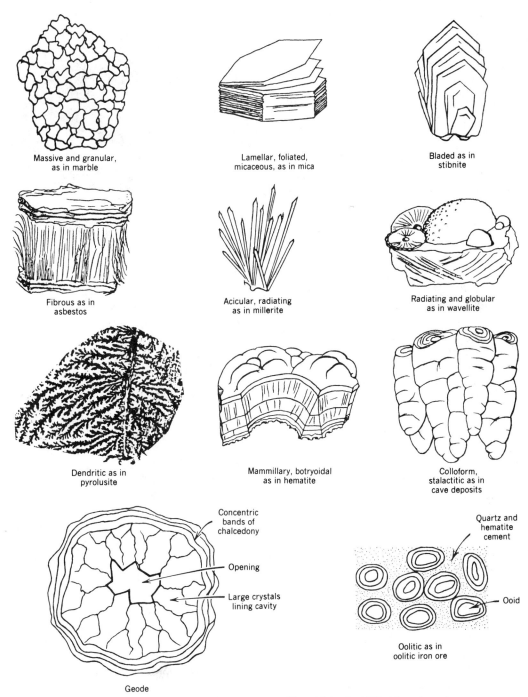

Massive and granular, as in marble

Lamellar, foliated, micaceous, as in mica

Bladed as in stibnite

Fibrous as in asbestos

Acicular, radiating as in millerite

Radiating and globular as in wavellite

Dendritic as in pyrolusite

Mammillary, botryoidal as in hematite

Colloform, stalactitic as in cave deposits

Concentric bands of chalcedony

Opening

Large crystals lining cavity

Geode

Quartz and hematite cement

Ooid

Oolitic as in oolitic iron ore

FIG. 6.1. Some common mineral habits and occurrences. (From Klein, C., 1989, *Minerals and Rocks: Exercises in Crystallography, Mineralogy, and Hand Specimen Petrology*, John Wiley & Sons, New York, 402 pp.)

(g) *Reniform.* Radiating individuals terminating in round kidney-shaped masses (Fig. 6.2).

(h) *Mammillary.* Large rounded masses resembling mammae, formed by radiating individuals.

(i) *Colloform.* Spherical forms composed of radiating individuals without regard to size; this includes botryoidal, reniform, and mammillary.

4. A mineral aggregate composed of scales or lamellae is described as:

(a) *Foliated.* Easily separable into plates or leaves.

(b) *Micaceous.* Similar to foliated, but splits into exceedingly thin sheets, as in the micas.

(c) *Lamellar* or *tabular.* Flat, platelike individuals superimposed upon and adhering to each other.

(d) *Plumose.* Fine sales with divergent or featherlike structure.

5. A mineral aggregate composed of equant grains is *granular*.

6. Miscellaneous terms:

(a) *Stalactitic.* Pendent cylinders or cones. Stalactites form by deposition from mineral-bearing waters dripping from the roofs of caverns.

(b) *Concentric.* More or less spherical layers superimposed upon one another about a common center.

(c) *Pisolitic.* Rounded masses about the size of peas.

(d) *Oölitic.* A mineral aggregate formed of small spheres resembling fish roe.

(e) *Banded.* A mineral in narrow bands of different color or texture.

(f) *Massive.* Compact material without form or distinguishing features.

(g) *Amygdaloidal.* A rock, such as basalt, containing almond-shaped nodules.

(h) *Geode.* A rock cavity lined by mineral matter but not wholly filled. Geodes may be banded, as in agate, due to successive depositions of material, and the inner surface is frequently covered with projecting crystals.

(i) *Concretion.* Masses formed by deposition of material about a nucleus. Some concretions are roughly spherical, whereas others assume a great variety of shapes.

CLEAVAGE, PARTING, AND FRACTURE

The properties of *cleavage*, *parting*, and *fracture* are the response of a crystalline material to an external force. When such a force is applied, a mineral is subjected to stress. If the internal structure of the crystalline substance is deformed, due to the stress, it is said to have undergone *strain*. As such, stress relates to the force applied, and strain to the resultant deformation. The strength of a crystalline material is a function of its bonding mechanism(s) and the presence (or absence) of structural defects. The type of bonding (see Chapter 4) is of major importance in a mineral's reaction to an applied force. If a mineral contains structural defects along specific planes or directions, it will tend to deform along such features more easily than a mineral with a truly perfect (or better-ordered) structure. If the strain in the mineral is so great as to exceed its overall strength, it will break. Many minerals have planar directions in their structure that are systemati-

FIG. 6.2. Reniform hematite, Cumbria, England. (Harvard Mineralogical Museum.)

cally weaker than other directions. This is the result of planes in the crystal structure that are joined by fewer bonds per unit volume than are other planes in the structure, or are joined by weaker bonds.

Cleavage

Cleavage is the tendency of minerals to break parallel to atomic planes that are identified by Miller indices, just as the faces of the external form of a crystal (see Chapter 2). Cleavage may be very well developed (perfect) in some crystals, as shown by the basal cleavage of micas, or it may be fairly obscure, as in beryl and apatite. In some minerals it is completely absent, as, for example, in quartz. Graphite, for example, has a well-developed platelike cleavage parallel to the basal plane, that is, parallel to {0001}. Within these cleavage plates there is a strong covalent bond among the carbon atoms, but across the plates there are weak van der Waals bonds giving rise to the cleavage (see Fig. 4.38). A weak bond is usually accompanied by a large interplanar spacing because the attractive force cannot hold the planes closely together. Diamond has only one bond type, the covalent bond, and its excellent cleavage parallel to the octahedron {111} takes place along atomic planes having the largest interplanar spacing. The relationship between internal atomic structure and resultant cleavage directions is well shown by the structure image of a pyroxene in Fig. 6.3. The planes of lowest bond density (i.e., the planes of lowest relative electron density) coincide with the two directions of the prism, {110}, the intersection of which is parallel to the *c* axis. The internal angle between these two cleavage directions is approximately 88°.

In describing a cleavage, its quality and crystallographic direction should be given. The quality is ex-

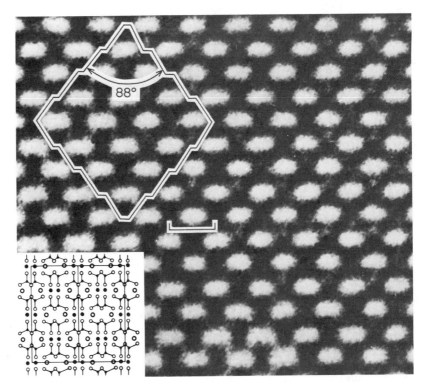

FIG. 6.3. High-resolution transmission electron microscope (HRTEM) image of orthopyroxene showing possible cleavage surfaces at 88° to each other. The length of the bar is 8.8 Å, which is the length of the *b* axis in orthopyroxene. The white regions in the image correspond to areas between M_2 sites in the pyroxene structure. These light regions have relatively fewer atoms than the rest of the structure and hence have a low electron density. The insert shows the pyroxene structure at the same scale as the structure image. (From Buseck, P. R. and Iijima, S., 1974, High resolution electron microscopy of silicates. *American Mineralogist*, v. 59, pp. 1–21.)

pressed as perfect, good, fair, and so forth. The direction is expressed by the name or indices of the form which the cleavage parallels, such as cubic {001} (Fig. 6.4), octahedral {111}, rhombohedral {10$\bar{1}$1}, prismatic {110}, or pinacoidal {001}. Cleavage is always consistent with the symmetry; thus, if one octahedral cleavage direction is developed, it implies that there must be three other symmetry related directions. If one dodecahedral cleavage direction is present, it similarly implies five other symmetry-related directions. Not all minerals show cleavage, and only a comparatively few show it in an eminent degree, but in these it serves as an outstanding diagnostic criterion.

Parting

When minerals break along planes of structural weakness, they have parting. The weakness may result from pressure or twinning or exsolution; and, because it is parallel to rational crystallographic planes, it resembles cleavage. However, parting, unlike cleavage, is not shown by all specimens but only by those that are

FIG. 6.4. Cleavage. *(a)* Cubic. *(b)* Octahedral. *(c)* Dodecahedral. *(d)* Rhombohedral. *(e)* Prismatic and pinacoidal. *(f)* Pinacoidal (basal).

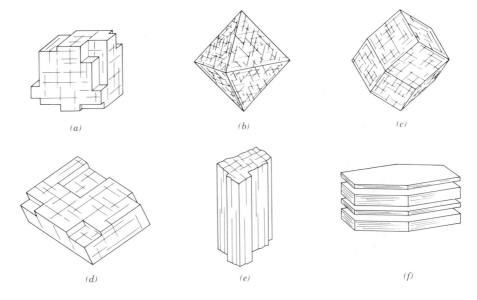

(a)   *(b)*   *(c)*

(d)   *(e)*   *(f)*

FIG. 6.5. (a) Basal parting, pyroxene. (b) Rhombohedral parting, corundum.

twinned or have been subjected to the proper pressure. Even in these specimens there are a limited number of planes in a given direction along which the mineral will break. For example, twinned crystals part along composition planes, but between these planes they fracture irregularly. Familiar examples of parting are found in the octahedral parting of magnetite, the basal parting of pyroxene, and the rhombohedral parting of corundum (see Figs. 6.5a and b).

Fracture

In some crystal structures the strength of the bonds is approximately the same in all directions. Breaking of such crystals generally will not follow a particular crystallographic direction. The way minerals break when they do not yield along cleavage or parting surfaces is their *fracture*. Fracture patterns can be distinctive and highly diagnostic in mineral identification. Different kinds of fracture are designated as follows:

   (a) *Conchoidal.* The smooth, curved fracture resembling the interior surface of a shell (see Fig. 6.6). This is most commonly observed in such substances as glass and quartz.

FIG. 6.6. Conchoidal fracture, obsidian.

   (b) *Fibrous* and *splintery*.
   (c) *Hackly.* Jagged fractures with sharp edges.
   (d) *Uneven* or *irregular*. Fractures producing rough and irregular surfaces.

HARDNESS

The resistance that a smooth surface of a mineral offers to scratching is its *hardness* (designated by **H**). The degree of hardness is determined by observing the comparative ease or difficulty with which one mineral is scratched by another, or by a file or knife. The hardness of a mineral might then be said to be its "scratchability." The evaluation of hardness is the assessment of the reaction of a crystal structure to stress without rupture (cleavage, parting, and fracture are various forms of rupture). In metallic bonded crystals that can flow plastically, scratching results in a groove. However, brittle materials with ionic and/or covalent bonds will react to a hardness test by microfracturing (rupture on a very fine scale). The effect of ionic size and charge in ionically bonded structures has been discussed in Chapter 4 (see Fig. 4.32). This illustrates how different chemical compounds with the same internal structure increase in hardness with decreasing ionic size and increasing ionic charge. In relating the hardness of a crystal structure to its bonding, it must be noted that the structure's overall strength is a composite of all of its bond types, whereas the hardness of that same structure is an expression of its weakest bonding. For example, in silicates, all of which are based on various arrangements of SiO_4 tetrahedra, the hardness ranges from 1, as in talc, to 7, as in quartz, and to 8, as in topaz. Such a variation suggests that hardness is not a function of Si-O bonding, but rather of the other bond types present in the structure. In talc the basal silicate layers are held together by weak van der Waals and/or hydrogen bonds; in quartz there is a very uniform bond strength within a relatively dense network of SiO_4 tetrahedra; and in topaz there are somewhat weaker Al-(F, OH) bonds.

   A series of ten common minerals was chosen by the Austrian mineralogist F. Mohs in 1824 as a scale, by comparison with which the relative hardness of any mineral can be told. The following minerals, arranged in order of increasing hardness, comprise what is known as the *Mohs scale of hardness*:

   1. Talc   6. Orthoclase
   2. Gypsum   7. Quartz
   3. Calcite   8. Topaz
   4. Fluorite   9. Corundum
   5. Apatite   10. Diamond

The above minerals are arranged in an order of increasing relative hardness. The hardness of these same minerals can be measured by more quantitative techniques than a scratch test, and this leads to an absolute hardness scale as shown in Fig. 6.7. The relative position of the minerals in the Mohs scale is preserved, but corundum, for example, is two times as hard as topaz and four times harder than quartz.

Talc, number 1 in the Mohs scale, has a structure made up of plates so weakly bound to one another that the pressure of the fingers is sufficient to slide one plate over the other. At the other end of the scale is diamond, with its constituent carbon atoms so firmly bound to each other that no other mineral can force them apart to cause a scratch.

In order to determine the relative hardness of any mineral in terms of this scale, it is necessary to find which of these minerals it can and which it cannot scratch. In making the determination, the following should be observed: Sometimes when one mineral is softer than another, portions of the first will leave a mark on the second that may be mistaken for a scratch. Such a mark can be rubbed off, whereas a true scratch will be permanent. The surfaces of some minerals are frequently altered to material that is much softer than the original mineral. A *fresh surface* of the specimen to be tested must therefore be used. The physical nature of a mineral may prevent a correct determination of its hardness. For instance, if a mineral is pulverulent, granular, or splintery, it may be broken down and apparently scratched by a mineral much softer than itself. It is always advisable when making the hardness test to confirm it by reversing the order of procedure; that is, do not only try to scratch mineral *A* by mineral *B*, but also try to scratch *B* by *A*.

The following materials serve in addition to the above scale: the hardness of the fingernail is a little over 2, a copper coin about 3, the steel of a pocket knife a little over 5, window glass $5\frac{1}{2}$, and the steel of a file $6\frac{1}{2}$. With a little practice, the hardness of minerals under 5 can be quickly estimated by the ease with which they can be scratched with a pocketknife.

Hardness, as we have seen (page 35), is a vectorial property. Thus crystals may show varying degrees of hardness, depending on the directions in which they are scratched. The directional hardness differences in most common minerals are so slight that, if they can be detected at all, it is only through the use of delicate instruments. Two exceptions are kyanite and calcite. In kyanite, $H = 5$ parallel to the length, but $H = 7$ across the length of the crystal. The hardness of calcite is 3 on all surfaces except $\{0001\}$; on this form it can be scratched by the fingernail and has a hardness of 2.

TENACITY

The resistance that a mineral offers to breaking, crushing, bending, or tearing—in short, its cohesiveness—is known as tenacity. The following terms are used to describe tenacity in minerals:

1. *Brittle.* A mineral that breaks and powders easily. This is characteristic of crystals with dominant ionic bonding.

2. *Malleable.* A mineral that can be hammered out into thin sheets.

3. *Sectile.* A mineral that can be cut into thin shavings with a knife.

4. *Ductile.* A mineral that can be drawn into a wire.

The characteristics described in 2, 3, and 4 (malleability, sectility, ductility) are diagnostic of materials held together by metallic bonding. The metallic bond conveys to crystalline substances the unique property of yielding to applied stress by plastic deformation. As shown in Fig. 4.36, metals are regarded as cations surrounded by a dense cloud of mobile electrons. These cations can, under stress, move past each other without setting up repulsive electrostatic forces. This atomic property is responsible for the physical behavior of metals under stress.

5. *Flexible.* A mineral that bends but does not resume its original shape when the pressure is released.

FIG. 6.7. Comparison of Mohs relative hardness scale and absolute measurements of hardness.

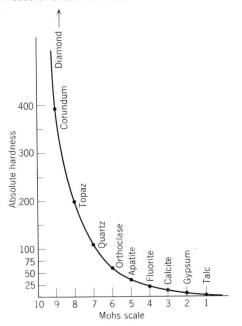

Cleavage sheets of chlorite and talc are flexible, but they do not snap back to their original position after having been bent. In other words, their deformation is permanent. The bonding between the OH-rich layers in these silicates is by a combination of van der Waals and hydrogen bonds, whereas the bonding within the tetrahedral (Si-Al-O) sheets is a mixture of covalent and ionic bonding. The flexibility of the sheets is the result of slippage along OH-layers in the structure (see Fig. 13.85).

6. *Elastic.* A mineral that, after being bent, will resume its original position upon the release of the pressure. Sheets of mica can be bent, and they will snap back into their original position after the bending has stopped. In contrast to the structures of talc and chlorite, the mica structure contains K^+-rich layers that exert a much stronger force on the sheets of Si–Al tetrahedra than do the hydrogen or van der Waals bonding mechanisms. The ionic bonding between K^+ ions and the Si–Al tetrahedral sheets is responsible for the elasticity of mica (see Figs. 13.83 and 13.84).

SPECIFIC GRAVITY

Specific gravity (G) or relative density[1] is a number that expresses the ratio between the weight of a substance and the weight of an equal volume of water at 4°C (this temperature is coincident with the maximum density of water). Thus a mineral with a specific gravity of 2 weighs twice as much as the same volume of water. The specific gravity of a mineral is frequently an important aid in its identification, particularly in working with fine crystals or gemstones, when other tests would injure the specimens.

The specific gravity of a crystalline substance depends on (1) the kind of atoms of which it is composed, and (2) the manner in which the atoms are packed together. In isostructural compounds in which the packing is constant, those with elements of higher atomic weight will usually have higher specific gravities. This is well illustrated by the orthorhombic carbonates listed in Table 6.1, in which the chief difference is in the cations.

In a solid solution series (see page 233), there is a continuous change in specific gravity (or density) with change in chemical composition. For example, the mineral olivine, $(Mg,Fe)_2SiO_4$, is a solid solution series between forsterite, Mg_2SiO_4 (**G** 3.3), and fay-

---

Table 6.1

SPECIFIC GRAVITY INCREASE WITH INCREASING ATOMIC WEIGHT OF CATION IN ORTHORHOMBIC CARBONATES

Mineral	Composition	Atomic Weight of Cation	Specific Gravity
Aragonite	$CaCO_3$	40.08	2.94
Strontianite	$SrCO_3$	87.62	3.78
Witherite	$BaCO_3$	137.34	4.31
Cerussite	$PbCO_3$	207.19	6.58

alite, Fe_2SiO_4 (**G** 4.4). Thus, from determination of specific gravity one can obtain a close approximation of the chemical composition of an Mg-Fe olivine (see Fig. 13.11b). A similar relationship in an amphibole series is shown in Fig. 6.8.

The influence of the packing of atoms on specific gravity is well illustrated in polymorphous compounds (see Table 3.7). In these compounds the composition remains constant, but the packing of the atoms varies. The most dramatic example is given by diamond and graphite, both elemental carbon. Diamond with specific gravity 3.5 has a closely packed structure, giving a high density of atoms per unit volume; whereas in graphite, with specific gravity 2.23, the layers of carbon atoms are loosely packed.

Average Specific Gravity

Most people from everyday experience have acquired a sense of relative weight even in regard to minerals. For example, ulexite (**G** 1.96) seems light, whereas barite (**G** 4.5) seems heavy for nonmetallic minerals. This means that one has developed an idea of an average specific gravity or a feeling of what a nonmetallic mineral of a given size should weigh. This average

FIG. 6.8. Variation of specific gravity with composition in the monoclinic cummingtonite-grunerite series ranging in composition from $Fe_2Mg_5Si_8O_{22}(OH)_2$ to $Fe_7Si_8O_{22}(OH)_2$. (After Klein, C., *American Mineralogist*, 1964.)

---

[1]Density and specific gravity are sometimes used interchangeably. However, density requires the citation of units, for example, grams per cubic centimeter or pounds per cubic foot.

specific gravity can be considered to be between 2.65 and 2.75. The reason for this is that the specific gravities of quartz (**G** 2.65), feldspar (**G** 2.60–2.75), and calcite (**G** 2.72), the most common and abundant nonmetallic minerals, fall mostly within this range. The same sense may be developed in regard to metallic minerals: graphite (**G** 2.23) seems light, whereas silver (**G** 10.5) seems heavy. The average specific gravity for metallic minerals is about 5.0, that of pyrite. Thus, with a little practice, one can, by merely lifting specimens, distinguish minerals that have comparatively small differences in specific gravity.

Determination of Specific Gravity

In order to determine specific gravity accurately, the mineral must be homogeneous and pure, requirements frequently difficult to fulfill. It must also be compact with no cracks or cavities within which bubbles or films of air could be imprisoned. For normal mineralogical work, the specimen should have a volume of about one cubic centimeter. If these conditions cannot be met, a specific gravity determination by any rapid and simple method means little.

The necessary steps in making an ordinary specific gravity determination are, briefly, as follows: the mineral is first weighed in air. Let this weight be represented by W_a. It is then immersed in water and weighed again. Under these conditions it weighs less, because in water it is buoyed up by a force equivalent to the weight of the water displaced. Let the weight in water be represented by W_w. Then $W_a - W_w$ equals the apparent loss of weight in water, or the weight of an equal volume of water. The expression $W_a/(W_a - W_w)$ will therefore yield a number which is the specific gravity.

Jolly Balance. Because specific gravity is merely a ratio, it is not necessary to determine the absolute weight of the specimen but merely values proportional to the weights in air and in water. This can be done by means of a *Jolly balance* (Fig. 6.9),[2] with which the data for making the calculations are obtained by the stretching of a spiral spring. In using the balance, a fragment is first placed on the upper scale pan and the elongation of the spring noted. This is proportional to the weight in air (W_a). The fragment is then transferred to the lower pan and immersed in water. The elongation of the spring is now proportional to the weight of the fragment in water (W_w).

A torsion balance was adapted by Harry Berman for obtaining specific gravities of small particles

[2]Manufactured by Eberbach and Son, Ann Arbor, Michigan.

FIG. 6.9. Jolly balance.

weighing less than 25 milligrams (Fig. 6.10).[3] To the advanced worker interested in accurate determinations this balance is particularly useful, because it is frequently possible to obtain only a tiny mineral fragment free from impurities. In using it, however, one must make a correction for temperature and use a liquid with a low surface tension.

Pycnometer. When a mineral cannot be obtained in a homogeneous mass large enough to permit use of one of the balance methods, the specific gravity of a powder or an aggregate of mineral fragments can be accurately obtained by means of a *pycnometer*. The pycnometer is a small bottle (Fig. 6.11) fitted with a ground-glass stopper through which a capillary opening has been drilled. In making a specific gravity determination, the dry bottle with stopper is first weighed empty (P). The mineral fragments are then introduced into the bottle and a second weighing (M) is made. ($M - P$) represents the weight of the sample in air. Subsequently the bottle (containing the mineral sample) is partially filled with distilled water and boiled for a few minutes to drive off any air bubbles. After cooling, the pycnometer is further filled with distilled water and weighed (S), care being taken that the water rises to the top of the capillary opening but that no excess water is present. The last weighing (W) is made after the bottle has been emptied and refilled with distilled water alone. In this last step, the pycnometer contains more water than in the previous

[3]This balance is distributed through Bethlehem Instrument Company, Bethlehem, Pennsylvania.

FIG. 6.10. Berman balance.

weighing; the volume of water added is equal to the aggregate volume of the grains comprising the sample. The specific gravity can be determined from:

$$\mathbf{G} = \frac{(M - P)}{W + (M - P) - S}$$

where $M - P$ = weight of sample,
$\quad\quad\quad\quad W$ = pycnometer + water content,
$\quad\quad\quad\quad S$ = sample + pycnometer
$\quad\quad\quad\quad\quad\quad$ + undisplaced water,
$W + (M - P) - S$ = weight of water displaced
$\quad\quad\quad\quad\quad\quad\quad\quad$ by sample.

FIG. 6.11. Pycnometer.

Heavy Liquids. Several liquids with relatively high densities are sometimes used in the determination of the specific gravity of minerals. The two liquids most easily used are bromoform (**G** 2.89) and methylene iodide (**G** 3.33).[4] These liquids are miscible with acetone (**G** 0.79), and thus, by mixing, a solution of any intermediate specific gravity may be obtained. A mineral grain is introduced into the heavy liquid and the liquid is diluted with acetone until the mineral neither rises nor sinks. The specific gravity of the liquid and the mineral are then the same, and that of the liquid may be quickly determined by means of a Westphal balance.

Heavy liquids are frequently used in the separation of grains from mixtures composed of several constituents. For example, a separation of the constituent mineral grains of a sand composed of quartz (**G** 2.65), tourmaline (**G** 3.20), and garnet (**G** 4.25) could be quickly made. In bromoform the quartz would float and the tourmaline and garnet would sink; they can be separated from quartz using a separatory tunnel. After removing and washing these "heavy minerals" in acetone, they could be separated from each other in methylene iodide; the tourmaline would float and the garnet would sink.

[4]Although bromoform and methylene iodide are miscible, *do not mix them*. The mixture will become black.

Calculation of Density. If one knows the number of the various kinds of atoms in the unit cell and the volume of the unit cell, the specific gravity can be calculated. The chemical formula of the mineral gives the proportions of the different atoms, but it does not provide us with a knowledge of the number of formula units per unit cell. This number, usually small, is indicated by Z. For example, in aragonite, $CaCO_3$, the ratio of the atoms is 1Ca : 1C : 3 oxygens, but there are four formula units per cell or 4Ca, 4C, 12 oxygens. The molecular weight, M, of $CaCO_3$ is 100.09; the molecular weight of the contents of the unit cell (Z = 4) is 4 × 100.09 = 400.36.

The volume of the unit cell, V, in the orthogonal crystal systems is found by multiplying the cell dimensions, as $a × b × c = V$. In the inclined systems the angles between cell edges must also be considered in obtaining the volume. Aragonite is orthorhombic with cell dimensions: $a = 4.96$ Å, $b = 7.97$, $c = 5.74$. Therefore, $V = 226.91$ Å^3.

Converting Å^3 to cm^3, we divide by $(10^8)^3 = 10^{24}$ or $V = 226.91 × 10^{-24}$ cm^3. Knowing the values M and V, the density, D, can be calculated using the formula:

$$D = \frac{Z × M}{N × V}$$

where N is Avogadro's number, $6.02338 × 10^{23}$. Substituting values for aragonite,

$$D = \frac{4 × 100.09}{6.02338 × 10^{23} × 226.91 × 10^{-24}}$$
$$= 2.93 \text{ gm/cm}^3$$

This value, 2.93, for the calculated density of aragonite is in excellent agreement with the best measured values, which are about 2.94 (see Tables 3.6 and 6.1).

In the study of new minerals, the numerical value of Z is commonly unknown. Hence it is necessary to make successive trials of the calculations given above, using different values of Z until the best possible agreement with the measured specific gravity is secured. Z is always an integer and generally small.

COLOR

Minerals possess many properties, of which color is usually the first and most easily observed. For many it is characteristic and serves as a distinguishing criterion; indeed most people will recognize a number of the gem minerals and gemstones shown in Plates I

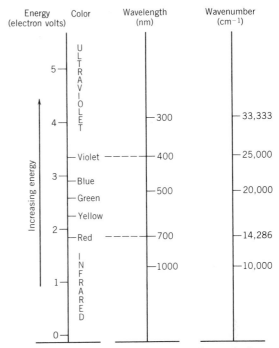

FIG. 6.12. The spectrum, with three ways of numerically specifying the colors. The wavelength scale is in nanometers (nm; 1 nm = 10 Å); the wavenumber scale expresses the number of wavelengths per unit length (cm).

to IV (Chapter 15) on the basis of their colors alone. Yet in many minerals color is one of the most changeable and unreliable diagnostic properties.

Color is the response of the eye to the visible light range of the electromagnetic spectrum (see Figs. 6.12 and 7.1). Visible light represents a range of wavelengths from about 350 to 750 nanometers (nm, 1 nm = 10 angstroms). The energy of light, as of all electromagnetic radiation, can be expressed as follows:

$$E = h\nu = \frac{hc}{\lambda} = hc\bar{\nu}$$

where E denotes energy, h is Planck's constant, c is the speed of light (a constant), ν is frequency, λ is the wavelength, and $\bar{\nu}$ is the wavenumber. In Fig. 6.12 the spectral range of visible light is defined in terms of energy, wavelength, and wavenumber scales. Wavenumber, which is the reciprocal of the wavelength, is directly proportional to energy.

When white light strikes the surface of a mineral, it may be transmitted, scattered, reflected, refracted, or absorbed (see Fig. 6.13). The processes of scattering and reflection are part of the property perceived as the *luster* of a material. If the light suffers no absorption, the mineral is colorless in reflected and transmitted light. Minerals are colored when certain wave-

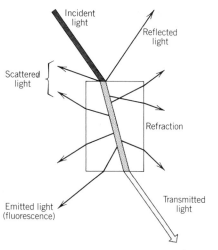

FIG. 6.13. Interaction of light with condensed matter causes reflection, refraction, scattering, and absorption. Some absorbed light can also be reemitted (usually at a longer wavelength) as fluorescence. (Redrawn after Nassau, K., 1980, The Causes of Color. *Scientific American*, v. 243, pp. 124–156.)

lengths of light are absorbed, and the perceived color results from the combination of those remaining wavelengths that reach the eye. The wavelengths that are absorbed by minerals can be quantitatively measured by a spectrometer. An example of an *absorption spectrum* for the mineral beryl is given in Fig. 6.14. The peaks in this pattern represent absorption of specific wavelengths of light, and are the result of the interaction of such wavelengths with ions, molecules, and bonds in the irradiated structure. The absorptions between 0.4 and 0.7 μm are caused by chromophoric transition metal ions such as Fe^{3+} and Cr^{3+} (*chromophore* is derived from the Greek and means color-causing). In the infrared region between 1 and 4.5 μm there are absorptions due to molecules such as H_2O and CO_2, which are interstitial in the hexagonal channels of the beryl structure (see

Fig. 5.16). Beyond 4.5 μm the absorptions are the result of vibrations of the crystal lattice, the so-called lattice modes.

As discussed in Chapter 4, the energies of electrons occur in discrete units, *quanta*. Furthermore, there are well-defined energy differences between these allowed energy levels. When electromagnetic radiation interacts with a material, those wavelengths whose energies correspond exactly to the energy differences between the electronic levels will be absorbed, resulting in electrons being excited from one level to another. In colored minerals, the energy differences between certain electron energy levels are in the range of energy of visible light. Thus, when white light shines on a mineral, certain wavelengths are absorbed (and therefore removed from the spectrum), causing excitation of electrons between these levels.

The electronic processes responsible for color in minerals can be classified as *crystal field transitions*, *molecular orbital (charge transfer) transitions*, and *color centers*. In our subsequent discussion of these phenomena we will draw extensively on the publications of Loeffler and Burns (1976) and Nassau (1978, 1980; see the end of this chapter for complete references).

Crystal Field Transitions

Crystal field transitions are electronic transitions between partially filled 3d orbitals of transition elements. Such transitions are most common in minerals containing the following transition elements: Ti, V, Cr, Mn, Fe, Co, Ni, and Cu. These elements belong to the first transition series with electronic configurations of the general form: $1s^2\ 2s^2\ 2p^6\ 3s^2\ 3p^6\ 3d^{10-n}\ 4s^{1-2}$, with partially filled 3d orbitals (see Tables 4.5 and 4.6). Of these elements Fe is most abundant in the Earth's crust, and for that reason is a dominant cause of color

FIG. 6.14. The visible and infrared spectrum of beryl. Peaks correspond to absorption bands. Absorption in the visible region (0.1 to 0.8 μm) is caused by chromophores Fe and Cr; absorption in the 0.8 to 6.5 μm region results from molecular vibrations; and absorption above 4.5 μm is the result of lattice vibrations. 1 micrometer (μm) = 1000 nanometers (nm). V = violet, R = red. (After Wood, D. L., and Nassau, K., 1968. The characterization of beryl and emerald by visible and infrared absorption spectroscopy. *The American Mineralogist*, v. 53, pp. 777–801.)

in minerals. The electrons in the partially filled $3d$ orbitals can be excited by quanta of energy from the visible spectrum; these electronic transitions are the basis for the production of color. This is in contrast to ionic compounds made up of ions with a noble gas configuration, which are commonly colorless. This is because the energy gap between an occupied p orbital and the next available unoccupied orbital is considerably greater than the energy of visible light.

Crystal field theory accounts for these electronic transitions between partially filled d orbitals. The negative charges of the coordinating anions create an electrical field about the central transition metal ion. This is known as the *crystal field*, which has a specific symmetry and shape as a result of the number of anions, their distances from the cation, and their charges. The five $3d$ orbitals of a transition metal cation have the same energy in the absence of neighboring ions and they have distinctive probability distributions for the electrons (see Fig. 4.10c). Two of these orbitals, $d_{x^2-y^2}$ and d_{z^2}, with their maximum electron densities along the x, y, and z coordinate axes, are referred to as the e_g set. The other three orbitals, d_{xy}, d_{yz}, and d_{xz}, have their greatest electron density in directions between the coordinate axes. These are referred to as the t_{2g} set. When a transition element ion is surrounded by a spherically symmetrical cloud of negative charge, the electron orbitals are the same as those of the free ion (with all orbitals of the same energy), but their overall energy levels will exceed those of the free ion because the spherical negative field will repel all electrons equally in these orbitals, which adds to their potential energy (see Figs. 6.15a and b).

When a transition metal ion is placed in a coordination site in a mineral, there will be a nonuniform crystal field interaction on the various d orbitals by the neighboring anions. If the coordination polyhedron about the cation is octahedral (see Fig. 6.15c) the electrostatic repulsion between the anion orbitals and the centrally located cation orbitals raises the energy level of the $d_{x^2-y^2}$ and d_{z^2} orbitals (whose lobes of electron density are *along* the axes, see Fig. 4.10) relative to the d_{xy}, d_{xz}, and d_{yz} orbitals (whose lobes of electron density are *between* the axes). This is known as *crystal field splitting*, meaning that the crystal field set up by the six surrounding anions splits the $3d$ energy levels of the central cation (Fig. 6.15d). In this illustration of crystal field splitting we have evaluated only what happens in an octahedral anion polyhedron. Minerals generally have several different coordination polyhedra; in these differing polyhedra, the energy level splitting of the transition metals' $3d$ orbitals will also be different. Furthermore, any distortion of the anionic coordination polyhedron about the central transition element will cause additional levels of splitting of $3d$ orbitals.

Let us now illustrate some of the above observations with respect to the color of three well-known minerals and their gem varieties: (1) peridot, a gem variety of olivine, with a yellow-green transmitted color (see Plate III, no. 1, Chapter 15); (2) chrysoberyl, with a characteristic pale yellow-green color (see Plate II, no. 8); and (3) almandine, a member of the garnet group, with a dark red transmitted color (see Plate III, no. 2).

Peridot is a member of the olivine series, $(Mg, Fe)_2SiO_4$. The olivine crystal structure consists of in-

FIG. 6.15. Schematic representation of the energy levels of the five $3d$ orbitals in transition metals. (a) A free ion without surrounding neighbors, as in a gaseous state. (b) The ion surrounded by a uniform and spherically distributed negative charge. (c) The transition metal ion surrounded by an octahedral field of negative charges (anions) experiences the energy separation (or crystal field splitting, Δ_0) of the $3d$ orbitals, as shown in (d). The five $3d$ orbitals are separated into a high-energy group (the e_g set) and a low-energy group (the t_{2g} set). (Adapted from Loeffler, B. M. and Burns, R. G., 1976. Shedding light on the color of gems and minerals. *American Scientist*, v. 64, pp. 636–647.)

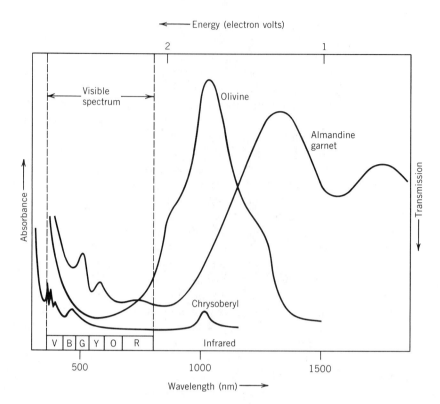

FIG. 6.16. Absorption spectra of two Fe^{2+}-bearing minerals, peridot, a gem variety of olivine, and almandine, a member of the garnet group. The spectrum of an Fe^{3+}-bearing mineral, chrysoberyl, is also shown. (From Loeffler, B. M. and Burns, R. G., 1976, Shedding light on the color of gems and minerals. *American Scientist*, v. 64, pp. 636–647. Reprinted by permission of *American Scientist*, journal of Sigma Xi, The Scientific Research Society.)

dependent SiO_4 tetrahedra linked to Mg^{2+} and Fe^{2+} in octahedral coordination (see Fig. 13.4). The Fe^{2+} ions are distributed among two slightly different octahedral sites (designated as *M1* and *M2*). If white light shines on a peridot and one measures the amount of light absorption as a function of wavelength, the optical spectrum shown in Fig. 6.16 is obtained. Wavelengths of light that correspond in energy exactly to the energy differences caused in the $3d$ orbitals by crystal field splitting are absorbed. This absorbance in peridot takes place mainly in the infrared region, with some extension into the visible range. The absorption of the red component of white light is responsible for the yellow-green (transmitted) color of peridot. As such, Fe^{2+} in 6-coordination produces a characteristic green transmitted color in minerals.

The oxidation state of the transition element also affects the transmitted color. The mineral chrysoberyl, Al_2BeO_4, may contain some Fe^{3+} in substitution for Al^{3+}. The crystal structure of chrysoberyl is very similar to that of olivine, with Al^{3+} (and small amounts of Fe^{3+}) in octahedral coordination. The optical absorption spectrum of chrysoberyl (see Fig. 6.16) is, however, very different from that of peridot. Chrysoberyl absorbs only weakly in the violet and blue region of the spectrum, giving the characteristic pale yellow color of Fe^{3+}-containing chrysoberyl. The differences in the absorption spectra (between peridot

and chrysoberyl) are the result of the differences in the electronic structures of Fe^{2+} and Fe^{3+}.

A change in coordination polyhedron about a transition element also affects the absorption spectrum and resultant transmitted color. In the structure of garnet (see Fig. 13.5), for example, independent SiO_4 tetrahedra are linked to trivalent cations in octahedral coordination and divalent cations in 8-fold coordination (distorted cubic). In a common variety of garnet, almandine, $Fe_3Al_2Si_3O_{12}$, Fe^{2+} is housed in the 8-coordinated site. The absorption spectrum of such an almandine is given in Fig. 6.16. Now the main absorption peaks occur entirely outside the visible region, with lesser absorption peaks in the region of violet-blue-green-yellow. The transmission color resulting from this is a deep red.

Two other minerals that occur as highly prized gems illustrate yet another factor contributing to color. These are (1) emerald, an emerald-green, Cr-containing variety of beryl, $Be_3Al_2Si_6O_{18}$ (see Plate I, nos. 2 and 5, Chapter 15), and (2) ruby, a ruby-red, Cr-containing variety of corundum, Al_2O_3 (see Plate I, numbers 4 and 7).

In both minerals, small amounts of Cr^{3+} replace Al^{3+} in 6-coordinated sites (somewhat distorted octahedral). In beryl, a ring silicate (see Fig. 5.16), the six oxygens about Al^{3+} (or Cr^{3+}) are shared with SiO_4 and BeO_4 tetrahedra; in corundum, an oxide consist-

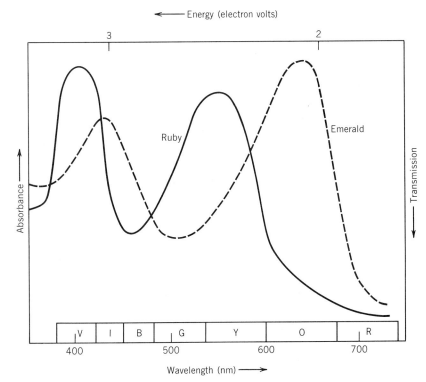

FIG. 6.17. Absorption spectra of emerald and ruby. In emerald, where the crystal field about Cr^{3+} is weaker, the absorption peaks are shifted to lower energy, producing transmission in green. In ruby, the absorption peaks are at higher energies, permitting transmission mainly in the blue and red regions. (From Loeffler, B. M. and Burns, R. G., 1976, Shedding light on the color of gems and minerals. *American Scientist*, v. 64, pp. 636–647. Reprinted by permission of *American Scientist*, journal of Sigma Xi, The Scientific Research Society.)

ing of hexagonal close-packed layers of oxygen, Al^{3+} (or Cr^{3+}) occupies interstices between these layers. In the silicate structure of beryl there is a covalent component to the bonding (see Fig. 4.35), whereas in corundum the bonding is more truly ionic. This results in a weaker crystal field around the Cr^{3+} in beryl than in corundum. These differences are clearly reflected in Fig. 6.17, which compares the absorption spectra of the two gems. In emerald, the absorption peaks are at lower energy than in ruby. In emerald, the absorptions are in the violet and blue, in the yellow, and in the orange and red. This gives transmission in the green and hence the emerald-green color. In ruby, there is absorption in the violet, green, and yellow, with transmission in blue and red. The overall red color of ruby is furthermore intensified by a characteristic fluorescence in the red. That is, not only does ruby absorb most wavelengths such that red is transmitted, it also emits red light by fluorescence (see page 268).

In summary, several factors which influence the transmitted color produced by crystal field interactions are:

1. the presence of a specific transition element,
2. its oxidation state (valence), which determines the number of electrons in $3d$ orbitals,
3. the geometry of the site in which the transition metal is housed, (octahedral, tetrahedral, etc.),

4. the strength of the crystal field (charges on anions, distortion of coordination polyhedra, etc.), and
5. the way in which the human eye interprets the pattern of transmitted wavelengths.

Table 6.2 is a summary of minerals whose color is the result of crystal field transitions.

Molecular Orbital Transitions

Molecular orbital transitions (also known as charge transfer transitions) occur in minerals when valence electrons transfer back and forth between adjacent ions. The electrons are contributed to *shared molecular orbitals* and as such are delocalized; that is, the valence electrons of a constituent atom are no longer in atomic orbitals centered on the atom. In such instances crystal field theory does not apply, but instead molecular orbital theory best describes the observed spectra.

Examples of molecular orbital transitions are found in many minerals. Of these, $Fe^{2+} \rightarrow Fe^{3+}$ and $Fe^{2+} \rightarrow Ti^{4+}$ are the most common metal-metal charge transfer transitions. In the $Fe^{2+} \rightarrow Fe^{3+}$ charge transfer transition, an electron is transferred from Fe^{2+} (in site A, making it Fe^{3+}) to Fe^{3+} (in site B, making it Fe^{2+}), such that $Fe^{2+}_{(A)} + Fe^{3+}_{(B)} \rightleftarrows Fe^{3+}_{(A)} + Fe^{2+}_{(B)}$. Energies of this reversible electron-hopping process generally correspond to wavelengths in visible light,

Table 6.2

EXAMPLES OF COMMON MINERALS WHOSE COLOR IS DUE TO THE INTERACTION OF TRANSITION ELEMENTS AND CRYSTAL FIELD TRANSITIONS*

Absorbing Ion	Mineral	Formula	Color
Cr^{3+}	Beryl (emerald)	$Be_3Al_2Si_6O_{18}$	Green
	Corundum (ruby)	Al_2O_3	Red
Mn^{3+}	Tourmaline (rubellite)	$Na(Li,Al)_3Al_6(BO_3)_3(Si_6O_{18})OH_4$	Pink
Mn^{2+}	Beryl (morganite)	$Be_3Al_2Si_6O_{18}$	Pink
	Spessartine garnet	$Mn_3Al_2(SiO_4)_3$	Yellow-orange
Fe^{3+}	Andradite garnet	$Ca_3Fe_2(SiO_4)_3$	Green
	Chrysoberyl	$BeAl_2O_4$	Yellow
Fe^{2+}	Olivine (peridot)	$(Mg,Fe)_2SiO_4$	Yellow-green
	Almandine garnet	$Fe_3Al_2(SiO_4)_3$	Dark red
Cu^{2+}	Turquoise	$CuAl_6(PO_4)_4(OH)_8 \cdot 5H_2O$	Light blue

*From Loeffler, B. M. and Burns, R. G., 1976, Shedding light on the color of gems and minerals. *American Scientist*, v. 64, pp. 636–647. Many of the minerals listed in the table are illustrated in Plates I through IV, Chapter 15.

and many minerals owe their intense blue color to such transitions. Examples are glaucophane (blue amphibole and crocidolite, blue amphibole asbestos), cordierite, kyanite (commonly blue), and sapphire (blue gem variety of corundum). The $Fe^{2+} \rightarrow Ti^{4+}$ charge transfer transition is also a large factor in the blue color of sapphire, Al_2O_3, which commonly contains small amounts of iron and titanium. Figure 6.18 shows the optical absorption spectrum of blue sapphire, in which the main absorbance peaks are identified as due to $Fe^{2+} \rightarrow Ti^{4+}$, and $Fe^{2+} \rightarrow Fe^{3+}$ charge transfers. The main transmission of light is in the blue range of the visible spectrum (see Plate I, nos. 3 and 6, Chapter 15 for color illustrations of sapphire). Table 6.3 is a summary of some minerals whose color is the result of molecular orbital transitions.

Color Centers

Coloration can also be caused by structural defects. This can be an excess electron that is unattached to any single atom and that is trapped at some structural defect, such as a missing ion or an interstitial impurity. A "hole," the absence of an electron, can have the same effect. These types are known as *color centers*, or *F* centers (from the German *Farbe*, color). The coloring mechanism in purple fluorite, CaF_2, is known to be the result of Frenkel defects (see Fig. 3.51) in the fluorite structure. Figure 6.19 is an illustration of the structure of fluorite in which an F^- ion is missing from its usual structural site. Such defects in the network of F^- ions can be the result of (1) high-energy radiation (e.g., X-rays) that displaced the F^- from its usual position to another one in the structure, (2) the growth of fluorite in a chemical environment with an excess of calcium, and (3) removal of some F from the crystal by the application of an electrical field. Because the overall structure must remain neutral, an electron usually occupies the empty position to produce an "electron color center," as in Fig. 6.19. Such an electron is not bound in place by a central nucleus, but by the electrical field (crystal field) of all of the surrounding ions. Within this field it can occupy a ground state and various excited states similar to those of the transition elements described above. The movement of electrons among these states can cause color and optical fluorescence. It should be noted that the original crystal structure of fluorite (without defects) can be restored by heating, whereupon the color fades.

FIG. 6.18. Optical absorption spectrum of sapphire, blue gem corundum (see Plate I, nos. 3 and 5, Chapter 15), and the molecular orbital transitions responsible for the absorption peaks ($Fe^{2+} \rightarrow Ti^{4+}$; $Fe^{2+} \rightarrow Fe^{3+}$; and $O^{2-} \rightarrow Fe^{3+}$ at the edge of the ultraviolet region). The only transmission in this spectrum is in the range of blue in the visible spectrum. (From Loeffler, B. M. and Burns, R. G., 1976, Shedding light on the color of gems and minerals. *American Scientist*, v. 64, pp. 636–647. Reprinted by permission of *American Scientist*, journal of Sigma Xi, The Scientific Research Society.)

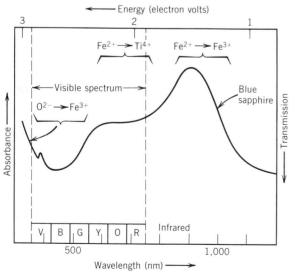

Table 6.3 **EXAMPLES OF SOME COMMON MINERALS WHOSE COLOR IS THE RESULT OF CHARGE-TRANSFER TRANSITIONS, DESCRIBED BY MOLECULAR ORBITAL THEORY***	**Ion Pair**	**Mineral**	**Formula**	**Color**
	$Fe^{2+} \rightarrow Fe^{3+}$	Beryl (aquamarine)	$Be_3Al_2Si_6O_{18}$	Blue-yellow
	$Fe^{2+} \rightarrow Fe^{3+}$	Cordierite	$(Mg,Fe)_2Al_4Si_5O_{18} \cdot nH_2O$	Blue
	$Fe^{2+} \rightarrow Ti^{4+}$	Corundum (sapphire)	Al_2O_3	Blue
	$Fe^{2+} \rightarrow Ti^{4+}$	Kyanite	Al_2SiO_5	Blue
	$O^{2-} \rightarrow Cr^{6+}$	Crocoite	$PbCrO_4$	Orange
	$O^{2-} \rightarrow Fe^{3+}$	Beryl (heliodore)	$Be_3Al_2Si_6O_{18}$	Yellow

*From Loeffer, B. M. and Burns, R. G., 1976, Shedding light on the color of gems and minerals. *American Scientist*, v. 64, pp. 636–647.

The smoky color of some quartz crystals is attributed to the occurrence of a "hole color center." In such quartz some Al^{3+} substitutes for Si^{4+}, and this substitution is coupled with some interstitial Na^+ or H^+ ions in order to maintain electrical neutrality. When this type of quartz with some Al substituting for Si in tetrahedral sites, is exposed to an intense X-ray or gamma ray beam for a few minutes, or when it has been exposed to low levels of radiation over geological periods, "hole color centers" are produced. The radiation expels one electron from a pair of electrons in an oxygen atom adjacent to an Al^{3+} ion, thereby leaving a single, unpaired electron in the orbital. This is illustrated schematically in Fig. 6.20. The missing electron is called a "hole," and the remaining unpaired electron has a set of excited states much like those of an excess electron, as described above. Table

6.4 lists some mineral examples in which coloration is due to color centers.

Other Causes of Color

Yet another coloring agent is the mechanical admixture of impurities, which can give a variety of colors to otherwise colorless minerals. Quartz may be green because of the presence of finely dispersed chlorite; calcite may be black, colored by manganese oxide or

FIG. 6.19. Schematic illustration of the structure of fluorite, CaF_2, in which an electron fills a vacancy created by a fluorine ion that was removed. Here a color center is the result of the electron taking the place of the dislodged ion. (Adapted from Nassau, K., The Causes of Color. *Scientific American*, v. 243, pp. 124–156. Copyright © 1980 by Scientific American, Inc. All rights reserved.)

FIG. 6.20. Schematic illustration of the quartz structure. (a) The normal structure of pure SiO_2. (b) The structure with some ionic substitution of Al^{3+} for Si^{4+}, coupled with introduction of H^+ into the structure, in order to retain overall neutrality. Radiation ejects one of a pair of electrons from an O^{2-} and leaves a "hole" color center of smoky quartz. (After Nassau, K., 1978, The origins of color in minerals. *American Mineralogist*, v. 63, pp. 219–229.)

(a)

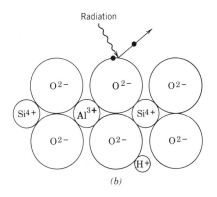

(b)

Table 6.4
EXAMPLES OF MINERALS IN WHICH THE COLORATION IS DUE TO COLOR CENTERS*

Mineral	Color
Amethyst, fluorite	Purple
Smoky quartz	Brown to black
Irradiated diamond	Green, yellow, brown, blue, pink
Natural and irradiated topaz	Blue
Halite	Blue and yellow

*From Nassau, K., 1978, The origins of color in minerals. *American Mineralogist*, v. 63, pp. 219–229.

carbon. Hematite, as the most common pigmenting impurity, imparts its red color to many minerals, including some feldspar and calcite and the fine-grained variety of quartz, jasper.

Streak

The color of a finely powdered mineral is known as its streak. Although the color of a mineral may vary, the streak is usually constant and thus is useful in mineral identification. The streak is determined by rubbing the mineral on a piece of unglazed porcelain, a *streak plate*. The streak plate has a hardness of about 7, and thus it cannot be used with minerals of greater hardness.

Luster

The term *luster* refers to the general appearance of a mineral surface in reflected light. There are two types of luster, *metallic* and *nonmetallic*, but there is no sharp division between them. Minerals with an intermediate luster are said to be *submetallic*.

A mineral having the brilliant appearance of a metal has a metallic luster. Such minerals are quite opaque to light and, as a result, give a black or very dark streak. Galena, pyrite, and chalcopyrite are common minerals with metallic luster.

In minerals with metallic bonding, the energy gaps between the ground states and the excited states of electrons are generally much smaller than those for ionic and covalently bonded structures. The energy of visible light is commonly much smaller than the energy gaps in ionic and covalent structures. In metallic compounds, however, there are a large number of excited states with energies that are available in the entire range of the visible spectrum. This means that any quantum of light that strikes the surface of a metallic or partially metallic bonded crystal is absorbed. Most of this absorbed energy is immediately re-emitted as

visible light. This results in materials with metallic luster reflecting light completely (see Fig. 6.13).

Minerals with a nonmetallic luster are, in general, light-colored and transmit light, if not through thick portions at least through thin edges. The streak of a nonmetallic mineral is either colorless or very light in color. The following terms are used to describe further the luster of nonmetallic minerals.

Vitreous. The luster of glass. Examples—quartz and tourmaline.

Resinous. Having the luster of resin. Examples—sphalerite and sulfur.

Pearly. An iridescent pearl-like luster. This is usually observed on mineral surfaces that are parallel to cleavage planes. Examples—basal plane of apophyllite and cleavage surface of talc.

Greasy. Appears as if covered with a thin layer of oil. This luster results from light scattered by a microscopically rough surface. Examples—nepheline and some speimens of sphalerite and massive quartz.

Silky. Silklike. It is caused by the reflection of light from a fine, fibrous parallel aggregate. Examples—fibrous gypsum, malachite, serpentine (chrysotile), and the silicified and oxidized product of crocidolite known as ''tiger's eye.''

Adamantine. An exceptionally brilliant luster like that of a diamond. It is due to a mineral's unusually high index of refraction (see page 290). Examples—the transparent lead-containing minerals such as cerussite and anglesite.

Play of Colors

Interference of light either at the surface or in the interior of a mineral may produce a series of colors as the angle of the incident light changes. The striking flashes of varied color against a white or black background, as seen in precious opal, is called *play of colors* (see Plate II, no. 7, Chapter 15). This phenomenon was originally thought to be the result of thin film interference, but electron microscopic study of opal has revealed that the underlying reason for the color play is the presence (in precious opal) of a regular three-dimensional array of equal-size spheres. These spheres consist of amorphous silica, SiO_2, with small amounts of water; they are cemented together by amorphous silica with a slightly different water content (see Fig. 6.21). In precious opal the uniformally packed spheres occur in patches (domains) ranging from less than a millimeter to more than a centimeter across. These regularly arrayed domains

FIG. 6.21. Scanning electron micrograph of an opal with chalky appearance, showing hexagonal closest packing of silica spheres (diameter of spheres is approximately 3000 Å). Because of the weak bonding between the spheres they are completely intact; in typical precious opal samples many of the spheres are cleaved. (Courtesy of Darragh, P. J., Gaskin, A. J. and Sanders, J. V., 1976, Opals. *Scientific American*, v. 234, no. 4, pp. 84–95.)

act as diffraction gratings for white light, and resolve white light into its spectral colors in accordance with the modified Bragg equation (see also Chapter 7):

$$n\lambda = \mu 2d \sin \theta$$

where n is a small number (1, 2, or 3) and is known as the order of reflection, λ is the wavelength of a specific spectral line, μ is the refractive index of opal (which must be considered because the process of diffraction takes place within the SiO_2 of the opal), d is the spacing between the spheres in the precious opal (in the sample in Fig. 6.21 this is about 3000 Å), and θ is the angle of incidence and reflection. The wavelength of the diffracted spectral color is determined by the value of d (interplanar spacing) and varies with the angle θ (see Fig. 6.22). Common opal lacks this

FIG. 6.22. The spectral colors of precious opal are the result of diffraction by regularly spaced lattice planes. These planes result from amorphous spheres in a regular close-packed array. The spacing of the lattice planes is shown as d. The λ of the diffracted spectral line is a function of d and the angle θ. (Redrawn after Darragh, P. J., Gaskin, A. J. and Sanders, J. V., 1976, Opals. *Scientific American*, v. 234, pp. 84–95. Copyright Scientific American Inc., George V. Kelvin.)

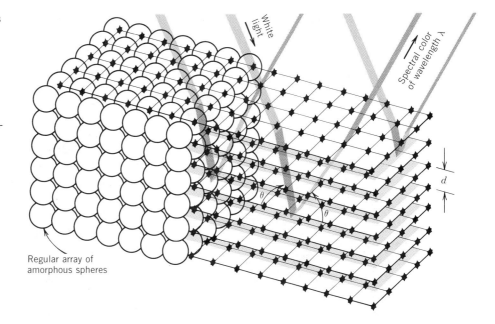

Regular array of amorphous spheres

FIG. 6.23. Microstructure in labradorite showing very fine, essentially parallel lamellae. These lamellae act as a diffraction grating for white light, producing spectral colors known as *labradorescence*. This photograph was taken with a transmission electron microscope. (From Champness, P. E. and Lorimer, G. W., 1976, Exsolution in silicates, chapter 4.1 in *Electron Microscopy in Mineralogy*, H. R. Wenk, ed. Springer-Verlag, New York.)

regular internal stacking of spheres, and the scattered white light produces a milky *opalescence*.

An internal iridescence is caused by light diffracted and reflected from closely spaced fractures, cleavage planes, twin lamellae, exsolution lamellae, or minute foreign inclusions in parallel orientation. Some specimens of labradorite show colors ranging from blue to green or yellow and red with changing thickness of lamellar pairs. This *iridescence*, also called *schiller* and *labradorescence*, is the result of light scattered by extremely fine, less than $\frac{1}{10}$ micron or thinner in width, exsolution lamellae in the range of An_{47} to An_{58} (see Fig. 6.23). The delicate bluish, grayish sheen of some albitic feldspars (peristerite), ranging in composition from An_2 to An_{16}, is caused by scatter from exsolution lamellae of An_0 and An_{25} composition.

A surface iridescence similar to that produced by soap bubbles or thin films of oil on water is caused by interference of light as it is reflected from thin surface films produced by oxidation or alteration. It is most commonly seen on metallic minerals, particularly hematite, bornite, limonite, and sphalerite.

CHATOYANCY AND ASTERISM

In reflected light some minerals have a silky appearance, which results from closely packed parallel fibers or from a parallel arrangement of inclusions or cavities. When a cabochon gemstone is cut from such a mineral, it shows a band of light at right angles to the length of the fibers or direction of the inclusions. This property, known as *chatoyancy*, is shown particularly well by "satin spar" gypsum, *cat's eye*, a gem variety of chrysoberyl, and *tiger's eye*, fibrous crocidolite replaced by quartz.

In some crystals, particularly those of the hexagonal system, inclusions may be arranged in three crystallographic directions at 120° to each other. A cabochon stone cut from such a crystal shows what might be called a triple chatoyancy, that is, one beam of light at right angles to each direction of inclusions producing a six-pointed star. The phenomenon, seen in star rubies and sapphires, is termed *asterism* (see Fig. 6.24) and is the result of scattering of light from inclusions of rutile arranged in three crystallographic directions. Some phlogopite mica containing rutile needles oriented in a pseudohexagonal pattern shows a striking asterism in transmitted light.

Luminescence

Any emission of light by a mineral that is not the direct result of incandescence is *luminescence*. This phenomenon may be brought about in several ways and is usually observed in minerals containing impurity ions called *activators*. Most luminescence is faint and can be seen only in the dark.

Fluorescence and Phosphorescence

Minerals that luminesce during exposure to ultraviolet light, X-rays, or cathode rays are *fluorescent*. If the

FIG. 6.24. Asterism in a sphere of rose quartz. Sphere diameter is 5.5 cm. The 6-rayed star is caused by microscopic needle-like inclusions of rutile (TiO_2) which are oriented in three directions (at 120° to each other) by the quartz structure. These inclusions reflect a spotlight source yielding the 6-rayed star. (Harvard Mineralogical Museum.)

luminescence continues after the exciting rays are cut off, the mineral is said to be *phosphorescent*. There is no sharp distinction between fluorescence and phosphorescence, because some minerals that appear only to fluoresce can be shown by refined methods to continue to glow for a small fraction of a second after the removal of the exciting radiation.

The cause of fluorescence is similar to the cause of color, and ions of the transition metals are effective activators. Electrons, excited by the invisible short radiation, are raised to higher energy levels. When they fall back to their initial (ground) state, they emit visible light of the same wavelength. However, these excited electrons may fall back to an energy level intermediate between their excited state and the ground state (Fig. 6.25). They then emit a photon of light of lower energy (longer wavelength) than that which provided the original excitation. If the original excitation is produced by ultraviolet (uv) light, the fluorescence is commonly in the visible range.

In phosphorescent minerals there is a time lag between the excitation of electrons to a higher energy level and their return to the ground state. Minerals vary in their ability to absorb uv light at a given wavelength. Thus some fluoresce only in shortwave uv, whereas others may fluoresce only in longwave uv, and still others will fluoresce under either wavelength of uv. The color of the emitted light varies considerably with the wavelengths or source of uv light.

Fluorescence is an unpredictable property, for some specimens of a mineral show it, whereas other apparently similar specimens, even from the same locality, do not. Thus, only some fluorite, the mineral from which the property receives its name, will fluoresce. Its usual blue fluorescence may result from the presence of organic material or rare earth ions. Other minerals that frequently but by no means invariably fluoresce are scheelite, willemite, calcite, eucryptite, scapolite, diamond, hyalite, and autunite. The pale blue fluorescence of most scheelite is ascribed to molybdenum substituting for tungsten. And the brilliant fluorescence of willemite and calcite from Franklin, New Jersey, is attributed to the presence of manganese.

FIG. 6.25. Schematic energy level diagram for the absorption (Abs) of ultraviolet radiation, and resulting fluorescence in the visible light region.

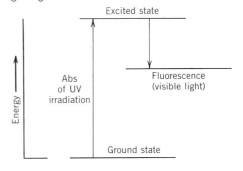

With the development of synthetic phosphors, fluorescence has become a commonly observed phenomenon in fluorescent lamps, paints, cloth, and tapes. The fluorescent property of minerals also has practical applications in prospecting and ore dressing. With a portable uv light one can at night detect scheelite in an outcrop, and underground the miner can quickly estimate the amount of scheelite on a freshly blasted surface. At Franklin, New Jersey, uv light has long been used to determine the amount of willemite that goes into the tailings. Eucryptite is an ore of lithium in the great pegmatite at Bikita, Zimbabwe. In white light it is indistinguishable from quartz, but under uv light it fluoresces a salmon pink and can be easily separated.

Thermoluminescence

This is the property of emitting visible light when a substance is heated to a temperature below that of red heat. It is best shown by nonmetallic minerals that contain impurity ions as activators. When a thermoluminescent mineral is heated, the initial visible light, usually faint, is given off at a temperature between 50° and 100°C, and light usually ceases to be emitted at temperatures higher than 475°C. For a long time, fluorite has been known to possess this property; the variety *chlorophane* was named because of the green light emitted. Other minerals that are commonly thermoluminescent are calcite, apatite, scapolite, lepidolite, and certain feldspars.

Triboluminescence

This is the property of becoming luminous on being crushed, scratched, or rubbed. Most minerals showing this property are nonmetallic and possess a good cleavage. Fluorite, sphalerite, and lepidolite may be triboluminescent; less commonly, so may pectolite, amblygonite, feldspar, and calcite.

ELECTRICAL PROPERTIES

The conduction of electricity in crystals is related to the type of bonding. Minerals with pure metallic bonding, such as the native metals, are excellent electrical *conductors*, whereas those in which the bonding is partially metallic, as in some sulfide minerals, are *semiconductors*. Ionically or covalently bonded minerals are usually *nonconductors*. For non-isometric minerals, electrical conductivity is a vectorial property varying with crystallographic direction. For ex-

ample, the hexagonal mineral graphite is a far better conductor at right angles to the c axis than parallel to it.

Piezoelectricity

Polar axes are present only in crystals that lack a center of symmetry. Of the 32 crystal classes, 21 have no center of symmetry (see Table 2.3), and of these all but one, the gyroidal class (432), has at least one polar axis with different crystal forms at opposite ends. If pressure is exerted at the ends of a polar axis, a flow of electrons toward one end produces a negative electrical charge, whereas a positive charge is induced at the opposite end. This is *piezoelectricity*, and any mineral crystallizing in one of the 20 classes with polar axes should show it. However, in some minerals the charge developed is too weak to be detected.

The property of piezoelectricity was first detected in quartz in 1881 by Pierre and Jacques Curie, but nearly 40 years passed before it was used in a practical way. Toward the end of World War I it was found that sound waves produced by a submarine could be detected by the piezoelectric current generated when they impinged on a submerged quartz plate. The device was developed too late to have great value during the war, but it pointed the way to other applications. In 1921 the piezoelectric property of quartz was first used to control radio frequencies, and since then millions of quartz plates have been used for this purpose. When subjected to an alternating current, a properly cut slice of quartz is mechanically deformed and vibrates by being flexed first one way and then the other; the thinner the slice, the greater the frequency of vibration. By placing a quartz plate in the electric field generated by a radio circuit, the frequency of transmission or reception is controlled when the frequency of the quartz coincides with the oscillations of the circuit. The tiny quartz plate used in digital and analog quartz watches serves the same function as quartz oscillators used to control radio frequencies. That is, it mechanically vibrates at a constant frequency which is a function of the plate's thickness and orientation; this quartz frequency controls accurately the radio frequency of the electronic circuit in the watch. This circuit counts the crystal frequency and provides the digital time display of the watch. Figure 6.26 is an illustration of the basic schematic of a liquid crystal watch for the display of hours and minutes. The quartz crystal controls an oscillator circuit, which in turn generates pulses of one second in time. These "second" pulses are counted to produce "minute" and "hour" pulses. Each of these pulses is decoded to

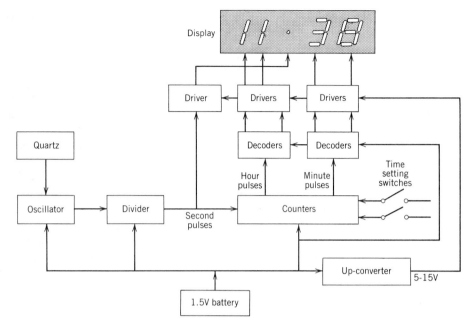

FIG. 6.26. Block diagram of a liquid crystal watch. (Redrawn after Burfoot, J. C. and Taylor, G. W., 1979, *Polar Dielectrics and Their Applications.* University of California Press, Berkeley, Calif. Copyright © 1979 Jack Burfoot and George Taylor.)

provide proper outputs for the digital watch display. Powered by a 1.5 V silver oxide battery, a quartz plate vibrates approximately 100,000 times per second. An inexpensive quartz watch today is more accurate than the best-made mechanical watch, and precision-manufactured quartz clocks are accurate to within one second per ten years.

The piezoelectric property of tourmaline has been known almost as long as that of quartz, but compared with quartz, tourmaline is a less effective radio oscillator and is rare in occurrence. Nevertheless, small amounts of it are used today in piezoelectric pressure gauges. In tourmaline, which is hexagonal, c is the polar axis. Plates cut normal to this direction will generate an electrical current when subjected to a transient pressure. The current generated is proportional to the area of the plate and to the pressure. Tourmaline gauges were developed to record the blast pressure of the first atomic bomb in 1945 and since then have been used by the United States with each atomic explosion. Lesser pressures also can be recorded by them, however, such as those generated by firing a rifle or by surf beating on a sea wall.

Pyroelectricity

Temperature changes in a crystal may cause the simultaneous development of positive and negative charges at opposite ends of a polar axis. This property of *pyroelectricity* is observed, as is piezoelectricity, only on crystals with polar axes. Crystals that belong to the ten crystal classes having a unique polar axis

are considered to show "true" or *primary* pyroelectricity. For example, tourmaline has a single polar axis, c, and falls within this group, whereas quartz with its three polar a axes does not. However, a temperature gradient in all crystals having polar axes such as quartz will produce a pyroelectric effect. In such crystals the polarization is the result of the deformation resulting from unequal thermal expansion that produces piezoelectric effects. If quartz is heated to about 100°C, on cooling it will develop positive charges at three alternate prismatic edges and negative charges at the three remaining edges. These charges have been called *secondary* pyroelectric polarization.

By means of single crystal X-ray photographs some minerals can be assigned to specific point or space groups, but for others X-ray data are ambiguous. For example, on the basis of X-ray photographs, a mineral might belong to either of the classes $2/m2/m2/m$ or $mm2$. If it could be shown to be piezoelectric or pyroelectric, it would definitely belong to $mm2$.

MAGNETIC PROPERTIES

Some minerals behave like magnets, whereas most do not. Such magnetic properties are the result of atomic properties that are specific to a number of elements. In Chapter 4 it was discussed that in order to specify the position of an electron in three-dimensional space, *three quantum numbers* are needed: *n*, the principal quantum number, l, the azimuthal (or orbital shape) quantum number, and m, the magnetic quantum number. In addition to these three quantum numbers, there is a fourth quantum number, the *spin quantum number*, s, which defines the spin of the electron in space (see Fig. 4.11). Because an electron can spin in only two directions, it has only two values, namely $+\frac{1}{2}$ and $-\frac{1}{2}$. A spinning electron behaves as a small magnet and will produce a magnetic field while moving around its orbit. This is analogous to the production of a magnetic field by an electrical current moving through a coiled wire. As such, each electron on an atom can be assigned values for four quantum numbers, n, l, m, and s, which determine the orbital in which the electron occurs and the direction of electron spin. There is, however, a restriction on the values these quantum numbers may have. This is known as the *Pauli exclusion principle*, which states that no two electrons in any one atom may have all four quantum numbers the same. This has the effect of limiting the number of electrons in any given orbital to two, and it also requires that the spins of the two electrons be in opposite directions (with s values of $+\frac{1}{2}$ and $-\frac{1}{2}$). This is summarized in Table 4.3.

The spin of the electron is mainly responsible for the magnetic properties of atoms and molecules. The spinning electron can be regarded as a minute magnet (or *magnetic dipole*) with a *magnetic moment* which is defined as a Bohr magneton, $\mu_\beta = 9.27 \times 10^{-24}$ Am^2, which is a product of the area about which the electron spins (in units of m^2) and the electron charge (in units of amperes, A). Because two electrons in the same orbital must have opposing spins (see above), with one having an "up" and the other a "down" orientation of its pole, this produces a zero net magnetic moment. Such materials are known as *diamagnetic* because they experience no attraction for a mag-

net. They are, in fact, slightly repelled by a magnetic field, a result of the behavior of the electron cloud of the atom, not of the electron spin. In these substances there are the same number of electrons of each spin ("up" and "down"), so that their magnetic effects cancel. Many common minerals have no magnetic response; they are diamagnetic. They are composed of elements with core electron configurations of the rare gases, or with completely filled d orbitals (see Table 4.5). A few examples of the many common diamagnetic minerals are calcite, $CaCO_3$, albite, $NaAlSi_3O_8$, quartz, SiO_2, and apatite, $Ca_5(PO_4)_3(F, Cl, OH)$.

The most important elements that produce magnetic moments are those with unshared electrons (not involved in bonding) in $3d$ orbitals of the first transition series, including Ti, V, Cr, Mn, Fe, Co, Ni, and Cu (Z 22 through 29; see Fig. 4.3 or Table 4.5). This includes some very common mineral constituents, namely Fe, Mn, Ti, and Cr. The magnetic moments of these transition elements are the result of the spin of single, unpaired electrons and are proportional to the number of such electrons. The distribution of electrons in the five $3d$ orbitals is summarized in *Hund's rule: Electrons entering a subshell with more than one orbital will be distributed over the available orbitals with their spins in the same direction.* This means that in going from top to bottom in Table 6.5, electrons are first added to each orbital with their spins in the same direction and then doubled up until all orbitals are filled. This build-up continues until all electrons are paired, in all five $3d$ orbitals in Zn. The table shows that Fe^{3+} and Mn^{2+}, each with five unpaired electrons, are among the most magnetic ions.

Although individual ions may be classified as more or less magnetic (as in Table 6.5), there is still the question of how such ions interact in crystal structures. If a structure has a *random arrangement of mag-*

Table 6.5

TRANSITION ELEMENTS (Z 21 THROUGH Z 30), THEIR COMMON IONS, NUMBER OF 3d ELECTRONS, ELECTRON SPIN DIRECTIONS, AND MAGNETIC MOMENT (EXPRESSED IN TERMS OF BOHR MAGNETONS, μ_β)	Elements	Ions	Spin Directions and Number of 3d Electrons for the Ions		Magnetic Moment
	Sc	Ti^{3+}, V^{4+}	↑ _ _ _ _	1	$1\mu_\beta$
	Ti	Ti^{2+}, V^{3+}	↑ ↑ _ _ _	2	$2\mu_\beta$
	V	V^{2+}, Cr^{3+}, Mn^{4+}	↑ ↑ ↑ _ _	3	$3\mu_\beta$
	Cr	Cr^{2+}, Mn^{3+}	↑ ↑ ↑ ↑ _	4	$4\mu_\beta$
	Mn	Mn^{2+}, Fe^{3+}	↑ ↑ ↑ ↑ ↑	5	$5\mu_\beta$
	Fe	Fe^{2+}, Co^{3+}	↑↓ ↑ ↑ ↑ ↑	6	$4\mu_\beta$
	Co	Co^{2+}	↑↓ ↑↓ ↑ ↑ ↑	7	$3\mu_\beta$
	Ni	Ni^{2+}	↑↓ ↑↓ ↑↓ ↑ ↑	8	$2\mu_\beta$
	Cu	Cu^{2+}	↑↓ ↑↓ ↑↓ ↑↓ ↑	9	$1\mu_\beta$
	Zn	Zn^{2+}, Cu^+	↑↓ ↑↓ ↑↓ ↑↓ ↑↓	10	0

netic dipoles (caused by specific constituent cations with unpaired spins, such as are listed in Table 6.5) it is said to be *paramagnetic*. When such a structure is placed into a magnetic field, the minute dipoles tend to align themselves with the external magnetic field. Thermal motion inside the structure, however, tends to randomize some of the dipole alignments. This results in only a small fraction of dipoles being aligned with the external magnetic field at any particular instant. As such, a paramagnetic material is drawn only weakly to an external magnetic field (shows low magnetic susceptibility). The magnetization imposed is also not permanent. Examples of two common minerals that show paramagnetic behavior are olivine $(Mg,Fe)_2SiO_4$, and augite $(Ca,Na)(Mg,Fe,Al)(Al,Si)_2O_6$.

Because of their different magnetic susceptibilities, minerals can be separated from each other by an electromagnet. Magnetic separation by a Franz Isodynamic Separator is a common procedure in the laboratory. Such a technique separates minerals that are paramagnetic from those that are diamagnetic. On a commercial scale, electromagnetic separation is used to separate ore minerals from gangue (waste).

Another property related to paramagnetism is called *ferromagnetism*, which is observed in metallic iron. In a paramagnetic material the magnetic dipoles are randomly oriented, but in a ferromagnetic substance these become aligned as a result of "exchange forces" which are the result of overlap of the orbitals of nearest neighbor atoms or ions (for further discussion of the exchange forces, see O'Reilly (1984); the complete reference is given at the end of this chapter). In a substance such as metallic iron, *domains* exist containing large numbers of paramagnetic atoms with their dipole moments well aligned. Ordinarily these domains are randomly oriented (see Fig. 6.27a) so that their net magnetic effect is zero. When such a material is placed in a magnetic field, the domains become aligned with the external field (Fig. 6.27b), and it responds with strong magnetic attraction. This interaction is much stronger than normally experienced in paramagnetic materials. When the external magnetic field is removed from a paramagnetic substance, the magnetic domains randomize and no permanent magnetism remains. However, in a ferromagnetic material, the domains tend to remain in the orientation imposed by the external magnetic field even in the absence of the field. For example, a nail can be magnetized simply by running a permanent magnet over it; this process aligns ("poles") the magnetic domains in the nail fairly permanently. When a ferromagnetic material in which permanent magnetism has been in-

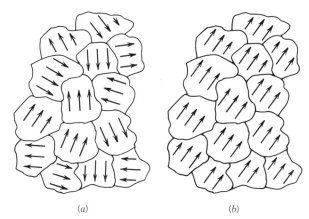

FIG. 6.27. Magnetic domains in a ferromagnetic solid. *(a)* Random domains when unmagnetized. *(b)* Parallel alignment of domains as a result of an external magnetic field.

duced is heated, the parallelism of the magnetic field is completely lost at the *Curie temperature*, above which it behaves paramagnetically. The Curie temperature for metallic iron is 770°C.

Yet another type of magnetism is known as *ferrimagnetism*, in which the ionic spin moments are antiparallel, instead of parallel as in ferromagnetism. In ferrimagnetic materials, the antiparallel spin moments are non-equal, and as such there are permanent magnetic domains (see Fig. 6.28a). Examples of ferrimagnetic minerals are members of the magnetite-ulvöspinel series, $Fe_3O_4 - Fe_2TiO_4$ (see spinel group, page 375), hematite-ilmenite solid solution members, $Fe_2O_3 - FeTiO_3$, and pyrrhotite, $Fe_{1-x}S$.

The distribution of magnetic dipoles in a ferrimagnetic material can be illustrated with reference to magnetite, a member of the spinel series. The formula of magnetite, Fe_3O_4, can be rewritten as $Fe^{3+}[Fe^{2+}Fe^{3+}]O_4$ in terms of a general XY_2O_4 formula for the spinel group. The spinel structure is based on cubic closest packing of oxygens with the cations in tetrahedral and octahedral interstices. The X cations occupy $\frac{1}{8}$ of the 64 ($= 8$) tetrahedral sites (per unit cell of spinel) and the Y cations occupy $\frac{1}{2}$ of the 32 ($= 16$) octahedral sites (per unit cell). The Fe^{3+} ions are therefore distributed in two different lattice sites, but with opposing magnetic spin directions. The Fe^{2+} ions (with lesser magnetic moment; see Table 6.5) are responsible for net unpaired spin and thus for the permanent magnetic domains in magnetite (see Fig. 6.28b). This net magnetization is considerably less than if the magnetic moments of all the cations were parallel, as in magnetized native iron, which is ferromagnetic. The Curie temperature of magnetite is 580°C, above which the magnetic ordering completely disappears.

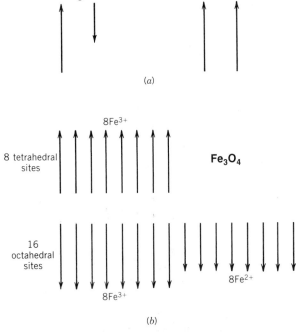

FIG. 6.28. *(a)* Schematic illustration of the spin alignments of dipoles in ferrimagnetic and ferromagnetic materials. In ferrimagnetic solids the ionic spins are antiparallel and their magnitudes are unequal. In ferromagnetic solids all spins are parallel and aligned in the same direction. *(b)* Schematic representation of the spin directions in the tetrahedral and octahedral sites of magnetite, Fe_3O_4. A net magnetic moment is due to the noncancellation of the dipole moments of Fe^{2+} ions.

The permanent magnetism of ferrimagnetic minerals in various rock types allows for the study of the ancient geomagnetic field of the Earth, known as *paleomagnetism*. The study of the natural remanent magnetization of rocks can yield a record of the Earth's magnetic field. For example, in the cooling of igneous rocks, from the temperature above to below the Curie temperature, the direction of the Earth's magnetic field is recorded in the orientation of the magnetic domains of ferrimagnetic minerals. *Lodestone*, a naturally occurring magnet of magnetite composition, is a ferrimagnetic substance in which all the net magnetic moments are strongly aligned ("poled"). The natural magnetism of lodestone is attributed to its having cooled from a melt (as part of an igneous rock) while under the influence of the Earth's magnetic field.

REFERENCES AND SUGGESTED READING

Banerjee, S. K., 1991, Magnetic properties of Fe-Ti oxides, in *Oxide Minerals: Petrologic and Magnetic Significance. Reviews in Mineralogy*, v. 25, pp. 107–128.

Keffer, F., 1967, The magnetic properties of materials. *Scientific American*, v. 217, pp. 222–238.

Loeffler, B. M. and Burns, R. G., 1976, Shedding light on the color of gems and minerals. *American Scientist*, v. 64, pp. 636–647.

Nassau, K., 1978, The origins of color in minerals. *American Mineralogist*, v. 63, pp. 219–229.

———, 1980, The causes of color. *Scientific American*, v. 243, pp. 124–156.

———, 1980, *Gems Made by Man*. Chilton Book Co., Radnor, Penn., 364 pp. [Specifically Chapter 26, "The origin of color in gemstones"]

———, 1983, *The Physics and Chemistry of Color: The Fifteen Causes of Color*. John Wiley & Sons, New York, 454 pp.

O'Reilly, W., 1984, *Rock and Mineral Magnetism*. Blackie, London, England, distributed by Chapman and Hall, New York, 220 pp.

Wood, D. L. and Nassau, K., 1968, The characterization of beryl and emerald by visible and infrared absorption spectroscopy. *American Mineralogist*, v. 53, pp. 777–801.

Zussman, J., 1977, *Physical Methods in Determinative Mineralogy*. Academic Press, New York, 720 pp.

CHAPTER 7

X-RAY CRYSTALLOGRAPHY

The application of X-rays to the study of crystals was the greatest single impetus ever given to crystallography. Prior to 1912, crystallographers had correctly deduced from cleavage, optical properties, and the regularity of external form that crystals had an orderly structure; but their thinking concerning the geometry of crystal structures had only the force of a hypothesis. X-rays have made it possible not only to measure the distance between successive atomic planes but also to locate the positions of the various atoms or ions within the crystal, thus enabling determinations of crystal structures.

X-rays were discovered accidently by Wilhelm Conrad Roentgen in 1895 while he was experimenting with the production of cathode rays in sealed discharge tubes wrapped in black paper. The electron stream in the discharge tube, striking the glass of the tube, produced low-intensity X-radiation, which caused some fluorescent material nearby to glow in the dark. Roentgen correctly inferred that a new type of radiation was being produced, fittingly called "X-radiation" because of the many mysteries connected with it. Roentgen was unsuccessful in his efforts to measure the wavelength of X-rays, and it was this unsolved problem that led to the discovery of the diffraction of X-rays by crystals.

The fact that most substances are more or less transparent to X-rays brought about their almost immediate use in hospitals for medical purposes in the location of fractures, foreign bodies, and diseased tissue in much the same manner in which they are used today.

It was not until 1912, 17 years after the discovery of X-radiation, that, at the suggestion of Max von Laue, X-rays were used to study crystals. The original experiments were carried out at the University of Munich where von Laue was a lecturer in Professor Sommerfeld's department. Sommerfeld was interested in the nature and excitation of X-rays and von Laue in interference phenomena. Also at the University of Munich was Paul Heinrich Groth, a leading crystallographer. With the gathering together of such a group of distinguished scientists having these special interests, the stage was set for the momentous discovery.

In 1912, Paul Ewald was working under Sommerfeld's direction on a doctoral dissertation involving the scattering of light waves on passing through a crystal. In thinking about this problem, von Laue raised the question: What would be the effect if it were possible to use electromagnetic waves having essentially the same wavelength as the interatomic distances in crystals? Would a crystal act as a three-dimensional diffraction grating forming spectra that could be recorded? If so, it would be possible to measure precisely the wavelength of the X-rays employed, assuming interatomic spacings of the crystal; or assuming a wavelength for the X-rays, measure the interplanar spacings of the crystal. Methods for making such an experiment were discussed, and Freidrich and Knipping, two research students, agreed to carry them out. Several experiments in which copper sulfate was used as the object crystal were unsuccessful. Finally they passed a narrow beam of X-rays through a cleavage plate of sphalerite, ZnS, allowing the beam to fall on a photographic plate. The developed plate showed a large number of small spots arranged in a geometrical pattern around the large spot produced by the direct X-ray beam. This pattern was shown to be identical with that predictable from the diffraction of X-rays by a regular arrangement of scattering points within the crystal. Thus, a single experiment demonstrated the regular, orderly arrangement of atomic particles within crystals and the agreement as to order of magnitude of the wavelength of X-rays with the interplanar spacing of crystals. Although largely replaced by more powerful means of X-ray investigation, this techinque, the Laue method, is still in use.

Within the next few years, great strides were made as a result of the work of the English physicists, William Henry Bragg and William Lawrence Bragg, father and son. In 1914 the first structure of a compound, halite, NaCl, was worked out by the Braggs; and in the years following many more structures were solved by them. The Braggs also greatly simplified von Laue's mathematical generalizations as to the geometry of X-ray diffraction and popularized the results of their research through their well-written books on the subject.

X-RAY SPECTRA

Electromagnetic waves form a continuous series varying in wavelength from long radio waves with wavelengths of the order of thousands of meters to cosmic radiation whose wavelengths are of the order of 10^{-12} meters (a millionth of a millionth of a meter!). All forms of electromagnetic radiation have certain properties in common such as propagation along straight lines at a speed of 300,000 km per second in a vacuum, reflection, refraction according to Snell's law, diffraction at edges and by slits or gratings and a relation between energy and wavelength given by the Einstein equation:

$$E = h\nu = hc/\lambda$$

where E is energy, ν frequency, c velocity of propagation, λ wavelength, and h Planck's constant. Thus, the shorter the wavelength the greater its energy and the greater its powers of penetration. X-rays occupy only a small portion of the spectrum, with wavelengths varying between slightly more than 100 Å and 0.02 Å (see Fig. 7.1). X-rays used in the investigation of crystals have wavelengths of the order of 1 Å. Visible light has wavelengths between 7200 and 4000 Å, more than 1000 times as great, and hence is less penetrating and energetic than X-radiation.

When electrons moving at high velocity strike the atoms of any element, as is the case in an X-ray tube where electrons bombard a target material, X-rays are produced. These X-rays result in two types of X-ray spectra—*continuous* and *characteristic* (Fig. 7.2).

In a modern X-ray tube there is nearly a complete vacuum. The tube is fitted with a tungsten filament as a cathode that provides the source of electrons. The anode consists of one of a number of metals such as Mo, Cu, or Fe, and acts as the target for the electrons. When the filament is heated by passage of a current, electrons are emitted which are accelerated toward the target anode by a high voltage applied across the tube. X-rays are generated when the electrons impact on the target (anode). The nature of the X-rays depends on the metal of the target and the applied voltage. No X-rays are produced until the voltage reaches a minimum value dependent on the target material. At that point a *continuous spectrum* is generated. With increasing potential, the intensity of all wavelengths increases, and the value of the minimum wavelength becomes progressively less (Fig. 7.2*a*). The *continuous spectrum*, also referred to as *white radiation*, is caused by the stepwise loss of energy of bombarding electrons in series of encounters with atoms of the target material. When an electron is instantaneously stopped, it loses all of its energy at once and the X-ray radiation emitted is that of the shortest wavelength (see Fig. 7.2*a*). The stepwise energy losses from a stream of electrons produce a continuous range of wavelengths that can be plotted as a smooth function of intensity against wavelength (Fig. 7.2*a*). The curve begins at the short wavelength limit, rises to a maximum, and extends toward infinity at very low intensity levels.

If the voltage across the X-ray tube is increased to a critical level, which is dependent on the element of the target, there becomes superimposed on the continuous spectrum a *line spectrum of characteristic radiation* peculiar to the target material. This characteristic radiation, many times more intense than the continuous spectrum, consists of several isolated wavelengths, as shown in Fig. 7.2*b* by β and α peaks.

The characteristic X-ray spectrum is produced when the bombarding electrons have sufficient energy to dislodge electrons from the inner electron

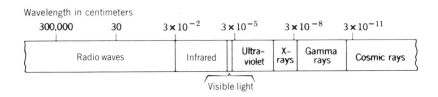

FIG. 7.1. The electromagnetic spectrum.

shells in the atoms of the target material. When these inner electrons are expelled, they leave vacancies that are filled by electrons from surrounding electron shells. The electron transitions, from outer to inner shells, are accompanied by the emission of X-radiation with specific wavelengths. Electron transitions from the L- to the K-shell produce $K\alpha$ radiation, and those from the M- to the K-shell cause $K\beta$ radiation (see Fig. 5.6). The $K\beta$ peak can be eliminated by an appropriate filter yielding essentially a single wavelength, which by analogy to monochromatic light is called *monochromatic X-radiation*. The $K\overline{\alpha}$ radiation consists of two peaks, $K\alpha_1$ and $K\alpha_2$, which are very close together in wavelength.

The wavelengths of the characteristic X-radiation emitted by various metals have been accurately determined. The $K\overline{\alpha}$ wavelengths (weighted averages of $K\alpha_1$ and $K\alpha_2$) of the most commonly used are:

	Å		Å
Molybdenum	0.7107	Cobalt	1.7902
Copper	1.5418	Iron	1.9373
		Chromium	2.2909

DIFFRACTION EFFECTS AND THE BRAGG EQUATION

Crystals consist of an ordered three-dimensional structure with characteristic periodicities, or *identity periods*, along the crystallographic axes. When an X-ray beam strikes such a three-dimensional arrangement, it causes electrons in its path to vibrate with a frequency of the incident X-radiation. These vibrating electrons absorb some of the X-ray energy and, acting as a source of new wave fronts, emit (scatter) this energy as X-radiation of the same frequency and wavelength. In general the scattered waves interfere destructively, but in some specific directions they reinforce one another, producing a cooperative scattering effect known as *diffraction*.

In a row of regularly spaced atoms that is bombarded by X-rays every atom can be considered the center of radiating, spherical wave shells (Fig. 7.3). When the scattered waves interfere constructively, they produce wave fronts that are *in phase*, and diffraction will occur. Figure 7.4 illustrates that rays 1 and 2 will be in phase only when distance AB represents an integral number of wavelengths, in other

FIG. 7.2. X-ray spectra. *(a)* Distribution of intensity with wavelength in the continuous X-ray spectrum of tungsten at various voltages. *(b)* Intensity curve showing characteristic wavelengths superimposed on the continuous X-ray spectrum of molybdenum. (After Ulrey, C. T., 1918, An experimental investigation of the energy in the continuous X-ray spectra of certain elements. *Phys. Reviews*, v. 11, pp. 401–410.)

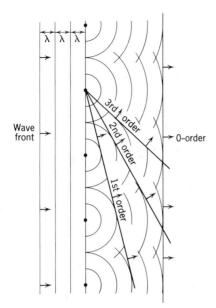

FIG. 7.3. Scattering of X-rays by a row of equally spaced, identical atoms.

words when $AB = n\lambda = c \cos \phi$ (where n represents whole numbers such as 0, 1, 2, 3, . . . , n). For a specific value of $n\lambda$, ϕ is constant, and the locus of all possible diffracted rays will be represented by a cone with the row of scattering points as the central axis. Because the scattered rays will be in phase for the same angle ϕ on the other side of the incident beam, there will be another similar but inverted cone on that side (see Fig. 7.5). The two cones with $n = 1$ would have ϕ (as in Fig. 7.4) as the angle between the cone axis and the outer surface of the cone. When $n = 0$, the cone becomes a plane that includes the incident beam. The greater the value of n the larger the value of $\cos \phi$, and hence the smaller the angle ϕ and the

narrower the cone. All have the same axis, however, and all have their vertices at the same point, the intersection of the incident beam and the row of atoms.

In a three-dimensional lattice there are three axial directions, each with its characteristic periodicity of scattering points and each capable of generating its own series of nested cones with characteristic apical angles. Diffraction cones from any three noncoplanar rows of scattering atoms may or may not intersect each other, but only when all three intersect in a common line is there a diffracted beam (Fig. 7.6). This direction (shown by the arrow in Fig. 7.6) represents the direction of the diffracted beam, which can be recorded on a film or registered electronically. The geometry of the three intersecting cones in Fig. 7.6 can be expressed by three independent equations (the Laue equations) in which the three cone angles (ϕ_1, ϕ_2, and ϕ_3) define a common direction along the path of the arrow (the common intersection of the three cones). In order to produce a diffraction effect (a spot on a photographic plate or film) these three geometric equations must be simultaneously satisfied. The three equations are named after Max von Laue, who originally formulated them.

Shortly after the publication of these equations, W. L. Bragg, working on X-ray diffraction in England, pointed out that, although X-rays are indeed diffracted by crystals, the diffracted X-rays act as though they were reflected from planes within the crystal. Unlike the reflection of light, however, X-rays are not "reflected" continuously from a given crystal plane. Using a given wavelength, λ, Bragg showed that a "reflection" takes place from a family of parallel planes

FIG. 7.5. Diffraction cones from row of atoms.

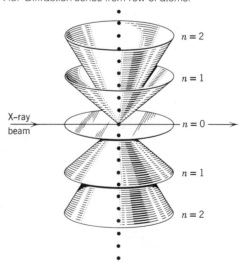

FIG. 7.4. Conditions for X-ray diffraction from a row of atoms.

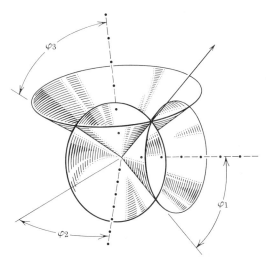

FIG. 7.6 Diffraction cones from three noncoplanar rows of scattering atoms, intersecting in a common line.

only under certain conditions. These conditions must satisfy the equation: $n\lambda = 2d \sin \theta$, where n is an integer $(1, 2, 3, \ldots, n)$, λ the wavelength, d the distance between successive parallel planes, and θ the angle of incidence and "reflection" of the X-ray beam from the given atomic plane. This equation, known as the *Bragg law*, expresses in a simpler manner the simultaneous fulfillment of the three Laue equations.

We have seen that the faces most likely to appear on crystals are those parallel to atomic planes having the greatest density of lattice nodes. Parallel to each face is a family of equispaced identical planes. When an X-ray beam strikes a crystal, it penetrates it, and the resulting diffraction effect is not from a single plane but from an almost infinite number of parallel planes, each contributing a small bit to the total diffraction maximum. In order that the diffraction effect ("reflection") be of sufficient intensity to be recorded, the individual "reflections" must be *in phase* with one another. The following conditions necessary for reinforcement were demonstrated by W. L. Bragg.

In Fig. 7.7 the lines p, p_1, and p_2 represent the traces of a family of atomic planes with spacing d. X-rays striking the outer plane pp would be reflected at the incident angle θ, whatever the value of θ. However, to reinforce one another in order to give a "reflection" that can be recorded, all "reflected" rays must be *in phase*. The path of the waves along DEF "reflected" at E is longer than the path of the waves along ABC "reflected" at B. If the two sets of waves are to be *in phase*, the path difference of ABC and DEF must be a whole number of wavelengths $(n\lambda)$. In Fig. 7.7 BG and BH are drawn perpendicular to AB and

BC, respectively, so that $AB = DG$ and $BC = HF$. To satisfy the condition that the two waves be in phase, $GE + EH$ must be equal to an integral number of wavelengths. BE is perpendicular to the lines p and p_1 and is equal to the interplanar spacing d. In ΔGBE, $d \sin \theta = GE$; and, in ΔHBE, $d \sin \theta = EH$. Thus for in phase "reflection" $GE + EH = 2d \sin \theta = n\lambda$.

This equation,

$$n\lambda = 2d \sin \theta,$$

is the *Bragg equation*. For a given interplanar spacing (d) and given λ, "reflections" (diffraction maxima) occur only at those angles of θ that satisfy the equation. Suppose, for example, a monochromatic X-ray beam is parallel to a cleavage plate of halite and the plate is supported in such a way that it can be rotated about an axis at right angles to the X-ray beam. As the halite is slowly rotated there is no "reflection" until the incident beam makes an angle θ that satisfies the Bragg equation, with $n = 1$. On continued rotation there are further "reflections" only when the equation is satisfied at certain θ angles with $n = 2, 3$, and so on. These are known as the first-, second-, third-order, etc., "reflections." These "reflections" are in actuality the diffraction effects that occur when the three diffraction cones about three noncoplanar rows of atoms intersect in a common direction (see Fig. 7.6).

Laue Method

In the *Laue method* a single crystal that remains stationary is used. A photographic plate or flat film enclosed in a light-tight envelope is placed a known distance, usually 5 cm, from the crystal. A beam of white X-radiation is passed through the crystal at right angles to the photographic plate. The direct beam causes a darkening at the center of the photograph, so a small disc of lead is usually placed in front of the film to intercept and absorb it. The angle of incidence, θ, be-

FIG. 7.7. Geometry of X-ray "reflection."

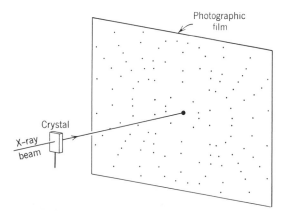

FIG. 7.8. Obtaining a Laue photograph with a stationary crystal.

tween the X-ray beam and the various atomic planes with their given interplanar spacings (d) within the crystal is fixed. However, because X-rays of a large range of wavelengths are present, the Bragg law, $n\lambda = 2d \sin \theta$, can be satisfied by each family of atomic planes, provided $(2d \sin \theta)/n$ is within the range of wavelengths. Around the central point of a Laue photograph are arranged diffraction spots, each one resulting from diffractions from a given series of atomic planes (Fig. 7.8).

The *Laue method*, although of great historical interest, has largely been replaced by other more powerful methods of X-ray crystal analysis. Its use today is primarily for alignment of large (X-ray opaque) crystals and for determining the presence of symmetry elements parallel to the X-ray beam. If a crystal is so oriented that the X-ray beam is parallel to a symmetry element, the arrangement of spots on the photograph reveals this symmetry. A Laue photograph of a mineral taken with the X-ray beam parallel to the 2-fold axis of a monoclinic crystal will show a 2-fold arrangement of spots; if the beam is parallel to a symmetry plane, the photograph shows a line of symmetry. A photograph of an orthorhombic crystal (symmetry $2/m2/m2/m$) with the beam parallel to a crystallographic axis shows a 2-fold arrangement as well as two lines of symmetry. Figure 7.9 shows a 4-fold arrangement of spots as given by vesuvianite with the X-ray beam parallel to the 4-fold rotation axis.

Unfortunately, X-rays do not differentiate between opposite ends of a polar axis and thus diffraction effects always introduce a center of symmetry.

Other Single-Crystal Methods

In the *Laue method*, discussed above, the single crystal under investigation is stationary in the X-ray beam and the flat film, behind the crystal, is also stationary.

The X-ray beam, however, is not filtered and therefore contains a large number of wavelengths. A monochromatized (single λ) X-ray beam, however, is most appropriate in single-crystal techniques where the investigator wishes to assign a Miller index to each diffraction spot. Because a stationary single-crystal method provides only a very limited number of diffraction spots on a film (because it gives only a small number of solutions to the Bragg equation, with λ = constant), several rotating-crystal methods were developed. The simplest of these is the *rotation method*, in which a single crystal is rotated about one of its principal crystallographic axes, inside a cylindrical camera of known diameter. Because it is commonly necessary to identify specific spots on the film with the atomic planes that gave rise to them, a modification of the rotation method led to the development of the *Weissenberg method*. In this technique the cylindrical camera is translated back and forth during a rotation of the crystal, and the resulting X-ray spots on the film are spread out in festoons. Although it is a laborious process, each spot can be identified with a specific Miller index.

Today the most commonly used single-crystal technique is that known as the *precession method*, developed in the 1930s by Professor Martin J. Buerger of the Massachusetts Institute of Technology (MIT). This method provides the X-ray information in a much more straightforward fashion than any of the other single-crystal techniques. In the precession method a flat film and crystal both move in a complex gyratory motion, mechanically compensating for distortions produced in the Weissenberg method. Thus a great amount of data is provided in a readily usuable form. The instrumentation is illustrated in Fig. 7.10 and an example of a precession photograph is given in Fig. 7.11. Further consideration of these single-crystal (film) methods is beyond the scope of this discussion, and the reader is referred to the references at the end of this chapter.

Although an enormous number of crystal structures have been solved on the basis of data obtained on film using single-crystal camera techniques, the modern approach to data acquisition is not by film technique, but by *single-crystal diffractometer*. In such instrumentation (see also "X-ray Powder Diffractometer," below) the intensity of X-ray reflections is not evaluated from the intensity of a spot on an X-ray film, but instead is measured by an electronic device, an X-ray counter (or detector). Such detectors greatly improve the accuracy of X-ray intensity measurements over those obtained by film techniques. Furthermore, in automated instrumentation, such detectors can measure large numbers of X-ray reflections with high

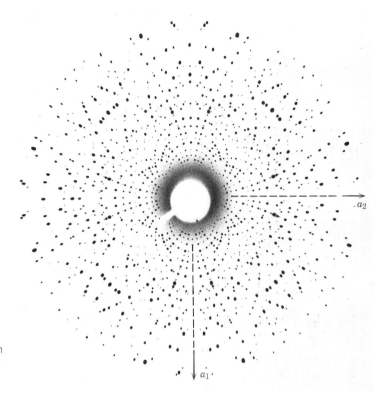

FIG. 7.9. Laue photograph of vesuvianite with point group symmetry $4/m2/m2/m$. The photograph was taken along the 4-fold rotation axis (c axis) of vesuvianite, thus revealing 4-fold symmetry and mirrors in the arrangement of diffraction spots. The axial directions, a_1 and a_2, were inked onto the photograph after it had been developed.

FIG. 7.10. A Buerger precession camera, model VIII, manufactured by the Charles Supper Co., Natick, Massachusetts. 1. Collimator through which the X-ray beam enters from the X-ray tube. 2. Goniometer head, which allows for the orientation of the crystal. 3. Film cassette. 4. Scale for setting the inclination angle, μ, which is responsible for the precession motion of crystal and film cassette. 5. Motor drive. (Courtesy of Charles Supper Co.)

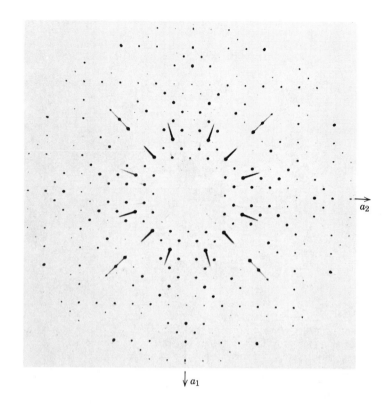

FIG. 7.11. Precession photograph of vesuvianite, with point group symmetry 4/m2/m2/m. The photo was taken along the 4-fold rotation (c) axis, thus revealing the 4-fold symmetry about the axis, as well as mirror planes. Compare with Fig. 7.9.

accuracy. The most commonly used automated technique in X-ray structure analysis is the *four-circle diffractometer*. The name four-circle arises from its possession of four arcs that are used to orient the single crystal so as to bring desired (atomic) planes into reflecting positions. An automated four-circle diffractometer is shown in Fig. 7.12. A discussion of the steps and parameters necessary in the determination of a crystal structure appears on page 147.

POWDER METHOD

The relative rarity of well-formed crystals and the difficulty of making the precise orientation required by single-crystal methods led to the discovery of the *powder method* of X-ray investigation. For the powder method, the specimen is ground as finely as possible and bonded together by an amorphous material, such as flexible collodion, into a needle-like spindle 0.2 and 0.3 mm in diameter. This spindle, the *powder mount*, consists ideally of crystalline particles in completely random orientation. To ensure randomness of orientation of these tiny particles with respect to the impinging X-ray beam, the mount is generally rotated in the path of the beam during the exposure.

When the beam of monochromatic X-rays strikes the mount, all possible diffractions take place simultaneously. If the orientation of the crystalline particles

in the mount is truly random, for each family of atomic planes with its characteristic interplanar spacing (d), there are many particles whose orientation is such that they make the proper θ angle with the incident beam to satisfy the *Bragg law*, $n\lambda = 2d \sin \theta$. The diffraction maxima from a given set of planes form cones with the incident beam as axis and the internal angle 4θ. Any set of atomic planes yields a series of nested cones corresponding to "reflections" of the first, second, third, and higher orders ($n = 1, 2, 3, \ldots$). Different families of planes with different interplanar spacings will satisfy the Bragg law at appropriate values of θ for different integral values of n, thus giving rise to separate sets of nested cones of "reflected" rays.

If the rays forming these cones are permitted to fall on a flat photographic plate at right angles to the incident beam, a series of concentric circles will result (Fig. 7.13). However, only "reflections" with small values of the angle 2θ can be recorded in this manner.

In order to record diffraction effects of 2θ up to 180°, the powder camera is used. This is a flat, disc-shaped box with an adjustable pin at the center for attachment of the mount. The cylindrical wall of the camera is pierced diametrically for a demountable slit system and opposing beam catcher. The light-tight lid may be removed to insert and remove the film, which, for the most popular type of camera, is a narrow strip about 14 inches long and 1 inch wide. Two holes are

(a)

(b)

FIG. 7.12. *(a)* Schematic illustration of the P4 single-crystal X-ray diffractometer manufactured by Siemens Industrial Automation, Inc. *(b)* Close-up of the four-circle goniometer for controlling the orientation of the single crystal (in center of photograph). To the right is the X-ray tube, and to the left is the X-ray detector, a scintillation counter. (Courtesy of Siemens Industrial Automation, Inc., Madison, Wis.)

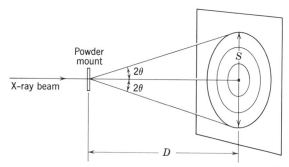

FIG. 7.13. X-ray diffraction from powder mount recorded on a flat plate.

FIG. 7.15. X-ray diffraction from powder mount recorded on cylindrical film.

punched in the film and so located that, when the film is fitted snugly to the inner curve of the camera, the collimating tube and beam catcher pass through the holes. This type of mounting is called the Straumanis method (Fig. 7.14).

A narrow beam of monochromatic X-rays is allowed to pass through the collimating slit and fall on the spindle-shaped mount, which is carefully centered on the short axis of the camera so that the mount remains in the X-ray beam as it rotates during the exposure. The undeviated beam passes through and around the mount and enters the lead-lined beam catcher. Under these conditions, the film intercepts the cones of diffracted rays along curved lines (Fig. 7.15). Because the axes of the cones coincide with the X-ray beam, for each cone there will be two curved lines on the film symmetrically disposed on each side of the slit through which the X-rays leave the camera. The angular distance between these two arcs is 4θ.

FIG. 7.14. Powder diffraction camera.

When the film is developed and laid flat, the arcs are seen to have their centers at the two holes in the film. Cones of low θ have their center at the exit hole through which the beam catcher passes through the film. Going out from this point, the arcs are of increasing radius until at $2\theta = 90°$ they are straight lines. Cones made by diffraction maxima with $2\theta > 90°$ curve in the opposite direction and are concentric about the hole through which the X-rays enter the camera. These are known as *back reflections*.

If a flat film is used and the distance D from the specimen to the film is known, it is possible to calculate θ by measuring S, the diameter of the rings. It may be seen in Fig. 7.13 that $\tan 2\theta = S/2D$. When a cylindrical camera is used, the distance S is measured with the film laid flat. Under these conditions, $S = R \times 4\theta$ in radians[1] or $\theta = S/4R$ in radians, where R is the radius of the camera and S is measured in the same units as R. Most powder cameras are constructed with a radius such that S, measured in millimeters on the flat film, can be converted easily to θ. For example, when the radius of the camera is 57.3 mm, the circumference is 360 mm. When such a camera is used, each millimeter measured along the film is equal to 1°. Hence, a distance S of 60 mm measured on the film is equal to $60° = 4\theta$, and θ is accordingly equal to 15°.

It is not possible to measure the symmetrical distance S for values of θ much in excess of 40° ($S = 160$ mm) on two-hole films of the Straumanis type. In order to obtain θ for lines of θ higher than about 40°, the center about which the low lines are concentric must first be found by measurement of a number of such lines. Then the distance $S/2$ may be measured. It must be borne in mind that, if the camera radius is not 57.3°, a correction factor must be applied. Thus, if a camera having a radius of 28.65 mm is used, the correction factor is $57.3/28.65 = 2$, and all S values measured must be divided by 2 to obtain θ in degrees.

[1]One radian = 57.3°.

The Straumanis method, the type of mounting described with two holes in the film, is now widely used. In older-type powder cameras the X-ray beam entered between the ends of the cylindrical film and left through a centered hole. During the process of developing, the film usually shrinks. With the Straumanis camera, the effective diameter of the film can be measured and compared with the true diameter. This provides a shrinkage factor that can be applied to film measurements in making calculations.

When the diffraction angle θ corresponding to a given line on a powder photograph has been determined, it is possible to calculate the interplanar spacing of the family of atomic planes that gave rise to this line by use of the Bragg equation $n\lambda = 2d \sin\theta$, or $d = n\lambda/(2\sin\theta)$. As it is usually impossible to tell the order of a given reflection, n in the formula above is generally given the value 1, and d is determined in every case as though the line were produced by a first-order "reflection." For substances crystallizing in the isometric and tetragonal systems, it is easy to index the lines of a powder photograph and thus determine cell dimensions. For crystals of the other systems, indexing is more difficult. However, for these crystals, if the unit cell dimensions are known, indexing is performed routinely with high-speed computer techniques.

The powder method finds its chief use in mineralogy as a determinative tool. One can use it for this purpose without knowing anything of the crystal structure or symmetry. Every crystalline substance produces its own powder pattern, which, because it is dependent on the internal structure, is characteristic of that substance. The powder photograph is often spoken of as the "fingerprint" of a mineral, because it differs from the powder pattern of every other mineral. Thus, if an unknown mineral is suspected of being the same as a known mineral powder, photographs of both are taken. If the photographs correspond exactly line for line, the two minerals are identical. Many organizations maintain extensive files of standard photographs of known minerals, and by comparison, an unknown mineral may be identified readily if there is some indication as to its probable nature.

One frequently has no clue to the identity of the unknown mineral, and a systematic comparison with the thousands of photographs in a reference file would be too time-consuming. When this happens, the investigator turns to the card or microfiche file of X-ray diffraction data prepared by the Joint Committee on Powder Diffraction Standards (JCPDS) (Fig. 7.16). The interplanar spacings for thousands of crystalline substances are recorded on cards or on microfiches. To use these systems, the investigator must calculate the interplanar spacings for the most prominent lines in the powder pattern of the unknown substance and estimate the relative intensities of the lines on a scale where the strongest line is taken as 100. Then he or she may seek a corresponding series of d's in the JCPDS file cards, which are arranged in order of d of the most intense line. Because many substances have intense lines corresponding to the same d value and because many factors may operate to change the relative intensity of the lines in a powder pattern, all substances are cross-indexed for their second and third most intense lines. After the most likely "suspects" have been selected from the file, comparison of the weaker "reflections" (which are also listed on the JCPDS card or microfiche) will speedily identify

FIG. 7.16. JCPDS card for quartz. The interplanar spacings (d), their relative intensities and Miller indices are given. At the top of card are given the three strongest lines and their relative intensities. The fourth d is of the greatest spacing.

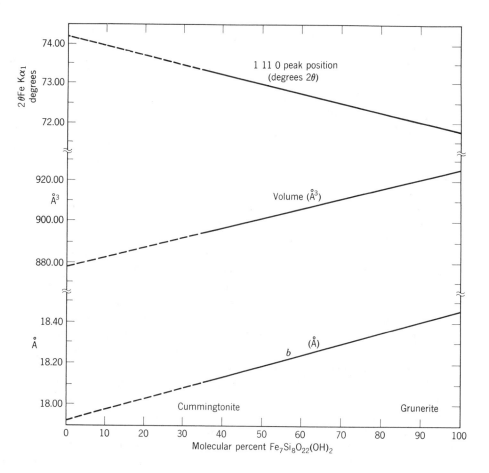

FIG. 7.17. Variation of b and V of the unit cell and the position of the 1,11,0 peak as a function of composition in the monoclinic cummingtonite-grunerite series with composition ranging from $Fe_2Mg_5Si_8O_{22}(OH)_2$ to $Fe_7Si_8O_{22}(OH)_2$. (After Klein, C., and Waldbaum, D. R., 1967, X-ray crystallographic properties of the cummingtonite-grunerite series. *Jour. Geology*, v. 75, pp. 379–392.)

the substance in most cases. In this way a completely unknown substance may generally be identified in a short time on a very small volume of sample.

The powder method is of wider usefulness, however, and there are several other applications in which it is of great value. Variations in chemical composition of a known substance involve the substitution of ions, generally of a somewhat different size, in specific sites in a crystal structure. As a result of this substitution the unit cell dimensions and hence the interplanar spacings are slightly changed, and the positions of the lines on the powder photograph corresponding to these interplanar spacings are shifted accordingly. By measuring these small shifts in position of the lines in powder patterns of substances of known structure, changes in chemical composition may often be accurately detected. Figure 7.17 illustrates a variation diagram that correlates unit cell dimensions and changes in the position of a specific diffraction maximum (1, 11, 0) with composition in the cummingtonite-grunerite series.

Further, the relative proportions of two or more known minerals in a mixture are often conveniently determined by the comparison of the intensities of the

same lines in photographs of prepared control samples of known composition.

X-Ray Powder Diffractometer

The usefulness of the powder method has been greatly increased and its field of application extended by the introduction of the X-ray *powder diffractometer*. This powerful research tool uses essentially monochromatic X-radiation and a finely powdered sample, as does the powder film method, but records the information about the "reflections" present as an inked trace on a printed strip chart, or as electronic counts (X-ray counts) that can be stored in a computer. The automated equipment supplied by one manufacturer is shown in Fig. 7.18.

The sample is prepared for powder diffractometer analysis by grinding to a fine powder, which is then spread uniformly over the surface of a glass slide, using a small amount of adhesive binder. The instrument is so constructed that this slide, when clamped in place, rotates in the path of a collimated X-ray beam while an X-ray detector, mounted on an arm, rotates about it to pick up the diffracted X-ray signals.

FIG. 7.18. Automated X-ray powder diffractometer with X-ray safety cabinet in the open position. Scintag XDS-2000 Diffraction System, with goniometer, solid-state detector, and single sample holder. 1. X-ray tube. 2. Holder for powdered sample. 3. X-ray detector. 4. Goniometer mechanism. 5. Controls for X-ray generator. (Courtesy of Scintag, Inc., Sunnyvale, Calif.)

When the instrument is set at zero position, the X-ray beam is parallel to the slide and passes directly into the X-ray detector. The slide mount and the counter are driven by a motor through separate gear trains so that, while the slide and specimen rotate through the angle θ, the detector rotates through 2θ.

If the specimen has been properly prepared, there will be thousands of tiny crystalline particles on the slide in random orientation. As in powder photography, all possible "reflections" from atomic planes take place simultaneously. Instead of recording all of them on a film at one time, however, the X-ray detector maintains the appropriate geometrical relationship to receive each diffraction maximum separately.

In operation, the sample, the X-ray detector, and the paper drive of the strip chart recorder are set in motion simultaneously. If an atomic plane has an interplaner spacing (d) such that a reflection occurs at $\theta = 20°$, there is no evidence of this reflection until the counting tube has been rotated through 2θ, or $40°$. At this point the diffracted beam enters the X-ray de-

tector, causing it to respond. The pulse thus generated is amplified and causes a deflection on the recording strip chart. Thus, as the X-ray detector scans, the strip chart records as peaks the diffraction maxima from the specimen. The angle 2θ at which the diffraction occurs may be read directly from the position of the peak on the strip chart. The heights of the peaks are directly proportional to the intensities of the diffraction effects.

The chart on which the record is drawn is divided in tenths of inches and moves at a constant speed, such as for example, 0.5 inch per minute. At this chart speed and a scanning speed of the X-ray detector of $1°$ per minute, 0.5 inch on the chart is equivalent to a 2θ of $1°$. The positions of the peaks on the chart can be read directly, and interplanar spacings giving rise to them can be determined by use of the equation $n\lambda = 2d \sin \theta$.

Although the powder diffractometer yields data similar to those derived from a powder photograph, it has definite advantages. The powder film method may require many hours of exposure plus the time necessary for developing, fixing, washing, and drying the film. In contrast, a diffractometer run can be made much more quickly, over a period of 10 to 30 minutes, depending on the scan speed of the diffractometer. Furthermore, it is commonly difficult to accurately estimate the intensity of lines on a powder photograph, whereas the height of peaks on a diffractometer chart can be determined graphically with much greater precision (the counts of intensities of X-ray peaks can also be stored electronically in an on-line computer, and these can be analyzed statistically as well). The powder photograph must be carefully measured to obtain 2θ values, whereas 2θ can be read directly from the diffractometer strip chart (or X-Y recorder chart). Figure 7.19 compares a powder diffractometer strip chart with a powder photograph of the same mineral.

Several student exercises using X-ray powder diffraction film and diffractometer techniques are given in Klein (1989; see reference list for complete reference).

Although, until recent years, most of the structural studies of crystalline materials were based on single-crystal X-ray and neutron diffraction experiments, a newly developed X-ray powder diffraction technique, the *Rietveld refinement method*, allows the extraction of structural information from powdered instead of single-crystal specimens. This is an especially important development for the determination of the crystal structures of minerals that are typically finely crystalline and are not found in well-developed single-crystals. Examples of such finely crystalline or poorly ordered minerals are clay minerals, manganese

FIG. 7.19. Comparison of diffractometer record and powder film of quartz. On the diffractometer recording are given the Miller indices of the crystal planes that gave rise to the various low-angle diffraction peaks. (Courtesy of Philips Electronic Instruments, Inc., Mahwah, N.J.)

and iron oxides and hydroxides, and some zeolites. The basic requirements of the Rietveld refinement are: (1) accurate powder diffraction intensity data measured at specific intervals of 2θ; (2) a basic understanding ("starting model") of the actual crystal structure of the material that is being studied; and (3) a quantitative understanding of shapes, width, and any systematic errors in the positions of X-ray peaks in the powder pattern. For further discussion of the Rietveld method see Post and Bish (1989; see reference list for complete reference).

REFERENCES AND SUGGESTED READING

Azaroff, L. V., 1968, *Elements of X-ray Crystallography.* McGraw-Hill Book Co., New York, 610 pp.

Azaroff, L. V. and M. J. Buerger, 1958, *The Powder Method in X-ray Crystallography.* McGraw-Hill Book Co., New York, 342 pp.

Bragg, W. L., 1949, *The Crystalline State: A General Survey.* G. Bell and Sons Ltd., London, 352 pp.

Buerger, M. J., 1964, *The Precession Method in X-ray Crystallography.* John Wiley & Sons, New York, 276 pp.

Cullity, B. D., 1978, *Elements of X-ray Diffraction*, 2nd ed. Addison-Wesley, Reading, Mass., 555 pp.

Klein, C., 1989, *Minerals and Rocks: Exercises in Crystallography, Mineralogy, and Hand Specimen Petrology.* John Wiley & Sons, New York, 402 pp.

Klug, H. P. and Alexander, L. E., 1974, *X-ray Diffraction Procedures for Polycrystalline and Amorphous Materials*, 2nd ed. John Wiley & Sons, New York, 966 pp.

Nuffield, E. W., 1966, *X-ray Diffraction Methods.* John Wiley & Sons, New York, 409 pp.

Post J. E. and Bish, D. L., 1989, Rietveld refinement of crystal structures using X-ray powder diffraction data, in *Modern Powder Diffraction*, Bish, D. L. and Post, J. E., eds. *Reviews in Mineralogy*, 20, pp. 277–308.

CHAPTER 8

OPTICAL PROPERTIES OF MINERALS

The optical properties are less easily determined than the other physical properties of minerals but are important in mineral characterization and identification. Because they are usually observed microscopically, only a very small amount of material is necessary. Most of the following discussion deals with the optical properties that can be determined in transmitted light and thus applies to nonopaque minerals. However, brief mention is made at the end of this chapter of the optical properties of opaque minerals.

NATURE OF LIGHT

To account for all the properties of light it is necessary to resort to two theories: the *wave theory* and the *corpuscular theory*. However, it is the wave theory that we shall consider in explaining the optical behavior of crystals. This theory assumes that visible light, as part of the electromagnetic spectrum, travels in straight lines with a transverse wave motion; that is, it vibrates at right angles to the direction of propagation. The wave motion is similar to that generated by dropping a pebble into still water with waves moving out from the central point. The water merely rises and falls, it is only the wave front that moves forward. The *wavelength* (λ) of such a wave motion is the distance between successive crests (or troughs); the *amplitude* is the displacement on either side of the position of equilibrium; the *frequency* is the number of waves per second passing a fixed point; and the *velocity* is the frequency multiplied by the wavelength. Similarly light waves (Fig. 8.1) have length, amplitude, frequency, and velocity but their transverse vibrations, perpendicular to the direction of propagation, take place in all possible directions.

Visible light occupies a very small portion of the electromagnetic spectrum (see Figs. 6.12 and 7.1). The wavelength determines the color and varies from slightly more than 7000 Å at the red end to about 4000 Å at the violet end. White light is composed of all wavelengths between these limits, whereas light of a single wavelength is called monochromatic.

Reflection and Refraction

When a light ray passes from a rare medium, such as air, into a denser medium, such as glass, part of it is reflected from the surface back into the air and part enters the glass (Fig. 8.2). The reflected ray obeys the laws of reflection, which state: (a) that the angle of incidence (*i*) equals the angle of reflection (*r'*), when both angles are measured from the surface normal, and (b) that the incident and reflected rays lie in the

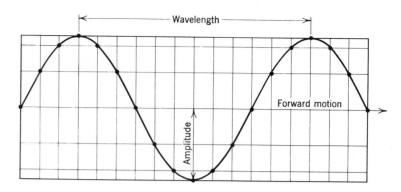

FIG. 8.1 Wave motion.

same plane. The light that passes into the glass travels with a lesser velocity than in air and no longer follows the path of the incident ray but is bent or *refracted*. The amount of bending depends on the obliquity of the incident ray and the relative velocity of light in the two media; the greater the angle of incidence and the greater the velocity difference, the greater the refraction.

Index of Refraction

The index of refraction, n, of a material can be expressed as the ratio between the velocity of light in the air (V) and its velocity in the denser material (v), that is, $n = V/v$. As a basis for comparison the velocity of light in air is considered equal to 1, $n = 1/v$, or the index of refraction is equal to the reciprocal of the velocity.[1]

The precise relationship of the angle of incidence (i) to the angle of refraction (r) is given by Snell's law, which states that for two media (e.g., in going from air into glass) the ratio of sin i : sin r is a constant. This is usually expressed as sin i/sin r = n, where the constant, n, is the *index of refraction*.

The velocity of light in glass is equal to frequency multiplied by wavelength; therefore, with fixed frequency, the longer the wavelength the greater the ve-

[1]As a standard for comparison the velocity of light in vacuum is taken as unity. Other light velocities are thus expressed in terms of $V = 1$. Because for air, $v = 0.9997$ (almost as great as in vacuum), it may also be considered unity.

locity. Red light with its longer wavelength has a greater velocity than violet light and because of the reciprocal relation between velocity and refractive index, n for red light is less than n for violet light (Fig. 8.3). A crystal thus has different refractive indices for different wavelengths of light. This phenomenon is known as *dispersion* and because of it, monochromatic light is used for accurate determination of refractive index.

TOTAL REFLECTION AND THE CRITICAL ANGLE

We have seen (Fig. 8.2) that light is reflected toward the normal when it passes from a medium with a lower refractive index to a medium with a higher refractive index. When the conditions are reversed, as in Fig. 8.4, and the light moves from the higher to the lower index medium, it is refracted away from the normal. In Fig. 8.4, assume that lines A, B, C, and so forth represent light rays moving through glass, and into air at point O. The greater the obliquity of the incident ray, the greater the angle of refraction. Finally, an angle of incidence is reached, as at ray D, for which the angle of refraction is 90°, and the ray then grazes the surface. The angle of incidence at which this takes place is known as the *critical angle*. Rays such as E and F, striking the interface at a greater angle, are totally reflected back into the higher index medium.

FIG. 8.2 Reflected and refracted light.

FIG. 8.3. Different refraction for different wavelengths of light.

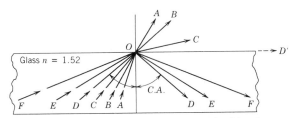

FIG. 8.4. Light rays moving through glass and striking the glass–air interface at angle *C.A.*, and greater angles, are totally reflected.

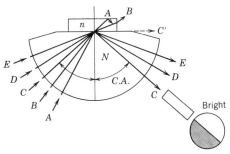

FIG. 8.5. Pulfrich refractometer and measurement of the critical angle *C.A. N = 1.90, C.A. = 50°, n = 1.455.*

The measurement of the critical angle is a quick and easy method of determining the refractive index of both liquids and solids. The instrument used is a refractometer, of which there are many types. A description of one of these, the Pulfrich refractometer, will suffice to illustrate the underlying principles. This instrument employs a polished hemisphere of high refractive index glass (Fig. 8.5). A crystal face or polished surface of the mineral is placed on the equatorial plane of the hemisphere and, depending on the angle of incidence, is either partly refracted through the unknown or totally reflected back through the hemisphere. If a telescope is placed in a position to receive the reflected rays, one can observe a sharp boundary between the portion of the field intensely illuminated by the totally reflected light and the remainder of the field. When the telescope is moved so that its cross hairs are precisely on the contact, the critical angle is read on a scale. Knowing this angle and the index of refraction of the hemisphere, *N*, one can calculate the index of refraction of the mineral: *n* mineral = sin critical angle × *N* hemisphere.

ISOTROPIC AND ANISOTROPIC CRYSTALS

For optical considerations all transparent substances can be divided into two groups: *isotropic* and *anisotropic*. The isotropic group includes such noncrystalline substances as gases, liquids, and glass, but it also includes crystals that belong to the isometric crystal system. In them light moves in all directions with equal velocity and hence each isotropic substance has a single refractive index. In anisotropic substances, which include all crystals except those of the isometric system, the velocity of light varies with crystallographic direction and thus there is a range of refractive index.

In general, light passing through an anisotropic crystal is broken into two polarized rays vibrating in mutually perpendicular planes. Thus for a given ori-

entation, a crystal has two indices of refraction, one associated with each polarized ray.

POLARIZED LIGHT

We have seen that light can be considered a wave motion with vibrations taking place in all directions at right angles to the direction of propagation. When the wave motion is confined to vibrations in a single plane, the light is said to be *plane polarized*. The three principal ways of polarizing light are by double refraction, absorption, and reflection.

Polarized Light by Double Refraction

It has been pointed out that when light passes through an anisotropic crystal it is divided into two polarized rays. The principle on which the first efficient polarizer was based was the elimination of one of these rays. The crystalline material used was the optically clear variety of calcite, Iceland spar, and the polarizer was called the *Nicol prism*, after the inventor William Nicol. Calcite has such a strong double refraction that each ray produces a separate image when an object is viewed through a cleavage fragment (Fig. 8.6*a*). In making the Nicol prism (Fig. 8.7), an elongated cleavage rhombohedron of calcite is sawed at a specified angle and the two halves are rejoined by cementing with Canada balsam. Faces are then ground at the ends of the prism to make angles of 90° with the cemented surface. On entering the prism, light is resolved into two rays *O* and *E*. Because of the greater refraction of the *O* ray it is totally reflected at the Canada balsam surface. The *E* ray with refractive index close to that of the balsam proceeds essentially undeviated through the prism and emerges as plane polarized light.

Polarized Light by Absorption

The polarized rays into which light is divided in anisotropic crystals may be differentially absorbed. If one

FIG. 8.6. *(a)* Calcite, viewed normal to the rhombohedron face, showing double refraction. The double repetition of "calcite" at the top of the photograph as seen through a face cut on the specimen parallel to the base. *(b)* Calcite showing no double refraction. Viewed parallel to the *c* axis. (Specimens from the Harvard Mineralogical Museum.)

ray suffers nearly complete absorption and the other very little, the emerging light will be plane polarized. This phenomenon is well illustrated by some tourmaline crystals (Fig. 8.8). Light passed through the crystal at right angles to [0001] emerges essentially plane polarized, with vibrations parallel to the *c* axis. The other ray, vibrating perpendicularly to it, is almost completely absorbed. When two crystals are placed at right angles one above the other, the polarized ray

emerging from one is absorbed by the other. Polarizing sheets, such as *Polaroid*, are made by aligning crystals on an acetate base. These crystals absorb very little light in one vibration direction, but are highly absorptive in the other. The light transmitted by the sheet is thus plane polarized. Because they are thin and can be made in large sheets, manufactured polarizing plates are extensively used in optical equipment, including many polarizing microscopes.

Polarized Light by Reflection

Light reflected from a smooth, nonmetallic surface is partially polarized with the vibration directions parallel to the reflecting surface. The extent of polarization depends on the angle of incidence (Fig. 8.9) and the index of refraction of the reflecting surface. It is

FIG. 8.7. Nicol prism.

FIG. 8.8. Polarized light by absorption. *(a)* Tourmaline. *(b) Polaroid.* Arrows indicate directions of maximum transmission; directions of maximum absorption are at right angles.

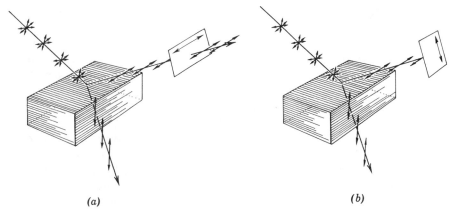

FIG. 8.9. Polarized light by reflection and refraction.

most nearly polarized when the angle between the reflected and refracted ray is 90° (Brewster's law). The fact that reflected light is polarized can be easily demonstrated by viewing it through a polarizing filter. When the vibration direction of the filter is parallel to the reflecting surface, the light passes through the filter with only slight reduction in intensity (Fig. 8.9*a*); when the filter is turned 90° only a small percentage of the light reaches the eye (Fig. 8.9*b*).

THE POLARIZING MICROSCOPE

The polarizing microscope is the most important instrument for determining the optical properties of crystals; with it more information can be obtained easily and quickly than with more specialized devices. Several manufacturers each make a number of models of polarizing microscopes that vary in complexity of design and sophistication and hence in price. A student model made by Nikon Inc. is illustrated in Fig. 8.10 with the essential parts named.

Although a polarizing microscope differs in detail from an ordinary compound microscope, its primary function is the same: to yield an enlarged image of an object placed on the stage. The magnification is produced by a combination of two sets of lenses: the objective and the ocular. The function of the objective lens, at the lower end of the microscope tube, is to produce an image that is sharp and clear. The ocular merely enlarges this image including any imperfection resulting from a poor quality objective. For mineralogical work it is desirable to have three objectives: low, medium, and high power. In Fig. 8.10 these are shown mounted on a revolving nosepiece and can be successively rotated into position. The magnification produced by an objective is usually indicated on its housing, such as $2\times$ (low), $10\times$ (medium), and $50\times$

(high). Oculars also have different magnifications such as $5\times$, $7\times$, $10\times$. The total magnification of the image can be determined by multiplying the magnification of the objective by that of the ocular as: $50\times \cdot 10\times = 500\times$. Although in routine work the

FIG. 8.10. Polarizing microscope, Labophot-Pol, manufactured by Nikon Inc. 1. Oculars. 2. Analyzer. 3. Slot for accessory plate. 4. Revolving nosepiece for objectives. 5. Objectives. 6. Rotating stage. 7. Lever for swinging in and out condenser lens. 8. Rotatable polarizer. 9. Vertical adjustment of stage, for focusing (knobs on both sides). 10. Field diaphragm. 11. Substage illuminator. 12. Intensity adjustment knob for illuminator. 13. On/off switch. (Courtesy of S & M Microscopes Inc., Colorado Springs, Col.)

three objectives are frequently interchanged, a single ocular usually suffices. The ocular assembly, which slips into the upper end of the microscope tube, carries cross hairs—one N–S (front-back) the other E–W. These enable one to locate under high power a particular mineral grain that has been brought to the center of the field under low power. They are also essential in aligning cleavage fragments for making angular measurements. A condenser is located below the stage. The upper lens of the condenser, used with high-power objectives, makes the light strongly converging and can be rotated easily into or out of the optical system. The iris diaphragm, also located below the stage, can be opened or closed to control the depth of focus and to regulate the intensity of light striking the object.

In addition to the lenses, condenser, and diaphragm mentioned above which are common to all compound microscopes, the polarizing microscope has several other features. The *polarizer* below the stage is a polarizing plate, or Nicol prism, that transmits plane polarized light vibrating in a N–S (front-back) direction. The *analyzer*, fitted in the tube above the stage, is a similar plate or prism that transmits light vibrating only in an E–W direction. The polarizer and analyzer are collectively called *polars*.[2] When both polars are in position, they are said to be crossed and, if no anisotropic crystal is between them, no light reaches the eye. The polarizer remains fixed but the analyzer can be removed from the optical path at will. The Bertrand lens is an accessory that is used to observe interference figures (see page 299). In working with crystals it is frequently necessary to change their orientation. This is accomplished by means of a rotating stage, whose axis of rotation is the same as the microscope axis.

Microscopic Examination of Minerals and Rocks

The polarizing microscope is also called the petrographic microscope because it is used in the study of rocks. In examining thin sections of rocks, the textural relationships are brought out and certain optical properties can be determined. It is equally effective in working with powdered mineral fragments. On such loose grains all the optical properties can be determined, and in most cases they characterize a mineral sufficiently to permit its identification.

---

[2]Before using a microscope the vibration directions of the polars should be checked; for, although the orientation is usually as given, in some microscopes the vibration directions of transmitted light are reversed; that is, the polarizer is E–W, the analyzer N–S.

The optimum size of mineral grains for examination with a polarizing microscope is minus 50 mesh–plus 100 mesh, but larger or smaller sizes may be used. To prepare a mount for examination (1) put a few mineral grains on an object glass, a slide 40 mm $\times$ 27 mm, (2) immerse the grains in a drop of liquid of known refractive index, and (3) place a cover glass on top of the liquid. When this type of mount is used the refractive indices of mineral grains are determined by the *immersion method*.

In using this method there should be available a series of calibrated liquids ranging in refractive index from 1.41 to 1.77, with a difference of 0.01 or less between adjacent liquids. These liquids cover the refractive index range of most of the common minerals. The immersion method is one of trial and error and involves comparing the refractive index of the unknown with that of a known liquid.

Isotropic Crystals and the Becke Line

Because light moves in all directions through an isotropic substance with equal velocities, there is no double refraction and only a single index of refraction. With a polarizing microscope, objects are always viewed in polarized light, which conventionally is vibrating N–S. If the object is an isotropic mineral, the light passes through it and continues to vibrate in the same plane. If the analyzer is inserted, darkness results for this polar permits light to pass only if it is vibrating in an E–W direction. Darkness remains as the position of the crystal is changed by rotating the microscope stage. This is a characteristic that distinguishes isotropic from anisotropic crystals.

Let us first consider how the single refractive index of isotropic substances is determined. The first mount may be made by using any liquid, but if the mineral is a complete unknown, it is wise to select a liquid near the middle of the range. When the grains are brought into sharp focus using a medium-power objective and plane polarized light, in all likelihood they will stand in relief; that is, they will be clearly discernible from the surrounding liquid. This is because the light is refracted as it passes from one medium to another of different refractive index. The farther apart the indices of refraction of mineral and liquid, the greater is the relief. But when the indices of the two are the same, there is no refraction at the interface, and the grains are essentially invisible. Relief shows that the index of refraction of the mineral is different from that of the liquid; but is it higher or lower? The answer to this important question can be found by means of the *Becke line* (Fig. 8.11). If the mineral grain is thrown slightly out of focus by raising the microscope tube (in most modern microscopes

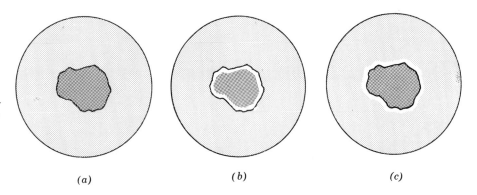

FIG. 8.11. The Becke Line. When thrown out of focus by raising of microscope tube (or lowering of stage), white line moves into medium of higher refractive index. *(a)* In focus. *(b) n* of grain > *n* of liquid. *(c) n* of grain < *n* of liquid.

(a)   *(b)*   *(c)*

this is accomplished by lowering the stage), a narrow line of light will form at its edge and move toward the medium of higher refractive index. Thus if the Becke line moves into the mineral grain, a new mount must be made using a liquid of higher refractive index. After several tries it may be found that there is no Becke line and the mineral grains are invisible in a given liquid. The index of refraction of the mineral is then the same as that of the calibrated liquid. More frequently, however, the refractive index of the mineral is found to be greater than that of one liquid but less than that of its next higher neighbor. In such cases it is necessary to interpolate. If the Becke line moving into the mineral in the lower index liquid is more intense than the line moving out of the mineral into the next higher index liquid, it can be assumed that the refractive index of the mineral is closer to that of the higher liquid. In this way it is usually possible to report a refractive index to ±0.003.

Frequently the Becke line can be sharpened by restricting the light by means of the substage diaphragm. Most liquids have a greater dispersion than minerals. Thus, if the refractive index of liquid and mineral are matched for a wavelength near the center of the spectrum, the mineral has a higher index than the liquid for red light but a lower index for violet light. This is evidenced, when observed in white light, by a reddish line moving into the grain while a bluish line moves out.

Aside from color, the single index of refraction is the only significant optical characteristic of isotropic minerals. It is, therefore, important in mineral identification to consider other properties such as cleavage, fracture, color, hardness, and specific gravity.

UNIAXIAL CRYSTALS

We have seen that light moves in all directions through an isotropic substance with equal velocity and vibrates in all directions at right angles to the direction of propagation. In hexagonal and tetragonal crystals, there is one and only one direction in which light moves in this way. This is parallel to the *c* axis, with vibrations in all directions in the basal plane. For this reason the *c* axis is called the *optic axis*, and hexagonal and tetragonal crystals are called optically *uniaxial*. This distinguishes them from orthorhombic, monoclinic, and triclinic crystals which have two optic axes and are called *biaxial*.

When light moves in uniaxial crystals in any direction other than parallel to the *c* axis, it is broken into two rays traveling with different velocities. One, the *ordinary ray*, vibrates in the basal plane; the other, the *extraordinary ray*, vibrates at right angles to it and thus in a plane that includes the *c* axis. Such a plane, of which there is an infinite number, is referred to as the *principal section*. The nature of these two rays can be brought out in the following way. Assume that the direction of the incident beam is varied to make all possible angles with the crystal axes, and that the distance traveled by the resulting rays in any instant can be measured. We would find that:

1. One ray, with waves always vibrating in the basal plane, traveled the same distance in the same time. Its surface can be represented by a sphere, and because it acts much as ordinary light does it is the *ordinary ray (O ray)*.

2. The other ray, with waves vibrating in the plane that includes the *c* axis, traveled in the same time different distances depending on the orientation of the incident beam. If the varying distances of this, the *extraordinary ray (E ray)* were plotted, they would outline an ellipsoid of revolution, with the optic axis the axis of revolution.

Uniaxial crystals are divided into two optical groups: positive and negative. They are *positive* if the *O* ray has the greater velocity, and *negative* if the *E* ray has the greater velocity. Cross sections of the ray velocity surfaces are shown in Fig. 8.12. Note that in both positive and negative crystals the *O* and *E* rays have the same velocity when traveling along the optic

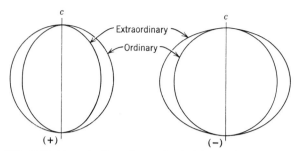

FIG. 8.12. Ray velocity surfaces of uniaxial crystals.

(c) axis. But the difference in their velocities becomes progressively greater as the direction of light propagation moves away from the optic axis, reaching a maximum at 90°.

Because the two rays have different velocities, there are two indices of refraction in uniaxial crystals. Each index is associated with a vibration direction. The index related to vibration along the ordinary ray is designated ω (omega), whereas that associated with the extraordinary ray is ε (epsilon) or ε′. In positive crystals the O ray has a greater velocity than the E ray, and ω is less than ε. But in negative crystals with the E ray having the greater velocity, ω is greater than ε. The two principal indices of refraction of a uniaxial crystal are ω and ε and the difference between them is the *birefringence*.

The *uniaxial indicatrix* is a geometrical figure that is helpful in visualizing the relation of the refractive indices and their vibration directions that are perpendicular to the direction of propagation of light through a crystal. For positive crystals the indicatrix is a prolate spheroid of revolution; for negative crystals it is an oblate spheroid of revolution (Fig. 8.13). In their construction the direction of radial lines is proportional to the refractive indices. First consider light moving parallel to the optic axis. It is not doubly refracted but moves through the crystal as the ordinary ray with waves vibrating in all directions in the basal plane.

FIG. 8.13. Optical indicatrix, uniaxial crystals. *(a)* Positive. *(b)* Negative.

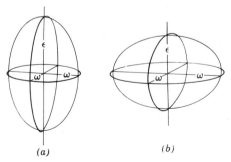

(a)   (b)

This is why light moving parallel to the c axis of calcite (Fig. 8.6b) produces a single image. There is a single refractive index for all these vibrations, proportional to the radius of the equatorial circle of the indicatrix. Now consider light traveling perpendicular to the optic axis. It is doubly refracted. The waves of the ordinary ray vibrate, as always, in the basal plane and the associated refractive index, ω, is again an equatorial radius of the indicatrix. The vibration direction of waves of the extraordinary ray must be at right angles to both the vibration direction of the ordinary waves and the direction of propagation. Thus, in this special case, it is parallel to the optic axis. The axis of revolution of the indicatrix is then proportional to ε, the greatest index in (+) crystals and the least in (−) crystals. It can be seen that light moving through a crystal in a random direction gives rise to two rays: (1) the O ray with waves vibrating in the basal section, the associated index, ω, and (2) the E ray with waves vibrating in the principal section in a direction at right angles to propagation. The length of the radial line along this vibration direction is ε′, a refractive index lying between ω and ε.

A study of the indicatrix shows that (1) ω can be determined on *any* crystal grain; and only ω can be measured when light moves parallel to the optic axis, (2) ε can be measured only when light moves normal to the c axis, and (3) a randomly oriented grain yields, in addition to ω, an index intermediate to ω and ε, called ε′. The less the angle between the direction of light propagation and the normal to the optic axis, the closer is the value ε′ to true ε.

Uniaxial Crystals Between Crossed Polars

Extinction

We have seen that because isotropic crystals remain dark in all positions between crossed polars, they can be distinguished from anisotropic crystals. However, there are special conditions under which uniaxial crystals present a dark field when viewed between crossed polars. One of these conditions is when light moves parallel to the optic axis. Moving in this direction, light from the polarizer passes through the crystal as through an isotropic substance and is completely cut out by the analyzer. The other special condition is when the vibration direction of light from the polarizer coincides exactly with one of the vibration directions of the crystal. In this situation, light passes through the crystal as either the O ray or the E ray to be completely eliminated by the analyzer, and the crystal is said to be at *extinction*. As the crystal is ro-

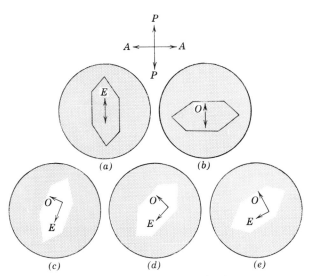

FIG. 8.14. Quartz crystal between crossed polars.

tated from this extinction position it becomes progressively lighter, reaching a maximum brightness at 45°. There are four extinction positions in a 360° rotation, one every 90°.

Interference

Let us consider how the crystal affects the behavior of polarized light as it is rotated from one extinction position to another. Figure 8.14 represents five positions of a tiny quartz crystal elongated on the c axis and lying on a prism face. In the diagrams it is assumed that light from the polarizer is moving upward, normal to the page, and vibrating in direction P–P. The vibration direction of the analyzer is A–A. The crystal in (a) is at an extinction position and light moves through it as the E ray vibrating parallel to the c axis. At 90°, position (b), the crystal is also at extinction with light moving through it as the O ray. When the crystal is turned as in (c), (d), and (e), polarized light entering it is resolved into two components. One moves through it as the O ray, vibrating in the basal plane; the other as the E ray, vibrating in the principal section. In (c) most light is transmitted as the E ray but in (e) it is transmitted mostly as the O ray. In (d), the 45° position, the amounts of light transmitted by the two rays are equal.

When these rays from the crystal enter the analyzer, each is broken up into an O and E ray conforming in vibration directions to those of the analyzer. Only the components of the rays vibrating in an E–W direction are permitted to pass. During their passage through the crystal the two rays travel with different velocities and thus on emerging there is a phase difference because one is ahead of the other. The amount it is ahead depends both on the difference in velocities and on the thickness of crystal traversed. Because both rays vibrate in the same plane of the analyzer, they interfere. For monochromatic light, if one ray is an integral number of wavelengths ($n\lambda$) behind the other, the interference results in darkness. On the other hand, if the path difference is $\lambda/2$, $3\lambda/2$, or in general $(2n - 1)\lambda/2$, the waves reinforce one another to produce maximum brightness.

Each wavelength has its own sets of critical conditions when interference produces darkness. Consequently, when white light is used, "darkness" for one wavelength means its elimination from the spectrum and its complementary color appears. The colors thus produced are called *interference colors*. There are different *orders* of interference depending on whether the color results from a path difference of 1λ, 2λ, 3λ, $\ldots$, $n\lambda$. These are called first-order, second-order, third-order, and so forth interference colors as indicated in Fig. 8.15. A color plate of this figure appears on the endpaper at the back of the book.

Interference colors depend on three factors: orientation, thickness, and birefringence. With a continuous change in direction of light, from parallel to perpendicular to the optic axis, there is a continuous increase in interference colors. For a given orientation, the thicker the crystal and the greater its birefringence the higher the order of interference color. If a crystal plate is of uniform thickness, as a cleavage flake may be or a grain in a rock thin section, it will show a single interference color. As seen in immersion liquids, grains commonly vary in thickness and a variation in interference colors reflects this irregularity.

Accessory Plates

The gypsum plate, mica plate, and quartz wedge are accessory plates used with the polarizing microscope; their function is to produce interference of known amounts and thus predetermined colors. They are all constructed so that the fast ray (the vibration direction of the lesser refractive index) is parallel to the long dimension. The *gypsum plate* is made by cleaving a gypsum crystal to such a thickness that in white light it produces a uniform red interference color: *red of the first order*. The *mica plate* is made with a thin mica flake, cleaved to a thickness that for yellow light it yields a path difference of a quarter of a wavelength. It is thus also called the *quarter wave plate*. The *quartz wedge* is an elongated wedge-shaped piece of quartz (see Fig. 8.16a) with the vibration direction of the fast ray (ω) parallel to its length and the slow ray (ϵ) across its length. As thicker portions of the wedge are placed in the optical path, the path difference of the rays pass-

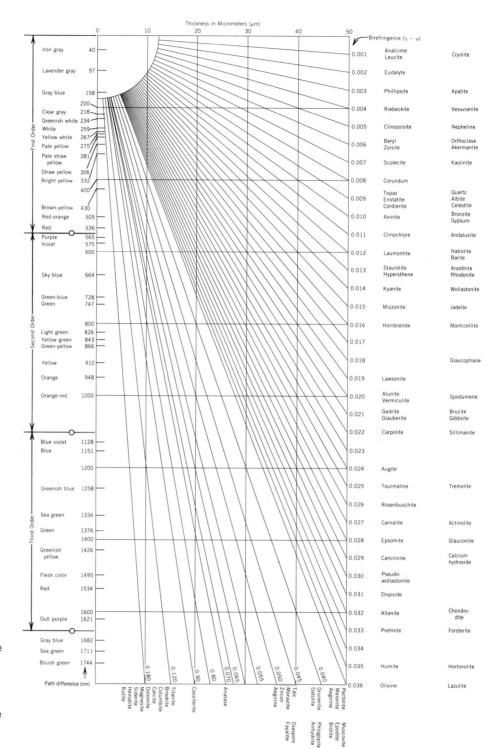

FIG. 8.15. Chart showing the relation of interference colors to thickness and birefringence and the birefringence of some common minerals. A color plate of this same figure is given on the back end-paper of this text.

ing through it also increases, producing a succession of interference colors. The number of orders depends on the wedge angle: the greater the angle the more orders per unit of length.

When the quartz wedge is viewed between crossed polars in monochromatic light, it is crossed by alternating dark and light bands: dark where the path difference is $n\lambda$ and brightest where the path difference is $(2n - 1)\lambda/2$, Fig. 8.16b. In white light a succession of interference colors is observed that resemble colors seen in thin oil films on water, Fig. 8.16c. The colors result from interference phenomena

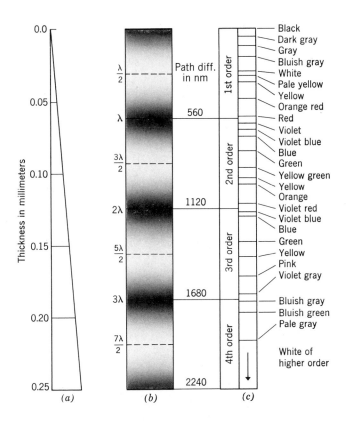

Path diff. in nm		Order	Color
			Black
			Dark gray
			Gray
$\lambda/2$		1st order	Bluish gray
			White
			Pale yellow
			Yellow
λ	560		Orange red
			Red
			Violet
			Violet blue
$3\lambda/2$		2nd order	Blue
			Green
			Yellow green
			Yellow
			Orange
2λ	1120		Violet red
			Violet blue
			Blue
$5\lambda/2$		3rd order	Green
			Yellow
			Pink
			Violet gray
3λ	1680		Bluish gray
			Bluish green
			Pale gray
$7\lambda/2$		4th order	
			White of higher order
	2240		

Thickness in millimeters: 0.0, 0.05, 0.10, 0.15, 0.20, 0.25

FIG. 8.16. Quartz wedge. *(a)* Cross section; *(b)* in monochromatic light, $\lambda = 560$ nm; *(c)* colors in white light.

described by Sir Isaac Newton and are called *Newton's colors*.

Uniaxial Crystals in Convergent Polarized Light

What are known as *interference figures* are seen when properly oriented crystal sections are examined in convergent polarized light. To see them, the polarizing microscope (usually used as an orthoscope) is converted to a conoscope by swinging in the upper substage condensing lens, so that the section can be observed in strongly converging light, using a high-power objective. The interference figure then appears as an image just above the upper lens of the objective and can be seen between crossed polars by removing the ocular and looking down the microscope tube. If the Bertrand lens, an accessory lens located above the analyzer, is inserted, an enlarged image of the figure can be seen through the ocular.

The principal interference figure of a uniaxial crystal, the optic axis figure, Fig. 8.17, is seen when one views the crystal parallel to the *c* axis. Only for the central rays from the converging lens is there now double refraction; the others, traversing the crystal in directions not parallel to the *c* axis, are resolved into *O* and *E* rays having increasing path difference as the obliquity to the *c* axis increases. The interference of these rays produces concentric circles of interference colors. The center is black with no interference but

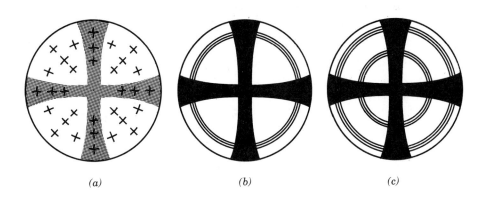

FIG. 8.17. Uniaxial optic axis interference figures. *(a)* Radial lines indicate vibration directions of *E* ray; tangential lines indicate vibration directions of *O* ray. *(b)* and *(c)* show isochromatic curves.

(a)   *(b)*   *(c)*

moving outward there is a progression from first-order to second-order to third-order, and so forth, interference colors. If the crystal section is of uniform thickness, no change will be noted as it is moved horizontally. If, however, the thickness varies, as in a wedge-shaped fragment, the positions of the colors change with horizontal movement. At the thin edge there may be only gray of the first-order, but as the crystal is moved so the light path through it becomes greater, all the first-order colors may appear. And with increasing crystal thickness, the path difference of the two rays may be great enough to yield second-, third-, and higher-order interference colors.

The reason for the black cross superimposed on the rings of interference colors is brought out in Fig. 8.17a. In this drawing the radial dashes indicate the vibration directions of the E ray and those at right angles the vibration directions of the O ray. It will be seen that where these vibration directions are parallel or nearly parallel to the vibration directions of the polarizer and analyzer no light passes and thus the formation of the dark cross.

Figure 8.17 illustrates a centered optic axis figure as obtained on a crystal plate whose c axis coincides with the axis of the microscope; as the stage is rotated, no movement of the figure is seen. If the optic axis of the crystal makes an angle with the axis of the microscope, the black cross is no longer symmetrically located in the field of view (Fig. 8.18). When the stage is rotated, the center of the cross moves in a circular path, but the bars of the cross remain parallel to the vibration directions of the polarizer and analyzer. Even if the inclination of the optic axis is so great that the center of the cross does not appear, on rotation of the crystal the bars move across the field, maintaining their parallelism to the vibration directions of the polars.

The *flash figure* is an interference figure produced by a uniaxial crystal when its optic axis is normal to the axis of the microscope; that is, a hexagonal or tetragonal crystal lying on a face in the prism zone. When the crystal is at an extinction position, the figure is an ill-defined cross occupying much of the field. On rotation of the stage, the cross breaks into two hyperbolas that rapidly leave the field in those quadrants containing the optic axis. The cross forms because the converging light is broken into O and E rays with vibration directions mostly parallel or nearly parallel to the vibration directions of the polarizer and analyzer. A centered flash figure not only indicates the vibration direction of the E ray, but assures one that in this direction a true value of ε can be obtained in plane polarized light.

Determination of Optic Sign

The mica plate, the gypsum plate, and the quartz wedge may be used with a uniaxial optic axis figure to determine the optic sign; that is, whether the crystal is positive or negative. They are inserted below the analyzer in a slot in the microscope tube so positioned that when the plates are in place, their vibration directions make angles of 45° with the vibration directions of the polars.

From the previous discussion we have learned that in the optic axis interference figure, the E ray vibrates radially and the O ray tangentially. By use of an accessory plate in which vibration directions of the slow and fast rays are known, one can tell whether the E ray of the crystal is slower (positive crystals) or faster (negative crystals) than the O ray and thus determine the optic sign. For most American-made equipment the vibration of the slow ray is at right angles to the length of the plate and is so marked on the metal carrier. However, before using an accessory the vibration directions should be checked. The principle in the use of all the plates is the same: to add or subtract from the path difference of the O and E rays of the crystal.

If the *mica plate* is superimposed on a uniaxial optic axis figure in which the ordinary ray is slow (negative crystal), the interference of the plate reinforces the interference colors in the SE and NW quadrants, causing them to shift slightly toward the center. At the same time subtraction causes the colors in the NE and SW quadrants to shift slightly away from the center. The most marked effect produced by the mica plate is the formation of two black spots near the center of the black cross in the quadrants where subtraction occurs (Fig. 8.19).

FIG. 8.18. Positions of off-centered uniaxial optic axis figure on clockwise rotation of microscope stage.

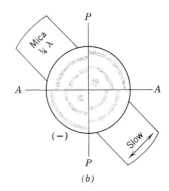

FIG. 8.19. Determination of optic sign with mica plate.

(a)   *(b)*

The *gypsum plate* is usually used to determine the optic sign when low-order interference colors or no colors at all are seen in the optic axis figure. It has the effect of superimposing red of the first-order on the interference figure. If the figure shows several orders of interference colors, one should consider the color effect on the grays of the first-order near the center. In the quadrants where there is addition, the red plus the gray gives blue; in the alternate quadrants the red minus the gray gives yellow. The arrangement of colors in positive crystals is: yellow SE–NW; blue NE–SW; and in negative crystals, yellow NE–SW, blue SE–NW. It is suggested that the student insert the colors in Fig. 8.20.

The *quartz wedge* is most effective in determining optic sign when high-order interference colors are present in the optic axis figure. The wedge is usually inserted with the thin edge first. If its retardation is added to that of the crystal, the interference colors in two opposite quadrants will increase progressively as the wedge moves through the microscope tube. If the retardation is subtracted from that produced by the crystal, the order of colors will decrease. Thus as the quartz wedge is slowly inserted over an optic axis figure of a negative crystal, the color bands in the SE–NW quadrants move toward the center and disappear. At the same time in the NE–SW quadrants the

colors move outward to the edge of the field. In a positive crystal similar phenomena are observed, but the colors move in the opposite directions, that is, away from the center in the SE–NW quadrants and toward the center in the NE–SW quadrants.

Sign of Elongation
Hexagonal and tetragonal crystals are frequently elongated on the c axis or have prismatic cleavage that permits them to break into splintery fragments also elongated parallel to c. If such an orientation is known, one can determine the optic sign by turning the elongated grain to the 45° position and inserting the gypsum plate. If the interference colors rise (gray of mineral plus first-order red equals blue), the slow ray of the gypsum has been superimposed on the slow ray of the mineral. If this is also the direction of elongation, it means the *E* ray is slow (ϵ is the higher refractive index) and the mineral has positive elongation and is optically positive. When the slow ray of the gypsum plate is parallel to the elongation of the mineral grain and the interference colors fall (gray of mineral minus first-order red equals yellow), the mineral has negative elongation and is optically negative (Fig. 8.21).

FIG. 8.20. Determination of optic sign with gypsum plate.

(a)   *(b)*

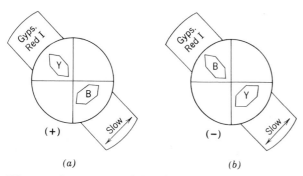

FIG. 8.21. Determination of sign of elongation with gypsum plate. In diagrams Y = yellow, B = blue. *(a)* Positive elongation, *(b)* negative elongation.

Very commonly the interference colors of small grains are grays of the first order. Thus, on superimposing red of the first order, addition gives a blue color and subtraction a yellow color.

Absorption and Dichroism

In the discussion of polarized light (page 291) it was pointed out that in some tourmaline the absorption of one ray is nearly complete but for the other ray it is negligible. Although less striking, many crystals show a similar phenomenon: more light is absorbed in one vibration direction than in the other. In tourmaline where absorption of the O ray is greatest, it is expressed as, absorption: $O > E$ or $\omega > \epsilon$. In other crystals certain wavelengths may be absorbed in one direction and the complementary colors are transmitted. Thus the crystal has different colors in different vibration directions and is said to be *dichroic*. Dichroism is expressed by giving the colors, for example, O or ω = yellow, E or ϵ = pink. Absorption is independent of other properties and is considered, as are refractive indices, a fundamental optical property of crystals.

BIAXIAL CRYSTALS

Orthorhombic, monoclinic, and triclinic crystals are called optically biaxial because they have two directions in which light travels with zero birefringence. In uniaxial crystals there is only one such direction.

Light moving through a biaxial crystal, except along an optic axis, travels as two rays with mutually perpendicular vibrations. The velocities of the rays differ from each other and change with changing crystallographic direction. The vibration directions of the fastest ray, X, and the slowest ray, Z, are at right angles to each other. The direction perpendicular to the

plane defined by X and Z is designated as Y. For biaxial crystals there are thus three indices of refraction resulting from rays vibrating in each of these principal optical directions. The numerical difference between the greatest and least refractive indices is the *birefringence*. Various letters and symbols have been used to designate the refractive indices, but the most generally accepted are the Greek letters as follows:

Index*		Direction	Ray Velocity
(alpha)	α lowest	X	highest
(beta)	β middle	Y	intermediate
(gamma)	γ highest	Z	lowest

*Other equivalent designations are: $\alpha = nX, n_x, N_x, N_p$; $\beta = nY, n_y, N_y, N_m$; $\gamma = nZ, n_z, N_z, N_g$.

The Biaxial Indicatrix

The biaxial indicatrix is a triaxial ellipsoid with its three axes the mutually perpendicular optical directions X, Y, and Z. The lengths of the semiaxes are proportional to the refractive indices: α along X, β along Y, and γ along Z. Figure 8.22 shows the three principal sections through the indicatrix; these are the planes XY, YZ, and XZ. They all are ellipses and each has the

FIG. 8.22. Principal sections through the biaxial indicatrix of a positive crystal.

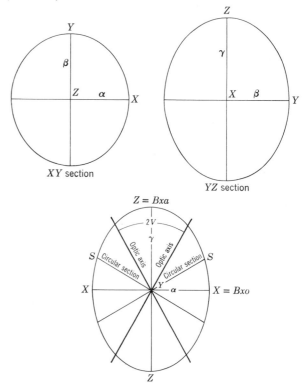

length of its semimajor and semiminor axes proportional to refractive indices as shown. Of most interest is the *XZ* section. With its semimajor axis proportional to γ and its semiminor axis proportional to α, there must be points on the ellipse between these extremes where the radius is proportional to the intermediate index, β. In Fig. 8.22 this radius is marked *S*. With two exceptions every section passing through the center of a triaxial ellipsoid is an ellipse. The exceptions are *circular sections* of which *S* is the radius. The two directions normal to these sections are the *optic axes*; and the *XZ* plane in which they lie is called the *optic plane*. The *Y* direction perpendicular to this plane is the *optic normal*. Light moving along the optic axes and vibrating in the circular sections shows no birefringence and gives the constant refractive index, β. The optic axis of a uniaxial crystal is analogous to these directions, because light moving parallel to it also vibrates in a circular section with constant refractive index.

With variation in refractive indices there is a corresponding variation in the axial lengths of the biaxial indicatrix. Some crystals are nearly uniaxial and in these the intermediate index, β, is very close to either α or γ. If β is close to α, the circular sections make only a small angle with the *XY* plane and the optic axes make the same angle with the *Z* direction. This is angle *V* and the angle between the two optic axes, known as the *optic angle*, is *2V*. The optic angle is always acute and because, in this case, it is bisected by *Z*, *Z* is called the *acute bisectrix* (Bxa); *X* is the *obtuse bisectrix* (Bxo) because it bisects the obtuse angle between the optic axes. When *Z* is the Bxa, the crystal is optically positive.

If β is closer to γ than to α, the acute angle between the optic axes is bisected by *X* and the obtuse angle bisected by *Z*. In this case, with *X* the Bxa, the crystal is negative. When β lies exactly halfway between α and γ, the optic angle is 90°.

The relation between the optic angle and the indices of refraction is expressed by formula (a) below. A close approximation to the optic angle can be made using formula (b).

$$\text{(a)} \quad \cos^2 V_x = \frac{\gamma^2 (\beta^2 - \alpha^2)}{\beta^2 (\gamma^2 - \alpha^2)}$$

$$\text{(b)} \quad \cos^2 V_x = \frac{\beta - \alpha}{\gamma - \alpha}$$

The error using the simplified formula increases with increase of both birefringence and *V* and always yields values for *V'* less than true *V*. It should be noted that

in using either formula *half* the optic angle is calculated, and that it is determined with *X* the bisectrix. Thus, when *V* < 45°, the crystal is negative but when *V* > 45°, the crystal is positive.

Biaxial Crystals in Convergent Polarized Light

Biaxial interference figures are obtained and observed in the same manner as uniaxial figures, that is, with converging light, high-power objective, and Bertrand lens. Although interference figures can be observed on random sections of biaxial crystals, the most symmetrical and informative are obtained on sections normal to the optical directions *X*, *Y*, and *Z* and to an optic axis.

The *acute bisectrix figure* is observed on a crystal plate cut normal to the acute bisectrix. If *2V* is very small, there are four positions during a 360° rotation at which the figure resembles the uniaxial optic axis figure. That is, a black cross is surrounded by circular bands of interference colors. However, as the stage is turned, the black cross breaks into two hyperbolas that have a slight but maximum separation at a 45° rotation; and the color bands, known as *isochromatic curves*, assume an oval shape. The hyperbolas are called *isogyres* and the dark spots, called *melatopes*, at their vertices in the 45° position result from light rays that traveled along the optic axes in the crystal. Thus with increasing optic angle the separation of the isogyres increases, and the isochromatic curves are arranged symmetrically about the melatopes as shown in Fig. 8.23. For most crystals when *2V* exceeds 60° the isogyres leave the field at the 45° position; the larger the optic angle, the faster they leave.

The portion of the interference figure occupied by the isogyres is dark, for here, light as it emerges from the section has vibration directions parallel to those of the polarizer and analyzer. The dark cross is thus present when the obtuse bisectrix and optic normal coincide with the vibration directions of the polars. The bar of the cross parallel to the optic plane is narrower and better defined than the other bar (Fig. 8.23). Because light travels along the optic axes with no birefringence, their points of emergence are, of course, dark in all positions of the figure.

The Apparent Optic Angle

The distance between the points of emergence of the optic axes is dependent not only on *2V* but on β as well. The refractive index of the crystals is β for light rays moving along the optic axes. These rays are refracted on leaving the crystal, giving an apparent optic

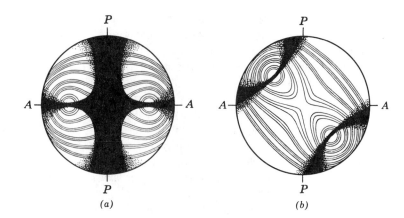

FIG. 8.23. Acute bisectrix interference figure.
(a) Parallel position. *(b)* 45° position.
(a)   (b)

angle, 2*E*, greater than the real angle, 2*V* (Fig. 8.24). The higher the β refractive index, the greater the refraction. Thus if two crystals have the same 2*V*, the one with the higher β index has the larger apparent angle and the farther apart the optic axes emerge in the interference figure.

The *optic axis figure* is observed on mineral grains cut normal to an optic axis. Such grains are easy to select for they remain essentially dark between crossed polars on complete rotation. The figure consists of a single isogyre at the center of which is the emergence of the optic axis. When the optic plane is parallel to the vibration direction of either polar, the isogyre crosses the center of the field as a straight bar. On rotation of the stage it swings across the field forming a hyperbola in the 45° position. In this position the figure can be pictured as half an acute bisectrix figure with the convex side of the isogyre pointing toward the acute bisectrix. As 2*V* increases, the curvature of the isogyre decreases and when 2*V* = 90°, the isogyre is straight (Fig. 8.25).

The *obtuse bisectrix figure* is obtained on a crystal section cut normal to the obtuse bisectrix. When the plane of the optic axes is parallel to the vibration direction of either polar, there is a black cross. On rotation of the stage the cross breaks into two isogyres that move rapidly out of the field in the direction of

the acute bisectrix. Although not as informative as an acute bisectrix figure, a centered obtuse bisectrix figure indicates that an accurate determination of β and of either α or γ can be made on the mineral section producing it.

The *optic normal figure* is obtained on sections cut parallel to the plane of the optic axes and resembles the flash figure of a uniaxial crystal. When the *X* and *Z* optical directions are parallel to the vibration directions of the polars, the figure is a poorly defined cross. On slight rotation of the stage it splits into hyperbolas that move rapidly out of the field in the quadrants containing the acute bisectrix. An optic normal figure is obtained on sections with maximum birefringence and indicates that α and γ can be determined on this section.

Determination of Optic Sign of a Biaxial Crystal

The optic sign of biaxial crystals can best be determined on acute bisectrix or optic axis figures with the aid of accessory plates. Let us assume that *a* in Fig. 8.26 represents an acute bisectrix figure of a negative crystal in the 45° position. By definition, *X* is the acute bisectrix and *Z* the obtuse bisectrix. *OP* is the trace of the optic plane and *Y* the vibration direction of β, at

FIG. 8.24. Relation of 2*V* to 2*E*.

FIG. 8.25. Curvature of isogyre in optic axis figure from 0° to 90° 2*V*.

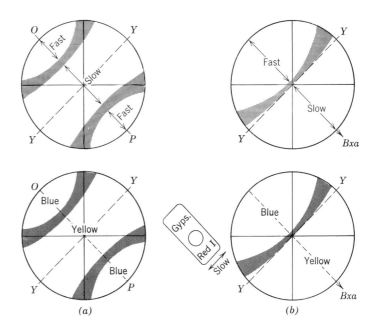

FIG. 8.26. Optic sign determination of negative crystal with gypsum plate. *(a)* Acute bisectrix figure. *(b)* Optic axis figure.

(a)   *(b)*

right angles. The velocity is constant for all rays moving along the optic axes, and for them the refractive index of the crystal is β, including those vibrating in the optic plane. Consider the velocities of other rays, vibrating in the optic plane. In a negative crystal, those emerging between the isogyres of an acute bisectrix figure have a lesser velocity, those emerging outside the isogyres have a greater velocity. If the gypsum plate is superimposed over such a figure, the slow ray of the plate combines with the fast ray of the crystal and subtraction of interference colors produces a yellow color on the convex side of the isogyres. On the concave sides of the isogyres a blue color is produced by addition, the slow ray of the plate over the slow ray of the crystal. In a positive crystal the reverse color effect is seen, for here Z is the acute bisectrix.

An optic axis figure in the 45° position can be used in a manner similar to the acute bisectrix figure in determining optic sign. Insertion of the gypsum plate yields for (−) crystal: convex side yellow, concave side blue (Fig. 8.26b), for (+) crystal: convex side blue, concave side yellow.

The quartz wedge may be used to determine optic sign if several isochromatic bands are present. As the wedge is inserted the colors move out in those portions of the figure where there is subtraction and move in where there is addition. In other words, in the areas that a gypsum plate would render blue, the colors move in; in those that it would render yellow, they move out.

There is a tendency for students learning optical techniques to search for interference figures in the initial immersion when, in all probability, the indices of the mineral are far removed from the refractive index of the liquid. It is time saving, particularly in working with biaxial minerals, to first match as closely as possible the refractive indices of the unknown with *n* of the liquid. An interference figure, then, in addition to the optic sign, will yield other useful information. Using a grain giving an optic axis figure, β can be compared with *n* of the liquid. On a grain yielding an acute bisectrix figure, β and either α or γ can be compared with the refractive index of the liquid. The indices of a grain showing highest interference colors are most likely to be close to α and γ. Therefore, in the same mount, one should check such a grain to estimate how far α and γ are from the refractive index of the liquid.

Optical Orientation in Biaxial Crystals

The orientation of the optical indicatrix is one of the fundamental optical properties. It is given by expressing the relationship of X, Y, and Z optical directions to the crystallographic axes, *a*, *b*, and *c*.

In *orthorhombic crystals* each of the crystallographic axes is coincident with one of the principal optical directions. For example, the optical orientation of anhydrite is: $X = c$, $Y = b$, $Z = a$. Usually the optical directions coinciding with only two axes are given, for this completely fixes the position of the indicatrix.

It is difficult or impossible to determine the axial directions on microscopic grains of some minerals. To do it one must use a fragment oriented by X-ray study or broken from a faced crystal. However, even in

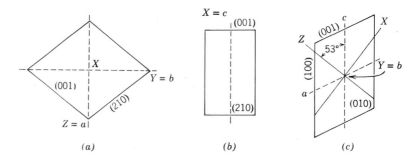

FIG. 8.27. Optical orientation. *(a)* Barite on {001} showing symmetrical extinction. *(b)* Barite on {210} showing parallel extinction. *(c)* Gypsum on {010} showing extinction angle.

small particles, orientation can be expressed relative to cleavages. Powdered fragments tend to lie on cleavages, which in orthorhombic crystals are commonly pinacoidal or prismatic. For example, barite has {001} and {210} cleavage and most grains lie on faces of these forms. Those lying on {001} will be diamond shaped (Fig. 8.27*a*) and have *symmetrical extinction.* That is, the extinction position makes equal angles with the bounding cleavage faces. Fragments lying on {210} will have *parallel extinction* (Fig. 8.27*b*). Symmetrical and parallel extinction are characteristic of orthorhombic crystals.

Barite is (+), $X = c$, $Y = b$, $Z = a$. Thus, grains lying on {001} yield a centered Bxo figure and one can determine β in the b direction and γ in the a direction. Grains showing parallel extinction do not give a centered interference figure, but α can be measured parallel to c.

In *monoclinic crystals* one of the principal optical directions (X, Y, or Z) of the indicatrix coincides with the b axis; the other two lie in the $a–c$ plane of the crystal. The orientation is given by stating which optical direction equals b and indicating the *extinction angle,* the angle between the c axis and one of the other optical directions. If the extinction lies between the $+$ ends of the a and c axes, the angle is positive; between $+c$ and $-a$, the angle is negative. In gypsum $Y = b$ and $Z \wedge c = 53°$. Thus a fragment lying on the {010} cleavage would yield an optic normal interference figure and α and γ could be determined; γ at the extinction position $+53°$; α at extinction position $-37°$ (see Fig. 8.27*c*). A grain lying on the {100} cleavage would show parallel extinction and β could be measured at right angles to the trace of {010}. Crystal fragments lying on any other face in the [001] zone, as on {110}, will show an extinction angle, but the angle to be recorded, usually that of *maximum extinction,* is observed on {010}. Parallel extinction indicates the grain is lying on a face in the [010] zone and the index of the ray vibrating parallel to b can be measured.

In *triclinic crystals* the optic indicatrix can occupy any position relative to the crystallographic axes. Thus

a complete optical orientation necessitates giving ϕ and ρ angles of the principal optical directions. But in most cases it suffices to give the extinction angles observed on grains lying on known cleavage faces.

Dispersion of the Optic Axes

We have seen that the refractive indices of a mineral vary with the wavelength of light. This dispersion (of the indices) means that, with a variation in the color of light, there is a variation in the indicatrix. The resulting change in the positions of the optic axes, and the accompanying change in $2V$, is known as dispersion of the optic axes. Instead of giving different values for $2V$ for different wavelengths of light, this phenomenon is usually expressed by stating whether $2V$ is greater or lesser for red light than for violet light.

Using white light, dispersion can be observed in acute bisectrix figures and optic axis figures and is evidenced as a red fringe on one side of an isogyre and a blue fringe on the other side. Let us assume that $2V$ is greater for red than for violet light. Red light moving along the "red" optic axis has zero path difference. Thus where this axis emerges, red has been removed from the white light and a blue color appears. Similarly, violet light has been removed at the point of emergence of the "violet" axis and a red color appears. In this case the red fringe would appear on the convex side of the isogyre, the blue fringe on the concave side and the dispersion is expressed as: $r > v$. If the positions of the color fringes were reversed, the dispersion would be: $r < v$. In most interference figures the color fringes are subtle and the isogyre is essentially black.

The foregoing explanation of dispersion is strictly true only for orthorhombic crystals, where the plane of the optic axis is an axial plane of the crystal and the acute bisectrix a crystal axis. In monoclinic crystals there is dispersion of the bisectrices as well as of the optic axes giving rise to three types of dispersion. The dispersion is called: *crossed* when Bxa $= b$; *horizontal* when Bxo $= b$; and *inclined* when the optic normal $= b$. The distribution of the color fringes in-

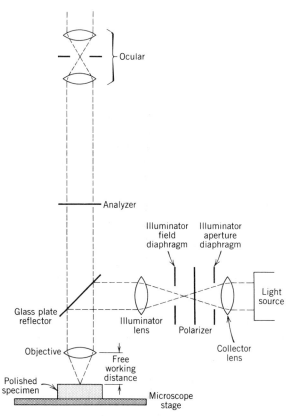

FIG. 8.28. Schematic cross section through a reflecting light microscope showing the light path and some of the components of the microscope.

dicating these special types of dispersion is rarely seen, and monoclinic dispersion is usually expressed as r > or < v.

Absorption and Pleochroism

The absorption of light in biaxial crystals may differ in the X, Y, and Z optical directions. If the difference is only in intensity and X has the greatest absorption and Z the least, it is expressed as X > Y > Z. If different wavelengths are absorbed in different directions, the mineral is said to be pleochroic and the color of the transmitted light is given. For example, in hypersthene the *pleochroism* is: X = brownish red, Y = reddish yellow, Z = green. The term *pleochroism* is commonly used to denote all differential absorption in both uniaxial and biaxial crystals.

OPTICAL PROPERTIES OF OPAQUE MINERALS

The discussion in this chapter has dealt entirely with the optical properties of nonopaque minerals. It should be mentioned, however, that opaque minerals

also possess optical properties by which they are characterized. Specially built microscopes are used for viewing polished sections of these minerals in reflected light. Because most of the ore minerals are opaque, the techniques for polished section study with the *ore microscope* (or *reflecting light microscope*) have been largely developed and used by the student of ores (see Fig. 8.28 for an illustration of the light path in a reflecting light microscope). In a polished section, color is the first and frequently the most important property observed, and the skilled microscopist is able to distinguish one mineral from another by a subtle color difference. Figure 8.29 illustrates the appearance of some opaque minerals in the ore microscope. Using polarized light many nonisometric minerals show a *bireflectance*, a property analogous to pleochroism in nonopaque minerals. That is, there is a change in brightness or color as the mineral is rotated on the microscope stage. Different minerals show this property in varying degrees. When viewed between crossed polars, nonisometric opaque minerals will show, as do nonopaque minerals, four positions of extinction in a 360° rotation. Isotropic minerals show neither bireflectance nor extinction positions and thus can be distinguished from anisotropic minerals.

The microscopic study of polished sections of opaque minerals gives important clues to the paragenesis and origin of mineral deposits. From the textural relationship of the minerals one can determine their order of deposition and their subsequent replace-

FIG. 8.29. Photomicrograph of a polished section of opaque minerals from Balmat, New York, as obtained with a reflecting light microscope. The minerals are: white, pyrite; light gray with triangular pits, galena; dark gray, sphalerite. (Courtesy of Charles Grocetti, Harvard University.)

0.10 mm

ment and exsolution phenomena. There is a large literature on ore microscopy, but even a brief discussion of the methods and techniques is beyond the scope of this book.

REFERENCES AND SUGGESTED READING

Nonopaque Minerals

Ehlers, E. G., 1987, *Optical Mineralogy*, v. 1, *Theory and Techniques*, v. 2, *Mineral Descriptions*. Blackwell Scientific Publications, Palo Alto, Calif., 158 pp. and 286 pp.

Heinrich, E. W., 1965, *Microscopic Identification of Minerals*. McGraw-Hill Book Co., New York, 414 pp.

Larsen, E. S. and Berman, H., 1934, *The Microscopic Determination of the Nonopaque Minerals*, 2nd ed. U.S. Geological Survey Bulletin no. 848. U.S. Government Printing Office, Washington, D.C., 266 pp.

MacKenzie, W. S. and Guilford, C., 1980, *Atlas of Rock-Forming Minerals in Thin Section*. John Wiley & Sons, New York, 98 pp.

Nesse, W. D., 1991, *Introduction to Optical Mineralogy*, 2nd ed. Oxford University Press, New York, 335 pp.

Phillips, W. R., 1971, *Minerals Optics: Principles and Techniques*. W. H. Freeman & Co., San Francisco, 249 pp.

Phillips, W. R. and Griffen, D. T., 1981, *Optical Mineralogy: The Nonopaque Minerals*. W. H. Freeman & Co., San Francisco, 677 pp.

Shelley, D., 1985, *Optical Mineralogy*, 2nd ed. Elsevier Science Publishers, New York, 321 pp.

Stoiber, R. E. and Morse, S. A., 1981, *Microscopic Identification of Crystals*, rev. ed. Robert E. Krieger Publishing Co., Melbourne, Fla., 286 pp.

Wahlstrom, E. E., 1979, *Optical Crystallography*, 5th ed. John Wiley & Sons, New York, 488 pp.

Opaque Minerals

Craig, J. R. and Vaughan, D. J., 1981, *Ore Microscopy and Ore Petrography*. John Wiley & Sons, New York, 406 pp.

Galopin, R. and Henry, N. F. M., 1972, *Microscopic Study of Opaque Minerals*. McCrone Research Associates Ltd., London, 322 pp.

Ramdohr, P., 1969, *The Ore Minerals and Their Intergrowths*. Pergamon Press, New York, 1174 pp.

Stanton, R. L., 1972, *Ore Petrology*. McGraw-Hill Book Co., New York, 697 pp.

Uytenbogaardt, W. and Burke, E. A. J., 1971, *Tables for Microscopic Identification of Ore Minerals*, 2nd ed. Elsevier-North Holland, New York, 430 pp.

CHAPTER 9

MINERAL STABILITY DIAGRAMS

The mineral definition given in Chapter 1 states that a mineral is a naturally occurring homogeneous *solid* with a definite (but generally not fixed) chemical composition and a *highly ordered atomic arrangement*. The atomic order (or long-range periodicity) distinguishes the *solid crystalline state* from gases, liquids, as well as glasses. In the *gaseous state* atoms and molecules move around freely and the structure of a gas can be considered chaotic. Although molecules or groups of molecules may be part of the molecular arrangement of a gas, such groupings are generally small in scale and are commonly in the process of formation or breaking apart. A gas has no unique volume and no boundaries except those imposed by the container in which it is put. The *liquid state* may be considered as intermediate between that of a solid and that of a gas. In a liquid the atoms and molecules are much more ordered than in a gas, and some *short-range order* may occur. Many of the molecules or molecular groups, however, will be moving from one position to another. A liquid always has a boundary surface, commonly in contact with a gas.

The behavior of solids, liquids, and gases under variable external conditions, such as those of temperature and pressure, is commonly expressed in what is known as a *phase diagram* (or a *stability diagram*). A *phase* is a homogeneous substance with a well-defined set of physical and chemical properties. As such the term *phase* can be used interchangeably with

the term *mineral* only if the mineral is homogeneous (i.e., exhibits no compositional variation). For example, low quartz (SiO_2) is a low-temperature phase in the chemical system $Si-O_2$ (or SiO_2); kyanite (Al_2SiO_5) is a high-pressure phase in the chemical system $Al_2O_3-SiO_2$ (or Al_2SiO_5). When a mineral exhibits solid solution, as in the complete solid solution series between forsterite (Mg_2SiO_4) and fayalite (Fe_2SiO_4), we speak of a *phase region*. A phase may be solid, liquid, or gaseous as in the case of H_2O with three distinct phases, ice, water, and steam.

PHASE DIAGRAM FOR H$_2$O

Figure 9.1 is a phase diagram for the chemical system H_2O; specifically it is a *pressure-temperature stability diagram*, commonly known as a *P-T diagram*. The diagram in Fig. 9.1a shows that at low temperatures and high pressure gas condenses to liquid, and that at even lower temperatures both gas and liquid give way to various polymorphs of ice. Along the various curves in this diagram two phases can *coexist stably* (*in equilibrium*, see below). For example, ice and water can coexist along the various *P-T* curves (freezing point curves) on the left side of the diagram. In Fig. 9.1b, at the point where all three curves meet, three phases, ice, water, and steam, can coexist; this is known as the *triple point, t*. Along the curve *t-c* water and steam

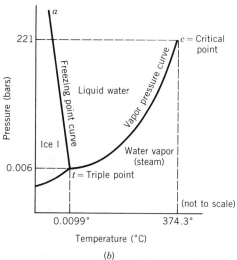

FIG. 9.1 *(a) P-T* diagram for H_2O. Six polymorphic types of ice are indicated by I, II, III, V, VI, and VII. (After Bridgeman, P. W., *Jour. Chemical Physics*, 1937, v. 5, p. 965, and *Phase Diagrams for Ceramists*, copyright © American Ceramic Society, Columbus, Ohio, 1964.) For clarity the water/water vapor curve has been offset slightly toward higher pressure. The shaded region in figure *(a)* is enlarged, but not to scale, in *(b)*.

can coexist but with increasing *P* and *T* (going in the upper right direction along the curve) the water phase becomes less dense (expands due to increasing temperature) and the steam phase more compressed (due to increasing *P*). At point *c* (*critical point*) the two phases become identical, hence indistinguishable. In *P-T* space, to the upper right of *c* we no longer speak of water or steam but we speak of a *supercritical* aqueous fluid.

STABILITY, ACTIVATION ENERGY, AND EQUILIBRIUM

Phase, or mineral stability, diagrams are very useful in providing a visual image of what mineral or group of minerals is stable with respect to some other mineral or mineral groups at a specific set of external (e.g., specified *P* and *T*) conditions. For example, in Fig.

9.1*a*, different polymorphs of ice are stable in different parts of the diagram. Each numbered ice field outlines the *P-T* space in which a specific polymorph of ice is stable. Ice VI has a much larger *P-T* region over which it is stable than, for example, ice III. Furthermore, one can say that ice VII is stable at the highest pressures and a range of low temperatures, whereas ice I is stable only over a range of relatively low *P* and *T* conditions. This leads to the conclusion that ice VII is unstable in the *P-T* space of ice I; the reverse statement is also true.

The concept of *stability* in a chemical system is related to the energy of the system (see discussion of Gibbs free energy, below), but it can be intuitively appreciated in terms of the situation of some mechanical blocks with respect to their resting surface. In Figure. 9.2 three of the same blocks are shown in different positions. In Fig. 9.2*a* the block is barely stable, even though the bottom edge on which it stands was

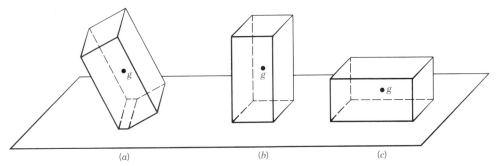

FIG. 9.2. Illustration of various degrees of stability for the same block in different orientations: *(a)* unstable, *(b)* metastable, and *(c)* stable (see text for discussion). The center of gravity is marked as *g*.

beveled off. This position is clearly *unstable*, because as a result of a slight displacement the block in (*a*) will change to either position (*b*) or (*c*). Note also that the center of gravity (*g*) of the block in (*a*) is in the highest position above the resting surface. The block in (*b*) is in a *metastable* condition, because if a reasonably large displacement is applied, it would very likely change to the position of the block in (*c*). If only a small disturbance is applied to the block in (*b*) it will remain in the same orientation. Note also that the center of gravity (*g*) of block (*b*) is lower than that in (*a*) but considerably higher than that in (*c*). In (*c*) the orientation of the block is said to be *stable* because even after a reasonably large disturbance (or displacement) it will return to the orientation in which it is shown. The center of gravity (*g*) is now at the lowest position of the three orientations. The physical displacements (or disturbances) applied to the various blocks can be expressed in energy terms, as the *activation energy*. The amount of energy necessary to change the position of block (*b*) to that of block (*c*) is the activation energy needed to cause the change in position; in a chemical system the activation energy is the energy needed to cause a chemical reaction to occur. In applying the concept of stability to phase diagrams, we are generally concerned with differences in stability (that is, differences in energy values) and not the absolute values. For example, with reference to Fig. 9.1, we would say that liquid water is unstable with respect to ice I, in the *P-T* of ice I stability.

The concept of *equilibrium* is related to time. If, for example, water and ice coexist together in constant amounts indefinitely, that is, no water is forming at the expense of ice, or vice versa (as along curve *a-t* in Fig. 9.1*b*), we can say that under these specific conditions water and ice are in *equilibrium*. In rocks, where the constituent minerals have coexisted since their formation, perhaps several million years ago, one cannot always conclude unambiguously whether the mineral constituents are in equilibrium with each other or not. If no *reaction rims* are observed between minerals that touch each other in the rock, we may assume that the minerals were in equilibrium at the time of formation; this assumption, however, must generally be supported by further detailed chemical and textural information. If, however, megascopically or microscopically visible rims exist between minerals, we may tend to conclude that some of the minerals were *not* in equilibrium with each other. For example, garnet may be separated by a chlorite rim from a coexisting biotite. Here we would conclude that the garnet and biotite were *not* in equilibrium with each other, as they are separated by a reaction product, chlorite. In experimental studies, the experimentalist will conclude that the phases under investigation are in equilibrium when no further change takes place between them during a certain time interval. This may range from a few hours, to several months, or even years, depending on the speed or sluggishness of the reactions studied, and on the patience of the investigator.

COMPONENTS

In Fig. 9.1 we have illustrated stability fields, in terms of bordering equilibrium curves, for phases in a system that can be described chemically by one compound component, H_2O. Phases in a system are described by independent chemical species known as *components*; a minimum number of chemical variables is generally chosen. This can be rephrased as *components are the smallest number of chemical entities necessary to define the compositions of all the phases in a system.* For example, in the case of H_2O, we chose the compound component H_2O, instead of defining the chemical system in terms of two components, H_2 and O_2. In the system Al_2SiO_5 (andalusite-sillimanite-kyanite), Al_2SiO_5 is generally chosen as the compound component although three elements, Al, Si, and O, or two oxide components, Al_2O_3 and SiO_2, could have been selected to define the system chemically. If one is interested in the stability relations of wollastonite, $CaSiO_3$, one might choose to represent one's findings in terms of the compound component, $CaSiO_3$. If, however, one wishes to know the stability fields of pyroxenes in the system $CaO-MgO-FeO-SiO_2$, one generally chooses the three compound components $CaSiO_3-MgSiO_3-FeSiO_3$ to define the system chemically (see Fig. 9.13a).

INTRODUCTORY THERMODYNAMICS

The phase diagram for H_2O, as given in Fig. 9.1, can be delineated by a large number of pressure-temperature experiments on H_2O and very careful characterization of the resultant reaction products. However, such a diagram can also be calculated on the basis of known parameters that relate to the various configurations of atoms, ions, and molecules as a function of physical and chemical conditions. This scientific method, which allows us to make quantitative assessments of mineral and other phase equilibria, is known as *thermodynamics*. Here we will give only a very cursory introduction to this subject, enough to aid in the understanding of some aspects of the phase diagrams that follow.

A fundamental and universal observation is that all organizations of matter drive toward a minimal energy state (or arrangement). Minerals and rocks tend toward the lowest energy state that is, the most stable state for their constituents.

The *first law of thermodynamics* states that "the internal energy (E) of an isolated system is constant," in which a system is defined as any part of the universe being considered (e.g., it may be a hand specimen of a specific rock type, a specific mineral assemblage, or a specific chemical mix in a platinum crucible ready for experimental study). In a *closed system* (one in which there is no addition or subtraction of material; the mass remains constant, but there can be a loss or gain of energy) the change in internal energy (the differential of E, dE) will be the difference between heat (a form of energy, defined as Q) added to the system (expressed as dQ) and work (another form of energy, defined as W) done by the system (expressed as dW). The first law can be stated as:

$$dE = dQ - dW$$

Because work (W) = force × distance, and because force = pressure (P) × surface area, $W = P \times$ sur-

face area × distance, or $W =$ pressure (P) × volume (V).

At constant pressure, and in the absence of electrical, etc., work, this results in $dW = PdV$. When this is substituted in the above equation it yields the most familiar form of the first law of thermodynamics as

$$dE = dQ - PdV$$

The effect of this equation on a mineral can be qualitatively assessed. When energy, as heat, is added to a mineral, the increase in internal energy (dE) of the mineral is proportional to, but less than, the heat added because part of the added energy is transformed into the work of thermal expansion of the mineral.

The *second law of thermodynamics* relates a change in thermal energy of a system (at constant pressure, P, and constant temperature, T) to a change in the degree of order (or disorder) in that system. *Entropy (S)* is a quantity that represents the degree of disorder in a system. As discussed in Chapter 3 (see "Order-Disorder Polymorphism"), the state of greatest order in a crystalline material is at the lowest temperature, with increasing disorder of atoms in the structure the result of increasing temperatures. Relating this to entropy, a rigorously ordered structure has a lower entropy than one that is disordered. The concept of entropy can be illustrated with respect to Fig. 9.1. When heat is added to ice, some of it will convert to water, which, because of its much less regularly ordered structure than ice, will have a higher entropy. As a consequence of the second law,

$$\frac{dQ}{T} = dS$$

where dQ is the absorption of a quantity of heat. The concept of entropy can be further illustrated with respect to the system H_2O, as shown in Fig. 9.1. Because entropy is related to the amount of disorder (or chaos) in a system, there will be increases in entropy in the following reactions in the system H_2O: (1) ice → vapor, (2) ice → liquid, and (3) liquid → vapor.

The third law of thermodynamics states that "at absolute zero (0 Kelvin, which is equivalent to $-273°C$), a crystalline structure approaches perfect order, and the entropy of such a perfect crystal is zero."

In the discussion of Fig. 5.22 we defined another important thermodynamic function, namely, the *Gibbs free energy, G*, as:

$$G = E + PV - TS$$

where E = internal energy, P = pressure, V = volume, T = temperature, and S = entropy. The term "free energy" expresses the energy in excess of the internal energy; this is the excess energy which is needed to drive a chemical reaction. When the Gibbs free energy equation is differentiated and combined with the equations for the first and second laws (given above) the following relation results:

$$dG = VdP - SdT.$$

For a system in equilibrium, at constant P and T, it can be shown that the Gibbs free energy (G) is a minimum. Furthermore, at equilibrium, the Gibbs free energies of the reactant (r) and product (p) are equal, that is $G_r = G_p$, and $dG = 0$.

The above equation of the Gibbs free energy can be differentiated with respect to T, at constant P, or with respect to P, at constant T. This results in two important relations that express the change in free energy (dG) with respect to pressure and temperature.

These are:

$$\left(\frac{\partial G}{\partial P}\right)_T = V \qquad \text{where } V = \text{volume}$$

and

$$\left(\frac{\partial G}{\partial T}\right)_P = -S \qquad \text{where } S = \text{entropy}$$

The first of these expressions tells us that dense phases (i.e., those with small volumes) are favored at high pressures, and the second shows that high entropy states (with greater atomic disorder) are favored at high temperatures. Basic aspects of the Gibbs free energy equation can be shown graphically. The equations given above state that dG, a change in the Gibbs free energy, is a function of only P (stated as dP) and of T (stated as dT). Because the three variables G, P, and T, are interrelated, the function G can be represented graphically in terms of two variables, P and T. This is shown in Fig. 9.3a for two minerals (or phases) marked A and B. Each phase has its distinct G surface.

FIG. 9.3. *(a)* Three-dimensional representation of G-T-P space, with free energy surfaces for two minerals, A and B. Where the two surfaces intersect (along the line of intersection) the free energies of the two minerals are equal. Two sections through this space are shown: *(b)* a G-T section and *(c)* a G-P section. *(d)* is a projection of the line of intersection from above onto the P-T plane.

(a)

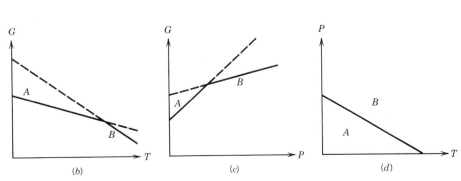

(b)   (c)   (d)

Where the two G surfaces intersect, the two minerals (or phases) are in equilibrium because the condition of $G_A = G_B$ is satisfied. Figures 9.3b and c illustrate two cross sections through the three-dimensional picture in a. Fig. 9.3b shows a cross-over of the entropy of B with respect to A (as a function of G and T at constant P). Figure 9.3c shows a cross-over in volume of B with respect to that of A (as a function G and P at constant T). Figure 9.3b is a cross section of the diagram in a at constant pressure; this is also known as an *isobaric* section. Figure 9.3c is a cross section at constant temperature, also known as an *isothermal* section. Figure 9.3d is the standard P-T diagram obtained from Fig. 9.3a by projection of the equilibrium curve (line of intersection) along which phases A and B coexist onto the basal plane of P-T. This shows that for any phases (minerals) to be at equilibrium, they must be at the same P and T.

Figure 9.1, for the system H_2O, illustrates the P-T regions in which the specifically labeled state (e.g., ice, or liquid, or vapor) has a lower free energy than other possible states. Along the curves (= phase boundaries) the adjoining states have equal free energies and are in equilibrium.

The last relationship that we will mention is the *Clapeyron equation*, which allows us to determine the P-T trajectory of equilibrium states (on a P-T diagram) as a function of entropy and volume changes. When ice and water are in equilibrium along a specific curve, as in Fig. 9.1, the changes in the Gibbs free energy along the equilibrium curve must be equal. This allows us to equate two Gibbs free energy expressions (that is, one for the reactant, ice, and one for the product, water). As such the one equation must equal the other, as follows: For the reactant (r):

$$dG_r = V_r dP - S_r dT$$

and for the product (p):

$$dG_p = V_p dP - S_p dT$$

At equilibrium:

$$V_r dP - S_r dT = V_p dP - S_p dT$$

Rearrangement results in:

$$(V_p - V_r)dP = (S_p - S_r)dT$$

which results in:

$$\frac{dP}{dT} = \frac{\Delta S}{\Delta V}$$

the *Clapeyron equation*. In this equation ΔS is the total entropy of the products minus the total entropy of the reactants. Similarly, ΔV is the total volume of the products minus the total volume of the reactants. The dP/dT function, as expressed on a P-T diagram, is a function of both changes in entropy and volume of the system.

Gibbs Phase Rule

The number of minerals that may coexist in equilibrium, be it in a mineral association (that is, a rock type) or in an experimentally produced reaction product, is not unlimited. This number is restricted by what is known as the *Gibbs phase rule*, which is commonly stated as

$$p + f = c + 2,$$

where p = the number of phases, f = the variance, or the number of degrees of freedom in a system, and c = the number of components.

The selection of the most appropriate component(s) to describe a specific chemical system was discussed above. Here it suffices to note that the Gibbs phase rule does not stipulate a particular set of components, but only the minimum necessary number.

The term f, degrees of freedom, is best discussed with respect to a specific phase diagram. For example, the phase diagram for the system H_2O (with H_2O the compound component) is given in Fig. 9.1. This diagram is subdivided into several stability regions for the various phases of H_2O. Along the curves that separate these regions, two phases are in equilibrium, at any pressure and temperature defined by a specific curve. The number of degrees of freedom (f) represents the minimum number of variables (in this case P and T) that must be fixed to define a particular condition of the system. At the triple point, t, in Fig. 9.1b three phases coexist (ice I, liquid water, and water vapor; note that there are numerous other triple points in Fig. 9.1a). Such a triple coexistence in Fig. 9.1b is possible only at the specific values of P and T noted on the axes of the figure. At this unique point specific P and T values define the point, and therefore this point is said to have zero degrees of freedom. The same answer may be derived by the use of the Gibbs phase rule:

$$p + f = c + 2$$

but solving for f,

$$f = c - p + 2$$

and with regard to the triple point in Fig. 9.1*b*, the number of components = 1, the number of phases = 3. Substituting these numbers into the above expression yields

$$f = 1 - 3 + 2 = 0$$

From this one can conclude that when three phases coexist at a triple point in a one-component system, $f = 0$, which is known as *invariant* (or an *invariant point*). Similar invariant points can be seen in the various one-component diagrams in Fig. 9.4.

Along any of the curves in Figs. 9.1 or 9.4, the coexistence of two phases (or minerals) along the curve does not uniquely define the *P* and *T* of the two-phase assemblage because there are a variety of *P* and *T* combinations at which the two phases coexist. Along any of the curves this can be done by either specifying the *P* or the *T* of the coexistence. For example, along the vapor pressure curve in Fig. 9.1*b*, it can be seen that if liquid and water coexist at 220 bars of pressure, the corresponding temperature is just below 374°C. This means that there is one independent and one dependent variable. Such a condition can be derived from the Gibbs phase rule by substituting $c = 1$ (H_2O is one component), $p = 2$ (two phases coexisting along the vapor pressure curve), which leads to

$$f = 1 - 2 + 2 = 1$$

known as *univariance*. Equilibrium curves, in one-component systems, such as in Figs. 9.1 and 9.4, are known as *univariant curves*, possessing one degree of freedom.

Within a phase region in a one-component system, one specified parameter does not lead to a unique answer for the other parameter (as in *P-T* diagrams; see Figs. 9.1 and 9.4). Within a phase region, a certain phase can exist over a range of *P* and *T* conditions. Using the Gibbs phase rule with respect to H_2O, $c = 1$, $p = 1$, leads to

$$f = 1 - 1 + 2 = 2$$

This result means that there are two degrees of freedom in a phase region. This is also known as a *divariant region*.

It is obvious that the minimum value for *f*, the degrees of freedom, = 0. When this is entered into the Gibbs phase rule,

$$f = c - p + 2$$

it results in $0 = c - p + 2$, or $c + 2 = p$. This means that there should not be more than $c + 2$ phases (or minerals) in a rock if equilibrium exists. When a very large number of minerals is found in a specific rock sample, one should suspect that the mineral assem-

blage may not have formed under equilibrium conditions.

EXAMPLES OF MINERAL STABILITY (PHASE) DIAGRAMS

In the following discussion and illustrations of some representative mineral stability diagrams, we will concentrate on those that represent conditions pertaining to the *solid state*. This decision is made because this is a mineralogy text in which almost all of our concerns center on solids. A few diagrams, used in igneous petrology, may involve a high-temperature liquid, and some, used in low-temperature processes (such as in chemical sedimentation in the ocean), may involve a low-temperature fluid phase.

One-Component Diagrams

Figure 9.4 shows four stability diagrams for various common minerals. The diagrams are defined by pressure and temperature axes (*P-T* diagrams) identical to those used in Fig. 9.1. Recalling our earlier discussion of *polymorphism* (page 153), it becomes clear that all four diagrams depict the stability fields of various polymorphic forms, in a specific chemical system, as a function of *P* and *T*. Figure 9.4*a* shows the stability fields of diamond, graphite, carbon III, and liquid for a system that is composed only of carbon (C). Diamond and graphite are the two common polymorphs of carbon. Diamond has a very large stability field in the high-pressure region of the diagram. The diamond field also extends to high temperature at certain pressures (up to about 4000 Kelvin at about 150 kilobars). Graphite is stable over a wide temperature range but only at relatively low pressures. Diamond has a much more densely packed structure than graphite, with a molecular volume that is about 36 percent smaller than that of graphite. As such, diamond would be expected to be the high-pressure phase, as shown in the stability diagram. The polymorphic reaction in transforming diamond to graphite, and vice versa, is reconstructive. In such reconstructions, chemical bonds must be broken, and the structure must be reassembled. This requires large amounts of energy (activation energy), and because of this, diamond is very stable under ordinary pressures and temperatures (atmospheric conditions) even though thermodynamically the diagram shows that graphite is the stable phase under atmospheric conditions. The conversion rate of diamond to graphite, at atmospheric conditions, must be infinitely slow, and cannot be detected.

Figure 9.4*b* is a stability diagram for the three polymorphic forms of Al_2SiO_5: andalusite, sillimanite,

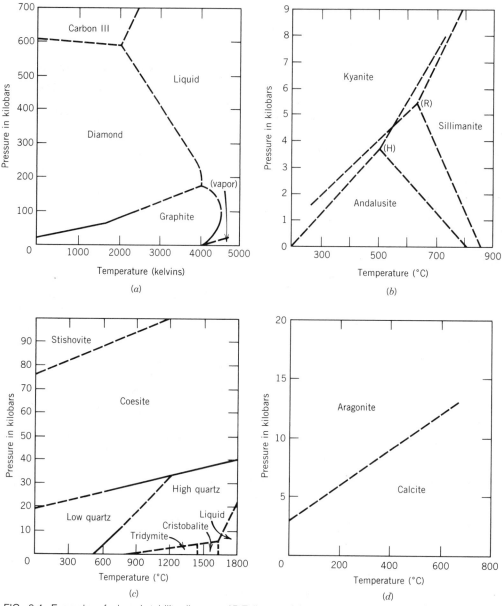

FIG. 9.4. Examples of mineral stability diagrams (*P-T* diagrams) for one-component systems. Each of these has been determined by experiments. When curves, or lines, are solid their locations have been determined with certainty; when they are dashed, their location is less well known. (*a*) The system C based on experimental data from various sources. (*b*) The system Al₂SiO₅ as based on results from two different experimental studies (R = Richardson et al., 1969; H = Holdaway, 1971; complete references are given at the end of this chapter). (*c*) The system SiO₂ from various sources. (*d*) The system CaCO₃. A kilobar = 1000 bars; 1 bar = 0.987 atmosphere.

and kyanite. This diagram is regarded as a one-component system because Al_2SiO_5 is the compound but single component that represents the composition of all three polymorphs (see "Components," page 311). Because kyanite occupies the high-pressure region of the *P-T* diagram, one would conclude kyanite to be the densest polymorph of Al_2SiO_5. This is so, as shown by the densities and structural studies of the three Al_2SiO_5 polymorphs. The three structures may

be represented as follows:

$$[Al]^6[Al]^6SiO_5—kyanite,$$
$$[Al]^5[Al]^6SiO_5—andalusite,$$
$$[Al]^4[Al]^6SiO_5—sillimanite,$$

where the square brackets with superscripts denote the coordination polyhedron about Al ($[Al]^6$ = octahedral; $[Al]^5$ = irregular 5-fold; $[Al]^4$ = tetrahedral).

It is clear from this notation that kyanite represents the closest atomic packing of the three polymorphs.

The diagram in Fig. 9.4b shows two independent sets of equilibrium boundaries that radiate from a triple point. These are marked (R) and (H) and represent two sets of independent experimental results on this system. Both results are shown because they illustrate the experimental difficulties in determining the exact equilibrium boundaries in a highly refractory system (refractory meaning that, in this chemical system, the transformation reactions are difficult to delineate because of the sluggishness of the reactions, causing metastability of a specific polymorph into another polymorphic phase region). Recent work by Bohlen et al. (1991) has largely confirmed the triple point location marked (H); this configuration will therefore be used in Chapters 13 and 14 as well.

In the case of the experimental study of the stability fields of the three polymorphs of Al_2SiO_5, very high purity minerals may be used for the experiments although most commonly the starting materials are high purity chemicals. For example, gem grade and inclusion-free kyanite may be found at increasing T to react to form sillimanite (see Fig. 9.4b). The position of the curve separating the kyanite and sillimanite fields is based on the first evidence of sillimanite forming at the expense of kyanite with increasing T and on evidence of the reverse reaction, namely sillimanite giving way to kyanite at decreasing T or increasing P. The determination of the beginning of such reactions is generally based on a combination of X-ray powder diffraction and optical microscopic techniques. Because of uncertainties in various aspects of the experimental techniques in locating specific reaction boundaries, the reaction curves, defined in P and T values, may represent relatively broad reaction zones, although they are commonly shown as narrow lines or curves.

Kyanite, or sillimanite, or andalusite, without any textural indication of reaction to another Al_2SiO_5 polymorph, can be found in metamorphic rocks of all geological ages, including early Precambrian. This observation signifies that the activation energy necessary to transform a high-temperature or high-pressure polymorph into the lower P-T polymorph stable at atmospheric conditions has not been provided. However, in many other metamorphic rocks that contain Al_2SiO_5 polymorphs, the following rimming reactions have been observed:

a center of kyanite rimmed by sillimanite;

a center of sillimanite rimmed by kyanite;

a center of sillimanite rimmed by andalusite;

a center of andalusite rimmed by kyanite.

All of the above textural occurrences delineate various P-T paths across the different equilibrium boundaries in Fig. 9.4b.

Figure 9.4c outlines the stability fields of phases in the system SiO_2. This diagram was also discussed in Chapter 3, under "Polymorphism" (Fig. 3.38). All the equilibrium boundaries between solids (except for the line separating high and low quartz) are boundaries of reconstructive polymorphic transformations. The low to high quartz boundary is a displacive transformation boundary. Quartz, as low quartz, is the SiO_2 phase in plutonic, metamorphic, and sedimentary rocks, reflecting their general temperature of origin below 1000°C. Tridymite and cristobalite are found in volcanic assemblages of many geological ages. This means that these minerals exist metastably for long geological time periods. In other volcanic occurrences, the original cristobalite or tridymite may have been converted to low quartz, but still preserving the crystal form of cristobalite or tridymite (these occurrences, therefore, represent pseudomorphs of low quartz after higher-temperature SiO_2 polymorphs). Such pseudomorphic occurrences suggest that the activation energy necessary for the reconstructive transformations (from cristobalite or tridymite to low quartz) was possibly provided by some reheating due to later metamorphism of the original volcanics. The high-pressure polymorphs of SiO_2, coesite and stishovite, have been found in meteorite impact craters. Coesite also occurs as inclusions in diamonds, and inside pyrope garnets in very pyrope-rich high-grade metamorphic rocks from Parigi, Northern Italy (Chopin, 1984; for complete reference see reference list at end of this chapter). Coesite has also been found in a xenolith nodule (a xenolith is a foreign inclusion) in a kimberlite pipe (see Fig. 4.52). This kimberlite pipe originated at a depth of about 170 to 200 km in the mantle, at pressures of approximately 60 to 70 kilobars (a pressure region in Figs. 9.4a and c where both diamond and coesite are stable). Coesite-diamond coexistences have also been reported from very high-pressure metamorphism of crustal rocks in eastern China (Shutong et al., 1992; for complete reference see reference list at end of this chapter). The coesite-diamond-jadeite (another high-pressure mineral) assemblage is considered the result of burial of the crust to great depths, metamorphism, and subsequent exhumation back toward the Earth's surface.

Figure 9.4d shows the stability diagram for $CaCO_3$, with two polymorphic forms, calcite and aragonite. By far the most common carbonate, be it sedimentary, metamorphic, or igneous in origin, is calcite. The phase diagram in Fig. 9.4c suggests that $CaCO_3$ that is formed at normal temperatures and pressures (at essentially atmospheric conditions) should be calcite. However, this is not what is observed. Primary precipitation of $CaCO_3$ from seawater is commonly in the form of aragonite, giving rise to aragonite muds. Many organisms, furthermore, build their shells of aragonite, and the main constituent of pearls is aragonite. These observations imply that aragonite is formed and precipitated as a metastable phase that, over time, will convert to calcite. This polymorphic transformation can be very slow, as shown by the preservation of aragonite in Pennsylvanian reefs, which are about 300 million years old. However, the aragonite to calcite polymorphic reaction is more easily accomplished than the transformations mentioned above for diamond-graphite or coesite-quartz. If energy is applied in the form of increased temperature, as a result of burial, aragonite will transform to calcite more quickly.

Two-Component Diagrams

Two-component stability diagrams are most commonly constructed with a horizontal composition bar and a vertical temperature axis. Such *temperature-composition (T-X) diagrams* may exhibit very different features as a result of: (1) complete solid solution between two end members, (2) partial solid solution between two end members as expressed by the presence of a miscibility gap, and (3) no solid solution between various mineral species that can be represented along the two-component composition bar.

Let us begin with some examples of two-component *T-X* diagrams in which there is complete solid solution over the whole temperature range of the diagram. This is shown by the olivine series, between Mg_2SiO_4 (forsterite) and Fe_2SiO_4 (fayalite), and by the high-temperature region in the plagioclase feldspar series ($NaAlSi_3O_8$ to $CaAl_2Si_2O_8$). Both diagrams, depicted in Figs. 9.5a and b, are the result of experimental studies at atmospheric pressure. That is, the pressure conditions of the experiments are fixed at 1 atmosphere (= 1.01325 bars). These two diagrams are also known as *liquidus diagrams* because they involve a liquid phase, or melt. Such liquidus diagrams are very instructive in evaluating the melting relations of igneous rock compositions, as well as for the study of a crystallization sequence from a melt. The upper curve in Fig. 9.5a is known as the *liquidus*, a line or surface along which compositions of melt are in equilibrium with a crystalline phase. The lower curve is known as *solidus*, a line or surface along which compositions of a crystalline phase are in equilibrium with the melt. Everything above the liquidus is liquid; everything below the solidus is solid. With the aid of

the schematic diagram in Fig. 9.5c the crystallization behavior of solids from a melt in such diagrams can be illustrated. This diagram shows that the pure end member A melts at T_A and pure end member B melts at T_B, but intermediate compositions (AB), which consist of a single phase (part of the solid solution series), melt through a range of temperatures intermediate between T_A and T_B. A melt of composition M at temperature T_B will be entirely melt. When it cools to T_1 it will start to crystallize out a member of the solid solution series AB with specific composition xA,yB. These crystals are enriched in the B component with respect to the melt composition M, and their growth will deplete the melt in component B. The melt composition will, as a result of this depletion, move along the liquidus curve toward A as indicated by the upper arrow. As a result of the continual lowering of the temperature, the solid phase of original composition xA,yB will react with the melt in the direction of the lower arrow along the solidus. As such, both the melt and crystalline products will increase in content of A with decreasing T and the ratio of solid to melt will increase. Finally, at T_2 the crystallized products have a composition which is that of the original melt M, and the amount of melt in equilibrium with the crystals will reach zero. Now, because only a solid phase remains (i.e., we are in the subsolidus region), with continued lowering of the temperature the composition of the crystalline product will remain constant at the bulk composition M of the original melt. The above observations apply to both diagrams, Figs. 9.5a and b.

Although the original equilibrium melting and crystallization experiments necessary to construct Fig. 9.5b indicated a complete solid solution series for plagioclase at high temperature, subsequent detailed single-crystal X-ray diffraction and transmission electron microscope studies have shown this to be untrue at lower temperatures. Figure 9.6a shows three broadly outlined miscibility gaps (see also page 236) in the lower temperature region of the plagioclase series. These regions have been named, in increasing An content, the *peristerite gap*, the *Bøggild intergrowth* region, and the *Huttenlocker intergrowth* region. These regions represent exsolution phenomena on an extremely small scale. Exsolution lamellae are responsible for the *iridescence* (see page 268) of moonstone, as a result of the presence of closely spaced lamellae of An_0 and An_{25} composition (peristerite gap). Similarly, the *schiller* or *labradorescence* of labradorite is caused by very fine exsolution lamellae of An_{47} to An_{58} composition (see Fig. 6.23) across the gap described as Bøggild intergrowth. A temperature-composition diagram that incorporates the data in

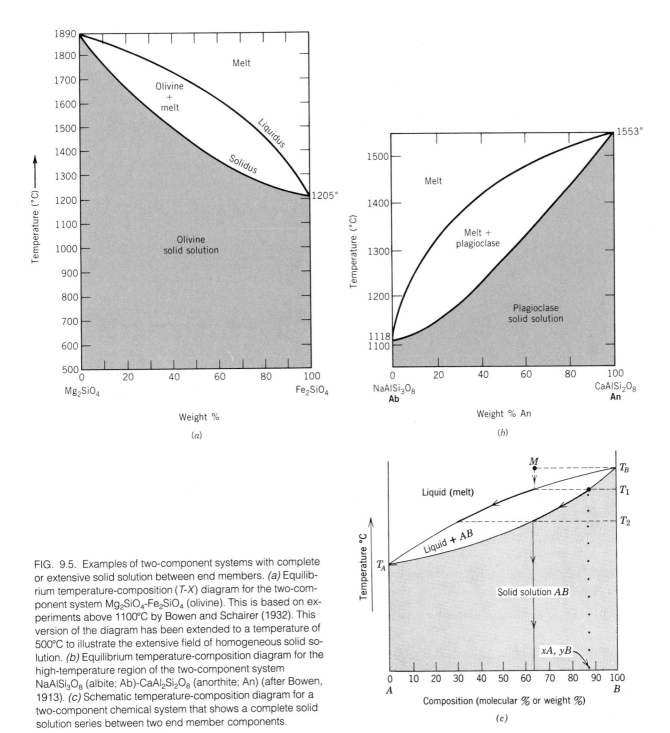

FIG. 9.5. Examples of two-component systems with complete or extensive solid solution between end members. *(a)* Equilibrium temperature-composition (*T-X*) diagram for the two-component system Mg_2SiO_4-Fe_2SiO_4 (olivine). This is based on experiments above 1100°C by Bowen and Schairer (1932). This version of the diagram has been extended to a temperature of 500°C to illustrate the extensive field of homogeneous solid solution. *(b)* Equilibrium temperature-composition diagram for the high-temperature region of the two-component system $NaAlSi_3O_8$ (albite; Ab)-$CaAl_2Si_2O_8$ (anorthite; An) (after Bowen, 1913). *(c)* Schematic temperature-composition diagram for a two-component chemical system that shows a complete solid solution series between two end member components.

Figs. 9.5*b* and 9.6*a* for the plagioclase series is given in Fig. 9.6*b*. This figure represents the most up-to-date interpretation of data showing three miscibility gaps at temperatures below about 800°C and an almost complete solid solution series above about 800°C, up to the solidus of the diagram.

The two-component system for the alkali feldspars series ($NaAlSi_3O_8$-$KAlSi_3O_8$) is an example of only very limited solid solution at low temperature with an almost complete range of solid solution at high temperature. This is shown in a *T-X* diagram in Fig. 9.7*a*. The lack of solid solution at low temperatures is mainly the result of the large difference in ionic sizes of Na^+ (1.18 Å in 8-coordination) and K^+ (1.51 Å in 8-coordination; see Table 4.8). Lamellar exsolution textures (see page 237) are common in

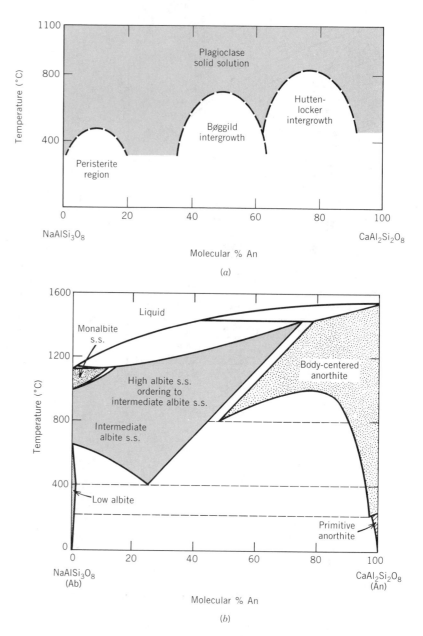

FIG. 9.6. (a) Approximate location of three miscibility gaps in the low-temperature region of the plagioclase feldspar series. (b) Schematic phase diagram for the plagioclase feldspar series showing a range of almost complete subsolidus solid solution at high temperatures, and the various miscibility regions at lower temperatures. (Simplified with permission after Smith, J. V. and Brown, W. L., 1988, *Feldspar Minerals*, v. 1, *Crystal Structures, Physical, Chemical, and Microtextural Properties.* Springer-Verlag, New York, Fig. 1.4, 828 pp.)

this system, i.e., perthite. The mineral names listed inside the diagram are the names for polymorphs of $NaAlSi_3O_8$ (monalbite = monoclinic albite; high, intermediate, and low albite) and of $KAlSi_3O_8$ (sanidine, orthoclase, and microcline).

It is of considerable geological interest to see what happens to the configuration of the diagram in Fig. 9.7a when H_2O (at high water pressure = P_{H_2O}) is added to the chemical system as in Fig. 9.7b. A high aqueous-fluid pressure considerably lowers the melting temperatures, while increased pressure applied to the system raises the maximum temperature of the miscibility gap such that now the solidus and the mis-

cibility gap intersect. Point *e*, the *eutectic point* (see further discussion below), is the lowest temperature point on the liquidus at which a unique melt of fixed composition is in equilibrium with two feldspar compositions (that is, a Na-rich feldspar solid solution and a K-rich feldspar solid solution). The crystallization of a melt composition at *e*, therefore, allows for the direct crystallization from the liquid of coexisting albite and K-spar, without the need for an exsolution process (in the solid state) as is required in the anhydrous system of Fig. 9.7a. Coarse-grained coexistences of albite and K-spar without exsolution textures are common in granites and pegmatites; such occurrences reflect

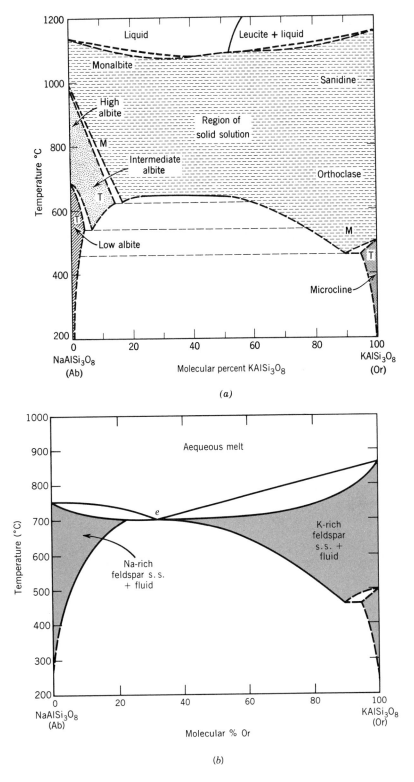

(a)

(b)

FIG. 9.7. (*a*) Schematic phase diagram for the system NaAlSi$_3$O$_8$ (Ab)-KAlSi$_3$O$_8$ (K-spar) showing a large miscibility gap at temperatures below approximately 650°C. *M* and *T* mean monoclinic and triclinic, respectively. (Modified with permission after Smith, J. V. and Brown, W. L., 1988, *Feldspar Minerals*, v. 1, *Crystal Structures, Physical, Chemical, and Microtextural Properties*. Springer-Verlag, New York, Fig. 1.2, 828 pp.) (*b*) Approximate and schematic phase diagram for the same chemical system, except for addition of H$_2$O to the melt, at 5 kilobars of P_{H_2O}. Now the solidus and miscibility gap intersect at point *e*, the eutectic point; see text for discussion. S.S. = solid solution. (Modified after Morse, S. A., 1970, Alkali feldspars with water at 5 kb pressure. *Journal of Petrology*, v. 11, pp. 221–251.)

crystallization of both minerals from a fluid. Perthitic intergrowths, however, represent exsolution in the solid state.

A very different temperature-composition diagram results when there is *no* solid solution between various minerals in a two-component system. Figure 9.8*a* is a two-component phase diagram with a *eutectic* relationship. Phases *A* and *B* are pure substances and as such there is no solid solution between them. For composition *A* the melting temperature is T_A; sim-

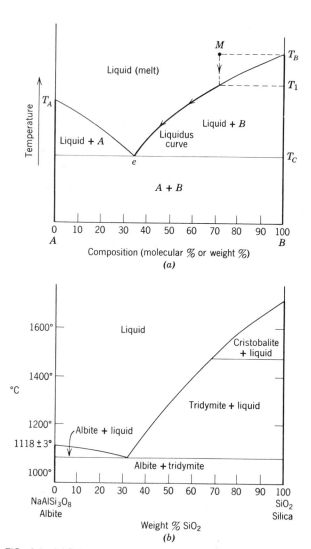

FIG. 9.8. *(a)* Schematic temperature-composition section showing eutectic crystallization of components *A* and *B*; both components are pure substances (no solid solution between them). *(b)* The system NaAlSi₃O₈(albite)-SiO₂. (After Schairer, J. F. and Bowen, N. L., 1956, *American Journal of Science*, v. 254, p. 161.)

FIG. 9.9. *(a)* Schematic temperature-composition section showing peritectic as well as eutectic crystallization involving three pure phases, *A, B,* and *C*. *(b)* The system Mg₂SiO₄ (forsterite)-SiO₂. (After Bowen, N. L. and Andersen, O., modified by Greig, J. W., 1964, *Phase Diagrams for Ceramists*. American Ceramic Society, Columbus, Ohio, p. 112.)

ilarly for composition *B* the melting temperature is T_B. The addition of some of composition *B* to a melt of *A* lowers the temperature of the liquid that can coexist with *A*, along the curve between T_A and *e* (*eutectic point*, which is the minimum temperature point of the liquid field). Similarly the melting temperature of the liquid that can coexist with *B* is lowered by the addition of some of *A*, as shown by the curve between T_B and *e*. The lowest temperature at which crystals and melt are in equilibrium is T_C, the temperature of the eutectic. Let us now look at the crystallization sequence of a melt of composition *M*. Upon lowering of the temperature of the melt at *M* (original temperature = T_B) the melt will start to crystallize some of

the pure substance *B* at T_1. The crystallization of *B* from the melt will continue along the *liquidus curve* (T_B to *e*), continually increasing the content of substance *A* in the melt. At the eutectic point (*e*) phase *A* will join *B* as a crystallization product. At this point there is no further change in composition of the melt because both A and B crystallize from the melt in the same proportions as are present in the melt. With continued crystallization the melt will disappear and the final crystalline products will be B and A in the proportions represented by the original bulk composition M. Fig. 9.8*b* is an illustration of a eutectic melting relationship in the high-temperature part of the system $NaAlSi_3O_8$-SiO_2.

Yet another type of *T-X* section through a liquidus diagram is shown in Fig. 9.9*a*. This diagram illustrates a *peritectic* as well as *eutectic* phenomenon in the crystallization sequence. A *peritectic point* (or *reaction point*) is an inflection point on the liquidus surface at which a unique melt (of specified composition) is in equilibrium with two crystalline phases. The diagram, in addition to phases A and B (which lack solid solution between them), shows the presence of a third phase of constant composition, C. Phases of composition A and B melt to liquid of their own compositions at T_A and T_B, respectively; this is known as *congruent melting*. Phase C, however, melts *incongruently* (which means that it reacts and decomposes to form another solid of another composition plus liquid), in this case to B and a melt of composition M_C at temperature T_C. If we look at the crystallization history of a melt of composition M (originally at temperature T_B) it can be cooled until it reaches the liquidus curve at T_1, at which point phase B begins to crystallize. With continued loss of heat the melt composition will move toward A along the liquidus curve with continued precipitation of phase B until point p (*peritectic point*) is reached. At the peritectic point the reaction: melt + $B \rightarrow C$ will take place, and will continue until one of the reactants (melt or B) is used up. Which of these is used up depends on the position of M with respect to the composition of the incongruently melting phase (C in this case). It is simple to see which must be used up first by projecting the composition of M down onto the composition bar, and seeing whether the bulk composition (M) lies between A and C, or C and B. With a continued decrease in temperature in the melt, more C will continue to crystallize and the temperature of the product phase C will fall until it finally reaches point e (eutectic). At this point C is joined by crystallization products of composition A. All of this discussion assumes equilibrium crystallization. Figure 9.9*b* illustrates these types of crystallization phenomena in the system Mg_2SiO_4-SiO_2. In mafic igneous rocks it is common to see early olivine crystals rimmed by later orthopyroxene as a result of the peritectic crystallization relationship between these two mineral groups. Such rimming of an earlier mineral by a later one is indicative of disequilibrium crystallization.

Another example of a temperature-composition diagram in a two-component system with a peritectic as well eutectic point is given in Fig. 9.10. This is for the system $KAlSi_2O_6$ (leucite)-SiO_2 (cristobalite or tridymite) with an intermediate composition, $KAlSi_3O_8$ (K-feldspar, which is sanidine under the high-temper-

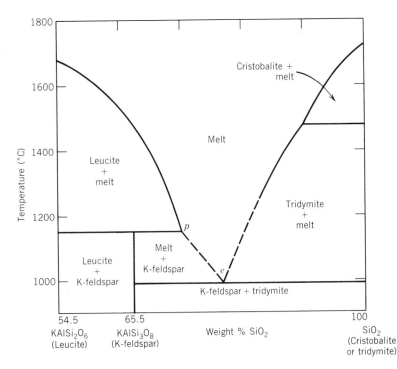

FIG. 9.10. Temperature-composition diagram for the system $KAlSi_2O_6$-SiO_2. The polymorph of $KAlSi_3O_8$ that is stable under these conditions is sanidine. *p* and *e* are peritectic and eutectic points, respectively. (From Schairer, J. F. and Bowen, N. L., 1947, The system anorthite-leucite-silica. *Commission Géologique de Finlande Bull.*, no. 140, pp. 67–87.)

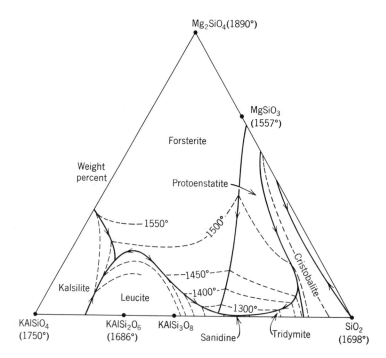

FIG. 9.11. Melting relations among minerals in the system SiO$_2$-KAlSiO$_4$-Mg$_2$SiO$_4$. Temperatures next to compositions refer to the melting temperatures of those compositions in °C. Protoenstatite instead of enstatite is produced in experiments of this type. The compositions of protoenstatite and enstatite are identical, but their structures differ in detail. (After Luth, W. C., 1967, *Journal of Petrology*, p. 373, reprinted by permission of Oxford University Press.)

ature conditions). There is no solid solution whatsoever between these three minerals. In this diagram, K-feldspar bears a reaction relation to the melt just as enstatite (and clinoenstatite) does in the system forsterite-SiO$_2$ (Fig. 9.9). From Figs. 9.9 and 9.10 it is clear that possible equilibrium coexistences are: forsterite-enstatite (or clinoenstatite); clinoenstatite-cristobalite; leucite-K-spar; and K-spar-tridymite. Nonequilibrium pairs are: forsterite-cristobalite, and leucite-tridymite.

Three- or More-Component Diagrams

Because most rock types (igneous, metamorphic, and sedimentary) consist on the order of six to ten, or more, chemical components, one-component and two-component stability diagrams generally have only limited applicability to more complex naturally occurring mineral assemblages. However, easy graphic representation of multicomponent chemical systems is limited mainly to three-component (triangular) diagrams. For the methods of plotting mineral formulas on triangular diagrams see page 245. One may also consider four-component systems with chemical components at each of the four corners of a tetrahedron, but such tetrahedral representations become graphically complex (see Figs. 14.3, 14.4, and 14.18a).

Triangular phase diagrams are commonly used in igneous petrology to represent the experimentally studied melting relations of igneous rock composi-

tions, as well as the crystallization sequence from a melt. An example of such a diagram is given in Fig. 9.11 in terms of three compound components SiO$_2$, KAlSiO$_4$, and Mg$_2$SiO$_4$. The contours in this diagram represent melting temperatures and are known as *isotherms*; the surface defined by these isotherms is the *liquidus surface*. The arrows along the boundaries of the various phase fields indicate crystallization paths, with decreasing temperature. The previous figure, 9.10, is a temperature-composition section along the lower edge of the triangle in Fig. 9.11, from leucite to SiO$_2$.

In the study of the mineralogy of assemblages and rock types, triangular diagrams that depict mineral stabilities below the liquidus and solidus surfaces—that is, *subsolidus* triangular phase diagrams—are most instructive. A schematic, composite liquidus-subsolidus diagram is shown in Fig. 9.12a for the triangular composition space of feldspar. This diagram is a composite of the more detailed information contained in several earlier *T-X* diagrams that relate to alkali feldspars and plagioclase (see Figs. 9.6 and 9.7). The bottom triangle is based on three major feldspar end members: KAlSi$_3$O$_8$-NaAlSi$_3$O$_8$-CaAl$_2$Si$_2$O$_8$. This compositional triangle alone is commonly used to depict the extent of solid solution in the feldspars. Figure 9.12b shows the experimentally determined extent of solid solution between end member compositions as a function of the temperature of the experiments at P_{H_2O} = 1 kilobar. This diagram illustrates that at the highest temperature (900°C) the feldspars show the most exten-

(a)

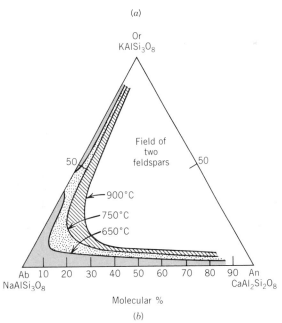

FIG. 9.12. (a) Highly schematic temperature-composition diagram for the three-component system $KAlSi_3O_8$ (K-feldspar)-$NaAlSi_3O_8$ (albite)-$CaAl_2Si_2O_8$ (anorthite), at water pressure of about 5 kilobars. Details of the interior are complex and have been omitted. The upper, contoured surface of the diagram is the liquidus surface. The three intergrowth regions, as a result of miscibility gaps, in the lower temperature range of the plagioclase series are: P = peristerite, B = Bøggild intergrowth, H = Huttenlocher intergrowth. Solid solution is shown by shading. Compare the two vertical sides of this diagram with more detailed information given in Figs. 9.6a and b, and 9.7a and b. (Adapted from Ribbe, P. H., 1987, Feldspars, in *McGraw-Hill Encyclopedia of Science and Technology*, 6th ed., v. 7, p. 45; reproduced with permission of McGraw-Hill.) (b) Experimentally determined extent of solid solution plotted on the triangular base of the diagram in (a), at P_{H2O} = 1 kilobar. (After Ribbe, P. H., 1975, The chemistry, structure, and nomenclature of feldspar. *Reviews of Mineralogy, Feldspar Mineralogy*, v. 2, pp. R1–R72, fig. R1.)

sive solid solution field. At the lowest temperature studied, 650°C, there is the least extent of solid solution. Indeed, along the join $KAlSi_3O_8$-$NaAlSi_3O_8$ there is a considerable break in the central part of the solid solution series at 650°C. This reflects the inter-

section of the miscibility gap in the system at about that temperature (see Fig. 9.7a, which is the *T-X* section for $KAlSi_3O_3$-$NaAlSi_3O_8$, without H_2O present in the system). The plagioclase join in Fig. 9.12b suggests that solid solution at 650°C is continuous between NaAlSi$_3$O$_8$-CaAl$_2$Si$_2$O$_8$. It is known (see Fig. 9.6) that three miscibility gaps exist in this series, below about 800°C. These gaps are not reflected in the experimental results of Fig. 9.12b, because the chemical analytical techniques (used to determine compositions of experimental products) cannot resolve the extremely fine-scale nature of the exsolution lamellae across the three gaps. In order to ascertain that the plagioclase grains obtained experimentally are homogeneous (or nonhomogeneous, that is, consisting of lamellar intergrowth), single-crystal X-ray or

transmission electron microscope techniques must be used, in addition to the commonly used technique of electron probe microanalysis (see Chapter 5).

Another very commonly used triangular diagram is one that represents the main compositional variation among members of the pyroxene group. This is shown in Fig. 9.13a as a function of three components, CaSiO$_3$ (wollastonite, a pyroxenoid), MgSiO$_3$ (enstatite), and FeSiO$_3$ (ferrosilite). The compositional extent between end members of various pyroxene series, in common igneous and metamorphic rocks, is

FIG. 9.13 (a) The extent of pyroxene solid solution in the system CaSiO$_3$-MgSiO$_3$-FeSiO$_3$. Representative tielines across the miscibility gap between augite and the more Mg-Fe-rich pyroxenes are shown. AA' locates the section across this diagram shown below. (b) A temperature-composition section across the diagram in (a) at about En$_{65}$Fs$_{35}$ (section position noted as AA'). Pigeonite is stable only at high-temperature conditions.

(a)

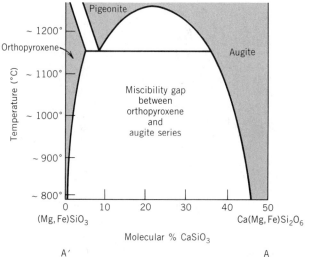

(b)

shown by shading. This diagram is the result of the compilation and graphic plotting of thousands of pyroxene analyses from the published literature. Such a compilation of analyses from natural occurrences (instead of carefully controlled high-temperature experiments on specific compositions, as in Fig. 9.12b) does not distinguish compositional regions that specifically represent high-temperature pyroxenes (as in basalt flows) from lower-temperature pyroxenes (as in gabbros). Such a diagram is useful only as a graphic representation of the average extent of solid solution of common pyroxenes. The open space between the augite region and that of pigeonite/orthopyroxene is a miscibility gap. The presence of this gap can be seen in a *T-X* section along A-A. This is shown in Fig. 9.13b. This subsolidus diagram (note that the solidus and/or liquidus surfaces are not even shown) shows the slight expansion of solid solution at elevated temperatures in orthopyroxene, and the much enlarged solid solution field of augite at high temperatures. It also demonstrates that pigeonite is specifically a high-temperature phase region that must invert at about 1150°C to an orthopyroxene matrix with augite exsolution lamellae (due to presence of the miscibility gap and the fact that the structure of pigeonite is different from that of orthopyroxene). Pigeonite occurs only in quickly quenched high-temperature basalts. If the basalt has cooled slowly the original pigeonite is now represented by orthopyroxene with augite exsolution lamellae.

Triangular composition diagrams similar to Fig. 9.13a are commonly used in the graphic representation of other mineral groups, such as members of the olivine (see Fig. 13.8) and amphibole groups (Fig. 13.65). Three-component triangular diagrams may also be used to illustrate the common coexistence of possible mineral pairs, or of groups of three minerals. Such diagrams are known as *assemblage diagrams*.

In the construction of such a diagram, one of the first steps is the listing of the minerals that compose a specific rock under study. A simple igneous granite may consist of orthoclase, albite, quartz, and biotite. The texture of this specific granite may indicate that all four minerals formed as crystallization products at about the same elevated temperature. A petrologist, therefore, might conclude that orthoclase-albite-quartz-biotite is the *mineral assemblage* (also referred to as the *mineral paragenesis*) of this granite. Although the term *assemblage* (or *paragenesis*) is often loosely used to include all the minerals that compose a rock, it should be restricted to those minerals in a rock that appear to have been *in equilibrium* (*the equilibrium mineral assemblage*). The above granite may be

found to contain vugs in outcrop which are lined by several clay minerals, bauxite as well as some limonite. Clearly the minerals in the vugs represent relatively low-temperature alteration products of the original much higher-temperature granite. The high-temperature granitic assemblage consists of orthoclase-albite-quartz-biotite, and a separate, later formed, low-temperature assemblage consists of clays-bauxite-limonite. In short, an assemblage consists of minerals that formed under the same, or very similar, conditions of pressure and temperature. In practice all minerals that coexist (physically touch each other) and show no alteration or rimming relations are commonly considered to constitute the assemblage. In a distinctly banded rock the mineral assemblages in the various bands are often very different from each other because the bands may represent major differences in bulk chemistries.

The evaluation of whether all minerals in a rock are in *equilibrium* with each other is generally not straightforward. If the texture does not show reaction rims and alterations, it is possible that the minerals are in equilibrium; nevertheless, further detailed chemical tests are often needed to define the equilibrium assemblage without ambiguity. Similarly, the evaluation of equilibrium in an experimental study is also not straightforward. However, in spite of the problems inherent in the assessment of an equilibrium assemblage, triangular mineral assemblage diagrams are commonly used to depict the observed pairs, and groupings of three possible minerals in a specific chemical system. Figure 9.14a shows commonly observed coexistences (as deduced from natural assemblage occurrences as well as synthetic studies) of minerals in the system SiO_2 (quartz, tridymite, cristobalite)-$NaAlSiO_4$(nepheline)-$KAlSiO_4$(kalsilite). This chemical system also includes the alkali feldspar series and leucite ($KAlSi_2O_6$). The diagram shows that there is complete solid solution in the alkali feldspar series and extensive solid solution in nepheline, kalsilite, and leucite. The diagram was determined experimentally and is drawn for high-temperature conditions of about 1000°C. The SiO_2 phase, as would be expected at this temperature, is tridymite. The assemblages depicted, therefore, are analogous to those found in high-temperature volcanic occurrences. Tielines connect mineral compositions that represent equilibrium coexistences. Triangles consist of three sets of tielines (outlining the triangle) depicting three-phase equilibrium coexistences. The diagram illustrates that feldspathoids do not coexist with tridymite because of the intervening alkali feldspar series. This specific triangular diagram is of importance in igneous petrology because it outlines mineral compositions

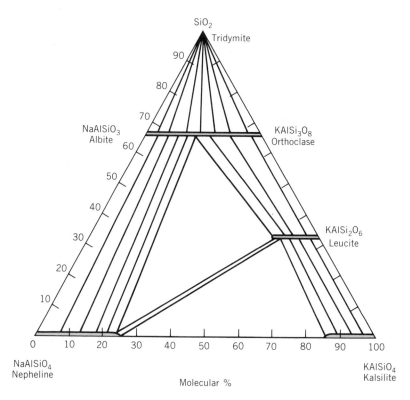

FIG. 9.14. The system SiO_2-$NaAlSiO_4$ (nepheline)-$KAlSiO_4$ (kalsilite) at approximately 1000°C and atmospheric pressure, as determined experimentally. The extent of solid solution for several phase regions is shown, as are tielines between coexisting pairs of minerals. Triangles represent three-phase coexistences. (Adapted from *Petrologic Phase Equilibria*, 21st ed., Fig. 4.39, by W. G. Ernst. Copyright © 1976 by W. H. Freeman and Company.)

in SiO_2-rich and SiO_2-poor rocks. All assemblages above the albite-orthoclase series contain an SiO_2 phase, whereas all assemblages below the alkali feldspar series are free of SiO_2 minerals (see Chapter 14 and Fig. 14.4 for further discussion).

Another illustration of an assemblage diagram for a carbonate system with very common carbonate minerals is shown in Fig. 9.15a. This diagram is based upon naturally occurring compositional ranges, and coexistences of metamorphic carbonates in a temperature range of about 400°C (the biotite zone of the greenschist facies). It shows a very small solid solution region for calcite, $CaCO_3$, an extensive (but not complete) series between dolomite, $CaMg(CO_3)_2$, and ankerite, $CaFe(CO_3)_2$, and a complete series between magnesite, $MgCO_3$, and siderite, $FeCO_3$. Common tielines between coexisting pairs are shown, as well as one three-phase coexistence. The tielines cross miscibility gaps, as illustrated in Fig. 9.15b in a *T-X* section along the magnesite-calcite edge of Fig. 9.15a.

Triangular assemblage diagrams are most commonly determined from experimental results, or from carefully studied and chemically analyzed natural assemblages. In order to delineate the extent of solid solution of any of the minerals in such diagrams, one of several chemical analytical techniques must be used (see Chapter 5). However, for mineral assemblages that represent a relatively simple chemical system, and where the minerals involved are relatively

easily identifiable, mineral assemblage diagrams can be constructed on the basis of hand specimen study of a large suite of assemblages that relate to the system in question. In hand specimen study no knowledge can be gained about solid solution; therefore, the mineral formula for each mineral is plotted as a point on the triangular diagram, without reference to any possible solid solution extent. Coexistences (and as such tielines) are based upon the visual observation of two or three minerals touching each other in the hand specimen. An example of such an assemblage diagram in the system Cu-Fe-S is shown in Fig. 9.16. Here all minerals are represented as points on the diagram, reflecting the fact that any possible solid solution within any of the minerals was ignored.

Until now we have discussed and illustrated triangular diagrams with only three components (one at each corner). Petrologists may also use triangular diagrams for assemblage representations in which two components are grouped at a specific corner, so as to represent more complex chemical systems that are more alike to those found in rocks. Figure 9.17 illustrates two assemblage diagrams in which two oxide components have been grouped together in the right-hand corner. This allows for a more complete representation of the mineral compositions in this instance. In Fig. 9.17 the basalt assemblage may have crystallized at a constant and relatively low *P* (e.g., 1 to 2 kilobars) and a temperature of somewhere between

(a)

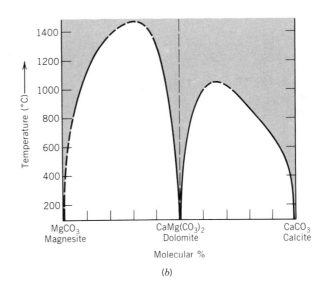

(b)

FIG. 9.15. (a) The system CaCO₃-MgCO₃-FeCO₃ showing extent of solid solution for phase regions and coexistences for natural assemblages metamorphosed to
about 400°C (biotite zone of greenschist facies). A three-
phase assemblage is shown by a triangle. In this study
the extent of solid solution is based on experimental data
as well as on analytical results of natural assemblages.
The composition of coexisting carbonates was based on
electron microprobe analyses of carbonates over a wide
compositional range in low- to medium-grade metamorphic rocks. The open regions across which tielines are
drawn are miscibility gaps. Extent of solid solution is
shaded. (b) Temperature-composition diagram for the
left-hand side of the triangle in a. This diagram shows
the presence of two miscibility gaps, one between calcite and dolomite and the other between dolomite and
magnesite. Extent of solid solution is shaded. Vertical
dashed line is ideal dolomite composition. (a and b
adapted from Anovitz, L. M. and Essene, E. J., 1987,
Phase equilibria in the system CaCO₃-MgCO₃-FeCO₃.
Journal of Petrology, v. 28, pp. 389–415.)

1000 and 1200°C. The chemically equivalent eclogite
assemblage formed at a much higher pressure, possibly over a range of 12 to 30 kilobars, and a temperature range of 400 to 800°C. In other words, the two
assemblage diagrams represent very different P and T
conditions for identical bulk chemical compositions.
In assemblage diagrams obtained from experimental
studies the temperatures and pressures can be closely
controlled and isobaric and isothermal phase diagrams can be obtained for much narrower ranges in

P and T than is generally possible for natural rock or
mineral systems. Examples of more complex assemblage representations are given in Chapter 14 as part
of the discussion of metamorphic rocks. Figure 14.18a
illustrates a commonly used graphic projection
scheme in which assemblages that are depicted in a
tetrahedral chemical system are projected onto a two-
dimensional surface. Because the chemistry of many
rocks and mineral associations is so complex that a
graphic representation of mineral assemblages is not

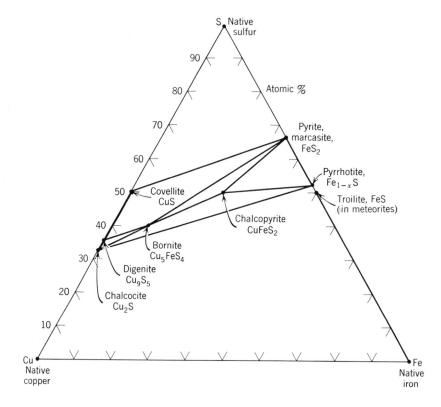

FIG. 9.16. Some of the most common sulfides represented in the Cu-Fe-S system. Many of these sulfides (e.g., bornite and chalcopyrite) show some solid solution of especially Cu and Fe; this is not shown in the diagram. Tielines connect commonly occurring pairs of minerals. Triangles indicate coexistences of three sulfides. The Fe-FeS coexistence is common in iron meteorites.

feasible, mathematical expressions are commonly used instead.

Diagrams for Mineral Reactions Involving H₂O or CO₂

Pressure-temperature diagrams are used to outline the stability field of a mineral that, along some *P-T* curve, gives way to another mineral by a chemical reaction.

The only *P-T* diagrams that we have discussed until now are those that have involved polymorphic reactions (see Figs. 9.1 and 9.4). However, experiments can also be done that delineate the stability fields of, for example, a hydrous mineral and its anhydrous (higher temperature) reaction products. Figure 9.18a shows the stability region of anthophyllite, $Mg_7Si_8O_{22}(OH)_2$, in the low-temperature part of the diagram and the stability of the higher-temperature

FIG. 9.17. Coexisting minerals in the system Al_2O_3-CaO-(MgO+FeO)-SiO_2-Na_2O. In *basalt* a common assemblage is plagioclase-clinopyroxene-orthopyroxene, as shown by the triangle of connecting tielines (the *x* represents a possible bulk composition of basalt with this mineral assemblage). In *eclogite*, of the same bulk composition, marked by *x*, the assemblage kyanite-omphacite-pyrope (as a component in almandine) occurs. Composition bars and omphacite field in the eclogite diagram outline approximate extent of solid solution. Eclogitic rocks may form in the mantle of the Earth from basaltic compositions.

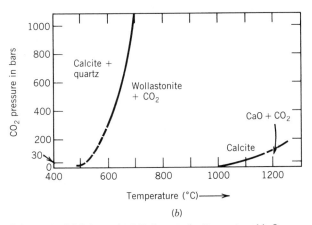

FIG. 9.18. *(a)* Schematic *P-T* diagram for the system MgO-SiO₂-H₂O. The stability fields of anthophyllite, Mg₇Si₈O₂₂(OH)₂, and reaction products, enstatite, MgSiO₃ + quartz + H₂O are shown. (After Greenwood, H. J., 1963, *Journal of Petrology*, v. 4, p. 325.) *(b)* *P-T* diagram for the system CaO-SiO₂-CO₂, showing the stability fields of calcite in the presence of SiO₂, and without SiO₂ in the assemblage. (Modified from *Petrologic Phase Equilibria*, 21st ed. Fig. 6.14a, by W. G. Ernst. Copyright © 1976 by W. H. Freeman and Company. Reprinted by permission.)

reaction products, enstatite (MgSiO₃), and quartz (SiO₂), and fluid. This chemical system can be defined as MgO-SiO₂-H₂O. The curve delineates the *P-T* region over which the breakdown of Mg-amphibole occurs, according to

$$Mg_7Si_8O_{22}(OH)_2 \rightleftarrows 7MgSiO_3 + SiO_2 + H_2O'$$

the H₂O enters the fluid phase produced in the reaction.

A pressure-temperature diagram involving mineral stabilities as a function of CO₂ is shown in Fig. 9.18b. In the system CaO-CO₂, calcite is stable to high temperatures, even at relatively low CO₂ pres-

sures (in Fig. 9.18b calcite is still stable at about 1000°C and P_{CO_2} pressure of less than 10 bars). At temperatures above the curve calcite decomposes as follows:

$$CaCO_3 \rightleftarrows CaO + CO'_2$$

However, the addition of SiO₂ (quartz) to the system, resulting in the system CaO-SiO₂-CO₂, produces a large stability field for wollastonite, CaSiO₃, on account of the reaction:

$$CaCO_3 + SiO_2 \rightleftarrows CaSiO_3 + CO'_2$$

Figure 9.18 illustrates that CaCO₃ decomposes at much lower temperatures in the presence of SiO₂ than in the simple system CaO-CO₂. The diagrams shown in Fig. 9.18 are commonly used in metamorphic petrology to evaluate mineral stabilities as a function of increasing temperature.

In the evaluation of the conditions that metamorphic rocks may have undergone, composite *P-T* diagrams with reaction curves for several minerals are commonly used. Figure 9.19 is such a diagram. The reaction curves for Al₂SiO₅ and for muscovite + quartz = K-feldspar + sillimanite + H₂O are especially relevant to metamorphic rocks with high Al₂O₃ contents relative to other components such as CaO, MgO, and FeO. Shales with abundant clay minerals have relatively high Al₂O₃ contents and during metamorphism mineral reactions take place in the shale, as well as recrystallization. The metamorphic equivalents of shales, known collectively as pelitic schists,

FIG. 9.19. Reaction curves for some common metamorphic minerals. For significance of shaded area, see text.

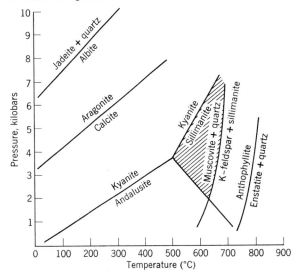

may contain abundant sillimanite, muscovite, and quartz. This type of schist, on the basis of the reaction curves, and stability fields of minerals in Fig. 9.19 may have equilibrated under *P-T* conditions as outlined by the shaded area in Fig. 9.19. In metamorphic rocks that may have undergone very high pressures, such as eclogites, the originally present albite may, during metamorphism, have reacted to form jadeite + quartz, and any original calcite may have transformed to aragonite. The presence of jadeite + quartz, for example, would indicate a minimum pressure during metamorphism of about 10 kilobars at 300°C (see upper left curve in Fig. 9.19).

Eh-pH Diagrams

Until now we have discussed various types of phase diagrams that are applicable to minerals that have formed in high-temperature and/or high-pressure environments, as may be the case for igneous and metamorphic rocks. At low temperatures (essentially room temperature; 25°C) and low pressure (atmospheric pressure), such as prevail under atmospheric conditions, it is often useful to express the stability fields of phases (or minerals) in terms of Eh (oxidation potential) and pH (the negative logarithm of the hydrogen ion concentration; it allows for the definition of acidic conditions with pH = 0–7, for basic conditions with pH = 7–14, and for neutral solutions at pH = 7). The construction of such diagrams on the basis of thermodynamic parameters is outlined in Garrels and Christ (1965; see reference list). Figure 9.20 illustrates the stability fields of some iron oxides and iron sulfides as calculated for atmospheric conditions (25°C, 1 atmosphere pressure). It shows at a glance that hematite is stable in an oxidizing region (high Eh). In hematite, $Fe_2^{3+}O_3$, all iron is present in the most oxidized, trivalent state. Magnetite, on the other hand, is stable under more reducing conditions, lower Eh. Magnetite, Fe_3O_4, which can be rewritten as $FeO \cdot Fe_2O_3$, consists of $\frac{1}{3}Fe^{2+}$ and $\frac{2}{3}Fe^{3+}$. The two sulfides, pyrite and pyrrhotite, occur under reducing conditions and pH values between 4 and 9. In both of these, the oxidation state of iron is Fe^{2+}. Along the lines that separate the various mineral fields, two of the minerals can coexist in equilibrium. For example, hematite and magnetite can coexist under conditions of variable pH and Eh (line *ab*, Fig. 9.20); indeed, hematite and magnetite commonly occur together in Precambrian sedimentary iron-formations. Diagrams of this type are especially useful in evaluating some of the physical chemical parameters that prevail during conditions of atmospheric weathering and of chemical sedimentation and diagenesis of water-laid

sediments (at essentially atmospheric pressure conditions and temperatures ranging from 25°C to about 100°C). The size of the sulfide fields (pyrite and pyrrhotite) depends on the amount of sulfur in solution. If sulfur were lower and the CO_2 content high there would be a stability field of siderite ($FeCO_3$) instead. On the basis of the frequent coexistence of hematite and magnetite in Proterozoic banded iron-formations one may be able to evaluate to some extent parameters such as Eh and pH of the original Proterozoic sedimentary basin.

Another example of an Eh-pH diagram is given in Fig. 9.21. This is for the stability fields of three manganese oxides, one manganese hydroxide, and one manganese sulfide. As in Fig. 9.20, the most oxidized oxide is stable at the highest Eh values. Pyrolusite, MnO_2, is the most oxidized naturally occurring oxide, with all Mn in the Mn^{4+} state. Manganite, Mn_2O_3, with all Mn in the Mn^{3+} state, is stable over a slightly less highly oxidizing Eh range. Hausmannite, Mn_3O_4,

FIG. 9.20. Stability of iron oxides and iron sulfides in water at 25°C and 1 atmosphere total pressure, with an activity of total dissolved sulfur = 10^{-6}. The pyrrhotite field is calculated on the basis of the formula FeS in this diagram; more correctly it would be $Fe_{1-x}S$. (Adapted from Garrels, R. M. and Christ, C. L., 1965, *Solutions, Minerals, and Equilibria.* Freeman, Cooper and Co., San Francisco, fig. 7.20.)

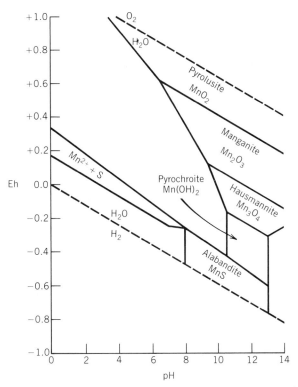

FIG. 9.21. Stability fields of manganese oxides, hydroxide, and sulfide in water at 25°C and 1 atmosphere total pressure, with activity of total dissolved sulfur = 10^{-1}. (Adapted from Runnells, D. D., 1962, in *Equilibrium Diagrams for Minerals at low Temperatures and Pressure*, Schmitt, H. H., ed. Geology Club of Harvard, Cambridge, Mass., 199 pp.)

which can be rewritten as $MnO \cdot Mn_2O_3$, contains $\frac{1}{3}Mn^{2+}$ and $\frac{2}{3}Mn^{3+}$. As a consequence, it is stable at Eh values below manganite. In $Mn(OH)_2$ and MnS, both of which show stability fields at the bottom of the Eh scale, Mn is present as Mn^{2+}. This type of diagram is useful not only in the prediction of what Mn-rich minerals may precipitate from an Mn-rich system at atmospheric conditions, but also for the interpretation of the conditions of formation of naturally occurring Mn compounds in manganese formations.

REFERENCES AND SUGGESTED READING

Bohlen, S. R., Montana, A. and Kerrick, D. M., 1991, Precise determinations of the equilibria kyanite ⇌ sillimanite and kyanite ⇌ andalusite and a revised triple point for

Al_2SiO_5 polymorphs. *American Mineralogist*, v. 76, pp. 677–680.

Bowen, N. L., 1913, The melting phenomena of the plagioclase feldspars. *American Journal Science*, v. 34, pp. 577–599.

Bowen, N. L. and Schairer, J. F., 1932, The system FeO-SiO$_2$. *American Journal of Science*, v. 29, pp. 197–217.

Brookins, D. G., 1988, *Eh-pH Diagrams for Geochemistry*. Springer-Verlag, New York, 174 pp.

Chopin, D., 1984, Coesite and pure pyrope of the Western Alps: A first record and some consequences. *Contributions to Mineralogy and Petrology*, v. 86, pp. 107–118.

Denbigh, K., 1981, *The Principles of Chemical Equilibrium*, 4th ed. Cambridge University Press, New York, 494 pp.

Ehlers, E. G., 1972, *The Interpretations of Geological Phase Diagrams*. W. H. Freeman and Co., San Francisco, 280 pp.

Ernst, W. G., 1976, *Petrologic Phase Equilibria*. W. H. Freeman and Co., San Francisco, 333 pp.

Garrels, R. M. and Christ, C. L., 1965, *Solutions, Minerals, and Equilibria*. Freeman, Cooper and Co., San Francisco, 450 pp.

Holdaway, M. J., 1971, Stability of andalusite and the aluminum silicate phase diagram. *American Journal of Science*, v. 271, pp. 97–131.

Kern, R. and Weisbrod, A., 1967, *Thermodynamics for Geologists*, Freeman, Cooper and Co., San Francisco, 304 pp.

Mitchell, R. H., 1986, *Kimberlites: Mineralogy, Geochemistry and Petrology*. Plenum Press, New York, 442 pp.

Nordstrom, D. K. and Munoz, J. L., 1985, *Geochemical Thermodynamics*. Benjamin/Cummings Publishing Co., Menlo Park, Calif., 477 pp.

Ribbe, P. H., 1975, The chemistry, structure, and nomenclature of feldspar. *Reviews of Mineralogy, Feldspar Mineralogy*, v. 2., pp. R1–R72.

Richardson, S. W., Gilbert, M. C. and Bell, P. M., 1969, Experimental determination of kyanite-andalusite and andalusite-sillimanite equilibria: The aluminum silicate triple point. *American Journal of Science*, v. 267, pp. 259–272.

Shutong, X., Okay, A. I., Shouyuan, J., Sengör, A. M. C., Wen S., Yican, L. and Laili, J., 1992, Diamond from the Dabie Shan metamorphic rocks and its implication for tectonic setting. *Science*, v. 256, pp. 80–82.

Smith, J. V. and Brown, W. L., 1988, *Feldspar Minerals*, v. 1, *Crystal Structures, Physical, Chemical, and Microtextural Properties*. Springer-Verlag, New York, 828 pp.

CHAPTER 10

SYSTEMATIC MINERALOGY
PART 1: NATIVE ELEMENTS, SULFIDES, AND SULFOSALTS

This chapter and three subsequent chapters include descriptions of all the common minerals and those rarer ones of most economic importance. The list, numbering about 200, is relatively small, because approximately 3500 minerals are recognized as valid species. The names and chemical compositions of some other minerals are also given.

The treatment of all minerals[1] follows a common scheme of presentation. The headings used and the data given under each are as follows:

Crystallography. Under this heading is given the following crystallographic information:

The crystal system and symbol of the crystal class.

Crystallographic features usually observed by inspection, such as habit, twinning, and crystal forms.

Other pertinent data are listed without subheadings. For example, consider the following data for cerussite:

Pmcn (space group), $a = 5.15$, $b = 8.47$, $c = 6.11$ Å (unit cell dimensions), $Z = 4$ (formula units per unit cell).

d's: 3.59(10), 3.50(4), 3.07(2), 2.49(3), 2.08(3) (strongest X-ray lines in angstrom units with relative intensities in parentheses).

Physical Properties. The physical properties: *Cleavage*, **H** (hardness), **G** (specific gravity), *Luster*, *Color*, and *Optics* (brief summary of optical data) are listed.

Composition and Structure. Chemical composition; commonly with the percentage of elements or oxides and the elements that may substitute for those given in the chemical formula. Brief description of the most important aspects of the crystal structure.

Diagnostic Features. The outstanding properties and tests that aid one in recognizing the mineral and distinguishing it from others.

Occurrence. A brief statement of the mode of occurrence and characteristic mineral associations is given.[2] The localities where a mineral is or has been found in notable amount or quality are mentioned. For minerals found abundantly the world over, emphasis is on North American localities.

Use. For a mineral of economic value, there is a brief statement of its uses.

Similar Species. The similarity of the species listed to the mineral whose description precedes may

---

[1]Mineral names are given in boldface capitals (e.g., **GOLD**) and smaller letters (e.g., **Acanthite**). Those in boldface capitals are considered most common or important.

[2]See Chapter 14 for a discussion of rock types, mineral associations, and some terms relating to ore deposits.

be either on the basis of chemical composition or crystal structure.

MINERAL CLASSIFICATION

Chemical composition has been the basis for the classification of minerals since the middle of the nineteenth century. According to this scheme, minerals are divided into classes depending on the dominant anion or anionic group (e.g., oxides, halides, sulfides, silicates, etc.). There are a number of reasons why this criterion is a valid basis for the broad framework of mineral classification. First, minerals having the same anion or anionic group dominant in their composition have unmistakable family resemblances, in general stronger and more clearly marked than those shared by minerals containing the same dominant cation. Thus the carbonates resemble each other more closely than do the minerals of copper. Second, minerals related by dominance of the same anion tend to occur together or in the same or similar geological environment. Thus the sulfides occur in close mutual association in deposits of vein or replacement type, whereas the silicates make up the great bulk of the rocks of the Earth's crust. Third, such a scheme of mineral classification agrees well with the current chemical practice in the naming and classification of inorganic compounds.

However, it was early recognized that chemistry alone does not adequately characterize a mineral. A full appreciation of the nature of minerals was to wait until X-rays were used to determine internal structures. It is now clear that mineral classification must be based on chemical composition *and* internal structure, because these together represent the essence of a mineral and determine its physical properties. Crystallochemical principles were first used by W. L. Bragg and V. M. Goldschmidt for silicate minerals. This large mineral group was divided into subclasses partially on the basis of chemical composition but principally in terms of internal structure. Within the silicate class, therefore, framework, chain and sheet silicate (etc.) subclasses exist on the basis of the structural linkage of SiO_4 tetrahedra. Such structural principles in combination with chemical composition provide a logical classification. It is this scheme of classification that is used in the subsequent sections on systematic mineralogy.

The broadest divisions of the classification used in this book (as based on C. Palache, H. Berman, and C. Frondel, *Dana's System of Mineralogy*, 7th ed., and H. Strunz, *Mineralogische Tabellen*, 5th ed.; see references) are:

1. Native elements	⎫
2. Sulfides	⎬ Chapter 10
3. Sulfosalts	⎭
4. Oxides	⎫
(a) Simple and multiple	⎬ Chapter 11
(b) Hydroxides	
5. Halides	⎭
6. Carbonates	⎫
7. Nitrates	⎬
8. Borates	⎬ Chapter 12
9. Phosphates	⎬
10. Sulfates	⎭
11. Tungstates	
12. Silicates	Chapter 13

The above classes are subdivided into *families* on the basis of chemical types, and the family in turn may be divided into *groups* on the basis of structural similarity. A group is made up of *species*, which may form *series* with each other. Species have the same structure but different chemistries. A *species* may be subdivided into chemical *varieties* by adjectival modifiers which reflect the presence of unusual amounts of chemical constituents. Examples of such adjectival modifiers are:

aluminian	:	Al-rich
calcian	:	Ca-rich
chromian	:	Cr-rich
ferroan	:	Fe^{2+}-rich
ferrian	:	Fe^{3+}-rich
magnesian	:	Mg-rich
manganoan	:	Mn-rich

and so on. Some examples are: manganoan aegirine, or ferrian diopside, or magnesian augite.

In each of the classes, the mineral with the highest ratio of metal to nonmetal is given first, followed by those containing progressively less metal. Because a relatively small number of minerals is described in this book, often only one member of a group or family is represented, and thus a rigorous adherence to division and subdivision is impractical.

The discussion of each of the chemical classes, families, or groups is preceded by an introduction regarding the most important aspects of the crystal chemistry and crystal structure.

NATIVE ELEMENTS

With the exception of the free gases of the atmosphere, only about 20 elements are found in the native state. These elements can be divided into (1) metals;

(2) semimetals; (3) nonmetals. The more common *native metals*, which display very simple structures, constitute three groups: the *gold group* (space group: *Fm3m*), gold, silver, copper, and lead, all of which are isostructural; the *platinum group* (space group: *Fm3m*), platinum, palladium, iridium and osmium, all of which are isostructural; and the *iron group*, iron and nickel-iron, of which pure Fe as well as kamacite have space group *Im3m* and the more Ni-rich variety of nickel-iron (taenite) *Fm3m*.

NATIVE ELEMENTS

Metals		Semimetals	
Gold group		Arsenic group	
Gold	Au	(Arsenic	As)
Silver	Ag	(Bismuth	Bi)
Copper	Cu	Nonmetals	
Platinum group		Sulfur	S
Platinum	Pt	Diamond	C
Iron group		Graphite	C
Iron	Fe		
(Kamacite	Fe, Ni)		
(Taenite	Fe, Ni)		

In addition, mercury, tantalum, tin, and zinc have been found. The *native semimetals* form two isostructural groups: arsenic, antimony, and bismuth (space group *R3m*), and the less common selenium and tellurium (space group *P3₁21*). The important *nonmetals* are sulfur and carbon in the form of diamond and graphite.

Native Metals

It is fitting that *systematic mineralogy* begin with a discussion of the gold group, for our knowledge of the properties and usefulness of metals arose from the chance discovery of nuggets and masses of these minerals. Many relatively advanced early cultures were restricted in their use of metal to that found in the native state.

The elements of the *gold group* belong to the same group in the periodic table of elements and, hence, their atoms have somewhat similar chemical properties; all are sufficiently inert to occur in an elemental state in nature. When uncombined with other elements, the atoms of these metals are united into crystal structures by the rather weak metallic bond. The minerals are isostructural and are built on the face-centered cubic lattice with atoms in 12-coordination (see Fig. 10.1a). Complete solid solution takes place between gold and silver, as these two elements have the same atomic radii (1.44 Å). Copper, because of its smaller atomic radius (1.28 Å), exhibits only limited solid solution in gold and silver. Conversely, native copper carries only traces of gold and silver in solid solution.

The similar properties of the members of this group arise from their common structure. All are rather soft, malleable, ductile, and sectile. All are excellent conductors of heat and electricity, display metallic luster and hackly fracture, and have rather low melting points. These are properties conferred by the metallic bond. All are isometric hexoctahedral and have high densities resulting from the cubic closest packing of their structures.

Those properties with respect to which the minerals of this group differ arise from the properties of the atoms of the individual elements. Thus, the yellow of gold, the red of copper, and the white of silver are atomic properties. The specific gravities likewise depend on atomic properties and show a rough proportionality to the atomic weights.

Although only platinum is discussed here, the *platinum group* also includes the rarer minerals palladium, platiniridium, and iridosmine. The last two are respectively alloys of iridium and platinum and iridium and osmium, with hexagonally close-packed structures and space group *P6₃/mmc*. Platinum and iridium, however, have cubic closest packed structures, similar to metals of the gold group, with space group *Fm3m*. The platinum metals are harder and have higher melting points than metals of the gold group.

Members of the *iron group* metals are isometric and include pure iron (Fe), which occurs only rarely on the Earth's surface, and two species of nickel-iron (*kamacite* and *taenite*), which are common in meteorites. Iron and nickel have almost identical atomic radii (1.26 Å and 1.25 Å, respectively), and thus nickel can and usually does substitute for some of the iron. Pure iron and kamacite, which contains up to about 5.5 weight percent Ni, show cubic close packing with space group *Im3m*. Taenite, which shows a range in Ni content from 27 to 65 weight percent, is cubic closest packed with space group *Fm3m* (see Figs. 10.1a and 10.1c). These two minerals are characteristic of iron meteorites and it is believed that Fe-Ni alloys of this type constitute a large part of the Earth's core.

GOLD—Au

Crystallography. Isometric; $4/m\bar{3}2/m$. Crystals are commonly octahedral (Fig. 10.2a), rarely showing the faces of the dodecahedron, cube, and trapezohedron {113}. Often in arborescent crystal groups with crystals elongated in the direction of a 3-fold symmetry axis, or flattened parallel to an octahedron face. Crystals are irregularly formed, passing into filiform, reticulated, and dendritic shapes (Fig. 10.2b).

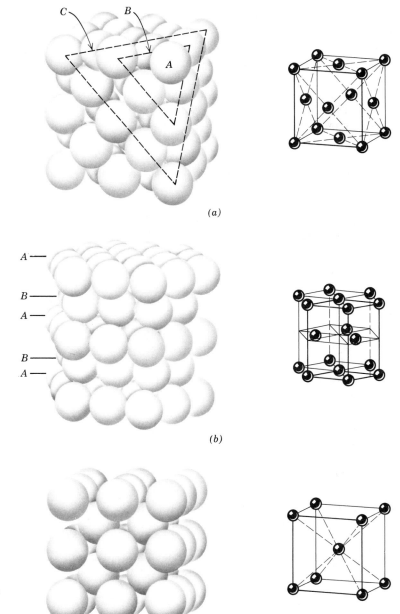

(a)

(b)

(c)

FIG. 10.1. *(a)* Close-packed model of cubic closest packing (*ABCABC* . . .) of equal spheres as shown by Cu, Au, Pt, and many other metals. Each metal atom is surrounded by 12 closest neighbors (see also Figs. 4.19 and 4.20). Close-packed layers are parallel to {111}. The face-centered cubic lattice (*F*) compatible with this packing sequence is illustrated on the right. *(b)* Close-packed model of hexagonal closest packing (*ABAB* . . .) of equal spheres as shown by Mg, Zn, and Cd. Each metal atom is surrounded by 12 closest neighbors (see also Figs. 4.19 and 4.20). This type of stacking leads to a hexagonal (*H*) lattice, as illustrated on the right, which can be reinterpreted as a rhombohedral (*R*) lattice (see Fig. 3.17; no. 11). *(c)* Close-packed model of body-centered cubic packing of equal spheres, as shown by Fe. Each sphere is surrounded by 8 closest neighbors. This packing is not as close as that exhibited by CCP and HCP (above). The body-centered (*I*) lattice compatible with this packing model is illustrated on the right.

Seldom shows crystal forms; usually in irregular plates, scales, or masses.

Fm3m; *a* = 4.079 Å; *Z* = 4. *d*'s: 2.36(10), 2.04(7), 1.443(6), 1.229(8), 0.785(5).

Physical Properties. H $2\frac{1}{2}$–3, **G** 19.3 when pure. The presence of other metals decreases the specific gravity, which may be as low as 15. *Fracture* hackly. Very malleable and ductile. Opaque. *Color* various shades of yellow, depending on the purity, becoming paler with increase of silver.

Composition and Structure. A complete solid solution series exists between Au and Ag, and most gold contains some Ag. California gold carries 10 to

FIG. 10.2. *(a)* Distorted gold octahedron. *(b)* Dendritic gold.

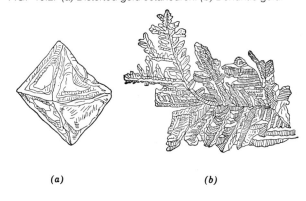

(a)   (b)

15% Ag. When Ag is present in amounts of 20% or greater, the alloy is known as *electrum*. Small amounts of Cu and Fe may be present as well as traces of Bi, Pb, Sn, Zn, and the platinum metals. The purity or *fineness* of gold is expressed in parts per 1000. Most gold contains about 10% of other metals and thus has a fineness of 900. The structure of gold is based on cubic closest packing of Au atoms (see Fig. 10.1a).

Diagnostic Features. Gold is distinguished from the yellow sulfides pyrite and chalcopyrite and from yellow flakes of altered micas by its sectility and its high specific gravity.

Occurrence. Gold has an average abundance of 0.004 ppm in the Earth's crust, and is therefore a rare element. It occurs in nature widely distributed in small amounts. It is found most commonly in veins that bear a genetic relation to silicic types of igneous rocks. Most gold occurs as the native metal. For a listing of some other, rare gold minerals see "Similar Species," below.

The chief sources of gold are hydrothermal gold-quartz veins where, together with pyrite and other sulfides, gold was deposited from ascending mineral-bearing solutions. Gold is also recovered as a by-product from sulfide deposits mined essentially for the base metals. The gold is merely mechanically mixed with the sulfides and is not in chemical substitution. In most of the veins, gold is so finely divided and uniformly distributed that its presence in the ore can only be detected by microscopic techniques.

When gold-bearing veins are weathered, the gold liberated either remains in the soil mantle as an eluvial deposit, or is washed into the neighboring streams to form a placer deposit. Because of its high specific gravity, gold works its way through the lighter sands and gravels to lodge behind irregularities or to be caught in crevices in the bedrock. The rounded or flattened nuggets of placer gold can be removed by panning, a process of washing away all but the heavy concentrate from which the gold can be easily separated. On a larger scale, gold-bearing sand is washed through sluices where the gold collects behind cross-bars or riffles and amalgamates with mercury placed behind the riffles. Most placer mining is today carried on with dredges, some of which are gigantic and can extract the gold from thousands of cubic yards of gravel a day.

Of the estimated world production of gold in 1981 of about 41 million ounces, over one-half came from the Republic of South Africa. The principal sources of South African gold are the Precambrian Witwatersrand conglomerate "the Rand," in the Transvaal and similar conglomerates in the Orange Free State. The "reefs" in these conglomerates in which the gold is concentrated are thought to be fossil placers. It is estimated that today about 20% of the world's gold production comes from the Commonwealth of Independent States (CIS), making this country the second largest producer. Gold is mined in the Ural Mountains, but it is believed that two-thirds of the production comes from placers, mostly in Siberia. China ranks third in gold production, followed by Canada, United States, and Brazil.

The most important gold-producing states in the United States, in decreasing importance, are: Nevada, South Dakota, Utah, Arizona, and Montana. The gold from Utah and Arizona is essentially a by-product of copper mining. For 100 years following the discovery of gold in the streams of California in 1848, that state was the leading producer. The most important gold-producing districts were those of the Mother Lode, gold-quartz veins along the western slope of the Sierra Nevada.

Use. Most of the existing gold is owned by various countries as bullion and is used for international settlements. An increasing amount in the form of medallions and small bars is used for investment purposes. Other uses are in jewelry, scientific instruments, electroplating, gold leaf, and dental appliances.

Similar Species. In addition to native gold there are 19 relatively rare to extremely uncommon minerals in which gold combines with other elements. The majority of these are tellurides. The most prominent examples are: *calaverite*, $AuTe_2$, associated with other tellurides at Cripple Creek, Colorado, and at Kalgoorlie, Western Australia; and *sylvanite*, $(Au,Ag)Te_2$, also found at Cripple Creek, Colorado, and at Kalgoorlie, Western Australia. Others are: *krennerite*, $(Au,Ag)Te_2$; *petzite*, Ag_3AuTe_2; *kostovite*, $CuAuTe_4$; *montbrayite*, $(Au,Sb)_2Te_3$; and *muthmannite*, $(Ag,Au)Te$. Other compounds are: *uytenbogaardtite*, Ag_3AuS_3; *fischesserite*, Ag_3AuSe_2; and *nagyagite*, $Pb_5Au(Te,Sb)_4S_{5-8}$. For a complete listing of gold-containing minerals see Wilson (1982).

SILVER—Ag

Crystallography. Isometric; $4/m\bar{3}2/m$. Crystals are commonly malformed and in branching, arborescent, or reticulated groups. Found usually in irregular masses, plates, and scales; in places as coarse or fine wire (Fig. 10.3).

$Fm3m$; $a = 4.09$ Å; $Z = 4$. d's: 2.34(10), 1.439(6), 1.228(8), 0.936(7), 0.934(8).

Physical Properties. H $2\frac{1}{2}$–3. G 10.5 when pure, 10–12 when impure. *Fracture* hackly. Malle-

FIG. 10.3. Silver, Kongsberg, Norway.

able and ductile. *Luster* metallic. *Color* and *streak* silver-white, often tarnished to brown or gray-black.

Composition and Structure. Native silver frequently contains alloyed Au, Hg, and Cu, more rarely traces of Pt, Sb, and Bi. *Amalgam* is a solid solution of Ag and Hg. The structure of silver is based on cubic closest packing of Ag atoms (see Fig. 10.1a).

Diagnostic Features. Silver can be distinguished by its malleability, color on a fresh surface, and high specific gravity.

Occurrence. Native silver is widely distributed in small amounts in the oxidized zone of ore deposits. However, native silver in larger deposits is the result of deposition from primary hydrothermal solutions. There are three types of primary deposits: (1) Associated with sulfides, zeolites, calcite, barite, fluorite, and quartz as typified by the occurrence in

Kongsberg, Norway. Mines at this locality were worked for several hundred years and produced magnificent specimens of wire silver. (2) With arsenides and sulfides of cobalt, nickel, and silver and with native bismuth. Such was the type at the old silver mines at Freiberg and Schneeberg in Saxony, Germany, and at Cobalt, Ontario. (3) With uraninite and cobalt-nickel minerals. Deposits at Joachimsthal, Bohemia, Czechoslovakia, and at Great Bear Lake, Northwest Territories, Canada, are of this type.

In the United States small amounts of primary native silver are associated with native copper on the Keweenaw Peninsula, Michigan. This has been the best U.S. source of crystallized silver. Elsewhere in the United States native silver is largely secondary. Most of the world's supply comes from silver sulfides and sulfosalts. Large producers of silver are Canada, Peru, Mexico, and the United States, with Mexico a very important producer historically.

Use. An ore of silver, although most of the world's supply comes from other minerals such as *acanthite*, Ag_2S, *proustite*, Ag_3AsS_3 and *pyrargyrite*, Ag_3SbS_3. Silver has many uses chief of which are in photographic film emulsions, plating, brazing alloys, tableware, and electronic equipment. Because of declining production and increased price, silver in coinage has been largely replaced by other metals such as nickel and copper.

COPPER—Cu

Crystallography. Isometric; $4/m\bar{3}2/m$. Tetrahexahedron faces are common, as well as the cube, dodecahedron, and octahedron. Crystals are usually malformed and in branching and arborescent groups (Figs. 10.4a and b). Usually occurs in irregular masses, plates, and scales, and in twisted and wire-like forms.

FIG. 10.4. *(a)* Dendritic copper, Broken Hill, New South Wales, Australia, *(b)* Native copper, Keweenaw Peninsula, Michigan. (Harvard Mineralogical Museum.)

(a)

(b)

Fm3m; $a = 3.615$ Å; $Z = 4$. *d's*: 2.09(10), 1.81(10), 1.28(10), 1.09(10), 1.05(8).

Physical Properties. **H** $2\frac{1}{2}$–3. **G** 8.9. Highly ductile and malleable. *Fracture* hackly. *Luster* metallic. *Color* copper-red on fresh surface, usually dark with dull luster because of tarnish.

Composition and Structure. Native copper often contains small amounts of Ag, Bi, Hg, As, and Sb. The structure of copper is based on cubic closest packing of Cu atoms (see Fig. 10.1*a*).

Diagnostic Features. Native copper can be recognized by its red color on fresh surfaces, hackly fracture, high specific gravity, and malleability.

Occurrence. Small amounts of native copper have been found at many localities in the oxidized zones of copper deposits associated with cuprite, malachite, and azurite.

Most primary deposits of native copper are associated with basaltic lavas, where deposition of copper resulted from the reaction of hydrothermal solutions with iron oxide minerals. The only major deposit of this type is on the Keweenaw Peninsula, Michigan, on the southern shore of Lake Superior. The rocks of the area are composed of Precambrian basic lava flows interbedded with conglomerates. The series, exposed for 100 miles along the length of the peninsula, dips to the northwest beneath Lake Superior to emerge on Isle Royale 50 miles away. Some copper is found in veins cutting these rocks, but it occurs principally either in the lava flows filling cavities or in the conglomerates filling interstices or replacing pebbles. Associated minerals are prehnite, datolite, epidote, calcite, and zeolites. Small amounts of native silver are present. Michigan copper had long been used by the American Indians, but it was not until 1840 that its source was located. Exploitation began shortly thereafter and during the next 75 years there was active mining through the length of the Keweenaw Peninsula. Most of the copper in the district was in small irregular grains, but notable large masses were found; one found in 1857 weighed 520 tons.

Sporadic occurrences of copper similar to the Lake Superior district have been found in the sandstone areas of the eastern United States, notably in New Jersey, and in the glacial draft overlying a similar area in Connecticut. In Bolivia at Corocoro, southwest of La Paz, there is a noted occurrence in sandstone. Native copper, sometimes as fine crystal groups, occurs in small amounts associated with the oxidized copper ores of Ajo, Bisbee, and Ray, Arizona, and the Chino Mine, Santa Rita, New Mexico.

Use. A minor ore of copper; copper sulfides are today the principal ores of the metal. The greatest use of copper is for electrical purposes, mostly as wire. It is also extensively used in alloys, such as brass (copper and zinc), bronze (copper and tin with some zinc), and German silver (copper, zinc, and nickel). These and many other minor uses make copper second only to iron as a metal essential to modern civilization.

PLATINUM—Pt

Crystallography. Isometric; $4/m\overline{3}2/m$. Cubic crystals are rare and commonly malformed. Usually found in small grains and scales. In some places it occurs in irregular masses and nuggets of larger size.

Fm3m; $a = 3.923$ Å; $Z = 4$. *d's*: 2.27(9), 1.180(10), 1.956(8), 1.384(8).

Physical Properties. **H** 4–$4\frac{1}{2}$ (unusually high for a metal). **G** 21.45 when pure, 14–19 when native. Malleable and ductile. *Color* steel-gray, with bright luster. Magnetic when rich in iron.

Composition and Structure. Platinum is usually alloyed with several percent Fe and with smaller amounts of Ir, Os, Rh, Pd; also Cu, Au, Ni. The structure of platinum is based on cubic closest packing of atoms (see Fig. 10.1*a*).

Diagnostic Features. Determined by its high specific gravity, malleability, and steel-gray color.

Occurrence. Most platinum occurs as the native metal in ultrabasic rocks, especially dunites, associated with olivine, chromite, pyroxene, and magnetite. It has been mined extensively in placers, which are usually close to the platinum-bearing igneous parent rock.

Platinum was first discovered in Colombia, South America. It was taken to Europe in 1735 where it received the name *platina* from the word *plata* (Spanish for silver) because of its resemblance to silver. A small amount of platinum is still produced in Colombia from placers in two districts near the Pacific Coast. In 1822 platinum was discovered in placers on the Upper Tura River on the eastern slope of the Ural Mountains in a large district surrounding Nizhniy Tagil, Russia, CIS. From that time until 1934 most of the world's supply of platinum came from placers of that district, which centered around the town of Nizhne Tagil. In 1934, because of the large amount of platinum recovered from the copper-nickel ore of Sudbury, Ontario, Canada, that area became the leading producer. In 1954 the Republic of South Africa moved into first place. Part of the South African production is a by-product from gold mining on the Rand, but the chief source is the Merensky Reef in the ultrabasic rocks of the Bushveld igneous complex. The Merensky Reef is a horizon of this layered intrusive about 12 inches thick and extending for many miles with a uniform platinum content of about one-half ounce per ton of ore. The chief mining operations are near Rustenburg in the Transvaal. It is estimated that today about half of the world's platinum-group metals comes from the CIS. In the United States small amounts are produced as a by-product of copper mining. In Canada most of the platinum comes from *sperrylite*, $PtAs_2$, which occurs in the copper-nickel ore in Sudbury, Ontario.

Use. Many of the uses of platinum depend on its high melting point (1755°C), resistance to chemical attack, and superior hardness. But its major use is as catalysts to control automobile exhaust emissions and in the chemical and petroleum industries. It is also used in dentistry, surgical instruments, jewelry, and electrical equipment.

Similar Species. *Iridium*; *iridosmine*, an alloy of iridium and osmium; and *palladium* are rare minerals of the platinum group associated with platinum.

Iron—Fe

Crystallography. Isometric; $4/m\bar{3}2/m$. Crystals are rare. Terrestrial: in blebs and large masses; meteoritic (*kamacite*): in plates and lamellar masses, and in regular intergrowth with nickel-iron (*taenite*).

Ordinary iron (α-Fe and kamacite) body-centered cubic, $Im3m$, $a = 2.86$ Å; $Z = 2$. d's: 2.02(9), 1.168(10), 1.430(7), 1.012(7). Nickel-iron (taenite), face-centered cubic, $Fm3m$, $a = 3.56$ Å, $Z = 4$. d's: 2.06(10), 1.073(4), 1.78(3), 1.27(2).

Physical Properties. (α-Fe). *Cleavage* {001} poor. **H** $4\frac{1}{2}$. **G** 7.3–7.9. *Fracture* hackly. Malleable. Opaque. *Luster* metallic. *Color* steel-gray to black. Strongly magnetic.

Composition and Structure. α-Fe always contains some Ni and frequently small amounts of Co, Cu, Mn, S, C. *Kamacite* contains approximately 5.5 weight percent Ni. *Taenite* is highly variable in Ni content and ranges from about 27 to 65 weight percent Ni. The structures of α-Fe and kamacite are based on body-centered cubic packing of atoms (see Fig. 10.1c), whereas the structure of taenite is based on face-centered cubic packing (see Fig. 10.1a).

Diagnostic Features. Iron can be recognized by its strong magnetism, its malleability, and the oxide coating usually on its surface.

Occurrence. Occurs sparingly as terrestrial iron but commonly in meteorites. Iron in the elemental (native) state is highly unstable in the oxidizing conditions in rocks of the upper crust and in the Earth's atmosphere. Iron is normally present as Fe^{2+} or Fe^{3+} in oxides such as magnetite, Fe_3O_4, or hematite,

Fe_2O_3, or as the hydroxide, goethite, FeO·OH. Terrestrial iron is regarded as a primary magmatic constituent or a secondary product formed by the reduction of iron compounds by assimilated carbonaceous material. The most important locality is Disko Island, Greenland, where fragments ranging from small grains to masses of many tons are included in basalts. Masses of nickel-iron have been found in Josephine and Jackson counties, Oregon.

Iron meteorites are composed largely of a regular intergrowth of *kamacite* and *taenite*. Evidence of the intergrowth, the Widmanstätten pattern (Fig. 10.5), is seen on polished and etched surfaces of such meteorites. Stony-iron meteorites consist of mixtures of silicates, kamacite, taenite and troilite (FeS).

Native Semimetals

The semimetals arsenic, antimony, and bismuth belong to an isostructural group with space group $R\bar{3}m$. Their structures, unlike those of the metals, cannot be represented as a simple packing of spheres, because each atom is somewhat closer to three of its neighbors than to the remainder of the surrounding atoms. In Fig. 10.6a, which represents the structure of arsenic (or antimony), the spheres intersect each other along the flat circular areas. The bonding between the four closest atoms, forming pyramidal groups (see Fig. 10.6b), is due to the covalent nature of these bonds. This covalency is related to the position of these semimetals in group V of the periodic table toward the electronegative end of a row. The relatively strong bonding to four closest neighbors results in a layered structure, as shown in Fig. 10.6b. These layers are parallel to {0001} and the weak bonding between them gives rise to a good cleavage. Members of this group have similar physical properties. They are rather brittle and much poorer conductors of heat and electricity than the native metals. These properties reflect a bond type intermediate between metallic and covalent; hence it is stronger and more directional than pure metallic in its properties, leading to lower symmetry.

Because native arsenic and bismuth as minerals are uncommon, there will be no detailed descriptions of these species.

Native Nonmetals

The structure of the nonmetals sulfur, diamond, and graphite are very different from those of the metals and semimetals. *Sulfur* ordinarily occurs in orthorhombic crystal form (with space group *Fddd*) in nature; two

FIG. 10.5. Widmanstätten pattern in the Edmonton (Kentucky) iron meteorite. The octahedral pattern is outlined by very coarse kamacite (which is visible) and very thin taenite lamellae (invisible). The irregular inclusion at the right center consists of troilite (FeS). (Smithsonian Astrophysical Observatory, courtesy of J. A. Wood; Harvard Mineralogical Museum.)

FIG. 10.6. (a) Close-packed model of the structures of As and Sb. Flat areas represent overlap between adjoining atoms. (b) Expanded view of the structure of As or Sb, showing corrugated layers of covalently bonded groups, parallel to {0001}.

(a)   (b)

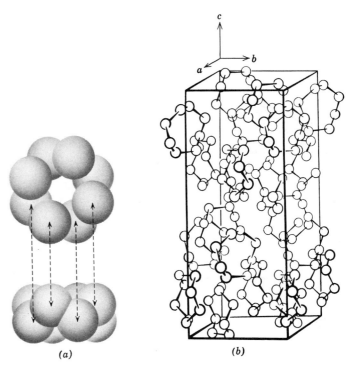

FIG. 10.7. *(a)* S$_8$ rings in orthorhombic sulfur as seen parallel (below) and perpendicular to the rings (above) *(b)* Unit cell of the structure of sulfur, showing the arrangement of S$_8$ rings.

(a)   (b)

monoclinic polymorphs are very rare in nature but are commonly produced synthetically. The orthorhombic structure of S is stable at atmospheric pressure below 95.5°C, and a monoclinic polymorph is stable from 95.5 to 119°C. This monoclinic polymorph melts above 119°C. The unit cell of the orthorhombic form of sulfur contains a very large number of S atoms, namely 128. The sulfur atoms are arranged in puckered rings of eight atoms which form closely bonded S$_8$ molecules (see Fig. 10.7*a*). These rings are bonded to each other by van der Waals forces, with a relatively large spacing between the rings (see Fig. 10.7*b*); 16 of such rings are present in the unit cell. Small amounts of selenium (atomic radius 1.16 Å) may substitute for S (atomic radius 1.04 Å) in the structure.

The structures of the two carbon polymorphs, *diamond* and *graphite*, are shown in Figs. 10.8*a* and *b*. Diamond has an exceptionally close-knit and strongly bonded structure in which each carbon atom is bound by powerful and highly directive covalent bonds to four carbon neighbors at the apices of a regular tetrahedron. The resulting structure, although strongly bonded, is not close packed, and only 34% of the available space is filled. The presence of rather widely spaced sheets of carbon atoms parallel to {111} (see Fig. 10.8*a*) accounts for prominent octahedral cleavage in diamond. The {111} sheets are planes of maximum atomic population.

The structure of graphite, illustrated in Fig. 10.8*b*, consists of six-membered rings in which each carbon atom has three near neighbors arranged at the apices of an equilateral triangle. Three of the four valence electrons in each carbon atom may be considered to be locked up in tight covalent bonds with its three close neighbors in the plane of the sheet. The fourth is free to wander over the surface of the sheet, creating a dispersed electrical charge which bestows on graphite its relatively high electrical conductivity. In contrast, diamond, in which all four valence electrons are locked up in covalent bonds, is among the best electrical insulators.

The sheets composing the graphite crystal are stacked in such a way that alternate sheets are in identical position, with the intervening sheet translated a distance of one-half the identity period in the plane of the sheets (see Fig. 10.8*b*). The distance between sheets is much greater than one atomic diameter, and van der Waals bonding forces perpendicular to the sheets are very weak. This wide separation and weak binding give rise to the perfect basal cleavage and easy gliding parallel to the sheets. Because of this open structure, only about 21% of the available space in graphite is filled, and the specific gravity is proportionately less than that of diamond. A photograph of the two polymorphs of carbon, diamond and graphite, is given in Fig. 10.9.

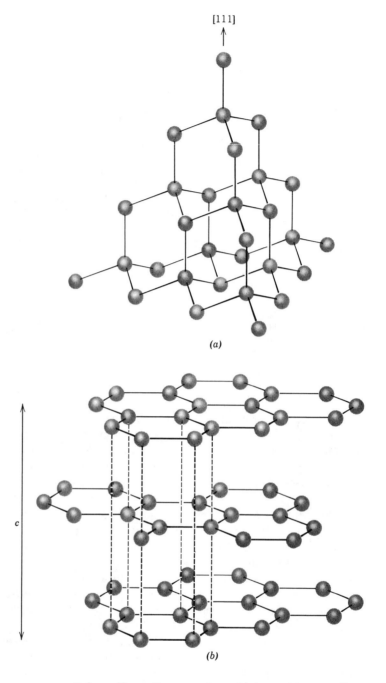

FIG. 10.8. *(a)* Partial representation of the structure of diamond. The horizontal plane is (111). *(b)* The structure of graphite with sheets // {0001}. Dashed vertical lines link atoms in successive sheets; these lines do not represent bonds.

SULFUR—S

Crystallography. Orthorhombic; $2/m2/m2/m$. Pyramidal habit common (Fig. 10.10a), often with two dipyramids, prism {011}, and base in combination (Figs. 10.10b and 10.11). Commonly found in irregular masses imperfectly crystallized. Also massive, reniform, stalactitic, as incrustations, earthy.

Fddd; $a = 10.47$, $b = 12.87$, $c = 24.49$ Å; $Z = 128$. *d*'s: 7.76(4), 5.75(5), 3.90(10), 3.48(4), 3.24(6).

Physical Properties. *Fracture* conchoidal to uneven. Brittle. **H** $1\frac{1}{2}$–$2\frac{1}{2}$. **G** 2.05–2.09. *Luster* resinous.

Color sulfur-yellow, varying with impurities to yellow shades of green, gray, and red. Transparent to translucent. *Optics*: (+); $\alpha = 1.957$, $\beta = 2.037$, $\gamma = 2.245$, 2V = 69°.

Sulfur is a poor conductor of heat. When a crystal is held in the hand close to the ear, it will be heard to crack. This is due to the expansion of the surface layers because of the heat from the hand, while the interior, because of the slow heat conductivity, is unaffected. Crystals of sulfur should therefore be handled with care.

FIG. 10.9. The two polymorphs of carbon: fibrous graphite from Buckingham, Quebec, Canada and an octahedral diamond from the Republic of South Africa. (Harvard Mineralogical Museum.)

Composition and Structure. Native sulfur may contain small amounts of selenium in substitution for S. The structure of the orthorhombic polymorph is given in Fig. 10.7. This structure consists of covalently bonded S_8 groups in the shape of puckered rings. These discrete and compact S_8 rings are relatively loosely bound to each other by van der Waals bonds. The unit cell of sulfur contains 128 atoms of sulfur distributed among 16 S_8 rings. Sulfur melts at 119°C to a liquid in which the covalently bonded S_8 rings retain their identity up to 160°C. The monoclinic polymorphs (α and β sulfur) are very rare in nature.

Diagnostic Features. Sulfur can be identified by its yellow color and the ease with which it burns; it ignites in a candle flame. The absence of a good cleavage distinguishes it from orpiment.

Occurrence. Sulfur often occurs at or near the crater rims of active or extinct volcanoes where it has been derived from the gases given off in fumaroles. These may furnish sulfur as a direct sublimation prod-

uct or by the incomplete oxidation of hydrogen sulfide gas. It is also formed from sulfates, by the action of sulfur-forming bacteria. Sulfur may be found in veins associated with metallic sulfides and formed by the oxidation of the sulfides. It is most commonly found in Tertiary sedimentary rocks associated with anhydrite, gypsum, and limestone; often in clay rocks; frequently with bituminous deposits. The large deposits near Girgenti, Sicily, are world famous for the fine crystals associated with celestite, gypsum, calcite, aragonite. Sulfur is also found associated with the volcanoes of Mexico, Hawaii, Japan, Argentina, and at Ollague, Chile, where it is mined at an elevation of 19,000 feet.

In the United States the most productive deposits are in Texas and Louisiana, where sulfur is associated with anhydrite, gypsum, and calcite in the cap rock of salt domes. There are 10 to 12 producing localities, but the largest are Boling Dome in Texas and Grand Ecaille, Plaquemines Parish, Louisiana. Sulfur is also recovered from salt domes in Mexico and offshore in the shallow waters of the Gulf of Mexico. Sulfur is obtained from these deposits by the Frasch method. Superheated water is pumped down to the sulfur horizon, where it melts the sulfur; compressed air then forces the sulfur to the surface.

About half of the world's sulfur is produced as the native element; the remainder is recovered as a by-product in smelting sulfide ores, from sour natural gas, and from pyrite.

Use. Sulfur is used primarily for the manufacture of sulfur compounds, such as sulfuric acid (H_2SO_4) and hydrogen sulfide (H_2S). Large quantities of elemental sulfur are used in insecticides, artificial fertilizers, and the vulcanization of rubber. Sulfur

FIG. 10.11. Sulfur crystals, Cinciana, Sicily, Italy. (Harvard Mineralogical Museum.)

FIG. 10.10. Sulfur crystals.

(a)   *(b)*

FIG. 10.12. Diamond crystals all showing a pronounced octahedral habit. (Courtesy of the Diamond Information Center, New York, N.Y.).

compounds are used in the manufacture of soaps, textiles, leather, paper, paints, dyes, and in the refining of petroleum.

DIAMOND—C

Crystallography. Isometric; $4/m\bar{3}2/m$. Crystals usually octahedral but may be cubic or dodecahedral. Curved faces, especially of the octahedron (see Fig. 10.12) and hexoctahedron are frequently observed (Figs. 10.13*a* and *b*; see also Plate I, no. 1, in Chapter 15). Crystals may be flattened on {111}. Twins on {111} (spinel law) are common, usually flattened parallel to the twin plane (Fig. 10.13*c*). *Bort*, a variety of diamond, has rounded forms and rough exterior resulting from a radial or cryptocrystalline aggregate. The term is also applied to badly colored or flawed diamonds without gem value.

*Fd*3*m*; *a* = 3.567 Å; *Z* = 8. *d*'s: 2.06(10) 1.26(8), 1.072(7), 0.813(6), 0.721(9).

Physical Properties. *Cleavage* {111} perfect. **H** 10 (hardest known mineral). **G** 3.52. *Luster* adamantine; uncut crystals have a characteristic greasy appearance. The very high refractive index, 2.42, and the strong dispersion of light account for the brilliancy and "fire" of the cut diamond. *Color* usually pale yellow or colorless; also pale shades of red, orange, green, blue, and brown. Deeper shades are rare. *Carbonado* or *carbon* is black or grayish-black bort. It is noncleavable, opaque, and less brittle than crystals.

Composition, Structure, and Synthesis. The composition is pure carbon. The structure of diamond is illustrated in Figs. 10.8*a* and 3.30. Figure 10.8*a* shows that each carbon is surrounded by four neighboring carbon atoms in tetrahedral coordination. This is the result of covalent bonding in which four valence

FIG. 10.13. Diamond crystals.

(a)   *(b)*   *(c)*

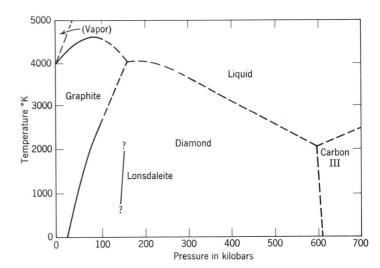

FIG. 10.14. Pressure-temperature phase diagram for carbon, based on experimental data from various sources.

electrons in each carbon fill the bonding orbitals of the four neighboring carbon atoms by electron sharing. Every carbon atom is linked in this way to four others, forming a continuous network. Figure 3.30c shows the face-centered arrangement of the carbon atoms in a cubic unit cell and Fig. 3.30d shows the location of some of the symmetry elements in the unit cell of diamond. Figure 3.30e illustrates the many complexities of the diamond space group, $F4_1/d\bar{3}2/m$ (which is abbreviated to $Fd3m$).

Diamond, as is to be expected from its high specific gravity and fairly close packing, is the high-pressure polymorph (see Fig. 10.14). At low pressures or temperatures it is unstable with respect to graphite and may be converted to graphite at moderate temperatures. The reason that diamond and graphite can coexist at room temperatures and pressures is because the reconstructive polymorphic reaction between the two minerals is very sluggish. In order to permit the change from graphite to diamond, extremely high temperature, that is, high activation energy, is needed to cause the carbon atoms in the graphite structure to break loose by thermal agitation, and to make them available for building the diamond structure. Such temperatures also increase the pressure required to bring about the polymorphic reaction.

The synthesis of diamond is based on the stability fields of the various carbon polymorphs as shown in Fig. 10.14. The first synthetic diamonds were made in 1953 by Dr. Erik Lundblad in the research laboratories of ASEA in Sweden. These resulted from experimental conditions of about 2400 K at about 76 kilobars using an apparatus that simultaneously exerted very high pressures and withstood extremely high temperatures. The ASEA technique was not patented at that time because the synthesis mechanism was not yet fully understood. In 1955 General Electric began the synthetic manufacture of small diamonds. World production of synthetic diamonds is now on the order of 300 million carats[3], or about three times the production of diamonds mined from the Earth. The synthetic material, which is identical to natural diamonds, offers the prospect of an assured and unlimited supply of diamonds. In recent years experimental techniques have produced diamonds as large as 11.1 ct. of good quality with some yellow coloration; however, the synthetic process is still too costly to be competitive with high-quality larger diamonds of natural origin.

The success of diamond synthesis encouraged experimentation with boron nitride, BN, whose structure is similar to that of graphite. At very high pressures and temperatures (on the order of 85 kilobars and 2100 K) it transforms, similar to the graphite-diamond transition, to cubic boron nitride (with the trade name of *Borazon*). This is extremely strong and hard, second only to diamond. It is manufactured in considerable quantities and is used mainly as an abrasive. This synthesis illustrates the application of crystal chemical concepts first evolved and tested in minerals.

After the initial synthesis of diamond, scientists at the General Electric Company continued experimentation. In 1967 they produced a hexagonal transparent polymorph of carbon with specific gravity and refractive indices close to those of diamond. Almost simultaneously a hexagonal diamond was found in meteoritic material from the Cañon Diablo, Arizona, meteorite crater. The name *lonsdaleite* is given to this

[3]One carat = 200 milligrams; a unit weight in gemstones.

naturally occurring hexagonal polymorph of diamond (see Fig. 10.14). Lonsdaleite is also found closely intergrown with polycrystalline diamond, *carbonado*, in *yakutiite*, named after its occurrence in alluvial deposits of northern Yakutia, Siberia. It is likely that yakutiite is the result of a meteorite impact.

Much recent research involving the synthesis of diamond is centered on the deposition of diamond from a carbonaceous gas at less than atmospheric pressure in order to obtain a thin film of polycrystalline diamond on substrates such as glass, silicon, tungsten, and molybdenum. This process is known as chemical vapor deposition, or CVD. These thin diamond films can provide protective coatings on other materials, and may improve the performance of cutting tools and semiconductor materials.

Diagnostic Features.   Diamond is distinguished from minerals which it resembles by its great hardness, adamantine luster, and cleavage.

Occurrence.   Diamonds have been discovered in many different localities but in only a few in notable amounts. Most commonly diamond is found in alluvial deposits, where it accumulates because of its inert chemical nature, its great hardness, and its fairly high specific gravity. In several countries in Africa and North America, and in China, Venezuela, Australia, India, and Siberia, diamonds have been found *in situ*. The rock in which they occur is called *kimberlite*, after the type locality of Kimberley, South Africa. Kimberlites are a group of volatile-rich (mainly CO_2), potassic ultrabasic rocks with a range of megacryst compositions set in a fine-grained groundmass (a megacryst is a megascopically visible crystal that stands out clearly from the finer groundmass, but whether or not it formed in the same kimberlite magma is uncertain). The megacryst assemblage may consist of ilmenite, pyrope garnet, olivine, clinopyroxene, phlogopite, enstatite, and chromite. The matrix mineralogy is complex by may include olivine, phlogopite, perovskite, spinel, and diopside. Although these intrusive bodies vary in size and shape, many are roughly circular with a pipelike shape and are referred to as "kimberlite" or "diamond pipes." In the deepest mines, in Kimberley, South Africa, the diameter of the pipes decreased with depth and mining stopped at a depth of 3500 feet, although the pipes continued. At the surface the kimberlite is weathered to a soft yellow rock, "yellow ground," that in depth gives way to a harder "blue ground." The ratio of diamonds to barren rock varies from one pipe to another. In the Kimberley mine it was 1:8,000,000, but in some it may be as high as 1:30,000,000.

Diamonds were first found in India, which remained virtually their only source until they were discovered in Brazil in 1725. The early diamonds came from stream gravels in southern and central India, and it is estimated that 12 million carats were produced from this area. Today the Indian production is only a few hundred carats a year.

Following their discovery in stream gravels in the state of Minas Gerais, Brazil, diamonds were also found in the states of Bahia, Goias, and Mato Grosso. Brazil's production today is chiefly from stream gravels. Extensive upland deposits of diamond-bearing gravels and clays are also worked. The black carbonado comes only from Bahia, Brazil.

In 1866 diamonds were discovered in the gravels of the Vaal River, South Africa, and in 1871 in the "yellow ground" of several pipes located near the present city of Kimberley. Although some diamonds are still recovered from gravels, the principal South African production is from kimberlite pipes. The deposits were originally worked as open pits, but as they deepened, underground methods were adopted. The world's largest and most productive diamond mine is the Premier, 24 miles east of Pretoria, South Africa. Since mining began there in 1903, nearly 30 million carats or six tons of diamonds have been produced. It was at the Premier Mine in 1905 that the world's largest diamond, the Cullinan, weighing 3106 carats, was found. Prospecting for diamonds has located in South Africa over 700 kimberlite pipes, dikes, and sills, most of which are barren. As recently as 1966 the Finsch Mine, 80 miles west of Kimberley, came into production. Even more recently diamonds have been found in kimberlite in Botswana and in northern South Africa, close to the border with Zimbabwe. The most productive pipe of all was found in about 1979 in Western Australia, and is now known as the Argyle Mine. This pipe consists of *lamproite*. Lamproite is a rock name that describes highly potassic and somewhat aluminum-poor igneous rocks with the following range of minerals: phlogopite, K-rich soda tremolite (also known as richterite), leucite, sanidine, diopside, and a variety of rare K-rich, Ba-rich, Ti-rich, and Zr-rich oxides and silicates.

In Cape Province, South Africa, on the desert coast just south of the mouth of the Orange River, terrace deposits containing high-quality stones were discovered in 1927; previously, in 1908, diamonds were found near Luderitz on the coast of Namibia. Elsewhere in Africa diamonds, mostly alluvial, have been found in a dozen different countries. The principal producers are Zaire, Botswana, Angola, Ghana, Sierra Leone, the Central African Republic, and Tanzania. Zaire is by far the largest producer and furnishes from kimberlites and minor placer deposits about 20% of the world's supply. These Zaire diamonds are mostly of industrial grade, with only 10% of gem quality. The Commonwealth of Independent States is also a major producer of diamonds from both pipes and placers. Important kimberlite localities in the CIS are: Mir kimberlite, Mirny, and Udachnaya kimberlite, Udachnaya, both in Yakutia; and Pomorskaya kimberlite, Arkhangelsk, Onega Peninsula, White Sea. Important alluvial deposits with diamonds in rivers and associated placers are found in the Urals and in Yakutia.

The world's main diamond producers, listed for 1988, with their natural rough diamond production recorded in millions of carats are:

Australia	35.0
Zaire	23.0
Botswana	15.0
CIS	12.0
South Africa	9.0
Namibia	0.9
South America	0.85

(Data from Wilks and Wilks, 1991)

Diamonds have been found sparingly in various parts of the United States. Small stones have occasionally been discovered in the stream sands along the eastern slope of the Appalachian Mountains from Virginia south to Georgia. Diamonds have also been reported from the gold sands of northern California and southern Oregon, and sporadic occurrences have been noted in the glacial drift in Wisconsin, Michigan, and Ohio. In 1906 diamonds were found in a lamproite pipe near Murfreesboro, Pike County, Arkansas. This locality, resembling the diamond pipes of South Africa, has yielded about 40,000 stones but is at present unproductive. In 1951 the old mine workings were opened to tourists who, for a fee, were permitted to look for diamonds. This area is now a state park.

Use. *In Industry.* Fragments of diamond crystals are used to cut glass. The fine power is employed in grinding and polishing diamonds and other gemstones. Wheels are impregnated with diamond powder for cutting rocks and other hard materials. Steel bits are set with diamonds, especially the cryptocrystalline variety, *carbonado*, for diamond drilling in exploratory mining work. Diamond is also used in wire drawing and in tools for the truing of grinding wheels.

Gemstones. The diamond is one of the most highly prized gemstones. Its value depends on its hardness, its brilliancy, which results from its high index of refraction, and its 'fire,'' resulting from its strong dispersion. In general, the most valuable are those flawless stones that are colorless or possess a "blue-white" color. A faint straw-yellow color, which diamond often shows, detracts from its value. Diamonds colored deep shades of yellow, red, green, or blue, known as fancy stones, are greatly prized and bring very high prices. Diamonds can be colored deep shades of green by irradiation with high-energy nuclear particles, neutrons, deuterons, and alpha particles, and blue by exposing them to fast-moving electrons. A stone colored green by irradiation can be made a deep yellow by proper heat treatment. These artificially colored stones are difficult to distinguish from those of natural color. See Chapter 15 for further discussion and illustrations of gem diamonds.

Name. The name diamond is a corruption of the Greek word *adamas*, meaning *invincible*. The luster term, *adamantine*, which describes the very high luster of minerals with a high index of refraction, such as of diamond and cerussite, is derived from *adamas*.

Similar Species. For a discussion of materials that are marketed as gem diamond imitations see Chapter 15 and specifically Table 15.2.

GRAPHITE—C

Crystallography. Hexagonal; $6/m2/m2m$. In tabular crystals of hexagonal outline with prominent basal plane. Distinct faces of other forms very rare. Triangular markings on the base are the result of gliding along an $\{hh\overline{2}hl\}$ pyramid. Usually in foliated or scaly masses, but may be radiated or granular.

$P6_3/mmc$. $a = 2.46$, $c = 6.74$ Å; $Z = 4$. d's: 3.36(10), 2.03(5), 1.675(8), 1.232(3), 1.158(5).

Physical Properties. *Cleavage* $\{0001\}$ perfect. **H** 1–2 (readily marks paper and soils the fingers). **G** 2.23. *Luster* metallic, sometimes dull earthy. *Color* and streak, black. Greasy feel. Folia flexible but not elastic (see Fig. 10.9).

Composition and Structure. Carbon. Some graphite impure with iron oxide, clay, or other minerals. The structure is illustrated in Fig. 10.8*b*.

Diagnostic Features. Graphite is recognized by its color, foliated nature, and greasy feel. Distinguished from molybdenite by its black color (molybdenite has a blue tone), and black streak on glazed porcelain.

Occurrence. Graphite most commonly occurs in metamorphic rocks such as crystalline limestones, schists, and gneisses. It may be found as large, crystalline plates or disseminated in small flakes in sufficient amount to form a considerable proportion of the rock. In these cases, it has probably been derived from carbonaceous material of organic origin that has been converted into graphite during metamorphism. Metamorphosed coal beds may be partially converted into graphite, as in the graphite coals of Rhode Island and in the coal fields of Sonora, Mexico. Graphite also occurs in hydrothermal veins associated with quartz, biotite, orthoclase, tourmaline, apatite, pyrite, and titanite, as in the deposits at Ticonderoga, New York. The graphite in these veins may have been formed from hydrocarbons introduced into them during the metamorphism of the region and derived from the surrounding carbon-bearing rocks. Graphite occurs occasionally as an original constituent of igneous rocks as in the basalts of Ovifak, Greenland, in a nepheline syenite in India, and in a graphite pegmatite in Maine. It is also found in some iron meteorites as graphite nodules.

The principal countries producing natural graphite are: China, the CIS, North and South Korea, India, and Mexico.

Synthetic. Graphite is manufactured on a large scale in electrical furnaces using anthracite coal or petroleum coke as the raw materials. The use of artificial graphite in the United States is considerably in excess of that of the natural mineral.

Use. Used in the manufacture of refractory crucibles for the steel, brass, and bronze industries. Flake graphite for crucibles comes mostly from Sri Lanka and the Malagasy Republic. Mixed with oil, graphite is used as a lubricant, and mixed with fine clay, it forms the "lead" of pencils. It is employed in the manufacture of protective paint for structural steel and is used in foundry facings, batteries, electrodes, generator brushes, and in electrotyping.

Name. Derived from the Greek word meaning *to write*, in allusion to its use in pencils.

SULFIDES

The sulfides form an important class of minerals that includes the majority of the ore minerals. With them are classed the similar but rarer sulfarsenides, arsenides, and tellurides.

Most of the sulfide minerals are opaque with distinctive colors and characteristically colored streaks. Those that are nonopaque, such as cinnabar, realgar, and orpiment, have extremely high refractive indices and transmit light only on thin edges.

The general formula for the sulfides is given as X_mZ_n in which X represents the metallic elements and Z the nonmetallic element. The general order of listing of the various minerals is in a decreasing ratio of $X:Z$.

The sulfides can be divided into small groups of similar structures but it is difficult to make broad generalizations about their structure. Regular octahedral or tetrahedral coordination about sulfur is found in many simple sulfides such as in galena, PbS, (with an NaCl-type structure), and in sphalerite, ZnS (see Fig. 10.15a). In more complex sulfides, as well as sul-

FIG. 10.15. The sphalerite structure and derivatives. *(a)* Sphalerite, ZnS. Compare this illustration with Fig. 4.44a. *(b)* Chalcopyrite, CuFeS$_2$. *(c)* Tetrahedrite, Cu$_{12}$Sb$_4$S$_{13}$. (After Wuensch, B.J., 1974, in *Sulfide Mineralogy, Reviews in Mineralogy*, v. 1, Mineralogical Society of America, Washington, D.C.).

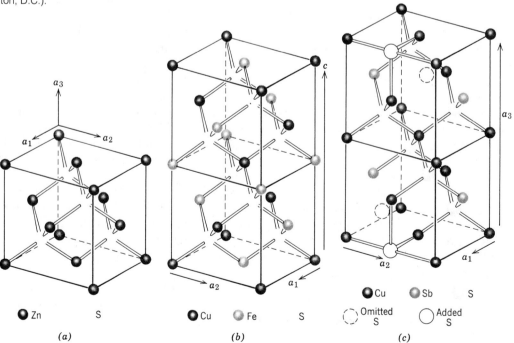

fosalts, distorted coordination polyhedra may be found (see tetrahedrite, Fig. 10.15c). Many of the sulfides have ionic and covalent bonding, whereas others, displaying most of the properties of metals, have metallic bonding characteristics. The structures and some aspects of the crystal chemistry of a few of the most common sulfides (e.g., sphalerite, chalcopyrite, pyrite, marcasite, and covellite) will be discussed.

ZnS occurs in two polymorphic forms: the sphalerite-type (Fig. 10.15a) and the wurtzite-type structures (Fig. 10.23). In both the sphalerite and wurtzite structures Zn is surrounded by four sulfurs in tetrahedral coordination, but in sphalerite the Zn atoms are arranged in a face-centered cubic lattice, whereas in wurtzite they are approximately in positions of hexagonal closet packing. The atomic arrangement in sphalerite is like that in diamond (see Figs. 10.8a and 4.44) in which the carbon has been replaced by equal amounts of Zn and S. Greenockite, CdS, a relatively rare sulfide, is isostructural with ZnS and occurs in both the sphalerite and wurtzite type structures. Natural sphalerite shows extensive Fe^{2+} and very limited Cd^{2+} substitution for Zn (see Table 5.3).

SULFIDES, SULFARSENIDES, AND ARSENIDES

Acanthite (Argentite)	Ag_2S	Cinnabar	HgS
Chalcocite	Cu_2S	Realgar	AsS
Bornite	Cu_5FeS_4	Orpiment	As_2S_3
Galena	PbS	Stibnite	Sb_2S_3
Sphalerite	ZnS	Pyrite	FeS_2
Chalcopyrite	$CuFeS_2$	Marcasite	FeS_2
Pyrrhotite	$Fe_{1-x}S$	Molybdenite	MoS_2
Nickeline	NiAs	Cobaltite	(Co, Fe)AsS
Millerite	NiS	Arsenopyrite	FeAsS
Pentlandite	$(Fe, Ni)_9S_8$	Skutterudite	$(Co, Ni)As_3$
Covellite	CuS		

Chalcopyrite, $CuFeS_2$, has a structure (Fig. 10.15b) that can be derived from the sphalerite structure by regularly substituting Cu and Fe ions for Zn in sphalerite; this leads to a doubling of the unit cell. Stannite, Cu_2FeSnS_4, has a structure based on sphalerite in which layers of ordered Fe and Sn alternate with layers of Cu (chalcopyrite can be rewritten as $Cu_2Fe_2S_4$; this illustrates chemically the close similarity to Cu_2FeSnS_4).

Covellite, CuS, is an example of a chemically very simple substance with a rather complex structure (see Fig. 10.16) in which part of the Cu is tetrahedrally coordinated by four S, and part coordinated by three S in the form of a triangle. The structure can, therefore, be viewed as made of sheets of CuS_3 triangles, between double layers of CuS_4 tetrahedra; covalent sulfur-sulfur bonds link the layers.

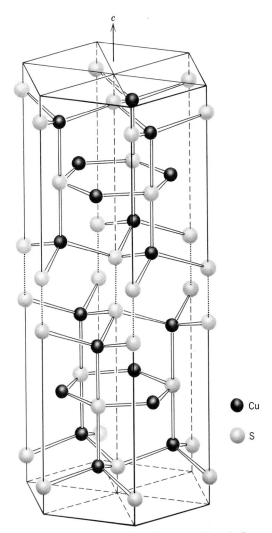

FIG. 10.16. Structure of covellite, CuS. The dotted lines indicate covalent S-S bonds. (After Wuensch, B. J., 1974, in *Sulfide Mineralogy, Reviews in Mineralogy*, v. 1, Mineralogical Society of America, Washington, D.C..)

Pyrite, FeS_2, has a cubic structure as shown in Fig. 10.17a. The structure contains covalently bonded S_2 pairs, which occupy the position of Cl in the NaCl structure type. The pyrite structure may be considered as derived from the NaCl structure in which Fe is found in the Na positions of NaCl. Another polymorphic form of FeS_2 is marcasite (see Fig. 10.17b) with an orthorhombic structure. This structure, as pyrite, contains closely spaced S_2 pairs. It is still unclear what the stability fields of the two polymorphs, pyrite and marcasite, are. From geological occurrences one would conclude that marcasite occurs over a range of low to medium temperature in sedimentary rocks and metalliferous veins. Pyrite, however, occurs also as a magmatic (high T) constituent.

The structure of arsenopyrite, FeAsS, can be derived from the marcasite (FeS_2 or FeSS) structure (see

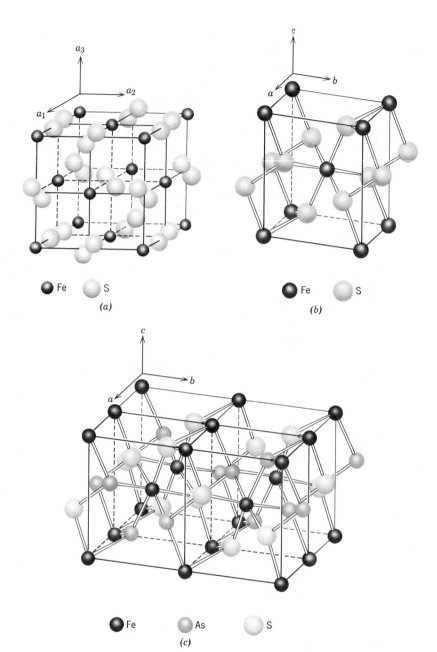

FIG. 10.17. *(a)* The structure of pyrite, as based on NaCl type structure (compare with Fig. 3.31). Note the closely spaced sulfurs in S$_2$ groups. *(b)* The structure of marcasite, showing coordination of ions to closest neighbors. *(c)* The structure of arsenopyrite, FeAsS, a derivative of the marcasite structure type. (*b* and *c* after Wuensch, B. J., 1974 in *Sulfide Mineralogy, Reviews in Mineralogy*, v. 1, Mineralogical Society of America, Washington, D.C.)

Fig. 10.17*c*) in which one S per unit formula is replaced by As. The general coordination of the atoms in marcasite and FeAsS is approximately the same in both structures.

Acanthite—Ag$_2$S

Crystallography. Monoclinic, $2/m$ below 173°C; isometric, $4/m\overline{3}2/m$ above 173°C. Crystals, twinned polymorphs of the high-temperature form, commonly show the cube, octahedron, and dodecahedron and frequently are arranged in branching or reticulated groups. Most commonly massive or as a coating. Historically acanthite has been referred to as argentite.

$P2_1/n$; $a = 4.23$, $b = 6.93$, $c = 7.86$ Å, $\beta = 99°35'$; $Z = 4$. d's: 2.60(10), 2.45(8), 2.38(5), 2.22(3), 2.09(4).

Physical Properties. **H** 2–2$\frac{1}{2}$. **G** 7.3. Very sectile; can be cut with a knife like lead. *Luster* metallic. *Color* black. *Streak* black, shining. Opaque. Bright on fresh surface but on exposure becomes dull black, owing to the formation of an earthy sulfide.

Composition and Structure. Ag 87.1, S 12.9%. Ag$_2$S has space group $P2_1/n$ below 173°C, and space group $Im3m$ above 173°C. On cooling Ag$_2$S from above 173°C the structure twins pervasively to produce apparently cubic crystals of twinned acanthite, known to mineralogists as *argentite*. However, it seems that acanthite is the only stable form of Ag_2S at ordinary temperatures.

Diagnostic Features. Acanthite can be distinguished by its color, sectility, and high specific gravity.

Occurrence. Acanthite is an important primary silver mineral found in veins associated with native silver, the ruby silvers, polybasite, stephanite, galena, and sphalerite. It may also be of secondary origin. It is found in microscopic inclusions in argentiferous galena. Acanthite is an important ore in the silver mines of Guanajuato and elsewhere in Mexico. Formerly important European localities are Freiberg, Saxony, Germany; Joachimsthal, Schemnitz, and Kremnitz, Czechoslovakia, and Kongsberg, Norway. In the United States it has been an important ore mineral in Nevada, notably at the Comstock Lode and at Tonopah. It is also found in the silver districts of Colorado, and in Montana at Butte associated with copper ores.

Use. An important ore of silver.

Name. The name acanthite comes from the Greek word meaning *thorn*, in allusion to the shape of the crystals. The name argentite comes from the Latin *argentum*, meaning silver.

CHALCOCITE—Cu_2S

Crystallography. Monoclinic, pseudo-orthorhombic; $2/m$ or m, (below 105°C); above 105°C, hexagonal. Crystals are uncommon, usually small and tabular with hexagonal outline; striated parallel to the a axis (Fig. 10.18). Commonly fine-grained and massive.

$P2_1/c$ or Pc; $a = 11.82$, $b = 27.05$, $c = 13.43$Å; $Z = 96$. d's: 3.39(3), 2.40(7), 1.969(8), 1.870(10), 1.695(4).

Physical Properties. *Cleavage* {110} poor. *Fracture* conchoidal. **H** $2\frac{1}{2}$–3. **G** 5.5–5.8. *Luster* metallic.

FIG. 10.18. Chalcocite crystals.

Imperfectly sectile. *Color* shining lead-gray, tarnishing to dull black on exposure. *Streak* grayish-black. Some chalcocite is soft and sooty.

Composition and Structure. Cu 79.8, S 20.2%. May contain small amounts of Ag and Fe. Below 105°C the structure is based on hexagonal closest packing of sulfur atoms with a monoclinic space group. Above 105°C it inverts to high chalcocite with space group $P6_3/mmc$.

Diagnostic Features. Chalcocite is distinguished by its lead-gray color and sectility.

Occurrence. Chalcocite is one of the most important copper-ore minerals. Fine crystals are rare but have been found in Cornwall, England, and Bristol, Connecticut. Chalcocite may occur as a primary mineral in veins with bornite, chalcopyrite, enargite, and pyrite. But its principal occurrence is as a supergene mineral in enriched zones of sulfide deposits. Under surface conditions the primary copper sulfides are oxidized; the soluble sulfates formed move downward, reacting with the primary minerals to form chalcocite and thus enriching the ore in copper. The water table is the lower limit of the zone of oxidation and here a "chalcocite blanket" may form (see Fig. 14.21). Many famous copper mines owe their greatness to this process of secondary enrichment, as: Rio Tinto, Spain; Ely, Nevada; Morenci, Miami, and Bisbee, Arizona; and Butte, Montana.

Much of the world's copper is today produced from what is called "porphyry copper" ore. In these deposits primary copper minerals disseminated through the rock, usually a porphyry, have been altered, at least in part, to chalcocite and thus enriched to form a workable ore body. The amount of copper in such deposits is small, rarely greater than 1 or 2% and may be as low as 0.40%.

Use. An important copper ore.

Name. From the Greek word *chalkos* meaning *copper*.

Similar Species. *Digenite*, Cu_9S_5, is blue to black, associated with chalcocite. *Djurleite*, $Cu_{31}S_6$, a black to dark gray mineral, occurs in many of the large porphyry copper deposits of the Western Cordillera.

BORNITE—Cu_5FeS_4

Crystallography. Tetragonal, $\bar{4}2m$, below 228°C; isometric, $4/m\bar{3}2/m$, above 228°C. The most common form in ores is tetragonal. Rarely in rough pseudocubic and less commonly dodecahedral and octahedral crystals. Usually massive.

$P\bar{4}2_1c$; $a = 10.94$, $c = 21.88$ Å; $Z = 16$. d's: 3.31(4), 3.18(6), 2.74(5), 2.50(4), 1.94(10).

$Fm3m$; $a = 5.50$ Å; $Z = 1$. d's: 3.17(5), 2.75(5), 1.94(10), 1.66(1).

Physical Properties. **H** 3. **G** 5.06–5.08. *Luster* metallic. *Color* brownish-bronze on fresh fracture but quickly tarnishing to variegated purple and blue (hence called *peacock ore*) and finally to almost black on exposure. *Streak* grayish-black.

Composition and Structure. Cu 63.3, Fe 11.2, S 25.5% for stoichiometric Cu_5FeS_4. Shows extensive solid solution within the Cu-Fe-S system. Its stoichiometric composition in terms of Cu-Fe-S is shown in Fig. 10.19, with the compositions of other common sulfides in this system. The structure of the high T polymorph of bornite is relatively complex with sulfur atoms in a face-centered cubic lattice and the Cu and Fe atoms in tetrahedral coordination to S. The structure of the low temperature form is derived from the high T form with the addition of defects. This defect structure gives rise to large variations in Cu, Fe, and S contents.

Diagnostic Features. Bornite is distinguished by its characteristic bronze color on the fresh fracture and by the purple tarnish.

Alteration. Bornite alters readily to chalcocite and covellite.

Occurrence. Bornite is a widely occurring copper ore usually found associated with other sulfides (chalcocite, chalcopyrite, covellite, pyrrhotite and pyrite; see Fig. 10.19) in hypogene deposits. It is less frequently found as a supergene mineral, in the upper, enriched zone of copper veins. It occurs disseminated in basic rocks, in contact metamorphic deposits, in replacement deposits, and in pegmatites. It is not as important an ore of copper as chalcocite and chalcopyrite.

Good crystals of bornite have been found associated with crystals of chalcocite at Bristol, Connecticut, and in Cornwall, England. Found in large masses in Chile, Peru, Bolivia, and Mexico. In the United States it has been an ore mineral at the Magma Mine, Arizona; Butte, Montana; Engels Mine, Plumas County, California; Halifax County, Virginia; and Superior, Arizona.

Use. An ore of copper.

Name. Bornite was named after the Austrian mineralogist Ignatius von Born (1742–1791).

GALENA—PbS

Crystallography. Isometric; $4/m\bar{3}2/m$. The most common form is the cube, sometimes truncated by the octahedron (Figs. 10.20 and 10.21). Dodecahedron and trisoctahedron rare.

$Fm3m$; $a = 5.936$Å; $Z = 4$. d's: 3.44(9), 2.97(10), 2.10(10), 1.780(9), 1.324 (10).

Physical Properties. *Cleavage* perfect {001}. **H** $2\frac{1}{2}$. **G** 7.4–7.6. *Luster* bright metallic. *Color* and *streak* lead-gray.

FIG. 10.19. Some of the most common sulfides represented in the Cu-Fe-S sysem. Many of these sulfides (e.g., bornite and chalcopyrite) show some solid solution of especially Cu and Fe; this is not shown in the diagram. Tielines connect commonly occurring pairs of minerals. Triangles indicate coexistences of three sulfides. The Fe-FeS coexistence is common in iron meteorites.

FIG. 10.20. Galena crystals.

Composition and Structure. Pb 86.6, S 13.4%. Silver is usually present as admixtures of silver minerals such as acanthite or tetrahedrite but also in solid solution. Inclusions probably also account for the small amounts of Zn, Cd, Sb, As, and Bi that may be present. Selenium may substitute for sulfur and a complete series of PbS-PbSe has been reported. Galena has an NaCl type of structure with Pb in place of Na and S in place of Cl.

Diagnostic Features. Galena can be easily recognized by its good cleavage, high specific gravity, softness, and lead-gray streak.

Alteration. By oxidation galena is converted into anglesite, $PbSO_4$, and cerussite, $PbCO_3$.

Occurrence. Galena is a very common metallic sulfide, found in veins associated with sphalerite, pyrite, marcasite, chalcopyrite, cerussite, anglesite, dolomite, calcite, quartz, barite, and fluorite. When found in hydrothermal veins galena is frequently associated with silver minerals; it often contains silver

FIG. 10.21. (a) Galena crystals, Wellington Mine, Breckenridge, Colorado. (b) Galena on sphalerite and dolomite crystals, Jasper County, Missouri. (Both specimens from the Harvard Mineralogical Museum).

(a)

(b)

itself and so becomes an important silver ore. A large part of the supply of lead comes as a secondary product from ores mined chiefly for their silver. In a second type of deposit typified by the lead-zinc ores of the Mississippi Valley, galena, associated with sphalerite, is found in veins, open space filling, or replacement bodies in limestones. These are low-temperature deposits, located at shallow depths, and usually contain little silver. Galena is also found in contact metamorphic deposits, in pegmatites, and as disseminations in sedimentary rocks.

Famous world localities are Freiberg, Saxony, and the Harz Mountains, Germany; Pribram, Bohemia, Czechoslovakia; Derbyshire and Cumbria, England; Sullivan Mine, British Columbia; and Broken Hill, Australia.

In the past, many districts in the United States were significant lead producers, but today only three are of importance. Chief of these is in southeast Missouri, which yields nearly 90% of the domestic production. Here galena is disseminated through limestone. A small production comes from the Coeur d'Alene district, Idaho, where galena is the chief ore mineral in the lead-silver veins, and from Colorado, where lead is recovered mostly as a by-product from the mining of silver and gold ores. Formerly, the Tri-State district of Missouri, Kansas, and Oklahoma was an important lead producer. In this region galena, associated with sphalerite, occurred in irregular veins and pockets in limestone and chert. Because the minerals could grow into open spaces, the Tri-State district is famous for its many beautiful crystallized specimens of galena, sphalerite, marcasite, chalcopyrite, calcite, and dolomite.

Use. Practically the only source of lead and an important ore of silver. The largest use of lead is in storage batteries, but nearly as much is consumed in making metal products such as pipe, sheets, and shot. Lead is converted into the oxides (*litharge*, PbO, and *minium*, Pb_3O_4) used in making glass and in giving a glaze to earthenware, and into white lead (the basic carbonate), the principal ingredient of many paints. However, this latter use is diminishing because of the poisonous nature of lead-based paints. Diminishing also is its use in gasoline antiknock additives because of environmental restrictions. Lead is a principal metal of several alloys as solder (lead and tin), type metal (lead and antimony), and low-melting alloys (lead, bismuth, and tin). Lead is used as shielding around radioactive materials.

Name. The name galena is derived from the Latin *galena*, a name originally given to lead ore.

Similar Species. *Altaite*, PbTe, and *alabandite*, MnS, like galena, have an NaCl type of structure.

 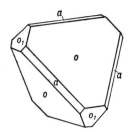

FIG. 10.22. Sphalerite crystals.

SPHALERITE—ZnS

Crystallography. Isometric; $\bar{4}3m$. Tetrahedron, dodecahedron, and cube common forms (Fig. 10.22), but the crystals, frequently highly complex and usually distorted or in rounded aggregates, often show polysynthetic twinning on {111}. Usually found in cleavable masses, coarse to fine granular. Compact, botryoidal, cryptocrystalline.

$F\bar{4}3m; a = 5.41$ Å; $Z = 4$. d's: 3.12(10), 1.910(8), 1.631(7), 1.240(4), 1.106(5).

Physical Properties. *Cleavage,* {011} perfect but some sphalerite is too fine grained to show cleavage. H $3\frac{1}{2}$–4. G 3.9–4.1. *Luster* nonmetallic and resinous to submetallic; also adamantine. *Color* colorless when pure, and green when nearly so. Commonly yellow, brown to black, darkening with increase in iron. Also red (ruby zinc). Transparent to translucent. *Streak* white to yellow and brown.

Composition and Structure. Zn 67, S 33% when pure. Nearly always contains some Fe with the amount of Fe dependent on the temperature and chemistry of the environment. If Fe is in excess as indicated by association with pyrrhotite the amount of FeS in sphalerite can reach 50 mole percent (see also Table 5.3). If sphalerite and pyrrhotite crystallize together, the amount of iron is an indication of the temperature of formation, and sphalerite can be used as a geological thermometer. Mn and Cd are usually present in small amounts in solid solution (see Table 5.3).

The structure of sphalerite is similar to that of diamond with one half of the carbon atoms replaced by Zn and the other half by S (see Figs. 4.44, 10.8a, and 10.15a). Sphalerite is considered the low-temperature cubic polymorph of ZnS, and wurtzite the high-temperature polymorph stable above 1020°C at 1 atmosphere pressure. The *wurtzite* polymorph (space group $P6_3mc$), with Zn atoms in hexagonal closest packing (see Fig. 10.23), shows a very large number of stacking sequences, referred to as wurtzite *polytypes*, which differ in the length of their c axes. For example the 4H polytype has a c axis of 12.46 Å and the 8H polytype a c axis of 24.96 Å (H = hexagonal).

FIG. 10.23. The wurtzite type polymorph of ZnS. See Fig. 10.15a for the sphalerite polymorph.

Diagnostic Features. Sphalerite can be recognized by its striking resinous luster and perfect cleavage. The dark varieties (black jack) can be told by the reddish-brown streak, always lighter than the massive mineral.

Occurrence. Sphalerite, the most important ore mineral of zinc, is extremely common. Its occurrence and mode of origin are similar to those of galena, with which it is commonly found. In the shallow seated lead-zinc deposits of the Tri-state district of Missouri, Kansas, and Oklahoma (now largely exhausted), these minerals are associated with marcasite, chalcopyrite, calcite, and dolomite. Sphalerite with only minor galena occurs in hydrothermal veins and replacement deposits associated with phyrrhotite, pyrite, and magnetite. Sphalerite is also found in veins in igneous rocks and in contact metamorphic deposits.

Zinc is mined in significant amounts in over 40 countries. Although in a few places the ore minerals are hemimorphite and smithsonite and, at Franklin and Sterling Hill, New Jersey, willemite, zincite, and franklinite, most of the world's zinc comes from sphalerite. The principal producing countries are: Canada, CIS, United States, Australia, Poland, Mexico, and Japan. In the United States nearly 60% of the zinc is produced east of the Mississippi River with Tennessee, New York, Pennsylvania, and New Jersey the principal producing states. In the western United States, Idaho, Colorado, and Utah are the chief producers. Elmwood, Tennessee, produces some outstanding specimen material.

Use. The most important ore of zinc. The chief uses for metallic zinc, or *spelter*, are in galvanizing iron; making brass, an alloy of copper and zinc; in electric batteries; and as sheet zinc. Zinc oxide, or zinc white, is used extensively for making paint. Zinc chloride is used as a preservative for wood. Zinc sulfate is used in dyeing and in medicine. Sphalerite also serves as the most important source of cadmium, indium, gallium, and germanium.

Name. Sphalerite comes from the Greek mean-

ing *treacherous*. It was called blende because, although often resembling galena, it yielded no lead; from the German word meaning *blind* or *deceiving*.

Similar Species. *Greenockite*, CdS, a rare mineral mined as a source of cadmium, is isostructural in two polymorphic forms with sphalerite and wurtzite. Cadmium is recovered from greenockite associated with zinc minerals, especially sphalerite.

CHALCOPYRITE—CuFeS$_2$

Crystallography. Tetragonal; $\overline{4}2m$. Commonly tetrahedral in aspect with the disphenoid p {112} dominant (see Figs. 10.24 and 10.25). Other forms shown in Fig. 10.25 are rare. Usually massive.

$I\overline{4}2d$; $a = 5.25$, $c = 10.32$ Å; $Z = 4$. d's: 3.03(10), 1.855(10), 1.586(10), 1.205(8), 1.074(8).

Physical Properties. H $3\frac{1}{2}$–4. G 4.1–4.3. *Luster* metallic. *Color* brass-yellow; often tarnished to bronze or iridescent. *Streak* greenish-black. Brittle.

Composition and Structure. Cu 34.6, Fe 30.4, S 35.0%. It deviates very little from ideal CuFeS$_2$. See Fig. 10.19 for chalcopyrite composition in the Cu-Fe-S system. Its structure can be regarded as a derivative of the sphalerite structure in which half the Zn is replaced by Cu and the other half by Fe. This leads to a doubling of the unit cell (see Fig. 10.15b).

Diagnostic Features. Recognized by its brass-yellow color and greenish-black streak. Distinguished from pyrite by being softer than steel and from gold by being brittle. Known as "fool's gold," a term also applied to pyrite.

Occurrence. Chalcopyrite is the most widely occurring copper mineral and one of the most important sources of that metal. Most sulfide ores contain some chalcopyrite but the most important economically are the hydrothermal vein and replacement deposits. In the low-temperature deposits as in the Tri-State district, it occurs as small crystals associated with galena, sphalerite, and dolomite. Associated with pyrrhotite and pentlandite, it is the chief copper mineral in the ores of Sudbury, Ontario, and similar high-temperature deposits. Chalcopyrite is the principal primary copper mineral in the "porphyry-copper" deposits. Also occurs as an original constituent of igneous rocks; in pegmatite dikes; in contact metamorphic deposits; and disseminated in schistose rocks. It may carry gold or silver and become an ore of those metals. Often in subordinate amount with large bodies of pyrite, making them serve as low-grade copper ores.

A few of the localities and countries in which chalcopyrite is the chief ore of copper are: Cornwall, England; Falun, Sweden; Schemnitz and Schlaggen-

FIG. 10.24. Chalcopyrite and quartz crystals, St. Agnes, Cornwall, England. (Harvard Mineralogical Museum.)

wald, Czechoslovakia; Freiberg, Saxony, Germany; Rio Tinto, Spain; South Africa; Zambia; and Chile. Found widely in the United States but usually with other copper minerals in equal or greater amount; found at Butte, Montana; Bingham, Utah; Jerome, Arizona; Ducktown, Tennessee; and various districts in California, Colorado, and New Mexico. In Canada the most important occurrences of chalcopyrite are at Sudbury, Ontario, and in the Rouyn district, Quebec.

Alteration. Chalcopyrite is the principal source of copper for the secondary minerals malachite, azurite, covellite, chalcocite, and cuprite. Concentrations of copper in the zone of supergene enrichment are often the result of such alteration and removal of copper in solution with its subsequent deposition.

Use. Important ore of copper.

FIG. 10.25. Chalcopyrite crystals.

Name. Derived from Greek word *chalkos* meaning *copper* and from *pyrites*.

Similar Species. *Stannite*, Cu_2FeSnS_4, tetragonal, is a rare mineral and a minor ore of tin. The crystal structure of stannite can be derived from that of chalcopyrite by substitution of Fe and Sn in place of part of the Cu in chalcopyrite, see page 350.

PYRRHOTITE—$Fe_{1-x}S$

Crystallography. Monoclinic; $2/m$, for low-temperature form stable below about 250°C; hexagonal, $6/m2/m2/m$, for high-temperature forms. Hexagonal crystals, usually tabular but in some cases pyramidal (Fig. 10.26), indicate formation as the high-temperature polymorph.

$A2/a$; $a = 12.78$, $b = 6.86$, $c = 11.90$, $\beta = 117°17'$; $Z = 4$. No specific d's are given here because there are several monoclinic forms.

$C6/mmc$; $a = 3.44$, $c = 5.73$ Å; $Z = 2$. d's: 2.97(6), 2.63(8), 1.06(10), 1.718(6), 1.045(8).

Physical Properties. **H** 4. **G** 4.58–4.65. *Luster* metallic. *Color* brownish-bronze. *Streak* black. Magnetic, but varying in intensity; the greater the amount of iron, the lesser the magnetism. Opaque.

Composition and Structure. Most pyrrhotites have a deficiency of iron with respect to sulfur, as

FIG. 10.26. Pyrrhotite crystal.

indicated by the formula $Fe_{1-x}S$, with x between 0 and 0.2. This is referred to as omission solid solution (see page 235 and Fig. 5.17). A complete solid solution series from FeS (with 50 atomic percent Fe) to pyrrhotite, with 44.9 atomic percent Fe, exists in the high-temperature part of the pyrrhotite stability field (from its melting point, at 1190°C, to about 400°C). This phase field has hexagonal symmetry. At lower temperatures (see Fig. 10.27), the pyrrhotite phase field narrows and the single, hexagonal solid solution field of pyrrhotite gives way to several hexagonal and monoclinic types. The monoclinic pyrrhotite compositions center at about Fe_7S_8. The monoclinic variety is stable from 0° to 254°C, where it inverts to the hexagonal type (see Fig. 10.27).

The structure of pyrrhotite is a complex derivative of the NiAs structure type. The sulfur atoms are arranged in approximate hexagonal closest packing. A large number of ordered structures exist for this nonstoichiometric sulfide. Several hexagonal structure types are stable over various temperature ranges in the field labeled "hexagonal varieties of pyrrhotite" in Fig. 10.27. The monoclinic structure is stable only be-

FIG. 10.27. The stability field of pyrrhotite, $Fe_{1-x}S$, in a temperature-composition section for the join FeS-FeS$_2$. The stability field of monoclinic pyrrhotite is fairly narrow and centers on Fe_7S_8. Its maximum temperature limit is 254°C. The hexagonal pyrrhotite field is narrow at 0° but widens all the way to the FeS axis at 140°C. The extent of solid solution is shown by shading. (Simplified after Kissin, S. A., 1974, in *Sulfide Mineralogy, Reviews in Mineralogy*, v. 1, Mineralogical Society of America, Washington, D. C., p. CS–25.)

low 254°C. The structural relations within the various polymorphs are complex. The mineral *troilite* is essentially FeS in composition.

Diagnostic Features. Recognized usually by its massive nature, bronze color, and magnetism.

Occurrence. Pyrrhotite is commonly associated with basic igneous rocks, particularly norites. It occurs in them as disseminated grains or, as at Sudbury, Ontario, as large masses associated with pentlandite, chalcopyrite, or other sulfides. At Sudbury vast tonnages of pyrrhotite are mined principally for the copper, nickel, and platinum that are extracted from associated minerals. Pyrrhotite is also found in contact metamorphic deposits, in vein deposits, and in pegmatites.

Large quantities are known in Finland, Norway, Sweden, and the CIS. Fine crystals are found in Santa Eulalia, Chihuahua, Mexico; in the Morro Velho gold mine, Nova Lima, Minas Gerais, Brazil; and in Trepča, Yugoslavia. In the United States considerable amounts are found at Ducktown, Tennessee.

Use. It is mined for its associated nickel, copper, and platinum. At Sudbury, Ontario, it is also a source of sulfur and an ore of iron.

Name. The name pyrrhotite comes from the Greek meaning *reddish*.

Similar Species. *Troilite*, FeS, common in iron meteorites.

Nickeline—NiAs

Crystallography. Hexagonal; $6/m2/m2/m$. Rarely in tabular crystals. Usually massive, reniform with columnar structure.

$C6/mmc$; $a = 3.61$, $c = 5.02$ Å; $Z = 2$. d's: 2.66(10), 1.961(9), 1.811(8), 1.328(3), 1.071(4).

Physical Properties. H $5-5\frac{1}{2}$. G 7.78. *Luster* metallic. *Color* pale copper-red (hence called *copper nickel*), with gray to blackish tarnish. *Streak* brownish-black. Opaque.

Composition and Structure. Ni 43.9, As 56.1%. Usually a little Fe, Co, and S. As frequently replaced in part by Sb. The structure of NiAs is based on hexagonal closest packing of arsenic atoms giving a sequence *ABABAB*. . . . The nickel atoms are coordinated to six closest As atoms.

Diagnostic Features. Characterized by its copper-red color. Gives nickel test with dimethylglyoxime.

Alteration. Quickly alters to *annabergite* (green nickel bloom, $Ni_3(AsO_4)_2 \cdot 8H_2O$) in moist atmosphere.

Occurrence. Nickeline with other nickel arsenides and sulfides, pyrrhotite, and chalcopyrite, frequently occurs in, or is associated with, norites. Also found in vein deposits with cobalt and silver minerals.

Found in Germany in the silver mines of Saxony, the Harz Mountains, in Hessen-Nassau; and at Cobalt, Ontario.

Use. A minor ore of nickel.

Name. The first name of this mineral, *kupfernickel*, gave the name *nickel* to the metal. It is now called nickeline, after its nickel content.

Similar Species. *Breithauptite*, NiSb, is isostructural with nickeline with similar occurrence and association. Breithauptite has a distinctive light copper-red color on fresh fracture.

Millerite—NiS

Crystallography. Hexagonal; $\bar{3}2/m$ (low-temperature polymorph stable below 379°C). Usually in hairlike tufts and radiating groups of slender to capillary crystals, Fig. 10.28. In velvety incrustations. Rarely in coarse, cleavable masses.

$R3m$; $a = 9.62$, $c = 3.16$ Å; $Z = 9$. d's: 4.77(8), 2.75(10), 2.50(6), 2.22(6), 1.859(10).

Physical Properties. *Cleavage* {$10\bar{1}1$}, {$01\bar{1}2$} good. **H** $3-3\frac{1}{2}$. **G** 5.5 ± 0.2. *Luster* metallic. *Color* pale brass-yellow; with a greenish tinge when in fine hairlike masses. *Streak* black, somewhat greenish.

Composition and Structure. Ni 64.7, S 35.3%. The low-temperature form, stable below 379°C, shows little if any solid solution. Ni and S are both in 5-coordination in the low-temperature form. The high-temperature form (stable above 379°C) has an NiAs type of structure and exhibits considerable metal deficiency, Ni$_{1-x}$S, similar to pyrrhotite.

Diagnostic Features. Characterized by its cap-

FIG. 10.28. Millerite in a calcite-lined vug, Keokuk, Iowa. (Harvard Mineralogical Museum.)

illary crystals and distinguished from minerals of similar color by nickel tests.

Occurrence. Millerite forms as a low-temperature mineral often in cavities and as an alteration of other nickel minerals, or as crystal inclusions in other minerals.

In the United States, it is found with hematite and ankerite at Antwerp, New York; in vugs and geodes in limestone in Harrodsburg, Indiana; Keokuk, Iowa; and Milwaukee, Wisconsin. In coarse, cleavable masses it is a major ore mineral at the Marbridge Mine, Lamotte Township, Quebec.

Use. A subordinate ore of nickel.

Name. In honor of the mineralogist, W. H. Miller (1801–1880), who first studied the crystals.

Pentlandite—(Fe,Ni)$_9$S$_8$

Crystallography. Isometric; $4/m\bar{3}2/m$. Massive, usually in granular aggregates with octahedral parting.

$Fm3m$; $a = 10.07$ Å; $Z = 4$. d's: 5.84(2), 3.04(6), 2.92(2), 2.31(3), 1.78(10).

Physical Properties. Parting on {111}. **H** $3\frac{1}{2}$–4. **G** 4.6–5.0. Brittle. *Luster* metallic. *Color* yellowish-bronze. *Streak* light bronze-brown. Opaque. Nonmagnetic.

Composition and Structure. (Fe,Ni)$_9$S$_8$. Usually the ratio of Fe : Ni is close to 1 : 1. Commonly contains small amounts of Co. A rather complicated face-centered cubic structure with the metal atoms in octahedral as well as tetrahedral coordination with sulfur, see Fig. 10.29. Pure (Fe,Ni)$_9$S$_8$, without Co, is stable up to 610°C in the Fe-Ni-S system. Pentlandite with as much as 40.8 weight percent Co is stable up to 746°C. Pentlandite commonly occurs as exsolution lamellae within pyrrhotite.

Diagnostic Features. Pentlandite closely resembles pyrrhotite in appearance but can be distinguished from it by octahedral parting and lack of magnetism. Gives nickel test with dimethyglyoxime.

Occurrence. Pentlandite usually occurs in basic igneous rocks where it is commonly associated with other nickel minerals and pyrrhotite, and chalcopyrite, and has probably accumulated by magmatic segregation.

Found at widely separated localities in small amounts but its chief occurrences are in Canada where, associated with pyrrhotite, it is the principal source of nickel at Sudbury, Ontario, and the Lynn Lake area, Manitoba. It is also an important ore mineral in similar deposits in the Petsamo district of Karelia in the CIS and in the Kimbalda area of Western Australia.

FIG. 10.29. The structure of pentlandite, $(Fe,Ni)_9S_8$, which consists of Fe and Ni in octahedral and tetrahedral coordination with S. The starlike clusters consist of eight linked $(Fe,Ni)S_4$ tetrahedra. The sulfur atoms are in cubic closest packing.

Use. The principal ore of nickel. The chief use of nickel is in steel. Nickel steel contains $2\frac{1}{2}$–$3\frac{1}{2}$% nickel, which greatly increases the strength and toughness of the alloy, so that lighter machines can be made without loss of strength. Nickel is also an essential constituent of stainless steel. The manufacture of Monel metal (68% Ni, 32% Cu) and Nichrome (38–85% Ni) consumes a large amount of the nickel produced. Other alloys are German silver (Ni, Zn and Cu); metal for coinage—the 5-cent coin of the United States is 25% Ni and 75% Cu; low-expansion metals for watch springs; and other instruments. Nickel is used in plating; although chromium now largely replaces it for the surface layer, nickel is used for a thicker underlayer.

Name. After J. B. Pentland, who first noted the mineral.

Covellite—CuS

Crystallography. Hexagonal; $6/m2/m2/m$. Rarely in tabular hexagonal crystals. Usually massive as coatings or disseminations through other copper minerals.

$P6_3/mmc$; $a = 3.80$, $c = 16.36$ Å; $Z = 6$. d's: 3.06(4), 2.83(6), 2.73(10), 1.899(8), 1.740(5).

Physical Properties. *Cleavage* {0001} perfect giving flexible plates. **H** $1\frac{1}{2}$–2. **G** 4.6–4.76. *Luster* metallic. *Color* indigo-blue or darker. *Streak* lead-gray to black. Often iridescent. Opaque.

Composition and Structure. Cu 66.4, S 33.6%. A small amount of Fe may be present. Covellite has a rather complex structure (see Fig. 10.16). One type of Cu atom is in tetrahedral coordination with S, with the tetrahedra sharing corners to form layers. A second type of Cu is in trigonal coordination with S to build planar layers. The excellent {0001} cleavage is parallel to this layer structure. Covellite is stable up to 507°C, its temperature of decomposition.

Diagnostic Features. Characterized by the indigo-blue color, micaceous cleavage yielding flexible plates, and association with other copper sulfides.

Occurrence. Covellite is not an abundant mineral but is found in most copper deposits as a supergene mineral, usually as a coating, in the zone of sulfide enrichment. It is associated with other copper minerals, principally chalcocite, chalcopyrite, bornite, and enargite, and is derived from them by alteration. Primary covellite is known but uncommon.

Found at Bor, Serbia, Yugoslavia; and Leogang, Austria. In large iridescent crystals from the Calabona Mine, Alghero, Sardinia, Italy. In the United States covellite is found in appreciable amounts and excellent crystals at Butte, Montana, and Summitville, Colorado. Formerly found at Kennecott, Alaska.

Use. A minor ore of copper.

Name. In honor of N. Covelli (1790–1829), the discoverer of the Vesuvian covellite.

CINNABAR—HgS

Crystallography. Hexagonal; 32 (low-temperature polymorph stable below approximately 344°C). High-temperature form, known as *metacinnabar*, isometric; $\overline{4}3m$. Crystals of cinnabar usually rhombohedral, often in penetration twins. Trapezohedral faces rare. Usually fine granular, massive; also earthy, as incrustations and disseminations through the rock. Metacinnabar crystals are tetrahedral, with rough faces; usually massive.

$P3_121$ or $P3_221$; $a = 4.146$, $c = 9.497$ Å; $Z = 3$. d's: 3.37(10), 3.16(8), 2.87(10), 2.07(8), 1.980(8).

$F\overline{4}3m$; $a = 5.852$; $Z = 4$. d's: 3.38(10), 2.93(3), 2.07(5), 1.764(4).

Physical Properties. *Cleavage* {10$\overline{1}$0} perfect. **H** $2\frac{1}{2}$. **G** 8.10. *Luster* adamantine when pure to dull earthy when impure. *Color* vermilion-red when pure to brownish-red when impure. *Streak* scarlet. Transparent to translucent. *Metacinnabar* has a metallic luster and grayish-black color. *Hepatic cinnabar* is an inflammable liver-brown variety of cinnabar with bituminous impurities; usually granular or compact, sometimes with a brownish streak.

Composition and Structure. Hg 86.2, S 13.8%, with small variations in Hg content. Traces of Se and Te may replace S. Frequently impure from admixture of clay, iron oxide, bitumen. The structure of cinnabar, which is different from that of any other sulfide, is based on infinite spiral Hg-S-Hg chains which extend along the c axis; it can be represented by one of two enantiomorphic space groups (see above). *Metacinnabar*, isometric, is stable above about 344°C. Some Hg deficiency is found in this polymorph and its composition may be represented as $Hg_{1-x}S$, with x ranging from 0 to 0.08.

Diagnostic Features. Recognized by its red color and scarlet streak, high specific gravity, and cleavage.

Occurrence. Cinnabar is the most important ore of mercury but is found in quantity at comparatively few localities. Occurs as impregnations and as vein fillings near recent volcanic rocks and hot springs and evidently deposited near the surface from solutions which were probably alkaline. Associated with pyrite, marcasite, stibnite, and sulfides of copper in a gangue of opal, chalcedony, quartz, barite, calcite, and fluorite.

The important localities for the occurrence of cinnabar are at Almaden, Spain; Idria, Yugoslavia; and in superb twinned crystals in the provinces of Kweichow and Hunan, China. In the United States deposits are in California at New Idria, and New Almaden.

Use. The only important source of mercury. The principal uses of mercury are in electrical apparatus, industrial control instruments, electrolytic preparation of chlorine and caustic soda, and in mildew proofing of paint. Other lesser but important uses are in dental preparations, scientific instruments, drugs, catalysts, and in agriculture. Formerly a major use of mercury was in the amalgamation process for recover-

ing gold and silver from their ores, but amalgamation has been essentially abandoned in favor of other methods of extraction.

Name. The name cinnabar is supposed to have come from India, where it is applied to a red resin.

REALGAR—AsS

Crystallography. Monoclinic; $2/m$. Found in short, vertically striated, prismatic crystals (Fig. 10.30). Frequently coarse to fine granular and often earthy and as an incrustation.

$P2_1/n$; $a = 9.29$, $b = 13.53$, $c = 6.57$ Å; $\beta = 106°33'$, $Z = 16$. d's: 5.40(10), 3.19(9), 2.94(8), 2.73(8), 2.49(5).

Physical Properties. *Cleavage* {010} good. **H** $1\frac{1}{2}$–2. **G** 3.48. Sectile. *Luster* resinous. *Color* and *streak* red to orange. Translucent to transparent.

Composition and Structure. As 70.1, S 29.9%. The realgar structure contains ringlike groups of As_4S_4, somewhat similar to the rings of S_8 in native sulfur. Each As is bonded covalently to another arsenic as well as to two sulfur atoms.

Diagnostic Features. Realgar is distinguished by its red color, resinous luster, orange-red streak, and almost invariable association with orpiment.

Alteration. On long exposure to light disintegrates to a reddish-yellow powder.

Occurrence. Realgar is found in veins of lead, silver, and gold ores associated with orpiment, other arsenic minerals, and stibnite. It also occurs as a volcanic sublimation product and as a deposit from hot springs.

Found in good crystals at Nagyág, Romania; Binnenthal, Switzerland; and Allchar, Macedonia. In the United States realgar is found at Mercur, Utah; at Manhattan, Nevada; and deposited from the geyser waters in the Norris Geyser Basin, Yellowstone National Park.

FIG. 10.30. Realgar.

Allchar, Macedonia

FIG. 10.31. Orpiment.

Use. Realgar was used in fireworks to give a brilliant white light when mixed with saltpeter and ignited. Today, artificial arsenic sulfide is used for this purpose. It was formerly used as a pigment.

Name. The name is derived from the Arabic, Rahj al ghar, *powder of the mine.*

ORPIMENT—As$_2$S$_3$

Crystallography. Monoclinic; $2/m$. Crystals small tabular or short prismatic (Fig. 10.31), and rarely distinct; many pseudo-orthorhombic. Usually in foliated or columnar masses.

$P2_1/n$; $a = 11.49$, $b = 9.59$, $c = 4.25$ Å, $\beta = 90°27'$; $Z = 4$. d's: 4.78(10), 2.785(4), 2.707(6), 2.446(6), 2.085(4).

Physical Properties. *Cleavage* {010} perfect; cleavage laminae flexible but not elastic. Sectile. **H** $1\frac{1}{2}$–2. **G** 3.49. *Luster* resinous, pearly on cleavage face. *Color* lemon-yellow. *Streak* pale yellow. Translucent.

Composition and Structure. As 61, S 39%. Contains up to 2.7% Sb. In the structure of orpiment trigonal pyramids of AsS$_3$ can be recognized that share corners to form six-membered rings. These rings are linked to produce a corrugated As$_2$S$_3$ layer structure. The bonds within the layers are essentially covalent, whereas those between the layers are van der Waals in nature. The perfect {010} cleavage is parallel to these layers.

Diagnostic Features. Characterized by its yellow color and foliated structure. Distinguished from sulfur by its perfect cleavage.

Occurrence. Orpiment is a rare mineral, associated usually with realgar and formed under similar conditions. Found in good crystals at several localities in Romania and Czechoslovakia. In the United States it occurs at Mercur, Utah, and at Manhattan, Nevada. Deposited with realgar from geyser waters in the Norris Geyser Basin, Yellowstone National Park.

Use. Used in dyeing and in a preparation for the removal of hair from skins. Artificial arsenic sulfide is largely used in place of the mineral. Both realgar and orpiment were formerly used as pigments but this use has been discontinued because of their poisonous nature.

Name. Derived from the Latin, *auripigmentum,* "golden paint," in allusion to its color and because the substance was supposed to contain gold.

STIBNITE—Sb$_2$S$_3$

Crystallography. Orthorhombic; $2/m2/m2/m$. Slender prismatic habit, prism zone vertically striated. Crystals often steeply terminated (Figs. 10.32 and 10.33) and sometimes curved or bent (Fig. 10.33). Often in radiating crystal groups or in bladed forms with prominent cleavage. Massive, coarse to fine granular.

Pbnm; $a = 11.22$, $b = 11.30$, $c = 3.84$ Å; $Z = 4$. d's: 5.07(4), 3.58(10), 2.76(3), 2.52(4), 1.933(5).

Physical Properties. *Cleavage* {010} perfect, showing striations parallel to [100]. **H** 2. **G** 4.52–4.62. *Luster* metallic, splendent on cleavage surfaces. *Color* and *streak* lead-gray to black. Opaque.

Composition and Structure. Sb 71.4, S 28.6%. May carry small amounts of Au, Ag, Fe, Pb, and Cu. The structure of stibnite is composed of zigzag chains of closely bonded Sb and S atoms that are parallel to the c axis. Sb-S distances in the chains range from 2.5Å to 3.1Å; these represent covalent linking. Distances between adjoining bands are larger, ranging from 3.2 to 3.6 Å (see Fig. 10.34). The long, striated prisms (//c) of stibnite are parallel to these structural chains. The excellent {010} cleavage occurs between Sb-S chains.

Diagnostic Features. Characterized by its easy fusibility, bladed habit, perfect cleavage in one direction, lead-gray color, and soft black streak. Fusible in a candle flame.

Occurrence. Stibnite is found in low-temperature hydrothermal veins or replacement deposits and in hot spring deposits. It is associated with other anti-

FIG. 10.32. Stibnite.

FIG. 10.33. Cluster of stibnite crystals, and a single, curved crystal. Ischinokowa, Japan. (Harvard Mineralogical Museum.)

FIG. 10.34. *(a)* The structure of stibnite, Sb_2S_3, projected on (001). The heights of the atoms are indicated by fractions along the *c* axis. The regions with the lined patterns represent the projections of zigzag chains parallel to the *c* axis. The interatomic distances within the chains are shown as ranging between 2.5 and 3.1 Å; those between chains range from 3.4 to 3.6 Å. (Redrawn after Hellner, E. and Leineweber, G., 1956, Über komplex zusammengesetzte sulfidische Erze. *Zeitschrift für Kristallographie*, v. 107, pp. 105–154.) *(b)* Representation of one of the zigzag chains in the structure of stibnite.

(a)   (b)

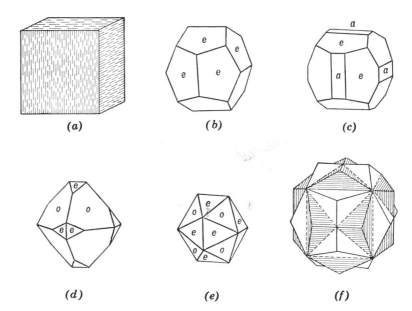

FIG. 10.35. Pyrite crystals. *(a)* Striated cube. *(b)* Pyritohedron {210}; *(c)* Cube and pyritohedron. *(d)* and *(e)* Octahedron and pyritohedron. *(f)* Twinned pyritohedrons, *iron cross*.

mony minerals that have formed as the product of its decomposition, and with galena, cinnabar, sphalerite, barite, realgar, orpiment, and gold.

There are numerous localities in Romania for well-crystallized stibnite, but the finest crystals have come from the province of Iyo, Island of Shikoku, Japan. The world's most important producing district is in the province of Hunan, China. Found in quantity at only a few localities in the United States, notably Manhattan, Nevada.

Use. The chief ore of antimony but much of the metal is produced as a by-product from smelting lead ores. Antimony trioxide is used as a pigment and for making glass.

Name. The name stibnite comes from an old Greek word that was applied to the mineral.

Similar Species. *Bismuthinite*, Bi_2S_3, is a rare mineral isostructural with stibnite and with similar physical properties.

PYRITE—FeS₂

Crystallography. Isometric; $2/m\bar{3}$. Frequently in crystals (Figs. 10.35, 10.36 and 10.37). The most common forms are the cube, the faces of which are usually striated, the pyritohedron, and the octahedron. Figure 10.35*f* shows a penetration twin, known as the *iron cross* with [001] the twin axis. Also massive, granular, reniform, globular, and stalactitic.

$Pa3$; $a = 5.42$ Å; $Z = 4$. *d*'s: 2.70(7), 2.42(6), 2.21(5), 1.917(4), 1.632(10).

Physical Properties. *Fracture* conchoidal. Brittle. **H** $6-6\frac{1}{2}$ (unusually hard for a sulfide). **G** 5.02. *Luster* metallic, splendent. *Color* pale brass-yellow; may

be darker because of tarnish. *Streak* greenish or brownish-black. Opaque. Paramagnetic.

Composition and Structure. Fe 46.6, S 53.4%. May contain small amounts of Ni and Co. Some analyses show considerable Ni, and a complete solid solution series exists between pyrite and *bravoite*, $(Fe,Ni)S_2$. Frequently carries minute quantities of Au and Cu as microscopic impurities. The structure of pyrite can be considered as a modified NaCl type of structure with Fe in the Na and S_2 in the Cl positions (Fig. 10.17*a*). FeS_2 occurs in two polymorphs, pyrite and *marcasite* (see page 351).

FIG. 10.36. Pyrite, Bingham, Utah. (Harvard Mineralogical Museum.)

0   2   4
cm

FIG. 10.37. Pyrite. (Harvard Mineralogical Museum.)

Diagnostic Features. Distinguished from chalcopyrite by its paler color and greater hardness, from gold by its brittleness and hardness, and from marcasite by its deeper color and crystal form.

Alteration. Pyrite is easily altered to oxides of iron, usually limonite. In general, however, it is much more stable than marcasite. Pseudomorphic crystals of limonite after pyrite are common. Pyrite veins are usually capped by a cellular deposit of limonite, termed *gossan*. Rocks that contain pyrite are unsuitable for structural purposes because the ready oxidation of pyrite would serve both to disintegrate the rock and to stain it with iron oxide.

Occurrence. Pyrite is the most common and widespread of the sulfide minerals and the worldwide occurrences of fine crystals are too many to enumerate. A few of the notable localities of crystals in the United States are: Chester, Vermont; Leadville and Central City, Colorado; and Bingham, Utah (Fig. 10.36). Fine crystals are found in La Libertad and Huanzala, Peru. Sculptural groups of crystals occur at Ambasaguas and Navajun, Spain.

Pyrite has formed at both high and low temperatures, but the largest masses probably at high temperature. It occurs as magmatic segregations, as an accessory mineral in igneous rock, and in contact metamorphic deposits and hydrothermal veins. Pyrite is a common mineral in sedimentary rocks, being both primary and secondary. It is associated with many minerals but found most frequently with chalcopyrite, sphalerite, and galena.

Large and extensively developed deposits occur at Rio Tinto and elsewhere in Spain and also in Portugal. Important deposits of pyrite in the United States are in Virginia, where it occurs in large lenticular masses that conform in position to the foliation of the enclosing schists; in St. Lawrence County, New York, at the Davis Mine, near Charlemont, Massachusetts; and in various places in California, Colorado, and Arizona.

Use. Pyrite is often mined for the gold or copper associated with it. Because of the large amount of sulfur present in the mineral, it is used as an iron ore only in those countries where oxide ores are not available. Its chief use is a source of sulfur for sulfuric acid and *copperas* (ferrous sulfate). Copperas is used in dyeing, in the manufacture of inks, as a preservative of wood, and as a disinfectant. Pyrite may be cut as a gemstone which is sold under the name of its polymorph, marcasite.

Name. The name *pyrite* is from a Greek word meaning *fire*, in allusion to the brilliant sparks emitted when struck by steel.

MARCASITE—FeS_2

Crystallography. Orthorhombic; $2/m2/m2/m$. Crystals commonly tabular {010}; less commonly prismatic [001] (Fig. 10.38). Often twinned, giving cockscomb and spear-shaped groups (Figs. 10.39 and 10.40). Usually in radiating forms. Often stalactitic, having an inner core with radiating structure and covered with irregular crystal groups. Also globular and reniform.

$Pmnn$; $a = 4.45$, $b = 5.42$, $c = 3.39$ Å; $Z = 2$. d's: 2.70(10), 2.41(6), 2.32(6), 1.911(5). 1.755(9).

Physical Properties. **H** $6-6\frac{1}{2}$. **G** 4.89. *Luster* metallic. *Color* pale bronze-yellow to almost white on fresh fracture, hence called *white iron pyrites*. Yellow to brown tarnish. *Streak* grayish-black. Opaque.

Composition and Structure. Of constant composition, FeS_2, dimorphous with pyrite. The marcasite structure is shown in Fig. 10.17b. The configuration of nearest neighbor atoms is the same as in pyrite. The

FIG. 10.38. Marcasite crystals.

FIG. 10.39. "Cockscomb" marcasite.

stability relationships of pyrite and marcasite are still unclear. Experimental evidence indicates that marcasite is metastable relative to pyrite and pyrrhotite above about 157°C. Geological occurrences of marcasite indicate a lower temperature stability range than for pyrite which may occur in magmatic segregations.

Diagnostic Features. Usually recognized and distinguished from pyrite by its pale yellow color, its crystals or its fibrous habit.

Alteration. Marcasite usually disintegrates more easily than pyrite with the formation of ferrous sulfate and sulfuric acid. The white powder that forms on marcasite is *melanterite*, $FeSO_4 \cdot 7H_2O$.

Occurrence. Marcasite is found in metalliferous veins, frequently with lead and zinc ores. It is less stable than pyrite, being easily decomposed, and is much less common. It is deposited at low temperatures from acid solutions and commonly formed under surface conditions as a supergene mineral. Marcasite most frequently occurs as replacement deposits in limestone, and often in concretions embedded in clays, marls, and shales.

Found abundantly in clay near Carlsbad and elsewhere in Czechoslovakia; in the chalk marl of Folkestone and Dover, England. In the United States marcasite is found with zinc and lead deposits of the Joplin, Missouri district; at Mineral Point, Wisconsin; and at Galena, Illinois.

Use. Marcasite is used to a slight extent as a source of sulfur.

Name. Derived from an Arabic word, at one time applied generally to pyrite.

MOLYBDENITE–MoS₂

Crystallography. Hexagonal; $6/m2/m2/m$. Crystals in hexagonal plates or short, slightly tapering prisms. Commonly foliated, massive, or in scales.

$P6_3/mmc$; $a = 3.16$, $c = 12.32$ Å; $Z = 2$. d's: 6.28(10), 2.28(9), 1.824(6), 1.578(4), 1.530(4).

Physical Properties. *Cleavage* {0001} perfect, laminae flexible but not elastic. Sectile. **H** $1-1\frac{1}{2}$. **G** 4.62–4.73. Greasy feel. *Luster* metallic. *Color* lead-gray. *Streak* grayish-black. Opaque.

Composition and Structure. Mo 59.9, S 40.1%, of essentially constant composition. In the structure of molybdenite a sheet of Mo atoms is sandwiched between two sheets of S atoms, the three sheets together forming a layered structure. The bond strengths within

FIG. 10.40. Marcasite, Ottawa County, Oklahoma. (Harvard Mineralogical Museum.)

0   2   4
cm

the layer are much stronger than between the layers, giving rise to the excellent {0001} cleavage. MoS_2 occurs as two polytypes of which one is hexagonal (2H) and the other rhombohedral (3R).

Diagnostic Features. Resembles graphite but is distinguished from it by higher specific gravity; by a blue tone to its color, whereas graphite has a brown tinge. On glazed porcelain, it gives a greenish streak, graphite a black streak.

Alteration. Molybdenite alters to yellow *ferrimolybdite*, $Fe_2(MoO_4)_3 \cdot 8H_2O$.

Occurrence. Molybdenite forms as an accessory mineral in certain granites; in pegmatites and aplites; also associated will porphyry copper deposits. Commonly in high-temperature vein deposits associated with cassiterite, scheelite, wolframite, and fluorite. Also in contact metamorphic deposits with lime silicates, scheelite, and chalcopyrite.

Exceptionally good crystals are found at Kingsgate, New South Wales, Australia. In the United States molybdenite is found in many localities; examples are Okanogan County, Washington, and Questa, New Mexico. From various pegmatites in Ontario, Canada. The bulk of the world's supply comes from Climax, Colorado, where molybdenite occurs in quartz veinlets in silicified granite with fluorite and topaz. Much molybdenum is produced at Bingham Canyon, Utah, as a by-product of the copper mining.

Use. The principal ore of molybdenum.

Name. The name molybdenite comes from the Greek word *molybdos* meaning *lead*.

COBALTITE—(Co,Fe)AsS

Crystallography. Orthorhombic; *mm*2. Pseudoisometric with forms that appear isometric, that is, cubes and pyritohedrons with the faces striated as in pyrite. Also granular.

$Pca2_1$; $a = 5.58$, $b = 5.58$, $c = 5.58$ Å; $Z = 4$. *d*'s: 2.77(8), 2.49(10), 2.27(9), 1.680(10), 1.490(8).

Physical Properties. *Cleavage* pseudocubic, perfect. Brittle. **H** $5\frac{1}{2}$. **G** 6.33. *Luster* metallic. *Color* silver-white, inclined to red. *Streak* grayish-black.

Composition and Structure. Usually contains considerable Fe (maximum about 10%) and lesser amounts of Ni. *Gersdorffite*, NiAsS, and cobaltite form a complete solid solution series, but intermediate compositions are rare. The structure of cobaltite is closely related to that of pyrite (Fig. 10.17a) in which half the S_2 pairs are replaced by As. Such a substitution causes a lowering of the symmetry of the structure of cobaltite as compared to that of pyrite. Natural cobaltites have a somewhat disordered distribution of As and S in the structure.

Diagnostic Features. Although in crystal form cobaltite resembles pyrite, it can be distinguished by its silver color and cleavage.

Occurrence. Cobaltite is usually found in high-temperature deposits, as disseminations in metamorphosed rocks, or in vein deposits with other cobalt and nickel minerals. Notable occurrences of cobaltite are at Tunaberg, Sweden, and Cobalt, Ontario. The largest producer of cobalt today is Zaire, where oxidized cobalt and copper ores are associated.

Use. An ore of cobalt.

Name. In allusion to its chemistry.

ARSENOPYRITE—FeAsS

Crystallography. Monoclinic; $2/m$, pseudo-orthorhombic. Crystals are commonly prismatic elongated on *c* and less commonly on *b* (Fig. 10.41). Twinning on {100} and {001} produces pseudo-orthorhombic crystals; on {110} generates contact or penetration twins; may be repeated, as in marcasite.

$P2_1c$; $a = 5.74$, $b = 5.67$, $c = 5.78$ Å, $\beta = 112°17'$; $Z = 4$. *d*'s: 2.68(10), 2.44(9), 2.418(9), 2.412(9), 1.814(9).

Physical Properties. *Cleavage* {101} poor. **H** $5\frac{1}{2}$–6. **G** 6.07. *Luster* metallic. *Color* silver-white. *Streak* black. Opaque.

Composition and Structure. Close to FeAsS with some variation in As and S contents ranging from $FeAs_{0.9}S_{1.1}$ to $FeAs_{1.1}S_{0.9}$. Cobalt may replace part of the Fe and a series extends to *glaucodot*, (Co,Fe)AsS. The structure of arsenopyrite is a derivative from the marcasite structure type (see Fig. 10.17c) in which one-half of the S is replaced by As.

Diagnostic Features. Distinguished from marcasite by its silver-white color. Its crystal form distinguishes it from skutterudite.

Occurrence. Arsenopyrite is the most common mineral containing arsenic. It occurs with tin and tungsten ores in high-temperature hydrothermal deposits, associated with silver and copper ores, galena,

FIG. 10.41. Aresnopyrite crystals.

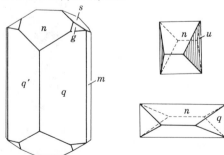

sphalerite, pyrite, chalcopyrite. Frequently associated with gold. Often found sparingly in pegmatites, in contact metamorphic deposits, disseminated in crystalline limestones.

Arsenopyrite is a widespread mineral and is found in considerable abundance in many localities, as at Freiberg and Munzig, Saxony, Germany; with tin ores in Cornwall, England; from Tavistock, Devonshire; in various places in Bolivia. In the United States it is associated with gold at Lead, South Dakota. Large quantities occur at Deloro, Ontario.

Use. The principal source of arsenic. Most of the arsenic produced is recovered in the form of the oxide, as a by-product in the smelting of arsenical ores for copper, gold, lead, and silver. Metallic arsenic is used in some alloys, particularly with lead in shot metal. Arsenic is used chiefly, however, in the form of white arsenic or arsenious oxide in medicine, insecticides, preservatives, pigments, and glass. Arsenic sulfides are used in paints and fireworks.

Name. Arsenopyrite is a contraction of the older term *arsenical pyrites*.

SKUTTERUDITE—(Co,Ni)As₃

Crystallography. Isometric; $2/m\overline{3}$. Common crystal forms are cube and octahedron, more rarely dodecahedron and pyritohedron. Usually massive, dense to granular.

$Im3$; $a = 8.21 - 8.29$ Å; $Z = 8$; d's: 2.61(10), 2.20(8), 1.841(9), 1.681(7), 1.616(9).

Physical Properties. H $5\frac{1}{2}$–6. G 6.5 ± 0.4. Brittle. *Luster* metallic. *Color* tin-white to silver-gray. *Streak* black. *Opaque.*

Composition and Structure. Essentially $(Co,Ni)As_3$ but Fe usually substitutes for some Ni or Co. The high nickel varieties are called *nickel skutterudite*, $(Ni,Co)As_3$. A remarkable feature of the structure of skutterudite is the square grouping of As as As_4. The unit cell contains 8(Co + Ni) atoms and six As_4 groups which leads to the formula $(Co,Ni)As_3$. *Smaltite*, $(Co,Ni)As_{3-x}$ and *chloanthite*, $(Ni,Co)As_{3-x}$ (with $x = 0.5 - 1.0$) have skutterudite-like structures but a deficiency of As.

Diagnostic Features. Tin-white to silver-gray color. Chemical tests may be necessary for identification.

Occurrence. Skutterudite is usually found with cobaltite and nickeline in veins formed at moderate temperature. Native silver, bismuth, arsenopyrite, and calcite are also commonly associated with it.

Notable localities are Skutterud, Norway; Annaberg and Schneeberg, in Saxony, Germany; and Cobalt, Ontario, where skutterudite is associated with sil-

ver ores. Well-crystallized material is found near Bou Azzer, Morocco.

Use. An ore of cobalt and nickel. Cobalt is chiefly used in alloys for making permanent magnets and high-speed tool steel. Cobalt oxide is used as blue pigment in pottery and glassware.

Name. Skutterudite from the locality, Skutterud, Norway.

Similar Species. *Linnaeite*, Co_3S_4 associated with cobalt and nickel minerals.

SULFOSALTS

The term *sulfosalt* was originally proposed to indicate that a compound was a salt of one of a series of acids in which sulfur had replaced the oxygen of an ordinary acid. Because such acids are purely hypothetical, it is somewhat misleading to endeavor to thus explain this class of minerals. Nevertheless, the term sulfosalt has been retained for indicating a certain type of unoxidized sulfur mineral that is structurally distinct from a sulfide.

The sulfosalts comprise a diverse and relatively large group of minerals with about 100 species. They differ from sulfides, sulfarsenides, and arsenides in that As, Sb, and rarely Bi play a role more or less like that of metals in the structure; in sulfarsenides and arsenides the semimetals take the place of sulfur in the structure. For example, in arsenopyrite, FeAsS (Fig. 10.17c), the As is substituted for S in a marcasite type of structure. In enargite, Cu_3AsS_4, on the other hand, As enters the metal position of the wurtzite type of structure and is coordinated to four neighboring S ions (see Fig. 10.42). Sulfosalts may be considered double sulfides. As such enargite, Cu_3AsS_4, may be considered $3Cu_2S \cdot As_2S_5$.

The sulfosalts usually occur as minor minerals in hydrothermal veins associated with the more common sulfides. With only rare exceptions they are compounds containing silver, copper, or lead but only a very few are abundant enough to have served as ores of these metals. Mention is made below of only five of the more important sulfosalts.

ENARGITE—Cu₃AsS₄

Crystallography. Orthorhombic; $mm2$. Crystals elongated parallel to c and vertically striated, also tabular parallel to {001}. Columnar, bladed, massive.

$Pnm2_1$; $a = 6.41$, $b = 7.42$, $c = 6.15$ Å; $Z = 2$. d's: 3.22(10), 2.87(8), 1.859(9), 1.731(6), 1.590(5).

Physical Properties. *Cleavage* {110} perfect,

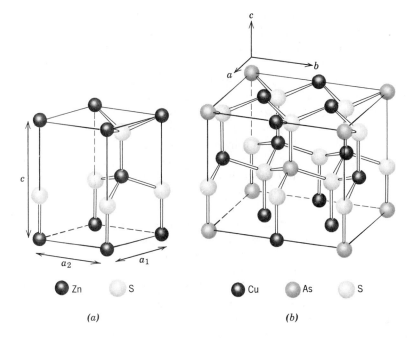

FIG. 10.42. *(a)* Wurtzite structure type of ZnS. *(b)* Enargite, Cu_3AsS_4, an orthorhombic, derivative of the wurtzite structure. (After Wuensch, B. J., 1974, in *Sulfide Mineralogy, Reviews in Mineralogy*, v. 1, Mineralogical Society of America, Washington, D. C.)

{100} and {010} distinct. **H** 3. **G** 4.45. *Luster* metallic. *Color* and *streak* grayish-black to iron-black. Opaque.

Composition and Structure. In Cu_3AsS_4: Cu 48.3, As 19.1, S 32.6%. Sb substitutes for As up to 6% by weight, and some Fe and Zn are usually present. The structure of enargite can be regarded as a derivative of the wurtzite structure (Fig. 10.42), in which $\frac{3}{4}$ of Zn is replaced by Cu and $\frac{1}{4}$ of Zn by As. In chemical terms Zn_4S_4 becomes Cu_3AsS_4. The low-temperature polymorph of Cu_3AsS_4 (stable below 320°C) is *luzonite* with a tetragonal structure ($I\bar{4}2m$). *Famatinite*, Cu_3SbS_4, the antimony analogue of enargite is isostructural with luzonite. Extensive solid solution exists in famatinite toward luzonite, and famatinite toward enargite.

Diagnostic Features. Characterized by its color and cleavage. Distinguished from stibnite by a test for Cu.

Occurrence. Enargite is a comparatively rare mineral, found in vein and replacement deposits formed at moderate temperatures associated with pyrite, sphalerite, bornite, galena, tetrahedrite, covellite, and chalcocite.

Notable localities: Morococha, Quiruvilca, and Cerro de Pasco, Peru; also from Chile and Argentina; Island of Luzon, Philippines. In the United States it was an important ore mineral at Butte, Montana, and to a lesser extent in the Tintic district, Utah. Occurs in the silver mines of the San Juan Mountains, Colorado.

Use. An ore of copper.

Name. From the Greek word *enarges* meaning *distinct*, in allusion to the cleavage.

Pyrargyrite—Ag_3SbS_3, Proustite—Ag_3AsS_3

These minerals are known as the *ruby silvers* and in places have been important ores. They are isostructural with similar crystals forms, physical properties, and occurrences but there is little solid solution between them. Ruby red, hexagonal ($3m$) prismatic crystals make handsome mineral specimens.

Tetrahedrite—$Cu_{12}Sb_4S_{13}$, Tennantite—$Cu_{12}As_4S_{13}$

These two isostructural minerals form a complete solid solution series. They are so similar in crystallographic and physical properties that it is impossible to distinguish them by inspection. They are isometric, $\bar{4}3m$, and frequently occur in tetrahedral crystals (Fig. 10.43). Fe, Zn and less commonly Ag, Pb, and Hg may

FIG. 10.43. Tetrahedrite crystals.

substitute for Cu. The argentiferous variety, *freibergite*, may contain 18% Ag and thus become a silver ore.

REFERENCES AND SUGGESTED READING

Anthony, J. W., Bideaux, R. A., Bladh, K. W. and Nichols, M. C., 1990, *Handbook of Mineralogy*, v. 1, *Elements, Sulfides, Sulfosalts.* Mineral Data Publications, Tucson, Ariz., 588 pp.

Bragg, L. and Claringbull, G. F., 1965, *Crystal Structures of Minerals*, G. Bell and Sons, Ltd., London, 409 pp.

Cabri, L. J., 1972, The mineralogy of the platinum-group elements. *Minerals Science and Engineering*, v. 4, pp. 3–29.

Deer, W. A., Howie, R. A. and Zussman, J., 1962, *Rock Forming Minerals*, v. 5, *Non-Silicates.* John Wiley & Sons, New York, 371 pp.

———, 1992, *An Introduction to the Rock Forming Minerals*, 2nd ed. John Wiley & Sons, New York, 691 pp.

Fleischer, M. and Mandarino, J. A., 1991, *Glossary of Mineral Species 1991.* Mineralogical Record Inc., Tucson, Ariz., 256 pp.

Nickel, E. H. and Nichols, M. C., 1991, *Mineral Reference Manual.* Van Nostrand Reinhold, New York, 250 pp.

Palache, C., Berman, H. and Frondel, C., 1944, *The System of Mineralogy*, 7th ed., v. 1, *Elements, Sulfides, Sulfosalts, Oxides.* John Wiley & Sons, New York, 843 pp.

Strunz, H., 1970, *Mineralogische Tabellen*, 5th ed. Akademische Verlagsgesellschaft, Geest und Portig K. G., Leipzig, 621 pp.

Sulfide Mineralogy. Reviews in Mineralogy, 1974, v. 1. Mineralogical Society of America, Washington, D.C.

Vaughan, D. J. and Craig, J. R., 1978, *Mineral Chemistry of Metal Sulfides.* Cambridge University Press, New York, 493 pp.

Wilks, J. and Wilks, E., 1991, *Properties and Applications of Diamond.* Butterworth-Heinemann Ltd., Oxford, 525 pp.

Wilson, W. E., 1982, The gold-containing minerals: A review. *Mineralogical Record*, v. 13, pp. 389–400.

CHAPTER 11

SYSTEMATIC MINERALOGY
PART II: OXIDES, HYDROXIDES, AND HALIDES

The *oxides* are a group of minerals that are relatively hard, dense, and refractory and generally occur as accessory minerals in igneous and metamorphic rocks and as resistant detrital grains in sediments. The *hydroxides*, on the other hand, tend to be lower in hardness and density and are found mainly as secondary alteration or weathering products. The *halides* comprise about 80 chemically related minerals with diverse structure types and of diverse geological origins. Here we will discuss only six of the most common halides.

OXIDES

The oxide minerals include those natural compounds in which oxygen is combined with one or more metals. They are here grouped as simple oxides and multiple oxides. The simple oxides, compounds of one metal and oxygen, are of several types with different $X : O$ ratios (the ratio of metal to oxygen) as X_2O, XO, X_2O_3. Although not described on the following pages, ice, H_2O (see Fig. 9.1), is a simple oxide of the X_2O type in which hydrogen is the cation. The most common of all oxides, SiO_2, quartz and its polymorphs, is not considered in this chapter, but instead is treated in Chapter 13 with the silicates because the structure of quartz (and its polymorphs) is most closely related to that of other Si-O compounds. The multiple oxides, XY_2O_4, have two nonequivalent metal atom sites (A and B). Within the oxide class are several minerals

of great economic importance. These include the chief ores or iron (hematite and magnetite), chromium (chromite), manganese (pyrolusite, as well as the hydroxides, manganite, and romanechite), tin (cassiterite), and uranium (uraninite).

The bond type in oxide structures is generally strongly ionic in contrast to the sulfide structures with ionic, covalent, as well as metallic bonding. One of the first structure determinations reported by Sir Lawrence Bragg in 1916 was that of cuprite, Cu_2O, in which the oxygen atoms are arranged at the corners and center of tetrahedral groups. The Cu atoms lie halfway between the oxygens (see Fig. 11.1) within the tetrahedral groups.

FIG. 11.1. Structure of cuprite. *(a)* Oxygen atoms at corners and center of tetrahedral group; Cu halfway between two oxygens. *(b)* Oxygens in a cubic lattice, as shown in a polyhedral representation of the cuprite structure.

- Cu
- Oxygen

(a)   *(b)*

The structures of the *hematite group* are based on hexagonal closest packing of oxygens with the cations in octahedral coordination between them (see Fig. 11.2a). As can be seen in a basal projection of this structure only $\frac{2}{3}$ of the octahedral spaces are occupied by Fe^{3+} or Al^{3+}. The presence of $\frac{1}{3}$ oxygen octahedra without central Al^{3+} or Fe^{3+} ions is related to the electrostatic valency (e.v.) or bond strength of the $Al^{3+} - O^{2-}$ and $Fe^{3+} - O^{2-}$ bonds. Because Al^{3+} is surrounded by six oxygens, the e.v. of each of the six Al-O bonds $= \frac{1}{2}$. Each oxygen is shared between four octahedra, which means that four bonds of e.v. $= \frac{1}{2}$

FIG. 11.2. Structures of the hematite group. *(a)* Basal sheet of octahedra in corundum, Al_2O_3, or hematite, Fe_2O_3, with one octahedron vacant for every two octahedra with Al or Fe^{3+} in center. *(b)* Vertical section through the corundum structure showing the locations of filled and empty octahedra. Note locations of Al^{3+} within octahedra. *(c)* Schematic vertical cross section through the ilmenite structure. This representation is the same as in Fig. 11.2b with oxygen positions and octahedral outlines eliminated; only cation locations are shown.

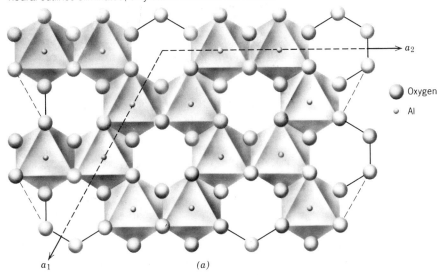

(a)

Vacant	○	○	Vacant
×	×	Vacant	×
○	Vacant	○	○
Vacant	×	×	Vacant
○	○	Vacant	○
×	Vacant	×	×

○ Fe
× Ti

(b)

(c)

OXIDES

X_2O and XO types	
Cuprite	Cu_2O
Zincite	ZnO

X_2O_3 type		XY_2O_4 type	
Hematite Group		**Spinel Group**	
Corundum	Al_2O_3	Spinel	$MgAl_2O_4$
Hematite	Fe_2O_3	Gahnite	$ZnAl_2O_4$
Ilmenite	$FeTiO_3$	Magnetite	Fe_3O_4
		Franklinite	(Zn,Fe,Mn)- $(Fe,Mn)_2O_4$
		Chromite	$FeCr_2O_4$

XO_2 type (excluding SiO_2)			
Rutile Group			
Rutile	TiO_2	Chrysoberyl	$BeAl_2O_4$
Pyrolusite	MnO_2	Columbite	(Fe,Mn)-
Cassiterite	SnO_2		$(Nb,Ta)_2O_6$
Uraninite	UO_2		

can radiate from an oxygen position. In the (0001) plane, as in Fig. 11.2a, this allows for only *two* Al-O bonds from each oxygen, as shown by the geometry of two octahedra sharing one oxygen corner. A vertical section through the corundum (or hematite) structure (Fig. 11.2b) shows the arrangement of the Al^{3+} (or Fe^{3+}) ions and of the omitted cations as well. In the vertical stacking of octahedra each octahedron shares a face between two adjoining layers. The cations, Al^{3+} or Fe^{3+}, within the octahedra that share faces will tend to move away from the shared face on account of the repulsive forces between them (see Fig. 11.2b). The corundum and hematite structures have space group $R\bar{3}m$. The ilmenite structure, in which Fe and Ti are arranged in alternating Fe-O and Ti-O layers (Fig. 11.2c), has a lower symmetry (space group $R\bar{3}$) because of the ordered Fe and Ti substitution for Al^{3+} (as in corundum) or Fe^{3+} (as in hematite).

The structure of periclase, MgO, is identical to that of NaCl with cubic space group $Fm3m$. In a sec-

tion parallel to {111} this structure appears very similar (see Fig. 11.3) to that of members of the *hematite group*. In such a section a "layer" of octahedra is seen without any cation omissions in the octahedral sites. With Mg in 6-coordination, the e.v. of each of the Mg-O bonds is $\frac{1}{3}$. This in turn allows for six bonds, each with e.v. $= \frac{1}{3}$, to radiate from every oxygen ion in the structure. In other words, each oxygen is shared between six Mg-O octahedra, and not between just four Al-O (or Fe-O) octahedra as in the structures of the *hematite group*. The MgO type of structure, therefore, shows no cation vacancies.

The structures of XO_2 type oxides that we will consider fall into two structure types. One is the structure typified by *rutile*, in which the cation is in 6-coordination with oxygen. The radius ratio, $R_X : R_O$, in this structure lies approximately between the limits of 0.732 and 0.414 (see Table 11.1). The other has the *fluorite* structure (see Fig. 11.38), in which each oxygen has four cation neighbors arranged about it at the apices of a more or less regular tetrahedron, whereas each cation has eight oxygens surrounding it at the corners of a cube. The oxides of 4-valent uranium, thorium, and cerium, of great interest because of their connection with nuclear chemistry, have this structure. In general, all dioxides in which the radius ratio of cation to oxygen lies within and close to the limits for 8-coordination (0.732–1) may be expected to have this structure and to be isometric hexoctahedral (see Table 11.1).

The members of the *rutile group* are isostructural with space group $P4_2/mnm$. (Rutile itself has two additional polymorphs, anatase and brookite, with slightly different structures.) In the rutile structure (see Figs. 11.4 and 4.46) Ti^{4+} is located at the center of oxygen octahedra the edges of which are shared forming chains parallel to the c axis. These chains are cross-linked by the sharing of octahedral corners (see Fig. 11.4b). Each oxygen is linked to three titanium ions and the e.v. of each of the Ti-O bonds $= \frac{2}{3}$. The

FIG. 11.3. Structure of periclase, MgO, as seen parallel to (111). Compare this with Fig. 11.2a as well as with Fig. 11.26, of brucite.

○ Oxygen

• Mg

Table 11.1

RADIUS RATIOS IN XO_2 TYPE OXIDES (R_O = 1.36 Å)

R_x	[C.N.]	R_x/R_O	Ion	Mineral	Structure Type
0.53	[6]	0.39	Mn^{4+}	Pyrolusite	Rutile
0.61	[6]	0.45	Ti^{4+}	Rutile	Rutile
0.69	[6]	0.51	Sn^{4+}	Cassiterite	Rutile
0.97	[8]	0.71	Ce^{4+}	Cerianite	Fluorite
1.00	[8]	0.74	U^{4+}	Uraninite	Fluorite
1.05	[8]	0.77	Th^{4+}	Thorianite	Fluorite

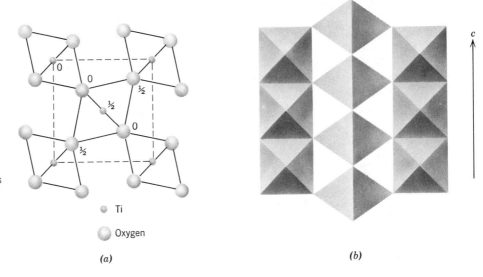

FIG. 11.4. Structure of rutile. (a) Projection on (001) showing location of Ti and O atoms and the outline of the unit cell (dashed lines). (b) Projection on (110) showing chains of octahedra parallel to c. Compare these illustrations with Fig. 4.46.

• Ti

Oxygen

(a)

(b)

prismatic habit of minerals of the rutile group is a reflection of the chainlike arrangement of octahedra.

The minerals of the *spinel group*, with the general formula XY_2O_4, are based on an arrangement of oxygens in approximate cubic closest packing along (111) planes in the structure (see Fig. 11.5). Several end members are listed in Table 11.2. The cations that are interstitial to the oxygen framework are in octahedral and tetrahedral coordination polyhedra with oxygen. In a unit cell of spinel, with edge length of approximately 8 Å, there are 32 possible octahedral sites and 64 possible tetrahedral sites; of these, 16 octahedral and 8 tetrahedral sites are occupied by cations. Occupied octahedra are joined along edges to form rows and planes parallel to {111} of the structure and occupied tetrahedra, with the apices along $\bar{3}$ axes, provide cross-links between layers of octahedra (see Fig. 11.5). A plan view of an oxygen layer parallel to (111), and its coordination with cations, is given in Fig. 11.6.

The general chemical formula of the spinel group is XY_2O_4 (or $X_8Y_{16}O_{32}$ per unit cell), where X and Y are various cations with variable valence. In magnetite, X = Fe^{2+} and Y = Fe^{3+}, whereas in ulvöspinel, X = Ti^{4+}, and Y = Fe^{2+}. There are two types of spinel

structures, and these are referred to as the "normal spinel" and the "inverse spinel" structures. In the normal spinel structure, the eight X cations occupy the 8 tetrahedral sites and the Y cations occupy the 16 octahedral sites, giving the formula $X_8Y_{16}O_{32}$ (which is

FIG. 11.5. The spinel (AB_2O_4) structure ($F4_1/d\bar{3}2/m$ = $Fd3m$) with alternating layers parallel to {111} of octahedral and octahedral-tetrahedral polyhedra, as based upon approximate cubic closest packing. (From Waychunas, G. A., 1991, Crystal chemistry of oxides and hydroxides, in *Oxide Minerals*, *Reviews in Mineralogy*, v. 25. Mineralogical Society of America, Washington, D.C.)

— A

— B

— C

— A

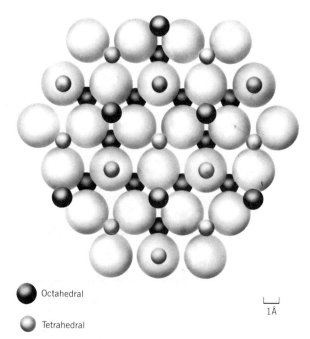

● Octahedral

◯ Tetrahedral

|___|
1Å

FIG. 11.6. An oxygen layer in the spinel structure, projected onto the (111) plane. See Fig. 11.5 for the position of such planes in the structure. The large circles are oxygen in approximate cubic closest packing. The cation layers on either side of the oxygen layer are shown as well. (Redrawn after Lindsley, D. H., 1976, The crystal chemistry and structure of oxide minerals as exemplified by the Fe-Ti oxides, in *Oxide Minerals, Reviews in Mineralogy,* v. 3. Mineralogical Society of America, Washington, D.C.)

deed the larger cation is in the smaller polyhedron, and vice versa. When crystal field stabilization energies are considered instead of the geometric aspects of the ions, it becomes clear that the larger cation may occupy tetrahedral sites.

The spinel structure has a coordination scheme similar to that of the silicates of the olivine series. The composition of members of the forsterite-fayalite series with a total solid solution series from Mg_2SiO_4 to Fe_2SiO_4 can be represented as $X_2^{2+}Y^{4+}O_4$. Although this is not the same as $X^{2+}Y_2^{3+}O_4$ as in spinel, in both cases the total cation charge is identical. If one compares the structure of an Fe-Mg olivine with that of a possible Fe-Mg spinel, one finds that the spinel structure is about 12% denser than the olivine structure for the same composition. This observation has led petrologists to suggest that olivine, which is thought to be abundant in the mantle, is converted to a spinel-type structure at great depth (see page 216). In 1969 an olivine, *ringwoodite*, from a stony meteorite, was found to have a spinel structure. This same olivine composition $(Fe_{0.7}Mg_{0.3})_2SiO_4$ had been previously synthesized as a spinel in the laboratory of A. E. Ringwood, Australian National University, Canberra, Australia.

CUPRITE—Cu₂O

Crystallography. Isometric; $4/m\bar{3}2/m$. Commonly occurs in crystals showing the cube, octahedron, and dodecahedron, frequently in combination (Fig. 11.7). Sometimes in elongated capillary crystals, known as "plush copper" or *chalcotrichite.*

$Pn3m$; $a = 4.27$ Å; $Z = 2$. *d*'s: 2.46(10), 2.13(6), 1.506(5), 1.284(4), 0.978(3).

Physical Properties. **H** $3\frac{1}{2}-4$. **G** 6.1. *Luster* metallic-adamantine in clear crystallized varieties. *Color* red of various shades; ruby-red in transparent crystals, called "ruby copper." *Streak* brownish-red.

Composition and Structure. Cu 88.8, O 11.2%. Usually pure, but FeO may be present as an impurity. The structure, as based on O atoms at the corners and centers of tetrahedral groups, is shown in Fig. 11.1.

equivalent to XY_2O_4). In the inverse spinel structure 8 of the 16 Y cations occupy the eight tetrahedral sites, resulting in the formula $Y(YX)O_4$ (see Table 11.2). Most naturally occurring spinels have cation distributions between the normal and inverse structure types.

The coordination polyhedra about the various cations in spinel are not what might be predicted on the basis of the ionic sizes of the cations. Because Mg^{2+} is larger than Al^{3+}, one would expect Mg to occur in octahedral and Al in tetrahedral coordination with oxygen. In the normal spinel structure, however, the general concepts of radius ratio do not apply; in-

Table 11.2				
END MEMBERS OF THE SPINEL GROUP (XY_2O_4)	**Normal Spinel Structure**		**Inverse Spinel Structure**	
	Spinel	$MgAl_2^{3+}O_4$	Magnetite	$Fe^{3+}(Fe^{2+}Fe^{3+})O_4$
	Hercynite	$FeAl_2^{3+}O_4$	Magnesioferrite	$Fe^{3+}(Mg^{2+}Fe^{3+})O_4$
	Gahnite	$ZnAl_2^{3+}O_4$	Jacobsite	$Fe^{3+}(Mn^{2+}Fe^{3+})O_4$
	Galaxite	$MnAl_2^{3+}O_4$	Ulvöspinel	$Fe^{2+}(Fe^{2+}Ti^{4+})O_4$
	Franklinite	$ZnFe_2^{3+}O_4$		
	Chromite	$Fe^{2+}Cr_2^{3+}O_4$		
	Magnesiochromite	$Mg^{2+}Al_2^{3+}O_4$		

FIG. 11.7. Cuprite crystals.

Diagnostic Features. Usually distinguished from other red minerals by its crystal form, high luster, streak, and association with limonite.

Occurrence. Cuprite is a supergene copper mineral and in places an important copper ore. It is found in the upper oxidized portions of copper veins, associated with limonite and secondary copper minerals such as native copper, malachite, azurite, and chrysocolla (see Figs. 14.21 and 14.22).

Noteworthy foreign countries where cuprite is an ore are Chile, Bolivia, Australia, and Zaire. Fine crystals have been found at Cornwall, England; Chessy, France; and the Onganja Mine and Tsumeb, Namibia. Found in the United States in excellent crystals in the copper deposits at Bisbee, Arizona and Chino at Santa Rita, New Mexico. Also found at Clifton and Morenci, Arizona.

Use. A minor ore of copper.

Name. Derived from the Latin *cuprum*, copper.

Similar Species. *Tenorite*, CuO, is a black supergene mineral.

Zincite—ZnO

Crystallography. Hexagonal; *6mm*. Crystals are rare, terminated at one end by faces of a steep pyramid and the other by a pedion (see Fig. 2.86). Usually massive with platy or granular appearance.

$P6_3mc; a = 3.25, c = 5.19$ Å; $Z = 2$. *d*'s: 2.83(7), 2.62(5), 2.49(10), 1.634(6), 1.486(6).

Physical Properties. *Cleavage* $\{10\bar{1}0\}$ perfect;

$\{0001\}$ parting. **H** 4. **G** 5.68. *Luster* subadamantine. *Color* deep red to orange-yellow. *Streak* orange-yellow. Translucent. *Optics*: (+); $\omega = 2.013$, $\epsilon = 2.029$.

Composition and Structure. Zn 80.3, O 19.7%. Mn^{2+} often present and probably colors the mineral; pure ZnO is white. The structure of zincite is similar to that of wurtzite (see Fig. 10.23). Zn is in hexagonal closest packing and each oxygen lies within a tetrahedral group of four Zn ions.

Diagnostic Features. Distinguished chiefly by its red color, orange-yellow streak, and the association with franklinite and willemite.

Occurrence. Zincite is confined almost exclusively to the zinc deposits at Franklin and Sterling Hill, New Jersey, where it is associated with franklinite and willemite in calcite, often in an intimate mixture. Reported only in small amounts from other localities.

Use. An ore of zinc, particularly used for the production of zinc white (zinc oxide).

Similar Species. *Periclase*, MgO, is another oxide of *XO* composition. The structure of MgO is of the NaCl type with space group *Fm3m*. Continuous octahedral sheets can be identified in the structure parallel to $\{111\}$, see Fig. 11.3. It is found in contact metamorphosed Mg-rich limestones by the reaction: $CaMg(CO_3)_2 \rightarrow CaCO_3 + MgO + CO_2$.

Hematite Group, $X_2O_3^1$

CORUNDUM—Al₂O₃

Crystallography. Hexagonal; $\bar{3}2/m$. Crystals commonly tabular on $\{0001\}$ or prismatic $\{11\bar{2}0\}$. Often tapering hexagonal dipyramids (Fig. 11.8), rounded into barrel shapes with deep horizontal striations. May show rhombohedral faces. Usually rudely crystallized or massive; coarse or fine granular. Polysynthetic twinning is common on $\{10\bar{1}1\}$ and $\{0001\}$.

---

[1]The long-established morphological unit with *c* one-half the length of *c* of the unit cell is retained here for members of this group; the indices of forms of these minerals are expressed accordingly.

FIG. 11.8. Corundum crystals.

(a)   *(b)*   *(c)*

$R\bar{3}c$; a = 4.76, c = 12.98 Å; Z = 6. d's: 2.54(6), 2.08(9), 1.738(5), 1.599(10), 1.374(7).

Physical Properties. Parting on {0001} and {10$\bar{1}$1}, the latter giving nearly cubic angles; more rarely prismatic parting. **H** 9. Corundum may alter to mica, and care should be exercised in obtaining a fresh surface for hardness test. **G** 4.02. *Luster* adamantine to vitreous. Transparent to translucent. *Color* usually some shade of brown, pink, or blue, but may be colorless or any color. *Ruby* is red gem corundum; *sapphire* is gem corundum of any other color. Rubies and sapphires having a stellate opalescence when viewed in the direction of the *c* crystal axis are termed *star ruby* or *star sapphire* (see Fig. 6.24). *Emery* is a black granular corundum intimately mixed with magnetite, hematite, or hercynite. *Optics*: ($-$), ω = 1.769, ϵ = 1.760.

Composition and Structure. Al 52.9, O 47.1%. Rubies contain trace amounts of Cr (ppm) as a coloring agent and sapphires are blue due to trace amounts of Fe and Ti (see page 260). The structure of corundum is illustrated in Fig. 11.2 and consists of oxygen in hexagonal closest packing and Al in octahedral coordination. Two-thirds of the octahedra are occupied by Al and $\frac{1}{3}$ are vacant (see also discussion on page 373).

Diagnostic Features. Characterized chiefly by its great hardness, high luster, specific gravity, and parting. Infusible.

Occurrence. Corundum is common as an accessory mineral in some metamorphic rocks, such as crystalline limestone, mica-schist, gneiss. Found also as an original constituent of silica deficient (undersaturated) igneous rocks such as syenites and nepheline syenites. May be found in large masses in the zone separating peridotites from adjacent country rocks. It is disseminated in small crystals through certain lamprophyric dikes and is found in large crystals in pegmatites. Found frequently in crystals and rolled pebbles in detrital soil and stream sands, where it has been preserved through its hardness and chemical inertness. Associated minerals are commonly chlorite, micas, olivine, serpentine, magnetite, spinel, kyanite, and diaspore.

The finest rubies have come from Burma; the most important locality is near Mogok, 90 miles north of Mandala. Some stones are found here in the metamorphosed limestone that underlies the area, but most of them have been recovered from the overlying soil and associated stream gravels. Recently rubies of fine quality occurring in situ have been mined in southern Kenya. Darker, poorer quality rubies have been found in alluvial deposits near Bangkok, Thailand, and at Battambang, Cambodia. In Sri Lanka rubies of relatively inferior grade, associated with more abundant sapphires and other gemstones, are recovered from stream gravels. In the United States a few rubies have been found associated with the large corundum deposits of North Carolina.

Sapphires are found in the alluvial deposits of Thailand, Sri Lanka, and Cambodia associated with rubies. The stones from Cambodia of a cornflower-blue color are most highly prized. Sapphires occur in Kashmir, India, and are found over an extensive area in central Queensland, Australia. In the United States small sapphires of fine color are found in various localities in Montana. They were first discovered in the river sands east of Helena during placer operations for gold, and later found embedded in the rock of a lamprophyre dike at Yogo Gulch.

Common abrasive corundum has in the past been mined in many countries, but today Zimbabwe, the Commonwealth of Independent States (CIS), and India are the only significant producers. At one time it was extensively mined in North Carolina, where it occurs in large masses at the edges of intruded olivine rock (dunite), and in Craigmont, Ontario, where, as a primary constituent of nepheline syenite, corundum makes up more than 10% of the rock mass.

Emery is found in large quantities on Cape Emeri on the island of Naxos, Greece, where mining continues today as it has for many centuries. In the United States emery was mined at Chester, Massachusetts and at Peekskill, New York.

Artificial. Artificial corundum is manufactured from bauxite on a large scale. This synthetic material, together with other manufactured abrasives, notably silicon carbide, has largely taken the place of natural corundum as an abrasive.

Synthetic rubies and sapphires, colored with small amounts of Cr and Ti, are synthesized by fusing alumina powder in an oxy-hydrogen flame which on cooling forms single crystal "boules." This, the Verneuil process, has been in use since 1902. In 1947 the Linde Air Products of the United States succeeded in synthesizing star rubies and star sapphires. This was accomplished by introducing titanium which, during proper heat treatment, exsolves as oriented rutile needles to produce the star. The artificial rival the natural stones in beauty, and it is difficult for the untrained person to distinguish them (see also Chapter 15).

Use. As a gemstone and abrasive. The deep red ruby is one of the most valuable of gems, second only to emerald. The blue sapphire is also valuable, and stones of other colors may command good prices (for color illustrations see Plate I in Chapter 15). Stones of gem quality are used as watch jewels and as bearings

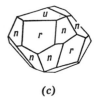

FIG. 11.9. Hematite crystals.   *(a)*   *(b)*   *(c)*

in scientific instruments. Corundum is used as an abrasive, either ground from the pure massive mineral or in its impure form as emery.

Name. Probably from Kauruntaka, the Indian name for the mineral.

HEMATITE—Fe₂O₃

Crystallography. Hexagonal; $\bar{3}2/m$. Crystals are usually thick to thin tabular on {0001}, basal planes often show triangular markings, and the edges of the plates may be beveled with rhombohedral forms (Fig. 11.9*b*). Thin plates may be grouped in rosettes (*iron roses*) (Fig. 11.10). More rarely crystals are distinctly rhombohedral, often with nearly cubic angles and may be polysynthetically twinned on {0001} and {10$\bar{1}$1}. Also in botryoidal to reniform shapes with radiating structure, *kidney ore* (see Fig. 6.2). May also be micaceous and foliated, *specular*. Usually earthy; called *martite* when in octahedral pseudomorphs after magnetite.

$R\bar{3}c$; $a = 5.04$, $c = 13.76$ Å; $Z = 6$. d's: 2.69(10), 2.52(8), 2.21(4), 1.843(6), 1.697(7).

Physical Properties. *Parting* on {10$\bar{1}$1} with nearly cubic angles, also on {0001}. **H** $5\frac{1}{2}$–$6\frac{1}{2}$, **G** 5.26 for crystals. *Luster* metallic in crystals and dull in

earthy varieties. *Color* reddish-brown to black. Red earthy variety is known as *red ocher*, platy and metallic variety as *specularite*. *Streak* light to dark red which becomes black on heating. Translucent.

Composition and Structure. Fe 70, O 30%; essentially a pure substance at ordinary temperatures, although small amounts of Mn and Ti have been reported. Forms a complete solid solution series with ilmenite above 950°C (see Figs. 11.11 and 11.21). The structure is similar to that of corundum (see Fig. 11.2 and page 373).

Diagnostic Features. Distinguished chiefly by its characteristic red streak.

Occurrence. Hematite is widely distributed in rocks of all ages and forms the most abundant and important ore of iron. It may occur as a sublimation product in connection with volcanic activities. Occurs in contact metamorphic deposits and as an accessory mineral in feldspathic igneous rocks such as

FIG. 11.10. Hematite, iron rose, St. Gotthard, Switzerland. (Harvard Mineralogical Museum.)

FIG. 11.11. Compositions of naturally occurring oxide minerals in the system FeO-Fe₂O₃-TiO₂. Solid lines (tielines) indicate common geological coexistences at relatively low temperatures. Dashed lines indicate complete solid solution between end members. Hematite-ilmenite is a complete series above 950°C. Magnetite-ulvöspinel is a complete series above about 600°C. Magnetite commonly contains ilmenite lamellae. It is possible that such intergrowths result from the oxidation of members of the magnetite-ulvöspinel series. Compare with Fig. 11.21.

granite. Found from microscopic scales to enormous masses in regionally metamorphosed rocks where it may have originated by the oxidation of limonite, siderite, or magnetite. It is found in red sandstones as the cementing material that binds the quartz grains together. The quantity of hematite in a deposit that can be mined economically must be measured in tens of millions of tons. Such major accumulations are largely sedimentary in origin and many of them, through leaching of associated silica by meteoric waters, have been enriched to high-grade (direct shipping) ores (over 50% Fe). Like limonite, it may be formed in irregular masses and beds as the result of the weathering and oxidation of iron-bearing rocks. The oölitic ores are of sedimentary origin and may occur in beds of considerable size.

Noteworthy localities for hematite crystals are on the island of Elba; St. Gotthard, Switzerland, in "iron roses"; in the lavas of Vesuvius; at Cleator Moor, Cumbria, England; and in Minas Gerais, Brazil.

In the United States the columnar and earthy varieties are found in enormous bedded, Precambrian deposits that have furnished a large proportion of the iron ore of the world. The chief iron ore districts of the United States are grouped around the southern and northwestern shores of Lake Superior in Michigan, Wisconsin, and Minnesota. The chief districts, which are spoken of as iron ranges, are, from east to west, the Marquette in northern Michigan; the Menominee in Michigan to the southwest of the Marquette; the Penokee-Gogebic in northern Wisconsin. In Minnesota the Mesabi, northwest of Duluth; the Vermilion, near the Canadian boundary; and the Cuyuna, southwest of the Mesabi. The iron ore of these different ranges varies from the hard specular variety to the soft, red earthy type. Of the several ranges the Mesabi is the largest, and since 1892 has yielded over 2.5 billion tons of high-grade ore, over twice the total production of all the other ranges.

Oölitic hematite is found in the United States in the rocks of the Clinton formation, which extends from central New York south along the line of the Appalachian Mountains to central Alabama. The most important deposits lie in eastern Tennessee and northern Alabama, near Birmingham. Hematite has been found at Iron Mountain and Pilot Knob in southeastern Missouri. Deposits of considerable importance are located in Wyoming in Laramie and Carbon counties.

Although the production of iron ore within the United States remains large, the rich deposits have been largely worked out. In the future much of the iron must come from low-grade deposits or must be imported. The low-grade, silica-rich iron formation from which the high-grade deposits have been derived is known as *taconite*, which contains about 25% iron. The iron reserves in taconite are far greater than were the original reserves of high-grade ore.

Exploration outside the United States has been successful in locating several ore bodies with many hundreds of millions of tons of high-grade ore. These are notably in Venezuela, Brazil, Canada, and Australia. Brazil's iron mountain, Itabira, is estimated to have 15 billion tons of very pure hematite. In 1947 Cerro Bolivar in Venezuela was discovered as an extremely rich deposit of hematite, and by 1954 ore was being shipped from there to the United States. In Canada major Precambrian iron ore deposits have been located in the Labrador Trough close to the boundary between Quebec and Labrador. A 350-mile railroad built into this previously inaccessible area began in 1954 to deliver iron ore to Seven Islands, a major port on the St. Lawrence River. Large Australian deposits are located in the Precambrian rocks of the Hamersley Range, Western Australia. The main producers of iron ore are Australia, the CIS, the United States, Brazil, Canada, China, and Sweden.

Use. Most important ore of iron for steel manufacture. Also used in pigments, red ocher, and as polishing powder. Black crystals may be cut as gems.

Name. Derived from a Greek word *haimatos* meaning *blood*, in allusion to the color of the powdered mineral.

ILMENITE—FeTiO$_3$

Crystallography. Hexagonal; $\bar{3}$. Crystals are usually thick tabular with prominent basal planes and small rhombohedral truncations. Often in thin plates. Usually massive, compact; also in grains or as sand.

$R\bar{3}$; $a = 5.09$, $c = 14.06$ Å; $Z = 6$. d's: 2.75(10), 2.54(7), 1.867(5), 1.726(8), 1.507(4).

Physical Properties. H $5\frac{1}{2}$–6. G 4.7. *Luster* metallic to submetallic. *Color* iron-black. *Streak* black to brownish-red. May be magnetic without heating. Opaque.

Composition and Structure. Fe 36.8, Ti 31.6, O 31.6% for stoichiometric FeTiO$_3$. The formula can be more realistically expressed as (Fe, Mg, Mn)TiO$_3$ with limited Mg and Mn substitution. Ilmenite may contain limited amounts of Fe$_2$O$_3$ (less than 6 weight percent) at ordinary temperatures. However, a complete solid solution exists between Fe$_2$O$_3$ and FeTiO$_3$ above 950°C (see Figs. 11.11 and 11.21). Related species are *geikielite*, MgTiO$_3$, and *pyrophanite*, MnTiO$_3$. The structure of ilminite is similar to that of corundum (Fig. 11.2b) with Fe and Ti ordered in alternate octahedrally coordinated layers perpendicular to the c axis (see Fig. 11.2c).

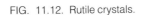

FIG. 11.12. Rutile crystals.

Diagnostic Features. Ilmenite can be distinguished from hematite by its streak and from magnetite by its lack of strong magnetism.

Occurrence. Ilmenite is a common accessory mineral in igneous rocks. It may be present in large masses in gabbros, diorites, and anorthosites as a product of magmatic segregation intimately associated with magnetite. It is also found in some pegmatites and vein deposits. As a constituent of black sands it is associated with magnetite, rutile, zircon, and monazite.

Found in large quantities at Krägerö and other localities in Norway; in Finland; and in crystals at Miask in the Ilmen Mountains, Russia, CIS. A large percentage of the production of ilmenite is recovered from beach sands, notably in Australia, Republic of South Africa, India, Brazil, and, in the United States from Florida. Until recently ilmenite was actively mined at Tahawas, New York. A large ilmenite-hematite deposit is mined at Allard Lake, Quebec, Canada. The world production of ilmenite is about 4 million tons.

Use. The major source of titanium. It is used principally in the manufacture of titanium dioxide for paint pigments, replacing older pigments, notably lead compounds. As the metal and in alloys, because of its high strength-to-weight ratio and high resistance to corrosion, titanium is used for aircraft and space vehicle construction in both frames and engines.

Name. From the Ilmen Mountains, Russia, CIS.

Similar Species. *Perovskite*, $CaTiO_3$, a pseudocubic titanium mineral found usually in nepheline syenites and carbonatites. *Pseudobrookite*, Fe_2TiO_5, and solid solutions toward *ferropseudobrookite*, $FeTi_2O_5$ (see Fig. 11.21), occur in igneous rocks, in kimberlites, and in meteorite-impacted basalts.

Rutile Group, XO_2

RUTILE—TiO_2

Crystallography. Tetragonal; $4/m2/m2/m$. Usually in prismatic crystals with dipyramid termi-

nations common and vertically striated prism faces. Frequently in elbow twins, often repeated (Figs. 11.12 and 11.13) with twin plane {011}. Crystals frequently slender acicular. Also compact, massive.

$P4_2/mnm$; $a = 4.59$, $c = 2.96$ Å; $Z = 2$. d's: 3.24(10), 2.49(5), 2.18(3), 1.687(7), 1.354(4).

Physical Properties. *Cleavage* {110} distinct. **H** $6-6\frac{1}{2}$. **G** 4.18–4.25. *Luster* adamantine to submetallic. *Color* red, reddish-brown to black. *Streak* pale brown. Usually subtranslucent, may be transparent. *Optics*: (+); $\omega = 2.612$, $\epsilon = 2.899$.

Composition and Structure. Ti 60, O 40%. Although rutile is essentially TiO_2, some analyses report considerable amounts of Fe^{2+}, Fe^{3+}, Nb, and Ta. When Fe^{2+} substitutes for Ti^{4+}, electrical neutrality is maintained by twice as much Nb^{5+} and Ta^{5+} entering the structure. This leads to the formulation: $Fe_x(Nb,Ta)_{2x}Ti_{1-3x}O_2$. The structure of rutile is given in Figs. 11.4 and 4.46 and described on page 374. TiO_2 occurs in two additional, relatively rare polymorphs, tetragonal *anatase* (space group $I4_1/amd$) and orthorhombic *brookite* (space group $Pbca$). In anatase TiO_6 octahedra share four edges, in brookite the octahedra share three edges, whereas in rutile they share only two edges. The stability fields of the three polymorphs

FIG. 11.13. Rutile. Elbow twin, Pfitch, Austria. Cyclic twin, Parksburg, Chester County, Pennsylvania. (Harvard Mineralogical Museum.)

are still not well defined. Rutile is generally considered to be a high T polymorph of TiO_2 and occurs in rocks which formed over a wide range of P and T. Its large stability range may suggest that the polymorph in which only two edges are shared between adjoining octahedra is indeed the most stable structure (see page 197).

Diagnostic Features. Characterized by its peculiar adamantine luster and red color. Lower specific gravity distinguishes it from cassiterite.

Occurrence. Rutile is found in granite, granite pegmatites, gneisses, mica schists, metamorphic limestone, and dolomite. It may be present as an accessory mineral in the rock or in quartz veins traversing it. Often occurs as slender crystals inside quartz and micas. Is found in considerable quantities in black sands associated with ilmenite, magnetite, zircon, and monazite.

Notable European localities are: Krägerö, Norway; Yrieix, near Limoges, France; in Switzerland; and the Tyrol. Rutile from beach sands of northern New South Wales and southern Queensland makes Australia the largest producer of rutile. In the United States remarkable crystals have come from Graves Mountain, Lincoln County, Georgia. Rutile is found in Alexander County, North Carolina, and at Magnet Cove, Arkansas. It has been mined in Amherst and Nelson counties, Virginia, and derived in commercial quantities from the black sands of northeastern Florida.

Artificial. Single crystals of rutile have been manufactured by the Verneuil process (see page 607). With the proper heat treatment they can be made transparent and nearly colorless, quite different from the natural mineral. Because of its high refractive index and dispersion, this synthetic material makes a beautiful cut gemstone with only a slight yellow tinge. It is sold under a variety of names, some of the better known are *titania*, *kenya gem*, and *miridis*.

Use. Most of the rutile produced is used as a coating of welding rods. Some titanium derived from rutile is used in alloys; for electrodes in arc lights; to give a yellow color to porcelain and false teeth. Manufactured oxide is used as a paint pigment (see ilmenite).

Name. From the Latin *rutilus*, red, an allusion to the color.

Similar Species. *Stishovite*, a very high-pressure polymorph of SiO_2, isostructural with rutile, is found rarely in meteorite impact craters. This dense form of SiO_2 is the only example of Si in 6-coordination rather than its common tetrahedral coordination. *Leucoxene*, a fine-grained, yellow to brown alteration product of minerals high in Ti (ilmenite, perovskite, titanite), consists mainly of rutile, less commonly anatase.

PYROLUSITE—MnO₂

Crystallography. Tetragonal; $4/m2/m2/m$. Rarely in well-developed crystals. Usually in radiating fibers or columns. Also granular massive; often in reniform coats and dendritic shapes (Fig. 11.14) finely intergrown with other Mn-oxides and hydroxides. Frequently pseudomorphous after manganite.

FIG. 11.14. Dendrites of pyrolusite on limestone, Sardinia, Italy. (Harvard Mineralogical Museum.)

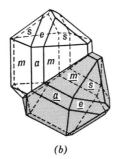

FIG. 11.15. Cassiterite crystals.

$P4_2/mnm$; $a = 4.39$, $c = 2.86$ Å; $Z = 2$. d's: 3.11(10), 2.40(5), 2.11(4), 1.623(7), 1.303(3).

Physical Properties. *Cleavage* {110} perfect. **H** 1–2 (often soiling the fingers). For coarsely crystalline *polianite* the hardness is 6–6½. **G** 4.75. *Luster* metallic. *Color* and *streak* iron-black. *Fracture* splintery. Opaque.

Composition and Structure. Mn 63.2, O 36.8%. Commonly contains some H_2O. The structure is the same as that of rutile (see Fig. 11.4) with Mn in 6-coordination with oxygen.

Diagnostic Features. Characterized by and distinguished from other manganese minerals by its black streak and low hardness.

Occurrence. Manganese is present in small amounts in most crystalline rocks. When dissolved from these rocks, it may be redeposited as various minerals, but chiefly as pyrolusite. Nodular deposits of pyrolusite are found in bogs, on lake bottoms, and on the floors of seas and oceans. Nests and beds of manganese ores are found enclosed in residual clays, derived from the decay of manganiferous limestones. Also found in veins with quartz and various metallic minerals.

Pyrolusite is the most common manganese ore and is widespread in its occurrence. The chief manganese-producing countries are the CIS, Republic of South Africa, Brazil, China, and India. The present production in the United States is negligible.

Use. Most important manganese ore. Manganese is used with iron in the manufacture of *spiegeleisen* and *ferromanganese*, employed in making steel. This is its principal use, for about 13.2 pounds of manganese are consumed in the production of one ton of steel. Also used in various alloys with copper, zinc, aluminum, tin, and lead. Pyrolusite is used as an oxidizer in the manufacture of chlorine, bromine, and oxygen; as a disinfectant in potassium permanganate; as a drier in paints; as a decolorizer of glass; and in electric dry-cells and batteries. Manganese is also used as a coloring material in bricks, pottery, and glass.

Name. *Pyrolusite* is derived from two Greek words, *pyros* meaning *fire* and *louo to wash*, because it is used to free glass through its oxidizing effect of the colors due to iron.

Similar Species. *Alabandite*, MnS, is comparatively rare, associated with other sulfides in veins. *Wad* is the name given to manganese ore composed of an impure mixture of hydrous manganese oxides.

CASSITERITE—SnO_2

Crystallography. Tetragonal; $4/m2/m2/m$. The common forms are the prisms {110} and {010} and the dipyramids {111} and {011} (Fig. 11.15a). Frequently in elbow-shaped twins with a characteristic notch, giving rise to the miner's term *visor tin* (Figs. 11.15b and 11.16); the twin plane is {011}. Usually massive granular; often in reniform shapes with radiating fibrous appearance, *wood tin*.

$P4_2/mnm$; $a = 4.73$, $c = 3.18$ Å; $Z = 2$. d's: 2.36(8), 2.64(7), 1.762(10), 1.672(4), 1.212(5).

Physical Properties. *Cleavage* {010} imperfect. **H** 6–7. **G** 6.8–7.1 (unusually high for a nonmetallic mineral). *Luster* adamantine to submetallic and dull. *Color* usually brown or black; rarely yellow or white.

FIG. 11.16. Twinned crystal of cassiterite, Schlaggenwald, Bohemia, Czechoslovakia. (Harvard Mineralogical Museum.)

Streak white. Translucent, rarely transparent. *Optics:* $(+)$; $\omega = 1.997$, $\epsilon = 2.093$.

Composition and Structure. Close to SnO_2 with Sn 78.6, O 21.4%. Small amounts of Fe^{3+} may be present and lesser amounts of Nb and Ta substitute for Sn. The structure is that of rutile (see Fig. 11.4 and page 374).

Diagnostic Features. Recognized by high specific gravity, adamantine luster, and light streak.

Occurrence. Cassiterite is widely distributed in small amounts but is produced on a commercial scale in only a few localities. It has been noted as an original constituent of igneous rocks and pegmatites, but it is more commonly found in high-temperature hydrothermal veins in or near granitic rocks. Tin veins usually have minerals that contain fluorine or boron, such as tourmaline, topaz, fluorite, and apatite, and the minerals of the wall rocks are commonly much altered. Frequently associated with wolframite, molybdenite and arsenopyrite. Cassiterite is commonly found in the form of rolled pebbles in placer deposits, *stream tin*.

Most of the world's supply of tin comes from alluvial deposits chiefly in Malaysia, the CIS, Indonesia, and Thailand. Bolivia is the only country where a significant production comes from vein deposits; but in the past the mines of Cornwall, England, were major producers. Fine specimens originate from Schlaggenwald, Czechoslovakia, from Cornwall, England, and from Araca, Bolivia. In the United States cassiterite is not found in sufficient quantities to warrant mining but is present in small amounts in numerous pegmatites. *Wood tin* is found in rhyolites in New Mexico and Mexico.

Use. Principal ore of tin. The chief use of tin was in the manufacture of *tin plate* and *tern plate* for food containers. Tern plate is made by applying a coating of tin and lead instead of pure tin. At present aluminum, glass, paper, plastic, and tin-free steel are replacing tin-plated containers. Tin is also used with lead in solders, in babbitt metal with antimony and copper, and in bronze and bell-metal with copper. "Phosphor bronze" contains 89% Cu, 10% Sn, and 1% P. Artificial tin oxide is a polishing powder.

Name. From the Greek word *kassiteros* meaning *tin*.

Similar Species. *Stannite*, Cu_2FeSnS_4, is structurally similar to chalcopyrite and sphalerite. It is a minor ore of tin.

URANINITE—UO_2

Crystallography. Isometric; $4/m\bar{3}2/m$. The rare crystals are usually octahedral with subordinate cube and dodecahedron faces. More commonly as the variety *pitchblende*: massive or botryoidal with a banded structure.

$Fm3m$; $a = 5.46$ Å; $Z = 4$. d's: 3.15(7), 1.926(6), 1.647(10), 1.255(5), 1.114(5).

Physical Properties. **H** $5\frac{1}{2}$. **G** 7.5–9.7 for crystals; 6.5–9 for pitchblende. The specific gravity decreases with oxidation of U^{4+} to U^{6+}. *Luster* submetallic to pitchlike, dull. *Color* black. *Streak* brownish-black.

Composition and Structure. Uraninite is always partially oxidized, and thus the actual composition lies between UO_2 and U_3O_8 ($= U^{4+}O_2 + 2U^{6+}O_3$). Th can substitute for U and a complete series between uraninite and *thorianite*, ThO_2, has been prepared synthetically. In addition to Th, analyses usually show the presence of small amounts of Pb, Ra, Ce, Y, N, He, and A. The lead occurs as one of two stable isotopes (^{206}Pb and ^{207}Pb) which result from the radioactive decay of uranium. For example: $^{238}U \rightarrow {}^{206}Pb + 8{}^4He$ and $^{235}U \rightarrow {}^{207}Pb + 7{}^4He$. Thorium decays as follows: $^{232}Th \rightarrow {}^{208}Pb + 6{}^4He$. In addition to ionized helium atoms (α particles), electrons (β particles) are emitted during the decay process. Because the radioactive disintegration proceeds at a uniform and known rate, the accumulation of both helium and lead can be used as a measure of the time elapsed since the mineral crystallized. For example, for the decay of ^{238}U, the *half-life*, T, which is the time needed to reduce the number of ^{238}U atoms by one-half, is 4.51×10^9 years (see Table 4.13). Both lead-uranium and helium-uranium ratios have been used by geologists to determine the age of rocks (see also page 218).

It was in uraninite that helium was first discovered on Earth, having been previously noted in the sun's spectrum. Also radium was discovered in uraninite. The structures of uraninite and thorianite are like that of fluorite (Fig. 11.38) in which the metal atom is at the center of eight oxygens at the corner of a cube, with each oxygen at the center of a tetrahedral group of metal atoms.

Diagnostic Features. Characterized chiefly by its pitchy luster, high specific gravity, color, and streak. Because of its radioactivity, uraninite, as well as other uranium compounds, can be detected in small amounts by Geiger and scintillation counters.

Occurrence. Uraninite occurs as a primary constituent of granitic rocks and pegmatites. It also is found in high-temperature hydrothermal veins associated with cassiterite, chalcopyrite, pyrite, and arsenopyrite as at Cornwall, England; or in medium-temperature veins with native silver and Co-Ni-As minerals as at Joachimsthal, Czechoslovakia, and

Great Bear Lake, Canada. The world's most productive uranium mine during and immediately following World War II was the Shinkolobwe Mine in Zaire. Here uraninite and secondary uranium minerals are in vein deposits associated with cobalt and copper minerals. The ores also carry significant amounts of Mo, W, Au, and Pt.

Wilberforce, Ontario, has been a famous Canadian locality, but the present major Canadian sources are the mines at Great Bear Lake, the Beaverlodge region, Saskatchewan; and the Blind River area, Ontario. At Blind River uraninite occurs as detrital grains in a Precambrian quartz conglomerate. In the same type of deposit it is found in the gold-bearing Witwatersrand conglomerates, Republic of South Africa.

In the United States uraninite was found long ago in isolated crystals in pegmatites at Middletown, Glastonbury, and Branchville, Connecticut, and the mica mines of Mitchell County, North Carolina. Following World War II, exploration has located many workable deposits of uraninite and associated uranium minerals on the Colorado Plateau in Arizona, Colorado, New Mexico, and Utah.

Use. Uraninite is the chief ore of uranium although other minerals are important sources of the element such as *carnotite* (page 439), *tyuyamunite* (page 439), *torbernite* (page 439), and *autunite* (page 438).

Uranium has assumed an important place among the elements because of its susceptibility to nuclear fission, a process by which the nuclei of uranium atoms are split apart with the generation of tremendous amounts of energy. This energy, first demonstrated in the atomic bomb, is now produced by nuclear-power reactors for generating electricity.

Uraninite is also the source of radium but contains it in very small amounts. Roughly 750 tons of ore must be mined in order to furnish 12 tons of concentrates; chemical treatment of these concentrates yields about 1 gram of a radium salt. In the form of various compounds uranium has a limited use in coloring glass and porcelain, in photography, and as a chemical reagent.

Name. Uraninite in allusion to the composition. **Similar Species.** *Thorianite*, ThO_2, is dark gray to black with submetallic luster. Found chiefly in pegmatites and as water-worn crystals in stream gravels. *Cerianite*, $(Ce,Th)O_2$, an extremely rare mineral, has, as does thorianite, the fluorite structure.

Spinel Group, XY_2O_4

The minerals of the *spinel group* show extensive solid solution between the various end-member composi-

tions listed in Table 11.2. There is, for example, extensive solid solution between magnetite, Fe_3O_4, ulvöspinel, Fe_2TiO_4, and a synthetic end-member composition, Mg_2TiO_4 (a magnesian titanate with inverse spinel structure). Furthermore, there are substitutions between chromite, $FeCr_2O_4$, and magnesiochromite, $MgCr_2O_4$; between spinel, $MgAl_2O_4$, and hercynite, $FeAl_2O_4$, and so on. The complexity of the chemical substitutions in this group makes it very difficult to use triangular composition diagrams for expression of the various solid solutions extents; instead, a "spinel prism" is used for such chemical representations. Figure 11.17 shows three such prisms: (a) for the compositional space between normal spinels at the base of the prism and inverse spinels, with $(Fe^{2+} + Ti^{4+})$ substitutions, at the top edge; (b) for the compositional space between normal spinels at the base of the prism and inverse spinels, with Fe^{3+} substitutions, along the upper edge; and (c) the general nomenclature for compositions in this prism.

Spinel—$MgAl_2O_4$

Crystallography. Isometric; $4/m\bar{3}2/m$. Usually in octahedral crystals or in twinned octahedrons (spinel twins) (Figs. 11.18a and b). Dodecahedron may be present as small truncations (Fig. 11.18c), but other forms rare. Also massive and as irregular grains.

$Fd3m$; $a = 8.10$ Å; $Z = 8$. d's: 2.44(9), 2.02(9), 1.552(9), 1.427(10), 1.053(10).

Physical Properties. H 8. G 3.5–4.1. G 3.55 for the composition as given. Nonmetallic. *Luster* vitreous. *Color* various: white, red, lavender, blue, green, brown, black. *Streak* white. Usually translucent, may be clear and transparent. *Optics*: $n = 1.718$.

Composition and Structure. MgO 28.2, Al_2O_3 71.8%. Fe^{2+}, Zn, and less commonly Mn^{2+} substitute for Mg in all proportions. Fe^{3+} and Cr may substitute in part for Al. The clear red, nearly pure magnesium spinel is known as *ruby spinel. Ferroan spinel*, intermediate between spinel and hercynite, $FeAl_2O_4$, is dark green to black, and *chromian spinel*, intermediate between hercynite and *chromite*, $FeCr_2O_4$, is yellowish to greenish-brown (see Fig. 11.17 for compositional definitions). The structure of spinel is illustrated in Figs. 11.5 and 11.6 and discussed on page 375.

Diagnostic Features. Recognized by its hardness (8), its octahedral crystals, and its vitreous luster. The iron spinel can be distinguished from magnetite by its nonmagnetic character and white streak.

Occurrence. Spinel is a common high-temperature mineral occurring in contact metamorphosed

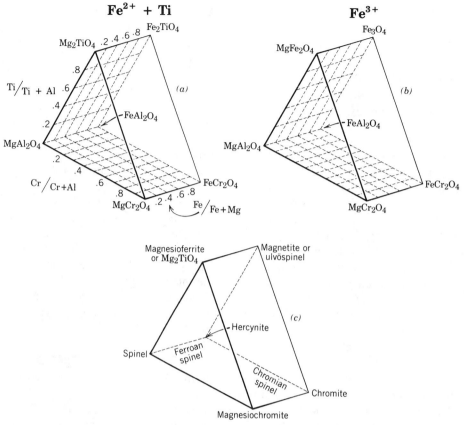

FIG. 11.17. End member compositions in the spinel group as represented in a spinel prism. *(a)* Compositions in this space range from those of normal spinels (in the base of the prism) to those with [Fe^{2+}(or Mg^{2+}) + Ti] end members along the upper edge (these are inverse spinels). *(b)* Compositions in this space range from those of the normal spinels (in the base) to those with Fe^{3+}-rich end members along the upper edge (these are inverse spinels). *(c)* Nomenclature for members of the spinel group as based on the chemical compositions represented in *(a)* and *(b)*.

limestones and metamorphic argillaceous rocks poor in SiO_2. Occurs also as an accessory mineral in many dark igneous rocks. In contact metamorphic rocks it is associated with phlogopite, pyrrhotite, chondrodite, and graphite. Found frequently as rolled pebbles in stream sands, where it has been preserved because of its resistant physical and chemical properties. The ruby spinels are found in this way, often associated with the gem corundum, in the sands of Sri Lanka, Thailand, Upper Burma, and Malagasy Republic. Or-

dinary spinel is found in various localities in New York and New Jersey.

Use. When transparent and finely colored, spinel is used as a gem (see Plate II, no. 10, Chapter 15). Usually red and known as *ruby spinel* or *balas ruby*. Some stones are blue. The largest cut stone known weighs about 80 carats. The stones are usually comparatively inexpensive.

Artificial. Synthetic spinel has been made by the Verneuil process (see corundum) in various colors

FIG. 11.18. Spinel crystals.

(a)

(b)

(c)

FIG. 11.19. Magnetite crystals.   (a)   (b)   (c)

rivaling the natural stones in beauty. Synthetic spinel is also used as a refractory.

Similar Species. *Hercynite*, $FeAl_2O_4$, is associated with corundum in some emery; also found with andalusite, sillimanite, and garnet. *Ferroan galaxite*, $(Fe^{2+},Mn)Al_2O_4$, an Fe^{2+}-rich variety of *galaxite*, $MnAl_2O_4$, occurs sporadically in nature.

Gahnite—ZnAl$_2$O$_4$

Crystallography. Isometric; $4/m\bar{3}2/m$. Commonly octahedral with faces striated parallel to the edge between the dodecahedron and octahedron. Less frequently showing well-developed dodecahedrons and cubes.

$Fd3m$; $a = 8.12$ Å; $Z = 8$. d's: 2.85(7), 2.44(10), 1.65(4), 1.48(6), 1.232(8).

Physical Properties. **H** $7\frac{1}{2}$–8. **G** 4.55. *Luster* vitreous. *Color* dark green. *Streak* grayish. Translucent. *Optics*: $n = 1.80$.

Composition and Structure. Fe^{2+} and Mn^{2+} may substitute for Zn; and Fe^{3+} for Al. Structure is that of spinel (see Figs. 11.5 and 11.6 and page 375).

Diagnostic Features. Characterized by crystal form (striated octahedrons) and hardness.

Occurrence. Gahnite is a rare mineral. It occurs in granitic pegmatites, in zinc deposits, and also as a metamorphic mineral in crystalline limestones. Found in large crystals at Bodenmais, Bavaria, Germany; in a talc schist near Falun, Sweden. In the United States found at Charlemont, Massachusetts and Franklin, New Jersey.

Name. After the Swedish chemist J. G. Gahn, the discoverer of manganese.

MAGNETITE—Fe$_3$O$_4$

Crystallography. Isometric; $4/m\bar{3}2/m$. Magnetite is frequently in octahedral crystals (Figs. 11.19a and 11.20), more rarely in dodecahedrons (Fig. 11.19b). Dodecahedrons may be striated parallel to

FIG. 11.20. Octahedral magnetite crystal, Binnenthal, Switzerland. (Harvard Mineralogical Museum.)

the intersection with the octahedron (Fig. 11.19c). Other forms rare. Usually granular massive, coarse-, or fine-grained.

$Fd3m$; $a = 8.40$ Å; $Z = 8$. d's: 2.96(6), 2.53(10), 1.611(8), 1.481(9), 1.094(8).

Physical Properties. Octahedral parting on some specimens. **H** 6. **G** 5.18. *Luster* metallic. *Color* iron-black. *Streak* black. Strongly magnetic; may act as a natural magnet, known as *lodestone*. Opaque.

Composition and Structure. Fe 72.4, O 27.6%. The composition of much magnetite corresponds closely to Fe_3O_4. However, analyses may show considerable percentages of Mg and Mn^{2+} substituting for Fe^{2+} and Al, Cr, Mn^{3+}, and Ti^{4+} substituting for Fe^{3+}. Above 600°C a complete solid solution series is possible between magnetite and *ulvöspinel*, Fe_2TiO_4 (see Figs. 11.11 and 11.21). The structure of magnetite is that of an *inverse spinel* (see page 375), which can be expressed by rewriting the formula as $Fe^{3+}(Fe^{2+},Fe^{3+})_2O_4$ (see Table 11.2).

Diagnostic Features. Characterized chiefly by its strong magnetism, its black color, and its hardness (6). Can be distinguished from magnetic franklinite by streak.

FIG. 11.21. Solid solution series and their extent among oxides of the system TiO_2-FeO-Fe_2O_3. The extent of solid solution between end member compositions is shown by solid bars, for mineral assemblages synthesized in the laboratory at about 600°C. At higher temperatures, the extent of solid solution will increase considerably in, for example, the ilmenite-hematite and pseudobrookite series. Lack of solid solution at about 600° C is shown by dashed lines. Compare with Fig. 11.11. (Compiled from Haggerty, S. E., 1976, Oxidation of opaque mineral oxides in basalts, and Opaque mineral oxides in terrestrial igneous rocks; articles in *Oxide Minerals, Reviews in Mineralogy,* v. 3. Mineralogical Society of America, Washington, D.C.)

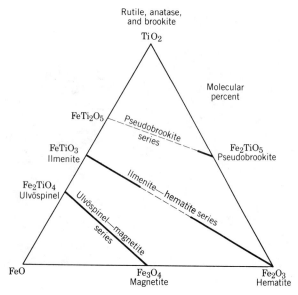

Occurrence. Magnetite is a common mineral found disseminated as an accessory through most igneous rocks. In certain types of rocks, through magmatic segregation, magnetite may become one of the chief constituents of the rock and may thus form large ore bodies. Such bodies are often highly titaniferous. Commonly associated with crystalline metamorphic rocks and may occur as large beds and lenses. It is a common constituent of sedimentary and metamorphic banded Precambrian iron-formations. In such occurrences it is of chemical, sedimentary origin. Found in the black sands of the seashore. Occurs as thin plates and dendritic growths between plates of mica. Often intimately associated with corundum, forming emery.

The largest magnetite deposits in the world are in northern Sweden at Kiruna and Gellivare associated with apatite, and are believed to have formed by magmatic segregation. Other important deposits are in Norway, Romania, and the Ukraine, CIS. The most powerful natural magnets are found in Siberia, in the Harz Mountains, on the Island of Elba, and in the Bushveld igneous complex, Republic of South Africa. The extensive Precambrian iron deposits in the Lake Superior region and in the Labrador Trough, Canada, carry considerable magnetite as an ore mineral.

The Precambrian iron-formations of the Lake Superior region, containing about 25% iron, mostly in the form of magnetite, are a major source of iron in the United States. This low-grade ore, *taconite*, has been actively mined and the magnetite separated magnetically from the waste material. Found as lodestone and in crystals at Magnet Cove, Arkansas.

Use. An important iron ore.

Name. Probably derived from the locality Magnesia, bordering on Macedonia. A fable told by Pliny ascribes its name to a shepherd named Magnes, who first discovered the mineral on Mount Ida by noting that the nails of his shoes and the iron ferrule of his staff adhered to the ground.

Similar Species. *Magnesioferrite,* $Fe^{3+}(Mg,Fe^{3+})_2O_4$ (or $MgFe_2O_4$) with an inverse spinel structure, is a rare mineral found chiefly in fumaroles. *Jacobsite,* $Fe^{3+}(Mn,Fe^{3+})_2O_4$ (or $MnFe_2O_4$), an inverse spinel and a rare mineral found at Långban, Sweden. *Ulvöspinel,* $Fe^{3+}(Ti,Fe^{2+})_2O_4$ (or Fe_2TiO_4), also with an inverse spinel structure, is not uncommonly present as exsolution blebs and lamellae within magnetite (see also Table 11.2).

Franklinite—(Zn,Fe,Mn) (Fe,Mn)$_2$O$_4$

Crystallography. Isometric; $4/m\overline{3}2/m$. Crystals octahedral (see Fig. 11.22) with dodecahedral trun-

FIG. 11.22. Franklinite, Franklin, New Jersey. (Harvard Mineralogical Museum.)

cations; often rounded. Also massive, coarse, or fine granular, in rounded grains.

Fd3m; a = 8.42 Å; *Z* = 8. *d*'s: 2.51(10), 1.610(10), 1.480(9), 1.278(6), 1.091(6).

Physical Properties. H 6. **G** 5.15. *Luster* metallic. *Color* iron-black. *Streak* reddish-brown to dark brown. Slightly magnetic.

Composition and Structure. Dominantly $ZnFe_2O_4$ but always with some substitution of Fe^{2+} and Mn^{2+} for Zn, and Mn^{3+} for Fe^{3+}. Analyses show a wide range in the proportions of the various elements. The structure of franklinite is that of a normal spinel (see page 375 and Figs. 11.5 and 11.6).

Diagnostic Features. Resembles magnetite but is only slightly magnetic and has a dark brown streak. Usually identified by its characteristic association with willemite and zincite.

Occurrence. Franklinite, with only minor exceptions, is confined to the zinc deposits at Franklin, New Jersey, where, enclosed in a granular limestone, it is associated with zincite and willemite.

Use. As an ore of zinc and manganese. The zinc is converted into zinc white, ZnO, and the residue is smelted to form an alloy of iron and manganese, *spiegeleisen*, used in the manufacture of steel.

Name. From Franklin, New Jersey.

CHROMITE—$FeCr_2O_4$

Crystallography. Isometric; $4/m\bar{3}2/m$. Habit octahedral but crystals small and rare. Commonly massive, granular to compact.

Fd3m; a = 8.36 Å; *Z* = 8. *d*'s: 4.83(4), 2.51(10), 2.08(5), 1.602(6), 1.473(8).

Physical Properties. H $5\frac{1}{2}$. **G** 4.6. *Luster* metallic to submetallic; frequently pitchy. *Color* iron-black to brownish-black. *Streak* dark brown. Subtranslucent. *Optics*: n = 2.16.

Composition and Structure. For $FeCr_2O_4$, FeO 32.0, Cr_2O_3 68.0%. Some Mg is always present substituting for Fe^{2+} and some Al and Fe^{3+} may substitute for chromium. There is extensive solid solution between chromite and magnesiochromite, $MgCr_2O_4$. The structure of chromite is that of a normal spinel (see page 375 and Figs. 11.5 and 11.6).

Diagnostic Features. The submetallic luster usually distinguishes chromite.

Occurrence. Chromite is a common constituent of peridotites and other ultrabasic rocks and of serpentines derived from them. One of the first minerals to separate from a cooling magma; large chromite ore deposits are thought to have been derived by such magmatic differentiation. For example, the Bushveld igneous complex of South Africa and the Great Dike of Zimbabwe contain many seams of chromite enclosed in pyroxenites. Associated with olivine, serpentine, and corundum.

The important countries for its production are: Republic of South Africa, CIS, Albania, and Zimbabwe. Chromite is found only sparingly in the United States, but in the past, Pennsylvania, Maryland, North Carolina, Wyoming, California, Alaska, and Oregon were small producers. During World War II, bands of low-grade chromite were mined in the Stillwater igneous complex, Montana.

Use. The only ore of chromium. Chromite ores are grouped into three categories—metallurgical, refractory, and chemical—on the basis of their chrome content and their Cr/Fe ratio. As a metal, chromium is used as a ferroalloy to give steel the combined properties of high hardness, great toughness, and resistance to chemical attack. Chromium is a major constituent in stainless steel. *Nichrome*, an alloy of Ni and Cr, is used for resistance in electrical heating equipment. Chromium is widely used in plating plumbing fixtures, automobile accessories, and so forth.

Because of its refractory character, chromite is made into bricks for the linings of metallurgical furnaces. The bricks are usually made of crude chromite and coal tar but sometimes of chromite with kaolin, bauxite, or other minerals. Chromium is a constituent of certain green, yellow, orange, and red pigments and in $K_2Cr_2O_7$ and $Na_2Cr_2O_7$, which are used as mordants to fix dyes.

Similar Species. *Magnesiochromite*, $MgCr_2O_4$, is in both occurrence and appearance similar to chromite.

Name. Named in allusion to the composition.

Chrysoberyl—BeAl$_2$O$_4$

Crystallography. Orthorhombic; $2/m2/m2/m$. Usually in crystals tabular on {001}, the faces of which are striated parallel to [100]. Commonly twinned on {130}, giving pseudohexagonal appearance (Fig. 11.23).

Pmnb; $a = 5.47$, $b = 9.39$, $c = 4.42$ Å; $Z = 4$.
d's: 3.24(8), 2.33(8), 2.57(8), 2.08(10), 1.61(10).

Physical Properties. *Cleavage* {110}. **H** $8\frac{1}{2}$. **G** 3.65–3.8. *Luster* vitreous. *Color* various shades of green, brown, yellow; may be red by transmitted light. *Optics*: (+), $\alpha = 1.746$, $\beta = 1.748$, $\gamma = 1.756$; $2V = 45°$; $X = c$, $Y = b$.

Alexandrite is a gem variety, emerald-green in daylight, but red by transmitted light and usually red by artificial light. *Cat's eye*, or *cymophane*, is a chatoyant variety which, when cut as an oval or round *cabochon gem*, shows a narrow band of light on its surface. The effect results from minute tubelike cavities or needlelike inclusions parallel to the c axis.

Composition and Structure. BeO 19.8, Al_2O_3 80.2%. Be 7.1%. From its formula, $BeAl_2O_4$, it would appear that chrysoberyl is a member of the spinel group. However, because of the small size of Be^{2+}, chrysoberyl has a structure of lower symmetry. It consists of oxygen atoms in hexagonal closest packing with Be in 4-coordination with O and Al in 6-coordination with O. Morphologically and structurally chrysoberyl is very similar to olivine $(Mg,Fe)_2SiO_4$. Be is in the same coordination as Si, and Al in the same coordination as Fe^{2+} or Mg. The hexagonal closest packed arrangement of O in chrysoberyl leads to pseudohexagonal lattice, angles, and twinning.

Diagnostic Features. Characterized by its high hardness, its yellowish to emerald-green color, and its twin crystals.

Occurrence. Chrysoberyl is a rare mineral. It occurs in granitic rocks, pegmatites, and mica schists. Frequently in river sands and gravels. The outstanding alluvial gem deposits are found in Brazil and Sri Lanka; the *alexandrite* variety comes from the Ural Mountains and Brazil. In the United States chrysoberyl of gem quality is rare, but it has been found in Oxford County and elsewhere in Maine; Haddam, Connecticut; and Greenfield, New York. Recently found in Colorado.

Use. As a gemstone. The ordinary yellowish-green stones are inexpensive; the varieties alexandrite and cat's eye are highly prized gems (see Chapter 15).

Name. *Chrysoberyl* means *golden beryl*. *Cymophane* is drived from two Greek words meaning *wave* and *to appear*, in allusion to the chatoyant effect. *Alexandrite* was named in honor of Alexander II of Russia.

Columbite-Tantalite— (Fe,Mn)Nb$_2$O$_6$-(Fe,Mn)Ta$_2$O$_6$

Crystallography. Orthorhombic; $2/m2/m2/m$. Commonly in crystals. The habit is short prismatic or thin tabular on {010}; often in square prisms because of prominent development of {100} and {010} (Fig. 11.24). Also in heart-shaped twins, twinned on {201}.

Pcan; $a = 5.74$, $b = 14.27$, $c = 5.09$ Å; $Z = 4$.
d's for columbite: 3.66(7), 2.97(10), 1.767(6), 1.735(7), 1.712(8).

Physical Properties. *Cleavage* {010} good. **H** 6. **G** 5.2–7.9, varying with the composition, increasing with rise in percentage of Ta_2O_5 (Fig. 11.25). *Luster* submetallic. *Color* iron-black, frequently iridescent. *Streak* dark red to black. Subtranslucent.

Composition and Structure. A complete solid solution exists from *columbite*, $(Fe,Mn)Nb_2O_6$, to *tantalite*, $(Fe,Mn)Ta_2O_6$. Often contains small

FIG. 11.23. Twinned chrysoberyl.

FIG. 11.24. Columbite-tantalite crystals.

amounts of Sn and W. A variety known as *mangano-tantalite* is essentially tantalite with Mn^{2+} substituting for most of the Fe^{2+}. In the structure octahedral chains of $(Mn,Fe)O_6$ and $(Ta,Nb)O_6$ exist in which octahedra join through sharing of edges. The chains are linked to each other by common apices.

Diagnostic Features. Recognized usually by its black color with lighter-colored streak, and high specific gravity. Distinguished from wolframite by lower specific gravity and less distinct cleavage.

Occurrence. The minerals occur in granitic rocks and pegmatites, associated with quartz, feldspar, mica, tourmaline, beryl, spodumene, cassiterite, wolframite, microlite, and monazite.

Notable localities for their occurrence are Bernie Lake, Manitoba, Canada; Zaire; Nigeria; Brazil; near Moss, Norway; Bodenmais, Bavaria, Germany; Ilmen Mountains, Russia, CIS; Western Australia (*mangano-tantalite*); and Malagasy Republic. In the United States it is found at Standish, Maine; Haddam, Middletown, and Branchville, Connecticut; in Amelia County, Virginia; Mitchell County, North Carolina; Black Hills, South Dakota; and near Canon City, Colorado.

Use. Source of tantalum and niobium. Because of its resistance to acid corrosion, tantalum is employed in chemical equipment, in surgery for skull plates and sutures, also in some tool steels and in electronic tubes. Niobium has its chief use in alloys in weldable high-speed steels, stainless steels, and alloys

resistant to high temperatures, such as used in the gas turbine of the aircraft industry.

Name. *Columbite* from Columbia, a name for America, whence the original specimen was obtained. *Tantalite* from the mythical Tantalus in allusion to the difficulty in dissolving in acid.

Similar Species. *Microlite*, $Ca_2Ta_2O_6(O,OH,F)$, is found in pegmatites; *pyrochlore*, $(Ca,Na)_2(Nb, Ta)_2O_6(O,OH,F)$, and *fergusonite*, $YNbO_4$, an oxide of niobium, tantalum, and rare earths (represented by *Y* in the fergusonite formula) are found associated with alkalic rocks.

HYDROXIDES

The hydroxide minerals to be considered in this section are:

HYDROXIDES

Brucite	$Mg(OH)_2$	
Manganite	$MnO(OH)$	
Romanechite	$BaMn^{2+}Mn_8^{4+}O_{16}(OH)_4$	
Goethite Group		
Diaspore		$\alpha\ AlO(OH)$
Goethite		$\alpha\ FeO(OH)$
Bauxite—mixture of diaspore, gibbsite, and boehmite		

FIG. 11.25. Columbite-tantalite. Variation of specific gravity with composition.

c

● OH

FIG. 11.26. Structure of brucite, composed of parallel layers of Mg^{2+} in octahedral coordination with $(OH)^-$. Large spacing along the *c* axis is caused by weak bonding between adjacent layers.

All structures in this group are characterized by the presence of the hydroxyl $(OH)^-$ group, or H_2O molecules. The presence of $(OH)^-$ groups causes the bond strengths in these structures generally to be much weaker than in the oxides.

The structure of *brucite* (see Fig. 11.26) consists of Mg^{2+} octahedrally coordinated to $(OH)^-$, with the octahedra sharing edges to form a layer. Because each $(OH)^-$ group is shared between three adjoining octahedra, the Mg^{2+} to $(OH)^-$ bond strengths have an e.v. $= \frac{1}{3}$. With three such bonds $(3 \times \frac{1}{3} = 1)$ the $(OH)^-$ group is neutralized. For this reason, the layers in the brucite structure are held together by only weak

bonds (compare Fig. 11.26 with Fig. 11.3). The structure of *gibbsite*, $Al(OH)_3$, is in principle identical to that of brucite except that, because of charge requirements, $\frac{1}{3}$ of the octahedrally coordinated cation positions are vacant (see Fig. 11.2a). A basal structural layer in corundum (as in Fig. 11.2) is structurally equivalent to the Al-OH sheets in gibbsite. The brucite structure type is referred to as *trioctahedral* (each $(OH)^-$ group is surrounded by three occupied octahedral positions) and the gibbsite structure type as *dioctahedral* (only two out of three octahedrally coordinated cation sites are filled). Such trioctahedral and dioctahedral layers are essential building units of the phyllosilicates (see page 498).

The structure of *diaspore* ($\alpha AlO(OH)$; space group *Pbnm*) is shown in Fig. 11.27. Oxygen and $(OH)^-$ groups are arranged in hexagonal closest packing with Al^{3+} in octahedral coordination between them. A chainlike pattern is produced by $Al(O,OH)_6$ octahedra extending along the *c* axis. The octahedra in each chain share edges, and the chains are joined to each other by adjoining apical oxygens. *Goethite*, $\alpha FeO(OH)$, is isostructural with diaspore. Both compositions, $AlO(OH)$ and $FeO(OH)$, occur in nature in two crystal forms, $\gamma AlO(OH)$, *boehmite*, and $\gamma FeO(OH)$, *lepidochrocite*, both with space group *Amam*, and show a somewhat different linkage of the octahedrally coordinated cations from that found in diaspore or goethite as seen in Fig. 11.28. The octahedra are linked by their apices to form chains; the chains, in turn, are joined by the sharing of octahedral edges, which results in corrugated sheets parallel to {010}. The sheets are weakly held together by hydrogen bonds between pairs of oxygens. Such bonding is represented as $(OH)^-$ groups in Fig. 11.28.

FIG. 11.27. The structure of diaspore, $\alpha AlO(OH)$, and goethite, $\alpha FeO(OH)$. The double chains of $AlO_3(OH)_3$ or $FeO_3(OH)_3$ octahedra run parallel to the *c* axis. Only (OH) groups are indicated; all unmarked apices of octahedra represent oxygen.

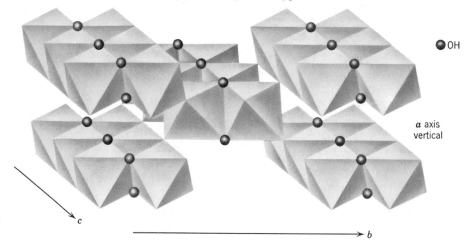

● OH

a axis vertical

c

b

FIG. 11.28. The structure of boehmite, $\gamma AlO(OH)$ and lepidocrocite, $\gamma FeO(OH)$, showing the arrangement of $AlO_4(OH)_2$ or $FeO_4(OH)_2$ octahedra. The corrugated sheets are parallel to (010). Only (OH) groups are indicated; all unmarked apices of octahedra represent oxygen.

Brucite—$Mg(OH)_2$

Crystallography. Hexagonal; $\bar{3}2/m$. Crystals usually tabular on {0001} and may show small rhombohedral truncations. Commonly foliated (Fig. 11.29), massive.

$C\bar{3}m; a = 3.13, c = 4.74$ Å; $Z = 1$. d's: 4.74(8), 2.37(10), 1.793(10), 1.372(7), 1.189(9).

Physical Properties. *Cleavage* {0001}) perfect. Folia flexible but not elastic. Sectile. **H** $2\frac{1}{2}$. **G** 2.39. *Luster* on base pearly, elsewhere vitreous to waxy.

FIG. 11.29. Brucite crystal, Wood's Mine, Texas. (Harvard Mineralogical Museum.)

Color white, gray, light green. Transparent to translucent. *Optics*: (+), ω = 1.566, ϵ = 1.581.

Composition and Structure. For $Mg(OH)_2$: MgO 69.0, H_2O 31.0%. Fe^{2+} and Mn^{2+} may substitute for Mg. The structure is illustrated in Fig. 11.26. The perfect {0001} cleavage is parallel to the octahedral sheets. Upon heating, brucite transforms into periclase (MgO).

Diagnostic Features. Recognized by its foliated nature, light color, and pearly luster on cleavage face. Distinguished from talc by its greater hardness and lack of greasy feel, and from mica by being inelastic.

Occurrence. Brucite is found associated with serpentine, dolomite, magnesite, and chromite; as an alteration product of periclase and magnesium silicates, especially serpentine. It is also found in crystalline limestone.

Notable foreign localities for its occurrence are at Unst, one of the Shetland Islands, Scotland, and Aosta, Italy. In the United States found at Tilly Foster Iron Mine, Brewster, New York; at Wood's Mine, Texas, Pennsylvania; and in Gabbs, Nevada.

Use. Brucite is used as a raw material for magnesia refractories and is a minor source of metallic magnesium.

Name. In honor of the early American mineralogist Archibald Bruce.

Similar Species. *Gibbsite*, $Al(OH)_3$, one of the three hydroxides of Al that are the main constituents of bauxite. The structure of gibbsite is like that of brucite, with one-third of the octahedral cation positions vacant (compare with Fig. 11.2a).

MANGANITE—MnO(OH)

Crystallography. Monoclinic; $2/m$ (pseudoorthorhombic). Crystals usually prismatic parallel to c and vertically striated (Figs. 11.30 and 11.31). Often columnar to coarse fibrous. Twinned on {011} as both contact and penetration twins.

FIG. 11.30. Manganite.

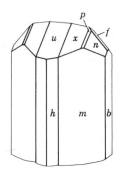

$B2_1/d$; a = 8.84, b = 5.23, c = 5.74 Å, β = 90°; Z = 8. d's: 3.38(10), 2.62(9), 2.41(6), 2.26(7), 1.661(9).

Physical Properties. Cleavage {010} perfect; {110} and {001} good. **H** 4. **G** 4.3. *Luster* metallic. *Color* steel-gray to iron-black. *Streak* dark brown. Opaque.

Composition and Structure. Mn 62.4, O 27.3, H_2O 10.3%. The structure consists of hexagonal closest packing of oxygen and $(OH)^-$ groups with Mn^{3+} in octahedral coordination with O^{2-} and $(OH)^-$. In this respect the structure is similar to that of diaspore; however the distribution of the cations is quite different.

Diagnostic Features. Recognized chiefly by its black color and prismatic crystals. Hardness (4) and brown streak distinguish it from pyrolusite.

Occurrence. Manganite is found associated with other manganese oxides in deposits formed by meteoric waters. Found often in low-temperature hydrothermal veins associated with barite, siderite, and calcite. It frequently alters to pyrolusite.

Occurs at Ilfeld, Harz Mountains, Germany, in fine crystals; also at Ilmenau, Thuringia, Germany, and Cornwall, England. In the United States at Negaunee, Michigan. In Nova Scotia, Canada.

Use. A minor ore of manganese.

Name. Named in allusion to the composition.

Romanechite—$BaMn^{2+}Mn_8^{4+}O_{16}(OH)_4$

Crystallography. Orthorhombic, 222. Massive, botryoidal, stalactitic (Fig. 11.32). Appears amorphous.

FIG. 11.31. Manganite, Ilfeld, Germany. (Harvard Mineralogical Museum.)

FIG. 11.32. Romanechite, Ironwood, Michigan. (Harvard Mineralogical Museum.)

$P222$; a = 9.45, b = 13.90, c = 5.72; Z = 2. d's: 3.46(7), 2.87(7), 2.41(7), 2.190(10), 1.820(7).

Physical Properties. **H** 5–6. **G** 3.7–4.7. *Luster* submetallic. *Color* black. *Streak* brownish-black. Opaque.

Composition and Structure. Small amounts of Mg, Ca, Ni, Co, Cu, and Si may be present. The structure is somewhat similar to that of rutile (Fig. 11.4) with complex chains of MnO,OH octahedra and large channels between adjoining chains. Both Ba and adsorbed H_2O are located in these channels. Upon heating to about 600°C, romanechite transforms into *hollandite*, $Ba_2Mn_8O_{16}$.

Diagnostic Features. Distinguished from the other manganese oxides by its greater hardness and botryoidal form, and from limonite by its black streak.

Occurrence. Romanechite is a secondary mineral; it occurs usually with pyrolusite, and its origin and associations are similar to those of that mineral.

Use. An ore of manganese. (See pyrolusite.)

Similar Species. Many of the hard botryoidal masses formerly called *psilomelane* are now known to be a mixture of several manganese oxides of which romanechite is a major constituent. Some of the other minerals commonly present in the mixture are *cryptomelane*, KMn_8O_{16}, *manjiroite*, $(Na,K)Mn_8O_{16}\cdot nH_2O$, and *todorokite* $(Mn,Ca,Mg)Mn_3O_7\cdot H_2O$. The presence of these and other minerals accounts for the oxides of Na, K, Ca, Co, Cu, Al, and Fe reported in chemical analyses of psilomelane.

Name. From the locality at Romanèche, France.

Diaspore—αAlO(OH)

Crystallography. Orthorhombic; $2/m2/m2/m$. Usually in thin crystals, tabular parallel to {010}; sometimes elongated on [001]. Bladed, foliated, massive, disseminated.

Pbnm; a = 4.41, b = 9.40, c = 2.84 Å; Z = 4. d's: 3.98(10), 2.31(8), 2.12(7), 2.07(7), 1.629(8).

Physical Properties. *Cleavage* {010} perfect. **H** $6\frac{1}{2}$–7. **G** 3.35–3.45. *Luster* vitreous except on cleavage face, where it is pearly. *Color* white, gray, yellowish, greenish. Transparent to translucent. *Optics*: (+), α = 1.702, β = 1.722, γ = 1.750; 2V = 85°; X = c, Y = b; $r < v$.

Composition and Structure. Al_2O_3 85, H_2O 15%. The structure is illustrated in Fig. 11.27 and consists of Al in 6-coordination with oxygen and OH^- forming $AlO_3(OH)_3$ octahedra. The structure of *boehmite*, $\gamma AlO(OH)$, is shown in Fig. 11.28.

Diagnostic Features. Characterized by its good cleavage, its bladed habit, and its high hardness.

Occurrence. Diaspore is commonly associated with corundum in emery rock, in dolomite, and in chlorite schist. In a fine-grained massive form it is a major constituent of much bauxite.

Notable localities are Mugla, Turkey; the island of Naxos, Greece; and Postmasburg, South Africa. In the United States it is found in Chester County, Pennsylvania; at Chester, Massachusetts; with alunite, forming rock masses at Mt. Robinson, Rosite Hills, Colorado. It is found abundantly in the bauxite and aluminous clays of Arkansas, Missouri, and elsewhere in the United States.

Use. As a refractory.

Name. Derived from a Greek word meaning to *scatter*, in allusion to its decrepitation when heated.

GOETHITE—αFeO(OH)

Crystallography. Orthorhombic; $2/m2/m2/m$. Rarely in distinct prismatic, vertically striated crystals (Fig. 11.33). Often flattened parallel to {010}. In acicular crystals. Also massive, reniform, stalactitic, in radiating fibrous aggregates (Fig. 11.34). Foliated. The so-called *bog ore* is generally loose and porous.

FIG. 11.33. Goethite.

FIG. 11.34. Goethite from an iron mine near Negaunee, Michigan. (Harvard Mineralogical Museum.)

Pbnm; *a* = 4.65, *b* = 10.02, *c* = 3.04 Å; *Z* = 4. *d*'s: 4.21(10), 2.69(8), 2.44(7), 2.18(4), 1.719(5).

Physical Properties. *Cleavage* {010} perfect. **H** 5–5½. **G** 4.37; may be as low as 3.3 for impure material. *Luster* adamantine to dull, silky in certain fine, scaly or fibrous varieties. *Color* yellowish-brown to dark brown. *Streak* yellowish-brown. Subtranslucent.

Composition and Structure. Fe 62.9, O 27.0, H_2O 10.1%. Mn is often present in amounts up to 5%. The massive varieties often contain adsorbed or capillary H_2O. Goethite is isostructural with diaspore (see Fig. 11.27). *Lepidocrosite*, $\gamma FeO(OH)$, a polymorph of goethite, is a platy mineral and is often associated with goethite; it is isostructural with boehmite (see Fig. 11.28).

Diagnostic Features. Distinguished from hematite by its streak.

Occurrence. Goethite is one of the commonest minerals and is typically formed under oxidizing conditions as a weathering product of iron-bearing minerals. It also forms as a direct inorganic or biogenic precipitate from water and is widespread as a deposit in bogs and springs. Goethite forms the gossan or "iron hat" over metalliferous veins. Large quantities of goethite have been found as residual lateritic mantles resulting from the weathering of serpentine.

Goethite in some localities constitutes an important ore of iron. It is the principal constituent of the valuable minette ores of Alsace-Lorraine, France. Other notable European localities are: Eiserfeld in Westphalia, Germany; Příbram, Bohemia, Czechoslovakia; and Cornwall, England. Large deposits of iron-rich laterites composed essentially of goethite are found in the Mayari and Moa districts of Cuba.

In the United States goethite is common in the Lake Superior hematite deposits and has been obtained in fine specimens at Negaunee, near Marquette, Michigan. Goethite is found in iron-bearing limestones along the Appalachian Mountains, from western Massachusetts as far south as Alabama. Such deposits are particularly important in Alabama, Georgia, Virginia, and Tennessee. Finely crystallized material occurs with smoky quartz and microcline in Colorado at Florissant and in the Pikes Peak region.

Use. An ore of iron.

Name. In honor of Goethe, the German poet.

Similar Species. *Limonite*, $FeO \cdot OH \cdot nH_2O$, is used mainly as a field term to refer to natural hydrous iron oxides of uncertain identity.

BAUXITE[2]—A Mixture of Diaspore, Gibbsite, and Boehmite

Crystallography. A mixture. Pisolitic, in round concretionary grains (Fig. 11.35); also massive, earthy, claylike.

Physical Properties. H 1–3. G 2–2.55. *Luster* dull to earthy. *Color* white, gray, yellow, red. Translucent.

Composition. A mixture of hydrous aluminum oxides in varying proportions. Some bauxites closely approach the composition of *gibbsite*, $Al(OH)_3$ (see brucite, "Similar Species"), but most are a mixture and usually contain some Fe. As a result, bauxite is not a mineral and should be used only as a rock name. The principal constituents of the rock bauxite are *gibbsite*, see Fig. 11.36; *boehmite*, $\gamma AlO(OH)$ (see Fig. 11.28), and *diaspore*, $\alpha AlO(OH)$ (see Fig. 11.27), any one of which may be dominant.

Diagnostic Features. Can usually be recognized by its pisolitic character.

Occurrence. Bauxite is of supergene origin, commonly produced under subtropical to tropical climatic conditions by prolonged weathering and leaching of silica from aluminum-bearing rocks. Also may be derived from the weathering of clay-bearing limestones. It has apparently originated as a colloidal precipitate. It may occur in place as a direct derivative of the original rock, or it may have been transported and deposited in a sedimentary formation. In the tropics

[2]Although bauxite is not a mineral species, it is described here because of its importance as the ore of aluminum.

FIG. 11.35. Pisolitic bauxite, Bauxite, Arkansas.

FIG. 11.36. Scanning electron micrograph of very small prismatic crystals of gibbsite (monoclinic) in bauxite from Surinam, South America. (Courtesy of H. H. Murray, Indiana University.)

deposits known as *laterites*, consisting largely of hyrous aluminum and ferric oxides, are found in the residual soils. These vary widely in composition and purity but many are valuable as sources of aluminum and iron.

Bauxite occurs over a large area in the south of France, an important district being at Baux, near Arles, France. The principal world producers are Surinam, Jamaica, and Guiana. Other major producing countries are Indonesia, the CIS, Australia, and Hungary. In the United States, the chief deposits are found in Arkansas, Georgia, and Alabama. In Arkansas bauxite has formed by the alteration of a nepheline syenite.

Use. The ore of aluminum. Eighty-five percent of the bauxite produced is consumed as aluminum ore. Because of its low density and great strength, aluminum has been adapted to many uses. Sheets, tubes, and castings of aluminum are used in automobiles, airplanes, and railway cars, where light weight is desirable. It is manufactured into cooking utensils, food containers, household appliances, and furniture. Aluminum is replacing copper to some extent in electrical transmission lines. Aluminum is alloyed with copper, magnesium, zinc, nickel, silicon, silver, and tin. Other uses are in paint, aluminum foil, and numerous salts.

The second largest use of bauxite is in the manufacture of Al_2O_3, which is used as an abrasive. It is also manufactured into aluminous refractories. Synthetic alumina is also used as the principal ingredient in heat-resistant porcelain such as spark plugs.

Name. From its occurrence at Baux, France.

HALIDES

The chemical class of halides is characterized by the dominance of the electronegative halogen ions, Cl^-, Br^-, F^-, and I^-. These ions are large, have a charge of only -1, and are easily polarized. When they combine with relatively large, weakly polarized cations of low valence, both cations and anions behave as almost perfectly spherical bodies. The packing of these spherical units leads to structures of the highest possible symmetry.

The structure of NaCl, shown in Figs. 11.37, 3.31, 3.36, and 4.42, was the first to be determined by X-ray diffraction techniques by W. H. and W. L. Bragg in 1913. The arrangement of the ions in the structure showed unambiguously that no molecules exist in the NaCl structure. Each cation and each anion is surrounded by 6 closest neighbors in octahedral coordination. Many halides of the *XZ* type crystallize with the *NaCl structure* (see Fig. 4.42); some mineral examples are, *sylvite*, KCl, *carobbiite*, KF and *chlorargyrite*, AgCl. Some *XZ* sulfides and oxides that have the NaCl-type structure are, *galena*, PbS, *alabandite*, MnS, and *periclase*, MgO. Several alkali halides such as CsCl, CsBr, and CsI, none of which is naturally occurring, do not have the NaCl type structure but crystallize with a geometric arrangement of 8 closest neighbors (cubic coordination) around the cation as well as the anion. This is known as the *CsCl-structure type* (see Fig. 4.43). The radius ratio $(R_X : R_Z)$ is the primary factor in determining which of the two struc-

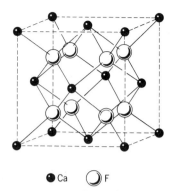

FIG. 11.38. The structure of fluorite. Compare this illustration with Fig. 4.45.

ture types is adopted by a given alkali halide of the *XZ* type.

The structures of several of the XZ_2 halides are identical to that of fluorite, CaF_2 (see Fig. 11.38), in which the Ca^{2+} ions are arranged at the corners and face centers of a cubic unit cell. F^- ions are tetrahedrally coordinated to four Ca^{2+}. Each Ca^{2+} is coordinated to eight F^- which surround it at the corners of a cube. The radius ratio ($R_{Ca} : R_F = 0.752$) leads to 8-coordination for Ca. Several oxides such as *uraninite*, UO_2, and *thorianite*, ThO_2, have a *fluorite type structure*.

Because the weak electrostatic charges are spread over the entire surface of the nearly spherical ions, the halides are the most perfect examples of pure ionic bonding. The isometric halides all have relatively low hardness and moderate to high melting points and are poor conductors of heat and electricity in the solid state. Such conduction of electricity as takes place does so by electrolysis, that is, by transport of charges by ions rather than by electrons. As the temperature increases and ions are liberated by thermal disorder, electrical conductivity increases rapidly, becoming excellent in the molten state. Advantage is taken of this conductivity of halide melts in the commercial methods for the preparation of sodium and chlorine by electrolysis of molten sodium chloride in the Downs cells, and in the Hall process for the electrolytic preparation of aluminum using molten cryolite. These properties are those conferred by the ionic bond.

When the halogen ions are combined with smaller and more strongly polarizing cations than those of the alkali metals, structures of lower symmetry result, and the bond has somewhat more covalent properties. In such structures, water or hydroxyl may enter as essential constituents, as in atacamite. Although there are over 85 species, we will consider only the following in detail:

FIG. 11.37. The structure of halite. Compare this illustration with Fig. 4.42.

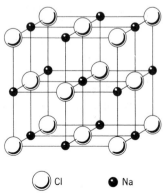

HALIDES

Halite	NaCl		Cryolite	Na_3AlF_6
Sylvite	KCl		Fluorite	CaF_2
Chlorargyrite	AgCl		Atacamite	$Cu_2Cl(OH)_3$

HALITE—NaCl

Crystallography. Isometric; $4/m\bar{3}2/m$. Habit cubic; other forms very rare. Some crystals hopper-shaped (Fig. 11.39). Found in crystals or granular crystalline masses showing cubic cleavage, known as *rock salt*. Also massive, granular to compact.

$Fm3m$; $a = 5.640$ Å; $Z = 4$. d's: 2.82(10), 1.99(4), 1.628(2), 1.261(2), 0.892(1).

Physical Properties. *Cleavage* {001} perfect. **H** $2\frac{1}{2}$. **G** 2.16. *Luster* transparent to translucent. *Color* colorless or white, or when impure may have shades of yellow, red, blue, purple. Salty taste. *Optics*: $n = 1.544$.

Composition and Structure. Na 39.3, Cl 60.7%. Commonly contains impurities, such as calcium and magnesium sulfates and calcium and magnesium chlorides. The structure of halite is illustrated in Figs. 11.37 and 4.42. This structure exists in a large number of *XZ* compounds with a radius ratio of between 0.41 and 0.73.

Diagnostic Features. Characterized by its cubic cleavage and taste, and distinguished from sylvite by less bitter taste.

Occurrence. Halite is a common mineral, occurring often in extensive beds and irregular masses, precipitated by evaporation with gypsum, sylvite, an-

FIG. 11.39. Halite, hopper-shaped crystals.

```
0    1    2
|____|____|
    cm
```

hydrite, and calcite. Halite is dissolved in the waters of salt springs, salt lakes, and the ocean. It is a major salt in playa deposits of enclosed basins.

The deposits of salt have been formed by the gradual evaporation and ultimate drying up of enclosed bodies of salt water (see Table 14.11 and related discussion). The salt beds formed in this way may have subsequently been covered by other sedimentary deposits and gradually buried beneath the rock strata formed on them. Salt beds range between a few feet to over 200 feet in thickness and it is estimated that some are buried beneath as much as 35,000 feet of overlying strata.

Extensive bedded deposits of salt are widely distributed throughout the world and are mined in many countries. Important production comes from China, the CIS, Great Britain, Germany, Canada, and Mexico.

The United States is the world's largest producer; salt in commercial amounts is, or has been, produced in every state, either from rock-salt deposits or by evaporation of saline waters. Thick beds of rock salt are found in New York State which extend through Ontario, Canada, into Michigan. Salt is recovered from these beds at many localities. Notable deposits are also found in Ohio, Kansas, and New Mexico, and in Canada in Nova Scotia and Saskatchewan. Salt is obtained by the evaporation of sea waters in California and Texas and from the waters of the Great Salt Lake in Utah.

Salt is also produced from *salt domes*, nearly vertical pipelike masses of salt that appear to have punched their way upward to the surface from an underlying salt bed. Anhydrite, gypsum, and native sulfur are commonly associated with salt domes. Geophysical prospecting for frequently associated petroleum has located several hundred salt domes along the Gulf coast of Louisiana and Texas and far out into the Gulf itself. Salt domes are also found in Germany, Romania, Spain, and Iran. In Iran, in an area of complete aridity, salt that has punched its way to the surface is not dissolved but moves down slope as a salt glacier.

Use. Halite finds its greatest use in the chemical industry, where it is the source of sodium and chlorine for the manufacture of hydrochloric acid and a large number of sodium compounds.

Salt is used extensively in the natural state in tanning hides, in fertilizers, in stock feeds, in salting icy highways, and as a weed killer. In addition to its familiar functions in the home, salt enters into the preparation of foods of many kinds, such as the preservation of butter, cheese, fish, and meat.

Name. Halite comes from the Greek word *halos* meaning *salt*.

SYLVITE—KCl

Crystallography. Isometric; $4/m\bar{3}2/m$. Cube and octahedron frequently in combination. Usually in granular crystalline masses showing cubic cleavage; compact.

$Fm3m$; $a = 6.293$ Å; $Z = 4$. d's: 3.15(10), 2.22(6), 1.816(2), 1.407(2), 1.282(4).

Physical Properties. Cleavage {001} perfect. **H** 2. **G** 1.99. Transparent when pure. *Color* colorless or white; also shades of blue, yellow, or red from impurities. Readily soluble in water. Salty taste but more bitter than halite. *Optics*: $n = 1.490$.

Composition and Structure. K 52.4, Cl 47.6%. May contain admixed NaCl. Sylvite has the NaCl structure (see Fig. 11.37) but because of the difference in the ionic radii of Na^+ (1.02 Å) and K^+ (1.38 Å) there is little solid solution between KCl and NaCl.

Diagnostic Features. Distinguished from halite by its more bitter taste.

Occurrence. Sylvite has the same origin, mode of occurrence, and associations as halite but is much rarer. It remains in the mother liquor after precipitation of halite and is one of the last salts to be precipitated (see evaporites, page 577).

It is found in quantity and frequently well crystallized, associated with the salt deposits at Stassfurt, Germany. In the United States it is found in large amount in the Permian salt deposits near Carlsbad, New Mexico, and in western Texas. More recently, deposits have been located in Utah. The most important world reserves are in Saskatchewan, Canada, where extensive bedded deposits have been found at depths greater than 3000 feet.

Use. The chief source of potassium compounds, which are principally used as fertilizers.

Name. Potassium chloride is the *sal digestivus Sylvii* of early chemistry, whence the name for the species.

Other Potassium Salts. Several other potassium minerals are commonly associated with sylvite and are found in Germany and Texas in sufficient amount to make them valuable as sources of potassium salts. These are: *carnallite*, $KMgCl_3 \cdot 6H_2O$, a usually massive to granular, generally light-colored mineral; *kainite*, $KMg(Cl,SO_4) \cdot 2\frac{3}{4}H_2O$ and *polyhalite*, $K_2Ca_2Mg(SO_4)_4 \cdot 2H_2O$.

CHLORARGYRITE—AgCl

Crystallography. Isometric; $4/m\bar{3}2/m$. Habit cubic but crystals rare. Usually massive, resembling wax; often in plates and crusts.

$Fm3m$; $a = 5.55$ Å; $Z = 4$. d's: 3.20(5), 2.80(10), 1.97(5), 1.67(2), 1.61(2).

Physical Properties. **H** 2–3. **G** 5.5 $\pm$. Sectile, can be cut with a knife; hornlike appearance, hence the name *horn silver*. Transparent to translucent. *Color* pearl-gray to colorless. Rapidly darkens to violet-brown on exposure to light. *Optics*: $n = 2.07$.

Composition and Structure. Ag 75.3, Cl 24.7%. A complete solid solution series exists between AgCl and *bromargyrite*, AgBr. Small amounts of F may be present in substitution for Cl or Br. Some specimens contain Hg. Chlorargyrite is isostructural with NaCl (see Fig. 11.37).

Diagnostic Features. Distinguished chiefly by its waxlike appearance and its sectility.

Occurrence. Chlorargyrite is an important supergene ore of silver found in the upper, enriched zone of silver deposits. It is found associated with native silver, cerussite, and secondary minerals in general.

Notable amounts have been found at Broken Hill, Australia; and in Peru, Chile, Bolivia, and Mexico. In the United States chlorargyrite was an important mineral in the mines at Leadville and elsewhere in Colorado, at the Comstock Lode in Nevada, and in crystals at the Poorman's Lode in Idaho.

Use. A silver ore.

Name. Chlorargyrite, from its composition.

Similar Species. Other closely related minerals that are less common but form under similar conditions, are *bromargyrite*, AgBr, and *iodian bromargyrite*, Ag(Cl,Br,I), isostructural with chlorargyrite; and *iodargyrite*, AgI, which is hexagonal.

CRYOLITE—Na$_3$AlF$_6$

Crystallography. Monoclinic; $2/m$. Prominent forms are {001} and {110}. Crystals rare, usually cubic in aspect, and in parallel groupings growing out of massive material. Usually massive.

$P2_1/n$; $a = 5.47$, $b = 5.62$, $c = 7.82$ Å, $\beta = 90°11'$; $Z = 2$. d's: 4.47(2), 3.87(2), 2.75(7), 2.33(4), 1.939(10).

Physical Properties. Parting on {110} and {001} produces cubical forms. **H** $2\frac{1}{2}$. **G** 2.95–3.0. *Luster* vitreous to greasy. *Color* colorless to snow-white. Transparent to translucent. *Optics*: (+), $\alpha = 1.338$, $\beta = 1.338$, $\gamma = 1.339$; $2V = 43°$; $X = b$, $Z \wedge c = -44°$, $r < v$. The low refractive index, near that of water, gives the mineral the appearance of watery snow or paraffin, and causes the powdered mineral to almost disappear when immersed in water.

Composition and Structure. Na 32.8, Al 12.8, F 54.4%. In the structure of cyrolite Al is octahedrally coordinated to six F^-. The Na^+ ions are also surrounded by six F^- ions, but in a somewhat less regular pattern. At high temperature (above 550°C) cryolite transforms to an isometric form with space group *Fm3m*.

Diagnostic Features. Characterized by pseudocubic parting, white color, and peculiar luster; and for the Greenland cryolite, the association of siderite, galena, and chalcopyrite.

Occurrence. The only important deposit of cryolite is at Ivigtut, on the west coast of Greenland. Here, in a large mass in granite, it is associated with siderite, galena, sphalerite, and chalcopyrite; and less commonly quartz, wolframite, fluorite, cassiterite, molybdenite, arsenopyrite, columbite. It is found at Miask, CIS; in the United States, at the foot of Pikes Peak, Colorado; and in crystals in Montreal, Quebec, Canada.

Use. Cryolite is used for the manufacture of sodium salts, of certain kinds of glass and porcelain, and as a flux for cleansing metal surfaces. It was early used as a source of aluminum. When bauxite became the ore of aluminum, cryolite was used as a flux in the electrolytic process. Today, with the essential exhaustion of the Ivigtut deposit, the sodium aluminum fluoride used in the aluminum industry is manufactured from fluorite.

Name. Name is derived from two Greek words, *kryos* meaning *frost* and *lithos* meaning *stone*, in allusion to its icy appearance.

FLUORITE—CaF$_2$

Crystallography. Isometric; $4/m\bar{3}2/m$. Usually in cubes, often as penetration twins twinned on [111] (Fig. 11.40a). Other forms are rare, but examples of all the forms of the hexoctahedral class have been observed; the tetrahexahedron (Fig. 11.40b) and hexoctahedron (Fig. 11.40c) are characteristic. Usually in crystals or in cleavable masses. Also massive; coarse or fine granular; columnar.

Fm3m; a = 5.46 Å; Z = 4. d's: 3.15(9), 1.931(10), 1.647(4), 1.366(1), 1.115(2).

Physical Properties. *Cleavage* {111} perfect. **H** 4. **G** 3.18. Transparent to translucent. *Luster* vitreous. *Color* varies widely; most commonly light green (see Plate IV, no. 1, Chapter 15), yellow, bluish-green, or purple; also colorless, white, rose, blue, brown. The color in some fluorite results from the presence of a hydrocarbon. A single crystal may show bands of varying colors; the massive variety is also often banded in color. The phenomenon of fluorescence (see page 268) received its name because it was early observed in some varieties of fluorite. *Optics*: n = 1.433.

Composition and Structure. Ca 51.3, F 48.7%. The rare earths, particularly Y and Ce, may substitute for Ca. The fluorite structure is shown in Figs. 11.38 and 4.45.

Diagnostic Features. Determined usually by its cubic crystals and octahedral cleavage; also vitreous luster and usually fine coloring, and by the fact that it can be scratched with a knife.

Occurrence. Fluorite is a common and widely distributed mineral. Usually found in hydrothermal veins in which it may be the chief mineral or as a gangue mineral with metallic ores, especially those of lead and silver. Common in vugs in dolomites and limestone and has been observed also as a minor accessory mineral in various igneous rocks and pegmatites. Associated with many different minerals, as calcite, dolomite, gypsum, celestite, barite, quartz, galena, sphalerite, cassiterite, topaz, tourmaline, and apatite.

Fluorite is found in quantity in England, chiefly from Cumbria, Derbyshire, and Durham; the first two localities are famous for their magnificent crystallized specimens. Found commonly in the mines of Saxony, Germany. Fine specimens come from the Alps. The large producers of commercial fluorite (fluorspar) are: Mongolia, the CIS, Republic of South Africa, China, Spain, and Thailand. Compared with these countries, the production in the United States is small, but there are significant amounts mined in southern Illinois near Rosiclare and Cave-in-Rock. Much of the fluorite at Cave-in-Rock is in coarse crystalline aggregates lining flat, open spaces, and thus the locality is the source of many beautifully crystallized specimens (Fig.

FIG. 11.40. Fluorite. *(a)* Penetration twin. *(b)* Cube and tetrahexahedron. *(c)* Cube and hexoctahedron.

(a)   *(b)*   *(c)*

FIG. 11.41. Yellow fluorite cubes coated by white quartz crystals, Cave-in-Rock, Harding County, Illinois. (Harvard Mineralogical Museum.)

11.41). Fluorite is also mined in small amounts in Nevada, Texas, and Utah.

Use. The bulk of the fluorite produced is used in the chemical industry (over 50%), mainly in the preparation of hydrofluoric acid, and as a flux in the making of steel (over 40%). Other uses are in the manufacture of glass, fiberglass, pottery, and enamel. Formerly used extensively as an ornamental material and for carving vases and dishes. Small amounts of fluorite are used for lenses and prisms in various optical systems, but most of the optical material is now made synthetically.

Name. From the Latin *fluere*, meaning to flow, because it melts more easily than other minerals with which it was (in the form of cut stones) confused.

Atacamite—Cu₂Cl(OH)₃

Crystallography. Orthorhombic; $2/m2/m2/m$. Commonly in slender prismatic crystals with vertical striations. Also tabular parallel to {010}. Usually in confused crystalline aggregates; fibrous; granular.

Pnam; $a = 6.02, b = 9.15, c = 6.85$ Å; $Z = 4$. *d*'s: 5.48(10), 5.03(7), 2.84(5), 2.78(5), 2.76(6).

Physical Properties. *Cleavage* {010} perfect. **H** $3-3\frac{1}{2}$. **G** 3.75–3.77. *Luster* adamantine to vitreous. *Color* various shades of green. Transparent to translucent. *Optics*: $(-)$, $\alpha = 1.831$, $\beta = 1.861$, $\gamma = 1.880$; $2V = 75°$; $r < v$. $X = b$, $Y = a$.

Composition and Structure. Cu 14.88, CuO 55.87, Cl 16.60, H₂O 12.65%. In the structure of atacamite part of the Cu atoms is in 6-coordination with five (OH) groups and one Cl. The remaining Cu is in 6-coordination with four (OH) groups and two Cl.

Diagnostic Features. Characterized by its green color and granular crystalline aggregates. Distinguished from malachite by its lack of effervescence in acid.

Occurrence. Atacamite is a comparatively rare copper mineral. Found originally as sand in the province of Atacama in Chile. Occurs in arid regions as a supergene mineral in the oxidized zone of copper deposits. It is associated with other secondary minerals in various localities in Chile, especially Chuquicamata, and in some of the copper districts of South Australia. In the United States occurs sparingly in the copper districts of Arizona.

Use. A minor ore of copper.

Name. From the province of Atacama, Chile.

REFERENCES AND SUGGESTED READING

Bragg, L. and Claringbull, G. F., 1965, *Crystal Structures of Minerals*. G. Bell and Sons, Ltd., London, 409 pp.

Deer, W. A., Howie, R. A. and Zussman, J., 1962, *Rock-Forming Minerals*, v. 5, Non-silicates. John Wiley & Sons, New York, 371 pp.

Evans, R. C., 1966, *An Introduction to Crystal Chemistry*, 2nd ed. Cambridge University Press, Cambridge, England, 410 pp.

Oxide Minerals. Reviews in Mineralogy, 1976, v. 3. Mineralogical Society of America, Washington, D.C.

Oxide Minerals: Petrologic and Magnetic Significance. Reviews in Mineralogy, 1991, v. 25, Lindsley, D. H., ed. Mineralogical Society of America, Washington, D.C., 509 pp.

Palache, C., Berman, H. and Frondel, C., 1944 and 1951, *The System of Mineralogy*, 7th ed., v. I and II. John Wiley & Sons, New York, 834 pp. and 1124 pp.

CHAPTER 12

SYSTEMATIC MINERALOGY

PART III: **CARBONATES**, NITRATES, BORATES, **SULFATES**, CHROMATES, TUNGSTATES, MOLYBDATES, **PHOSPHATES**, ARSENATES, AND VANADATES

In this chapter are discussed representative species of the carbonates, nitrates, borates, sulfates, chromates, tungstates, molybdates, phosphates, arsenates, and vanadates. These groups together contain over 1000 species but because most of them are uncommon minerals, only a small number will be considered.

These chemically diverse minerals are treated together because most of them contain *anionic complexes*, which can be recognized as strongly bonded units in their structures. Examples of such anionic complexes are: $(CO_3)^{2-}$ in carbonates, $(NO_3)^{1-}$ in nitrates, $(PO_4)^{3-}$ in phosphates, $(SO_4)^{2-}$ in sulfates, $(CrO_4)^{1-}$ in chromates, $(WO_4)^{2-}$ in tungstates, and $(AsO_4)^{3-}$ in arsenates. The bond strengths within such anionic complexes are always stronger than those between the anionic complex and other ions of the structure; these compounds are therefore referred to as *anisodesmic* (see page 199). For example, in the carbonates, radius ratio considerations predict three closest oxygen neighbors about carbon. This arrangement is triangular with carbon at the center and oxygen at each of the corners of the triangle (see Fig. 12.1); the (NO_3) group is also triangular. The e.v. of the bonds between carbon and each of the three closest oxygens is $\frac{1}{3} \times 4 = 1\frac{1}{3}$. This means that each oxygen has a residual charge of e.v. $= \frac{2}{3}$ for bonding other ions in the carbonate structure. Figure 12.1 illustrates the anisodesmic character of structures with tetrahedral (PO_4) and (SO_4) anionic groups as well.

Borates with triangular (BO_3) groups and silicates and tetrahedral (SiO_4) groups are examples of *mesodesmic* bonding (see page 200).

CARBONATES

The anionic $(CO_3)^{2-}$ complexes of carbonates are strongly bonded units and do not share oxygens with each other (as noted above, the residual e.v. of $\frac{2}{3}$ does not allow this). The triangular carbonate groups are the basic building units of all carbonate minerals and are largely responsible for the properties peculiar to the group.

Although the bond between the central carbon and its coordinated oxygens in the (CO_3) group is strong, it is not as strong as the covalent bond in CO_2. In the presence of hydrogen ion, the carbonate group becomes unstable and breaks down to yield CO_2 and water, according to $2H^+ + CO_3 \rightarrow H_2O + CO_2$. This reaction is the cause of the familiar "fizz" test with acid, which is widely used in the identification of carbonates.

The important anhydrous carbonates fall into three structurally different groups: the *calcite group*, the *aragonite group*, and the *dolomite group*. Aside from the minerals in these groups, the hydrous copper carbonates, azurite and malachite, are the only important carbonates.

CARBONATES

Calcite Group		**Aragonite Group**	
(Hexagonal; $R\bar{3}c$)		(Orthorhombic; $Pmcn$)	
Calcite	$CaCO_3$	Aragonite	$CaCO_3$
Magnesite	$MgCO_3$	Witherite	$BaCO_3$
Siderite	$FeCO_3$	Strontianite	$SrCO_3$
Rhodochrosite	$MnCO_3$	Cerussite	$PbCO_3$
Smithsonite	$ZnCO_3$		

		Dolomite Group	
		(Hexagonal; $R\bar{3}$)	
		Dolomite	$CaMg(CO_3)_2$
		Ankerite	$CaFe(CO_3)_2$

Monoclinic Carbonates with (OH)

Malachite	$Cu_2CO_3(OH)_2$
Azurite	$Cu_3(CO_3)_2(OH)_2$

FIG. 12.1. *(a), (b), (c)* Examples of anionic complexes, their bond strengths between the central cation and oxygen, and the residual charges on the oxygens. *(d)* The tetrahedral (SiO_4) group in which the e.v.'s between oxygen and the central cation are the same as the residual charge on the oxygen ($=1$).

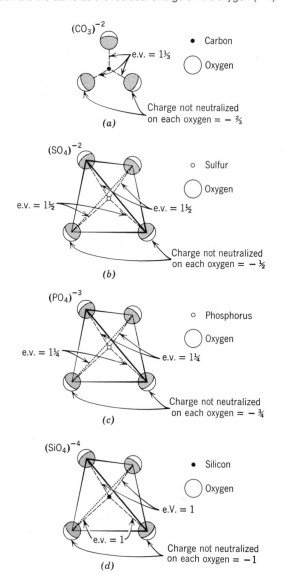

Calcite Group

The above five members of the calcite group are isostructural with space group $R\bar{3}c$. The structure of calcite, one of the earliest to be analyzed by X-rays by W. L. Bragg in 1914 (see Fig. 12.2a), can be thought of as a derivative of the NaCl structure in which triangular (CO_3) groups replace the spherical Cl, and Ca is in place of Na. The triangular shape of the (CO_3) groups causes the resulting structure to be rhombohedral instead of isometric as in NaCl. The (CO_3) groups lie in planes at right angles to the 3-fold (c) axis (Fig. 12.2) and the Ca ions, in alternate planes, are in 6-coordination with oxygens of the (CO_3) groups. Each oxygen is coordinated to two Ca ions as well as to a carbon ion at the center of the (CO_3) group.

FIG. 12.2. *(a)* Structure of calcite, $CaCO_3$. *(b)* The relation of the steep, true unit cell to the cleavage rhombohedron, which is face-centered. A hexagonal cell (rhomb-based prism) is also shown.

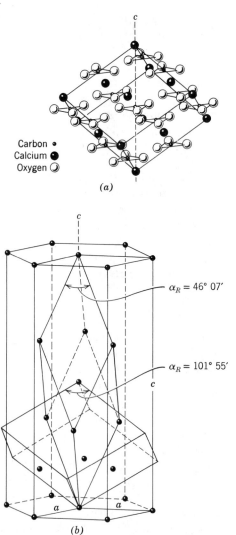

Calcite shows perfect rhombohedral cleavage to which, traditionally, the indices of {10$\overline{1}$1} have been assigned; axial ratios have been expressed accordingly. In the morphological descriptions and indexing of forms of the calcite and dolomite group minerals that follow, this convention has been preserved. However, X-ray structural determinations have shown that this rhombohedron does not correspond to the correct unit cell and that the simplest unit cell is a much steeper rhombohedron (Fig. 12.2b). Therefore, the structural axial ratios differ from the morphological. It should be noted that the radius ratio of Ca:O (= 0.714) in $CaCO_3$ is so close to the limiting value between 6- and 8-coordination (0.732) that $CaCO_3$ can occur in two structure types: *calcite*, with 6-coordination of Ca to O and *aragonite*, with 9-coordination of Ca to O (see aragonite, page 411).

CALCITE—CaCO₃

Crystallography. Hexagonal; $\overline{3}2/m$. Crystals are extremely varied in habit and often highly complex. Over 300 different forms have been described (Fig. 12.3). Three important habits exist: (1) prismatic, in long or short prisms, in which the prism faces are prominent, with base or rhombohedral terminations;

(2) rhombohedral, in which rhombohedral forms predominate; the unit (cleavage) form *r* is not common; (3) scalenohedral, in which scalenohedrons predominate, often with prism faces and rhombohedral truncations. The most common scalenohedron is {21$\overline{3}$1}. All possible combinations and variations of these types are found.

Twinning with the twin plane, {01$\overline{1}$2}, very common (Fig. 12.4); often produces twinning lamellae that may, as in crystalline limestones, be of secondary origin. This twinning may be produced artificially (see page 105). Twins with {0001}, the twin plane common. Calcite is usually in crystals or in coarse- to fine-grained aggregates. Also fine-grained to compact, earthy, and stalactitic.

$R\overline{3}c$. Hexagonal cell, a = 4.99, c = 17.06 Å; Z = 6; rhombohedral cell, a = 6.37 Å, α (rhombohedral angle) = 46°05'; Z = 2. d's: 3.04(10), 2.29(2), 2.10(2), 1.913(2), 1.875(2).

Physical Properties. *Cleavage* {10$\overline{1}$1} perfect (cleavage angle = 74°55'). Parting along twin lamellae on {01$\overline{1}$2}. **H** 3 on cleavage, 2$\frac{1}{2}$ on base. **G** 2.71. *Luster* vitreous to earthy. Color usually white to colorless, but may be variously tinted, gray, red, green, blue, yellow; also when impure, brown to black. Transparent to translucent. The chemically pure and

FIG. 12.3. Calcite crystals. Forms. *c*{0001}, *m*{10$\overline{1}$0}, *e*{01$\overline{1}$2}, *r*{10$\overline{1}$1}, *f*{02$\overline{2}$1}, *v*{21$\overline{3}$1}.

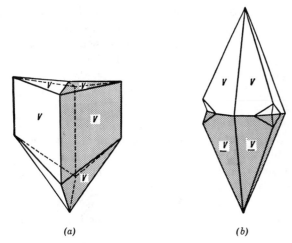

FIG. 12.4. Twinned calcite crystals. Twin planes: *(a)* {01$\bar{1}$2}. *(b)* {0001}.

optically clear, colorless variety is known as *Iceland spar* because of its occurrence in Iceland. *Optics*: (−); $\omega = 1.658, \epsilon = 1.486$.

Composition and Structure. Most calcites tend to be relatively close to pure $CaCO_3$ with CaO 56.0 and CO_2 44.0%. Mn^{2+}, Fe^{2+}, and Mg may substitute for Ca and a complete solid solution series extends to rhodochrosite, $MnCO_3$, above 550°C; a very partial series, with up to 5 weight percent FeO in calcite, exists between calcite and siderite, $FeCO_3$. Some in-

organic calcites may contain from 0 to about 2 weight percent MgO. Calcites in the hard parts of living organisms, however, may show a range of $MgCO_3$ of 2 to 16 molecular percent. See Fig. 12.5 for solid solution series in the system $CaO-MgO-FeO-CO_2$. The structure of calcite is shown in Fig. 12.2a and discussed on page 404.

Diagnostic Features. Fragments effervesce readily in cold dilute HCl. Characterized by its hardness (3), rhombohedral cleavage, light color, vitreous luster. Distinguished from dolomite by the fact that coarse fragments of calcite effervesce freely in cold HCl and distinguished from aragonite by lower specific gravity and rhombohedral cleavage.

Occurrence. As a *rock-forming mineral:* Calcite is one of the most common and widespread minerals. It occurs in extensive sedimentary rock masses in which it is the predominant mineral; in *limestones*, it is essentially the only mineral present. Crystalline, metamorphosed limestones are *marbles*. *Chalk* is a fine-grained pulverulent deposit of calcium carbonate. Calcite is an important constituent of calcareous marls and calcareous sandstones. Limestone has, in great part, been formed by the deposition on a sea bottom of great thicknesses of calcareous material in the form of shells and skeletons of sea animals. A smaller proportion of these rocks has been formed directly by precipitation of calcium carbonate.

FIG. 12.5. Carbonates and the extent of their solid solution in the system CaO-MgO-FeO-CO_2. The extent of solid solution series in this diagram is based upon chemical analyses of carbonates in metamorphic rocks that have been metamorphosed to about 400°C (biotite zone of greenschist facies). Tielines connect commonly coexisting carbonate species. Calcite-dolomite coexistences are common in Mg-containing limestones; ankerite-siderite coexistences are found in banded iron-formations. (Adapted from Anovitz, L. M. and Essence, E. J., 1987, Phase equilibria in the system $CaCO_3-MgCO_3-FeCO_3$. *Journal of Petrology*, v. 28, pp. 389–415.)

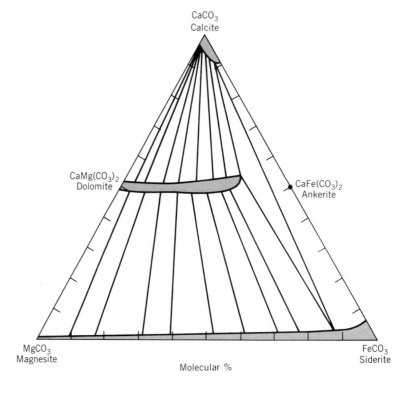

As cave deposits, etc.: Waters carrying calcium carbonate in solution and evaporating in limestone caves often deposit calcite as stalactites, stalagmites, and incrustations. Such deposits, usually semitranslucent and of light yellow colors, are often beautiful and spectacular. An example is Carlsbad Caverns, New Mexico. Both hot and cold calcareous spring water may form cellular deposits of calcite known as *travertine*, or *tufa*, around their mouths. The deposit at Mammoth Hot Springs, Yellowstone Park, is more spectacular than most, but of similar origin. *Onyx marble* is banded calcite and/or aragonite used for decorative purposes. Because much of this material comes from Baja California, Mexico, it is also called Mexican onyx.

Siliceous calcites: Calcite crystals may enclose considerable amounts of quartz sand (up to 60%) and form what are known as sandstone crystals. Such occurrences are found at Fontainebleau, France (Fontainebleau limestone), and in the Bad Lands, South Dakota.

Calcite occurs as a primary mineral in some igneous rocks such as carbonatites and nepheline syenites. It is a late crystallization product in the cavities in lavas. It is also a common mineral in hydrothermal veins associated with sulfide ores.

It is impossible to specify all the important districts for the occurrence of calcite in its various forms. Some of the more notable classic localities in which finely crystallized calcite is found are as follows: Andreasberg in the Harz Mountains, Germany; in Cumbria (Fig. 12.6), and Lancashire, England; Ice-land; and Guanajuato, Mexico. In the United States at Joplin, Missouri and the Lake Superior copper district.

Use. The most important use for calcite is for the manufacture of cements and lime for mortars. Limestone is the chief raw material, which when heated to about 900°C forms *quicklime*, CaO, by the reaction: $CaCO_3 \rightarrow CaO + CO_2'$. The CaO, when mixed with water, forms one or several CaO-hydrates (slaked lime), swells, gives off much heat, and hardens or, as commonly termed, "sets." Quicklime when mixed with sand forms common mortar.

The greatest consumption of limestone is in the manufacture of cements. The type known as Portland cement is most widely produced. It is composed of about 75% calcium carbonate (limestone) with the remainder essentially silica and alumina. Small amounts of magnesium carbonate and iron oxide are also present. In some limestone, known as *cement rocks*, the correct proportions of silica and alumina are present as impurities. In others these oxides are contributed by clay or shale mixed with the limestone before "burning." When water is mixed with cement, hydrous calcium silicates and calcium aluminates are formed.

Limestone is a raw material for the chemical industry, and finely crushed is used as a soil conditioner, for whiting and whitewash. Great quantities are quarried each year as a flux for smelting various metallic ores, as an aggregate in concrete, and as road material. A fine-grained limestone is used in lithography.

Calcite in several forms is used in the building

FIG. 12.6. Group of calcite crystals, Cumbria, England. (Harvard Mineralogical Museum.)

industry. Limestone and marble as dimension stone are used both for construction purposes and for decorative exterior facings. Polished slabs of travertine and Mexican onyx are commonly used as ornamental stone for interiors. Indiana is the chief source of building limestone (the Salem limestone) in the United States, with the most productive quarries in Lawrence and Monroe counties. Many of the federal buildings in Washington, D.C., have been constructed from this limestone. The most important marble quarries are in Vermont, New York, Georgia, and Tennessee.

Iceland spar, named for its occurrence in Iceland, is valuable for various optical instruments; its best known use was in the form of the Nicol prism to produce polarized light, prior to the use of Polaroid plates.

Name. From the Latin word *calx*, meaning burnt lime.

MAGNESITE—MgCO₃

Crystallography. Hexagonal; $\bar{3}2/m$. Crystals rhombohedral, $\{10\bar{1}1\}$, but rare. Usually cryptocrystalline in white, compact, earthy masses, less frequently in cleavable granular masses, coarse to fine.

$R\bar{3}c$. Hexagonal cell, $a = 4.63$, $c = 15.02$ Å; $Z = 6$; rhombohedral cell, $a = 5.62$ Å, α (rhombohedral angle) $= 48°10'$, $Z = 2$. *d*'s: 2.74(10), 2.50(2), 2.10(4), 1.939(1), 1.700(3).

Physical Properties. *Cleavage* $\{10\bar{1}1\}$ perfect. **H** $3\frac{1}{2}$–5. **G** 3.0–3.2. *Luster* vitreous. *Color* white, gray, yellow, brown. Transparent to translucent. *Optics:* (−); $\omega = 1.700$, $\epsilon = 1.509$.

Composition and Structure. MgO 47.8, CO₂ 52.2%. Fe^{2+} substitutes for Mg and a complete series extends to siderite (see Figs. 12.5 and 12.7). Small amounts of Ca and Mn may be present. Magnesite is isostructural with calcite (see Fig. 12.2).

Diagnostic Features. Cleavable varieties are distinguished from dolomite by higher specific gravity and absence of abundant calcium. The white massive variety resembles chert and is distinguished from it by inferior hardness. Scarcely acted upon by cold HCl, but dissolves with effervescence in hot HCl.

Occurrence. Magnesite commonly occurs in veins and irregular masses derived from the alteration of Mg-rich metamorphic and igneous rocks (serpentinites and peridotites) through the action of waters containing carbonic acid. Such magnesites are compact, cryptocrystalline, and often contain opaline silica. Beds of crystalline cleavable magnesite are (1) of metamorphic origin associated with talc schists, chlorite schists, and mica schists, and (2) of sedimen-

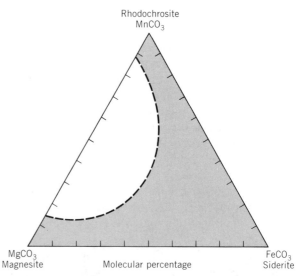

FIG. 12.7. The approximate extent of solid solution for carbonates in the three-component system MgCO₃-FeCO₃-MnCO₃. The lack of solid solution between MnCO₃ and MgCO₃ may be the result of a considerable difference in the size of the cationic radii, or it may reflect a lack of carbonate-rich rock types that bridge the gap between MnCO₃ and MgCO₃. (From Essene, E. J., 1983, Solid solutions and solvi among metamorphic carbonates with applications to geologic thermobarometry, in *Carbonates: Mineralogy and Chemistry. Reviews in Mineralogy*, v. 11, pp. 77–96.)

tary origin, formed as a primary precipitate or as a replacement of limestones by Mg-containing solutions, dolomite being formed as an intermediate product.

Notable deposits of the sedimentary type of magnesite are in China; at Satka in the Ural Mountains, CIS; and at Styria, Austria. The most famous deposit of the cryptocrystalline type is on the Island of Euboea, Greece. The best crystals originate from veins in Oberdorf, Austria, and Bahia, Brazil.

In the United States the compact variety is found in irregular masses in serpentine in the Coast Range, California. The sedimentary type is mined at Chewelah in Stevens County, Washington, and in the Paradise Range, Nye County, Nevada. There are numerous minor localities in the eastern United States in which magnesite is associated with serpentine, talc, or dolomite rocks.

Use. Dead-burned magnesite, MgO, that is, magnesite that has been calcined at a high temperature and contains less than 1% CO₂, is used in manufacturing bricks for furnace linings. Magnesite is the source of magnesia for industrial chemicals. It has also been used as an ore of metallic Mg, but at present the entire production of Mg comes from brines and seawater.

Name. Magnesite is named in allusion to the composition.

SIDERITE—FeCO$_3$

Crystallography. Hexagonal; $\bar{3}2/m$. Crystals usually unit rhombohedrons, frequently with curved faces. In globular concretions. Usually cleavable granular. May be botryoidal, compact, and earthy.

$R\bar{3}c$. Hexagonal cell, $a = 4.72$, $c = 15.45$ Å; $Z = 6$; rhombohedral cell, $a = 5.83$ Å, $\alpha = 47°45'$, $Z = 2$. $d's$: 3.59(6), 2.79(10), 2.13(6), 1.963(6), 1.73(8).

Physical Properties. Cleavage $\{10\bar{1}1\}$ perfect. **H** $3\frac{1}{2}$–4. **G** 3.96 for pure FeCO$_3$, but decreases with presence of Mn^{2+} and Mg. *Luster* vitreous. *Color* usually light to dark brown. Transparent to translucent. *Optics:* $(-)$; $\omega = 1.875$, $\epsilon = 1.633$.

Composition and Structure. For pure FeCO$_3$, FeO 62.1, CO$_2$ 37.9%. Fe 48.2%. Mn^{2+} and Mg substitute for Fe^{2+} and complete series extend to rhodochrosite and magnesite (see Figs. 12.5 and 12.7). The substitution of Ca for Fe^{2+} is limited due to the large difference in size of the two ions. Siderite is isostructural with calcite (see Fig. 12.2 and page 404).

Diagnostic Features. Distinguished from other carbonates by its color and high specific gravity, and from sphalerite by its rhombohedral cleavage. Soluble in hot HCl with effervescence.

Alteration. Pseudomorphs of limonite after siderite are common.

Occurrence. Siderite is frequently found as *clay ironstone*, impure by admixture with clay materials, in concretions with concentric layers. As *black-band ore* it is found, contaminated by carbonaceous material, in extensive stratified formations lying in shales and commonly associated with coal measures. These ores have been mined extensively in Great Britain in the past, but at present are mined only in North Staffordshire and Scotland. Clay ironstone is also abundant in the coal measures of western Pennsylvania and eastern Ohio, but it is not used to any great extent as an ore. Siderite is also formed by the replacement action of Fe-rich solutions upon limestones, and if such occurrences are extensive, they may be of economic value. The most notable deposit of this type is in Styria, Austria, where siderite is mined on a large scale. Siderite, in its crystallized form, is a common vein mineral associated with various metallic ores containing silver minerals, pyrite, chalcopyrite, tetrahedrite, and galena. When siderite predominates in such veins, it may be mined, as in southern Westphalia, Germany. Siderite is also a common constituent of banded Precambrian iron deposits, as in the Lake Superior region. A famous classic locality is Cornwall, England. Modern localities are the Morro Velho Gold Mine, Nova Lima, Brazil and Llallagua, Bolivia.

Use. An ore of iron. Important in Great Britain and Austria, but unimportant elsewhere.

Name. From the Greek word meaning *iron*. The name *spherosiderite* of the concretionary variety was shortened to siderite to apply to the entire species. *Chalybite*, used by some mineralogists, was derived from the Chalybes, ancient iron workers, who lived by the Black Sea.

RHODOCHROSITE—MnCO$_3$

Crystallography. Hexagonal; $\bar{3}2/m$. Only rarely in crystals of the unit rhombohedron; frequently with curved faces. Usually cleavable, massive; granular to compact.

$R\bar{3}c$. Hexagonal cell, $a = 4.78$, $c = 15.67$ Å; $Z = 6$; rhombohedral cell, $a = 5.85$ Å, $\alpha = 47°46'$, $Z = 2$. $d's$: 3.66(4), 2.84(10), 2.17(3), 1.770(3), 1.763(3).

Physical Properties. *Cleavage* $\{10\bar{1}1\}$ perfect. **H** $3\frac{1}{2}$–4. **G** 3.5–3.7. *Luster* vitreous. *Color* usually some shade of rose-red; may be light pink to dark brown. *Streak* white. Transparent to translucent. *Optics:* $(-)$; $\omega = 1.816$, $\epsilon = 1.597$.

Composition and Structure. For pure MnCO$_3$, MnO 61.7, CO$_2$ 38.3%. Fe^{2+} substitutes for Mn^{2+} forming a complete solid solution series between rhodochrosite and siderite (see Fig. 12.7). Ca^{2+} shows some substitution for Mn^{2+}. The occurrence of *kutnahorite*, CaMn(CO$_3$)$_2$, with an ordered structure of the dolomite type, suggests that only limited solid solution occurs, at ordinary temperatures, between CaCO$_3$ and MnCO$_3$. Mg may also substitute for Mn but the MnCO$_3$-MgCO$_3$ series is incomplete. Considerable amounts of Zn may substitute for Mn (see smithsonite). Rhodochrosite is isostructural with calcite (see Fig. 12.2 and page 404).

Diagnostic Features. Characterized by its pink color and rhombohedral cleavage; the hardness (4) distinguishes it from rhodonite, MnSiO$_3$, with hardness of 6. Infusible. Soluble in hot HCl with effervescence.

Occurrence. Rhodochrosite is a comparatively rare mineral, occurring in hydrothermal veins with ores of silver, lead, and copper, and with other manganese minerals. Beautiful banded rhodochrosite is mined for ornamental and decorative purposes at Capillitas, Catamarca, Argentina (see Fig. 12.8). Excellent crystals are found in the Kalahari manganese

FIG. 12.8. Rhodochrosite, Capillitas, Catamarca, Argentina. (Harvard Mineralogical Museum.)

region of Cape Province, Republic of South Africa, and at Pasto Bueno, Peru. In the United States it is found at Butte, Montana, where it has been mined as a manganese ore. In good crystals at Alicante, Lake County; Alma Park County; and elsewhere in Colorado.

Use. A minor ore of manganese. Small amounts used for ornamental purposes.

Name. Derived from two Greek words meaning *rose* and *color*, in allusion to its rose-pink color.

SMITHSONITE—ZnCO₃

Crystallography. Hexagonal; $\bar{3}2/m$. Rarely in small rhombohedral or scalenohedral crystals. Usually reniform, botryoidal (Fig. 12.9), or stalactitic, and in crystalline incrustations or in honeycombed masses known as *dry-bone ore*. Also granular to earthy.

$R\bar{3}c$. Hexagonal cell, $a = 4.66$, $c = 15.02$ Å; $Z = 6$; rhombohedral cell, $a = 5.63$ Å, $\alpha = 48°20'$, $Z = 2$. d's: 2.75(10), 3.55(5), 2.33(3), 1.946(3), 1.703(4).

Physical Properties. *Cleavage* $\{10\bar{1}1\}$ perfect. **H** 4–4½. **G** 4.30–4.45. *Luster* vitreous. *Color* usually dirty brown. May be colorless, white, green, blue, or pink. The yellow variety contains Cd and is known as *turkey-fat ore*. Streak white. Translucent. *Optics:* (−); $\omega = 1.850$, $\epsilon = 1.623$.

Composition and Structure. For pure $ZnCO_3$, ZnO 64.8, CO_2 35.2%. Considerable Fe^{2+} may substitute for Zn, but there appears to be a gap in the $ZnCO_3$-$FeCO_3$ series. Mn^{2+} is generally present in only a few percent, but the occurrence of a zincian rhodochrosite with Zn : Mn = 1 : 1.2 suggests there

may be a complete series between $ZnCO_3$ and $MnCO_3$. Ca and Mg are present in amounts of only a few weight percent. Small amounts of Co are found in a pink, and small amounts of Cu in a blue-green variety of smithsonite. Smithsonite is isostructural with calcite (see Fig. 12.2 and page 404).

Diagnostic Features. Soluble in cold HCl with effervescence. Distinguished by its effervescence in acids, tests for zinc, its hardness, and its high specific gravity.

Occurrence. Smithsonite is a zinc ore of supergene origin, usually found with zinc deposits in limestones. Associated with sphalerite, galena, hemimorphite, cerussite, calcite, and limonite. Often found in pseudomorphs after calcite. Smithsonite is

FIG. 12.9. Smithsonite, Kelly, New Mexico. (Harvard Mineralogical Museum.)

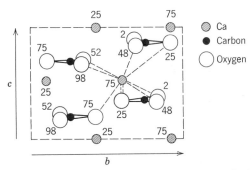

FIG. 12.10. The structure of aragonite, $CaCO_3$, as projected on (100). Oxygens that would normally superimpose have been made visible by some displacement. Numbers represent heights of atomic positions above the plane of origin, marked with respect to *a*. Dashed rectangle outlines unit cell. Ca-O bonds are also shown.

found in places in translucent green or greenish-blue material which is used for ornamental purposes. Laurium, Greece, is noted for this ornamental smithsonite, and Sardinia, Italy, for yellow stalactites with concentric banding. Fine crystallized specimens have come from the Broken Hill Mine, Zambia, and from Tsumeb, Namibia. In the United States smithsonite occurs as an ore in the zinc deposits of Leadville, Colorado; Arkansas and Wisconsin. Fine greenish-blue material has been found at the Kelly mine, Magdalena district, New Mexico.

Use. An ore of zinc. A minor use is for ornamental purposes.

Name. Named in honor of James Smithson (1754–1829), who founded the Smithsonian Institution in Washington, D.C. English mineralogists formerly called the mineral *calamine*.

Similar Species. *Hydrozincite*, $Zn_5(CO_3)_2(OH)_6$, occurs as a secondary mineral in zinc deposits.

Aragonite Group

When the (CO_3) group is combined with large divalent cations (ionic radii greater than 1.0 Å), the radius ratios generally do not permit stable 6-coordination and orthorhombic structures result. This is the *aragonite structure type* (Fig. 12.10 and Table 3.6) with space group *Pmcn*. $CaCO_3$ occurs in both the *calcite* and *aragonite structure types* because although Ca is somewhat large for 6-coordination (calcite), it is relatively small, at room temperature, for 9-coordination (aragonite); calcite is the stable form of $CaCO_3$ at room temperature (see Fig. 12.11). Carbonates with larger cations such as $BaCO_3$, $SrCO_3$, and $PbCO_3$, however, have the aragonite structure, stable at room temperature. In the aragonite structure the (CO_3) groups as in calcite, lie perpendicular to the *c* axis,

but in two structural planes, with the (CO_3) triangular groups of one plane pointing in opposite directions to those of the other. In calcite all (CO_3) groups lie in a single structural plane and point in the same direction (Fig. 12.2a). Each Ca is surrounded by nine closest oxygens. The cations have an arrangement in the structure approximating hexagonal closest packing, which gives rise to marked pseudohexagonal symmetry. This is reflected in both the crystal angles and in the pseudohexagonal twinning which is characteristic of all members of the group.

Solid solution within the aragonite group is somewhat more limited than in the calcite group, and it is interesting to note that Ca and Ba, respectively the smallest and largest ions in the group, form an ordered compound *barytocalcite*, $BaCa(CO_3)_2$, analogous to the occurrence of dolomite, $CaMg(CO_3)_2$, in the system $CaCO_3$-$MgCO_3$. The differences in physical properties of the minerals of the aragonite group are conferred largely by the cations. Thus the specific gravity is roughly proportional to the atomic weight of the metal ions (Table 6.1).

ARAGONITE–$CaCO_3$

Crystallography. Orthorhombic; $2/m2/m2/m$. Three habits of crystallization are common. (1) Acicular pyramidal; consisting of a vertical prism terminated by a combination of a very steep dipyramid and {110} prism (Fig. 12.12a). Usually in radiating groups of large to very small crystals (see Fig. 12.13). (2) Tabular; consisting of prominent {010} modified by {110} and a low prism, k{011} (Fig. 12.12b); often twinned on {110} (Fig. 12.12c). (3) In pseudohexagonal twins (Fig. 12.12d) showing a hexagonal-like prism terminated by a basal plane. These are formed by an intergrowth of three individuals twinned on {110} with {001} planes in common. The cyclic twins are distinguished from true hexago-

FIG. 12.11. The approximate location of the experimentally determined stability fields of calcite and aragonite.

FIG. 12.12. *(a)* and *(b)* Aragonite crystals. *(c)* and *(d)* Aragonite twins on {110} with a repeated twin producing a pseudohexagonal outline as in *(d)*.

(a)   *(b)*   *(c)*   *(d)*

nal forms by noting that the basal surface is striated in three different directions and that, because the prism angle of the simple crystals is not exactly 60°, the composite prism faces for the twin will often show slight reentrant angles. Also found in reniform, columnar, and stalactitic aggregates.

Pmcn; a = 4.96, *b* = 7.97, *c* = 5.74 Å; *Z* = 4. *d*'s: 3.04(9), 2.71(6), 2.36(7), 1.975(10), 1.880(8).

Physical Properties. *Cleavage* {010} distinct, {110} poor. *Luster* vitreous. *Color* colorless, white, pale yellow, and variously tinted. Transparent to translucent. **H** $3\frac{1}{2}$–4. **G** 2.94 (harder and higher specific gravity than calcite). *Optics:* (−); α = 1.530, β = 1.680, γ = 1.685; 2V = 18°; *X* = *c*, *Y* = *a*. *r* < *v* weak.

Composition, Structure, and Synthesis. Most aragonite is relatively pure $CaCO_3$. Small amounts of Sr and Pb substitute for Ca. The structure of aragonite is given in Fig. 12.10 and discussed on page 411. Calcite can be transformed into the aragonite structure by

extensive grinding with a mortar and pestle. This structural transformation can be determined only by X-ray diffraction techniques. The phase diagram for the polymorphs of $CaCO_3$ is given in Fig. 12.11. Aragonite, with a somewhat denser structure than calcite, is stable on the high *P*, low *T* side of the curve relating the two polymorphs. If the temperature of metamorphism in a metamorphic terrane can be arrived at independently, this curve allows for the estimation of a maximum or minimum pressure of metamorphism in calcite- or aragonite-bearing rocks.

Diagnostic Features. Effervesces in cold HCl. Distinguished from calcite by its higher specific gravity and lack of rhombohedral cleavage. Cleavage fragments of columnar calcite are terminated by a cross cleavage that is lacking in aragonite. Distinguished from witherite and strontianite by lower specific gravity.

Alteration. Pseudomorphs of calcite after aragonite are common. $CaCO_3$ secreted by mollusks as

FIG. 12.13. Aragonite, Cleator Moor, Cumbria, England. (Harvard Mineralogical Museum.)

aragonite is usually changed to calcite on the outside of the shell.

Occurrence. Aragonite is less stable than calcite under atmospheric conditions and much less common. It is precipitated in a narrow range of physicochemical conditions represented by low temperature, near-surface deposits. Experiments have shown that carbonated waters containing calcium more often deposit aragonite when they are warm and calcite when they are cold. The pearly layer of many shells and the pearl itself is aragonite. Aragonite is deposited by hot springs; found associated with beds of gypsum and deposits of iron ore where it may occur in forms resembling coral, and is called *flos ferri*. It is found as fibrous crusts on serpentine and in amygdaloidal cavities in basalt. The occurrence of aragonite in some metamorphic rocks as in the Franciscan Formation in California and in blue schists of New Zealand is the result of recrystallization at high pressures and relatively low temperature.

Notable localities for the various crystalline types are as follows: pseudohexagonal twinned crystals are found in Aragon, Spain; and at Girgenti, Sicily, Italy, associated with native sulfur. Prismatic crystals were found near Bílina, Czechoslovakia. The acicular type is found at Alston Moor and Cleator Moor, Cumbria, England. Flos ferri is found in the iron mines of Styria, Austria. Some *onyx marble* from Baja California, Mexico, is aragonite. In the United States pseudohexagonal twins are found at Lake Arthur, New Mexico. Flos ferri occurs in the Organ Mountains, New Mexico, and at Bisbee, Arizona.

Name. From Aragon, Spain, where the pseudohexagonal twins were first recognized.

WITHERITE—$BaCO_3$

Crystallography. Orthorhombic; $2/m2/m2/m$. Crystals always twinned on {110} forming pseudohexagonal dipyramids by the intergrowth of three individuals (Fig. 12.14). Crystals often deeply striated

FIG. 12.14. Witherite.

horizontally and by a series of reentrant angles have the appearance of one pyramid capping another. Also botryoidal to globular; columnar or granular.

$Pmcn$; $a = 5.31$, $b = 8.90$, $c = 6.43$ Å; $Z = 4$. d's: 3.72(10), 2.63(6), 2.14(5), 2.03(5), 1.94(5).

Physical Properties. *Cleavage* {010} distinct, {110} poor. **H** $3\frac{1}{2}$. **G** 4.31. *Luster* vitreous. *Color* colorless, white, gray. Translucent. *Optics:* (−), α = 1.529, β = 1.676, γ = 1.677; 2V = 16°; X = c, Y = b; r > v weak.

Composition and Structure. BaO 77.7, CO_2 22.3%. Small amounts of Sr and Ca may substitute for Ba. Witherite is isostructural with aragonite (see Fig. 12.10 and page 411).

Diagnostic Features. Soluble in cold HCl with effervescence. Witherite is characterized by high specific gravity. It is distinguished from barite by its effervescence in acid.

Occurrence. Witherite is a comparatively rare mineral, most frequently found in veins associated with galena. Found in England in fine crystals near Hexham in Northumberland and Alston Moor in Cumbria. In the United States found with fluorite at the Minerva Mine, Cave-in-Rock, Illinois.

Use. A minor source of barium.

Name. In honor of D. W. Withering (1741–1799), who discovered and first analyzed the mineral.

STRONTIANITE—$SrCO_3$

Crystallography. Orthorhombic; $2/m2/m2/m$. Crystals usually acicular, radiating like type 1 under aragonite. Twinning on {110} frequent, giving pseudohexagonal appearance. Also columnar, fibrous, and granular.

$Pmcn$; $a = 5.11$, $b = 8.41$, $c = 6.03$ Å; $Z = 4$. d's: 3.47(10), 2.42(5), 2.02(7), 1.876(5), 1.794(7).

Physical Properties *Cleavage* {110} good. **H** $3\frac{1}{2}$–4. **G** 3.78. *Luster* vitreous. *Color* white, gray, yellow, green. Transparent to translucent. *Optics:* (−), α = 1.520, β = 1.667, γ = 1.669; 2V = 7°; X = c, Y = b. r < v weak.

Composition and Structure. SrO 70.2, CO_2 29.8% for pure $SrCO_3$. Ca may be present in substitution for Sr to a maximum of about 25 atomic percent. Strontianite is isostructural with aragonite (see Fig. 12.10 and page 411).

Diagnostic Features. Characterized by high specific gravity and effervescence in HCl. Can be distinguished from celestite by poorer cleavage and effervescence in acid.

Occurrence. Strontianite is a low-temperature hydrothermal mineral associated with barite, celes-

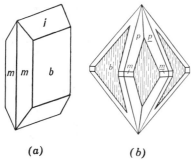

(a)   *(b)*

FIG. 12.15. Cerussite.

tite, and calcite in veins in limestone or marl, and less frequently in igneous rocks and as a gangue mineral in sulfide veins. Occurs in commercial deposits in Westphalia, Germany; Spain; Mexico; and England. Fine specimens are found in Oberdorf, Austria. Uncommon in the United States; found with fluorite at Cave-in-Rock, Illinois.

Use. Source of strontium. Strontium has no great commercial application; used in fireworks, red flares, military rockets, in the separation of sugar from molasses, and in various strontium compounds.

Name. From Strontian in Argyllshire, Scotland, where it was originally found.

CERUSSITE—PbCO₃

Crystallography. Orthorhombic; $2/m2/m2/m$. Crystals of varied habit common and show many forms. Often tabular on {010} (Fig. 12.15a). May form reticulated groups with the plates crossing each other at 60° angles, see Fig. 12.16; frequently in pseudo-hexagonal twins with deep reentrant angles in the vertical zone (Fig. 12.15b). Also in granular crystalline aggregates; fibrous; compact; earthy.

Pmcn; $a = 5.19$, $b = 8.44$, $c = 6.15$ Å; $Z = 4$. $d's$: 3.59(10), 3.50(4), 3.07(2), 2.49(3), 2.08(3).

Physical Properties. *Cleavage* {110} good, {021} fair. **H** $3-3\frac{1}{2}$. **G** 6.58. *Luster* adamantine. *Color* colorless, white, or gray. Transparent to subtranslucent. *Optics:* $(-)$; $\alpha = 1.804$, $\beta = 2.077$, $\gamma = 2.079$; $2V = 9°$, $X = c$, $Y = b$, $r > v$.

Composition and Structure. Most cerussite is very close in composition to PbCO₃, with PbO 83.5 and CO₂ 16.5%. It is isostructural with aragonite (see Fig. 12.10 and page 411).

Diagnostic Features. Recognized by its high specific gravity, white color, and adamantine luster. Crystal form and effervescence in nitric acid serve to distinguish it from anglesite.

Occurrence. Cerussite is an important and widely distributed supergene lead ore formed by the action of carbonated waters on galena. Associated with the primary minerals galena and sphalerite, and various secondary minerals such as anglesite, pyromorphite, smithsonite, and limonite.

Notable localities for its occurrence are Mies, Bohemia, Czechoslovakia; on the island of Sardinia, Italy; at Tsumeb, Namibia; Touissit, Morocco; and Broken Hill, New South Wales, Australia. In the United States found in various districts in Arizona;

FIG. 12.16. Cerussite, Tsumeb, Namibia. (Harvard Mineralogical Museum.)

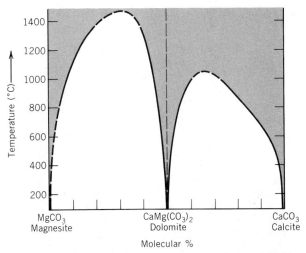

FIG. 12.17. CaCO$_3$-MgCO$_3$ system at CO$_2$ pressures sufficient to prevent decomposition of the carbonates. Vertical dashed line is ideal dolomite composition. Solid solution is shown by shading. (Adapted from Anovitz, L. M. and Essene, E. J., 1987, Phase equilibria in the system CaCO$_3$-MgCO$_3$-FeCO$_3$. *Journal of Petrology*, v. 28, pp. 389–415.)

from the Organ Mountains, New Mexico; and in the Coeur d'Alene district in Idaho.

Use. An important ore of lead.

Name. From the Latin word meaning *white lead*.

Similar Species. *Phosgenite*, Pb$_2$CO$_3$Cl$_2$ (tetragonal), is a rare carbonate.

Dolomite Group

The dolomite group includes *dolomite*, CaMg(CO$_3$)$_2$, *ankerite*, CaFe(CO$_3$)$_2$, and *kutnahorite*, CaMn(CO$_3$)$_2$. These three carbonates are isostructural with space group R$\overline{3}$. The structure of dolomite is similar to that of calcite but with Ca and Mg layers alternating along the c axis. The large difference in size of the Ca^{2+} and Mg^{2+} ions (33%) causes *cation ordering* with the two cations in specific, and separate levels in the structure. With the nonequivalence of Ca and Mg layers, the 2-fold rotation axes of calcite do not exist and the symmetry is reduced to that of the rhombohedral class, $\overline{3}$. The composition of dolomite is intermediate between CaCO$_3$ and MgCO$_3$ with Ca : Mg = 1 : 1.

The occurrence of this ordered compound, however, does not imply that solid solution exists between CaCO$_3$ and MgCO$_3$ (see Fig. 12.5). In the dolomite structure, especially at low temperatures, each of the two divalent cations occupies a structurally distinct position. At higher temperatures (above 700°C) dolomite shows small deviations from the composition with Ca : Mg = 1 : 1 as shown in Fig. 12.17. This diagram also shows that at elevated temperatures calcite coexisting with dolomite becomes more magnesian. Similarly dolomite in dolomite-calcite pairs becomes somewhat more calcic. At temperatures above about 1000° to 1100°C a complete solid solution exists between calcite and dolomite, but not between dolomite and magnesite. The compositions of coexisting dolomite and calcite have been used for an estimate of temperatures of crystallization of rocks containing both carbonates, using the temperature scale in Fig. 12.17.

DOLOMITE—CaMg(CO$_3$)$_2$; ANKERITE—CaFe(CO$_3$)$_2$

Crystallography. For *dolomite*. Hexagonal; $\overline{3}$. Crystals are usually the unit rhombohedron (Fig. 12.18a), more rarely a steep rhombohedron and base (Fig. 12.18c). Faces are often curved, some so acutely as to form "saddle-shaped" crystals (Figs. 12.18b and 12.19). Other forms rare. In coarse, granular, cleavable masses to fine-grained and compact. Twinning on {0001} common; lamellar twinning on {02$\overline{2}$1}. *Ankerite* generally does not occur in well-formed crystals. When it does, the crystals are similar to those of dolomite.

$R\overline{3}$; $a = 4.84$, $c = 15.95$ Å; $Z = 3$. d's : 2.88(10), 2.19(4), 2.01(3), 1.800(1), 1.780(1).

Physical Properties. *Cleavage* {10$\overline{1}$1} perfect. **H** 3$\frac{1}{2}$–4. **G** 2.85. *Luster* vitreous; pearly in some varieties, *pearl spar*. *Color* commonly some shade of pink, flesh color; may be colorless, white, gray, green, brown, or black. Transparent to translucent. *Optics*: (−); $\omega = 1.681$, $\epsilon = 1.500$. With increasing substitution of Fe (toward ankerite composition) **G** and indices of refraction increase. Ankerite is typically yel-

FIG. 12.18. Dolomite.

(a)   (b)   (c)

FIG. 12.19. Dolomite, St. Eustace, Quebec, Canada. (Harvard Mineralogical Museum.)

lowish-white but, due to oxidation of the iron, may appear yellowish-brown.

Composition and Structure. *Dolomite*: CaO 30.4, MgO 21.7, CO_2 47.9%. *Ankerite*: CaO 25.9, FeO 33.3, CO_2 40.8%. Naturally occurring dolomite deviates somewhat from Ca : Mg = 1 : 1 (see Fig. 12.17) with the Ca : Mg ratio ranging from 58 : 42 to $47\frac{1}{2} : 52\frac{1}{2}$. A complete solid solution series extends to *ankerite* and there is also a complete solid solution toward *kutnahorite*, $CaMn(CO_3)_2$ (see Fig. 12.20). Members of the dolomite group are isostructural. The structure of dolomite is discussed on page 415.

FIG. 12.20. The approximate extent of solid solution between $CaMn(CO_3)_2$-$CaMg(CO_3)_2$-$CaFe(CO_3)_2$. (From Essene, E. J., 1983, Solid solutions and solvi among metamorphic carbonates with applications to geologic thermobarometry, in *Carbonates: Mineralogy and Chemistry. Reviews in Mineralogy*, v. 11, pp. 77–96.)

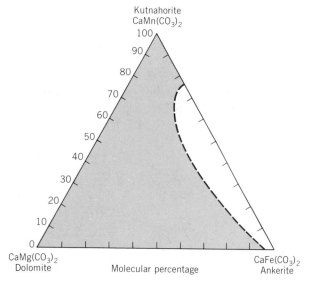

Kutnahorite
$CaMn(CO_3)_2$

$CaMg(CO_3)_2$
Dolomite

Molecular percentage

$CaFe(CO_3)_2$
Ankerite

Diagnostic Features. For *dolomite*. In cold, dilute HCl large fragments are slowly attacked but are soluble with effervescence only in hot HCl; the powdered mineral is readily soluble in cold acid. Crystallized variety told by its curved rhombohedral crystals and usually by its flesh-pink color. The massive rock variety is distinguished from limestone by the less vigorous reaction with HCl. *Ankerite* is similar to dolomite in its physical and chemical properties except for its color which ranges from yellow-brown to brown.

Occurrence. *Dolomite* is found in many parts of the world chiefly as extensive sedimentary strata, and the crystallized equivalent, dolomitic marble. Dolomite as a rock mass is generally thought to be secondary in origin, formed from ordinary limestone by the replacement of some of the Ca by Mg. The replacement may be only partial and thus most dolomite rocks (*dolostones*) are mixtures of dolomite and calcite. The mineral occurs also as a hydrothermal vein mineral, chiefly in the lead and zinc veins that traverse limestone, associated with fluorite, calcite, barite, and siderite. *Ankerite* is a common carbonate in Precambrian iron-formations.

Dolomite is abundant in the Dolomite region of southern Tyrol; in crystals from the Binnenthal, Switzerland; Traversella in Piedmont, Italy; northern England; and Guanajuato, Mexico. In the United States it is found as masses of sedimentary rock and gray to pink saddle-shaped crystals in vugs of rocks of many of the middle-western states, especially in the Joplin, Missouri, district.

Use of Dolomite. As a building and ornamental stone. For the manufacture of certain cements. For the manufacture of magnesia used in the preparation of refractory linings of the converters in the basic steel process. Dolomite is a potential ore of metallic Mg.

FIG. 12.21. Malachite, cut and polished.

Name. *Dolomite* in honor of the French chemist, Dolomieu (1750–1801). *Ankerite* after M. J. Anker (1772–1843), Austrian mineralogist.

Similar Species. *Kutnahorite*, $CaMn(CO_3)_2$, is isostructural with dolomite.

MALACHITE—$Cu_2CO_3(OH)_2$

Crystallography. Monoclinic; $2/m$. Crystals slender prismatic but seldom distinct. Crystals may be pseudomorphous after azurite. Usually in radiating fibers forming botryoidal or stalactitic masses (Fig. 12.21). Often granular or earthy.

$P2_1/a$; $a = 9.48$, $b = 12.03$, $c = 3.21$ Å, $\beta = 98°44'$; $Z = 4$. d's: 6.00(6), 5.06(8), 3.69(9), 2.86(10), 2.52(6).

Physical Properties. *Cleavage* $\{\overline{2}01\}$ perfect but rarely seen. **H** $3\frac{1}{2}$–4. **G** 3.9–4.03. *Luster* adamantine to vitreous in crystals; often silky in fibrous varieties; dull in earthy type. *Color* bright green. *Steak* pale green. Translucent. *Optics:* $(-)$, $\alpha = 1.655$, $\beta = 1.875$, $\gamma = 1.909$; $2V = 43°$; $Y = b$, $Z \wedge c = 24°$; pleochroism: X colorless, Y yellow green, Z deep green.

Composition and Structure. CuO 71.9, CO_2 19.9, H_2O 8.2%. Cu 57.4%. Cu^{2+} is octahedrally coordinated by O^{2-} and $(OH)^{1-}$ in $CuO_2(OH)_4$ and $CuO_4(OH)_2$ octahedra. These octahedra are linked along the edges forming chains that run parallel to the c axis. The chains are cross-linked by triangular $(CO_3)^{2-}$ groups.

Diagnostic Features. Soluble in HCl with effervescence, yielding a green solution. Recognized by its bright green color and botryoidal forms, and distinguished from other green copper minerals by its effervescence in acid.

Occurrence. Malachite is a widely distributed supergene copper mineral found in the oxidized portions of copper veins associated with azurite, cuprite, native copper, iron oxides (see Figs. 14.21 and 14.22, p. 590). It usually occurs in copper deposits associated with limestone.

Notable localities for its occurrence are at Nizhne Tagil in the Ural Mountains, CIS; pseudomorphous after cuprite at Chessy, near Lyons, France, and associated with azurite; from Tsumeb, Namibia; Katanga, Zaire; and Broken Hill, New South Wales, Australia. In the United States, formerly an important copper ore in the southwestern copper districts; at Bisbee, Morenci, and other localities in Arizona.

Use. A minor ore of copper. It is used extensively as an ornamental and gem mineral. In the nineteenth century Russia was the principal source, but today most of the material comes from Zaire.

Name. Derived from the Greek word *malache* for *mallows*, in allusion to its green color.

AZURITE—$Cu_3(CO_3)_2(OH)_2$

Crystallography. Monoclinic; $2/m$. Habit varies (Fig. 12.22); crystals frequently complex and malformed. Also in radiating spherical groups.

$P2_1c$; $a = 4.97$, $b = 5.84$, $c = 10.29$ Å, $\beta = 92°24'$; $Z = 2$. d's: 5.15(7), 3.66(4), 3.53(10), 2.52(6), 2.34(3).

Physical Properties. *Cleavage* $\{011\}$ perfect, $\{100\}$ fair. **H** $3\frac{1}{2}$–4. **G** 3.77. *Luster* vitreous. *Color* intense azure-blue. Transparent to translucent. *Optics:* $(+)$, $\alpha = 1.730$, $\beta = 1.758$, $\gamma = 1.838$; $2V = 67°$; $X = b$, $Z \wedge c = -13°$; pleochroic in blue $Z > Y > X$; $r > V$.

Composition and Structure. CuO 69.2, CO_2 25.6, H_2O 5.2%. Cu 55.3%. The structure of azurite contains Cu^{2+} ions in square, coplanar groups with 2 O^{2-} and $2(OH)^{1-}$. These square groups are linked

FIG. 12.22. Azurite crystals.

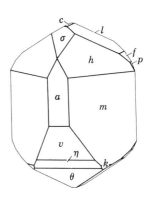

into chains parallel to the *b* axis. Each $(OH)^{1-}$ group is shared by 3 Cu^{2+} and each oxygen of the triangular (CO_3) group is bonded to one copper.

Diagnostic Features. Characterized chiefly by its azure-blue color and effervescence in HCl.

Alteration. Pseudomorphs of malachite after azurite commonly observed; less common after cuprite.

Occurrence. Azurite is less common than malachite but has the same origin and associations. It was found in fine crystals at Chessy, near Lyons, France; Tsumeb, Namibia; Touissit, Morocco; and Broken Hill, New South Wales, Australia. In the United States at the Copper Queen and other mines, Bisbee and Morenci, Arizona. Widely distributed with copper ores.

Use. A minor ore of copper.

Name. Named in allusion to its color.

Rare Hydrous Carbonates. *Aurichalcite*, $(Zn,Cu)_5(CO_3)_2(OH)_3$ pale green to blue in orthorhombic acicular crystals. *Gaylussite*, $Na_2Ca(CO_3)_2 \cdot 5H_2O$, monoclinic and *trona*, $Na_3(CO_3)(HCO_3) \cdot 2H_2O$, monoclinic, are both found in saline-lake deposits.

NITRATES

The minerals in this group are structurally similar to the carbonates with planar, triangular $(NO_3)^{1-}$ groups, much like the $(CO_3)^{2-}$ group. Like C in the (CO_3) group, the highly charged and highly polarizing N^{5+} ion binds its three coordinated oxygens into a close-knit group in which the strength of the oxygen-nitrogen bond (e.v. $= 1\frac{2}{3}$) is greater than any other possible bond in the structure. Because of the greater strength of this N—O bond as compared with the C—O bond, nitrates are less readily decomposed by acids than carbonates. There are eight nitrate minerals but, with the exception of nitratite and niter, they are very rare.

Nitratite—NaNO₃

In nitratite $(NO_3)^{1-}$ groups combine in one-to-one proportion with monovalent Na^+ in 6-coordination forming a structure analogous to that of calcite (see Fig. 12.2). Nitratite and calcite are thus isostructural with the same crystallography and cleavage. Because of the lesser cationic charge, nitratite is softer (**H** 1–2) and melts at a low temperature and, because of the lower atomic weight of sodium, has a lower specific gravity (**G** 2.29). Nitratite is soluble in water and is

thus found only in arid regions. The most notable occurrence is in northern Chile where it is recovered as a source of nitrogen.

Niter (Saltpeter)—KNO₃

Niter is isostructural with aragonite (see Fig. 12.10) with similar crystallography and analogous pseudo-hexagonal twinning on {110}. Like nitratite it fuses easily and is soluble in water. Niter is less common than nitratite but has been recovered from the soils in several countries as a source of nitrogen for fertilizer.

BORATES

Within the borate group of minerals BO_3 units are capable of polymerization (similar to the polymerization of SiO_4 tetrahedral groups in the silicates) in the form of chains, sheets, and isolated multiple groups (Fig. 12.23). This is possible because the small B^{3+} ion, which generally coordinates three oxygens in a triangular group, has bond strengths to each O with an e.v. $= 1$; this is exactly half the bonding energy of the oxygen ion. This permits a single oxygen to be shared between two boron ions linking the BO_3 triangles into expanded structural units (double triangles, triple rings, sheets, and chains). Because the triangular coordination of BO_3 is close to the upper stability limit of 3-coordination, boron is found also in 4-coordination in tetrahedral groups. In addition to BO_3 and BO_4 groups, natural borates may contain complex ionic groups such as $[B_3O_3(OH)_5]^{2-}$ (Fig. 12.23e) that consist of one triangle and two tetrahedra. In the structure of *colemanite*, $CaB_3O_4(OH)_3 \cdot H_2O$, complex infinite chains of tetrahedra and triangles occur, and in *borax*, $Na_2B_4O_5(OH)_4 \cdot 8H_2O$, a complex ion, $[B_4O_5(OH)_4]^{2-}$ consisting of two tetrahedra and two triangles is found (Fig. 12.23f). Borates can be classified on the basis of the structural anionic linking (or lack thereof) as insular (independent single or double BO_3 or BO_4 groups), chain, sheet, and framework structures.

Although it is possible to prepare a three-dimensional boxwork made up of BO_3 triangles only and having the composition B_2O_3, such a configuration has a very low stability and disorders readily, yielding a glass. Because of the tendency to form somewhat disordered networks of BO_3 triangles, boron is regarded as a "network-former" in glass manufacture and is used in the preparation of special glasses of light weight and high transparency to energetic radiation.

FIG. 12.23. Independent BO_3 triangles (a), multiple groups (b) and (c), and chains (d) in borates. Also complex triple and quadruple groups (e) and (f). The triple group in (e) consists of two $BO_2(OH)_2$ tetrahedra and a BO_2OH triangle. The group in (f) contains two BO_3OH tetrahedra and two BO_2OH triangles; this type of group is present in the structure of borax (see Fig. 12.27.)

Over 100 borate minerals are known but only the four most common are considered here:

Kernite $Na_2B_4O_6(OH)_2 \cdot 3H_2O$
Borax $Na_2B_4O_5(OH)_4 \cdot 8H_2O$
Ulexite $NaCaB_5O_6(OH)_6 \cdot 5H_2O$
Colemanite $CaB_3O_4(OH)_3 \cdot H_2O$

KERNITE—$Na_2B_4O_6(OH)_2 \cdot 3H_2O$

Crystallography. Monoclinic; $2/m$. Rarely in crystals; usually in coarse, cleavable aggregates.

$P2/a$; $a = 15.68$, $b = 9.09$, $c = 7.02$ Å; $\beta = 108°52'$; $Z = 4$. d's: 7.41(10), 6.63(8), 3.70(3), 3.25(2), 2.88(2).

Physical Properties. *Cleavage* {001} and {100} perfect. Cleavage fragments are thus elongated parallel to the b crystallographic axis, see Fig. 12.24. The angle between the cleavages: $(001) \wedge (100) = 71°8'$. **H** 3. **G** 1.95. *Luster* vitreous to pearly. *Color* colorless to white; colorless specimens become chalky white on long exposure to the air owing to a surface film of

tincalconite, $Na_2B_4O_5(OH)_4 \cdot 3H_2O$. *Optics:* $(-)$; $\alpha = 1.454$, $\beta = 1.472$, $\gamma = 1.488$; $2V = 70°$. $Z = b$, $X \wedge c = 71°$. $r > v$.

Composition and Structure. Na_2O 22.7, B_2O_3 51.0, H_2O 26.3%. The structure of kernite contains complex chains, parallel to the b axis, of composition $[B_4O_6(OH)_2]^{2-}$. These chains consist of BO_4 tetrahedra, joined at the vertices (as in the pyroxenes of the silicates) with remaining vertices connected to BO_3 triangles. The chains are linked to each other by Na^+ (in 5-coordination) and hydroxyl-oxygen bonds.

Diagnostic Features. Characterized by long, splintery cleavage fragments and low specific gravity. Slowly soluble in cold water.

Occurrence. The original locality and principal occurrence of kernite is in the Mohave Desert at Boron, California. Here, associated with borax, colemanite, and ulexite in a bedded series of Tertiary clays, it is present by the million tons. This deposit of sodium borates is 4 miles long, 1 mile wide, and as much as 250 feet thick, and lies from 150 to 1000 feet beneath the surface. The kernite occurs near the bottom of the deposit and is believed to have formed from

FIG. 12.24. Kernite showing typically excellent {100} and {001} cleavage. Scale = 1 cm. From Boron, Kern County, California.

FIG. 12.25. Borax.

borax by recrystallization caused by increased temperature and pressure. Kernite is also found associated with borax at Tincalayu, Argentina.

Use. Boron compounds are used in the manufacture of glass, especially in glass wool for insulation purposes. They are also used in soap, in porcelain enamels for coating metal surfaces, and in the preparation of fertilizers and herbicides.

FIG. 12.26. Borax crystals, altered to white, chalky tincalconite. Baker Mine, Kramer borate district, California. (Courtesy of Richard C. Erd, U.S. Geological Survey.)

Name. From Kern County, California, where the mineral is found.

BORAX—Na$_2$B$_4$O$_5$(OH)$_4$·8H$_2$O

Crystallography. Monoclinic; 2/*m*. Commonly in prismatic crystals (Fig. 12.25). Also as massive, cellular material or encrustations.

C2/c; *a* = 11.86, *b* = 10.67, *c* = 12.20 Å; β = 106°; Z = 4. *d*'s: 5.7(2), 4.86(5), 3.96(4), 2.84(5), 2.57(10).

Physical Properties. *Cleavage* {100} perfect. **H** 2–2$\frac{1}{2}$. **G** 1.7 ±. *Luster* vitreous. *Color* colorless or white. Translucent. Sweetish-alkaline taste. Clear crystals effloresce and turn white with the formation of *tincalconite*, Na$_2$B$_4$O$_5$(OH)$_4$·3H$_2$O, see Fig. 12.26. *Optics*: (−); α = 1.447, β = 1.469, γ = 1.472; 2V = 40°. X = *b*, Z ∧ *c* = −56°. *r* > *v*.

Composition and Structure. Na$_2$O 16.2, B$_2$O$_3$ 36.6, H$_2$O 47.2%. The structure of borax is shown in Fig. 12.27. It consists of [BO$_3$OH] tetrahedra and [BO$_2$OH] triangles that are joined parallel to the *c* crystallographic axis. The Na$^+$ is in 6-coordination with H$_2$O molecules, producing Na(H$_2$O)$_6$ octahedra

that have common edges, forming ribbons parallel to the *c* crystallographic axis. The boron groups are linked to the Na-H$_2$O ribbons by van der Waals and hydrogen bonding.

Diagnostic Features. Characterized by it crystals and tests for boron. Readily soluble in water.

Occurrence. Borax is the most widespread of the borate minerals. It is formed on the evaporation of enclosed lakes and as an efflorescence on the surface in arid regions. Deposits in Tibet furnished the first borax, under the name of *tincal*, to reach western civilization. In the United States it was first found in Lake County, California, and later in the desert region of southeastern California, in Death Valley, Inyo County, and in San Bernardino County. Associated with kernite, it is the principal mineral mined from the bedded deposits at Boron, California. It is also obtained commercially from the brines of Searles Lake at Trona, California.

Although the United States is the largest producer of borax, the world's most extensive deposits of borate minerals are in Turkey, notably in the Kirka area. Associated with other borates, borax is mined in the Andes Mountains in contiguous parts of Argentina,

FIG. 12.27. The structure of borax as seen on (100).

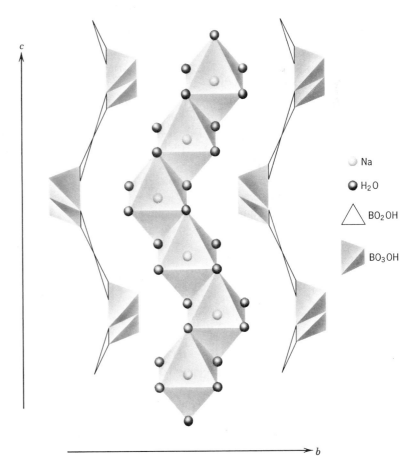

Na

H$_2$O

BO$_2$OH

BO$_3$OH

c

b

Bolivia, and Chile. Minerals commonly associated with borax are halite, gypsum, ulexite, colemanite, and various rare borates.

Use. Although boron is obtained from several minerals, it is usually converted to borax, the principal commercial product. There are many uses of borax. Its greatest use is in the manufacture of glass fibers for insulation and textiles. It is also used in soaps and detergents; as an antiseptic and preservative; in medicine; as a solvent for metallic oxides in soldering and welding; and as a flux in various smelting and laboratory operations. Elemental boron is used as a deoxidizer and alloy in nonferrous metals; and as a neutron absorber in shields for atomic reactors. Boron is used in rocket fuels and as an additive in motor fuel. Boron carbide, harder than corundum, is used as an abrasive.

Name. Borax comes from an Arabic name for the substance.

Ulexite—NaCaB$_5$O$_6$OH$_6$·5H$_2$O

Crystallography. Triclinic; $\overline{1}$. Usually in rounded masses of loose texture, consisting of fine fibers which are acicular or capillary crystals, "cotton-balls." Rarely in closely packed parallel fibers showing fiber-optical properties and called "television rock."

$P\overline{1}$; $a = 8.73$, $b = 12.75$, $c = 6.70$ Å; $\alpha = 90°16'$, $\beta = 109°8'$, $\gamma = 105°7'$; $Z = 2$. d's: 12.3(10), 7.89(7), 6.60(7), 4.19(7), 2.67(7).

Physical Properties. *Cleavage* {010} perfect. **H** $2\frac{1}{2}$; the aggregate has an apparent hardness of 1. **G** 1.96. *Luster* silky. *Color* white. Tasteless. *Optics:* (+); $\alpha = 1.491$, $\beta = 1.504$, $\gamma = 1.520$; $2V = 73°$.

Composition and Structure. Na$_2$O 7.7, CaO 13.8, B$_2$O$_3$ 43.0, H$_2$O 35.5%. The structure contains large anionic groups of composition [B$_5$O$_6$(OH)$_6$]$^{3-}$. Each Ca^{2+} cation is surrounded by an irregular polyhedron of five oxygens, three (OH) groups and two H$_2$O molecules. Each Na$^+$ cation is octahedrally coordinated by (OH) groups and H$_2$O molecules. Neighboring octahedra share edges to make continuous chains parallel to c.

Diagnostic Features. The soft "cotton-balls" with silky luster are characteristic of ulexite.

Occurrence. Ulexite crystallizes in arid regions from brines that have concentrated in enclosed basins, as in playa lakes. Usually associated with borax. It occurs abundantly in the dry basins of northern Chile and in Argentina. In the United States it has been found abundantly in certain of the enclosed basins of Nevada and California and with colemanite in bedded Tertiary deposits.

FIG. 12.28. Colemanite.

Use. A source of borax.

Name. In honor of the German chemist, G. L. Ulex (1811–1883), who discovered the mineral.

Colemanite—CaB$_3$O$_4$(OH)$_3$·H$_2$O

Crystallography. Monoclinic; $2/m$. Commonly in short prismatic crystals, highly modified (see Fig. 12.28). Cleavable massive to granular and compact.

$P2_1/a$; $a = 8.74$, $b = 11.26$, $c = 6.10$ Å, $\beta = 110°7'$; $Z = 4$. d's: 5.64(5), 4.00(4), 3.85(5), 3.13(10), 2.55(5).

Physical Properties. *Cleavage* {010} perfect. **H** $4-4\frac{1}{2}$. **G** 2.42. *Luster* vitreous. *Color* colorless to white. Transparent to translucent. *Optics:* (+); $\alpha = 1.586$, $\beta = 1.592$, $\gamma = 1.614$; $2V = 55°$; $X = b$, $Z \wedge c = 84°$. $r > v$.

Composition and Structure. CaO 27.2, B$_2$O$_3$ 50.9, H$_2$O 21.9%. The structure contains infinite chains with composition [B$_3$O$_4$(OH)$_2$]$^{2-}$ that are parallel to the a axis. The chains contain BO$_3$ triangles and BO$_2$(OH)$_2$ and BO$_3$(OH) tetrahedra. Interspersed between the chains are Ca ions and H$_2$O molecules.

Diagnostic Features. Characterized by one direction of highly perfect cleavage and exfoliation on heating.

Occurrence. Colemanite deposits are interstratified with lake bed deposits of Tertiary age. Ulexite and borax are usually associated, and colemanite is believed to have originated by their alteration. Found in California in Los Angeles, Ventura, San Bernardino, and Inyo counties, and in Nevada in the Muddy Mountains and White Basin, Clark County. Also occurs extensively in Argentina and Turkey.

Use. A source of borax that, at the time of the discovery of kernite, yielded over half of the world's supply.

Name. In honor of William T. Coleman, merchant of San Francisco, who marketed the product of the colemanite mines.

Similar Species. Other borates, locally abundant are: *boracite*, Mg$_3$ClB$_7$O$_{13}$; *hydroboracite*, CaMgB$_6$O$_8$(OH)$_6$·3H$_2$O and *inyoite*, CaB$_3$O$_3$(OH)$_5$·4H$_2$O.

FIG. 12.29. The structure of barite, BaSO$_4$, as projected on (010). The dashed lines outline the unit cell. The oxygen atoms, which would normally superimpose, have been displaced from their positions in this projection.

FIG. 12.30. The structure of anhydrite, CaSO$_4$.

Sinhalite, Mg(Al,Fe)BO$_4$ is a rare gem mineral found in the gravels of Sri Lanka and Burma. Before its description as a new mineral in 1952, cut stones had been misidentified as brown peridot.

SULFATES AND CHROMATES

In the discussion of the structures of sulfide minerals we have seen that sulfur occurs as the large, divalent sulfide anion. This ion results from the filling by captured electrons of the two vacancies in the outer, or valence, electron shell. The six electrons normally present in this shell may be lost, giving rise to a small, highly charged and highly polarizing positive ion (radius = 0.12 Å). It occurs in tetrahedral coordination with surrounding oxygens. The sulfur to oxygen bond in such an ionic group is very strong (e.v. = $1\frac{1}{2}$; see Fig. 12.1b) and covalent in its properties and produces tightly bound groups that are not capable of sharing oxygens. These anionic (SO$_4$)$^{2-}$ groups are the fundamental structure units of the sulfate minerals.

The most important and common of the anhydrous sulfates are members of the *barite group* (space

group *Pnma*), with large divalent cations coordinated with the sulfate ion. Part of the structure of *barite*, BaSO$_4$, is illustrated in Fig. 12.29. Each barium ion is coordinated to 12 oxygen ions belonging to seven different (SO$_4$) groups. The barite type structure is also found in manganates (with MnO$_4$ tetrahedral groups) and chromates (with CrO$_4$ tetrahedral groups) which contain large cations.

Anhydrite, CaSO$_4$, because of the smaller size of Ca^{2+} compared to Ba^{2+}, has a very different structure from that of barite (see Fig. 12.30). Each Ca^{2+} is coordinated to eight nearest neighbor oxygens from the tetrahedral (SO$_4$) groups.

Of the hydrous sulfates, *gypsum*, CaSO$_4$·2H$_2$O, is the most important and abundant. The structure of gypsum is illustrated in Fig. 12.31. Gypsum is monoclinic with space group $C2/c$. The structure consists of layers parallel to {010} of (SO$_4$)$^{2-}$ groups strongly bonded to Ca^{2+}. Successive layers of this type are separated by sheets of H$_2$O molecules. The bonds between H$_2$O molecules in neighboring sheets are weak, which explains the excellent {010} cleavage in

FIG. 12.31. The structure of gypsum, CaSO$_4$·2H$_2$O, projected on (001). Perfect (010) cleavage is indicated by dotted lines. Dashed lines outline unit cell. The 8-coordination (six oxygens and two water molecules) is indicated for one Ca^{2+}.

FIG. 12.32. Barite, Cumbria, England. (Harvard Mineralogical Museum.)

gypsum. Loss of the water molecules causes collapse of the structure and transformation into a metastable polymorph of anhydrite ($\gamma CaSO_4$) with a large decrease in specific volume and loss of the perfect cleavage.

A large number of minerals belong to this class but only a few of them are common. The class can be divided into (1) the anhydrous sulfates and (2) the hydrous and basic sulfates.

SULFATES AND CHROMATE

Anhydrous Sulfates and Chromate

Barite group
Barite	$BaSO_4$
Celestite	$SrSO_4$
Anglesite	$PbSO_4$
Anhydrite	$CaSO_4$
Crocoite	$PbCrO_4$

Hydrous and Basic Sulfates
Gypsum	$CaSO_4 \cdot 2H_2O$
Antlerite	$Cu_3SO_4(OH)_4$
Alunite	$KAl_3(SO_4)_2(OH)_6$

Barite Group

The sulfates of Ba, Sr, and Pb form an isostructural group with space group *Pnma*. They have closely related crystal constants and similar habits. The members of the group are: *barite, celestite, and anglesite.*

BARITE—BaSO₄

Crystallography. Orthorhombic; $2/m2/m2/m$. Crystals usually tabular on {001}; often diamond-shaped (Fig. 12.32). Both {0*kl*} and {*h0l*} prisms are usually present, either beveling the corners of the diamond-shaped crystals or the edges of the tables and forming rectangular prismatic crystals elongated parallel to either *a* or *b* (Fig. 12.33). Crystals may be very complex. Frequently in divergent groups of tabular crystals forming "crested barite" or "barite roses," see Fig. 12.34. Also coarsely laminated; granular, earthy.

Pnma; a = 8.87, *b* = 5.45, *c* = 7.14 Å; *Z* = 4. *d*'s: 3.90(6), 3.44(10), 3.32(7), 3.10(10), 2.12(8).

Physical Properties. *Cleavage* {001} perfect, {210} less perfect. **H** $3-3\frac{1}{2}$. **G** 4.5 (heavy for a nonmetallic mineral). *Luster* vitreous; on some specimens pearly on base. *Color* colorless, white, and light shades of blue, yellow, red (see Plate IV, no. 11, Chapter 15). Transparent to translucent. *Optics:* (+); α = 1.636, β = 1.637, γ = 1.648; 2V = 37°. *X* = *c*, *Y* = *b*. *r* < *v*.

Composition and Structure. BaO 65.7, SO₃ 34.3% for pure barite. Sr substitutes for Ba and a complete solid solution series extends to celestite, but most material is near one end or the other of the series. A small amount of Pb may substitute for Ba. The struc-

FIG. 12.33. Barite crystals.

FIG. 12.34. Crested barite, Cumbria, England.
(Harvard Mineralogical Museum.)

ture of barite is illustrated in Fig. 12.29 and discussed on page 423.

Diagnostic Features. Recognized by its high specific gravity and characteristic cleavage and crystals.

Occurrence. Barite is a common mineral of wide distribution. It occurs usually as a gangue mineral in hydrothermal veins, associated with ores of silver, lead, copper, cobalt, manganese, and antimony. It is found in veins in limestone with calcite, or as residual masses in clay overlying limestone. Also in sandstone with copper ores. In places acts as a cement in sandstone. Deposited occasionally as a sinter by waters from hot springs.

Notable localities for the occurrence of barite crystals are in Westmoreland, and Cumbria, England; Felsobanya and other localities, Romania; and Saxony, Germany. Massive barite, occurring usually as veins, nests, and irregular bodies in limestones, has been quarried in the United States in Georgia, Tennessee, Missouri, and Arkansas.

Use. More than 80% of the barite produced is used in oil- and gas-well drilling as "heavy mud" to aid in the support of drill rods, and to help prevent blowing out of gas. Barite is the chief source of Ba for chemicals. A major use of barium is in *lithopone*, a combination of barium sulfide and zinc sulfate that forms an intimate mixture of zinc sulfide and barium sulfate. Lithopone is used in the paint industry and to a lesser extent in floor coverings and textiles. Precipitated barium sulfate, "blanc fixe," is employed as a filler in paper and cloth, in cosmetics, as a paint pigment, and for barium meals in radiology.

Name. From the Greek word *barys* meaning *heavy*, an allusion to its high specific gravity.

CELESTITE—SrSO$_4$

Crystallography. Orthorhombic; $2/m2/m2/m$. Crystals closely resemble those of barite. Commonly tabular parallel to $\{001\}$ or prismatic parallel to a or b with prominent development of $\{0kl\}$ and $\{h0l\}$ prisms. Crystals elongated parallel to a are frequently terminated by nearly equal development of faces of $d\{101\}$ and $m\{210\}$ (Fig. 12.35). Also radiating fibrous; granular.

Pnma; a = 8.38, *b* = 5.37, *c* = 6.85 Å; *Z* = 4. *d*'s: 3.30(10), 3.18(6), 2.97(10), 2.73(6), 2.04(6).

Physical Properties. *Cleavage* $\{001\}$ perfect, $\{210\}$ good. **H** $3-3\frac{1}{2}$. **G** 3.95–3.97. *Luster* vitreous to pearly. *Color* colorless, white, often faintly blue or red. Transparent to translucent. *Optics:* (+); α = 1.622, β = 1.624, γ = 1.631; 2V = 50°; X = c, Y = b. r < v.

Composition and Structure. SrO 56.4, SO$_3$ 43.6% for pure celestite. Ba substitutes for Sr and a complete solid solution series exists between celestite and barite. At ordinary temperatures only a limited series exists between anhydrite, CaSO$_4$, and SrSO$_4$. Celestite is isostructural with barite (see Fig. 12.29).

Diagnostic Features. Closely resembles barite but is differentiated by lower specific gravity and by chemical test for strontium.

Occurrence. Celestite is found usually disseminated through limestone or sandstone, or in nests and lining cavities in such rocks. Associated with calcite, dolomite, gypsum, halite, sulfur, fluorite. Also found as a gangue mineral in lead veins.

Notable localities for its occurrence are with the sulfur deposits of Silicy; and Yate, Gloucestershire, England. Found in the United States at Clay Center, Ohio, and elsewhere in northwestern and southeastern Michigan; Lampasas, Texas; and with colemanite in Inyo County, California.

Use. Used in the preparation of strontium nitrate for fireworks and tracer bullets, and other strontium salts used in the refining of beet sugar.

Name. Derived from the Latin *caelestis* meaning *heavenly*, in allusion to the faint blue color of the first specimens described.

FIG. 12.35. Celestite crystals.

FIG. 12.36. Anglesite.

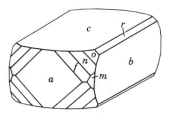

FIG. 12.37. Anhydrite.

ANGLESITE—PbSO$_4$

Crystallography. Orthorhombic; $2/m2/m2/m$. Frequently in crystals with habit often similar to that of barite but more varied. Crystals may be prismatic parallel to any one of the crystal axes and frequently show many forms, with a complex development (Fig. 12.36). Also massive, granular to compact. Frequently earthy, in concentric layers that may have an unaltered core of galena.

Pnma; a = 8.47, *b* = 5.39, *c* = 6.94 Å; *Z* = 4. *d*'s: 4.26 (9), 3.81(6), 3.33(9), 3.22(7), 3.00(10).

Physical Properties. *Cleavage* {001} good, {210} imperfect. Fracture conchoidal. **H** 3.0. **G** 6.2–6.4 (unusually high). *Luster* adamantine when crystalline, dull when earthy. *Color* colorless, white, gray, pale shades of yellow. May be colored dark gray by impurities. Transparent to translucent. *Optics:* (+); α = 1.877, β = 1.883, γ = 1.894; 2V = 75°; X = c, Y = b. r < v.

Composition and Structure. PbO 73.6, SO$_3$ 26.4%. Anglesite is isostructural with barite (see Fig. 12.29 and page 423).

Diagnostic Features. Recognized by its high specific gravity, its adamantine luster, and frequently by its association with galena. Distinguished from cerussite in that it will not effervesce in nitric acid.

Occurrence. Anglesite is a common supergene mineral found in the oxidized portions of lead deposits. It is formed through the oxidation of galena, either directly, as is shown by concentric layers of anglesite surrounding a core of galena, or by solution and subsequent deposition and recrystallization. Anglesite is commonly associated with galena, cerussite, sphalerite, smithsonite, hemimorphite, and iron oxides.

Notable localities for its occurrence are Monte Poni, Sardinia, Italy; Island of Anglesey, Wales; Derbyshire, England; and Leadhills, Scotland. From Sidi-Amorben-Salem, Tunisia; Tsumeb, Namibia; Broken Hill, New South Wales, and Dundas, Tasmania, Australia; and Touissit, Morocco. It is found in crystals embedded in sulfur at Los Lamentos, Chihuahua, Mexico. Occurs in the United States at Phoenixville, Pennsylvania; Tintic district, Utah; and Coeur d'Alene district, Idaho.

Use. A minor ore of lead.

Name. Named from the original locality on the Island of Anglesey.

ANHYDRITE—CaSO$_4$

Crystallography. Orthorhombic. $2/m2/m2/m$. Crystals rare; when observed are thick tabular on {010}, {100}, or {001} (see Fig. 12.37), also prismatic parallel to *b*. Usually massive or in crystalline masses resembling an isometric mineral with cubic cleavage. Also fibrous, granular, massive.

Amma; a = 6.95, *b* = 6.96, *c* = 6.21 Å; *Z* = 4. *d*'s: 3.50(10), 2.85(3), 2.33(2), 2.08(2), 1.869(2).

Physical Properties. *Cleavage* {010} perfect, {100} nearly perfect, {001} good. **H** 3–3½. **G** 2.89–2.98. *Luster* vitreous to pearly on cleavage. *Color* colorless to bluish or violet. Also may be white or tinged with rose, brown, or red. *Optics:* (+); α = 1.570, β = 1.575, γ = 1.614; 2V = 44°; X = b, Y = a; r < v.

Composition and Structure. CaO 41.2, SO$_3$ 58.8%. The structure illustrated in Fig. 12.30 is very different from the barite type. In anhydrite Ca is in 8-coordination with oxygen from SO$_4$ groups, whereas in barite Ca is in 12-coordination with oxygen. A metastable polymorph of anhydrite (γ CaSO$_4$) is hexagonal and is formed as the result of slow dehydration of gypsum (see Fig. 12.40).

Diagnostic Features. Anhydrite is characterized by its three cleavages at right angles. It is distinguished from calcite by its higher specific gravity and from gypsum by its greater hardness. Some massive varieties are difficult to recognize, and one should test for sulfate.

Alteration. Anhydrite by hydration changes to gypsum with an increase in volume, and in places large masses have been thus altered.

Occurrence. Anhydrite occurs in much the same manner as gypsum and is often associated with that mineral but is not nearly as common. Found in beds associated with salt deposits in the cap rock of salt domes, and in limestones. Found in some amygdaloidal cavities in basalt.

Notable foreign localities are: Wieliczka, Poland; Aussee, Styria and Hall near Innsbruck, Tyrol, Austria; and Bex, Switzerland. In the United States found in Lockport, New York; West Paterson, New Jersey; New Mexico; and Texas. Found in large beds in Nova Scotia.

Use. Ground anhydrite is used as a soil conditioner and to a minor extent as a setting retardant in Portland cement. In Great Britain and Germany it has been used as a source of sulfur for the production of sulfuric acid.

Name. Anhydrite is from the Greek meaning *without water*, in contrast to the more common hydrous calcium sulfate, gypsum.

Crocoite—PbCrO₄

Crystallography. Monoclinic; $2/m$. Commonly in slender prismatic crystals, vertically striated, and columnar aggregates. Also granular.

$P2_1/n$; $a = 7.11$, $b = 7.41$, $c = 6.81$ Å, $\beta = 102°33'$; $Z = 4$. d's: 4.96(3), 2.85(3), 2.33(2), 2.08(2), 1.869(2).

Physical Properties. *Cleavage* {110} imperfect. **H** $2\frac{1}{2}$–3. **G** 5.9–6.1. *Luster* adamantine. *Color* bright hyacinth-red. *Streak* orange-yellow. Translucent.

Composition and Structure. PbO 68.9, CrO₃ 31.1%. Crocoite is isostructural with *monazite*, CePO₄, with Pb in 9-coordination with oxygen, linking six CrO₄ tetrahedra.

Diagnostic Features. Crocoite is characterized by its color, high luster, and high specific gravity. It may be confused with wulfenite, PbMoO₄, but can be distinguished from it by its redder color, lower specific gravity, and crystal form.

Occurrence. Crocoite is a rare mineral found in the oxidized zones of lead deposits in those regions where lead veins have traversed rocks containing chromite. Associated with pyromorphite, cerussite, and wulfenite. Notable localities are: Dundas, Tasmania and Beresovsk near Sverdlovsk, Ural Mountains, CIS. In the United States it is found in small quantities in the Vulture district, Arizona.

Use. Not abundant enough to be of commercial value, but of historic interest, because the element chromium was first discovered in crocoite.

Name. From the Greek meaning *saffron*, in allusion to the color.

GYPSUM—CaSO₄·2H₂O

Crystallography. Monoclinic; $2/m$. Crystals are of simple habit (Figs. 12.38a and 12.39); tabular on {010}; diamond-shaped, with edges beveled by {120}

and {$\bar{1}11$}. Other forms rare. Twinning common, on {100} (Fig. 12.38b), often resulting in swallowtail twins.

$C2/c$; $a = 6.28$, $b = 15.15$, $c = 5.67$ Å, $\beta = 114°12'$; $Z = 4$. d's: 7.56(10), 4.27(5), 3.06(6), 2.87(2), 2.68(3).

Physical Properties. *Cleavage* {010} perfect yielding thin folia; {100}, with conchoidal surface; {011}, with fibrous fracture. **H** 2. **G** 2.32. *Luster* usually vitreous; also pearly and silky. *Color* colorless, white, gray; various shades of yellow, red, brown, from impurities. Transparent to translucent.

Satin spar is a fibrous gypsum with silky luster. *Alabaster* is the fine-grained massive variety. *Selenite* is a variety that yields broad colorless and transparent cleavage folia. *Optics:* (+); $\alpha = 1.520$, $\beta = 1.523$, $\gamma = 1.530$; $2V = 58°$. $Y = b$, $X \wedge c = -37°$; $r > v$.

Composition and Structure. CaO 32.6, SO₃ 46.5, H₂O 20.9%. As the result of dehydration of gypsum, CaSO₄·2H₂O, several phases may form: γ CaSO₄, when all H₂O is lost and a metastable phase CaSO₄·$\frac{1}{2}$H₂O. During the dehydration process the first $1\frac{1}{2}$ molecules of H₂O in gypsum are lost relatively continuously between 0° and about 65°C (see Fig. 12.40) perhaps with only slight changes in the gypsum structure. At about 70°C the remaining $\frac{1}{2}$H₂O molecule in CaSO₄·$\frac{1}{2}$H₂O is still retained relatively strongly, but at about 95°C this is lost and the structure transforms to

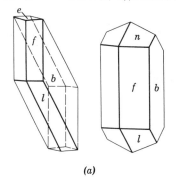

FIG. 12.38. *(a)* Gypsum crystals. *(b)* Gypsum twin.

(a)

(b)

FIG. 12.39. Gypsum crystals, Ellsworth, Ohio. (Harvard Mineralogical Museum.)

that of a polymorph of anhydrite. The structure of gypsum is given in Fig. 12.31 and discussed on page 423.

Diagnostic Features. Characterized by its softness and its three unequal cleavages.

Occurrence. Gypsum is a common mineral widely distributed in sedimentary rocks, often as thick beds. If frequently occurs interstratified with limestones and shales and is usually found as a layer underlying beds of rock salt, having been deposited there as one of the first minerals to crystallize on the evaporation of saltwaters. May recrystallize in veins, forming *satin spar*. Occurs also as lenticular bodies or scattered crystals in clays and shales. Frequently formed by the alteration of anhydrite and under these circumstances may show folding because of increased volume. Found in volcanic regions, especially where limestones have been acted upon by sul-

fur vapors. Also common as a gangue mineral in metallic veins. Associated with many different minerals, the more common being halite, anhydrite, dolomite, calcite, sulfur, pyrite, and quartz.

Gypsum is the most common sulfate, and extensive deposits are found in many localities through the world. The principal world producers are the United States, Canada, France, Japan, and Iran. In the United States commercial deposits are found in many states, but the chief producers are located in California, Iowa, New York, Texas, and Oklahoma. Gypsum is found in large deposits in Arizona and New Mexico in the form of windblown sand. Spectacular crystals originate from several mines in Mexico, especially Naica and Chihuahua.

Use. Gypsum is used chiefly for the production of plaster of Paris. In the manufacture of the material, the gypsum is ground and then heated until about 75% of the water has been driven off, producing the substance $CaSO_4 \cdot \frac{1}{2}H_2O$. This material, when mixed with water, slowly absorbs the water, crystallizes and thus hardens or "sets." Plaster of Paris is used extensively for "staff," the material from which temporary exposition buildings are built, for gypsum lath, wallboard, and for molds and casts of all kinds. Gypsum is employed in making adamant plaster for interior use. Serves as a soil conditioner, "land plaster," for fertilizer. Uncalcined gypsum is used as a retarder in Portland cement. Satin spar and alabaster are cut and polished for various ornamental purposes but are restricted in their use because of their softness.

Name. From the Greek name for the mineral, applied especially to the calcined mineral.

FIG. 12.40. Dehydration curve of gypsum showing formation of metastable $CaSO_4 \cdot \frac{1}{2}H_2O$ at about 65°C. At about 100°C the γ polymorph of $CaSO_4$ is formed.

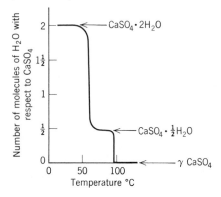

Antlerite—$Cu_3SO_4(OH)_4$

Crystallography. Orthorhombic; $2/m2/m2/m$. Crystals usually tabular on {010} (Fig. 12.41*b*). May be in slender prismatic crystals vertically striated. Also in parallel aggregates, reniform, massive.

Pnam; $a = 8.24$, $b = 11.99$, $c = 6.03$ Å; $Z = 4$. *d's*: 4.86(10), 3.60(8), 3.40(3), 2.68(6), 2.57(6).

Physical Properties. *Cleavage* {010} perfect. **H** $3\frac{1}{2}$–4. **G** 3.9±. *Luster* vitreous. *Color* emerald to blackish-green. *Streak* pale green. Transparent to translucent. *Optics:* (+); $\alpha = 1.726$, $\beta = 1.738$, $\gamma = 1.789$; 2V = 53°; X = *b*, Y = *a*; pleochroism X yellow-green, Y blue-green, Z green.

Composition and Structure. CuO 67.3, SO_3 22.5, H_2O 10.2%. The structure of antlerite contains Cu in two types of octahedral coordination: $CuO(OH)_5$ and $CuO_3(OH)_3$. These octahedra are linked with the SO_4 tetrahedra.

Diagnostic Features. Antlerite is characterized by its green color, {010} cleavage, and association. Lack of effervescence in HCl distinguishes it from malachite. It is commonly associated with atacamite and brochantite and one must use optical or chemical properties to distinguish it from these minerals.

Occurrence. Antlerite is found in the oxidized portions of copper veins, especially in arid regions. It was formerly considered a rare mineral, but in 1925 it was recognized as the chief ore mineral at Chuquicamata, Chile, the world's largest copper mine. It may form directly as a secondary mineral on chalcocite, or the copper may go into solution and later be deposited as antlerite, filling cracks. In the United States it is found at Bisbee, Arizona.

Use. An ore of copper.

Name. From the Antler Mine, Arizona, from which locality it was original described.

Similar Species. *Brochantite*, $Cu_4SO_4(OH)_6$, is similar in all its properties to antlerite, but, although more widespread, it is nowhere abundant. Until 1925 it was considered to be the chief ore mineral at Chuquicamata, Chile. Another rare hydrous copper sulfate is *chalcanthite*, $CuSO_4 \cdot 5H_2O$, which occurs abundantly at Chuquicamata and other arid areas in Chile. *Epsomite*, $MgSO_4 \cdot 7H_2O$, occurs as an efflorescence on the rocks in mine workings and on the walls of caves. *Melanterite*, $FeSO_4 \cdot 7H_2O$, is also a secondary mineral found as an efflorescence on mine timbers.

Alunite—$KAl_3(SO_4)_2(OH)_6$

Crystallography. Hexagonal; 3*m*. Crystals usually a combination of positive and negative trigonal pyramids resembling rhombohedrons with nearly cubic angles (90°50′). May be tabular on {0001}. Commonly massive or disseminated.

R3m; $a = 6.97$, $c = 17.38$ Å; $Z = 3$. *d's*: 4.94(5), 2.98(10), 2.29(5), 1.89(6), 1.74(5).

Physical Properties. *Cleavage* {0001} imperfect. **H** 4. **G** 2.6–2.8. *Luster* vitreous to pearly in crystals, earthy in massive material. *Color* white, gray, or reddish. Transparent to translucent. *Optics:* (+); $\omega = 1.572$, $\epsilon = 1.592$.

Composition and Structure. K_2O 11.4, Al_2O_3 37.0, SO_3 38.6, H_2O 13.0%. Na may replace K at least up to Na : K = 7 : 4. When Na exceeds K in the mineral it is called *natroalunite*, $(Na,K)Al_3(SO_4)_2(OH)_6$. In the structure of alunite K and Al are in 6-coordination with respect to O and (OH). The SO_4 groups share some of their oxygens with the K-containing octahedra.

Diagnostic Features. Alunite is usually massive and difficult to distinguish from rocks such as limestone and dolomite, and other massive minerals such as anhydrite and granular magnesite.

Occurrence. Alunite, also called *alumstone*, is usually formed by sulfuric acid solutions acting on rocks rich in potash feldspar, and in some places large masses have thus been formed. Found in smaller amounts about volcanic fumaroles. In the United States it is found at Red Mountain in the San Juan district, Colorado; Goldfield, Nevada; and Marysvale, Utah.

Use. In the production of alum. At Marysvale, Utah, alunite has been mined and treated in such a way as to recover potassium and aluminum.

Name. From the Latin meaning alum.

Similar Species. *Jarosite*, $KFe_3(SO_4)_2(OH)_6$, the Fe^{3+} analogue of alunite, is a secondary mineral found as crusts and coatings on ferruginous ores.

TUNGSTATES AND MOLYBDATES

W^{6+} and Mo^{6+} are considerably larger than S^{6+} and P^{5+}. Hence, when these ions enter into anisodesmic

FIG. 12.41. Antlerite crystals.

(a)   *(b)*

FIG. 12.42. Structure of scheelite, CaWO₄.

ionic groups with oxygen, the four coordinated oxygen ions do not occupy the apices of a regular tetrahedron, as is the case in the sulfates and phosphates, but form a somewhat flattened grouping of square outline. Although W (at. wt. 184) has a much greater atomic weight than Mo(96), both belong to the same family of the periodic table and, because of the lanthanide contraction, have the same ionic radii. As a result, each may freely substitute for the other as the coordinating cation in the warped tetrahedral oxygen groupings. In nature, however, the operation of the processes of geochemical differentiation often separates these elements, and it is not uncommon to find primary tungstates almost wholly free of Mo and vice versa. In secondary minerals, the two elements are more commonly in solid solution with each other.

The minerals of this chemical class fall mainly into two isostructural groups. The *wolframite group* contains compounds with fairly small divalent cations such as Fe^{2+}, Mn^{2+}, Mg, Ni, and Co in 6-coordination with (MoO_4). Complete solid solution occurs between Fe^{2+} and Mn^{2+} in the minerals of this group.

The *scheelite group* contains compounds of larger ions such as Ca^{2+} and Pb^{2+} in 8-coordination with $(WO_4)^{2-}$ and $(MoO_4)^{2-}$ groups. The structure of scheelite (see Fig. 12.42) is close to that of anhydrite and zircon, $ZrSiO_4$, but differs from these in the manner of linking of the CaO_8 polyhedra. The (WO_4) tetrahedra are somewhat flattened along the c axis and join edges with CaO_8. W and Mo may substitute for each other, forming partial series between *scheelite*,

$CaWO_4$, and *powellite*, $CaMoO_4$; and *stolzite*, $PbWO_4$, and *wulfenite*, $PbMoO_4$. The substitution of Ca and Pb for one another forms partial series between scheelite and stolzite and between powellite and wulfenite. The tungstates and molybdates comprise a group of nine minerals of which we will discuss only three: wolframite, scheelite, and wulfenite.

WOLFRAMITE—(Fe,Mn)WO₄

Crystallography. Monoclinic; $2/m$. Crystals commonly tabular parallel to {100} (Fig. 12.43) giving bladed habit, with faces striated parallel to c. In bladed, lamellar, or columnar aggregates. Massive granular.

$P2/c$; $a = 4.79$, $b = 5.74$, $c = 4.99$ Å; $\beta = 90°28'$; $Z = 2$. d's: 4.76(5), 3.74(5), 2.95(10), 2.48(6), 1.716(5). Cell parameters are slightly less for *ferberite*, $FeWO_4$, and greater for *huebnerite*, $MnWO_4$.

Physical Properties. *Cleavage* {010} perfect. **H** $4-4\frac{1}{2}$. **G** 7.0–7.5, higher with higher Fe content. *Luster* submetallic to resinous. *Color* black in ferberite to brown in huebnerite. *Streak* from nearly black to brown.

Composition and Structure. Fe^{2+} and Mn^{2+} substitute for each other in all proportions and a complete solid solution series exists between *ferberite*, $FeWO_4$, and *huebnerite*, $MnWO_4$. The percentage of WO_3 is 76.3 in ferberite and 76.6 in huebnerite. The structure of wolframite consists of distorted tetrahedral (WO_4) groups and octahedral $(Fe,Mn)O_6$ groups. From the interatomic distances around W, however, one may also conclude that W is in distorted octahedral coordination.

Diagnostic Features. The dark color, one direction of perfect cleavage, and high specific gravity serve to distinguish wolframite from other minerals.

Occurrence. Wolframite is a comparatively rare mineral found usually in pegmatites and high-temperature quartz veins associated with granites. More rarely in sulfide veins. Minerals commonly associated include cassiterite, scheelite, bismuth, quartz, pyrite, galena, sphalerite, and arsenopyrite. In

FIG. 12.43. Wolframite.

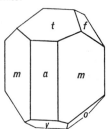

some veins wolframite may be the only metallic mineral present.

Found in fine crystals from Schlaggenwald and Zinnwald, Bohemia, Czechoslovakia, and in the various tin districts of Saxony, Germany, and Cornwall, England. Important producing countries are China, the CIS, Korea, Thailand, Bolivia, and Australia. Wolframite occurs in the United States in the Black Hills, South Dakota. Ferberite has been mined extensively in Boulder County, Colorado. Huebernite is found near Silverton, Colorado; Mammoth district, Nevada; and Black Hills, South Dakota. More than 90% of the world's estimated tungsten resources are outside the United States, with almost 60% in southeastern China.

Use. Chief ore of tungsten. Tungsten is used as hardening metal in the manufacture of high-speed tool steel, valves, springs, chisels, files, and so on. Its high melting point (3410°C) requires a special chemical process for reduction of the metal, which is produced in the form of a powder. By powder-metallurgy, pure metal products such as lamp filaments are fabricated. A large use for tungsten is in the manufacture of carbides, harder than any natural abrasives (other than diamond), that are used for cutting tools, rock bits, and hard facings. Sodium tungstate is used in fireproofing cloth and as a mordant in dyeing.

Name. Wolframite is derived from an old word of German origin.

SCHEELITE—CaWO$_4$

Crystallography. Tetragonal; $4/m$. Crystals usually simple dipyramids, {011} (Fig. 12.44). The dipyramid, {112}, closely resembles the octahedron in angles. Also massive granular.

$I 4/a; a = 5.25, c = 11.40$ Å; $Z = 4$. d's: 4.77(7), 3.11(10), 1.94(8), 1.596(9), 1.558(7).

Physical Properties. *Cleavage* {101}, distinct. **H** $4\frac{1}{2}$–5. **G** 5.9–6.1 (unusually high for a nonmetallic mineral). *Luster* vitreous to adamantine. *Color* white, yellow, green, brown. Translucent; some specimens transparent (see Plate IV, no. 14, Chapter 15). Most scheelite will fluoresce with bluish white color in short ultraviolet radiation. *Optics:* (+); $\omega = 1.920$, $\epsilon = 1.934$.

Composition and Structure. CaO 19.4, WO$_3$ 80.6%. Mo can substitute for W and a partial series extends to *powellite*, CaMoO$_4$. The structure of scheelite (Fig. 12.42) consists of flattened (WO$_4$) tetrahedra and CaO$_8$ polyhedra.

Diagnostic Features. Recognized by its high specific gravity, crystal form, and fluorescence in short ultraviolet light. The test for tungsten may be necessary for positive identification.

FIG. 12.44. Scheelite.

Occurrence. Scheelite is found in granite pegmatites, contact metamorphic deposits, and high-temperature hydrothermal veins associated with granitic rocks. Associated with cassiterite, topaz, fluorite, apatite, molybdenite, and wolframite. Occurs with tin in deposits of Bohemia, Czechoslovakia, Saxony, Germany, and Cornwall, England; and in quantity in New South Wales and Queensland, Australia. In the United States scheelite is mined near Mill City and Mina, Nevada; near Atolia, San Bernardino County, California; and in lesser amounts in Arizona, Utah, and Colorado.

Use. An ore of tungsten. Wolframite furnishes most of the world's supply of tungsten, but scheelite is more important in the United States. Transparent crystals may be cut as faceted gems.

Name. After K. W. Scheele (1742–1786), the discoverer of tungsten.

Wulfenite—PbMoO$_4$

Crystallography. Tetragonal; 4 or $4/m$. Crystals usually square, tabular in habit with prominent {001}; Figs. 12.45a and 12.46. Some crystals very thin. More rarely pyramidal (Fig. 12.45b). Some crystals may be twinned on {001} giving them a dipyramidal habit.

$I4_1/a; a = 5.42, c = 12.10$ Å; $Z = 4$. d's: 3.24(10), 3.03(2), 2.72(2), 2.02(3), 1.653(3).

Physical Properties. *Cleavage* {011} distinct. **H** 3. **G** 6.8 ±. *Luster* vitreous to adamantine. *Color* yellow, orange, red, gray, white. *Streak* white. Trans-

FIG. 12.45. Wulfenite crystals.

(a)

(b)

FIG. 12.46. Wulfenite, Stephenson-Bennett Mine, Dona Ana County, New Mexico. (Harvard Mineralogical Museum.)

parent to subtranslucent. Piezoelectricity suggests symmetry 4 rather than $4/m$ as determined from structure. *Optics:* $(-)$; $\omega = 2.404$, $\epsilon = 2.283$.

Composition and Structure. PbO 60.8, MoO_3 39.2%. Ca may substitute for Pb, indicating at least a partial series to *powellite*, $Ca(Mo,W)O_4$. Wulfenite is isostructural with scheelite (see Fig. 12.42).

Diagnostic Features. Wulfenite is characterized by its square tabular crystals, orange to yellow color, high luster, and association with other lead minerals. Distinguished from crocoite by test for Mo.

Occurrence. Wulfenite is found in the oxidized portions of lead veins with other secondary lead minerals, especially cerussite, vanadinite, and pyromorphite. Found in the United States in several places in the Southwest. Found in beautiful crystals at the Red Cloud, Glove, and Mammoth mines in Arizona; also at the Bennett mine in New Mexico; and in Schwarzenbach, Yugoslavia.

Use. A minor source of molybdenum. Molybdenite, MoS_2, is the chief ore.

Name. After X. F. Wulfen, Austrian mineralogist.

PHOSPHATES, ARSENATES, AND VANADATES

P^{5+} is only slightly larger than S^{6+} and, hence, like sulfur, forms a tetrahedral anionic $(PO_4)^{3-}$ group with oxygen (see Fig. 12.1). All phosphates contain this phosphate anionic complex as the fundamental building unit. Similar tetrahedral units, $(AsO_4)^{3-}$ and

$(VO_4)^{3-}$ occur in arsenates and vanadates. P^{5+}, As^{5+}, and V^{5+} may substitute for each other in the anionic groups. This type of substitution is best shown in the pyromorphite series of the apatite group. *Pyromorphite*, $Pb_5(PO_4)_3Cl$, *mimetite*, $Pb_5(AsO_4)_3Cl$, and *vanadinite*, $Pb_5(VO_4)_3Cl$, are isostructural, and all gradations of composition between the end members exist.

The structure of *apatite*, $Ca_5(PO_4)_3(OH,F,Cl)$, which is the most important and abundant phosphate, is illustrated in Fig. 12.47. The oxygens of the (PO_4) groups are linked to Ca in two different structural sites.

FIG. 12.47. Structure of fluorapatite, $Ca_5(PO_4)_3F$, projected on the (0001) plane. The dashed parallelogram outlines the base of the unit cell. The tetrahedral (PO_4) groups, triangular coordination of F to Ca, and examples of the two types of coordination about Ca are shown.

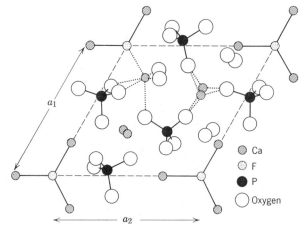

In one site Ca is in irregular 9-coordination, and in the other in irregular 8-coordination. Each fluorine (or Cl or OH) lies in a triangle with three calciums. Apatite shows extensive solid solution with respect to anions as well as cations. (PO_4) may be substituted for by (AsO_4) or (VO_4) as noted above, but also in part by tetrahedal (CO_3OH) groups, giving rise to *carbonate-apatite*, $Ca_5F(PO_4,CO_3OH)_3$. Small amounts of (SiO_4) and (SO_4) may also be present in substitution for (PO_4); these types of substitution must be coupled with other cation substitutions in apatite in order to retain the electrical neutrality of the structure (see page 233). F may be replaced by (OH) or Cl producing *hydroxylapatite*, $Ca_5(PO_4)_3(OH)$ and *chlorapatite*, $Ca_5(PO_4)_3Cl$. Mn^{2+} and Sr^{2+} may substitute for Ca. These varied ionic substitutions are typical of the phosphates which generally have rather complicated structures.

This mineral class, composed mostly of phosphates, is very large but most of its members are so rare that they need not be mentioned here. Of the list of phosphates, arsenates, and vanadates given below only apatite can be considered as common.

PHOSPHATES, ARSENATES, AND VANADATES

Triphylite—	$Li(Fe,Mn)PO_4$
Lithiophilite	$Li(Mn,Fe)PO_4$
Monazite	$(Ce,La,Y,Th)PO_4$
Apatite Group	
Apatite	$Ca_5(PO_4)_3(F,Cl,OH)$
Pyromorphite	$Pb_5(PO_4)_3Cl$
Vanadinite	$Pb_5(VO_4)_3Cl$
Erythrite	$Co_3(AsO_4)_2 \cdot 8H_2O$
Amblygonite	$LiAlPO_4F$
Lazulite—	$(Mg,Fe)Al_2(PO_4)_2(OH)_2$
Scorzalite	$(Fe,Mg)Al_2(PO_4)_2(OH)_2$
Wavellite	$Al_3(PO_4)_2(OH)_3 \cdot 5H_2O$
Turquoise	$CuAl_6(PO_4)_4(OH)_8 \cdot 4H_2O$
Autunite	$Ca(UO_2)_2(PO_4)_2 \cdot 10-12H_2O$
Carnotite	$K_2(UO_2)_2(VO_4)_2 \cdot 3H_2O$

Triphylite—$Li(Fe,Mn)PO_4$
Lithiophilite—$Li(Mn,Fe)PO_4$

Crystallography. Orthorhombic; $2/m2/m2/m$. Crystals rare. Commonly in cleavable masses. Also compact.

Pmcn. For triphylite: $a = 6.01$, $b = 4.86$, $c = 10.36$ Å; $Z = 4$. d's: 4.29(8), 3.51(9), 3.03(9), 2.54(10), 1.75(5).

Physical Properties. *Cleavage* {001} nearly perfect, {010} imperfect. **H** $4\frac{1}{2}$–5. **G** 3.42–3.56 increasing with Fe content. *Luster* vitreous to resinous. *Color* bluish-gray in triphylite to salmon-pink or clove-brown in lithiophilite. May be stained black by man-

ganese oxide. Translucent. *Optics:* (+); $\alpha = 1.669$–1.694, $\beta = 1.673$–1.695, $\gamma = 1.682$–1.700; $2V = 0°$–55°, $X = c$, $Y = a$, $Z = b$. Indices increase with Fe content.

Composition and Structure. A complete Fe^{2+}-Mn^{2+} series exists between two essentially pure end members. In the structure of members of this series Li and (Mn,Fe) are in 6-coordination. These octahedra are linked along their edges into zigzag chains which are connected by (PO_4) tetrahedra.

Diagnostic Features. Characterized by two cleavages at right angles, resinous luster, and association.

Occurrence. Triphylite and lithiophilite are pegmatite minerals associated with other phosphates, spodumene, and beryl. Notable localities are in Bavaria, Germany, and Finland. Lithiophilite is found at several localities in Argentina. In the United States triphylite is found in large crystals at the Palermo Mine, North Grothon, New Hampshire, and elsewhere in New England, and in the Black Hills, South Dakota. Lithiophilite is found at Branchville, Connecticut.

Name. Triphylite from the Greek words meaning *three* and *family* in allusion to containing three cations. Lithiophilite from the Greek words meaning *lithium* and *friend*.

Monazite—$(Ce,La,Y,Th)PO_4$

Crystallography. Monoclinic; $2/m$. Crystals rare and usually small, often flattened on {100}, or elongated on b. Usually in granular masses, frequently as sand.

$P2_1/n$; $a = 6.79$, $b = 7.01$, $c = 6.46$ Å; $\beta = 103°38'$; $Z = 4$. d's: 4.17(3), 3.30(5), 3.09(10), 2.99(2), 2.87(7).

Physical Properties. *Cleavage* {100} poor. Parting {001}. **H** 5–$5\frac{1}{2}$. **G** 4.6–5.4. *Luster* resinous. *Color* yellowish to reddish-brown. Translucent. *Optics:* (+); $\alpha = 1.785$–1.800, $\beta = 1.787$–1.801, $\gamma = 1.840$–1.850; $2V = 10°$–20°; $X = b$, $Z \wedge c = 2°$–6°.

Composition and Structure. A phosphate of the rare-earth metals essentially $(Ce,La,Y,Th)PO_4$. Th content ranges from a few to 20% ThO_2. Si is often present up to several percent SiO_2. The Si has been ascribed to admixture of *thorite*, $ThSiO_4$, but may be in part due to substitution of (SiO_4) for (PO_4). In the structure of monazite the rare earths are in 9-coodination with oxygen, linking six PO_4 tetrahedra. It is isostructural with *crocoite*, $PbCrO_4$ (see page 427).

Diagnostic Features. Radioactive. Large specimens may be distinguished from zircon by crystal form and inferior hardness, and from titanite by crystal form and higher specific gravity. On doubtful specimens it is usually wise to make a chemical phosphate test.

Occurrence. Monazite is a comparatively rare mineral occurring as an accessory in granites, gneisses, aplites, and pegmatites, and as rolled grains in the sands derived from the decomposition of such rocks. It is concentrated in sands because of its resistance to chemical attack and its high specific gravity, and is thus associated with other resistant and heavy minerals such as magnetite, ilmenite, rutile, and zircon.

The bulk of the world's supply of monazite comes from beach sands in Brazil, India, and Australia. A dikelike body of massive granular monazite was mined near Van Rhynsdorp, Cape Province, South Africa. Found in the United States in North Carolina, both in gneisses and in the stream sands; and in the beach sands of Florida.

Use. Monazite is the chief source of thorium oxide, which it contains in amounts varying between 1 and 20%; commercial monazite usually contains between 3 and 9%. Thorium oxide is used in the manufacture of mantles for incandescent gas lights.

Thorium is a radioactive element and is receiving considerable attention as a source of atomic energy. A commercial thorium-fueled reactor is in operation at Fort St. Vrain, Colorado.

Name. The name *monazite* is derived from the Greek word *monachos* meaning *solitary*, in allusion to the rarity of the mineral.

Apatite Group

APATITE—Ca$_5$(PO$_4$)$_3$(F,Cl,OH)

Crystallography. Hexagonal; $6/m$. Commonly occurs in crystals of long prismatic habit, see Fig. 12.48; some short prismatic or tabular. Usually terminated by prominent dipyramid, $\{10\bar{1}1\}$, and frequently a basal plane. Some crystals show faces of a hexagonal dipyramid (μ, Fig. 12.49c) which reveals the true symmetry. Also in massive granular to compact masses.

$P6_3/m$; $a = 9.39$, $c = 6.89$ Å; $Z = 2$. d's: 2.80(10), 2.77(4), 2.70(6), 1.84(6), 1.745(3).

Physical Properties. *Cleavage* $\{0001\}$ poor. **H** 5 (can just be scratched by a knife). **G** 3.15–3.20. *Luster* vitreous to subresinous. *Color* usually some shade of green or brown; also blue, violet, colorless. Transparent to translucent. *Optics:* $(-)$; $\omega = 1.633$, $\epsilon = 1.630$ (fluorapatite).

Composition and Structure. Ca$_5$(PO$_4$)$_3$F, *fluorapatite* is most common; more rarely Ca$_5$(PO$_4$)$_3$Cl, *chlorapatite*, and Ca$_5$(PO$_4$)$_3$(OH), *hydroxylapatite*. F, Cl, and OH can substitute for each other, giving com-

FIG. 12.48. Apatite in marble, Renfrew County, Ontario, Canada.

plete series. (CO$_3$, OH) may substitute for (PO$_4$) giving *carbonate-apatite*. The (PO$_4$) group can be partially replaced by (SO$_4$) as well as (SiO$_4$). The P^{5+} to S^{6+} replacement is compensated by the coupled substitution of Ca^{2+} by Na$^+$. Furthermore the P^{5+} to S^{6+} replacement may be balanced by substitution of Si^{4+} for P^{5+}. Mn and Sr can substitute in part for Ca (see page 433). The structure of apatite is illustrated in Fig. 12.47.

Collophane. The name collophane has been given to the massive, cryptocrystalline types of apatite that constitute the bulk of phosphate rock and fossil bone. X-ray study shows that collophane is essentially apatite and does not warrant designation as a separate species. In its physical appearance, collophane is usually dense and massive with a concretionary or colloform structure. It is usually impure and contains small amounts of calcium carbonate.

Diagnostic Features. Apatite is usually recognized by its crystals, color, and hardness. Distinguished from beryl by the prominent pyramidal terminations of its crystals and by its being softer than a knife blade.

Occurrence. Apatite is widely disseminated as an accessory constituent in all classes of rocks—igneous, sedimentary, and metamorphic. It is also found in pegmatites and other veins, probably of hydrothermal origin. Found in titaniferous magnetite bodies. Occasionally concentrated into large deposits or veins associated with alkalic rocks. Phosphate materials of bones and teeth are members of the apatite group.

Apatite occurs in large amounts along the south-

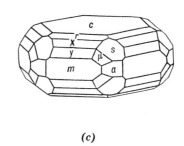

FIG. 12.49. Apatite crystals.   *(a)*   *(b)*   *(c)*

ern coast of Norway, between Langesund and Arendal, where it is found in veins and pockets associated with gabbro. It is distributed through the magnetite iron ore at Kiruna, Sweden, and Tahawas, New York. In Ontario and Quebec, Canada, large apatite crystals were formerly mined. The world's largest deposit of apatite is located on the Kola Peninsula, near Kirovsk, Karelia, CIS. Here apatite in granular aggregates intimately associated with nepheline and titanite, is found in a great lens between two types of alkalic rocks.

Finely crystallized apatite in pegmatites and veins occurs at various localities in the Tyrol, Austria; in Switzerland; in Panasquiera, Portugal; and Jumilla, Spain. In the United States at Auburn, Maine; St. Lawrence County, New York; Alexander County, North Carolina; and San Diego County, California.

The variety collophane is an important constituent of the rock *phosphorite or phosphate rock*. Bone is calcium phosphate, and large bodies of phosphorite are derived from the accumulation of animal remains as well as from chemical precipitation from seawater. Commercial deposits of phosphorite are found in northern France, Belgium, Spain, and especially in northern Africa in Tunisia, Algeria, and Morocco. In the United States, high-grade phosphate deposits are found in Tennessee and in Wyoming and Idaho. Deposits of "pebble" phosphate are found at intervals all along the Atlantic coast from North Carolina to Florida. The most productive deposits in the United States are in Florida.

Use. Crystallized apatite has been used extensively as a source of phosphate for fertilizer, but today only the deposits on the Kola Peninsula are of importance and phosphorite deposits supply most of the phosphate for fertilizer. The calcium phosphate is treated with sulfuric acid and changed to superphosphate to render it more soluble in the dilute acids that exist in the soil. Transparent varieties of apatite of fine color are occasionally used for gems (see Plate IV, no. 8 Chapter 15). The mineral is too soft, however, to allow its extensive use for this purpose.

Name. From the Greek word *apate* meaning

deceipt, because the gem varieties were confused with other minerals.

Pyromorphite—$Pb_5(PO_4)_3Cl$

Crystallography. Hexagonal; $6/m$. Crystals usually prismatic with basal plane (Fig. 12.50). Rarely shows pyramid truncations. Often in rounded barrel-shaped forms. Sometimes cavernous, the crystals being hollow prisms. Also in parallel groups. Frequently globular, reniform, fibrous, and granular.

$P6_3m$; $a = 9.97, c = 7.32$ Å; $Z = 2$. d's: 4.31(6), 4.09(9), 2.95(10), 2.05(8), 1.94(7).

Physical Properties. **H** $3\frac{1}{2}$–4. **G** 7.04. *Luster* resinous to adamantine. *Color* usually various shades of green, brown, yellow; rarely orange-yellow, gray, white. Subtransparent to translucent. *Optics:* $(-)$, $\omega = 2.058$, $\epsilon = 2.048$.

Composition and Structure. For pure $Pb_5(PO_4)_3Cl$, PbO 82.2, P_2O_5 15.7, Cl 2.6%. (AsO_4) substitutes for (PO_4) and a complete series extends to *mimetite*, $Pb_5(AsO_4)_3Cl$. Ca may substitute in part for Pb. Isostructural with apatite (see Fig. 12.47).

Diagnostic Features. Pyromorphite is characterized by its crystal form, high luster, and high specific gravity.

Occurrence. Pyromorphite is a supergene mineral found in the oxidized portions of lead veins, associated with other oxidized lead and zinc minerals.

FIG. 12.50. Pyromorphite.

Notable localities for its occurrence are the lead mines at Ems in Nassau and at Zschopau, Saxony, Germany; Příbram, Bohemia, Czechoslovakia; Beresovsk, Ural Mountains; in Cumbria, England and at Leadhills, Scotland. In the United States found at Phoenixville, Pennsylvania, and the Coeur d'Alene district, Idaho.

Use. A subordinate ore of lead.

Name. Derived from two Greek words meaning *fire* and *form*, in allusion to the apparent crystalline form it assumes on cooling from fusion.

Similar Species. *Mimetite*, $Pb_5(AsO_4)_3Cl$, is isostructural with pyromorphite and is similar in appearance, occurrence and most of its physical and chemical properties.

Vanadinite—$Pb_5(VO_4)_3Cl$

Crystallography. Hexagonal; $6/m$. Most commonly occurs in prismatic crystals with $\{10\bar{1}0\}$ and $\{0001\}$. May have small pyramidal faces, rarely the hexagonal dipyramid. In rounded crystals; in some cases cavernous. Also in globular forms. As incrustations.

$P6_3/m$; $a = 10.33$, $c = 7.35$ Å; $Z = 2$. d's: 4.47(3), 4.22(4), 3.38(6), 3.07(9), 2.99(10).

Physical Properties. **H** 3. **G** 6.9. *Luster* resinous to adamantine. *Color* ruby-red, orange-red, brown, yellow. Transparent to translucent. *Optics:* $(-)$; $\omega = 2.25–2.42$, $\epsilon = 2.20–2.35$. Indices lowered by substitution of As or P for V.

Composition. PbO 78.7, V_2O_5 19.4, Cl 2.5%. PO_4 and AsO_4 may substitute in small amounts for VO_4. In the variety *endlichite*, intermediate between vanadinite and mimetite, the proportion of V_2O_5 to As_2O_5 is nearly 1 : 1. Small amounts of Ca, Zn, and Cu substitute for Pb. Vanadinite is isostructural with apatite (see Fig. 12.47).

Diagnostic Features. Characterized by crystal form, high luster, and high specific gravity; distinguished from pyromorphite and mimetite by color.

Occurrence. Vanadinite is a rare secondary mineral found in the oxidized portion of lead veins associated with other secondary lead minerals. Found in fine crystals near Oudjda, Morocco, and Grootfontein, Namibia. In the United States it occurs in various districts in Arizona and New Mexico.

Use. Source of vanadium and minor ore of lead. Vanadium is obtained chiefly from other ores, such as *patronite*, VS_4; the vanadate *carnotite*, $K_2(UO_2)_2(VO_4)_2 \cdot 3H_2O$; and a vanadium mica, *roscoelite*, $KV_2(AlSi_3O_{10})(OH)_2$. Vanadium is used chiefly as a steel-hardening metal. Metavanadic acid,

HVO_3, is a yellow pigment, known as vanadium bronze. Vanadium oxide is a mordant in dyeing.

Name. In allusion to the composition.

Erythrite—$Co_3(AsO_4)_2 \cdot 8H_2O$

Crystallography. Monoclinic; $2/m$. Crystals prismatic and vertically striated. Usually as crusts in globular and reiniform shapes. Also pulverulent and earthy.

$C2/m$; $a = 10.20$, $b = 13.37$, $c = 4.74$ Å; $\beta = 105°$; $Z = 2$. d's: 6.65(10), 3.34(1), 3.22(1), 2.70(1), 2.32(1).

Physical Properties. *Cleavage* {010} perfect. **H** $1\frac{1}{2}$–$2\frac{1}{2}$. **G** 3.06. *Luster* adamantine to vitreous, pearly on cleavage. *Color* crimson to pink. Translucent. *Optics:* $(-)$; $\alpha = 1.626$, $\beta = 1.661$, $\gamma = 1.699$; $2V = 90°\pm$; $X = b$, $Z \wedge c = 31°$; $r > v$. Pleochroism X pink, Y violet, Z red.

Composition and Structure. CoO 37.5, As_2O_5 38.4, H_2O 24.1%. Ni substitutes for Co to form a complete series to *annabergite*, $Ni_3(AsO_4)_2 \cdot 8H_2O$. *Annabergite*, or *nickel bloom*, is light green in color. The structure of erythrite is of a layer type with strong bonds between (AsO_4) tetrahedra and $Co(O,H_2O)$ octahedra that are linked by common vertices. The layers parallel to (010) are held together by weak residual bonds.

Diagnostic Features. The association of erythrite with other cobalt minerals and its pink color are usually sufficient to distinguish it from all other minerals.

Occurrence. Erythrite is a rare secondary mineral. In pink crusts known as *cobalt bloom* it occurs as an alteration product of cobalt arsenides. It is rarely present in large amounts and usually forms as crusts or fine aggregates filling cracks. Notable localities are at Schneeberg, Saxony, Germany, and Bon Azzer, Morocco.

Use. Although erythrite has no economic importance, it is used by the prospector as a guide to other cobalt minerals and associated native silver.

Name. From the Greek word *erythros* meaning *red*.

Similar Species. *Vivianite*, $Fe_3(PO_4)_2 \cdot 8H_2O$, a rare mineral which is a weathering product of primary Fe-Mn phosphates in pegmatites.

AMBLYGONITE—$LiAlFPO_4$

Crystallography. Triclinic: $\bar{1}$. Usually occurs in coarse, cleavable masses. Crystals are rare, equant, and usually rough when large. Frequently twinned on $\{\bar{1}\bar{1}1\}$.

$P\bar{1}$; $a = 5.19$, $b = 7.12$, $c = 5.04$ Å; $\alpha = 112°02'$, $\beta = 97°50'$, $\gamma = 68°8'$; $Z = 2$. d's: 4.64(10), 3.15(10), 2.93(10), 2.39(5), 2.11(4).

Physical Properties. *Cleavage* {100} perfect, {110} good, {0$\bar{1}$1} distinct. **H** 6. **G** 3.0–3.1. *Luster* vitreous, pearly on {100} cleavage. *Color* white to pale green or blue, rarely yellow (see Plate IV, no. 4, Chapter 15). Translucent. *Optics:* usually (−); $\alpha = 1.58$–1.60, $\beta = 1.50$–1.62, $\gamma = 1.60$–1.63; 2V = 50–90°; $r > v$. Indices increase with increase in (OH) in substitution for F.

Composition and Structure. Li_2O 10.1, Al_2O_3 34.4, F 12.9, P_2O_5 47.9%. Na substitutes for Li; (OH) substitutes for F and probably forms a complete series. When OH > F, the mineral is known as *montebrasite*. In the structure of amblygonite AlO_6 octahedra and PO_4 tetrahedra are linked by vertices; Li is in 5-coordination and lies between PO_4 tetrahedra and the nearest Al octahedra. The structure is fairly compact as reflected in the relatively high density.

Diagnostic Features. Cleavage fragments may be confused with feldspar, but are distinguished by cleavage angles.

Occurrence. Amblygonite is a rare mineral found in granite pegmatites with spodumene, tourmaline, lepidolite, and apatite. Found at Montebras, France. In the United States it occurs at Hebron, Paris, Auburn, and Peru, Maine; Portland, Connecticut, and the Black Hills, South Dakota.

Use. A source of lithium.

Name. From the two Greek words *amblys* meaning *blunt* and *gonia* meaning *angle*, in allusion to the angle between the cleavages.

Similar Species. *Beryllonite*, $NaBePO_4$, colorless and *brazilianite*, $NaAl_3(PO_4)_2(OH)_4$, colorless to yellow, are rare gem minerals found in pegmatites.

Lazulite—$(Mg,Fe)Al_2(PO_4)_2(OH)_2$; Scorzalite—$(Fe,Mg)Al_2(PO_4)_2(OH)_2$

Crystallography. Monoclinic; $2/m$. Crystals showing steep {hkl} prisms rare. Usually massive, granular to compact.

$P2_1/c$; $a = 7.12$, $b = 7.26$, $c = 7.24$ Å; $\beta = 118°55'$; $Z = 2$. d's: 6.15(8), 3.23(8), 3.20(7), 3.14(10), 3.07(10).

Physical Properties. *Cleavage* {110} indistinct. **H** 5–5½. **G** 3.0–3.1. *Luster* vitreous. *Color* azure-blue. Translucent. *Optics:* (−); $\alpha = 1.604$–1.639, $\beta = 1.626$–1.670, $\gamma = 1.637$–1.680, 2V = 60°; $Y = b$, $X \wedge c = 10°$; $r < v$. Absorption $X < Y < Z$. Indices increase with increasing Fe^{2+} content.

Composition and Structure. Probably a complete solid solution series exists from lazulite to scorzalite with the substitution of Fe^{2+} for Mg. In the structure $(Mg,Fe)(O,OH)_6$ octahedra are linked by edges and faces with $Al(O,OH)_6$ octahedra to form groups. These groups are joined to each other and to PO_4 tetrahedra.

Diagnostic Features. If crystals are lacking, lazulite is difficult to distinguish from other blue minerals without optical or chemical tests.

Occurrence. The members of the lazulite-scorzalite series are rare minerals found in high-grade metamorphic rocks and in pegmatites. They are usually associated with kyanite, andalusite, corundum, rutile, sillimanite, and garnet. Notable localities are Salzburg, Austria; Krieglach, Styria; and Horrsjoberg, Sweden. In the United States they are found with corundum on Crowder's Mountain, Gaston Country, North Carolina; with rutile on Graves Mountain, Lincoln County, Georgia; and with andalusite in the White Mountains, Inyo County, California.

Use. A minor gemstone.

Name. Lazulite derived from an Arabic word meaning *heaven*, in allusion to the color of the mineral. Scorzalite after E. P. Scorza, Brazilian mineralogist.

Wavellite—$Al_3(PO_4)_2(OH)_3 \cdot 5H_2O$

Crystallography. Orthorhombic; $2/m2/m2/m$. Single crystals rare. Usually in radiating, spherulitic (Fig. 12.51) and globular aggregates.

Pcmn; $a = 9.62$, $b = 17.34$, $c = 6.99$ Å; $Z = 4$. d's: 8.39(10), 5.64(6), 3.44(8), 3.20(8), 2.56(8).

Physical Properties. *Cleavage* {110} and {101} good. **H** 3½–4. **G** 2.36. *Luster* vitreous. *Color* white, yellow, green, and brown. Translucent. *Optics:* (+); $\alpha = 1.525$, $\beta = 1.535$, $\gamma = 1.550$; 2V = 70°. $X = b$, $Y = a$; $r > v$.

Composition and Structure. Al_2O_3 38.0, P_2O_5 35.2, H_2O 26.8%. F may substitute for OH. Details of the structure are uncertain. $Al(O,OH)_6$ octahedra are linked by common vertices to PO_4 tetrahedra.

Diagnostic Features. Almost invariably in radiating globular aggregates.

Occurrence. Wavellite is a secondary mineral found in small amounts in crevices in aluminous, low-grade metamorphic rocks and in limonite and phosphorite deposits. Although it occurs in many localities, only rarely is it found in quantity. It is abundant in the tin veins of Llallagua, Bolivia. In the United States wavellite occurs in a number of localities in Arkansas.

FIG. 12.51. Wavellite, Hot Springs, Arkansas. (Harvard Mineralogical Museum.)

Name. After Dr. William Wavel, who discovered the mineral.

Turquoise—$CuAl_6(PO_4)_4(OH)_8 \cdot 5H_2O$

Crystallography. Triclinic; $\bar{1}$. Rarely in minute crystals, usually cryptocrystalline. Massive compact, reniform, stalactitic. In thin seams, incrustations, and disseminated grains.

$P\bar{1}$; $a = 7.48$, $b = 9.95$, $c = 7.69$; $\alpha = 111°39'$, $\beta = 115°23'$, $\gamma = 69°26'$; $Z = 1$. d's: 6.17(7), 4.80(6), 3.68(10), 3.44(7), 3.28(7).

Physical Properties. *Cleavage* {001} perfect, {010} good (rarely seen). **H** 6. **G** 2.6–2.8. *Luster* wax-like. *Color* blue, bluish-green, green (see Plate III, no. 7, Chapter 15). Transmits light on thin edges. *Optics:* (+); $\alpha = 1.61$, $\beta = 1.62$, $\gamma = 1.65$; $2V = 40°$; $r < v$ strong.

Composition and Structure. Fe^{3+} substitutes for Al and a complete series exists between turquoise and *chalcosiderite*, $CuFe_6^{3+}(PO_4)_4(OH)_8 \cdot 4H_2O$. The structure consists of PO_4 tetrahedra and (Al,Fe^{3+}) octahedra linked by common vertices. Fairly large holes in the structure contain Cu which is coordinated to 4(OH) and 2 H_2O molecules.

Diagnostic Features. Turquoise can be recognized by its color. It is harder than chrysocolla, the only common mineral which it resembles. There is much imitation material on the market (see Chapter 15).

Occurrence. Turquoise is a secondary mineral usually found in the form of small veins and stringers traversing more or less decomposed volcanic rocks in arid regions. The famous Persian deposits are found in trachyte near Nishapur in the province of Khorasan, Iran. In the United States it is found in a much altered trachytic rock in the Los Cerillos Mountains, near Santa Fe, and elsewhere in New Mexico. Turquoise is also found in Arizona, Nevada, and California. Small crystals have been found at Lunch Station, Virginia.

Use. As a gemstone. It is always cut in round or oval forms. Much cut turquoise is veined with the various gangue materials, and such stones are sold under the name of *turquoise matrix*.

Name. Turquoise is French and means *Turkish*, the original stones having come into Europe from the Persian locality through Turkey.

Similar Species. *Variscite*, $Al(PO_4) \cdot 2H_2O$, is a massive, bluish-green mineral somewhat resembling turquoise and used as a gem material. It has been found in nodules in a large deposit at Fairfield, Utah.

Autunite—$Ca(UO_2)_2(PO_4)_2 \cdot 10{-}12H_2O$

Crystallography. Tetragonal; $4/m2/m2/m$. Crystals tabular on {001}; subparallel growths are common; also foliated and scaly aggregates.

$I4/mmm$; $a = 7.00$, $c = 20.67$ Å; $Z = 2$. d's: 10.33(10), 4.96(8), 3.59(7), 3.49(7), 3.33(7).

Physical Properties. *Cleavage* {001} perfect. **H** 2–2½. **G** 3.1–3.2. *Luster* vitreous, pearly on {001}. *Color* lemon yellow to pale green. *Streak* yellow. In ultraviolet light fluoresces strongly yellow-green. *Optics:* (−); $\omega = 1.577$, $\epsilon = 1.553$. Pleochroism E pale yellow, O dark yellow.

Composition and Structure. Small amounts of Ba and Mg may substitute for Ca. The H_2O content apparently ranges from 10 to 12 H_2O. On drying and slight heating autunite passes reversibly to *meta-autunite* I (a tetragonal phase with 6½–2½ H_2O). On continued heating to about 80°C this passes irreversibly to *meta-autunite* II (an orthorhombic phase with 0–6 H_2O). Neither meta-I nor the meta-II hydrate occurs as a primary phase in nature. The structure of autunite consists of (PO_4) tetrahedra and $(UO_2)O_4$ polyhedra, which are joined into tetragonal corrugated layers of composition $UO_2(PO_4)$ parallel to {001}. These layers are held together by weak hydrogen bonds to H_2O molecules.

Diagnostic Features. Autunite is characterized by yellow-green tetragonal plates and strong fluorescence in ultraviolet light. Radioactive.

Occurrence. Autunite is a secondary mineral found chiefly in the zone of oxidation and weathering derived from the alteration of uraninite or other uranium minerals. Notable localities are near Autun, France; Sabugal and Vizeu, Portugal; Johanngeorgenstadt district and Falkenstein, Germany; Cornwall, England; and the Katanga district of Zaire. In the United States autunite is found in many pegmatites, notably at the Ruggles Mine, Grafton Center, New Hampshire; Black Hills, South Dakota; and Spruce Pine, Mitchell County, North Carolina. The finest specimens have come from the Daybreak Mine near Spokane, Washington.

Use. An ore of uranium (see uraninite, page 384).

Name. From Autun, France.

Similar Species. *Torbernite*, $Cu(UO_2)_2(PO_4)_2 \cdot 8-12H_2O$, is isostructural with autunite and has similar properties, but there is no evidence of a solid solution series. Color green, nonfluorescent. Associated with autunite.

Carnotite—$K_2(UO_2)_2(VO_4)_2 \cdot 3H_2O$

Crystallography. Monoclinic; $2/m$. Only rarely in imperfect microscopic crystals flattened on {001} or elongated on b. Usually found as a powder or as loosely coherent aggregates; disseminated.

$P2_1a$; $a = 10.47$, $b = 8.41$, $c = 6.91$ Å; $\beta = 103°40'$; $Z = 2$. d's: 6.56(10), 4.25(3), 3.53(5), 3.25(3), 3.12(7).

Physical Properties. *Cleavage* {001} perfect. *Hardness* unknown, but soft. **G** 4.7–5. *Luster* dull or earthy. *Color* bright yellow to greenish-yellow. *Optics:* $(-)$; $\alpha = 1.75$, $\beta = 1.93$, $\gamma = 1.95$; $2V = 40° \pm$, $Y = b$, $X = c$. $r < v$. Indices increase with loss of water.

Composition and Structure. The water content of carnotite is partly zeolitic and varies with the humidity at ordinary temperatures; the $3H_2O$ is for fully hydrated material. Small amounts of Ca, Ba, Mg, Fe, and Na have been reported. The structure of carnotite consists of a layer pattern that is the result of strong bonds between the (VO_4) groups and the $UO_2(O_5)$ polyhedra. These layers, of composition $(UO_2)_2(VO_4)_2$, are held together by hydroxyl-hydrogen bonds to water molecules and also by K, Ca, and Ba between the layers.

Diagnostic Features. Carnotite is characterized by its yellow color, pulverulent nature, radioactivity, and occurrence. Unlike many secondary uranium minerals, carnotite will not fluoresce in ultraviolet light.

Occurrence. Carnotite is of secondary origin, and its formation is usually ascribed to the action of meteoric waters on preexisting uranium and vanadium minerals. It has a strong pigmenting power and when present in a sandstone in amounts even less than 1% will color the rock yellow. It is found principally in the plateau region of southwestern Colorado and in adjoining districts of Utah where it occurs disseminated in a cross-bedded sandstone. Concentrations of relatively pure carnotite are found around petrified tree trunks.

Use. Carnotite is an ore of vanadium and, in the United States, a principal ore of uranium.

Name. After Marie-Adolphe Carnot (1839–1920), French mining engineer and chemist.

Similar Species. *Tyuyamunite*, $Ca(UO_2)_2(VO_4)_2 \cdot 5-8\frac{1}{2}H_2O$, is the calcium analogue of carnotite and similar in physical properties except for a slightly more greenish color and yellow-green fluorescence. It is found in almost all carnotite deposits. Named from Tyuya Muyum, southeastern Turkistan, CIS, where it is mined as a uranium ore.

REFERENCES AND SUGGESTED READING

Aristarain, L. F. and Hurlbut, C. S., Jr., 1972, Boron minerals and deposits. *Mineralogical Record*, v. 3. pp. 165–172, 213–220.

Carbonates. Reviews in Mineralogy, 1983. v. 11. Mineralogical Society of America, Washington, D.C., 394 pp.

Deer, W. A., Howie, R. A. and Zussman, J., 1962, *Rock-Forming Minerals*, v. 5, Non-silicates. John Wiley & Sons, New York, 371 pp.

Morgan, V. and Erd, R. C., 1969, Minerals of the Kramer borate district, California. *Mineral Information Service* (California Division of Mines and Geology), v. 22, pp. 143–153, 165–172.

Palache, C., Berman, H. and Frondel, C., 1951, *The System of Mineralogy*, 7th ed., v. II. John Wiley & Sons, New York, 1124 pp.

Phosphate, Potash, and Sulfur—a special issue, 1979, *Economic Geology*, v. 74, pp. 191–496.

Strunz, H., 1970, *Mineralogische Tabellen*, 5th ed. Akademische Verlagsgesellschaft, Leipzig, 621 pp.

CHAPTER 13

SYSTEMATIC MINERALOGY
PART IV: SILICATES

The silicate mineral class is of greater importance than any other, for about 25% of the known minerals and nearly 40% of the common ones are silicates. With a few minor exceptions all the igneous rock-forming minerals are silicates, and they thus constitute well over 90% of the Earth's crust (see Fig. 13.1).

When the average weight percentages of the eight most common elements in the Earth's crust are recalculated on the basis of atomic percent (see Fig. 5.1), we find that out of every 100 atoms 62.5 are O, 21.2 are Si, and 6.5 are Al. Fe, Mg, Ca, Na, and K each account for about two to three more atoms. With the possible exception of Ti, all other elements are present in insignificant amounts in the upper levels of the Earth's crust (see Table 5.1). When we recalculate the atomic percentages of the eight most abundant elements in terms of volume percentages (see Fig. 5.1, last column) we find that the Earth's crust can be regarded as a packing of oxygen ions, with interstitial metal ions, such as Si^{4+}, Al^{3+}, Fe^{2+}, Ca^{2+}, Na^+, K^+, and so forth.

The dominant minerals of the crust are thus shown to be silicates, with oxides and other oxygen compounds such as carbonates in subordinate amounts. Of the different assemblages of silicate minerals that characterize igneous, sedimentary, and metamorphic rocks, ore veins, pegmatites, weathered rocks, and soils, each has the potential to tell us something of the environment in which it was formed.

We have a further deep and compelling reason to study the silicates. The soil from which our food is ultimately drawn is made up in large part of silicates. The brick, stone, concrete, and glass used in the construction of our buildings either are silicates or largely derived from silicates. Even with the coming of the space age, we need not fear obsolescence of our studies of the silicates, but rather an enlargement of their scope, because we now know that the moon and the four terrestrial planets of our solar system have rocky crusts made of silicates and oxides much like those of Earth.

The fundamental unit on which the structure of all silicates is based consists of four O^{2-} at the apices of a regular tetrahedron surrounding and coordinated by one Si^{4+} at the center (Figs. 12.1d and 13.2). The powerful bond that unites the oxygen and silicon ions is the cement that holds the Earth's crust together. This bond may be estimated by use of Pauling's electronegativity concept (see Fig. 4.35) as 50% ionic and 50% covalent. That is, although the bond arises in part from the attraction of oppositely charged ions, it also involves sharing of electrons and interpenetration of the electronic clouds of the ions involved. The bond is strongly localized in the vicinity of these shared electrons.

Although electron sharing is present in the Si-O bond, the total bonding energy of Si^{4+} is still distributed equally among its four closest oxygen neighbors.

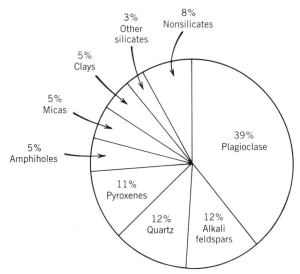

FIG. 13.1. Estimated volume percentages for the common minerals in the Earth's crust, inclusive of continental and oceanic crust. Ninety-two percent are silicates. (From Ronov, A. B. and Yaroshevsky, A. A., 1969, Chemical composition of the Earth's crust. American Geophysical Union Monograph no. 13, p. 50.)

Hence, the strength of any single Si-O bond is equal to just one-half the total bonding energy available in the oxygen ion. Each O^{2-} has, therefore, the potentiality of bonding to another silicon ion and entering into another tetrahedral grouping, thus uniting the tetrahedral groups through the shared (or *bridging*) oxygen. Such linking of tetrahedra is often referred to as *polymerization*, a term borrowed from organic chemistry, and the capacity for polymerization is the origin of the great variety of silicate structures. In no case, however, are three or even two oxygens shared between two adjacent tetrahedra in nature. Such sharing would place two highly charged Si^{4+} ions close together and the repulsion between them would render the structure unstable.

The sharing of oxygens may involve one, two, three, or all four of the oxygen ions in the tetrahedron, giving rise to a diversity of structural configurations. Figure 13.3 illustrates the various ways in which SiO_4

FIG. 13.2. Close packing representation of SiO_4 tetrahedron.

tetrahedra can be combined. Silicates with independent tetrahedral SiO_4 groups (in which the tetrahedra are not linked to each other) are known as *nesosilicates* (from the Greek *nesos*, meaning island) or *orthosilicates* (from the Greek *orthos*, meaning normal). Silicates in which two SiO_4 groups are linked, giving rise to Si_2O_7 groups, are classed as *sorosilicates* (from the Greek *soros*, meaning heap) or *disilicates* (in reference to the double tetrahedral groupings). If more than two tetrahedra are linked, closed ringlike structures are formed of a general composition Si_xO_{3x}. Fourfold tetrahedral rings have composition Si_4O_{12}. This group is known as the *ring silicates*, or the *cyclosilicates* (from the Greek *kyklos*, meaning circle). Tetrahedra may also be joined to form infinite single chains with a unit composition Si_2O_6 (or SiO_3). Infinite double chains give a ratio of Si : O = 4 : 11, resulting in Si_4O_{11} (or Si_8O_{22}). Both of these types of *chain silicates* are also known as *inosilicates* (from the Greek *inos*, meaning thread). When three of the oxygens of a tetrahedron are shared between adjoining tetrahedra, infinitely extending flat sheets are formed of unit composition Si_2O_5. Such *sheet silicates* are also referred to as *phyllosilicates* (from the Greek *phyllon*, meaning leaf). When all four oxygens of a SiO_4 tetrahedron are shared by adjoining tetrahedra, a three-dimensional network of unit composition SiO_2 results. These *framework silicates* are also known as *tectosilicates* (from the Greek word *tecton*, meaning builder).

In the subsequent treatment of silicates in this book, we will use the above structural classification (as illustrated in Fig. 13.3) of silicates. There are, however, alternate classifications such as those proposed by Liebau (1985) and Zoltai (1960) (see references at the end of this chapter).

Next to O and Si, the most important constituent of the crust is Al. Al^{3+} has a radius of 0.39 Å and thus the radius ratio Al : O = 0.286, corresponding to 4-coordination with oxygen. However, the radius ratio is sufficiently close to the upper limit for 4-coordination so that 6-coordination is also possible. It is this capacity for playing a double role in silicate minerals that gives Al^{3+} its outstanding significance in the crystal chemistry of the silicates. When Al coordinates four O's arranged at the apices of a regular tetrahedron, the resultant grouping occupies approximately the same space as a silicon-oxygen tetrahedron and may link with silicon tetrahedra in polymerized groupings. On the other hand, Al^{3+} in 6-coordination may serve to link the tetrahedral groupings through simple ionic bonds, weaker than those that unite the ions in the tetrahedra. It is thus possible to have Al in silicate structures both in the tetrahedral sites, substituting for Si, and in the octa-

Class	Arrangement of SiO$_4$ tetrahedra (central Si^{4+} not shown)	Unit composition	Mineral example
Nesosilicates		$(SiO_4)^{4-}$	Olivine, $(Mg, Fe)_2 SiO_4$
Sorosilicates		$(Si_2O_7)^{6-}$	Hemimorphite, $Zn_4Si_2O_7(OH) \cdot H_2O$
Cyclosilicates		$(Si_6O_{18})^{12-}$	Beryl, $Be_3Al_2Si_6O_{18}$
Inosilicates (single chain)		$(Si_2O_6)^{4-}$	Pyroxene e.g. Enstatite, $MgSiO_3$

FIG. 13.3. Silicate classification.

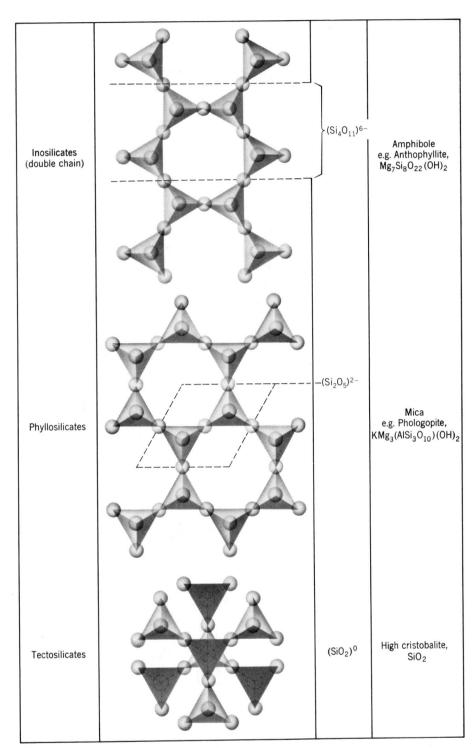

Inosilicates (double chain)

$(Si_4O_{11})^{6-}$

Amphibole e.g. Anthophyllite, $Mg_7Si_8O_{22}(OH)_2$

Phyllosilicates

$-(Si_2O_5)^{2-}$

Mica e.g. Phologopite, $KMg_3(AlSi_3O_{10})(OH)_2$

Tectosilicates

$(SiO_2)^0$

High cristobalite, SiO_2

FIG. 13.3. *(continued)*

hedral sites with 6-coordination, involved in solid solution relations with elements such as Mg and Fe^{2+}.

Mg, Fe^{2+}, Fe^{3+}, Mn^{2+}, Al^{3+}, and Ti^{4+} all tend to occur in silicate structures in 6-coordination with oxygen (see Table 13.1). Although divalent, trivalent, and tetravalant ions are included here, all have about the same space requirements and about the same radius ratio relations with oxygen, and, hence, tend to occupy the same type of atomic site. Solid solution relations between ions of different charge are possible through the mechanism of coupled substitution (see page 233).

The larger and more weakly charged cations, Ca^{2+} and Na^+, of ionic radii 1.12 Å and 1.16 Å, respectively, generally enter sites having 8-coordination with respect to oxygen. Although the sizes of these two ions are very similar, their charges are different. This charge difference is compensated for by coupled substitution such as in the plagioclase feldspars where $Na^+ + Si^{4+}$ substitute for $Ca^{2+} + Al^{3+}$ in order to keep the overall structure neutral.

The largest ions common in silicate structures are those of K, Rb, Ba, and the rarer alkalis and alkali earths. These ions generally do not enter readily into Na or Ca sites and are found in high-coordination-number sites of unique type. Hence, solid solution between these ions and the other common ions is limited.

Ionic substitution is generally common and extensive between elements that are grouped together in Table 13.1. The usual role played by the commonest elements permits us to write a general formula for all silicates:

$$X_m Y_n (Z_p O_q) W_r,$$

where X represents large, weakly charged cations in 8- or higher coordination with oxygen; Y represents medium-sized, two to four valent ions in 6-coordination; Z represents small, highly charged ions in tetrahedral coordination; O is oxygen; and W represents additional anionic groups such as $(OH)^-$ or anions such as Cl^- or F^-. The ratio $p : q$ depends on the degree of polymerization of the silicate framework, and the other subscript variables, m, n, and r, depend on the need for electrical neutrality. Any common silicate may be expressed by suitable substitution in this general formula.

The subsequent treatment of individual silicates will be on the basis of subclasses that reflect their internal structure (neso-, soro-, cyclosilicates, etc.) and their chemical composition. Such a scheme is the basis of *Mineralogische Tabellen*, 1970, 5th ed., by Hugo Strunz.

Table 13.1

COORDINATION OF COMMON ELEMENTS IN SILICATES, ARRANGED IN DECREASING IONIC SIZE*

	Ion	Coordination Number with Oxygen	Ionic Radius Å
X	K^+	8–12	1.51 (8)–1.64(12)
X	Na^+	8–6 ⎫ cubic to octahedral	1.18 (8)–1.02 (6)
	Ca^{2+}	8–6 ⎭	1.12 (8)–1.00 (6)
Y	Mn^{2+}	6 ⎫	0.83 (6)
	Fe^{2+}	6 ⎪	0.78 (6)
	Mg^{2+}	6 ⎬ octahedral	0.72 (6)
	Fe^{3+}	6 ⎪	0.65 (6)
	Ti^{4+}	6 ⎪	0.61 (6)
	Al^{3+}	6 ⎭	0.54 (6)
Z	Al^{3+}	4 ⎫ tetrahedral	0.39 (4)
	Si^{4+}	4 ⎭	0.26 (4)

*See Table 4.8 for a listing of ionic radii.

NESOSILICATES

In the nesosilicates the SiO_4 tetrahedra are isolated (Fig. 13.3) and bound to each other only by ionic bonds from interstitial cations. Their structures depend chiefly on the size and charge of the interstitial cations. The atomic packing of the nesosilicate structures is generally dense, causing the minerals of this group to have relatively high specific gravity and hardness. Because the SiO_4 tetrahedra are independent and not linked into chains or sheets, for example, the crystal habit of the nesosilicates is generally equidimensional and pronounced cleavage directions are absent. Although Al^{3+} substitutes commonly and easily in the Si position of silicates, the amount of Al substitution in SiO_4 tetrahedra in nesosilicates is generally low.

Very common members, especially in high-temperature igneous rocks, of the nesosilicate group are *forsterite*, Mg_2SiO_4, and *fayalite*, Fe_2SiO_4, end members of the $(Mg,Fe)_2SiO_4$ olivine series. The structure of olivine, which is shown in Fig. 13.4, can be viewed as consisting of layers parallel to {100}. These layers consist of octahedra cross-linked by independent SiO_4 tetrahedra. The octahedrally coordinated sites are known as *M1* and *M2* with *M1* distorted and *M2* somewhat more regular. In the $(Mg,Fe)_2SiO_4$ olivines Mg and Fe^{2+} occupy the *M1* and *M2* sites without any specific preference for either site. In the calcic olivines, however (e.g., *monticellite*, $CaMgSiO_4$), Ca enters into the *M2* site and Mg into *M1*.

Garnets are another group of very common nesosilicate minerals, especially in metamorphic rocks.

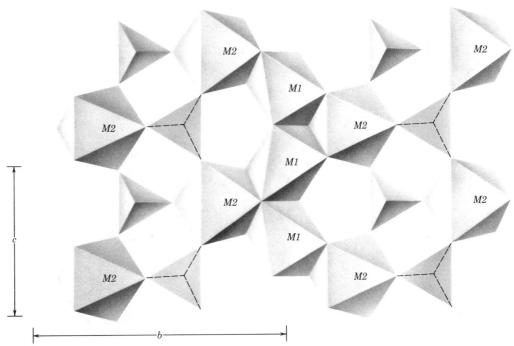

FIG. 13.4. Portion of the structure of olivine projected on (100). *M1* and *M2* are octahedral sites. The *M1* site is most distorted and the *M2* site is somewhat more regular. Extensive edge sharing among the polyhedra causes these distortions because shared edges are shortened as cations repel each other; see also page 197. (Redrawn after Papike, J. J. and Cameron, M., 1976, Crystal chemistry of silicate minerals of geophysical interest. *Reviews of Geophysics and Space Physics*, v. 14, pp. 37–80.)

Their structural formula may be represented as $A_3B_2(SiO_4)_3$, where A and B refer respectively to 8- and 6-coordinated cationic sites. The A sites are occupied by rather large divalent cations, whereas the B sites house smaller trivalent cations. Because of these size considerations in the filling of the A sites we may expect a fairly well-defined division of the garnets into those with Ca and those with easily interchangeable divalent ions such as Mg^{2+}, Fe^{2+}, Mn^{2+}. Similarly, because of the limited substitution possible in the B sites, we may expect a separation of garnets into Al^{3+}, Fe^{3+}, and Cr^{3+} bearing. These two trends are well marked and have given rise to a grouping of garnets into two series: *pyralspite* (Ca absent in A; B = Al) and *ugrandite* (A = Ca).

Pyralspite		Ugrandite	
Pyrope	$Mg_3Al_2Si_3O_{12}$	*Uvarovite*	$Ca_3Cr_2Si_3O_{12}$
Almandine	$Fe_3Al_2Si_3O_{12}$	*Grossular*	$Ca_3Al_2Si_3O_{12}$
Spessartine	$Mn_3Al_2Si_3O_{12}$	*Andradite*	$Ca_3Fe_2^{3+}Si_3O_{12}$

This classification serves as an excellent mnemonic aid for the names and formulas. Another grouping on the basis of the ions in the B site yields three unequal groups:

Aluminum Garnets	Ferri-Garnet	Chrome-Garnet
Pyrope	Andradite	Uvarovite
Almandine		
Spessartine		
Grossular		

Hydroxyl, as tetrahedral $(OH)_4$ groups, may substitute to a limited extent for SiO_4 tetrahedra in hydrogarnets such as *hydrogrossular*, $Ca_3Al_2Si_2O_8$-$(SiO_4)_{1-m}(OH)_{4m}$, with m ranging from 0 to 1. Ti^{4+} may enter into the B sites concomitant with replacement of Ca by Na in the A sites, producing the black *melanite*.

The structure of garnet, which is illustrated in Fig. 13.5, consists of alternating SiO_4 tetrahedra and BO_6 octahedra that share corners to form a continuous three-dimensional network. The A sites are surrounded by 8 oxygens in irregular coordination polyhedra.

The aluminosilicates of the nesosilicate group, kyanite, sillimanite, and andalusite, are commonly found in medium- to high-grade metamorphic rocks of Al-rich bulk composition. All three minerals are polymorphs of Al_2SiO_5, which may be stated structurally as $Al^{[4-6]}Al^{[6]}SiO_5$ (digits in square brackets

FIG. 13.5. Portion of the garnet structure projected on (001). Tetrahedra, octahedra and 8 coordination polyhedra (triangular dodecahedra, drawn as distorted cubes) are shown. (After Novak G. A. and Gibbs, G. V., 1971. The crystal chemistry of the silicate garnets, *American Mineralogist*, v. 56, pp. 791–825.)

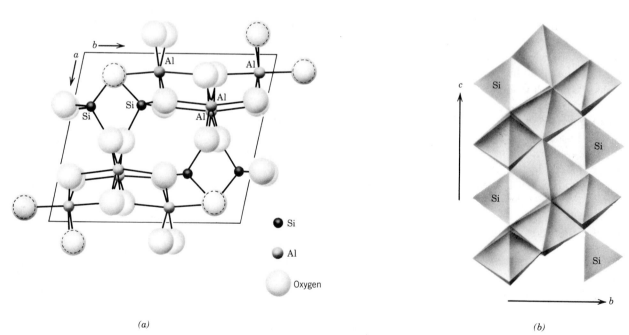

(a)   *(b)*

FIG. 13.6. The structure of kyanite. *(a)* Projected on (001) and *(b)* showing zigzag chains of edge-sharing octahedra parallel to the *c* axis. (Redrawn after Burnham, C. W., 1963, Refinement of the crystal structure of kyanite, *Zeitschrift für Kristallographie*, v. 118, pp. 337–360; and Winter, J. K. and Ghose, S., 1979, Thermal expansion and high-temperature crystal chemistry of the Al_2SiO_5 poly-morphs, *American Mineralogist*, v. 64, pp. 573–586; see also *Orthosilicates, Reviews in Mineralogy*, 1980, Mineralogical Society of America, Washington, D.C.)

indicate coordination number). Chains of edge-sharing octahedra, parallel to the c axis, are characteristic of all three structures. In *kyanite*, $Al^{[6]}Al^{[6]}SiO_5$, the triclinic polymorph, with space group $P\bar{1}$, all of the Al is octahedrally coordinated. It occurs as octahedral chains parallel to c, and as isolated Al octahedra (see Fig. 13.6). In *sillimanite*, $Al^{[4]}Al^{[6]}SiO_5$, an orthorhombic polymorph, with space group *Pbnm*, the octahedrally coordinated Al is found in octahedral chains (see Fig. 13.7a) and adjacent tetrahedral chains consist of alternating tetrahedral AlO_4 and SiO_4 groups. In *andalusite*, $Al^{[5]}Al^{[6]}SiO_5$, another orthorhombic polymorph, with space group *Pbnm*, half the Al is found in octahedral chains and the other half occurs in 5-coordinated polyhedra (see Fig. 13.7b) which are linked by SiO_4 tetrahedra.

The following minerals in the nesosilicate group will be discussed in detail:

NESOSILICATES

Phenacite Group

Phenacite	Be_2SiO_4
Willemite	Zn_2SiO_4

Olivine Group

Forsterite	Mg_2SiO_4
Fayalite	Fe_2SiO_4

Garnet Group $A_3B_2(SiO_4)_3$

Pyrope	Mg_3Al_2	Uvarovite	Ca_3Cr_2
Almandine	Fe_3Al_2	Grossular	Ca_3Al_2
Spessartine	Mn_3Al_2	Andradite	$Ca_3Fe_2^{3+}$

Zircon Group

Zircon	$ZrSiO_4$

Al_2SiO_5 Group

Andalusite ⎫	
Sillimanite ⎬	Al_2SiO_5
Kyanite ⎭	
Topaz	$Al_2SiO_4(F,OH)_2$
Staurolite	$FeAl_9O_6(SiO_4)_4(O,OH)_2$

Humite Group

Chondrodite	$Mg_5(SiO_4)_2(OH,F)_2$
Datolite	$CaB(SiO_4)(OH)$
Titanite	$CaTiO(SiO_4)$
Chloritoid	$(Fe,Mg)_2Al_4O_2(SiO_4)_2(OH)_4$

Phenacite Group

Phenacite—Be_2SiO_4

Crystallography. Hexagonal; $\bar{3}$. Crystals usually flat rhombohedral or short prismatic. Often with complex development. Frequently twinned on $\{10\bar{1}0\}$.

$R\bar{3}$; $a = 12.45$, $c = 8.23$ Å; $Z = 18$. d's: 3.58(6), 2.51(8), 2.35(6), 2.18(8), 1.258(10).

Physical Properties. *Cleavage* $\{11\bar{2}0\}$ imperfect. **H** $7\frac{1}{2}$–8. **G** 2.97–3.00. *Luster* vitreous. *Color* colorless, white. Transparent to translucent. *Optics:* (+); $\omega = 1.654$, $\epsilon = 1.670$.

Composition and Structure. BeO 45.6, SiO_2 54.4%. The structure consists of SiO_4 and BeO_4 tetrahedra with each oxygen linked to two Be and one Si at the corners of an equilateral triangle.

Diagnostic Features. Characterized by its crystal form and great hardness.

Occurrence. Phenacite is a rare pegmatite mineral associated with topaz, chrysoberyl, beryl, and apatite. Fine crystals are found at the emerald mines in the Ural Mountains, Commonwealth of Independent States (CIS), and in Minas Gerais, Brazil. In the United States found at Mount Antero, Colorado.

Use. Occasionally cut as a gemstone.

Name. From the Greek *phenakos* meaning a *deceiver*, in allusion to its resemblance to quartz.

WILLEMITE—Zn_2SiO_4

Crystallography. Hexagonal; $\bar{3}$. In hexagonal prisms with rhombohedral terminations. Usually massive to granular. Rarely in crystals.

$R\bar{3}$; $a = 13.96$, $c = 9.34$ Å; $Z = 18$. d's: 2.84(8), 2.63(9), 2.32(8), 1.849(8), 1.423(10).

Physical Properties. *Cleavage* $\{0001\}$ good. **H** $5\frac{1}{2}$. **G** 3.9–4.2. *Luster* vitreous to resinous. *Color* yellow-green, flesh-red, and brown; white when pure. Transparent to translucent. Most willemite from Franklin, New Jersey, fluoresces under ultraviolet light. *Optics:* (+); $\omega = 1.691$, $\epsilon = 1.719$.

Composition and Structure. ZnO 73.0, SiO_2 27.0%. Mn^{2+} often replaces a considerable part of the Zn (manganiferous variety called *troostite*); Fe^{2+} may also be present in small amount. Willemite is isostructural with phenacite (see above), with SiO_4 and ZnO_4 tetrahedra. Because Zn^{2+} (radius = 0.60 Å) is much larger than Be^{2+} (radius = 0.27 Å) the structure of willemite is much expanded over that of phenacite.

Diagnostic Features. Willemite from Franklin, New Jersey, can usually be recognized by its association with franklinite and zincite.

Occurrence. Willemite is found in crystalline limestone and may be the result of metamorphism of earlier hemimorphite or smithsonite. It is also found sparingly as a secondary mineral in the oxidized zone of zinc deposits.

Found at Altenberg, near Moresnet, Belgium; Tsumeb, Namibia. The most important locality is in the United States at Franklin, New Jersey, where willemite occurs associated with franklinite and zincite

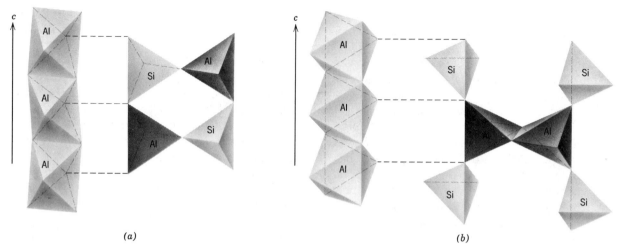

FIG. 13.7. *(a)* Projection of the sillimanite structure showing octahedral chains parallel to the *c* axis. (Redrawn after Burnham, C. W., 1963, Refinement of the crystal structure of sillimanite, *Zeitschrift für Kristallographie*, pp. 127–148.) *(b)* Projection of the andalusite structure showing octahedral chains parallel to the *c* axis, and the presence of AlO_5 polyhedra between SiO_4 tetrahedra. (Redrawn after Burnham, C. W. and Buerger, M. J., 1961, Refinement of the crystal structure of andalusite, *Zeitschrift für Kristallographie*, pp. 269–290.)

and as grains imbedded in calcite. It has also been found at the Merritt mine, New Mexico, and Tiger, Arizona.

Use. A valuable zinc ore at Franklin, New Jersey. Occasionally used as a gem.

Name. In honor of the King of the Netherlands, Willem I (1553–1584).

Olivine Group

The composition of the majority of olivines can be represented in the system $CaO-MgO-FeO-SiO_2$ (Fig. 13.8). The most common series in this system is from *forsterite*, Mg_2SiO_4, to *fayalite*, Fe_2SiO_4. Relatively rare olivines occur also along the *monticellite*, $CaMgSiO_4$, to *kirschteinite*, $CaFe^{2+}SiO_4$, join. Very little, if any, solid solution exists between these two series. Mn^{2+} may substitute for Fe^{2+}, forming a relatively rare series between fayalite and *tephroite*, Mn_2SiO_4. The structure of olivine, with fairly similar *M1* and *M2* octahedral sites, and independent SiO_4 tetrahedra, is shown in Fig. 13.4.

Members of the forsterite-fayalite series are common as primary crystallization products in Fe- and Mg-rich, silica-poor melts (magmas). *Dunites* and *peridotites* are pure olivine and olivine plus pyroxene rocks, respectively. Olivine concentrations in igneous rocks may result from the accumulation of olivine crystals, under the influence of gravity, during the cooling stages of a magma. Members of the forsterite-fayalite series are highly refractory, as can be seen from Fig. 13.9 (forsterite melting point = 1890°C and

fayalite melting point = 1205°C). This diagram represents a complete solid solution series without a maximum or minimum (see also Fig. 9.5 and related discussion). When a melt with composition x (50 weight percent Fe_2SiO_4) is cooled to the liquidus curve, olivine crystals of composition x_1 will start to form. The liquid, as a result of the Mg-rich olivine crystallization, will become more Fe-rich, as shown by the upper arrow. This, in turn, will cause a more Fe-rich olivine to crystallize, as shown by x_2, until finally all liquid is used up. At this point, under equilibrium conditions, the final composition to crystallize is of composition

FIG. 13.8. Olivine compositions in the system Ca_2SiO_4-Mg_2SiO_4-Fe_2SiO_4.

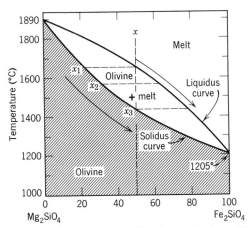

FIG. 13.9. Temperature-composition diagram for the system Mg_2SiO_4-Fe_2SiO_4 at atmospheric pressure (see also Fig. 9.5 and related discussion).

x_3, which is the same as the original x. Chemical zoning in olivines with Mg-rich cores and more Fe-rich rims may be found in high-temperature igneous rocks. Such rimming is the result of nonequilibrium crystallization in the direction of the arrows in Fig. 13.9.

OLIVINE—$(Mg,Fe)_2SiO_4$

Crystallography. Orthorhombic; $2/m2/m2/m$. Crystals are usually a combination of the three prisms, the three pinacoids, and a dipyramid. Often flattened parallel to either {100} or {010} (see Fig. 13.10). Usually appears as embedded grains or in granular masses.

Pbnm; $Z = 4$. Mg_2SiO_4: $a = 4.75$, $b = 10.20$, $c = 5.98$ Å. Fe_2SiO_4: $a = 4.82$, $b = 10.48$, $c = 6.09$ Å. d's for common olivine: 4.29(10), 2.41(8), 2.24(7), 1.734(8), 1.498(7). d_{130} can be used to determine compositions in this series (see Fig. 13.11a).

Physical Properties. *Fracture* conchoidal. **H** $6\frac{1}{2}$–7. **G** 3.27–4.37, increasing with increase in Fe content (see Fig. 13.11b). *Luster* vitreous. *Color* pale yellow-green to olive-green in forsterite; darker, brownish-green with increasing Fe^{2+}. Transparent to translucent. *Optics*: forsterite (+), others (−). For $Mg_{1.6}Fe_{0.4}SiO_4$: $\alpha = 1.674$, $\beta = 1.692$, $\gamma = 1.712$, $2V = 87°$; $X = b$, $Z = a$ (see Fig. 13.11b).

FIG. 13.10. Olivine.

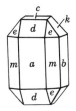

Composition and Structure. A complete solid solution exists from *forsterite*, Mg_2SiO_4, to *fayalite*, Fe_2SiO_4 (see Fig. 13.8). The more common olivines are richer in Mg than in Fe^{2+}. An example of the recalculation of an olivine analysis is given in Table 5.4. Compositions intermediate to the end members, forsterite (Fo) and fayalite (Fa), are commonly expressed as Fo_xFa_y, for example, $Fo_{60}Fa_{40}$, which is shortened to Fo_{60}. The structure of olivine is given in Fig. 13.4 and discussed on page 444. An HRTEM structure image is given in Fig. 13.12. Under very high pressures the olivine structure transforms to a spinel structure with Si in tetrahedral sites and Mg and Fe^{2+} in the octahedral sites of the spinel structure (see Fig. 13.13). The spinel structure is more dense than the olivine structure. Olivine is thought to be abundant in the mantle and occurs as a spinel-type structure at such great depths (see Fig. 4.51 and related text).

Diagnostic Features. Distinguished usually by its glassy luster, conchoidal fracture, green color, and granular nature.

Occurrence. Olivine is a rather common rock-forming mineral, varying in amount from that of an accessory to that of a main constituent. It is found principally in the dark-colored igneous rocks such as gabbro, peridotite, and basalt (see Table 14.4). In these rock types it coexists with plagioclase and pyroxenes. The rock, *dunite*, is made up almost wholly of olivine. Forsterite is not stable in the presence of free SiO_2 and will react with it to form pyroxene, according to:

$$Mg_2SiO_4 + SiO_2 \rightarrow 2MgSiO_3$$
forsterite   quartz   enstatite

Found as glassy grains in stony and stony-iron meteorites. Occasionally in crystalline dolomitic limestones where it formed by the reaction:

$$2CaMg(CO_3)_2 + SiO_2 \rightarrow$$
dolomite   quartz
$$Mg_2SiO_4 + 2CaCO_3 + 2CO_2$$
forsterite   calcite

Associated often with pyroxene, plagioclase, magnetite, corundum, chromite, and serpentine.

The transparent gem variety is known as *peridot* (see Plate III, no. 1, Chapter 15). It was used as a gem in ancient times in the East, but the exact locality for the stones is not known. At present peridot is found in Burma, and in rounded grains associated with pyrope garnet in the surface gravels of Arizona and New Mexico, but the best quality material comes from Zebirget, an island in the Red Sea. Crystals of olivine are

FIG. 13.11. *(a)* Relationship of the interplanar spacing of 130 ($= d_{130}$) and composition for the forsterite-fayalite series. (From Deer, W. A., Howie, R. A. and Zussman, J., 1982, *Orthosilicates*, vol. 1A, John Wiley and Sons, and Longman Group Ltd., London, England.) *(b)* Variations of refractive indices and specific gravity with composition.

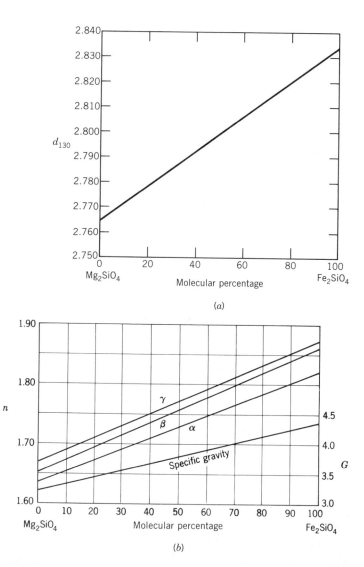

(a)

(b)

FIG. 13.12. High-resolution transmission electron microscope (HRTEM) image of an *a–c* section of olivine, of composition $(MG_{1.76} Fe_{0.24})SiO_4$. A unit cell is outlined. This image shows the highly regular and homogeneous nature of the structure of this olivine crystal. Compare with the radiation-damaged olivine in Fig. 3.48 and the olivine to clay alteration in Fig. 13.98 (Courtesy of L. M. Wang, University of New Mexico).

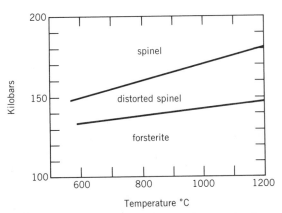

Kilobars

Temperature °C

FIG. 13.13. The stability fields of various polymorphs of Mg$_2$SiO$_4$: forsterite, distorted spinel, and spinel. (From Suito, K., 1972, Phase transformations of pure Mg$_2$SiO$_4$ into a spinel structure under high pressures and temperatures. *Journal of Physics of the Earth*, v. 20, pp. 225–243.)

found in the lavas of Vesuvius. Larger crystals, altered to serpentine, come from Snarum, Norway. Olivine occurs in granular masses in volcanic bombs in the Eifel district, Germany, and in Arizona. Dunite rocks are found at Dun Mountain, New Zealand, and with the corundum deposits of North Carolina.

Alteration. Very readily altered to *serpentine* minerals such as *antigorite*, Mg$_6$Si$_4$O$_{10}$(OH)$_8$. Magnesite, MgCO$_3$ and iron oxides form at the same time as a result of the alteration. Olivines in metamorphosed basic igneous rocks commonly show *coronas*, which are concentric rims consisting of pyroxene and amphibole. Such coronas are the result of the instability of the high temperature olivine in a lower temperature, H$_2$O-containing environment.

Use. As the clear green variety, *peridot*, it is used as a gem. Olivine is mined as refractory sand for the casting industry and for the manufacture of refractory bricks.

Name. Olivine derives its name from the usual olive-green color. *Chrysolite* is a synonym for olivine. Peridot is an old name for the species.

Similar Species. Other rarer members of the olivine group are: *monticellite*, CaMgSiO$_4$, a high-temperature contact metamorphic mineral in siliceous dolomitic limestones; and *tephroite*, Mn$_2$SiO$_4$.

Garnet Group

The garnet group includes a series of isostructural species with space group *Ia3d*; they crystallize in the hexoctahedral class of the isometric system and are similar in crystal habit. The structural arrangement (see Fig. 13.5) is such that the atomic population of the {100} and {111} families of planes is much depleted. As a result the cube and octahedon, common on most isometric hexoctahedral crystals, are rarely found on garnets.

Crystallography. Isometric; $4/m\overline{3}2/m$. Common forms are dodecahedron *d* (Fig. 13.14) and trapezohedron, *n* (Fig. 13.14*b*), often in combination (Figs. 13.14*c*, *d*, and 13.15). Hexoctahedrons are observed occasionally (Fig. 13.14*e*). Other forms are rare. Usually distinctly crystallized; also appears in rounded grains; massive granular, coarse or fine.

Ia3d; cell edge (see Table 13.2); $Z = 8$. *d*'s for pyrope: 2.89(8), 2.58(9), 1.598(9), 1.542(10), 1.070(8).

FIG. 13.14. Garnet crystals.

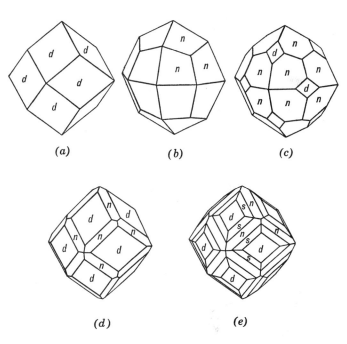

(a)   (b)   (c)

(d)   (e)

FIG. 13.15. Garnet crystals in chlorite schist.

Physical Properties. H $6\frac{1}{2}$–$7\frac{1}{2}$. G 3.5–4.3, varying with composition (see Table 13.2). *Luster* vitreous to resinous. *Color* varying with composition (see below); most commonly red, also brown, yellow, white, green, black. *Streak* white. Transparent to translucent.

Composition and Structure. Garnet compositions can be expressed by the general structural formula $A_3B_2(SiO_4)_3$ where the A site houses Ca, Mg, Fe^{2+} or Mn^{2+} and the B site incorporates Al, Fe^{3+}, and Cr^{3+} (see Fig. 13.5 and page 445). The formulas of the main species are given in Table 13.2, with the refractive indices, specific gravity, and unit cell edges for the end member compositions. There is extensive substitution among the *pyralspite* group and also among the *ugrandite* group but relatively little solid solution between these two major categories (see Fig. 13.16). Rarely are the pure end members found. Hydrous garnets such as *hydrogrossular*, may contain up to 8.5% H_2O. This water, in the form of $(OH)_4^{4-}$ groups, substitutes probably for (SiO_4) tetrahedra in the structure, assuming the replacement $Si^{4+} \rightleftarrows 4H^+$.

Experimental studies at high pressures and temperatures show that the garnet structure is stable in the conditions of the Earth's mantle. It appears that pyrope type compositions are most likely in the P and T regime of the Earth's mantle.

Pyrope, Mg_3Al_2. Some Ca and Fe^{2+} usually present. Color deep red to nearly black. Often transparent and then used as a gem. Name derived from Greek meaning *firelike*. *Rhodolite* is the name given to a pale rose-red or purple garnet, corresponding in composition to two parts of pyrope and one almandine.

Almandine, Fe_3Al_2. Fe^{3+} may replace Al, and Mg may replace Fe^{2+}. Color fine deep red, transparent in precious garnet; brownish-red, translucent in common garnet. Name derived from Alabanda in Asia Minor, where in ancient times garnets were cut and polished.

Spessartine, Mn_3Al_2. Fe^{2+} usually replaces some Mn^{2+} and Fe^{3+} some Al. Color brownish to red. Name derived from the locality Spessart, Germany.

Grossular, Ca_3Al_3 (*essonite, cinnamon stone*).

					Unit Cell
Species	Composition	n	G		Edge (Å)
Pyrope	$Mg_3Al_2Si_3O_{12}$	1.714	3.58		11.46
Almandine	$Fe_3Al_2Si_3O_{12}$	1.830	4.32		11.53
Spessartine	$Mn_3Al_2Si_3O_{12}$	1.800	4.19		11.62
Grossular	$Ca_3Al_2Si_3O_{12}$	1.734	3.59		11.85
Andradite	$Ca_3Fe_2Si_3O_{12}$	1.887	3.86		12.05
Uvarovite	$Ca_3Cr_2Si_3O_{12}$	1.868	3.90		12.00
Hydrogrossular	$Ca_3Al_2Si_2O_8(SiO_4)_{1-m}$ $(OH)_{4m}; m = 0$–1	1.734 to 1.675	3.59 to 3.13		11.85 to 12.16

Table 13.2
CHEMICAL COMPOSITIONS AND PHYSICAL PROPERTIES OF GARNETS

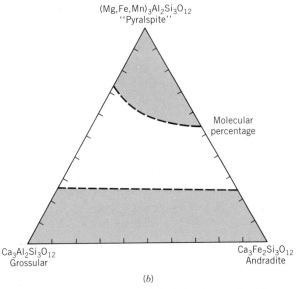

FIG. 13.16. *(a)* The extent of solid solution in garnets of the *pyralspite* group: pyrope-almandine-spessartine. *(b)* There is only limited solid solution between members of the *pyralspite* and *ugrandite* groups; ugrandite represents uvavorite-grossular-andradite.

Often contains Fe^{2+} replacing Ca, and Fe^{3+} replacing Al. Color white, green, yellow, cinnamon-brown, pale red. Name derived from the botanical name for gooseberry, in allusion to the light green color of the original grossular.

Andradite, $Ca_3Fe_2^{3+}$. Common garnet in part. Al may replace Fe^{3+}; Fe^{2+}, Mn^{2+}, and Mg may replace Ca. Color various shades of yellow, green, brown to black. *Demantoid* is a green variety with a brilliant luster, used as a gem. Named after the Portuguese mineralogist, d'Andrada.

Uvarovite, $Ca_3Cr_2^{3+}$. Color emerald-green. Named after Count Uvarov.

Diagnostic Features. Garnets are usually recognized by their characteristic isometric crystals, their hardness, and their color. Specific gravity, refractive index, and unit cell dimension taken together serve to distinguish members of the group.

Occurrence. Garnet is a common and widely distributed mineral, occurring abundantly in some metamorphic rocks (see Fig. 13.15) and as an accessory constituent in some igneous rocks. Its most characteristic occurrence is in mica schists, hornblende schists, and gneisses. It is frequently used as an index mineral in the delineation of isograds in metamorphic rocks (see Chapter 14). Found in pegmatite dikes, more rarely in granitic rocks. *Pyrope* occurs in ultrabasic rocks such a peridotites or kimberlites and in serpentines derived from them. The garnets, which in eclogites coexist with pyroxenes and kyanite, vary in composition from pyrope to almandine. *Almandine* is the common garnet in metamorphic rocks, resulting from the regional metamorphism of argillaceous sediments. It is also a widespread detrital garnet in sedimentary rocks. *Spessartine* occurs in skarn deposits and in Mn-rich assemblages containing rhodonite, Mn-oxides, and so forth. *Grossular* is found chiefly as a product of contact or regional metamorphism of impure limestones. *Andradite* occurs in geological environments similar to that of grossularite. It may be the result of metamorphism of impure siliceous limestone, by the reaction:

$$3CaCO_3 + 3SiO_2 + Fe_2O_3 \rightarrow$$
calcite   quartz   hematite

$$Ca_3Fe_2Si_3O_{12} + 3\ CO_2'$$
andradite

Melanite, a black variety of andradite, occurs in alkaline igneous rocks. *Uvarovite* is the rarest of this group of garnets and is found in serpentine associated with chromite. The best known locality is Outokumpu, Finland.

Pyrope of gem quality is found associated with clear grains of olivine (peridot) in the surface sands near Fort Defiance, close to the New Mexico–Arizona line. A locality near Meronitz, Bohemia, Czechoslovakia, is famous for pyrope gems. Almandine, of gem quality, is found in northern India, Sri Lanka, and Brazil. Fine crystals, although for the most part too opaque for cutting, are found in a mica schist on the Stikine River, Alaska. Grossular is used only a little in jewelry, but essonite or cinnamon stones of good size and color are found in Sri Lanka.

FIG. 13.17. Zircon crystals.

Alteration. Garnet often alters to other minerals, particularly talc, serpentine, and chlorite.

Use. All species, except uvarovite, are cut as gemstones (see Plate III, no. 2, Chapter 15). The most valued is a green andradite, known as *demantoid*, which comes from the Ural Mountains, CIS. At Gore Mountain, New York, large crystals of almandine in an amphibolite are mined. The unusual angular fractures and high hardness of the garnets make them desirable for a variety of abrasive purposes including garnet paper.

Name. *Garnet* is derived from the Latin *granatus*, meaning like a grain.

ZIRCON—ZrSiO₄

Crystallography. Tetragonal; $4/m2/m2/m$. Crystals usually show a simple combination of $a\{010\}$ and $e\{011\}$, but $m\{110\}$ and a ditegragonal dipyramid also observed (Fig. 13.17). Usually in crystals; also in irregular grains.

$I4/amd$; $a = 6.60$, c 5.98 Å; $Z = 4$. d's: 4.41(7), 3.29(10), 2.52(8), 1.710(9).

Physical Properties. *Cleavage* $\{010\}$ poor. **H** $7\frac{1}{2}$. **G** 4.68. *Luster* adamantine. *Color* commonly some shade of brown; also colorless, gray, green, red. *Streak* uncolored. Usually translucent; in some cases transparent. *Optics:* (+), $\omega = 1.923$–1.960, $\epsilon = 1.968$–2.015. Metamict zircon, $n = 1.78$.

Composition and Structure. For ZrSiO₄, ZrO₂ 67.2, SiO₂ 32.8%. Zircon always contains some hafnium. Although the amount is usually small (1 to 4%), analyses have reported up to 24% HfO₂. The structure of zircon is shown in Fig. 13.18. Zr is in 8-coordination with oxygen in the form of distorted cubelike polyhedra. The eight oxygens surrounding each Zr belong to six different SiO₄ tetrahedra. Although the structure of zircon is resistant to normal chemical attack, it is often in a metamict state (see page 159). This is caused by the structural damage

from Th and U, which are present in small amounts in many zircons. The "self-irradiation" from the decay of U and Th produces as a final result an isotropic glass with a reduction in density of about 16% and a lowering of refractive index (see Figs. 3.47 and 3.48).

Diagnostic Features. Usually recognized by its characteristic crystals, color, luster, hardness, and high specific gravity.

Occurrence. Zircon is a common and widely distributed accessory mineral in all types of igneous rocks. It is especially common in the more silicic types such as granite, granodiorite, syenite, and monzonite, and very common in nepheline syenite. Found also in crystalline limestone, gneisses, and schists. Because zircon is a stable chemical compound, it is a common accessory mineral in many sediments. Its characteristics in the heavy mineral fraction of sandstones are often useful in the evaluation of provenance. Zircon is frequently found as rounded grains in stream and beach sands, often with gold. Zircon has been produced from beach sands in Australia, Brazil, and Florida.

Gem zircons are found in the stream sands at Matura, Sri Lanka, and in the gold gravels in the Ural Mountains, CIS, and Australia. In large crystals from the Malagasy Republic. Found in the nepheline syenites of Norway. Found in the United States in Orange and St. Lawrence counties, New York; in considerable quantity in the sands of Henderson and Buncombe counties, North Carolina. Large crystals have been found in Renfrew County, Ontario, Canada.

Use. When transparent it serves as a gemstone. It is colorless in some specimens, but more often of a

FIG. 13.18. Structure of zircon, ZrSiO₄, consisting of independent SiO₄ tetrahedra and distorted ZrO₈ cubes. (Redrawn from Zoltai, T., *Polyhedral Structure Models.* University of Minnesota, Minneapolis.)

c

brownish or red-orange color. Blue is not a natural color for zircon but is produced by heat treatment. The colorless, yellowish, or smoky stones are called *jargon*, because although resembling diamond they have little value. Serves as the source of zirconium oxide, which is one of the most refractory substances known. Platinum, which fuses at 1755°C, can be melted in crucibles of zirconium oxide.

Overshadowing its other uses since 1945 is the use of zircon as the source of metallic zirconium. Pure zirconium metal is used in the construction of nuclear reactors. Its low neutron-absorption cross section, coupled with retention of strength at high temperatures and excellent corrosion resistance, makes it a most desirable metal for this purpose.

Mineral grains of zircon in igneous, metamorphic, or sedimentary rocks, are commonly used for radioactive age determinations.

Name. The name is very old and is believed to be derived from Persian words *zar*, gold, and *gun*, color.

Similar Species. *Thorite*, $ThSiO_4$, is isostructural with zircon. It is usually reddish-brown to black, hydrated, and radioactive.

Al_2SiO_5 Group

The three polymorphs of Al_2SiO_5 are: *andalusite*, $Al^{[6]}Al^{[5]}SiO_5$ (space group *Pnnm*), *sillimanite*, $Al^{[6]}Al^{[4]}SiO_5$ (space group *Pbnm*), and *kyanite*, $Al^{[6]}Al^{[6]}SiO_5$ (space group $P\bar{1}$). The structures of the three polymorphs have been discussed on pages 445–447 and their structures are shown in Figs. 13.6 and 13.7. All three minerals may be found in metamorphosed aluminous rocks, such as pelitic schists. The stability relations of the three polymorphs have been determined experimentally as shown in Fig. 13.19. A knowledge of these stability fields is of great value in the study of regionally and contact metamorphosed terranes (see also pages 584–587). Rocks of appropriate compositions tend to form sillimanite in high-temperature, regionally metamorphosed areas. Andalusite is frequently found in contact metamorphosed aureoles, and kyanite is located in metamorphic areas that have undergone considerable pressure. All three polymorphs are fairly easily recognized, in hand specimen, in medium- to coarse-grained rocks. The ease of recognition of these minerals and their presence in pelitic schists have led to their use as index minerals in the defining of metamorphic zones, as a function of temperature and pressure. Sillimanite can be used to define metamorphic zones that have been subjected to temperature of

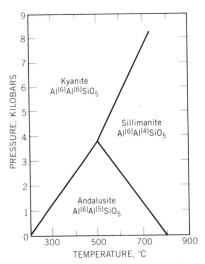

FIG. 13.19. Experimentally determined stability fields for the polymorphs of Al_2SiO_5. (From Holdaway, M. J., 1971, Stability of andalusite and the aluminum silicate stability diagram, *Amer. Jour. of Science*, v. 271, pp. 97–131; see also *Orthosilicates, Reviews in Mineralogy*, 1980, Mineralogical Society of America, Washington, D.C., p. 190.)

about 500°C and up (see Fig. 13.19). The metamorphic zones are named after the index mineral present, for example, sillimanite zone, kyanite zone. The boundaries between different zones are referred to as *isograds* (see page 587). It should be noted that kyanite, which has **G** = 3.55 to 3.66 (as compared with **G** = 3.23 for sillimanite and **G** = 3.16–3.20 for andalusite) has the densest structure of the three polymorphs and its stability field is located in the direction of increasing pressure (Fig. 13.19). Andalusite has the largest specific volume and is, therefore, the polymorph stable at the lowest pressure conditions.

ANDALUSITE—Al_2SiO_5

Crystallography. Orthorhombic; $2/m2/m2/m$. Usually occurs in coarse, nearly square prisms terminated by {001}.

Pnnm; $a = 7.78$, $b = 7.92$, $c = 5.57$ Å; $Z = 4$.
d's: 4.53(10), 3.96(8), 2.76(9), 2.17(10), 1.46(10).

Physical Properties. *Cleavage* {110} good. **H** $7\frac{1}{2}$. **G** 3.16–3.20. *Luster* vitreous. *Color* flesh-red, reddish-brown, olive-green. The variety *chiastolite* has dark-colored carbonaceous inclusions arranged in a regular manner forming a cruciform design (Fig. 13.20). Transparent to translucent. *Optics:* (−), α = 1.632, β = 1.638, γ = 1.643; 2V = 85°; X = c, Z = a. In some crystals strong pleochroism; X red, Y and Z green to colorless; r > v.

Composition and Structure. Al_2O_3 63.2, SiO_2

FIG. 13.20. Successive sections through a chiastolite crystal.

36.8%. The structure consists of chains of AlO_6 octahedra parallel to c, cross-linked by SiO_4 tetrahedra and AlO_5 polyhedra (see Fig. 13.7b).

Diagnostic Features. Characterized by the nearly square prism and hardness. Chiastolite is readily recognized by the symmetrically arranged inclusions.

Alteration. Pseudomorphs of fine-grained muscovite (sericite) after andalusite are common.

Occurrence. Andalusite is formed typically in the contact aureoles of igneous intrusions in argillaceous rocks. Here it commonly coexists with cordierite. Andalusite can be found in association with kyanite, or sillimanite, or both in regionally metamorphosed terrane. Such occurrences may reflect variations in P and T during metamorphism, as well as the sluggishness of the reactions in the system Al_2SiO_5 (see page 317).

Notable localities are in Andalusia, Spain; the Austrian Tyrol; in gem crystals and water-worn pebbles from Minas Gerais, Brazil. Crystals of chiastolite are found at Bimbowrie, South Australia. In the United States found in the White Mountains near Laws, California, and in Delaware County, Pennsylvania. Chiastolite is found at Westford, Lancaster, and Sterling, Massachusetts, and also in California.

Use. Andalusite has been mined in large quantities in California for use in the manufacture of spark plugs and other porcelains of a highly refractory nature. When transparent it may serve as a gemstone.

Name. From Andalusia, a province of Spain.

SILLIMANITE—Al_2SiO_5

Crystallography. Orthorhombic; $2/m2/m2/m$. Occurs in long, slender crystals without distinct terminations; often in parallel groups; frequently fibrous and called *fibrolite*.

$Pbnm$; $a = 7.44$, $b = 7.60$, $c = 5.75$ Å; $Z = 4$. d's: 3.32(10), 2.49(7), 2.16(8), 1.677(7), 1.579(7).

Physical Properties. *Cleavage* {010} perfect. **H** 6–7. **G** 3.23. *Luster* vitreous. *Color* brown, pale green, white. Transparent to translucent. *Optics:* (+), $\alpha = 1.657$, $\beta = 1.658$, $\gamma = 1.677$; 2V = 20°, $X = b$, $Z = c$; $r > v$.

Composition and Structure. Oxide components as in andalusite. The structure consist of AlO_6 chains parallel to the c axis and linking SiO_4 and AlO_4 tetrahedra which alternate along the c direction (see Fig. 13.7a).

Diagnostic Features. Characterized by slender crystals with one direction of cleavage.

Occurrence. Sillimanite occurs as a constituent of high-temperature metamorphosed argillaceous rocks. In contact metamorphosed rocks it may occur in sillimanite-cordierite gneisses or sillimanite-biotite hornfels. In regionally metamorphosed rocks it is found, for example, in quartz-muscovite-biotite-oligoclase-almandine-sillimanite schists (see also Table 14.14). In silica-poor rocks it may be associated with corundum.

Notable localities for its occurrence are Maldau, Bohemia, Czechoslovakia; Fassa, Austrian Tyrol; Bodenmais, Bavaria; and Freiberg, Saxony, Germany; and waterworn masses in diamantiferous sands from Minas Gerais, Brazil. In the United States found at Worcester, Massachusetts; at Norwich and Willimantic, Connecticut; and New Hampshire.

Name. In honor of Benjamin Silliman (1779–1864), professor of chemistry at Yale University.

Similar Species. *Mullite*, a nonstoichiometric compound of approximate composition $Al_6Si_3O_{15}$, is rare as a mineral but common in artificial Al_2O_3-SiO_2 systems at high temperature. *Dumortierite*, Al_7O_3-$(BO_3)(SiO_4)_3$, has been used in the manufacture of high-grade porcelain.

KYANITE—Al_2SiO_5

Crystallography. Triclinic; $\bar{1}$. Usually in long, tabular crystals, rarely terminated. In bladed aggregates.

$P\bar{1}$; $a = 7.10$, $b = 7.74$, $c = 5.57$ Å; $\alpha = 90°6'$, $\beta = 101°2'$, $\gamma = 105°45'$; $Z = 4$. d's: 3.18(10), 2.52(4), 2.35(4), 1.93(5), 1.37(8).

Physical Properties. *Cleavage* {100} perfect. **H** 5 parallel to length of crystals, 7 at right angles to this direction. **G** 3.55–3.66. *Luster* vitreous to pearly. *Color* usually blue, often of darker shade toward the center of the crystal. Also, in some cases, white, gray, or green. Color may be in irregular streaks and patches. *Optics:* (−); $\alpha = 1.712$, $\beta = 1.720$, $\gamma = 1.728$; 2V = 82°; $r > v$.

Composition and Structure. Oxide components as in andalusite. The structure consists of AlO_6 octahedral chains parallel to c and AlO_6 octahedra and SiO_4 tetrahedra between the chains (see Fig. 13.6).

Diagnostic Features. Characterized by its bladed crystals, good cleavage, blue color, and different hardness in different directions.

FIG. 13.21. Bladed kyanite crystals and prismatic staurolite (dark) in mica schist, St. Gotthard, Switzerland. (Harvard Mineralogical Museum.)

Occurrence.   Kyanite is typically a result of regional metamorphism of pelitic rocks and is often associated with garnet, staurolite, and corundum. It also occurs in some eclogites (rocks with omphacite-type pyroxenes associated with pyrope-rich garnets) and in garnet-omphacite-kyanite occurrences in kimberlite pipes. Both of these rock types reflect high to very high pressures of origin. Crystals of exceptional quality are found at St. Gotthard, Switzerland (Fig. 13.21); in the Austrian Tyrol; and at Pontivy and Morbihan, France. Commercial deposits are located in India, Kenya, and in the United States in North Carolina and Georgia.

Use.   Kyanite is used as is andalusite in the manufacture of spark plugs and other high refractory porcelains. Transparent crystals may be cut as gemstones.

Name.   Derived from the Greek word *kyanos* meaning *blue.*

TOPAZ—Al$_2$SiO$_4$(F,OH)$_2$

Crystallography.   Orthorhombic, $2/m2/m2/m$. Commonly in prismatic crystals terminated by dipyramids, {0kl} and {h0l} prisms and basal pinacoid (Fig. 13.22). Vertical prism faces frequently striated. Usually in crystals (Fig 13.23) but also in crystalline masses; granular, coarse or fine.

Pbnm; $a = 4.65$, $b = 8.80$, $c = 8.40$ Å; $Z = 4$. *d*'s: 3.20(9), 2.96(10), 2.07(9), 1.65(9), 1.403(10).

Physical Properties.   *Cleavage* {001} perfect. **H** 8. **G** 3.4–3.6. *Luster* vitreous. *Color* colorless, yellow, pink, wine-yellow, bluish, greenish. Transparent to translucent. *Optics*: (+); $\alpha = 1.606$–1.629, $\beta = 1.609$–1.631, $\gamma = 1.616$–1.638; $2V = 48°$–68°; $X = a$, $Y = b$; $r > v$.

Composition and Structure.   Al$_2$SiO$_4$(OH)$_2$ contains Al$_2$O$_3$ 56.6, SiO$_2$ 33.4 and H$_2$O 10.0%. Most of the (OH)$^-$ is generally replaced by F$^-$, the maximum fluorine content being 20.7%. The structure of topaz consists of chains parallel to the c axis of AlO$_4$F$_2$ octahedra which are cross-linked by independent SiO$_4$ tetrahedra. This arrangement is morphologically expressed by the prismatic [001] habit. The perfect {001} cleavage of topaz passes through the structure without breaking Si-O bonds; only Al-O and Al-F bonds are broken. The structure is relatively dense because it is based on closest packing of oxygens and fluorine; this packing scheme is neither cubic nor hexagonal, but more complex, *ABAC*.

Diagnostic Features.   Recognized chiefly by its crystals, basal cleavage, hardness (8), and high specific gravity.

Occurrence.   Topaz is a mineral formed by

FIG. 13.22. Topaz crystals.

FIG. 13.23. Topaz crystals on feldspar, Siberia, CIS. (Harvard Mineralogical Museum.)

fluorine-bearing vapors given off during the last stages of the solidification of siliceous igneous rocks. Found in cavities in rhyolite lavas and granite; a characteristic mineral in pegmatites, especially in those carrying tin. Associated with tourmaline, cassiterite, apatite, and fluorite; also with beryl, quartz, mica, and feldspar. Found in some localities as rolled pebbles in stream sands.

Notable localities for its occurrence are in the CIS in the Nerchinsk district of Siberia in large wine-yellow crystals, and in Mursinsk, Ural Mountains, in pale blue crystals; in Saxony, Germany, from various tin localities; Omi and Mino provinces, Japan; and San Luis Potosi, Mexico. Minas Gerais, Brazil, has long been the principal source of yellow gem-quality topaz. In the 1940s several well-formed, large crystals of colorless topaz were found there; the largest weighs 596 pounds. In the United States found at Pikes Peak, near Florissant and Nathrop, Colorado; Thomas Range, Utah; Streeter, Texas; San Diego County, California; Stoneham and Topsham, Maine; Amelia, Virginia; and Jefferson, South Carolina.

Use. As a gemstone. Frequently sold as "precious topaz" to distinguish it from citrine quartz, commonly called topaz. The color of the stones varies, being colorless, wine-yellow, golden brown, pale blue, and pink (see Plate III, no. 3, Chapter 15). The color of most deep blue topaz has been induced by irradiation plus heat treatment of colorless material (see Plate III, no. 4, Chapter 15).

Name. Derived from Topazion, the name of an island in the Red Sea, but originally probably applied to some other species.

Similar Species. *Danburite*, $Ca(B_2Si_2O_8)$, a member of the tectosilicates, shows crystal form and physical properties that are very similar to those of topaz. It can be distinguished from topaz by testing for boron.

STAUROLITE—$Fe_2^{2+}Al_9O_6(SiO_4)_4(O,OH)_2$

Crystallography. Monoclinic; $2/m$ (pseudo-orthorhombic). Prismatic crystals with common forms {110}, {010}, {001}, and {101} (Fig. 13.24a). Cruciform

FIG. 13.24. Staurolite.   *(a)*   *(b)*   *(c)*

twins very common, of two types: (1) with twin plane {031} in which the two individuals cross at nearly 90° (Fig. 13.24*b*); (2) with twin plane {231} in which they cross at nearly 60° (Fig. 13.24*c*). In some cases both types are combined in one twin group. Usually in crystals; rarely massive.

$C2/m; a = 7.83, b = 16.62, c = 5.65$ Å, $\beta = 90°; Z = 4.$ *d*'s: 3.01(8), 2.38(10), 1.974(9), 1.516(5), 1.396(10).

Physical Properties. **H** $7-7\frac{1}{2}$. **G** 3.65–3.75. *Luster* resinous to vitreous when fresh; dull to earthy when altered or impure. *Color* red-brown to brownish-black. Translucent. *Optics:* (+), $\alpha = 1.739–1.747, \beta = 1.745–1.753, \gamma = 1.752–1.761;$ $2V = 82°–88°; X = b$ (colorless), $Y = a$ (pale yellow), $Z = c$ (deep yellow); $r > v$.

Composition and Structure. FeO 16.7, Al_2O_3 53.3, SiO_2 27.9, H_2O 2.0% for the pure Fe end member; however, Mg, replacing Fe^{2+}, and Fe^{3+}, replacing Al, are generally both present in amounts of a few percent. The structure closely resembles that of kyanite with layers of $4Al_2SiO_5$ composition (AlO_6 octahedra in chains parallel to the *c* axis) alternating with layers of $Fe_2AlO_3(OH)_2$ composition along [010]. The ideal formula, on the basis of 24(O,OH), should contain one OH; however, this OH content is generally exceeded.

Diagnostic Features. Recognized by its characteristic crystals and twins. Distinguished from andalusite by its obtuse prism.

Occurrence. Staurolite is formed during regional metamorphism of aluminum-rich rocks and is found in schists and gneisses. Often associated with almandine garnet and kyanite in high-grade metamorphic rocks. May grow on kyanite in parallel orientation. In somewhat lower-grade metamorphic rocks it may occur with chloritoid. It is commonly used as in index mineral in medium-grade metamorphic rocks (see Chapter 14).

Notable localities are Monte Campione, Switzerland, and in large twin crystals in Brittany, France, and Scotland. In the United States found at Windham, Maine; Franconia and Lisbon, New Hampshire;

also in North Carolina, Georgia, Tennessee, Virginia, New Mexico, and Montana.

Use. In North Carolina the right angle twins are sold as amulets under the name "fairy stone," but most of the crosses offered for sale are imitations carved from a fine-grained rock and dyed.

Name. Derived from the Greek word *stauros* meaning *cross*, in allusion to its cruciform twins.

Humite Group

The humite group includes the following four members:

Norbergite	$Mg_3(SiO_4)(F,OH)_2$
Chondrodite	$Mg_5(SiO_4)_2(F,OH)_2$
Humite	$Mg_7(SiO_4)_3(F,OH)_2$
Clinohumite	$Mg_9(SiO_4)_4(F,OH)_2$

All four species have similar structures, chemistry, and physical properties. The structure of the members of the humite group is closely related to that of olivine with layers parallel to {100} which have the atomic arrangement of olivine, and alternating layers of $Mg(OH,F)_2$ composition. For example, the norbergite composition can be rewritten to express this layering as $Mg_2SiO_4 \cdot Mg(OH,F)_2$. The replacement of F by OH is extensive, but hydroxyl end members are unknown. The members of this group are rather restricted in their occurrence and are found mainly in metamorphosed and metasomatised limestones and dolomites, and skarns associated with ore deposits. Here we will discuss in detail only chondrodite, the most common member of the humite group.

Chondrodite—$Mg_5(SiO_4)_2(F,OH)_2$

Crystallography. Monoclinic; 2/m. Crystals are frequently complex with many forms. Usually in isolated grains. Also massive.

$P2_1/c; a = 7.89, b = 4.74, c = 10.29, \beta = 109°;$ $Z = 2.$ *d*'s: 3.02(5), 2.76(4), 2.51(5), 2.26(10), 1.740(7).

Physical Properties. **H** 6–6$\frac{1}{2}$. **G** 3.1–3.2. *Luster* vitreous to resinous. *Color* light yellow to red. Translucent. *Optics*: (+); α = 1.592–1.615, β = 1.602–1.627, γ = 1.621–1.646; 2V = 71°–85°; Z = b, X $\wedge$ c = 25°; r > v.

Composition and Structure. The composition of chondrodite can be rewritten as $2Mg_2SiO_4 \cdot Mg(OH,F)_2$ to reflect the olivine-type and brucite-like layering in the structure. The two main substitutions are Mg by Fe^{2+} and F by OH. The Fe substitution is limited to about 6 weight percent FeO.

Diagnostic Features. Characterized by its light yellow to red color and its mineral associations in crystalline limestone. The members of the humite group cannot be distinguished from one another without optical tests.

Occurrence. Chondrodite occurs most commonly in metamorphosed dolomitic limestones. The mineral association, including phlogopite, spinel, pyrrhotite, and graphite, is highly characteristic. In skarn deposits it is found with wollastonite, forsterite, and monticellite. Noteworthy localities of chondrodite are Monte Somma, Italy; Paragas, Finland; and Kafveltorp, Sweden. In the United States it was common at the Tilly Foster magnetite deposit near Brewster, New York.

Name. Chondrodite is from the Greek *chondros* meaning *a grain*; alluding to its occurrence as isolated grains. Humite is named in honor of Sir Abraham Hume.

DATOLITE—CaB(SiO$_4$)(OH)

Crystallography. Monoclinic; 2/m. Crystals usually nearly equidimensional in the three axial directions and often complex in development (Fig. 13.25). Usually in crystals. Also coarse to finely granular. Compact and massive, resembling unglazed porcelain.

$P2_1/a$; a = 4.83, b = 7.64, c = 9.66, β = 90°9'; Z = 4. d's: 3.76(5), 3.40(3), 3.11(10), 2.86(7), 2.19(6).

Physical Properties. **H** 5–5$\frac{1}{2}$. **G** 2.8–3.0. *Luster* vitreous. *Color* white, often with faint greenish tinge. Transparent to translucent. *Optics*: (−); α = 1.624, β = 1.652, γ = 1.668; 2V = 74°; Y = b, Z = c; r > v.

Composition and Structure. CaO 35.0, B_2O_3 21.8, SiO_2 37.6, H_2O 5.6%. The structure can be regarded as consisting of layers parallel to {100} of independent SiO_4 and $B(O,OH)_4$ tetrahedra. The layers are bonded by Ca in 8-coordination (6 oxygens and 2 OH groups). Despite the layerlike structure datolite has no conspicuous cleavage.

Diagnostic Features. Characterized by its

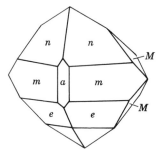

FIG. 13.25. Datolite.

glassy luster, pale green color, and its crystals with many faces.

Occurrence. Datolite is a secondary mineral found usually in cavities in basaltic lavas and similar rocks. Associated with zeolites, prehnite, apophyllite, and calcite. Notable foreign localities are Andreasberg, Harz Mountains, Germany; in Italy near Bologna, and from Seiser Alpe and Theiso, Trentino; and Arendal, Norway. In the United States it occurs in the trap rocks of Massachusetts, Connecticut, and New Jersey, particularly at Westfield, Massachusetts, and Bergen Hill, New Jersey. Masses of porcelain-like datolite, some white and some red from disseminated native copper, were found in the Lake Superior copper deposits.

Name. Derived from a Greek word meaning *to divide*, in allusion to the granular character of a massive variety.

TITANITE *(sphene)*—CaTiO(SiO$_4$)

Crystallography. Monoclinic; 2/m. Wedge-shaped crystals common resulting from a combination of {001}, {110}, and {111} (Fig. 13.26). May be lamellar or massive.

C2/c, a = 6.56, b = 8.72, c = 7.44 Å, β = 119°43'; Z = 4. d's: 3.23(10), 2.99(9), 2.60(9), 2.27(3), 2.06(4).

Physical Properties. *Cleavage* {110} distinct. Parting on {100} may be present. **H** 5–5$\frac{1}{2}$. **G** 3.4–3.55. *Luster* resinous to adamantine. *Color* gray, brown,

FIG. 13.26. Titanite crystals.

green, yellow, black. Transparent to translucent. *Optics*: (+); $\alpha = 1.900$, $\beta = 1.907$, $\gamma = 2.034$; $2V = 27°$; $Y = b$, $Z \wedge c = 51°$; $r > v$.

Composition and Structure. CaO 28.6, TiO_2 40.8, SiO_2 30.6%. Small amounts of rare earths, Fe, Al, Mn, Mg, and Zr may be present. The structure of titanite is shown in Fig. 13.27. It consists of corner-sharing TiO_6 octahedra that form kinked chains parallel to the *a* axis. These chains are cross-linked by isolated SiO_4 tetrahedra. This tetrahedral-octahedral framework produces cavities that house Ca in 7-coordination.

Diagnostic Features. Characterized by its wedge-shaped crystals and high luster. Hardness is less than that of staurolite and greater than that of sphalerite.

Occurrence. Titanite in small crystals is a common accessory mineral in granites, granodiorites, diorites, syenites, and nepheline syenites. Found in crystals of considerable size in metamorphic gneisses, chlorite schists, and crystalline limestone. Also found with iron ores, pyroxene, amphibole, scapolite, zircon, apatite, feldspar, and quartz.

The most notable occurrence of titanite is on the Kola Peninsula, CIS, where it is associated with apatite and nepheline syenite. It is mined there as a source of titanium. It is found in crystals in Tavetsch, Binnental, and St. Gotthard, Switzerland; Zillertal, Tyrol; Ala, Piemonte, and Vesuvius, Italy; and Arendal, Norway. In the United States in Diana, Rossie, Fine, Pitcairn, and Edenville, New York; and Riverside, California. Also in various places in Ontario and Quebec, Canada.

Use. As a source of titanium oxide for use as a paint pigment. A minor gemstone.

Name. In reference to its titanium content. The element titanium was named after the Titans, the mythical first sons of the Earth.

Similar Species. *Benitoite*, $BaTiSi_3O_9$, a cyclosilicate, occurs in association with *neptunite*, $KNa_2Li(Fe,Mn)_2Ti_2O(Si_4O_{11})_2$, a complex inosilicate, in San Benito, California. *Astrophyllite, aenigmatite, lamprophyllite, ramsayite,* and *fersmannite* are rare Ti-bearing silicates found associated with alkalic rocks. *Benitoite* is a minor gemstone.

Chloritoid—$(Fe,Mg)_2Al_4O_2(SiO_4)_2(OH)_4$

Crystallography. Generally monoclinic, $2/m$; may also be triclinic. Seldom in distinct tabular crystals, usually coarsely foliated, massive. Also in thin scales or small plates.

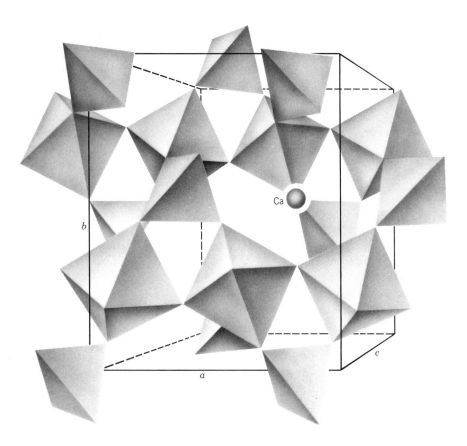

FIG. 13.27. Crystal structure of titanite (sphene) showing Si tetrahedra, Ti octahedra, and a 7-coordinated Ca site. The axes outline a unit cell. (From Papike, J. J., 1987, Chemistry of the rock-forming silicates: Ortho, ring, and single-chain structures. *Reviews of Geophysics*, v. 25, pp. 1483–1526.)

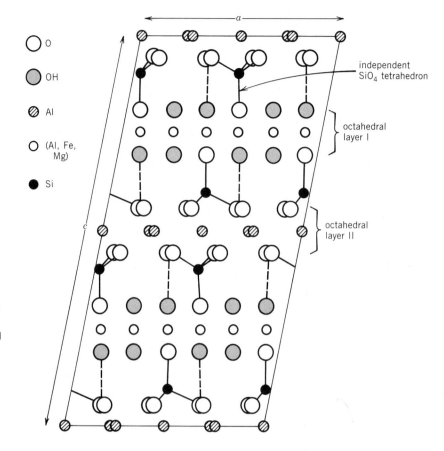

FIG. 13.28. The structure of monoclinic chloritoid projected on (010). Octahedral layer (I) has approximate composition $[Fe_2^{2+}AlO_2(OH)_4]^{1-}$ with some Mg^{2+} and Mn^{2+} substituting for Fe^{2+}; octahedral layer (II) has the composition of $[Al_3O_8]^{7-}$. Isolated tetrahedra link these layers. (From Harrison, F. W. and Brindley, G. W., 1957, The crystal structure of chloritoid. *Acta Crystallographica*, v. 10, pp. 77–82.)

$C2/c$; $a = 9.52$, $b = 5.47$, $c = 18.19$; $\beta = 101°39'$; $Z = 4$. d's: 4.50(10), 4.45(10), 3.25(6), 2.97(8), 2.70(7), 2.46(9).

Physical Properties. *Cleavage* {001} good (less so than in micas), producing brittle flakes. **H** $6\frac{1}{2}$ (much harder than chlorite). **G** 3.5–3.8. Luster pearly. *Color* dark green, greenish-gray, often grass-green in very thin plates. *Streak* colorless. *Optics*: (+); $\alpha = 1.713$–1.730, $\beta = 1.719$–1.734, $\gamma = 1.723$–1.740; $2V = 36°$–$60°$; $X = b$, $Z \wedge c = 2°$–$30°$. Pleochroism strong: Z yellow-green, Y indigo-blue, X olive-green.

Composition and Structure. For the Fe^{2+} end member, SiO_2 23.8, Al_2O_3 40.5, FeO 28.5, and H_2O 7.2%. All chloritoids show some Mg replacing Fe^{2+} but this substitution is limited. Fe^{3+} substitutes to some extent for Al. The structure of chloritoid is illustrated in Fig. 13.28. It can be described as consisting of two close-packed octahedral layers; one of a bucite-like sheet with composition $[(Fe,Mg)_2$-$AlO_2(OH)_4]^{1-}$ and another of a corundum-type composition, $[Al_3O_8]^{7-}$. These sheets alternate in the direction of the c axis. Independent SiO_4 tetrahedra link the brucite- and corundum-like sheets together. The SiO_4 tetrahedra, therefore, do not occur in continuous sheets, as in the micas and chlorite. Synthetic studies

show that chloritoid gives way to staurolite at elevated temperature, according to the reaction: chloritoid + andalusite $\rightarrow$ staurolite + quartz + fluid.

Diagnostic Features. Chloritoid is often difficult to distinguish from chlorite, with which it is generally associated. Optical study is necessary for unambiguous identification.

Occurrence. Chloritoid is a relatively common constituent of low- to medium-grade regionally metamorphosed iron-rich pelitic rocks. It generally occurs as porphyroblasts in association with muscovite, chlorite, staurolite, garnet, and kyanite. The original chloritoid was described from Kosoibrod, in the Ural Mountains, CIS. It is most commonly found in greenschist facies rocks; it is not uncommon in Vermont, for example, and northern Michigan.

Name. Chloritoid is named for its superficial resemblance to chlorite.

SOROSILICATES

The sorosilicates are characterized by isolated, double tetrahedral groups formed by two SiO_4 tetrahedra sharing a single apical oxygen (Fig. 13.29). The resulting ratio of silicon to oxygen is 2 : 7. Over 70 min-

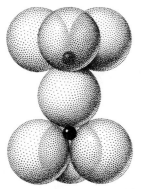

FIG. 13.29. Close-packed representation of Si_2O_7 group.

erals are known in this group, but most of them are rare. We will discuss only the following six species, of which members of the *epidote group* and *vesuvianite* are most important.

SOROSILICATES AND MIXED NESO-SOROSILICATES

Hemimorphite	$Zn_4(Si_2O_7)(OH)_2 \cdot H_2O$
Lawsonite	$CaAl_2(Si_2O_7)(OH)_2 \cdot H_2O$
Epidote group	
Clinozoisite	$Ca_2Al_3O(SiO_4)(Si_2O_7)(OH)$
Epidote	$Ca_2(Fe^{3+},Al)Al_2O(SiO_4)(Si_2O_7)(OH)$
Allanite	$X_2Y_3O(SiO_4)(Si_2O_7)(OH)$
Vesuvianite	$Ca_{10}(Mg,Fe)_2Al_4(SiO_4)_5(Si_2O_7)_2(OH)_4$

FIG. 13.30. Schematic representation of the structure of epidote, projected on (010). Dashed lines outline unit cell.

The structure of epidote (Fig. 13.30) contains both independent SiO_4 tetrahedra as well as Si_2O_7 groups. Chains of AlO_6 and $AlO_4(OH)_2$ octahedra, which share edges, run parallel to the *b* axis. (Similar octahedral chains are also present in the three polymorphs of Al_2SiO_5 and in staurolite.) An additional octahedral position occurs outside the chains; this site is occupied by Al in clinozoisite and by Fe^{3+} and Al in epidote. The chains are linked by independent SiO_4 and Si_2O_7 groups. Ca is in irregular 8-coordination with oxygen. The site in which Ca^{2+} is housed may in part be filled by Na^+. The octahedral site outside the chains may house, in addition to Al, Fe^{3+}, Mn^{3+} and more rarely Mn^{2+}.

All members of the epidote group are isostructural and form monoclinic crystals characteristically elongate on *b*. Orthorhombic *zoisite* has a structure that may be derived from that of its monoclinic polymorph, *clinozoisite*, by a twinlike doubling of the cell along the *a* axis, such that *a* of zoisite $= 2$ *a* sin β of clinozoisite (or epidote) (see Fig. 13.31). The structure of clinozoisite is the same as that of epidote, with all octahedral positions occupied by Al. *Allanite* may be derived from the epidote structure by replacing some of the Ca^{2+} by rare earths and adjusting the electrostatic charge balance by replacing some of the Fe^{3+} by Fe^{2+}. As a result of the very similar structures of the members of the epidote group, the various ionic substitutions provide the major variables. These are:

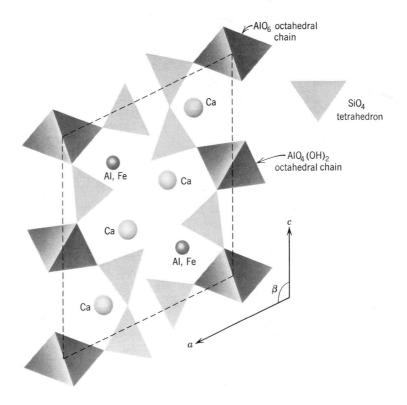

	Ions in Ca Site	Ions in Al Site Outside Chains
Clinozoisite	Ca^{2+}	Al^{3+}
Epidote	Ca^{2+}	Fe^{3+}, Al^{3+}
Piemontite	Ca^{2+}	$Mn^{3+}, Fe^{3+}, Al^{3+}$
Allanite	$Ca^{2+}, Ce^{3+}, La^{3+},$ Na^+	$Fe^{3+}, Fe^{2+}, Mg^{2+},$ Al^{3+}

FIG. 13.32. Hemimorphite.

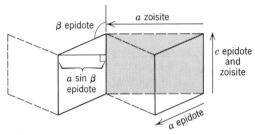

FIG. 13.31. Relation between unit cells of monoclinic epidote and orthorhombic zoisite, as projected on (010). The monoclinic unit cell outlined in solid can be related to its dashed equivalent by a mirror reflection (twinning). The orthorhombic unit cell is shown by shading.

HEMIMORPHITE—$Zn_4(Si_2O_7)(OH)_2 \cdot H_2O$

Crystallography. Orthorhombic; *mm*2. Crystals usually tabular parallel to {010}. They show prism faces and are terminated above usually by a combination of domes and pedion, and below by a pyramid (Fig. 13.32), forming polar crystals. Crystals often divergent, giving rounded groups with slight reentrant notches between the individual crystals, forming knuckle or coxcomb masses. Also mammillary, stalactitic, massive, and granular.

*Imm*2, $a = 8.38$, $b = 10.72$, $c = 5.12$ Å; $Z = 2$. d's: 6.60(9), 5.36(6), 3.30(8), 3.10(10), 2.56(5).

Physical Properties. *Cleavage* {110} perfect. **H** $4\frac{1}{2}$–5. **G** 3.4–3.5. *Luster* vitreous. *Color* white, in some cases with faint bluish or greenish shade; also yellow to brown. Transparent to translucent. Strongly pyroelectric and piezoelectric. *Optics*: (+); $\alpha = 1.614$, $\beta = 1.617$, $\gamma = 1.636$; $2V = 46°$; $X = b$, $Z = c$; $r > v$.

Composition and Structure. ZnO 67.5, SiO_2 25.0, H_2O 7.5%. Small amounts of Fe and Al may be present. The structure contains Si_2O_7 groups linked by $ZnO_3(OH)$ tetrahedra. The tetrahedra in the Si_2O_7 groups have their bases parallel to {001} and their apices all point the same way along the c direction. This orientation causes the polar character of the structure. Each (OH) group is bound to two Zn^{2+} ions. H_2O molecules lie in holes between the tetrahedra. The H_2O molecules are lost continuously by heating hemimorphite up to 500°C. At this temperature all H_2O molecules are driven off and only (OH) groups

are retained. These can be driven off only at much higher temperatures with destruction of the crystal structure.

Diagnostic Features. Characterized by the grouping of crystals. Resembles prehnite but has a higher specific gravity.

Occurrence. Hemimorphite is a secondary mineral found in the oxidized portions of zinc deposits, associated with smithsonite, sphalerite, cerussite, anglesite, and galena.

A notable locality for its occurrence is in Chihuahua, Mexico. In the United States it is found at Sterling Hill, Ogdensburg, New Jersey; Friedensville, Pennsylvania; Wythe County, Virginia; with the zinc deposits of southwestern Missouri; Leadville, Colorado; Organ Mountains, New Mexico; and Elkhorn Mountains, Montana.

Use. An ore of zinc.

Name. From the hemimorphic character of the crystals. The mineral was formerly called *calamine*.

Lawsonite—$CaAl_2(Si_2O_7)(OH)_2 \cdot H_2O$

Crystallography. Orthorhombic; 222. Usually in tabular or prismatic crystals. Frequently twinned polysynthetically on {110}.

$C222_1$; $a = 8.79$, $b = 5.84$, $c = 13.12$ Å; $Z = 4$. d's: 4.17(5), 3.65(6), 2.72(10), 2.62(7), 2.13(7).

Physical Properties. *Cleavage* {010} and {110} good. **H** 8. **G** 3.09. *Color* colorless, pale blue to bluish-gray. *Luster* vitreous to greasy. Translucent. *Optics*: (+); $\alpha = 1.665$, $\beta = 1.674$, $\gamma = 1.684$; $2V = 84°$; $X = a$, $Z = c$; $r > v$.

Composition and Structure. The composition of lawsonite is the same as that of anorthite, $CaAl_2Si_2O_8$, + H_2O. The structure consists of (AlO, OH) octahedra linked by Si_2O_7 groups. Ca^{2+} and H_2O molecules are located between these polyhedra.

Diagnostic Features. Lawsonite is characterized by its high hardness.

Occurrence. Lawsonite is a typical mineral of the glaucophane schist facies associated with chlorite, epidote, titanite, glaucophane, garnet, and quartz. The type locality is on the Tiburon Peninsula, San

FIG. 13.33. Epidote crystals.

Francisco Bay, California. It is a common constituent of gneisses and schists formed under low temperature and high pressure.

Name. In honor of Professor Andrew Lawson of the University of California.

Similar Species. *Ilvaite,* $CaFe_2^{2+}Fe^{3+}O(Si_2O_7)(OH)$, is related to lawsonite with a similar although not identical structure.

Epidote Group

The structure and crystal chemistry of the epidote group are discussed on page 463 (see also Figs. 13.30 and 13.31).

CLINOZOISITE—$Ca_2Al_3O(SiO_4)(Si_2O_7)(OH)$
EPIDOTE—$Ca_2(Al,Fe)Al_2O(SiO_4)(Si_2O_7)(OH)$

Crystallography. Monoclinic; $2/m$. Crystals are usually elongated parallel to *b* with a prominent development of the faces of the [010] zone, giving them a prismatic aspect (Figs. 13.33 and 13.34). Striated parallel to *b*. Twinning on {100} common. Usually coarse to fine granular; also fibrous.

$P2_1/m$; $a = 8.98$, $b = 5.64$, $c = 10.22$ Å, $\beta = 115°24'$; $Z = 2$. *d's:* 5.02(4), 2.90(10), 2.86(6), 2.53(6), 2.40(7).

Physical Properties. *Cleavage* {001} perfect and {100} imperfect. **H** 6–7. **G** 3.25–3.45. *Luster* vitreous. *Color* epidote: pistachio-green to yellowish-green to black; clinozoisite: pale green to gray. Transparent to translucent. *Optics.* Refractive indices and birefringence increase with iron content. Clinozoisite: (+): $\alpha = 1.670–1.715$, $\beta = 1.674–1.725$, $\gamma = 1.690–1.734$; $2V = 14°–90°$; $Y = b, X \wedge c = -2°$ to $-7°$; $r < v$. Epidote: (−), $\alpha = 1.715–1.751$, $\beta = 1.725–1.784$, $\gamma = 1.734–1.797$; $2V = 64°–90°$; $Y = b, X \wedge c = 1°$ to $-5°$; $r > v$. Absorption $Y > Z > X$. Transparent crystals may show strong absorption in ordinary light.

Composition and Structure. A complete solid solution series extends from clinozoisite (Al : Fe^{3+} = 3 : 0) to epidote (Al : Fe^{3+} = 2 : 1). *Piemontite,* $Ca_2Mn^{3+}Al_2O(SiO_4)(Si_2O_7)(OH)$, is isostructural with epidote and clinzoisite but contains mainly

Mn^{3+} instead of Fe^{3+} or Al^{3+} in the Al site outside the chains of the epidote structure. The structure of epidote has been discussed on page 463 and is illustrated in Fig. 13.30.

Diagnostic Features. Epidote is characterized by its peculiar green color and one perfect cleavage.

Occurrence. Epidote forms under conditions of regional metamorphism of the epidote-amphibolite facies (for discussion of metamorphic facies see Chapter 14). Characteristic associations of actinolite-albite-epidote-chlorite occur in the upper part of the greenschist facies. Epidote forms also during retrograde metamorphism and forms as a reaction product of plagioclase, pyroxene, and amphibole. Epidote is common in metamorphosed limestones with calcium-rich garnets, diopside, vesuvianite, and calcite. *Epidotization* is a low-temperature metasomatism and is found in veins and joint fillings in some granitic rocks.

Epidote is a widespread mineral. Notable localities for its occurrence in fine crystals are Knappenwand, Untersulzbachthal, Salzburg, Austria (Fig. 13.34*a*); Bourg d'Oisans, Isère, France; and the Ala Valley and Traversella, Piemonte, Italy. In the United States found at Riverside, California, and on Prince of Wales Island, Alaska (Fig. 13.34*b*).

Use. Sometimes cut as a gem.

Name. Epidote from the Greek meaning *increase,* because the base of the vertical prism has one

FIG. 13.34. *(a)* Epidote, Knappenwand, Austria. (Harvard Mineralogical Museum.) *(continued)*

FIG. 13.34. *(continued) (b)* Epidote with quartz crystals. Prince of Wales Island, Alaska.

side longer than the other. Zoisite was named after Baron von Zois and piemontite after the locality, Piemonte, Italy.

Similar Species. *Zoisite* is an orthorhombic polymorph (space group *Pnmc*) of clinozoisite. It is similar in appearance and occurrence to clinozoisite but is less common. In 1967 gem quality, blue-colored crystals were found in Tanzania. This variety is known as *tanzanite*.

Allanite—
$(Ca,Ce)_2(Fe^{2+},Fe^{3+})Al_2O(SiO_4)(Si_2O_7)(OH)$

Crystallography. Monoclinic; $2/m$. Habit of crystals similar to epidote. Commonly massive and in embedded grains.

$P2_1/m$; $a = 8.98$, $b = 5.75$, $c = 10.23$ Å, $\beta = 115°0'$; $Z = 2$. d's: 3.57(6), 2.94(10), 2.74(8), 2.65(6), 2.14(4).

Physical Properties. **H** $5\frac{1}{2}$–6. **G** 3.5–4.2. *Luster* submetallic to pitchy and resinous. *Color* brown to pitch-black. Often coated with a yellow-brown alteration product. Subtranslucent, will transmit light on thin edges. Slightly radioactive. *Optics*: $(-)$, usually, with $2V = 40°$–90°; when $(+)$, $2V = 60°$–90°, $\alpha = 1.690$–1.791, $\beta = 1.700$–1.815, $\gamma = 1.706$–1.828; $Y = b$, $X \wedge c\ 1°$–40°. Metamict allanites are isotropic with $n = 1.54$–1.72.

Composition and Structure. Of variable composition with Ce, La, Th, and Na in partial subsitution for Ca, and Fe^{2+}, Fe^{3+}, Mn^{3+}, and Mg in partial substitution for some of the Al. The structure of well-crystallized allanite is the same as that of epidote (see Fig. 13.30). Allanite is commonly found in a metamict state as the result of "self-irradiation" by radioactive constituents in the original mineral. Total destruction

of the structure leads to a glassy product that adsorbs considerable H_2O (see Fig. 3.47 and related discussion).

Diagnostic Features. Characterized by its black color, pitchy luster, and association with granitic rocks.

Occurrence. Allanite occurs as a minor accessory constituent in many igneous rocks, such as granite, syenite, diorite, and pegmatites. Frequently associated with epidote.

Notable localities are at Miask, Ural Mountains, CIS; Greenland; Falun and Ytterby, Sweden; and the Malagasy Republic. In the United States allanite is found at Franklin, New Jersey, and Barringer Hill, Texas.

Name. In honor of Thomas Allan, who first observed the mineral. *Orthite* is sometimes used as a synonym.

VESUVIANITE (*idocrase*)—
$Ca_{10}(Mg,Fe)_2Al_4(SiO_4)_5(Si_2O_7)_2(OH)_4$

Crystallography. Tetragonal; $4/m2/m2/m$. Crystals prismatic often vertically striated. Common forms are {110}, {010}, and {001} (Fig. 13.35). Fre-

FIG. 13.35. Vesuvianite.

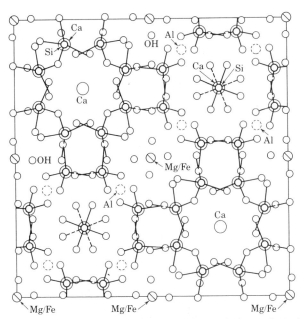

FIG. 13.36. The structure of a unit cell of vesuvianite, projected on (001). In order to compare this structural projection with the projected structure image in Fig. 13.37, one must note that regions of high atomic potential are dark areas in Fig. 13.37, whereas white regions represent parts of the structure with low electron density. The two vertical channels, with Ca at their centers, correspond with the darkest regions in the structure image of Fig. 13.37.

quently found in crystals, but striated columnar aggregates more common. Also granular, massive.

P4nnc; *a* = 15.66, *c* = 11.85 Å; *Z* = 4. *d*'s: 2.95(4), 2.75(10), 2.59(8), 2.45(5), 1.621(6).

Physical Properties. *Cleavage* {010} poor. **H** $6\frac{1}{2}$. **G** 3.35–3.45. *Luster* vitreous to resinous. *Color* usually green or brown; also yellow, blue, red. Subtransparent to translucent. *Streak* white. *Optics*: (−); ω = 1.703–1.752, ϵ = 1.700–1.746.

Composition and Structure. There is some substitution of Na for Ca; Mn^{2+} for Mg; Fe^{3+} and Ti for Al; and F for (OH). B and Be have been reported in some varieties. The structure of vesuvianite appears to be closely related to that of grossular garnet. Some parts of the structure are common to both minerals. Isolated SiO_4 tetrahedra as well as Si_2O_7 groups occur. Three-fourths of the Ca is in 8-coordination and one-fourth in 6-coordination with oxygen. The Al and Fe (and Mg) are in octahedral coordination with oxygen. A projection of a unit cell of the structure is given in Fig. 13.36. A structure image (obtained by high-resolution transmission electron microscopy) of this rather complicated silicate is given in Fig. 13.37. In this photograph a unit cell is outlined, which is equivalent to the structure shown in Fig. 13.36. Further-

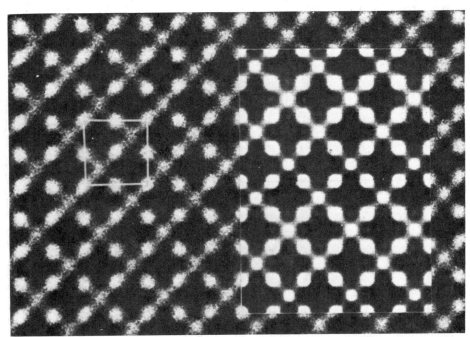

FIG. 13.37. High-resolution transmission electron microscope (HRTEM) image of vesuvianite. The edge of the unit cell, as outlined by a white square, is 15 Å. A computer-calculated image, which simulates the experimental image, is outlined by a rectangular area in the right-hand part of the illustration. The white areas are regions of low electron density in the vesuvianite structure. (From Buseck, P. R., 1978, Computed crystal structure images for high resolution electron microscopy. *Nature*, v. 274, pp. 322–324; reprinted by permission from *Nature*. Copyright © 1978, Macmillan Journals Limited.)

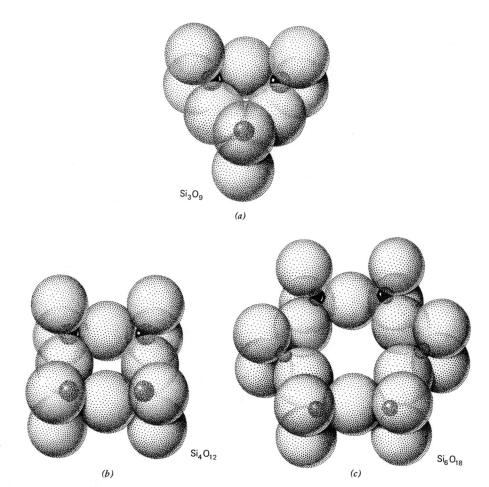

FIG. 13.38. Close-packed representation of ring structures in the cyclosilicates.

Si_3O_9

(a)

Si_4O_{12}

(b)

Si_6O_{18}

(c)

more, part of the photograph contains a computer-simulated image of the structure. Such calculated images are very helpful to the investigator, because they aid in the understanding of features of the experimentally obtained structure image.

Diagnostic Features. Brown tetragonal prisms and striated columnar masses are characteristic of vesuvianite.

Occurrence. Vesuvianite is usually formed as the result of contact metamorphism of impure limestones. Associated with other contact minerals, such as grossular and andradite garnet, wollastonite, and diopside. Originally discovered in the ancient ejections of Vesuvius and in the dolomitic blocks of Monte Somma.

Important localities are Zermatt, Switzerland; in Italy, at Ala, Piemonte, at Monzoni, Trentino and at Vesuvius; Achmatovsk, Ural Mountains, and River Vilui, Siberia, CIS; and Morelos, Mexico. In the United States it is found at Sanford, Maine; near Olmsteadville, New York; Franklin, New Jersey; Magnet Cove, Arkansas; and Crestmore, California. Found in many contact metamorphic deposits in the western United

States. A compact green variety resembling jade found in Siskiyou, Fresno, and Tulare counties, California, is known as *californite*. In Quebec, Canada, found at Litchfield, Pontiac County; at Templeton, Ottawa County; and at Asbestos.

Use. The green, massive variety californite is used as a jade substitute. Transparent crystals may be cut as faceted gems.

Name. From the locality Mount Vesuvius.

CYCLOSILICATES

The cyclosilicates contain rings of linked SiO_4 tetrahedra having a ratio of $Si : O = 1 : 3$. Three possible closed cyclic configurations of this kind may exist as shown in Fig. 13.38. The simplest is the Si_3O_9 ring, represented among minerals only by the rare titanosilicate *benitoite*, $BaTiSi_3O_9$. The Si_4O_{12} ring occurs only in a few very rare silicates. An example is *papagoite*, $Ca_2Cu_2Al_2Si_4O_{12}(OH)_6$. *Axinite* was originally classified on the basis of Si_4O_{12} rings in its structure. More recent data, however, have shown it to be

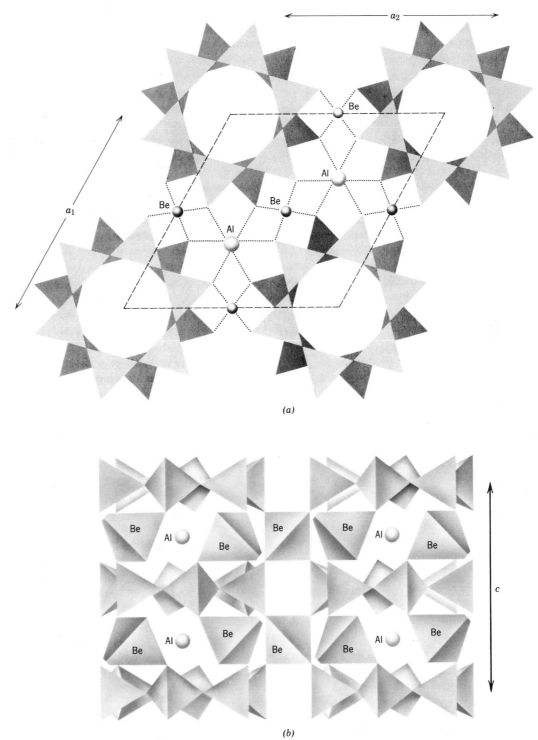

FIG. 13.39. The structure of beryl. *(a)* Projected onto (0001). Dashed lines outline unit cell. *(b)* View of the beryl structure with *c* axis vertical.

made of more complex $B_2Si_8O_{30}$ groups; see description below. The Si_6O_{18} ring, however, is the basic framework of the structures of the common and important minerals, *beryl*, $Be_3Al_2Si_6O_{18}$, and *tourmaline*. In the structure of beryl (Figs. 13.39*a* and *b*) Si_6O_{18}

rings are arranged in layers parallel to {0001}. Sheets of Be and Al ions lie between the layers of rings. Be in 4-coordination and Al in 6-coordination tie the layers together horizontally and vertically. The silicon-oxygen rings are so arranged as to be nonpolar; that

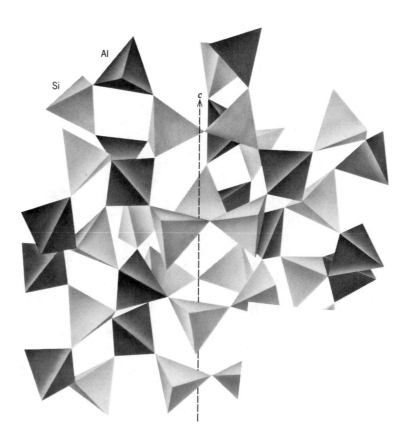

FIG. 13.40. Idealized drawing of the $(Al_4Si_5O_{18})$ framework in low cordierite. The octahedral coordination of Mg and Fe^{2+} is not shown. (After Gibbs, G. V., 1966, The polymorphism of cordierite I: the crystal structure of low cordierite, *American Mineralogist*, v. 51, pp. 1068–1087.)

is, a mirror plane can be pictured passing through the tetrahedra in the plane of the ring (Fig. 13.39*b*). Rings are positioned one above the other in the basal sheets so that the central holes correspond, forming prominent channels parallel to the *c* axis. In these channels a wide variety of ions, neutral atoms, and molecules, can be housed. $(OH)^-$, H_2O, F, He, and ions of Rb, Cs, Na, and K are housed in beryl in this way. With monovalent alkalis such as Na^+ in the channels, the overall charge of the structure is neutralized by the substitution 2 alkalis^{+1} $\leftrightharpoons$ Be^{2+}, or by 3 alkalis^{+1} + 1 Al^{3+} $\leftrightharpoons$ 3 Be^{2+}. Although beryl is here included as a member of the cyclosilicates on the basis of the Si_6O_{18} rings, consideration of the presence of the BeO_4 tetrahedra (see Fig. 13.39*b*) as an equally essential part of the structure reveals an overall three-dimensional network for this structure. For this reason, beryl can also be classified as a member of the framework (or tecto) silicates. For further discussion of the beryl structure and space group, see Fig. 3.29 and related text.

Cordierite, $(Mg,Fe)_2Al_4Si_5O_{18} \cdot nH_2O$, has a high-temperature polymorph, *indialite*, which is isostructural with beryl in which the Al is randomly distributed in the $(Si,Al)_6O_{18}$ rings. The common, lower-temperature form, *cordierite*, is orthorhombic

(pseudohexagonal) in which two of the tetrahedra in the 6-fold ring are occupied by Al, producing an ordered structure (see Fig. 13.40). Al occupies the Be sites and Mg and Fe^{2+} the octahedral Al sites of the beryl structure. H_2O molecules may reside in the channels of the structure. It has been pointed out by G. V. Gibbs (*American Mineralogist*, 1966, pp. 1068–1088) that cordierite should be classified with the tectosilicates rather than the cyclosilicates, because Al- and Si-tetrahedra are in perfect alternation in all directions of the structure except for two SiO_4

FIG. 13.41. Part of the structure of tourmaline projected on (0001).

FIG. 13.42. Axinite.

tetrahedra that share a common oxygen in the six-membered ring.

The structure of *tourmaline*, with a complex chemical composition, had long remained a mystery. It has been shown to be built on Si_6O_{18} rings along the center of which Na^+ and $(OH)^-$ alternate (see Fig. 13.41). The Si_6O_{18} rings in tourmaline are polar; that is, the net strength of bonds to one side of the ring is not the same as the strength of bonds extending to the other, looking first in one direction, then in the other along the *c* axis. Interlayered with the rings are sheets of triangular BO_3 groups. $(Li,Mg,Al)O_4(OH)_2$ octahedral groups link the Si_6O_{18} rings and BO_3 groups together. The columns of Si_6O_{18} rings are linked to each other by $(Al,Fe,Mn)O_5(OH)$ groups. The composition of tourmaline is very complex, but the following generalizations apply: Ca may substitute for Na; Al may be replaced by Fe^{3+}, Mn^{3+}; and Mg may be replaced by Fe^{2+}, Mn^{2+}, Al^{3+}, and Li^+. The varieties are determined by the relative proportions of the cations, and ionic substitution follows the usual pattern, with extensive mutual substitution of Mg by Fe^{2+} and Mn^{2+}, and Na^+ by Ca^{2+}, with concomitant coupled subsitution to maintain electrical neutrality. We will discuss in detail the following silicates:

Axinite	$(Ca,Fe^{2+},Mn)_3Al_2BSi_4O_{16}H$
Beryl	$Be_3Al_2(Si_6O_{18})$
Cordierite	$(Mg,Fe)_2Al_4Si_5O_{18}·nH_2O$
Tourmaline	$(Na,Ca)(Li,Mg,Al)(Al,Fe,Mn)_6$-
	$(BO_3)_3(Si_6O_{18})(OH)_4$

Axinite—$(Ca,Fe^{2+},Mn)_3Al_2BSi_4O_{16}H$

Crystallography. Triclinic; $\bar{1}$. Crystals usually thin with sharp edges but varied in habit (Fig. 13.42). Frequently in crystals and crystalline aggregates; also massive, lamellar to granular.

$P\bar{1}$; $a = 7.15$, $b = 9.16$, $c = 8.96$ Å, $\alpha = 88°04'$, $\beta = 81°36'$, $\gamma = 77°42'$; $Z = 2$. *d*'s: 6.30(7), 3.46(8), 3.28(6), 3.16(9), 2.81(10).

Physical Properties. *Cleavage* {100} distinct. **H** $6\frac{1}{2}$–7. **G** 3.27–3.35. *Luster* vitreous. *Color* clove-brown, violet, gray, green, yellow. Transparent to translucent. *Optics*: (−); $\alpha = 1.674$–1.693, $\beta = 1.681$–1.701, $\gamma = 1.684$–1.704; $2V = 63°80'$; $r < v$.

Composition and Structure. A considerable range exists in composition with varying amounts of Ca, Mn, and Fe. Some Mg may be present. The complex structure of axinite was originally regarded as made up of Si_4O_{12} rings and BO_3 triangles and (OH) groups. Such Si_4O_{12} rings would classify axinite as one of the cyclosilicates. More recent structure analysis, however, has shown that the axinite structure is best viewed as being made up $B_2Si_8O_{30}$ groups, in which BO_4 tetrahedra share three corners each, linking together four $S_{12}O_7$ groups. Zoltai (1960) would classify this structure under "complex silicate groups."

Diagnostic Features. Characterized by the triclinic crystals with very acute angles.

Occurrence. Axinite occurs in cavities in granite, and in the contact zones surrounding granitic instrusions. Notable localities for its occurrence are Bourg d'Oisans, Isère, France; various points in Switzerland; St. Just, Cornwall, England; and Obira, Japan. In the United States at Luning, Nevada, and a yellow manganous species at Franklin, New Jersey.

Use. A minor gem.

Name. Derived from a Greek word meaning axe, in allusion to the wedgelike shape of the crystals.

BERYL—$Be_3Al_2(Si_6O_{18})$

Crystallography Hexagonal; $6/m2/m2/m$. Strong prismatic habit. Frequently vertically striated and grooved. Cesium beryl frequently flattened on {0001}. Forms usually present consist only of {10$\bar{1}$0} and {0001} (Fig. 13.43*a*). Pyramidal forms are rare (Figs. 13.43*b* and 13.44). Crystals frequently of considerable size with rough faces. At Albany, Maine, a tapering crystal 27 feet long weighed over 25 tons.

$P6/mcc$; $a = 9.23$, $c = 9.19$ Å; $Z = 2$. *d*'s: 7.98(9), 4.60(5), 3.99(5), 3.25(10), 2.87(10).

Physical Properties. *Cleavage* {0001} imper-

FIG. 13.43. Beryl crystals.

(a)

(b)

FIG. 13.44. Transparent, light green beryl crystal (length 12.5 cm), from Perm, Ural Mountains, Russia, CIS. (Harvard Mineralogical Museum.)

fect. **H** $7\frac{1}{2}$–8. **G** 2.65–2.8. *Luster* vitreous. *Color* commonly bluish-green or light yellow, may be deep emerald-green, gold-yellow, pink, white, or colorless (see Plate I, nos. 2 and 5, and Plate III, no. 8, Chapter 15). Transparent to translucent. Frequently the larger, coarser crystals show a mottled appearance due to the alternation of clear transparent spots with cloudy portions.

Color serves as the basis for several variety names of gem beryl. *Aquamarine* is the pale greenish-blue transparent variety. *Morganite*, or rose beryl, is pale pink to deep rose. *Emerald* is the deep green, transparent beryl. *Golden beryl* is a clear golden-yellow variety. *Optics*: (−); ω = 1.560–1.602, ϵ = 1.557–1.599.

Composition and Structure. BeO 14.0, Al_2O_3 19.0, SiO_2 67.0 are the theoretical percentages of the oxides in the formula. However, the presence of alkalis (Na and Rb) and Li may considerably reduce the percentage of BeO. Small and variable amounts of

H_2O and CO_2 are housed interstitially in the large, hexagonal, vertical channels (see page 469). The structure of beryl is illustrated in Fig. 13.39 and discussed on page 469. See also Fig. 3.29 and related text.

Diagnostic Features. Recognized usually by its hexagonal crystal form and color. Distinguished from apatite by greater hardness and from quartz by higher specific gravity.

Occurrence. Beryl, although containing the rare element Be, is rather common and widely distributed. It occurs usually in granitic rocks, or in pegmatites. It is also found in mica schists and associated with tin ores. The world's finest emeralds are found in Colombia in a dark bituminous limestone; the most notable localities are Muzo, Cosquez, and Chivor. Emeralds of good quality are found in mica schists in the Transvaal, South Africa; Sandawana, Zimbabwe; and near Sverdlovsk, CIS. Rather pale emeralds have been found in small amount in Alexander County, North Carolina, associated with the green variety of spodumene, *hiddenite*. Beryl of the lighter aquamarine color is much more common and is found in gem quality in many countries.

The world's major source of gem beryl, other than emerald, is Brazil. Fine specimens come from several localities, but the most important are in the state of Minas Gerais. Other important localities are in the Ural Mountains, CIS, Malagasy Republic, and Namibia. In the United States gem beryl, chiefly *aquamarine*, has been found in various places in New England, Idaho, North Carolina, and Colorado. The most important localities for *morganite* are in the Pala and Mesa Grande districts, San Diego County, California. Common beryl, as an ore of beryllium, is produced in small amounts in many countries but the principal producers are the CIS, Brazil, and the United States.

Use. As a gemstone the emerald (see Plate I, nos. 2 and 5, Chapter 15), ranks as one of the most valuable of stones, and may have a greater value than the diamond. Beryl is also the principal source of beryllium, a lightweight metal similar to aluminum in many of its properties. A major use of beryllium is as an alloy with copper. Beryllium greatly increases the hardness, tensile strength, and fatigue resistance of copper.

Name. The name *beryl* is of ancient origin, derived from the Greek word which was applied to green gemstones.

Similar Species. *Euclase*, $BeAl(SiO_4)(OH)$, and *gadolinite*, $YFe^{2+}Be_2(SiO_4)_2O_2$, are rare beryllium silicates. Blue *euclase* is sometimes cut as a gemstone.

Cordierite—(Mg,Fe)$_2$Al$_4$Si$_5$O$_{18}$·nH$_2$O

Crystallography. Orthorhombic: $2/m2/m2/m$. Crystals are usually short prismatic, pseudohexagonal twins twinned on {110}. Also found as imbedded grains and massive.

Cccm; a = 17.13, *b* = 9.80, *c* = 9.35 Å; *Z* = 4. *d*'s: 8.54(10), 4.06(8), 3.43(8), 3.13(7), 3.03(8).

Physical Properties. *Cleavage* {010} poor. **H** 7–7½. **G** 2.60–2.66. *Luster* vitreous. *Color* various shades of blue to bluish-gray. Transparent to translucent. *Optics*: usually (−), may be (+). Indices increasing with Fe content. α = 1.522–1.558, β = 1.524–1.574, γ = 1.527–1.578. 2V = 0°–90°. *X* = *c*, *Y* = *a*; *r* < *v*. Pleochroism: cordierite is sometimes called *dichroite* because of pleochroism. Fe rich varieties: *X* colorless, *Y* and *Z* violet.

Composition and Structure. Although some substitution of Mg by Fe^{2+} occurs, most cordierites are magnesium-rich. Mn may replace part of the Mg. The Al content of cordierite shows little variation. Most analyses show appreciable but variable H$_2$O which is probably located in the large channels parallel to *c*. Small amounts of Na and K may be similarly housed. The structure of the low-temperature form, also known as low cordierite, is shown in Fig. 13.40 and is discussed on page 470. A high-temperature polymorph, *indialite*, with random distribution of Al in the (AlSi)$_6$O$_{18}$ ring, is isostructural with beryl and has space group *P6/mcc*.

Diagnostic Features. Cordierite resembles quartz and is distinguished from it with difficulty. Distinguished from corundum by lower hardness. Pleochroism is characteristic if observed.

Alteration. Commonly altered to some form of mica, chlorite, or talc and is then various shades of grayish-green.

Occurrence. Cordierite is a common constituent of contact and regionally metamorphosed argillaceous rocks. It is especially common in hornfels produced by contact metamorphism of pelitic rocks. Common assemblages are sillimanite-cordierite-spinel, and cordierite-spinel-plagioclase-orthopyroxene. Cordierite is also found in regionally metamorphosed cordierite-garnet-sillimanite gneisses. Cordierite-anthophyllite coexistences have been described from several localities. It occurs also in norites resulting from the incorporation of argillaceous material in gabbroic magmas. It is found in some granites and pegmatites. Gem material has come from Sri Lanka.

Use. Transparent cordierite of good color is used as a gem known by jewelers as *iolite* or *dichroite*.

Name. After the French geologist P. L. A. Cordier (1777–1861). *Iolite* is sometimes used as a synonym.

TOURMALINE—(Na,Ca)(Li,Mg,Al)-(Al,Fe,Mn)$_6$(BO$_3$)$_3$(Si$_6$O$_{18}$)(OH)$_4$

Crystallography. Hexagonal; 3*m*. Usually in prismatic crystals with a prominent trigonal prism and subordinate hexagonal prism, {11$\bar{2}$0}, vertically striated. The prism faces may round into each other giving the crystals a cross section like a spherical triangle. When doubly terminated, crystals usually show different forms at the opposite ends of the vertical axis (Fig. 13.45). May be massive, compact; also coarse to fine columnar, either radiating or parallel.

R3m; a = 15.95, *c* = 7.24 Å; *Z* = 3. *d*'s: 4.24(7), 4.00(7), 3.51(7), 2.98(9), 2.58(10).

Physical Properties. **H** 7–7½. **G** 3.0–3.25. *Luster* vitreous to resinous. *Color* varied, depending on the composition. *Fracture* conchoidal.

Black, Fe-bearing tourmaline (*schorl*) is most common; brown tourmaline (*dravite*) contains Mg. The rarer Li-bearing varieties (*elbaite*, containing Na and *liddicoatite*, containing Ca) are light colored in fine shades of green (*verdelite*), yellow, red-pink (*rubellite*), and blue (*indicolite*). Rarely white or colorless *achroite*. A single crystal may show several different colors arranged either in concentric envelopes about the *c* axis or in layers transverse to the length. Strongly pyroelectric and piezoelectric. *Optics*: (−); ω = 1.635–1.675, ε = 1.610–1.650. Some varieties strongly pleochroic, *O* > *E*.

Composition and Structure. A complex silicate of B and Al (see Fig. 13.41) with the following substitutions: Ca for Na along the centers of the ring channels; Mg and Al for Li in 6-coordination between Si$_6$O$_{18}$ rings and BO$_3$ groups; Fe^{3+} and Mn^{3+} for Al in polyhedra that link the Si$_6$O$_{18}$ rings. The structure is discussed on page 471.

Diagnostic Features. Usually recognized by the characteristic rounded triangular cross section of the crystals and conchoidal fracture. Distinguished from hornblende by absence of cleavage.

FIG. 13.45. Tourmaline.

FIG. 13.46. Tourmaline crystals with quartz and cleavelandite, Pala, California. (Harvard Mineralogical Museum.)

Occurrence. The most common and characteristic occurrence of tourmaline is in granite pegmatites and in the rocks immediately surrounding them. It is found also as an accessory mineral in igneous rocks and metamorphic rocks. Most pegmatitic tourmaline is black and is associated with the common pegmatite minerals, microcline, albite, quartz, and muscovite. Pegmatites are also the home of the light-colored lithium-bearing tourmalines frequently associated with lepidolite, beryl, apatite, fluorite, and rarer minerals. The brown Mg-rich tourmaline is found in crystalline limestones.

Famous localities for the occurrence of the gem tourmalines are the Island of Elba, Italy; the state of Minas Gerais, Brazil; Ural Mountains near Sverdlovsk, CIS; and the Malagasy Republic. In the United States found at Paris and Auburn, Maine; and Mesa Grande, Pala (Fig. 13.46), Rincon, and Ramona in San Diego County, California. Brown crystals are found near Gouverneur, New York, and fine black crystals at Pierrepont, New York.

Use. Tourmaline forms one of the most beautiful of the gemstones (see Plate II, no. 11, Chapter 15). The color of the stones varies, the principal shades being olive-green, pink to red, and blue. Sometimes a stone is so cut as to show different colors in different parts. The green-colored stones are usually known by the mineral name, tourmaline, or as *Brazil-ian emeralds*. The red or pink stones are known as *rubellite*, and the rarer dark blue stones are called *indicolite*.

Because of its strong piezoelectric property, tourmaline is used in the manufacture of pressure gauges to measure transient blast pressures (for piezoelectricity, see page 270).

Name. *Tourmaline* comes from *turamali*, a name given to the early gems from Sri Lanka.

INOSILICATES

SiO_4 tetrahedra may link into chains by sharing oxygens (Fig. 13.3). Such simple chains may be joined side by side by further sharing of oxygens in alternate tetrahedra to form bands or double chains (Fig. 13.3). In the simple chain structure, two of the four oxygens in each SiO_4 tetrahedron are shared, giving a ratio of $Si : O = 1 : 3$. In the band structure half of the tetrahedra share three oxygens, the other half share two oxygens yielding a ratio of $Si : O = 4 : 11$.

Included in the inosilicates are two important rock-forming groups of minerals: the *pyroxenes* as single chain members and the *amphiboles* as double chain members. Many similarities exist between the two groups in crystallographic, physical, and chemical properties. Although most pyroxenes and amphiboles are monoclinic, both groups have orthorhombic members. Also in both the repeat distance along the chains, that is, the c dimension of the unit cell, is approximately 5.2 Å. The a cell dimensions are also analogous but, because of the double chain, the b dimension of amphiboles is roughly twice that of the corresponding pyroxenes.

The same cations are present in both groups; but the amphiboles are characterized by the presence of (OH), which is lacking in pyroxenes. Although the color, luster, and hardness of analogous species are similar, the (OH) in amphiboles gives them, in general, slightly lower specific gravity and refractive indices than their pyroxene counterparts. Furthermore, the crystals have somewhat different habits. Pyroxenes commonly occur in stout prisms, whereas amphiboles tend to form more elongated crystals, often acicular. Their cleavages are distinctly different and can be directly related to the underlying chain structure (see Figs. 13.50 and 13.68).

Pyroxenes crystallize at higher temperatures than their amphibole analogues and hence are generally formed early in a cooling igneous melt (see Fig. 14.1) and occur also in high-temperature metamorphic rocks rich in Mg and Fe. If water is present in the melt or as a metamorphic fluid, the early-formed pyroxene may react with the liquid at lower temperatures to form amphibole. Under prograde metamorphic con-

ditions amphiboles commonly react to form pyroxenes, and under retrograde metamorphic conditions pyroxenes commonly give way to amphiboles (see Fig. 13.54).

Pyroxene Group

The chemical composition of pyroxenes can be expressed by a general formula as XYZ_2O_6, where X represents Na^+, Ca^{2+}, Mn^{2+}, Fe^{2+}, Mg^{2+}, and Li^+ in the $M2$ crystallographic site; Y represents Mn^{2+}, Fe^{2+}, Mg^{2+}, Fe^{3+}, Al^{3+}, Cr^{3+}, and Ti^{4+} in the $M1$ site; and Z represents Si^{4+} and Al^{3+} in the tetrahedral sites of the chain. (It should be noted that the X cations in general are larger than the Y cations, in accordance with the cation size requirements of the sites $M2$ and $M1$.) The pyroxenes can be divided into several groups, the most common of which can be represented as part of the chemical system $CaSiO_3$ (wollastonite, a pyroxenoid)–$MgSiO_3$(enstatite)–$FeSiO_3$ (ferrosilite). The trapezium part of this system includes members of the common series *diopside*, $CaMgSi_2O_6$-*hedenbergite*, $CaFeSi_2O_6$, and of the series *enstatite-ferrosilite* (see Fig. 13.47a). Only the names for end member compositions of the common pyroxenes are shown in Fig. 13.47a. In prior years, intermediate compositions have been given various species names, such as bronzite and hypersthene (for members of the

FIG. 13.47. *(a)* Pyroxene compositions in the system $CaSiO_3$-$MgSiO_3$-$FeSiO_3$. General compositional fields are outlined. Representative tielines across the miscibility gap between augite and more Mg-Fe-rich pyroxenes are shown. The "augite" field is labeled with quotation marks because all augite compositions contain considerable Al which is not considered in this triangular composition diagram.

(continued)

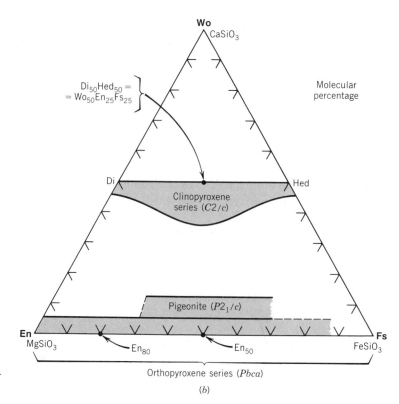

FIG. 13.47. *(continued)* *(b)* Nomenclature of pyroxenes in the system Wo (wollastonite), En (enstatite), and Fs (ferrosilite) as derived from chemical compositional information. Members of the orthopyroxene series range from En_{100} to En_0 (which is equivalent to Fs_0 to Fs_{100}). Any other, more general compositions are expressed in molecular percentages of Wo, En, and Fs (for example, Wo_{50} En_{25} Fs_{25}). The space groups of the various solid solution series are shown.

orthopyroxene series) and salite and ferrosalite (for members of the diopside-hedenbergite series). Modern nomenclature rulings (Morimoto et al., 1988; for complete reference see end of this chapter) disallow such species designations. The end member names, as shown in Fig. 13.47a, apply from the end member composition to the 50 molecular percentage point (the halfway point) in the diopside-hedenbergite and enstatite-ferrosilite series. If chemical analyses are available for intermediate compositions they can be expressed in terms of molecular percentages of the end members (see Fig. 13.47b). For example, pure enstatite can be stated as En_{100}. A member of the two-component orthopyroxene series composed of 80 molecular percent enstatite and 20 molecular percent ferrosilite can be listed as En_{80}. Similarly, the compositions of the two-component diopside-hedenbergite series can be expressed by molecular percentages that appear as subscripts to Di and Hed, e.g., $Di_{50}Hed_{50}$. Any more general chemical composition in the trapezium part of Fig. 13.47b can be expressed in terms of molecular percentages of the three end members, Wo (for wollastonite), En (for enstatite), and Fs (for ferrosilite). An example, shown in Fig. 13.47b, would be $Wo_{50}En_{25}Fs_{25}$. Table 5.5 illustrates the recalculation of a pyroxene analysis in terms of Wo-En-Fs end member components, and Fig. 5.25 (and related text) explains the procedure for plotting chemical compositions on triangular diagrams.

Compositionally *augite* is closely related to members of the diopside-hedenbergite series but with some substitution of, for example, Na for Ca in *M2*, Al for Mg(or Fe^{2+}) in *M1*, and Al for Si. *Pigeonite* represents a field of Mg-Fe solid solutions with a Ca-content somewhat larger than in the *enstatite-ferrosilite* series which represents the composition field of the orthopyroxenes. Sodium-containing pyroxenes are *aegirine*, $NaFe^{3+}Si_2O_6$, and *jadeite*, $NaAlSi_2O_6$. Aegirine and augite represent a complete solid solution series as shown by members of intermediate composition, *aegirine-augite*. *Omphacite* represents a solid solution series between augite and jadeite. *Spodumene*, $LiAlSi_2O_6$, is a relatively rare pyroxene found in Li-rich pegmatites.

The pyroxene structure is based on single SiO_3 chains that run parallel to the c axis. Figure 13.48 (see also Fig. 13.3) illustrates this tetrahedral chain as well as the double octahedral chain to which it is bonded. The structure contains two types of cation sites, labeled *M1* and *M2*. The *M1* site is a relatively regular octahedron, but, especially in the monoclinic pyroxenes, the *M2* site is an irregular polyhedron of 8-coordination (in orthorhombic pyroxenes with Mg in the *M2* site this polyhedron is closer to a regular octahedron). Figure 13.49a shows the pyroxene structure and the distribution of cation sites as seen in a direction parallel to the c axis. The cations in the *M1* sites are all coordinated by oxygens of two opposing

SiO₃ chains

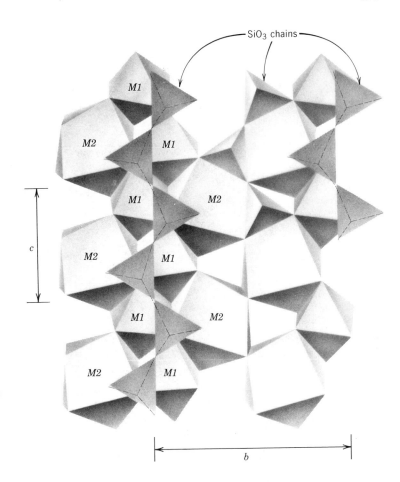

FIG. 13.48. The structure of jadeite, NaAlSi₂O₆, a monoclinic pyroxene in an approximate projection onto (100). The *M2* site is occupied by Na⁺, the *M1* site by Al³⁺. (After Prewitt, C. T. and Burnham, C. W., 1966, The crystal structure of jadeite, NaAlSi₂O₆, *American Mineralogist*, v. 51, pp. 956–975.)

SiO₃ chains and as such produce a tetrahedral-octahedral-tetrahedral ("*t-o-t*") strip. The coordination of cations in the *M2* position, however, is such that several of these *t-o-t* strips are cross-linked. These *t-o-t* strips are often schematically represented as in Fig. 13.49*b*; this in turn shows the relationship of the *t-o-t* strips to the cleavage angles in pyroxenes. Figure 13.50 is a direct structure image of the features shown in Fig. 13.49.

The majority of pyroxenes can be assigned to one of three space groups, two monoclinic (*C2/c* and *P2₁/c*) and one orthorhombic (*Pbca*) (see Fig. 13.47*b*).

FIG. 13.49. (a) Schematic projection of the monoclinic pyroxene structure on a plane perpendicular to the *c* axis. (b) Control of cleavage angles by *t-o-t* strips (also referred to as "I-beams") in the pyroxene structure, as compared with naturally occurring pyroxene cleavage.

(a)   (b)

FIG. 13.50. High-resolution transmission electron microscope (HRTEM) image of an *a-b* section of enstatite. The white regions correspond to areas between *M2* sites in the structure. A unit cell and cleavage surfaces are outlined. The insert shows the enstatite structure at the same scale as the structure image. (From Buseck, P. R. and Iijima, S., 1974, High resolution electron microscopy of silicates. *American Mineralogist*, v. 59, pp. 1–21.)

The $C2/c$ structure is found in most of the common clinopyroxenes such as diopside, $CaMgSi_2O_6$, jadeite, $NaAlSi_2O_6$, and augite. This structure is illustrated in Fig. 13.48. See Fig. 3.28 for an illustration of the $C2/c$ space group. The *M1* site is generally occupied by cations that are smaller than those of the *M2* site. For example, in the diopside-hedenbergite series, $CaMgSi_2O_6$-$CaFeSi_2O_6$, the *M1* site is occupied by Mg and Fe^{2+} in random distribution, whereas the *M2* is occupied solely by the larger Ca^{2+} ion, in 8-coordination. The *M2* site, however, can also house Mn^{2+}, Fe^{2+}, Mg, or Li, in which case the coordination is 6-fold. The $P2_1/c$ space group is found in pigeonite (see Fig. 13.47) which may be represented as $Ca_{0.25}(Mg,Fe)_{1.75}Si_2O_6$ in which the *M2* site is occupied by all of the Ca in the formula as well as additional Mg and Fe so as to make the composition of the site $[Ca_{0.25}(Fe + Mg)_{0.75}]$. The *M2* site, as present in the clinopyroxenes with space group $C2/c$, is too large to accommodate these smaller cations, so that *M2* is somewhat smaller in pigeonite (hence, this site has irregular 7-coordination) due to a slight shift in the pigeonite structure as compared with the more regular diopside type structure. The orthorhombic *Pbca* structure is that found in the orthopyroxene series which contains virtually no Ca. The Mg and Fe^{2+} ions are distributed among *M1* and *M2* with the larger cation (Fe^{2+}) showing strong preference for the somewhat larger and distorted *M2* site. The coordination of both the *M1* and *M2* sites in orthopyroxenes is 6-fold.

The unit cells of the orthorhombic pyroxenes are related to the monoclinic unit cells by a twinlike mirror across {100} accompanied by an approximate doubling of the *a* cell dimension (e.g., *a* of enstatite ≈ $2\,a\sin\beta$ of diopside; see Fig. 13.51a). Figures 13.51b and c show schematically the development of monoclinic and orthorhombic pyroxene structures built up from *t-o-t* strips. The compositions represented by the orthopyroxene series may in rare occurrences be found in a monoclinic form, known as the *clinoenstatite-clinoferrosilite series*.

We will discuss in detail the following common pyroxenes:

PYROXENES

Enstatite-Ferrosilite Series

Enstatite	$MgSiO_3$
Ferrosilite	$FeSiO_3$
Pigeonite	$Ca_{0.25}(Mg,Fe)_{1.75}Si_2O_6$

Diopside-Hedenbergite Series

Diopside	$CaMgSi_2O_6$
Hedenbergite	$CaFeSi_2O_6$
Augite	$XY(Z_2O_6)$

Sodium Pyroxene Group

Jadeite	$NaAlSi_2O_6$
Aegirine	$NaFe^{3+}Si_2O_6$
Spodumene	$LiAlSi_2O_6$

ENSTATITE—$MgSiO_3$-$(Mg,Fe)SiO_3$ FERROSILITE—$(Fe,Mg)SiO_3$

Crystallography. Orthorhombic; $2/m2/m2/m$. Prismatic habit, crystals rare. Usually massive, fibrous, or lamellar.

Pbca; $a = 18.22$, $b = 8.81$, $c = 5.17$ Å, for pure enstatite; $Z = 8$. *d*'s: 3.17(10), 2.94(4), 2.87(9), 2.53(4), 2.49(5). Unit cell parameters increase with Fe content (see Fig. 13.52).

Physical Properties. *Cleavage* {210} good. Because of the doubling of the *a* dimension in orthorhombic pyroxenes, the cleavage form is {210} rather than {110} as in monoclinic pyroxenes. Frequently

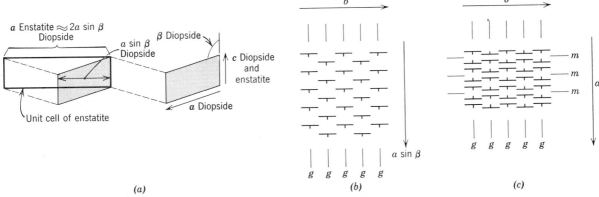

FIG. 13.51. *(a)* Relationship between unit cells of clinopyroxene (e.g., diopside) and orthopyroxene (e.g., enstatite) as projected on (010). The monoclinic unit cell outlined by shading can be related to the larger orthorhombic unit cell by a mirror reflection (compare with Fig. 13.72). *(b)* Schematic representation of a possible monoclinic pyroxene (looking down the *c* axis) with the *t-o-t* strips (or "I-beams") represented by strike and dip symbols; *g*'s indicate glide planes (with a translation component parallel to *c*). *(c)* Schematic representation of a possible orthorhombic pyroxene; *m*'s represent mirrors. (*b* and *c* after J. B. Thompson, Jr., Harvard University, pers. comm.)

good parting on {100}, less common on {001}. **H** $5\frac{1}{2}$–6 and **G** 3.2–3.6 for pure enstatite; increasing with Fe content. *Luster* vitreous to pearly on cleavage surfaces; En_{80} has submetallic, bronzelike luster. *Color* grayish, yellowish, or greenish-white to olive-green and brown. Translucent. *Optics*: enstatite (+); En_{90} to En_{10} (−). α = 1.653–1.710, β = 1.653–1.728, γ = 1.663–1.725, for En_{100} to En_{50}. $X = b$, $Z = c$. Indices increase with Fe content (see Fig. 13.53); in ferrosilite β = 1.765.

Composition and Structure. Fe^{2+} may substitute for Mg in all proportions up to nearly 90% $FeSiO_3$. However, in the more common orthopyroxenes the ratio of Fe : Mg rarely exceeds 1:1. Pure enstatite contains MgO 40.0, SiO_2 60.0%. The maximum amount of CaO in orthopyroxenes generally does not exceed 1.5 weight percent. The nomenclature for the orthopyroxenes is shown in Fig. 13.47; chemical compositions are generally expressed in terms of molecular percentages, for example, $En_{40}Fs_{60}$. The pure end member $FeSiO_3$, *ferrosilite*, is rarely found in nature because in most geologically observed pressure and temperature ranges the compositionally equivalent assemblage Fe_2SiO_4 (fayalite) + SiO_2 is more stable; all other varieties of the orthopyroxene series are found. The structure of members of the or-

FIG. 13.52. Variation in unit cell parameters as a function of composition in the orthopyroxene series. (From Deer, W. A., Howie, R. A. and Zussman, J., 1978, *Single-Chain Silicates*, 2nd ed., v. 2A. John Wiley & Sons, New York, and Longman Group Ltd., London, England.)

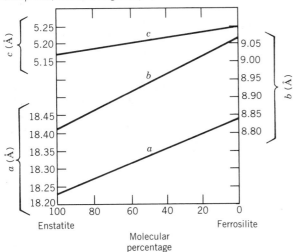

FIG. 13.53. Variation in two refractive indices (α and γ) and specific gravity as a function of composition in the orthopyroxene series. (From Deer, W. A., Howie, R. A. and Zussman, J., 1978, *Single-Chain Silicates*, 2nd ed., v. 2A. John Wiley and Sons, New York, and Longman Group Ltd., London, England.)

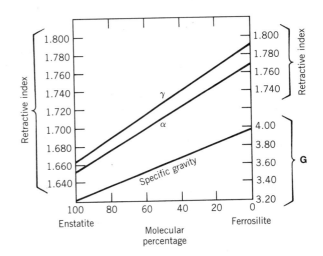

thopyroxene series can be considered to consist of the monoclinic *t-o-t* strips twinned along {100} so as to essentially double the *a* dimension of orthopyroxenes as compared to the *a* of clinopyroxenes. In the *Pbca* structure of orthopyroxenes Fe^{2+} shows a strong preference for the *M2* crystallographic site. Compositions between $MgSiO_3$-$FeSiO_3$ may also occur as members of the monoclinic series, *clinoenstatite-clinoferrosilite*, with space group $P2_1/c$. Experimental results of the stability fields of enstatite versus clinoenstatite are controversial. The common occurrence of orthopyroxenes versus the rare occurrence of clinopyroxenes in the series $MgSiO_3$-$FeSiO_3$ may indicate that the orthorhombic series is generally more stable, and at lower temperatures, than the monoclinic series.

Diagnostic Features. Usually recognized by their color, cleavage, and unusual luster. Varieties high in iron are black and difficult to distinguish from augite without optical tests.

Occurrence. Mg-rich orthopyroxene is a common constituent of peridotites, gabbros, norites, and basalts and is commonly associated with Ca-clinopyroxenes (e.g., augite), olivine, and plagioclase. It may be the major constituent of pyroxenites. Orthopyroxenes may also be found in metamorphic rocks, some types of which are of high *T* and high *P* origin, such as in the granulite facies. Iron-rich members of the orthopyroxene series are common in metamorphosed iron-formation in association with grunerite. In all such occurrences orthopyroxenes commonly coexist with clinopyroxenes because of a large miscibility gap between the two groups (see Fig. 13.47a). Orthopyroxenes frequently show exsolved lamellae of a Ca-rich clinopyroxene. Enstatite as well as clinoenstatite and clinohypersthene occur in both iron and stony meteorites. In prograde metamorphic rocks orthopyroxenes form commonly at the expense of Mg-Fe amphibole (e.g., anthophyllite) and in retrograde metamorphic rocks orthopyroxenes may give way to Mg-Fe amphiboles (see Fig. 13.54).

Enstatite is found in the United States at the Tilly Foster Mine, Brewster, New York, and at Edwards, St. Lawrence County, New York; at Texas, Pennsylvania; Bare Hills, near Baltimore, Maryland; and Webster, North Carolina. Hypersthene occurs in New York in the norites of the Cortland region, on the Hudson River, and in the Adirondack region. Fe-rich orthopyroxenes are common constituents of the metamorphosed Lake Superior and Labrador Trough iron-formations.

Use. A minor gemstone.

Name. Enstatite is named from the Greek word *enstates* meaning *opponent* because of its refractory nature. Hypersthene is named from two Greek words

FIG. 13.54. Schematic *P-T* diagram for the stability fields of anthophyllite, $Mg_7Si_8O_{22}(OH)_2$, and reaction products, enstatite, $MgSiO_3$, + quartz + H_2O. (After Greenwood, H. J., 1963, *Journal of Petrology*, v. 4, p. 325.)

meaning *very* and *strong* because its hardness is greater than that of hornblende.

Pigeonite ~ $Ca_{0.25}(Mg,Fe)_{1.75}Si_2O_6$

Crystallography. Monoclinic; 2/*m*. Very rarely as well-formed phenocrysts with a prismatic habit parallel to *c*.

$P2_1/c$; $a = 9.71$, $b = 8.96$, $c = 5.25$, $\beta = 108°33'$; $Z = 8$. *d*'s: 3.21(8), 3.02(10), 2.908(8), 2.904(10), 2.578(6).

Physical Properties. *Cleavage* {110} good; parting on {100} may be present. **H** 6. **G** 3.30–3.46. *Color* brown, greenish-brown, to black. *Optics*: (+); $\alpha = 1.682$–1.722, $\beta = 1.684$–1.722, $\gamma = 1.704$–1.752, increasing with Fe^{2+}. Two orientations occur: (1) with $Y = b$, and (2) more common, $X = b$. $Z \wedge c = 37°$–$44°$, 2V = 0°–30°.

Composition and Structure. Pigeonites are calcium-poor monoclinic pyroxenes that contain between 5 to 15 molecular percent of the $CaSiO_3$ component (see Fig. 13.47; field just above orthopyroxene base). The crystal structure of pigeonite is similar to that of diopside with all of the Ca and additional Fe and Mg in the *M2* site, and the remaining Mg and Fe in *M1*. Fe shows a strong preference for the *M2* sites. Pigeonite is stable at high temperatures in igneous rocks and inverts commonly at lower temperatures to orthopyroxene with augite-type exsolution lamellae. A temperature-composition stability diagram is given in Fig. 13.55.

Diagnostic Features. Can be distinguished from other pyroxenes only by optical or X-ray techniques. Augite 2V > 39°; pigeonite 2V < 32°.

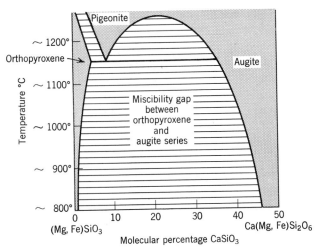

FIG. 13.55. Schematic *T*-composition section across the Wo-En-Fs diagram shown in Fig. 13.47*a*. The section is at about En$_{65}$ Fs$_{35}$.

Occurrence. Pigeonite is common in high-temperature, rapidly cooled lavas and in some intrusives such as diabases. It is present as phenocrysts in some volcanic rocks, but is not known from metamorphic rocks. If pigeonite formed in an igneous rock that cooled slowly, it may have exsolved augite lamellae on {001} and may subsequently have inverted to orthopyroxene (see Fig. 13.55) through a reconstructive transformation. At even lower temperatures the orthopyroxene may have developed augite exsolution along {100} planes.

Name. After the locality, Pigeon Cove, Minnesota.

DIOPSIDE—CaMgSi$_2$O$_6$
HEDENBERGITE—Ca,FeSi$_2$O$_6$
AUGITE—(Ca,Na)(Mg,Fe,Al)(Si,Al)$_2$O$_6$

Diopside and hedenbergite form a complete solid solution series with physical and optical properties varying linearly with composition. Augite is a clino-

pyroxene in which some Na substitutes for Ca, some Al substitutes for both Mg (or Fe) and Si, and in which Fe and Mg contents are higher than in diopside or hedenbergite (see Fig. 13.47). Although the crystal constants vary slightly from one member to another, a single description suffices for all.

Crystallography. Monoclinic; $2/m$. In prismatic crystals showing square or eight-sided cross section (Fig. 13.56). Also granular massive, columnar, and lamellar. Frequently twinned polysynthetically on {001}; less commonly twinned on {100}.

$C2/c$; $a = 9.73$, $b = 8.91$, $c = 5.25$ Å; $\beta = 105°50'$; $Z = 4$. d's: 3.23(8), 2.98(10), 2.94(7), 2.53(4), 1.748(4). See Fig. 3.28 for an illustration of the dinopyroxene structure and related $C2/c$ space group.

Physical Properties. *Cleavage* {110}, at 87° and 93°, imperfect. Frequently parting on {001}, and less commonly on {100} in the variety *diallage*. **H** 5–6. **G** 3.2–3.3. *Luster* vitreous. *Color* white to light green in diopside; deepens with increase of Fe. Augite is black. Transparent to translucent. *Optics*: (+); $\alpha = 1.67–1.74$, $\beta = 1.67–1.74$, $\gamma = 1.70–1.76$ (see Fig. 13.57). $2V = 55°–65°$; $Y = b$, $Z \wedge c = 39°–48°$; $r > v$. Darker members show pleochroism; X pale green, Y yellow-green, Z dark green.

Composition and Structure. In the diopside-hedenbergite series Mg and Fe^{2+} substitute for each other in all proportions. In the majority of analyses of members of this series the Al$_2$O$_3$ content varies be-

FIG. 13.57. Approximate trends of two refractive indices (α and γ) as a function of composition in the diopside-hedenbergite series. There is considerable scatter in the original data from which these lines were obtained. (From Deer, W. A., Howie, R. A. and Zussman, J., 1978, *Single-Chain Silicates*, v. 2A. John Wiley & Sons, New York, and Longman Group Ltd., London, England.)

FIG. 13.56. Augite crystals.

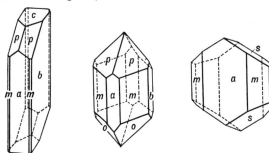

tween 1 and 3 weight percent. In augite, in addition to varying Mg and Fe^{2+} contents, Al substitutes for both Mg (or Fe^{2+}) and Si. Mn, Fe^{3+}, Ti, and Na may also be present. A complete series exists toward *aegirine-augite*, $(Na,Ca) (Fe^{3+},Fe^{2+},Mg,Al)Si_2O_6$, by the replacement $Ca(Mg,Fe^{2+}) \leftrightarrows NaFe^{3+}$. Compositional zoning is commonly found in igneous augites with the cores rich in the augite component and the rims tending toward aegirine-augite. Chemical analyses are generally recalculated in terms of molecular percentages of wollastonite (Wo), enstatite (En), and ferrosilite (Fs) and are expressed as $Wo_xEn_yFs_z$, where *x*, *y*, and *z* are molecular percentages (see Table 5.5). The structures of diopside, hedenbergite, and augite are all based on space group $C2/c$. Their structure is discussed on page 476, and a monoclinic structure is illustrated in Fig. 13.48. The Ca ions in the *M2* site are in 8-coordination, whereas the ions in *M1* are in 6-coordination.

Diagnostic Features. Characterized by crystal form and imperfect prismatic cleavage at 87° and 93°.

Occurrence. *Diopside* and *hedenbergite* are common in metamorphic rocks. Diopside, in association with forsterite and calcite, infrequently with monticellite, is the result of thermal metamorphism of siliceous, Mg-rich limestones, or dolomites. For example:

$$CaMg(CO_3)_2 + 2SiO_2 \rightarrow CaMgSi_2O_6 + 2CO_2$$
$$\text{dolomite} \qquad \text{quartz} \qquad \text{diopside}$$

Other associations include tremolite, scapolite, vesuvianite, garnet, and titanite. Hedenbergite occurs in more Fe-rich metamorphic rocks. Diopside and hedenbergite are also known as products of igneous crystallization. Early formed Ca-rich clinopyroxenes may be very close to diopside in composition, whereas the latest stages of crystallization may be represented by hedenbergite compositions, due to enrichment of the residual magma in Fe. The Skaergaard intrusion, East Greenland, contains late, fine-grained hedenbergite interstitial to earlier formed, coarser grained diopside and augite grains.

Fine crystals have been found in the Ural Mountains, CIS; Austrian Tyrol; Binnenthal, Switzerland; and Piemonte, Italy. At Nordmark, Sweden, fine crystals range between diopside and hedenbergite. In the United States they are found at Canaan, Connecticut; and DeKalb Junction and Gouverneur, New York.

Augite is the most common pyroxene and an important rock-forming mineral. It is found chiefly in the dark-colored igneous rocks, such as basaltic lavas and intrusives, gabbros, peridotites, and andesites.

Chemically zoned augites are common in quickly cooled rocks such as the lunar basalts. The clinopyroxene crystallization history is very well documented for many lunar basalts and gabbros as well as for the Skaergaard intrusion, East Greenland, for example; early formed crystals are more magnesian than later pyroxene grains. Fine crystals of augite have been found in the lavas of Vesuvius and at Val di Fassa, Trentino, Italy; and at Bilin, Czechoslovakia.

Use. Transparent varieties of diopside have been cut and used as gemstones.

Name. Diopside is from two Greek words meaning *double* and *appearance*, because the vertical prism zone can apparently be oriented in two ways. Hedenbergite is named after M. A. Ludwig Hedenberg, the Swedish chemist who discovered and described it. Augite comes from a Greek word meaning *luster*. The name pyroxene, *stranger to fire*, is a misnomer and was given to the mineral because it was thought that it did not occur in igneous rocks.

Jadeite—$NaAlSi_2O_6$

Crystallography. Monoclinic; $2/m$. Rarely in isolated crystals. Usually granular in compact, massive aggregates.

$C2/c$; $a = 9.50$, $b = 8.61$, $c = 5.24$ Å; $\beta = 107°26'$; $Z = 4$. *d*'s: 3.27(3), 3.10(3), 2.92(8), 2.83(10), 2.42(9).

Physical Properties. *Cleavage* {110} at angles of 87° and 93°. Extremely tough and difficult to break. **H** $6\frac{1}{2}$–7. **G** 3.3–3.5. *Color* apple-green to emerald-green, white. May be white with spots of green. *Luster* vitreous, pearly on cleavage surfaces. *Optics:* (+); $\alpha = 1.654$, $\beta = 1.659$, $\gamma = 1.667$; $2V = 70°$; $X = b$, $Z \wedge c = 34°$; $r > v$.

Composition and Structure. Na_2O 15.4, Al_2O_3 25.2, SiO_2 59.4 for pure end member. There is no replacement of Si by Al in jadeite and Fe^{3+} substitution for Al is very limited. Jadeite has a composition that is intermediate between that of nepheline, $NaAlSiO_4$, and albite, $NaAlSi_3O_8$, but does not form under the normal crystallization conditions of these two minerals. However, under high pressures (10 to 25 kilobars) and elevated temperatures (between 600° and 1000°C) jadeite forms by:

$$NaAlSiO_4 + NaAlSi_3O_8 \leftrightarrows 2\ NaAlSi_2O_6$$
$$\text{nepheline} \qquad \text{albite} \qquad \text{jadeite}$$

Similarly, jadeite forms at high pressures at the expense of albite alone according to the reaction:

$$NaAlSi_3O_8 \leftrightarrows NaAlSi_2O_6 + SiO_2$$

(see Fig. 13.58). The structure of jadeite is shown in Fig. 13.48.

Diagnostic Features. Characterized by its green color and tough aggregates of compact fibers. On polished surfaces nephrite has an oily luster, jadeite is vitreous.

Occurrence. Jadeite is found only in metamorphic rocks. Laboratory experiments have shown that high pressure and only relatively low temperatures are necessary for the formation of jadeite. Such occurrences are found near the margins of the continental crust as in the Alps, California, and Japan. In the Fransiscan Formation of California, jadeitic pyroxene is associated with glaucophane, aragonite, muscovite, lawsonite, and quartz.

Use. Jadeite has long been highly prized in the Orient, especially in China, where it is worked into ornaments and utensils of great variety and beauty. It was also used by primitive people for various weapons and implements.

Name. The term *jade* includes both jadeite and the amphibole, nephrite.

Similar Species. *Omphacite* is a bright green variety of augite-type composition rich in $NaAlSi_2O_6$, with space group $C2/c$ or $P2/n$. It occurs in eclogites which contain omphacite and pyrope-rich garnet as main constituents. These rocks are generally considered to be the result of high pressures and high temperatures of metamorphism. Eclogites are chemically equivalent to basalts, but contain denser minerals (see Fig. 9.17). Omphacite is also found in kimberlites. The name *omphacite* is from the Greek, *omphax*, an unripe grape, in allusion to its characteristic color.

FIG. 13.58. Experimentally determined stability fields of albite and jadeite + quartz. Reaction curve is only approximately located as shown by dashes.

AEGIRINE—NaFe^{3+}Si$_2$O$_6$

Crystallography. Monoclinic; $2/m$. Crystals slender prismatic with steep terminations. Often in fibrous aggregates. Faces often imperfect.

$C2/c$; $a = 9.66$, $b = 8.79$, $c = 5.26$ Å; $\beta = 107°20'$; $Z = 4$. d's: 6.5(4), 4.43(4), 2.99(10), 2.91(4), 2.54(6).

Physical Properties. *Cleavage* {110} imperfect at angles of 87° and 93°. **H** $6-6\frac{1}{2}$. **G** 3.40–3.55. *Luster* vitreous. *Color* brown or green. Translucent. *Optics* for aegirine: $(-)$; $\alpha = 1.776$, $\beta = 1.819$, $\gamma = 1.836$; $2V = 60°$; $Y = b$, $Z \wedge c = 8°$; $r > v$. Aegirine-augite: $(+)$ with lower indices and pleochroism in greens and brown.

Composition and Structure. Although the term *aegirine* is the recommended name for $NaFe^{3+}Si_2O_6$, both *aegirine* and *acmite* have been used for pyroxenes of this composition. Aegirines show a wide range of composition but in most species the replacement is according to $NaFe^{3+} \rightleftarrows Ca(Mg,Fe^{2+})$, which causes a complete series to augite, with an intermediate member known as *aegirine-augite*. Compositional zoning is very common in aegirine and compositions trending toward augite. Earlier crystallized material is richer in augite (in cores) and rims tend to be enriched in the aegirine component. The structure of aegirine is similar to that of other $C2/c$ pyroxenes (see Fig. 13.48).

Diagnostic Features. The slender prismatic crystals, brown to green color, and association are characteristic. However, it is not easily distinguished without optical tests.

Occurrence. Aegirine is a comparatively rare rock-forming mineral found chiefly in igneous rocks rich in Na and poor in SiO_2 such as nepheline syenite and phonolite. Associated with orthoclase, feldspathoids, augite, and soda-rich amphiboles. It is also found in some metamorphic rocks associated with glaucophane or riebeckite. It occurs in the nepheline syenites and related rocks of Norway, southern Greenland, and Kola Peninsula, the CIS. In the United States it is found in fine crystals at Magnet Cove, Arkansas. In Canada at Mount St. Hilaire, Quebec.

Name. After Aegir, the Scandinavian sea god, the mineral being first reported from Norway.

Spodumene—LiAlSi$_2$O$_6$

Crystallography. Monoclinic; $2/m$. Crystals prismatic, frequently flattened on {100}. Deeply striated vertically. Crystals usually coarse with roughened faces; some very large. Occurs also in cleavable masses. Twinning on {100} common.

$C2/c$; $a = 9.52$, $b = 8.32$, $c = 5.25$ Å; $\beta = 110°28'$; $Z = 4$. d's: 4.38(5), 4.21(6), 2.93(10), 2.80(8), 2.45(6).

Physical Properties. *Cleavage* {110} perfect at angles of 87° and 93°. Usually a well-developed parting on {100}. **H** $6\frac{1}{2}$–7. **G** 3.15–3.20. *Luster* vitreous. *Color* white, gray, pink, yellow, green. Transparent to translucent. *Optics:* (+); $\alpha = 1.660$, $\beta = 1.666$, $\gamma = 1.676$; $2V = 58°$; $Y = b$, $Z \wedge c = 24°$. $r < v$. Absorption: $X > Y > Z$. The clear lilac-colored variety is called *kunzite*, and the clear emerald-green variety *hiddenite*.

Composition and Structure. Li_2O 8.0, Al_2O_3 27.4, SiO_2 64.6%. A small amount of Na usually substitutes for Li. The structure of spodumene is the same as that of other $C2/c$ pyroxenes. The cell volume of spodumene is smaller than that of diopside, for example, because the larger Ca and Mg ions are substituted for by smaller Li and Al. This reduction in ionic sizes causes a somewhat closer packing of the SiO_3 chains.

Diagnostic Features. Characterized by its prismatic cleavage and {100} parting. The angle formed by one cleavage direction and the {100} parting resembles the cleavage angle of tremolite.

Alteration. Spodumene very easily alters to other species, becoming dull. The alteration products include clay minerals, albite, *eucryptite*, $LiAlSiO_4$, muscovite, and microcline.

Occurrence. Spodumene is found almost exclusively in lithium-rich pegmatites. Although it is a comparatively rare mineral, it occasionally occurs in very large crystals. At the Etta Mine, Black Hills, South Dakota, crystals measuring as much as 40 feet in length and weighing many tons have been found. It formerly was mined as the chief source of lithium, but today other minerals such as petalite, lepidolite, and amblygonite are of equal or greater importance. Also much of the lithium of commerce is extracted as Li_2CO_3 from brines. The major producers of lithium are the CIS, China, and Zimbabwe.

The principal countries for the production of gem spodumene are Brazil and Afghanistan. In the United States, beautiful crystals of kunzite have come from California, notably the Pala district in San Diego County. Hiddenite comes from Stony Point, Alexander County, North Carolina.

Use. As a gemstone and as a source of lithium. A major use of lithium is in grease to help it retain its lubricating properties over a wide range of temperatures. It is also used in aluminum, ceramics, storage batteries, air conditioning, and as a welding flux.

Names. *Spodumene* comes from a Greek word

meaning *ash-colored*. *Hiddenite* is named for W. E. Hidden; *kunzite*, for G. F. Kunz.

Similar Species. *Eucryptite*, $LiAlSiO_4$, is a major source of lithium at Bikita, Zimbabwe.

Pyroxenoid Group

There are a number of silicate minerals that have, as do the pyroxenes, a ratio of Si : O = 1 : 3, but with structures that are not identical to those of the pyroxenes. Both pyroxene and pyroxenoid structures contain octahedrally coordinated cations between SiO_3 chains but in pyroxenoids the geometry of the chains is not of the simple, infinitely extending type with a repeat distance of about 5.2 Å along the direction of the chain (see Fig. 13.59). In wollastonite, $CaSiO_3$, the smallest repeat of the chain consists of three twisted tetrahedra with a repeat distance of 7.1 Å (see Fig. 13.60), and in rhodonite, $MnSiO_3$, the unit repeat is built of five twisted tetrahedra with a repeat distance of 12.5 Å. Because of the lower symmetry of the chains (as compared with the pyroxene chain) the structures of pyroxenoids are triclinic. The chain structure in the pyroxenoids is expressed by their generally splintery cleavages and sometimes fibrous habit. We will discuss in detail the following three members of the pyroxenoid group:

Wollastonite	$CaSiO_3$
Rhodonite	$MnSiO_3$
Pectolite	$Ca_2NaH(SiO_3)_3$

WOLLASTONITE—$CaSiO_3$

Crystallography. Triclinic; $\bar{1}$. Rarely in tabular crystals with either {001} or {100} prominent. Commonly massive, cleavable to fibrous; also compact. *Pseudowollastonite*, $CaSiO_3$, is a polymorphic form stable above 1120°C; it is triclinic, pseudohexagonal, with different properties from wollastonite.

$P\bar{1}$; $a = 7.94$, $b = 7.32$, $c = 7.07$; $\alpha = 90°2'$, $\beta = 95°22'$, $\gamma = 103°26'$; $Z = 6$. d's: 3.83(8), 3.52(8), 3.31(8), 2.97(10), 2.47(6).

Physical Properties. *Cleavage* {100} and {001} perfect, {$\bar{1}$01} good giving splintery fragments elongated on b. **H** 5–$5\frac{1}{2}$. **G** 2.8–2.9. *Luster* vitreous, pearly on cleavage surfaces. May be silky when fibrous. *Color* colorless, white, or gray. Translucent. *Optics:* (−); $\alpha = 1.620$, $\beta = 1.632$, $\gamma = 1.634$; $2V = 40°$; Y near b, $X \wedge c = 32°$.

Composition and Structure. CaO 48.3, SiO_2 51.7% for pure $CaSiO_3$. Most analyses are very close to the pure end member in composition, although

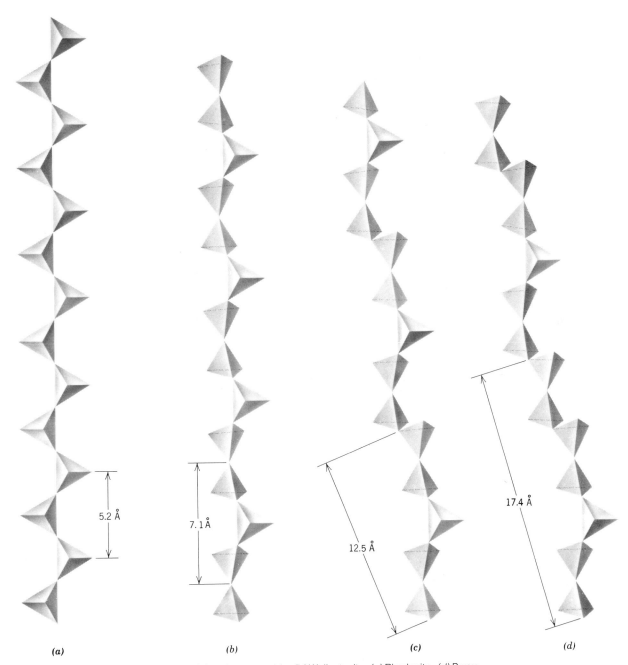

FIG. 13.59. SiO$_3$ chains in pyroxene *(a)* and pyroxenoids. *(b)* Wollastonite. *(c)* Rhodonite. *(d)* Pyroxmangite. (After Liebau, F., 1959, Über die Kristallstruktur des Pyroxmangits (Mn, Fe, Ca, Mg) SiO$_3$, *Acta Crystallographica*, v. 12, pp. 177–181.)

considerable amounts of Fe and Mn, and lesser Mg, may replace Ca. The structure of wollastonite consists of infinite chains, parallel to the *c* axis, with a unit repeat of three twisted tetrahedra (see Figs. 13.59*b* and 13.60). Ca is in irregular octahedral coordination and links the SiO$_3$ chains. *Pseudowollastonite,* stable above 1120°C, has space group *P*$\bar{1}$ but a much larger unit cell (*Z* = 24 as compared to *Z* = 6 for wollas-

tonite). The basic structure of pseudowollastonite is very similar to that of wollastonite.

Diagnostic Features. Characterized by its two perfect cleavages of about 84°. It resembles tremolite but is distinguished from it by the cleavage angle.

Occurrence. Wollastonite occurs chiefly as a contact metamorphic mineral in crystalline limestones, and forms by the reaction:

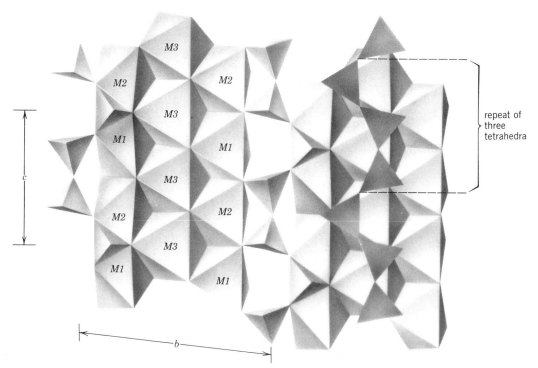

FIG. 13.60. The crystal structure of triclinic wollastonite. The tetrahedral silicate chain has a repeat of three tetrahedra. The Ca^{2+} ions are located in octahedral bands that are parallel to c and are three octahedra wide. Octahedral sites are identified as *M1*, *M2*, and *M3*. (Adopted from Papike, J. J., 1987, Chemistry of the rock-forming silicates: Ortho, ring, and single-chain structures. *Reviews of Geophysics and Space Physics*, v. 25, pp. 1483–1526.)

$$CaCO_3 \quad + SiO_2 \quad \rightarrow CaSiO_3 \quad + CO_2$$

calcite   quartz   wollastonite

It is associated with calcite, diopside, andradite, grossular, tremolite, plagioclase feldspar, vesuvianite, and epidote. During progressive metamorphism of siliceous dolomites the following approximate sequence of mineral formation is often found, beginning with the lowest-temperature product: talc-tremolite-diopside-forsterite-wollastonite-periclase-monticellite (see Fig. 14.16).

In places it may be so plentiful as to constitute the chief mineral of the rock mass. Such wollastonite rocks are found in the Black Forest, Germany; in Brittany, France; in Willsboro, New York; in California; and in Mexico. Crystals of the mineral are found at Csiklova in Romania; the Harz Mountains, Germany; and Chiapas, Mexico. In the United States found in New York at Diana, Lewis County, and St. Lawrence County. In California at Crestmore, Riverside County.

Use. Wollastonite is mined in those places where it constitutes a major portion of the rock mass and is used in the manufacture of tile.

Name. In honor of the English chemist, W. H. Wollaston (1766–1828).

RHODONITE—MnSiO₃

Crystallography. Triclinic; $\bar{1}$. Crystals commonly tabular parallel to {001} (Fig. 13.61); often rough with rounded edges. Commonly massive, cleavable to compact; in embedded grains.

$P\bar{1}$; a = 7.79, b = 12.47, c = 6.75 Å; α = 85°10', β = 94°4', γ = 111° 29'; Z = 10. d's: 4.78(4), 3.15(5), 3.09(3), 2.98(8), 2.93(9), 2.76(10).

Physical Properties. *Cleavage* {110} and {1$\bar{1}$0} perfect. **H** 5½–6. **G** 3.4–3.7. *Luster* vitreous. *Color* rose-red, pink, brown; frequently with black exterior of manganese oxide. Transparent to translucent. *Optics:* (+); α = 1.716–1.733, β = 1.720–1.737, γ = 1.728–1.747; 2V = 60°–75°, $r < v$.

Composition and Structure. Rhodonite is never pure MnSiO₃ but always contains some Ca, with a

FIG. 13.61. Rhodonite.

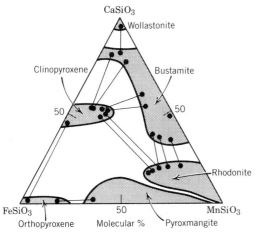

FIG. 13.62. Extent of compositional fields for pyroxenoid and pyroxene compositions in the system $CaSiO_3$-$FeSiO_3$-$MnSiO_3$. The temperature and pressure conditions applicable to this diagram are about 600°C and 6 kilobars, respectively. Straight line segments (tielines) join naturally occurring mineral pairs. (Redrawn from Brown, P. E., Essene, E. J. and Peacor, D. R., 1980, Phase relations inferred from field data for Mn pyroxenes and pyroxenoids. *Contributions to Mineralogy and Petrology*, v. 74, pp. 417–425; see also *Pyroxenes, Reviews in Mineralogy*, complete reference at end of chapter.)

maximum $CaSiO_3$ content of about 20 molecular percent. Fe^{2+} may replace Mn up to as much as 14 weight percent FeO (see Fig. 13.62). Zn may be present, and Zn-rich varieties are known as *fowlerite*. The structure of rhodonite consists of SiO_3 chains, parallel to the c axis, with a unit repeat of five twisted tetrahedra (see Fig. 13.59c). Layers of cations alternate with the chains. The structure is similar to that of wollastonite (see Fig. 13.60), and pyroxmangite, $(Mn,Fe)SiO_3$.

Diagnostic Features. Characterized by its pink color and near 90° cleavages. Distinguished from rhodochrosite by its greater hardness.

Occurrence. Rhodonite occurs in manganese deposits and manganese-rich iron formations, as a result of metamorphic and commonly associated metasomatic activity. It may form from rhodochrosite by the reaction:

$$MnCO_3 + SiO_2 \rightarrow MnSiO_3 + CO_2$$
rhodochrosite   quartz   rhodonite

Rhodonite is found at Långban, Sweden, with other manganese minerals and iron ore; in large masses near Sverdlovsk in the Ural Mountains, CIS; and at Broken Hill, New South Wales, Australia. In the United States rhodonite occurs in good-sized crystals in crystalline limestone with franklinite, willemite,

zincite, and so forth, at Franklin, New Jersey (Fig. 13.63).

Use. Some rhodonite is polished for use as an ornamental stone. This material is obtained chiefly from the Ural Mountains, CIS, and Australia.

Name. Derived from the Greek word for *a rose*, in allusion to the color.

Similar Species. *Pyroxmangite*, $(Mn,Fe)SiO_3$, is structurally very similar to rhodonite, but with a unit repeat of seven tetrahedra in the SiO_3 chain (see Fig. 13.59d). *Pyroxferroite*, $Ca_{0.15}Fe_{0.85}SiO_3$, isostructural with pyroxmangite, is a relatively common mineral in lunar lavas. *Bustamite*, $(Mn,Ca,Fe)SiO_3$, is very similar in structure to wollastonite. All of these pyroxenoids have extensive fields of solid solution, as shown in Fig. 13.62. *Tephroite*, Mn_2SiO_4, a red to gray mineral associated with rhodonite, is isostructural with olivine.

Pectolite—$Ca_2NaH(SiO_3)_3$

Crystallography. Triclinic; $\bar{1}$. Crystals elongated parallel to the b axis. Usually in aggregates of acicular crystals. Frequently radiating, with fibrous appearance (Fig. 13.64). In compact masses.

$P\bar{1}$; $a = 7.99$, $b = 7.04$, $c = 7.02$ Å; $\alpha = 90°31'$, $\beta = 95°11'$, $\gamma = 102°28'$; $Z = 2$. d's: 3.28(7), 3.08(9), 2.89(10), 2.31(7), 2.28(7).

Physical Properties. *Cleavage* {001} and {100} perfect. **H** 5. **G** 2.8±. *Luster* vitreous to silky. *Color* colorless, white, or gray. Transparent. *Optics:* (+); $\alpha = 1.595$, $\beta = 1.604$, $\gamma = 1.633$; $2V = 60°$; $Z \approx b$, $X \wedge c = 19°$.

FIG. 13.63. Rhodonite, Franklin, New Jersey. (Harvard Mineralogical Museum.)

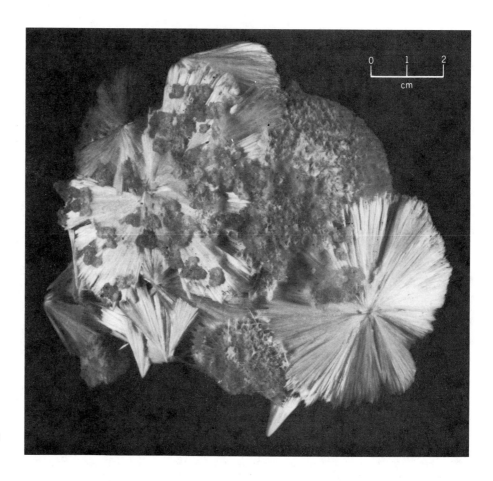

FIG. 13.64. Pectolite, Paterson, New Jersey. (Harvard Mineralogical Museum.)

Composition and Structure.   CaO 33.8, Na_2O 9.3, SiO_2 54.2, H_2O 2.7%. In some pectolite Mn^{2+} substitutes for Ca. The structure of pectolite contains SiO_3 chains, parallel to the *b* axis, the unit repeat of which consists of three twisted tetrahedra (see Fig. 13.59*b*) similar to that found in wollastonite. The Ca ions are in octahedral coordination, and Na is present in very distorted octahedral coordination.

Diagnostic Features.   Characterized by two directions of perfect cleavage, yielding sharp, acicular fragments that will puncture the skin if not handled carefully. Resembles wollastonite.

Occurrence.   Pectolite is a secondary mineral similar in its occurrence to the zeolites. Found lining cavities in basalt, associated with various zeolites, prehnite, calcite, and so forth. Found at Bergen Hill and West Paterson, New Jersey. In Canada, at Asbestos and Mount St. Hilaire, Quebec.

Name.   From the Greek word meaning *compact*, in allusion to its habits.

Amphibole Group

The chemical composition of members of the amphibole group can be expressed by the general formula $W_{0-1}X_2Y_5Z_8O_{22}(OH,F)_2$, where *W* represents Na^+ and K^+ in the *A* site, *X* denotes Ca^{2+}, Na^+, Mn^{2+}, Fe^{2+}, Mg^{2+}, and Li^+ in the *M4* sites, *Y* represents Mn^{2+}, Fe^{2+}, Mg^{2+}, Fe^{3+}, Al^{3+}, and Ti^{4+} in the *M1*, *M2*, and *M3* sites, and *Z* refers to Si^{4+} and Al^{3+} in the tetrahedral sites. Essentially complete ionic substitution may take place between Na and Ca and among Mg, Fe^{2+}, and Mn^{2+}. There is limited substitution between Fe^{3+} and Al and between Ti and other *Y*-type ions; and partial substitution of Al for Si in the tetrahedral sites of the double chains. Partial substitution of F and O for OH in the hydroxyl sites is also common. Several common series of the amphibole group can be represented compositionally in the chemical system $Mg_7Si_8O_{22}(OH)_2$ (anthophyllite)–$Fe_7Si_8O_{22}(OH)_2$ (grunerite)–"$Ca_7Si_8O_{22}(OH)_2$," a hypothetical end member composition, as in Fig. 13.65. This illustration is very similar to the Wo-En-Fs diagram for the pyroxene group (Fig. 13.47). A complete series exists from *tremolite*, $Ca_2Mg_5Si_8O_{22}(OH)_2$, to *ferroactinolite*, $Ca_2Fe_5Si_8O_{22}(OH)_2$. The commonly occurring *actinolite* composition is a generally Mg-rich member of the tremolite-ferroactinolite series. The compositional range from $Mg_7Si_8O_{22}(OH)_2$ to about $Fe_2Mg_5Si_8O_{22}(OH)_2$ is represented by the orthorhombic species, *anthophyl-*

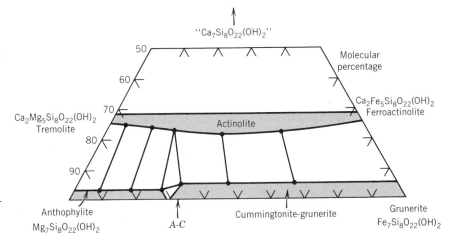

FIG. 13.65. Amphibole compositions in the system $Mg_7Si_8O_{22}(OH)_2$-$Fe_7Si_8O_{22}(OH)_2$-"$Ca_7Si_8O_{22}(OH)_2$". General compositional fields are outlined. Representative tielines across miscibility gaps are shown. Compare with Fig. 13.47. The composition marked A-C refers to an anthophyllite-cummingtonite intergrowth illustrated in Fig. 13.72.

lite. The *cummingtonite-grunerite* series is monoclinic and extends from about $Fe_2Mg_5Si_8O_{22}(OH)_2$ to $Fe_7Si_8O_{22}(OH)_2$. Miscibility gaps are present between anthophyllite and the tremolite-actinolite series, as well as between the members of the cummingtonite-grunerite series and the calcic amphiboles. These gaps are reflected in the common occurrence of anthophyllite-tremolite and grunerite-actinolite pairs, as well as exsolution textures between members of the Mg-Fe and the Ca-containing amphiboles of Fig. 13.65. No Ca-amphibole compositions are possible above the 2/7th line (representing two Ca out of a total of seven X + Y cations) because Ca can be housed only in the two M4 sites of the amphibole structure. *Hornblende* may be regarded as a tremolite-ferroactinolite type composition with additional partial substitution of Na in the A and M4 sites, Mn, Fe^{3+},

and Ti^{4+} for Y cations, and Al for Si in the tetrahedral sites, leading to a very complex general formula. Sodium-containing amphiboles are represented by members of the *glaucophane*, $Na_2Mg_3Al_2Si_8O_{22}(OH)_2$,–*riebeckite*, $Na_2Fe_3^{2+}Fe_2^{3+}$-$Si_8O_{22}(OH)_2$, series. *Arfvedsonite*, $NaNa_2Fe_4^{2+}Fe^{3+}$-$Si_8O_{22}(OH)_2$, contains additional Na in the A site of the structure. Table 13.3 compares the composition of some of the common members of the pyroxene and amphibole groups, and Table 5.6 illustrates the recalculation of an amphibole analysis.

The amphibole structure is based on double Si_4O_{11} chains that run parallel to the c axis. Figure 13.66 (see also Fig. 13.3) illustrates this chain as well as the octahedral strip to which it is bonded. The structure contains several cation sites, labeled A, M4, M3, M2, M1, as well as the tetrahedral sites in the

	Pyroxenes			Amphiboles		
Atomic Sites		**Name**		**Atomic Sites**		**Name**
M2	M1		A	M4	(M1 + M2 + M3)	
Mg	Mg	Enstatite	□*	Mg	Mg	Anthophyllite
		other members of	□	Fe	Mg	Cummingtonite
Fe	Mg	the orthopyroxene series	□	Fe	Fe	Grunerite
Ca	Mg	Diopside	□	Ca	Mg	Tremolite
Ca	Fe	Hedenbergite	□	Ca	Fe	Ferroactinolite
Ca	Mn	Johannsenite				
Ca, Na	Mg, Fe, Mn, Al, Fe^{3+}, Ti	Augite	□	Ca, Na	Mg, Fe^{2+}, Mn, Al, Fe^{3+}, Ti	Hornblende
Na	Al	Jadeite	□	Na	Mg, Al	Glaucophane
Na	Fe^{3+}	Aegirine	□	Na	Fe^{2+}, Fe^{3+}	Riebeckite
			Na	Na	Fe^{2+}, Fe^{3+}	Afrvedsonite
Li	Al	Spodumene	□	Li	Mg, Fe^{3+} Al, Fe^{2+}	Holmquistite

Table 13.3
IONS IN COMMON PYROXENES AND AMPHIBOLES

*□ represents a vacant atomic site.

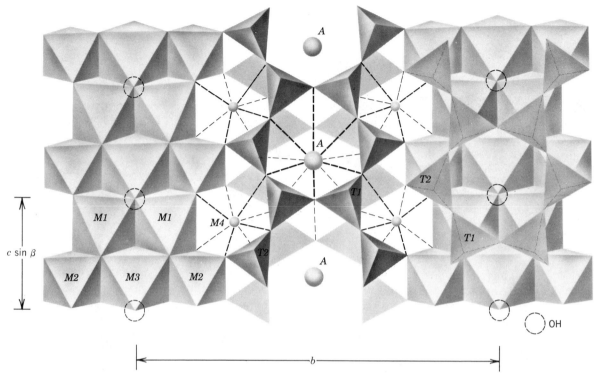

FIG. 13.66. Crystal structure of a monoclinic (*C2/m*) amphibole projected down the *a* axis. (OH) groups are located in the center of the large holes of the rings in the chains. *M1, M2,* and *M3* sites house *Y* cations and are 6-coordinated. The *M4* site houses larger *X* cations in 6- to 8-coordination. The ion in the *A* site, between the backs of double chains is 10- to 12-coordinated. (After Papike, J. J. et al., 1969, Mineralogical Society of America Special Paper no. 2, p. 120.)

chains. The *A* site has 10- to 12-coordination with oxygen and (OH) and houses mainly Na, and at times small amounts of K. The *M4* site has 6- to 8-coordination and houses *X* type cations (see Table 13.3). The *M1, M2,* and *M3* octahedra accommodate *Y* type cations and share edges to form octahedral bands parallel to *c. M1* and *M3* are coordinated by four oxygens and two (OH,F) groups, whereas *M2* is coordinated by six oxygens. Figure 13.67a shows the monoclinic amphibole structure and the distribution of cation sites as seen in a direction parallel to the *c* axis. The *t-o-t* strips (Fig. 13.67b) are approximately twice as

FIG. 13.67. *(a)* Schematic projection of the monoclinic amphibole structure on a plane perpendicular to the *c* axis (after Colville et al., 1966, *American Mineralogist*, v. 51, p. 1739). Compare with Fig. 13.49a. *(b)* Control of cleavage angles by *t-o-t* strips (also referred to as "I-beams") in the amphibole structure, as compared with naturally occurring cleavage angles.

FIG. 13.68. High-resolution transmission electron microscope (HRTEM) image of an *a-b* section of hornblende. The white regions correspond to the *A* sites. A unit cell and cleavage surfaces are outlined. The insert shows the hornblende structure at the same scale as the structure image. (From Buseck, P. R. and Iijima, S., 1974, High resolution electron microscopy of silicates. *American Mineralogist*, v. 59, pp. 1–21.)

wide (in the *b* direction) as the equivalent *t-o-t* strips in pyroxenes (see Fig. 13.49*b*) because of the doubling of the chain width in amphiboles. This wider geometry causes the typical 56° and 124° cleavage angles as shown in Fig. 13.67*b*. Figure 13.68 is a direct structure image of the features shown in Fig. 13.67.

The majority of amphiboles can be assigned to one of three space groups, two monoclinic (*C2/m* and *P2₁/m*) and one orthorhombic (*Pnma*). The *C2/m* structure is found in all of the common clinoamphiboles such as tremolite, $Ca_2Mg_5Si_8O_{22}(OH)_2$, and hornblende. This type of structure is illustrated in Fig. 13.66. The *P2₁/m* space group is found in some Mg-rich cummingtonites, close to (Mg, Fe)$_2$Mg$_5$Si$_8$O$_{22}$(OH)$_2$ in composition, and results because the *M4* site is somewhat collapsed to house the relatively small Mg and Fe^{2+} ions. The orthorhombic *Pnma* structure is found in members of the anthophyllite, $Mg_7Si_8O_{22}(OH)_2$ to $Fe_2Mg_5Si_8O_{22}(OH)_2$, series and in gedrite, an Al and Na-containing anthophyllite, and in holmquistite. The presence of cations of small size in *M4*, *M3*, *M2*, and *M1* leads to an orthorhombic rather than a monoclinic structure. The unit cells of orthorhombic amphiboles are related to the monoclinic unit cells by a twinlike mirror across (100) accompanied by an approximate doubling of the *a* cell dimension (e.g., *a* of anthophyllite ≈ 2 *a* sin β of cummingtonite). This is illustrated schematically and also with a high-resolution transmission electron microscope (HRTEM) image in Fig. 13.72. This relationship between the structures of clinoamphiboles and orthoamphiboles is identical to that for clinopyroxenes and orthopyroxenes (see Fig. 13.51).

The presence of (OH) groups in the structure of amphiboles causes a decrease in their thermal stabilities as compared with the more refractory pyroxenes. This causes amphiboles to decompose to anhydrous minerals (often pyroxenes) at elevated temperatures below the melting point (see Figs. 13.54 and 13.76).

In 1977 several new, ordered structures that are closely related to amphiboles were reported by D. R. Veblen (*Science*, 1977, v. 198, pp. 359–365). These are referred to as *biopyriboles*, a term derived from *bio*tite (mica), *pyr*oxene, and amph*ibole*. This collective term reflects the close architectural connection between numbers of the pyroxene, amphibole, and layer silicate groups. Indeed, two of the new structures discovered, *jimthompsonite*, $(Mg,Fe)_{10}Si_{12}O_{32}(OH)_4$, and *chesterite*, $(Mg,Fe)_{17}Si_{20}O_{54}(OH)_6$, have chain widths and chain repeat sequences that can be interpreted in terms of single chain repeats (SiO_3 chains as in pyroxenes), double chain repeats (Si_4O_{11}(OH) chains as in amphiboles), and triple chain repeats. These triple chain repeats are wider than those known in pyroxenes and amphiboles, and their increased width is suggestive of mica structures with infinitely extending sheets. Schematic illustrations of the pyroxene, amphibole, and new biopyribole structures are given in Fig. 13.69 in terms of I-beams. A direct image of some of these same structures, by high-resolution transmission electron microscopy (HRTEM), is given in Fig. 13.70.

The new ordered structure types were found in a fine intergrowth with anthophyllite, and are considered low-temperature alteration (and hydration) products of original enstatite and anthophyllite. They appear to be intermediate stages of structural devel-

FIG. 13.69. Schematic illustration of "I-beams" projected onto the (001) plane in orthopyroxene, orthoamphibole, jimthompsonite, and chesterite. The digits mean: 1—single chain width as in pyroxene, 2—double chain width as in amphibole, and 3—triple chain width. (Redrawn after Veblen, D. R., Buseck, P. R. and Burnham, C. W., 1977, Asbestiform chain silicates: New minerals and structural groups. *Science*, v. 198, pp. 359–365.)

opment between pyroxenes (anhydrous, high-temperature minerals) and phyllosilicates (hydrous, lower-temperature minerals; see also Fig. 13.93). Many of the intermediate reaction products, however, turn out to be disordered rather than ordered structures. Figure 13.71 is an example of a random intergrowth of double, triple, and quadruple chain widths in a material whose precursor was an amphibole, with only double chain widths, prior to alteration. Solid state reactions, possibly as a result of low-temperature metamorphism, facilitated the wide chain construction.

We will discuss in detail the following common amphiboles:

AMPHIBOLES

Anthophyllite	$(Mg,Fe)_7Si_8O_{22}(OH)_2$
Cummingtonite Series	
Cummingtonite	$Fe_2Mg_5Si_8O_{22}(OH)_2$
Grunerite	$Fe_7Si_8O_{22}(OH)_2$
Tremolite Series	
Tremolite	$Ca_2Mg_5Si_8O_{22}(OH)_2$
Actinolite	$Ca_2(Mg,Fe)_5Si_8O_{22}(OH)_2$
Hornblende	$X_{2-3}Y_5Z_8O_{22}(OH)_2$
Sodium Amphibole Group	
Glaucophane	$Na_2Mg_3Al_2Si_8O_{22}(OH)_2$
Riebeckite	$Na_2Fe_3^{2+}Fe_2^{3+}Si_8O_{22}(OH)_2$

FIG. 13.70. High-resolution transmission electron microscope (HRTEM) images of *(a)* anthophyllite, *(b)* jimthompsonite, and *(c)* chesterite. In each image the *c* axis of the mineral is coincident with the vertical striping. In anthophyllite the 2's locate double Si_4O_{11} chains. In jimthompsonite the 3's refer to triple chains, and in chesterite the alternating 2's and 3's refer to alternating double and triple chains. The inset in each image is an electron diffraction pattern; this is used mainly by the microscopist for crystallographic orientation. (From Veblen, D. R., Buseck, P. R. and Burnham, C. W., 1977, Asbestiform chain silicates: New minerals and structural groups. *Science*, v. 198, pp. 359–365; copyright © 1977 by the AAAS. See also Veblen, D. R., 1981, Non-classical pyriboles and polysomatic reactions, *in Amphiboles and other hydrous pyriboles*, *Reviews in Mineralogy*, v. 9A, Mineralogical Soc. of America, pp. 189–236.)

FIG. 13.71. High-resolution transmission electron microscope (HRTEM) image of disordered chain silicate, with chains of variable width aligned perpendicular to the plane of the photograph. The black regions represent the chains, and the white spots represent regions of low electron density between chains. The 2, 3, and 4 refer to double, triple, and quadruple chains, respectively. The precursor of this silicate material was an amphibole with double chains only. (From Buseck P. R. and Veblen D. R., 1978, Trace elements, crystal defects, and high resolution microscopy. *Geochimica et Cosmochimica Acta,* v. 42, pp. 669–678; see also Buseck P. R., 1983. Electron microscopy of minerals. *American Scientist,* v. 71, pp. 175–185.) Compare this illustration with Fig. 3.53.

Anthophyllite—$(Mg,Fe)_7Si_8O_{22}(OH)_2$

Crystallography. Orthorhombic; $2/m2/m2/m$. Rarely in distinct crystals. Commonly lamellar or fibrous.

Pnma; a = 18.56, *b* = 18.08, *c* = 5.28 Å; *Z* = 4. *d*'s: 8.26(6), 3.65(4), 3.24(6), 3.05(10), 2.84(4).

Physical Properties. *Cleavage* {210} perfect; $(210) \wedge (2\bar{1}0) = 55°$. **H** $5\frac{1}{2}$–6. **G** 2.85–3.2. *Luster* vitreous. *Color* gray to various shades of green and brown and beige. *Optics:* (−); α = 1.60–1.69, β = 1.61–1.71, γ = 1.62–1.72. 2V = 70°–100°; *X* = *a*, *Y* = *b*. Absorption *Z* > *Y* and *X*. Indices increase with Fe content.

Composition and Structure. Anthophyllite is part of a solid solution series from $Mg_7Si_8O_{22}(OH)_2$ to approximately $Fe_2Mg_5Si_8O_{22}(OH)_2$; at higher Fe contents the monoclinic cummingtonite structure results. *Gedrite* is an Al and Na-containing variety of anthophyllite, with an end member composition approximating $Na_{0.5}(Mg,Fe^{2+})_2(Mg,Fe^{2+})_{3.5}$-$(Al,Fe^{3+})_{1.5}Si_6Al_2O_{22}(OH)_2$, with Na in the *A* site of the structure. At moderate temperatures a miscibility gap exists between anthophyllite and gedrite as shown by coexisting anthophyllite and gedrite grains. The structures of anthophyllite and gedrite are similar,

both with orthorhombic space group *Pnma*. The relationship of the unit cell of orthoamphibole to that of clinoamphibole is shown in Fig. 13.72.

Diagnostic Features. Characterized by its clove-brown color but unless in crystals cannot be distinguished from other amphiboles such as cummingtonite or grunerite without optical or X-ray tests.

FIG. 13.72. *(a)* Relationship between unit cells of clinoamphiboles (e.g., cummingtonite or tremolite) and orthoamphibole (anthophyllite). The monoclinic unit cell (outlined by shading) can be repeated by a mirror reflection, which results in a zigzag pattern. This repeat pattern is described by the larger orthorhombic unit cell outline. This relationship is identical in the pyroxenes (see Fig. 13.51). *(b)* High-resolution transmission electron microscope (HRTEM) image provides a direct picture of the relationships shown in *(a)*. A cummingtonite (Cum) lamella (about 140 Å thick) is set in a matrix of anthophyllite (Ant). The stacking of the unit cells in cummingtonite is consistently in the same direction. However, the stacking of the unit cells in anthophyllite is the result of a mirror reflection parallel to (100), resulting in the zigzag pattern. Both amphiboles have approximately the same composition: $Na_{0.1}Fe_{2.2}Mg_{4.5}Al_{0.1}Si_8O_{22}(OH)_2$. In other words, the structure image is one of two polymorphs. The value of (Mg/Fe $\times$ 100)% for this formula is 67%, which plots it at the 67% molecular percentage point along the bottom edge of the triangle in Fig. 13.65, marked *A-C*. (Micrograph courtesy of Eugene A. Smelnik, Princeton University.)

(a)

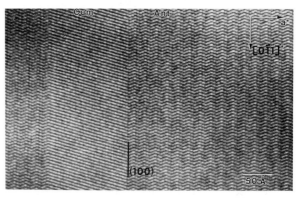

(b)

Occurrence. Anthophyllite is a metamorphic product of Mg-rich rocks such as ultrabasic igneous rocks and impure dolomitic shales. It is common in cordierite-bearing gneisses and schists. It may also form as a retrograde product rimming relict orthopyroxenes and olivine (see Fig. 13.54). Occurs at Kongsberg, Norway, and in many localities in southern Greenland. In the United States it is found at several localities in Pennsylvania, in southwestern New Hampshire and central Massachusetts, in the Gravelly Range and Tobacco Root Mountains of southwestern Montana, and at Franklin, North Carolina.

Name. From the Latin *anthophyllum*, meaning *clove*, in allusion to the clove-brown color.

CUMMINGTONITE—$(Mg,Fe)_7Si_8O_{22}(OH)_2$
GRUNERITE—$Fe_7Si_8O_{22}(OH)_2$

Crystallography. Monoclinic; $2/m$. Rarely in distinct crystals. Commonly fibrous or lamellar, often radiated.

$C2/m$; for grunerite, $a = 9.59$, $b = 18.44$, $c = 5.34$, $\beta = 102°0'$; cell lengths decrease with increasing Mg (see Fig. 13.73); $Z = 2$. d's: 9.21(5), 8.33(10), 3.07(8), 2.76(9), 2.51(6).

Physical Properties. *Cleavage* {110} perfect. **H** $5\frac{1}{2}$–6. **G** 3.1–3.6. *Luster* silky; fibrous. *Color* various shades of light brown. Translucent; will transmit light on thin edges. *Optics*: (−) for grunerite; (+) for cummingtonite; $\alpha = 1.65$–1.69; $\beta = 1.67$–1.71; $\gamma = 1.69$–1.73; 2V large; $Y = b$, $Z \wedge c = 13°$–$20°$ (see Fig. 13.74). $r < v$ for cummingtonite; $r > v$ for grunerite. Essentially nonpleochroic.

Composition and Structure. The cummingtonite-grunerite series extends from approximately $Fe_2Mg_5Si_8O_{22}(OH)_2$ to the end member $Fe_7Si_8O_{22}(OH)_2$. Members with Mg < Fe (atomic percentage) are referred to as cummingtonite; those with Fe > Mg as grunerite (in the literature this division is often taken at 30 atomic percent Fe). Mn^{2+} can be present to as much as $Mn_2Mg_5Si_8O_{22}(OH)_2$, with probably most of the Mn in the *M4* structure site. Al_2O_3 and CaO range up to a maximum of 0.4 and 0.9 weight percent, respectively. The structure (see Fig. 13.66) of the majority of members of the cummingtonite-grunerite series is $C2/m$, like that of tremolite; however, some Mg-rich cummingtonites have $P2_1/m$ symmetry.

Diagnostic Features. Characterized by light brown color and needlelike, often radiating habit. May be impossible to distinguish from anthophyllite or gedrite without optical or X-ray tests.

Occurrence. Cummingtonite is a constituent of regionally metamorphosed rocks and occurs in am-

phibolites. It commonly coexists with hornblende or actinolite (see tielines in Fig. 13.65). Mg-rich cummingtonite can also coexist with anthophyllite, because of a small miscibility gap between anthophyllite and the Mg-rich part of the cummingtonite-grunerite series (see Fig. 13.72b). Cummingtonite phenocrysts have been reported in some igneous rocks such as dacites. Mn-rich varieties occur in metamorphosed manganese-rich units. Grunerites are characteristic of metamorphosed iron-formations in the Lake Superior region and the Labrador Trough. Upon prograde metamorphism cummingtonite and grunerite give way to members of the orthopyroxene or olivine series.

Use. *Amosite*, a very rare asbestiform variety of

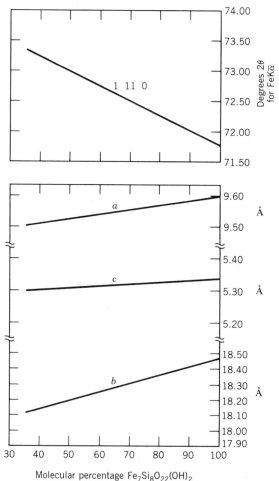

FIG. 13.73. Variation in the unit cell parameters *a*, *b*, and *c* as a function of composition in the cummingtonite-grunerite series. The topmost line relates the angular (2θ) position of the 1, 11, 0 X-ray diffraction peak (as recorded on a diffractometer pattern, or X-ray photograph; see Chapter 7) as a function of composition in the cummingtonite-grunerite series (using $FeK\overline{\alpha}$ radiation). (From Klein, C. and Waldbaum, D. R., 1967, X-ray crystallographic properties of the cummingtonite-grunerite series. *Journal of Geology*, v. 75, pp. 379–392.)

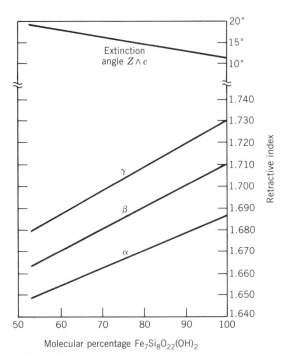

FIG. 13.74. Variation in refractive indices and extinction angle for members of the monoclinic cummingtonite-grunerite series. (From Klein, C., 1964, Cummingtonite-grunerite series: A chemical, optical and X-ray study. *American Mineralogist*, v. 49, pp. 963–982.)

grunerite, is mined only in the eastern part of the Transvaal Province of South Africa (see Fig. 13.75). The varietal name is derived from the word *Amosa*, an acronym for the company "Asbestos Mines of South Africa." Very few data are as yet available on the possible health hazards of this asbestos type (see Ross, 1981 and 1982; complete references at end of chapter). For comparison and more extensive discussion of asbestos, see *chrysotile* and *crocidolite*.

Names. Cummingtonite is named after Cummington, Massachusetts, and grunerite after Grüner, nineteenth century.

TREMOLITE—Ca$_2$Mg$_5$Si$_8$O$_{22}$(OH)$_2$
ACTINOLITE—Ca$_2$(Mg,Fe)$_5$Si$_8$O$_{22}$(OH)$_2$

Crystallography. Monoclinic; $2/m$. Crystals usually prismatic. The termination of the crystals is almost always formed by the two faces of a low {0kl} prism (similar to hornblende, see Fig. 13.77). Tremolite is often bladed and frequently in radiating columnar aggregates. In some cases in silky fibers. Coarse to fine granular. Compact.

$C2/m$; $a = 9.84$, $b = 18.05$, $c = 5.28$; $\beta = 104°42'$; $Z = 2$. d's: 8.38(10), 3.27(8), 3.12(10), 2.81(5), 2.71(9).

Physical Properties. *Cleavage* {110} perfect; (110) $\wedge$ (1$\bar{1}$0) = 56° often yielding a splintery surface. **H** 5–6. **G** 3.0–3.3 *Luster* vitreous; often with silky sheen in the prism zone. *Color* varying from white to green in actinolite. The color deepens and the specific gravity increases with increase in Fe content. Transparent to translucent. A felted aggregate of tremolite fibers goes under the name of *mountain leather* or *mountain cork*. A tough, compact variety that supplies much of the material known as *jade* is called *nephrite* (see Jadeite). *Optics:* (−); $\alpha = 1.60–1.68$, $\beta = 1.61–1.69$, $\gamma = 1.63–1.70$; 2V = 80°; Y = b, Z $\wedge$ c = 15°. Indices increase with iron content.

Composition and Structure. A complete solid solution exists from tremolite to ferroactinolite, Ca$_2$Fe$_5$Si$_8$O$_{22}$(OH)$_2$ (see Fig. 13.65). The name *actinolite* is used for intermediate members. In high-temperature metamorphic and in igneous occurrences a complete series exists from the tremolite-ferroactinolite series to hornblende as a result of a wide range of Al substitution for Si, and concomitant cation replacements elsewhere in the structure. The names *tremolite* and *ferroactinolite* are generally used to refer to the Al-poor members of this large compositional range. In lower-temperature occurrences tremolite (or actinolite) may coexist with hornblende as the result of a miscibility gap between tremolite (or

FIG. 13.75. Amosite, an asbestiform variety of the amphibole grunerite (in the trade this is also known as "brown asbestos"), Penge, Transvaal Province, Republic of South Africa (compare with Figs. 13.78 and 13.96).

actinolite) and hornblende. The structures of the members of the tremolite-ferroactinolite series are of the $C2/m$ type (see Figs. 13.66, 13.67, and page 491).

Diagnostic Features. Characterized by slender prisms and good prismatic cleavage. Distinguished from pyroxenes by the cleavage angle and from hornblende by lighter color.

Occurrence. Tremolite is most frequently found in metamorphosed dolomitic limestones where it forms according to:

$$5CaMg(CO_3)_2 + 8SiO_2 + H_2O$$

dolomite   quartz

$$\rightarrow Ca_2Mg_5Si_8O_{22}(OH)_2 + 3CaCO_3 + 7CO_2$$

tremolite   calcite

At higher temperatures tremolite is unstable and yields to diopside (see Fig. 13.76). Actinolite is a characteristic mineral of the greenschist facies of metamorphism. It occurs also in glaucophane schists in coexistence with quartz, epidote, glaucophane, pumpellyite, and lawsonite.

Notable localities for crystals of tremolite are: Ticino, Switzerland; in the Tyrol; and in Piemonte, Italy. In the United States from Russell, Gouverneur, Amity, Pierrepont, DeKalb, and Edwards, New York. Crystals of actinolite are found at Greiner, Zillerthal, Tyrol, Austria. The original jade was probably nephrite that came from the Khotan and Yarkand Mountains of Chinese Turkestan. Nephrite has been found in New Zealand, Mexico, and Central America; and, in the United States, near Lander, Wyoming.

Use. The compact variety, *nephrite*, is used as an ornamental and gem material.

FIG. 13.76. Comparison of the thermal stability limits of tremolite and ferroactinolite. (After Ernst, W. G., 1968, *Amphiboles.* Springer-Verlag, New York, p. 58.)

 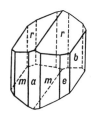

FIG. 13.77. Hornblende crystals.

Name. Tremolite is derived from the Tremola Valley near St. Gotthard, Switzerland. Actinolite comes from two Greek words meaning *a ray* and *stone*, in allusion to its frequently somewhat radiated habit.

HORNBLENDE—
$(Ca,Na)_{2-3}(Mg,Fe,Al)_5Si_6(Si,Al)_2O_{22}(OH)_2$

Crystallography. Monoclinic; $2/m$. Crystals prismatic, usually terminated by {011}. The vertical prism zone shows, in addition to the prism faces, usually {010}, and more rarely {100} (see Fig. 13.77). May be columnar or fibrous; coarse- to fine-grained.

$C2/m$; $a = 9.87$, $b = 18.01$, $c = 5.33$ Å; $\beta = 105°44'$; $Z = 2$. d's: 3.38(9), 3.29(7), 3.09(10), 2.70(10), 2.59(7).

Physical Properties. *Cleavage* {110} perfect. **H** 5–6. **G** 3.0–3.4. *Luster* vitreous; fibrous; fibrous varieties often silky. *Color* various shades of dark green to black. Translucent; will transmit light on thin edges. *Optics:* (−); $\alpha = 1.61$–1.71, $\beta = 1.62$–1.72, $\gamma = 1.63$–1.73; $2V = 30°$–90°; $Y = b$, $Z \wedge c = 15°$–34°. $r > v$ usually; may be $r < v$. Pleochroism X yellow-green, Y olive-green, Z deep green.

Composition and Structure. What passes under the name of hornblende is in reality a complex and large compositional range with variations in the ratios Ca/Na, Mg/Fe^{2+}, Al/Fe^{3+}, Al/Si, and OH/F. A generalized formula for a "common" hornblende is given above, and an example of the recalculation of an amphibole analysis is given in Table 5.6. At elevated temperatures hornblende shows a complete series toward tremolite-ferroactinolite, but at lower temperatures a miscibility gap exists. In some volcanic rocks oxy-hornblendes are found with considerable Fe^{3+}/Fe^{2+} ratios and low OH contents; an example composition is $NaCa_2Fe_4^{2+}Fe^{3+}(AlSi_7)O_{23}(OH)$. The structure of hornblende is similar to that of $C2/m$ tremolite (see Figs. 13.66 and 13.67). Because of the miscibility gap between Ca-amphiboles and Mg-Fe amphiboles, hornblende often contains grunerite exsolution lamellae (see Fig. 5.19).

Diagnostic Features. Crystal form and cleavage angles serve to distinguish hornblende from dark pyroxenes. Usually distinguished from other amphiboles by its dark color.

Occurrence. Hornblende is an important and widely distributed rock-forming mineral, occurring in both igneous and metamorphic rocks; it is particularly characteristic of some medium-grade metamorphic rocks known as amphibolites in which hornblende and associated plagioclase are the major constituents. It characteristically alters from pyroxene both during the late magmatic stages of crystallization of igneous rocks and during metamorphism. Hornblende is a common constituent of syenites and diorites.

Name. From an old German word for any dark prismatic mineral occurring in ores but containing no recoverable metal.

Similar Species. Because of the large chemical variation in the hornblende series, many names have been proposed for members on the basis of chemical composition and physical and optical properties. The most common names on the basis of chemical composition are: *edenite, pargasite, hastingsite,* and *tschermakite.*

Glaucophane—$Na_2Mg_3Al_2Si_8O_{22}(OH)_2$
Riebeckite—$Na_2Fe_3^{2+}Fe_2^{3+}Si_8O_{22}(OH)_2$

Crystallography. Monoclinic; $2/m$. In slender acicular crystals; frequently aggregated; riebeckite sometimes asbestiform.

$C2/m$; $a = 9.58$, $b = 17.80$, $c = 5.30$ Å; $\beta = 103°48'$; $Z = 2$. d's: 8.42(10); 4.52(5), 3.43(6), 3.09(8), 2.72(10).

Physical Properties. *Cleavage* {110} perfect. **H** 6. **G** 3.1–3.4. *Luster* vitreous. *Color* blue to lavender-blue to black, darker with increasing iron content. *Streak* white to light blue. Translucent. *Optics:* (−); $\alpha = 1.61–1.70$, $\beta = 1.62–1.71$, $\gamma = 1.63–1.72$; 2V = 40°–90°; $Y = b$, $Z \wedge c = 8°$. Pleochroism in blue $X < Y < Z$.

Composition and Structure. Very few glaucophanes are close to the end member composition, $Na_2Mg_3Al_2Si_8O_{22}(OH)_2$, because of some substitution by Fe^{2+} for Mg and Fe^{3+} for Al. Riebeckite analyses do not conform well with the end member formulation because the sum of $X + Y$ cations is frequently larger than 7, with Na entering the normally vacant A site. A partial series exists between glaucophane and riebeckite, with intermediate compositions known as *crossite.* The structures of both sodic amphiboles are similar to that of $C2/m$ tremolite (see Fig. 13.66).

Diagnostic Features. Glaucophane and riebeckite are characterized by their generally fibrous habit and blue color.

Occurrence. *Glaucophane* is found only in metamorphic rocks such as schists, eclogite, and marble. The occurrences of glaucophane reflect low-temperature, relatively high-pressure metamorphic conditions, in association with jadeite, lawsonite, and aragonite. It is a major constituent of glaucophane schists in the Franciscan Formation of California. It is also found in metamorphic rocks from Mexico, Japan, Taiwan, Indonesia, and eastern Australia. *Riebeckite* occurs most commonly in igneous rocks such as granites, syenites, nepheline syenites, and related pegmatites. It is a conspicuous mineral in the granite of Quincy, Massachusetts. It is present in some schists of regional metamorphic origin. The asbestiform variety of riebeckite is known as *crocidolite* (see Fig. 13.78).

Use. Crocidolite makes up about 4% of the total world production of asbestos. All of the present production comes from mines in Cape Province, South Africa. Extensive reserves of crocidolite also occur in the Hamersley Range of western Australia. However, no mining of crocidolite has taken place there since 1966. The crocidolite is closely interbanded with Precambrian banded iron-formation sequences, both in South Africa and western Australia. Medical studies have shown (see Ross, 1981 and

FIG. 13.78. Crocidolite, an asbestiform variety of the amphibole riebeckite (in the trade this is also known as "blue asbestos"), Kuruman, Cape Province, Republic of South Africa (compare with Figs. 13.75 and 13.96).

1982; also Mossman and Gee, 1989, and Mossman et al., 1990; complete references are given at the end of this chapter) that crocidolite is a much greater health hazard than chrysotile asbestos. For comparison, see discussions of *chrysotile* and *amosite*.

In many places in South Africa there has been a pseudomorphous replacement of oxidized crocidolite by quartz. The preservation of the fibrous texture makes it an attractive ornamental material with a chatoyancy. This is widely used for jewlery under the name of *tiger's eye*.

Name. Glaucophane is from the two Greek words meaning *bluish* and *to appear*. Riebeckite is in honor of E. Riebeck.

Similar Species. *Arfvedsonite*, $Na_3Fe_4^{2+}Fe^{3+}Si_8O_{22}(OH)_2$, is a deep green amphibole similar to riebeckite in occurrence. It is commonly associated with aegirine, or aegirine-augite. In the structure of arfvedsonite the *A* site is completely or almost completely filled by Na (see Table 13.3).

PHYLLOSILICATES

As the derivation of the name of this important group implies (Greek: *phyllon*, leaf), most of its many members have a platy or flaky habit and one prominent cleavage. They are generally soft, of relatively low specific gravity, and may show flexibility or even elasticity of the cleavage lamellae. All these characterizing peculiarities arise from the dominance in the structure of the indefinitely extended sheet of SiO_4 tetrahedra. In this sheet (see Fig. 13.3) three of the four oxygens in each SiO_4 tetrahedron are shared with neighboring tetrahedra, leading to a ratio of Si : O = 2 : 5. Each sheet, if undistorted, has 6-fold symmetry.

Most of the members of the phyllosilicates are hydroxyl bearing with the (OH) group located in the center of the 6-fold rings of tetrahedra, at the same height as the unshared apical oxygens in the SiO_4 tetrahedra (see Fig. 13.79). When ions, external to the Si_2O_5 sheet, are bonded to the sheets, they are coordinated to two oxygens and one OH, as shown in Fig. 13.79. The size of the triangle between two oxygens and one (OH) is just about the same as (but not identical to, see page 504) the triangular face of an XO_6 octahedron (with *X* commonly Mg or Al). This means that it is possible to bond to a regular net of apical oxygens and (OH) groups of $(Si_2O_5OH)^{3-}$ composition a sheet of regular octahedra, in which each octahedron is tilted onto one of its triangular sides (Fig. 13.80). When such tetrahedral and octahedral sheets are joined, one obtains the general geometry of the *antigorite* or *kaolinite* structures (Fig. 13.81a).

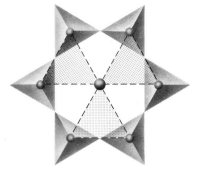

● Apical oxygens
● OH group

FIG. 13.79. Undistorted, hexagonal ring in Si_2O_5 sheet showing location of apical oxygens and (OH) group. In an ideal sheet structure the size of the triangles (outlined by shading) is the same as the size of triangular faces of XO_6 octahedra.

The cations in the octahedral layer may be divalent or trivalent. When the cations are divalent, for example, Mg or Fe^{2+}, the layer will have the geometry of that of brucite (Fig. 11.26) in which each cation site is occupied. In such a layer six bonds, each with e.v. = $\frac{2}{6} = \frac{1}{3}$, originate from the Mg^{2+} ion. Three such bonds radiate from each oxygen, or (OH) group, thus neutralizing one-half the charge of the oxygen but all of the OH. A layer in which each oxygen or (OH) group is surrounded by three cations, as in the brucite, $Mg(OH)_2$, structure, is known as *trioctahedral*. When the cations in the octahedral layer are trivalent, charge balance is maintained when one out of every three cations sites is unoccupied (see Fig. 13.81b). This is the case in the gibbsite, $Al(OH)_3$, and corundum structures (Fig. 11.2; see page 373). Such a layer structure in which each oxygen or (OH) group is surrounded only by two cations is known as *dioctahedral*. On the basis of the chemistry and geometry of the octahedral layers, the phyllosilicates are divided into two major groups: *trioctahedral* and *dioctahedral*.

It was pointed out above that the geometries of an average Si_2O_5 sheet and a layer of XO_6 octahedra are generally compatible such that one can be bonded

FIG. 13.80. Infinitely extending sheet of XO_6 octahedra. All octahedra lie on triangular faces.

● Oxygen
● Most commonly Mg or Al

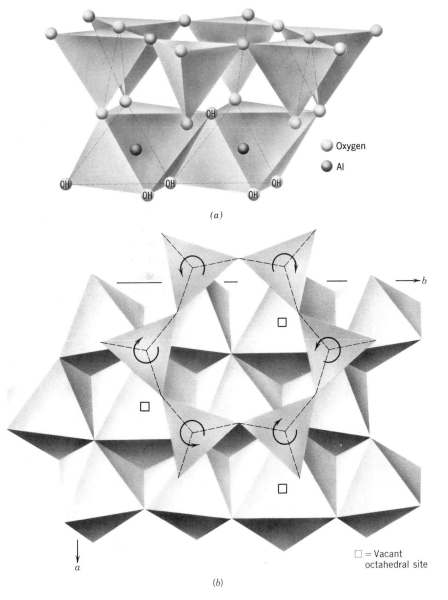

(a)

Oxygen
Al

FIG. 13.81. *(a)* Diagrammatic sketch of the structure of kaolinite with a tetrahedral sheet bonded on one side by an octahedral layer. (From Grim, R. E., 1968, *Clay Mineralogy*. McGraw-Hill Book Co., New York.) *(b)* Plan view of the structure of muscovite in which the overall dimensions of an octahedral sheet are somewhat smaller than those of an idealized tetrahedral sheet. To accommodate this slight misfit, the SiO_4 tetrahedra in the tetrahedral sheet are rotated (in opposite directions as shown by the arrows) about axes normal to the sheet. This changes the shape of the six-membered tetrahedral rings from hexagonal to trigonal. The octahedra marked with a square are vacant (see discussion of dioctahedral micas). (From Bailey, S. W., 1984, Crystal structure of the true micas, *in Micas, Reviews in Mineralogy*, v. 13, pp. 13–60, Mineralogical Society of America, Washington, D.C.)

□ = Vacant octahedral site

(b)

to the other. Let us look at this in terms of the chemistries of the octahedral and tetrahedral layers. Brucite, $Mg(OH)_2$, consists of two (OH) planes between which Mg is coordinated in octahedra. We can state this symbolically as $Mg_3 \frac{(OH)_3}{(OH)_3}$. If we replace two of the (OH) groups on one side of a brucite layer by two apical oxygens of an Si_2O_5 sheet, we obtain $Mg_3 \frac{Si_2O_5(OH)}{(OH)_3}$. This means that the other side of the brucite structure is not attached to an Si_2O_5 sheet as in Fig. 13.81a. This is known as the *antigorite*, $Mg_3Si_2O_5(OH)_4$, structure. The equivalent structure with a dioctahedral sheet is *kaolinite*, $Al_2Si_2O_5(OH)_4$. In short, the antigorite and kaolinite structures are built of one tetrahedral ("*t*") and one octahedral ("*o*")

sheet, giving "*t-o*" layers (see Fig. 13.84). Such *t-o* layers are electrically neutral and are bonded to one another by weak van der Waals bonds. As noted above, in the antigorite and kaolinite structures only one side of the octahedral sheet is coordinated to a tetrahedral sheet. However, we can derive further members of the phyllosilicate group by joining tetrahedral sheets on both sides of the *o* sheet. This leads to *t-o-t* type layers (Fig. 13.82) as in *talc*, $Mg_3Si_4O_{10}(OH)_2$, and *pyrophyllite*, $Al_2Si_4O_{10}(OH)_2$. We may again begin with brucite, $Mg_3 \frac{(OH)_3}{(OH)_3}$, and replace two (OH) groups in both the upper and lower (OH) planes by two oxygens of Si_2O_5 sheets, giving rise to $Mg_3 \frac{Si_2O_5(OH)}{Si_2O_5(OH)}$ or $Mg_3Si_4O_{10}(OH)_2$,

FIG. 13.82. Diagrammatic sketch of a *t-o-t* type layer as in pyrophyllite. (After Grim, 1968.)

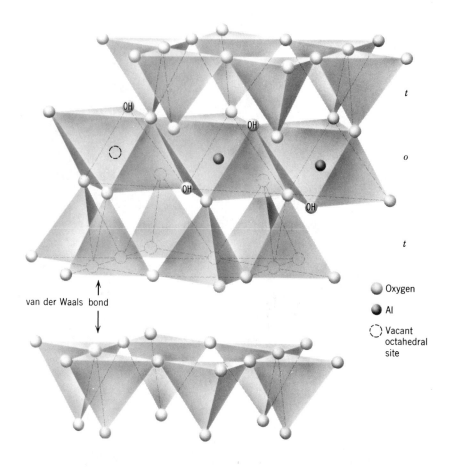

van der Waals bond

○ Oxygen

● Al

◌ Vacant octahedral site

FIG. 13.83. Diagrammatic sketch of the muscovite structure. (After Grim, 1968.)

○ Oxygen

● Al

○ K

talc. Similarly, gibbsite, $Al_2 \dfrac{(OH)_3}{(OH)_3}$, would result in $Al_2 \dfrac{Si_2O_5(OH)}{Si_2O_5(OH)}$ by replacement of two out of three (OH) groups in the (OH) planes by oxygens of Si_2O_5 sheets; this leads to pyrophyllite, $Al_2Si_4O_{10}(OH)_2$ (see Figs. 13.82 and 13.84). The *t-o-t* layers are electrically neutral and form stable structural units that are joined to each other by van der Waals bonds. Because this is a very weak bond, we may expect such structures to have excellent cleavage, easy gliding, and greasy feel, as indeed found in the minerals talc and pyrophyllite.

We can carry the process of evolution of phyllosilicate structures one step farther by substituting Al for some of the Si in the tetrahedral sites of the Si_2O_5 sheets. Because Al is trivalent, whereas Si is tetravalent, each substitution of this kind causes a free electrical charge to appear on the surface of the *t-o-t* layer. If Al substitutes for every fourth Si, a charge of significant magnitude is produced to bind univalent cations (*interlayer cations*), in regular 12-coordination to *t-o-t* layers (see Fig. 13.84). By virtue of these *t-o-t*-interlayer cation bonds, the structure is more firmly held together, the ease of gliding is diminished, the hardness is increased, and slippery feel is lost. The resulting mineral structures are those of the true *micas* (see Fig. 13.81b, 13.83, and 13.84). In the trioctahedral group of micas the univalent interlayer cation is K^+; in the dioctahedral group the cation can be K^+, as in muscovite, or Na^+, as in paragonite. It is easy to remember the formulas of the micas if it is recalled that *one* of the aluminums belongs in the tetrahedral sheet and the formulas are written accordingly. Thus,

Trioctahedral	$KMg_3(AlSi_3O_{10})(OH)_2$	Phlogopite
Dioctahedral	$KAl_2(AlSi_3O_{10})(OH)_2$	Muscovite

If half the Si in the tetrahedral sites of the Si_2O_5 sheets is replaced by Al, two charges per *t-o-t* layer become available for binding an interlayer cation. Such ions as Ca^{2+}, and, to a smaller extent, Ba^{2+}, may enter between two layers of the mica structure (see Fig. 13.106). The interlayer cations are held by ionic bonds so strong that the quality of the cleavage is diminished, the hardness increased, the flexibility of the layers is almost wholly lost, and the density is increased. The resulting minerals are the *brittle micas*, typified by

Trioctahedral	$CaMg_3(Al_2Si_2O_{10})(OH)_2$	Xanthophyllite
Dioctahedral	$CaAl_2(Al_2Si_2O_{10})(OH)_2$	Margarite

There is little solid solution between members of the dioctahedral and trioctahedral groups, although there may be extensive and substantially complete ionic substitution of Fe^{2+} for Mg, of Fe^{3+} for Al, and of Ca for Na in appropriate sites. Ba may replace K, Cr may substitute for Al, and F may replace OH to a limited extent. Mn, Ti, and Cs are rarer constituents of some micas. The lithium micas are structurally distinct from muscovite and biotite because of the small lithium ion.

Additional members of the phyllosilicate group can be developed. The important group of *chlorites* may be viewed as consisting of two layers of talc (or pyrophyllite) separated by a brucite- (or gibbsite-)like octahedral layer (see Figs. 13.85 and 13.106). This leads to a formula such as $Mg_3Si_4O_{10}(OH)_2 \cdot Mg_3(OH)_6$. However, in most chlorites Al, Fe^{2+}, and Fe^{3+} substitute for Mg in octahedral sites both in the talc and brucite-like sheets, and Al substitutes for Si in the tetrahedral sites. A more general chlorite formula would be:

$$(Mg,Fe^{2+},Fe^{3+},Al)_3(Al,Si)_4O_{10}(OH)_2 - (Mg,Fe^{2+},Fe^{3+},Al)_3(OH)_6$$

The various members of the chorite group differ from one another in amounts of substitution and the manner in which the successive octahedral and tetrahedral layers are stacked along the c axis.

The *vermiculite* minerals may be derived from the talc structure by interlamination of water molecules in definite sheets 4.98 Å thick, which is about the thickness of two water molecules (see Fig. 13.86). An example of a specific vermiculite formula would be $Mg_3(Si,Al)_4(OH)_2 \cdot 4.5H_2O[Mg]_{0.35}$ in which [Mg] represents exchangeable ions in the structure. The presence of exchangeable ions located between layers of H_2O molecules, and the ability of the structure to retain variable amounts of H_2O, are of great importance in agriculture. When the vermiculite structure is saturated with H_2O, the basal spacing is approximately 14.8 Å. This water can be gradually extracted as shown by a discontinuous collapse sequence along the c axis, leading to a basal spacing of about 9.0 Å for a vermiculite without interlayer water. The structure of the *montmorillonite* (or *smectite*) group may be derived from the pyrophyllite structure by insertion of sheets of molecular water containing exchangeable cations between the pyrophyllite *t-o-t* layers, leading to a structure that is essentially identical to that of vermiculite. The members of the vermiculite and montmorillonite groups exhibit an unrivalled capacity for swelling when wetted because of their ability to incorporate large amounts of interlayer water.

If occasional random substitution of aluminum for silicon in the tetrahedral sites of pyrophyllite sheets takes place, there may not be enough of an aggregate

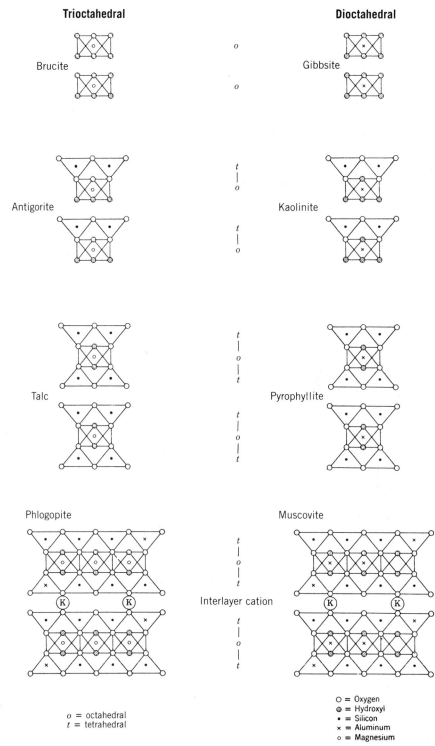

FIG. 13.84. Schematic development of some of the phyllosilicate structures (compare with Fig. 13.106).

o = octahedral
t = tetrahedral

O = Oxygen
⊘ = Hydroxyl
• = Silicon
× = Aluminum
o = Magnesium

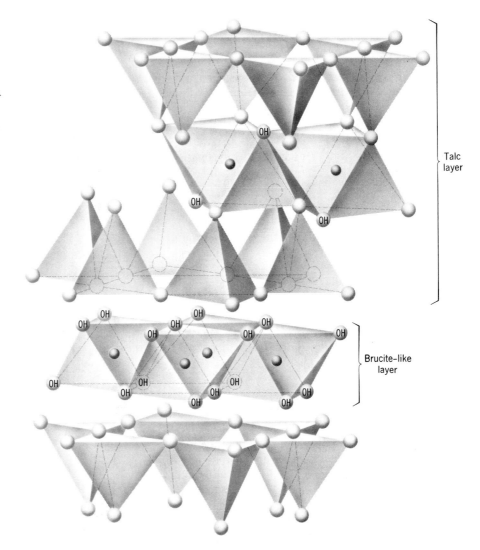

FIG. 13.85. Diagrammatic sketch of the structure of chlorite. (After Grim, 1968.)

Talc layer

Brucite-like layer

FIG. 13.86. Diagrammatic sketch of the vermiculite structure, showing layers of water. (After Grim, 1968.)

Tetrahedral sheet

Oriented water molecules

Tetrahedral sheet

charge on the *t-o-t* layers to produce an ordered mica structure with all possible interlayer cation sites filled. Locally, however, occasional cation sites may be occupied, leading to properties intermediate between those of clays and micas. Introduction of some molecular water may further complicate this picture. K-rich minerals of this type, intermediate between montmorillonite clays and the true micas, are referred to as *illite* or *hydromica*.

In our previous discussion of the derivation of various types of structures among the phyllosilicates (see Figs. 13.84 and 13.106) we suggested that there is a good geometric fit between an octahedral brucite layer and a tetrahedral Si_2O_5 sheet. In detail, however, one finds that there is considerable mismatch between a brucite layer and an undistorted Si_2O_5 sheet with hexagonal rings. The misfit is due to the fact that the edges of an $Mg(OH)_6$ octahedron in the brucite layer are somewhat greater than the distances between apical oxygens in the Si_2O_5 or $(Si,Al)_2O_5$ layer; this means that the geometry shown in Fig. 13.79 is an oversimplification. In the case of the *serpentine minerals*, *antigorite* and *chrysotile*, both $Mg_3Si_2O_5(OH)_4$, this misfit is compensated for by a bending (stretching of the distance between apical oxygens) of the tetrahedral layer so as to provide a better fit with the adjoining octahedral brucite layer (see Fig. 13.87). In the variety *antigorite*, the bending is not continuous but occurs as corrugations. In the fibrous variety, *chrysotile*, the mismatch is resolved by a continuous bending of the structure into cylindrical tubes (see Figs. 13.88 and 13.89). These high-magnification electron micro-

FIG. 13.88. Electron micrograph showing tubular fibers of chrysotile from Globe, Arizona, Magnification 35,000 ×. (After Nagy, B., and Faust, G. T., 1956, Serpentines: natural mixtures of chrysotile and antigorite, *American Mineralogist*, v. 41, pp. 817–838.)

FIG. 13.87. *(a)* Diagrammatic sketch of the antigorite structure as viewed along the *b* axis, showing curved *t* and *o* layers. *(b)* Highly schematic representation of the possible curvature in the chrysotile structure.

☐ Brucite layer
△△ Tetrahedral layer

(a)

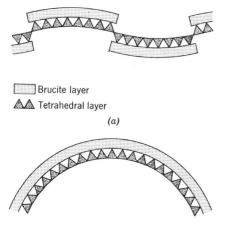

(b)

graphs reveal the existence of tubelike structures for the layers that occur in essentially parallel sheets in the kaolinite group.

In our discussion of the phyllosilicates we have pointed out the existence of hexagonal rings in the undistorted Si_2O_5 sheet. In order to join the *t* sheet to a brucite *o* layer, or to an *o-t* sequence so as to produce *t-o* or *t-o-t* type layers, the octahedral sheet is staggered with respect to the tetrahedral sheets (see any of the phyllosilicate illustrations). This stagger reduces the symmetry of the overall structure to monoclinic (see Fig. 13.90), even though the tetrahedral sheets contain hexagonal holes. Most of the phyllosilicates, therefore, have monoclinic structures, some are triclinic, and a few orthorhombic or trigonal.

Because of the hexagonal to trigonal symmetry about the (OH) group in the $Si_2O_5(OH)$ tetrahedral sheets, there are three alternate directions (see Fig. 13.91) in which $Si_2O_5(OH)$ sheets can be stacked. These three directions can be represented by three

FIG. 13.89. High-resolution transmission electron microscope (HRTEM) image of an intergrowth of planar and coiled structures in serpentine. The planar serpentine (S) is seen ending in rolls of *chrysotile*. Talc (T) is interlayered with the planar serpentine, *lizardite*. The two types of serpentine are formed against grains of amphibole (Amph). (From Veblen, D. R. and Buseck, P. R., 1979, Serpentine minerals: Intergrowths and new combination structures. *Science*, v. 206, pp. 1398–1400; copyright © 1979 by the AAAS. See also Buseck, P. R., 1983. Electron microscopy of minerals. *American Scientist*, v. 71, pp. 175–185.)

vectors (1, 2, and 3; and negative directions $\bar{1}$, $\bar{2}$, and $\bar{3}$) at 120° to each other. If the stacking of $Si_2O_5(OH)$ sheets is always in the same direction (see Fig. 13.91*b*), the resulting structure will have monoclinic symmetry, with space group $C2/m$; this is referred to as the 1*M* (*M* = monoclinic) *polytype* with stacking sequence [1]. If the stacking sequence of $Si_2O_5(OH)$ sheets consists of alternating 1 and $\bar{1}$ directions (in opposing directions along the same vector), a structure results which is best described as orthorhombic with space group $Ccm2_1$. This stacking sequence can be expressed as [1 $\bar{1}$] and the polytype is known as 2*O* (*O* = orthorhombic). If the stacking sequence consists of two vectors at 120° to each other, that is, [1 2], another monoclinic polytype results; this is known as 2*M*$_1$ with space group $C2/c$ (see Fig. 13.91*c*). When all three vectors come into play, such as in [1 2 3], a trigonal polytype, 3*T* (*T* = trigonal), results which is compatible with either of two enantiomorphic space groups: $P3_112$ or $P3_212$. Additional *polytype* stacking sequences are possible, but most micas that commonly show *polytypism* belong to 1*M*($C2/m$), 2*M*$_1$($C2/c$) or 3*T*($P3_112$) polytypes. Polytypism (see also page 158) is found in many layer silicates, among them serpentine, mica, and chlorite. Figure 13.92(*a*) illustrates two different polytypes for planar serpentine structures, and Fig. 13.92(*b*) shows two different polytypes of mica.

The great importance of the phyllosilicates lies in part in the fact that the products of rock weathering, and hence the constituents of soils, are mostly of this structural type. The release and retention of plant foods, the stockpiling of water in the soil from wet to dry seasons, and the accessibility of the soil to atmospheric gases and organisms depend in large part on the properties of the sheet silicates.

Geologically, the phyllosilicates are of great significance. The micas are the chief minerals of schists and are widespread in igneous rocks. They form at

FIG. 13.90. Schematic projection along the *b* axis of the mica structure. Note stagger of *t* and *o* sheets with respect to each other. Unit cell is outlined by dashed lines.

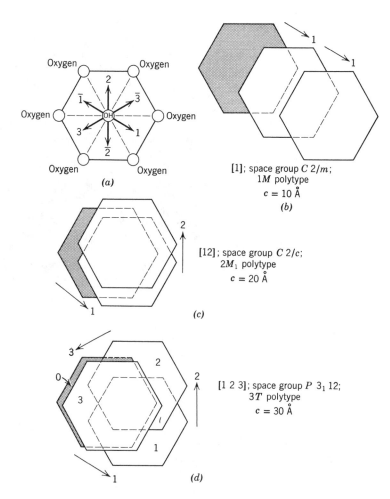

FIG. 13.91. Schematic illustration of some possible stacking polymorphs (*polytypes*) in mica. *(a)* Three vectorial directions for possible location of the (OH) group in an $Si_2O_5(OH)$ sheet which is stacked above or below the hexagonal ring shown. *(b)* Stacking of $Si_2O_5(OH)$ sheets in the same direction. *(c)* Stacking in 2 directions at 120° to each other. *(d)* Stacking in three directions at 120° to each other. Here, sheet 3 should lie directly above sheet 0 but is offset slightly for illustrative purposes.

lower temperatures than amphiboles or pyroxenes and frequently are formed as replacements of earlier minerals as a result of hydrothermal alteration. The chemical changes and structural transformations that take place in such alteration reactions are complex. Figure 13.93 is a schematic representation of the various reaction paths an originally high-temperature pyroxene (of igneous origin) might take during lower-temperature reactions (such as alteration by hydrothermal solutions). The lower-temperature end products, marked on the far right of the diagram, are layer silicates. The intermediate reaction products are mainly amphiboles, but also various pyriboles (see Figs. 13.69 and 13.70 and related text). Figure 13.94 gives two high-resolution transmission electron microscope (HRTEM) structure images for igneous pyroxenes from the Palisades Sill, New Jersey. Figure 13.94a shows the presence of a triple chain pyribole between a relict (unaltered) augite and sheet silicate which is mainly talc. The pyribole material appears to be an intermediate product in the reaction sequence: pyroxene → triple chain silicate → talc (see Fig.

13.93). In Figure 13.94b talc appears to have directly replaced orthopyroxene along a fracture.

We will discuss in detail the following common phyllosilicates:

PHYLLOSILICATES

Serpentine Group	
Antigorite ⎫	$Mg_3Si_2O_5(OH)_4$
Chrysotile ⎭	
Clay Mineral Group	
Kaolinite	$Al_2Si_2O_5(OH)_4$
Talc	$Mg_3Si_4O_{10}(OH)_2$
Pyrophyllite	$Al_2Si_4O_{10}(OH)_2$
Mica Group	
Muscovite	$KAl_2(AlSi_3O_{10})(OH)_2$
Phlogopite	$KMg_3(AlSi_3O_{10})(OH)_2$
Biotite	$K(Mg,Fe)_3(AlSi_3O_{10})(OH)_2$
Lepidolite	$K(Li,Al)_{2-3}(AlSi_3O_{10})(OH)_2$
Margarite	$CaAl_2(Al_2Si_2O_{10})(OH)_2$
Chlorite Group	
Chlorite	$(Mg,Fe)_3(Si,Al)_4O_{10}(OH)_2(Mg,Fe)_3(OH)_6$

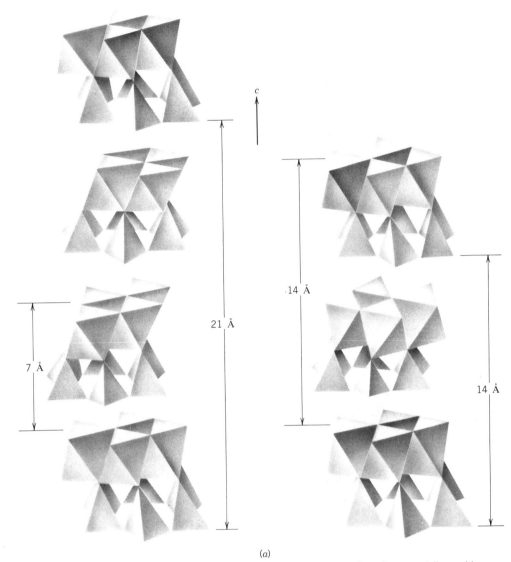

(a)

FIG. 13.92. (a) Two different polytypes (different stackings of t-o sheets along the c crystallographic axis) for planar serpentine.

(continued)

and three minerals closely related to the above phyllosilicate groups:

Apophyllite	$KCa_4(Si_4O_{10})_2F \cdot 8H_2O$
Prehnite	$Ca_2Al(AlSi_3O_{10})(OH)_2$
Chrysocolla	$Cu_4H_4Si_4O_{10}(OH)_8$

Serpentine Group

ANTIGORITE, LIZARDITE AND CHRYSOTILE—$Mg_3Si_2O_5(OH)_4$

Crystallography. Monoclinic; $2/m$. Crystals, except as pseudomorphs, unknown. Serpentine oc-curs in three common polymorphs: *antigorite, lizardite,* and *chrysotile.* Antigorite and lizardite are commonly massive and fine-grained; chrysotile is fibrous (Fig. 13.95). Orthorhombic polymorphs of both antigorite and chrysotile are known.

Cm, C2, or C2/m. For antigorite: $a = 5.30$, $b = 9.20$, $c = 7.46$ Å; $\beta = 91°24'$; $Z = 2$. d's: 7.30(10), 3.63(8), 2.52(2), 2.42(2), 2.19(1). For chrysotile: $a = 5.34$, $b = 9.25$, $c = 14.65$ Å; $\beta = 93°16'$; $Z = 4$.

Physical Properties. **H** 3–5, usually 4. **G** 2.5–2.6. *Luster* greasy, waxlike in the massive varieties, silky when fibrous. *Color* often variegated, showing mottling in lighter and darker shades of green. Translucent. *Optics:* $(-)$; *Chrysotile:* $\alpha =$

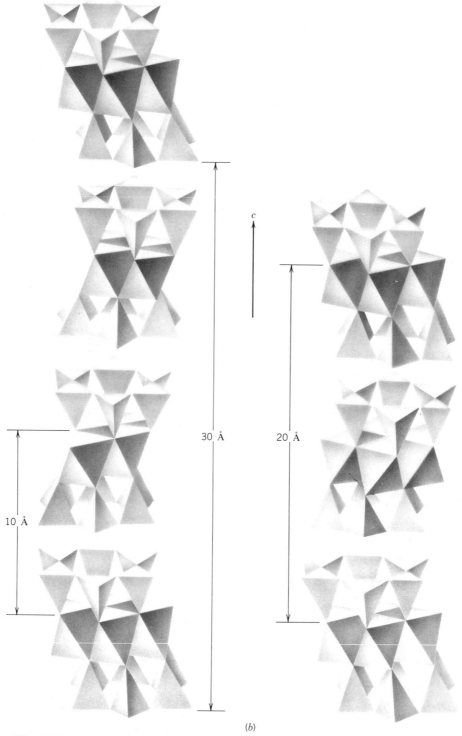

(b)

FIG. 13.92. *(continued) (b)* Two different polytypes for mica. (*a* and *b* from Bailey, S. W., 1988, X-ray diffraction identification of the polytypes of mica, serpentine, and chlorite. *Clays and Clay Minerals*, v. 36, pp. 193–213.)

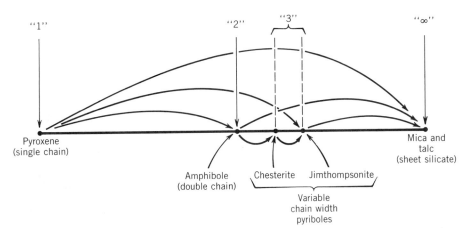

FIG. 13.93. The many possible reaction paths that lead from a high temperature pyroxene to lower temperature sheet silicates. Ch and Jt are pyriboles, chesterite and jimthompsonite, respectively, "1," "2," and "3" refer to chain width: "1" in pyroxenes, "2" in amphiboles, and "3" in biopyriboles with variable chain widths; "∞" means infinite chain width, resulting in sheets (see also Figs. 13.69 and 13.70). (From Veblen, D. R. and Buseck, P. R., 1981, Hydrous pyriboles and sheet silicates in pyroxenes and uralites: Intergrowth microstructures and reaction mechanisms. *American Mineralogist*, v. 66, pp. 1107–1134.)

1.532–1.549 (across fiber), $\gamma = 1.545$–1.556 (parallel to fiber length). *Antigorite:* nearly isotropic, $n = 1.55$–1.56.

The characteristic morphology of all asbestos minerals, in their natural form, is a parallel-sided fiber with a length to width ratio (referred to as an *aspect ratio*) in excess of 100 : 1 (Champness, P. E., Cliff, G. and Lorimer, G. W., 1976, The identification of as-

bestos, *Journal of Microscopy*, v. 108, pp. 231–249). Chrysotile asbestos occurs in very narrow tubular fibers that are often hollow (see Figs. 13.88 and 13.89). Other asbestos minerals are *crocidolite* (Fig. 13.78), an asbestiform variety of the amphibole riebeckite, and *amosite* (Fig. 13.75), an asbestiform variety of the amphibole grunerite.

Composition and Structure. Close to Mg_3Si_2-

FIG. 13.94. High-resolution transmission electron microscope (HRTEM) images of altered pyroxenes. *(a)* A complex intergrowth of high-temperature pyroxene and alteration products of talc and triple-chain silicate ("3" = pyribole). *(b)* Direct replacement of original orthopyroxene by later talc along a fracture. (From Veblen, D. R. and Buseck, P. R., 1981, Hydrous pyriboles and sheet silicates in pyroxenes and uralites: Intergrowth microstructures and reaction mechanisms. *American Mineralogist*, v. 66, pp. 1107–1134.)

(a)

(b)

FIG. 13.95. Veins of chrysotile asbestos in serpentine. Thetford, Quebec, Canada.

$O_5(OH)_4$ with MgO 43.0, SiO_2 44.1, and H_2O 12.9%. Fe and Ni may substitute to some extent for Mg and Al for some Si. The idealized structure of antigorite is shown in Fig. 13.84 in terms of trioctahedral *t-o* layers, in analogy with the dioctahedral kaolinite structure. Because of a misfit between the *t* and *o* layers the actual antigorite structure consists of finite and corrugated layers parallel to {001} (see Fig. 13.87*a* and page 504). In chrysotile this same mismatch is responsible for the curvature of the *t-o* layers forming cylindrical tubes (see Fig. 13.87*b* and page 504). Much of the matrix material of serpentine, known as *lizardite*, is extremely fine-grained and has a platy morphology as seen under the high magnification of the electron microscope (see Fig. 13.89).

Diagnostic Features. Recognized by its variegated green color and its greasy luster or by its fibrous nature.

Occurrence. Serpentine is a common mineral and widely distributed: usually as an alteration of magnesium silicates, especially olivine, pyroxene, and amphibole. From forsterite it may form by the reaction:

$$2Mg_2SiO_4 + 3H_2O \rightarrow Mg_3Si_2O_5(OH)_4 + Mg(OH)_2$$
forsterite serpentine brucite

Frequently associated with magnesite, chromite, and magnetite. Found in both igneous and metamorphic rocks, frequently in disseminated particles, in places in such quantity as to make up practically the entire rock mass. Serpentine, as a rock name, is applied to such rock masses made up mostly of a mixture of fine-grained antigorite and lizardite. The asbestiform variety, *chrysotile*, as the principal asbestos mineral, is mined extensively. More than half of the world's production comes from deposits in the CIS, just east of the Ural Mountains. Although for many years asbestos has been mined from large deposits in southeastern Quebec, making Canada the leading world producer, it is now second to the CIS; South Africa is ranked third in world production of asbestos minerals. In the United States there is a small production of chrysotile from the Coalinga area, San Benito Mountains, California.

Use. Massive serpentine, which is translucent and of a light to dark green color, is often used as an ornamental stone and may be valuable as building material. Mixed with white marble and showing beautiful variegated coloring, it is called *verde antique* marble. A transparent yellow-green variety, *bowenite*, is used as a jade substitute.

The use of chrysotile, as asbestos, depends on its fibrous, flexible nature, which allows it to be made into felt and woven into cloth and other fabrics, and upon its incombustibility and slow conductivity of heat. Asbestos products, therefore, are used for fireproofing and as an insulation material against heat and electricity.

FIG. 13.96. Chrysotile, asbesti-form variety of serpentine, Thetford, Quebec, Canada (compare with varieties of amphibole asbestos, Figs. 13.75 and 13.78).

About 90% of the past and 95% of the present world production of asbestos was or is the chrysotile form (Fig. 13.96). This is known as "white asbestos." The remaining 5% of world production consists of crocidolite (the asbestiform variety of the amphibole riebeckite), known as "blue asbestos," and amosite (the asbestiform variety of the amphibole grunerite), known as "brown asbestos." Other types of asbestos are of little economic importance.

Over the last several years there has been much attention given, in newspapers and on television, to the health hazards posed by asbestos. In such reports, no distinction is made between the various asbestos types, and furthermore, the health hazards that have been assessed in the dusty work places of earlier years (occupational settings such as sites of asbestos mining and asbestos board manufacturing) are linearly extrapolated to those in nonoccupational settings (e.g., school rooms and office buildings with asbestos ceiling and floor tiles). Medical studies on the pathogenicity of the different forms of asbestos (Mosmann et al., 1990; Mosmann and Gee, 1989; Ross, 1981, 1984; for complete references see the end of this chapter) show that "blue asbestos" (crocidolite, the asbestiform variety of the amphibole riebeckite) poses a much greater health hazard (in *occupational settings*) than chrysotile. Occupational exposure to asbestos can cause the following types of disorders: asbestosis, lung cancer, mesothelioma (cancer of the plural and peritoneal membranes), and benign changes in the pleura (Mossman et al., 1990). Crocidolite fibers appear to be most pathogenic, especially with respect to mesothelioma. Smoking is a strong contributor to the incidence of lung cancer (Ross, 1984).

The main asbestos type used in U.S. buildings is chrysotile. Available data do not support the concept that low-level (*nonoccupational*) exposure to asbestos is a health hazard in buildings and schools (Mossman et al., 1990). Furthermore, a low-level background of fibers is present globally, in the air and water, as a result of natural weathering processes of rocks with fibrous minerals. Abelson (1990; for complete reference see end of this chapter) stated: "We live on a planet on which there is an abundance of serpentine- and amphibole-containing rocks. Natural processes have been releasing fibers throughout Earth history, we breathe about 1 million fibers per year."

Name. Serpentine refers to the green, serpent-like cloudings of the massive variety. Antigorite after Antigorio in Italy and chrysotile from the Greek words for *golden* and *fiber*.

Similar Species. *Greenalite*, $(Fe,Mg)_3Si_2O_5(OH)_4$, the Fe-rich analogue of antigorite, is fairly common in unmetamorphosed Precambrian iron-formations. *Garnierite*, $(Ni,Mg)_3Si_2O_5(OH)_4$, is an apple-green Ni-bearing serpentine formed as an alteration product of Ni-rich peridotites. It is mined as a nickel ore in New Caledonia, the CIS, and Australia. In the United States it is found at Riddle, Oregon. *Sepiolite* (meerschaum), a hydrous magnesium silicate, is a claylike secondary mineral associated with serpentine. It is used in the manufacture of meerschaum pipes.

Clay Mineral Group

Clay is a rock term, and like most rocks it is made up of a number of different minerals in varying proportions. Clay also carries the implication of very small particle size. Usually the term *clay* is used in reference to fine-grained, earthy material that becomes plastic when mixed with a small amount of water. With the use of X-ray techniques, clays have been shown to consist mainly of a group of crystalline substances known as the *clay minerals*. They are all essentially hydrous aluminum silicates. In some, Mg or Fe substitute in part for Al and alkalies or alkaline earths may be present as essential constituents. Although a clay may be made up of a single clay mineral, there are usually several mixed with other minerals such as feldspar, quartz, carbonates, and micas.

KAOLINITE—$Al_2Si_2O_5(OH)_4$

Crystallography. Triclinic; $\overline{1}$. In very minute, thin, rhombic or hexagonal plates (Fig. 13.97). Usually in claylike masses, either compact or friable.

$P\overline{1}; a = 5.14. b = 8.93, c = 7.37$ Å; $\alpha = 91°48'$; $\beta = 104°30', \gamma = 90°; Z = 2. d's: 7.15(10), 3.57(10), 2.55(8), 2.49(9), 2.33(10)$.

Physical Properties. *Cleavage* {001} perfect. **H** 2. **G** 2.6. *Luster* usually dull earthy; crystal plates pearly. *Color* white. Often variously colored by impurities. Usually unctuous and plastic. *Optics:* (−); $\alpha = 1.553–1.565, \beta = 1.559–1.569, \gamma = 1.560–1.570; 2V = 24°–50°. r > v$.

Composition and Structure. Kaolinite shows little compositional variation; for $Al_2Si_2O_5(OH)_4$, Al_2O_3 39.5, SiO_2 46.5, H_2O 14.0%. The dioctahedral structure of kaolinite is shown schematically in Figs. 13.81*a* and 13.84. It consists of an Si_2O_5 sheet bonded to a gibbsite layer. The clay minerals *dickite* and *nacrite* are chemically identical to kaolinite but show a *t-o* stacking different from that in kaolinite.

Diagnostic Features. Recognized usually by its claylike character, but without X-ray tests it is impossible to distinguish from the other clay minerals of similar composition which collectively make up *kaolin*.

Occurrence. Kaolinite is a common mineral, the chief constituent of kaolin or clay. It is always a secondary mineral formed by weathering or hydrothermal alteration of aluminum silicates, particularly feldspar. It is found mixed with feldspar in rocks that are undergoing alteration; in places it forms entire deposits where such alteration has been carried to completion. As one of the common products of the decomposition of rocks it is found in soils, and transported by water it is deposited, mixed with quartz and other materials, in lakes, and so on, in the form of beds of clay. Good pseudomorphs after potassium feldspar occur in Cornwall, England.

Use. Clay is one of the most important of the natural industrial substances. It is available in every country of the world and is commercially produced in nearly every state in the United States. Many and

FIG. 13.97. Transmission electron micrograph of well-crystallized kaolinite from Georgia. (Courtesy of Dr. Kenneth M. Towe, Smithsonian Institution, Washington, D.C.)

0.5μ

varied products are made from it which include common brick, paving brick, drain tile, and sewer pipe. The commercial users of clay recognize many different kinds having slightly different properties, each of which is best suited for a particular purpose. High-grade clay, which is known as *china clay* or *kaolin*, has many uses in addition to the manufacture of china and pottery. Its largest is as a filler in paper, but it is also used in the rubber industry and in the manufacture of refractories.

The chief value of clay for ceramic products lies in the fact that when wet it can be easily molded into any desired shape, and when it is heated, part of the combined water is driven off, producing a hard, durable substance.

Name. Kaolinite is derived from *kaolin*, which is a corruption of the Chinese *kauling*, meaning high ridge, the name of a hill near Jauchu Fa, where the material is obtained.

Similar Species. *Dickite* and *nacrite* are of the same composition as kaolinite but differ somewhat in their structure; they are less important constituents of clay deposits.

Anauxite is also assigned to the kaolinite group but has a higher Si/Al ratio than kaolinite. *Halloysite* has two forms: one with kaolinite composition, $Al_2Si_2O_5(OH)_4$, the other with composition $Al_2Si_2O_5(OH)_4 \cdot 2H_2O$. The second type dehydrates to the first with loss of interlayer water molecules.

The *montmorillonite* group comprises a number of clay minerals composed of *t-o-t* layers of both dioctahedral and trioctahedral type. The outstanding characteristic of members of this group is their capacity to absorb water molecules between the sheets, thus producing marked expansion of the structure (see Fig. 13.86 for the similar structure of vermiculite). The dioctahedral members are *montmorillonite*, *beidellite*, and *nontronite*; the trioctahedral members are *hectorite* and *saponite*.

Montmorillonite is the dominant clay mineral in *bentonite*, altered volcanic ash (see also Fig. 13.98). Bentonite has the unusual property of expanding several times its original volume when placed in water. This property gives rise to interesting industrial uses. Most important is as a drilling mud in which the montmorillonite is used to give the fluid a viscosity several times that of water. It is also used for stopping leakage in soil, rocks, and dams.

Illite is a general term for the micalike clay minerals. The illites differ from the micas in having less

FIG. 13.98. Transmission electron micrograph of olivine that (through low-temperature hydrothermal alteration and weathering) has developed etch pits and channels lined with clay minerals (of the montmorillonite type). The white areas are very small void spaces that would have been completely filled if the montmorillonite had been completely hydrated. (From Banfield, J. F., Veblen, D. R. and Jones, B. F., 1990, Transmission electron microscopy of subsolidus oxidation and weathering of olivine. *Contributions to Mineralogy and Petrology*, v. 106, pp. 110–123.)

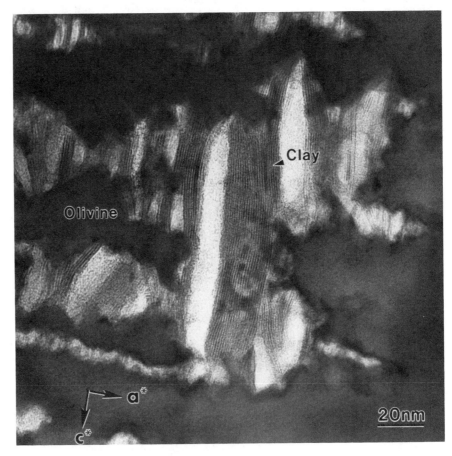

substitution of Al for Si, in containing more water, and in having K partly replaced by Ca and Mg. Illite is the chief constituent in many shales.

TALC—$Mg_3Si_4O_{10}(OH)_2$

Crystallography. Monoclinic; $2/m$. Crystals rare. Usually tabular with rhombic or hexagonal outline. Foliated and in radiating foliated groups. When compact and massive, known as *steatite* or *soapstone*.

$C2/c$; $a = 5.27$, $b = 9.12$, $c = 18.85$ Å; $β = 100°0'$; $Z = 4$. d's: 9.34(10), 4.66(9), 3.12(10), 2.48(7), 1.870(4).

Physical Properties. *Cleavage* {001} perfect. Thin folia somewhat flexible but not elastic. Sectile. **H** 1 (will make a mark on cloth). **G** 2.7–2.8. *Luster* pearly to greasy. *Color* apple-green, gray, white, or silver-white; in soapstone often dark gray or green. Translucent. Greasy feel. *Optics:* (−); $α = 1.539$, $β = 1.589$, $γ = 1.589$; $2V = 6°–30°$. $Z = b$, $X ⊥$ {001}; $r > v$.

Composition and Structure. There is little variation in the chemical composition of most talc; pure talc contains MgO 31.7, SiO_2 63.5, H_2O 4.8%. Small amounts of Al or Ti may substitute for Si and Fe may replace some of the Mg. The Fe-rich analogue, *minnesotaite*, $Fe_3Si_4O_{10}(OH)_2$, is common in low-grade metamorphic Precambrian iron deposits. It is probable that a complete solid solution exists between talc and minnesotaite. The trioctahedral structure of talc is similar to that of dioctahedral pyrophyllite and consists of essentially neutral *t-o-t* layers held together by weak residual bonds (see Fig. 13.84).

Diagnostic Features. Characterized by its micaceous habit, cleavage, softness, and greasy feel.

Occurrence. Talc is a secondary mineral formed by the alteration of magnesium silicates, such as olivine, pyroxenes (see Figs. 13.93 and 13.94), and amphiboles, and may be found as pseudomorphs after these minerals. Characteristically in low-grade metamorphic rocks, where, in massive form, *soapstone*, it may make up nearly the entire rock mass. It may also occur as a prominent constituent in schistose rocks, as in the talc schist. Figure 13.99 gives a *P-T* stability diagram for talc and related minerals in the system MgO-SiO_2-H_2O.

In the United States many talc or soapstone quarries are located along the line of the Appalachian Mountains from Vermont to Georgia. The major producing states are California, North Carolina, Texas, and Georgia.

Use. Most of the talc and soapstone produced is used in powdered form as an ingredient in paint, ceramics, rubber, insecticides, roofing, paper, and foundry facings. The most familiar use is in talcum powder. Talc is also used as an ornamental material for carving small objects.

Name. The name *talc* is of ancient and doubtful origin, probably derived from the Arabic, *talk*.

Similar Species. *Minnesotaite*, $Fe_3Si_4O_{10}(OH)_2$, may be isostructural with talc.

Pyrophyllite—$Al_2Si_4O_{10}(OH)_2$

Crystallography. Monoclinic; $2/m$. Not in distinct crystals. Foliated, in some cases in radiating la-

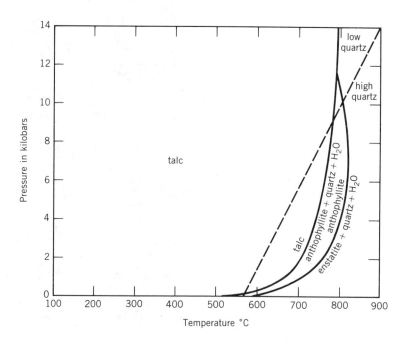

FIG. 13.99. Pressure-temperature diagram for the stability field of talc in the system MgO-SiO_2-H_2O. Talc is stable over a wide temperature range, from 100°C to between 500° and 700°C, depending on the pressure. At high temperature talc reacts to form anthophyllite, and at even higher temperature anthophyllite reacts to form enstatite. Compare with Fig. 13.93. The low–high quartz polymorphic transition is also shown. (Adapted from Evans, B. W. and Guggenheim, S., 1988, Talc, pyrophyllite, and related minerals, in *Hydrous Phyllosilicates, Reviews in Mineralogy*, v. 19, Mineralogical Soc. of America, Washington, D.C., pp. 225–294.)

mellar aggregates. Also granular to compact. Identical with talc in appearance.

$C2/c$; a = 5.15, b = 8.92, c = 18.59 Å; β = 99°55'; Z = 4. d's: 9.21(6), 4.58(5), 4.40(2), 3.08(10), 2.44(2).

Physical Properties. *Cleavage* {001} perfect. Folia somewhat flexible but not elastic. **H** 1–2 (will make a mark on cloth). **G** 2.8 *Luster* pearly to greasy. *Color* white, apple-green, gray, brown. Translucent, will transmit light on thin edges. *Optics:* (−); α = 1.552, β = 1.588, γ = 1.600; 2V = 57°; X ⊥ {001}; r > v.

Composition and Structure. Pyrophyllite shows little deviation from the ideal formula; Al_2O_3 28.3, SiO_2 66.7, H_2O 5.0%. The dioctahedral structure of pyrophyllite consists of essentially neutral *t-o-t* layers (Figs. 13.82 and 13.84) held together by weak residual bonds.

Diagnostic Features. Characterized chiefly by its micaceous habit, cleavage, and greasy feel. X-ray diffraction techniques are needed for positive identification.

Occurrence. Pyrophyllite is a comparatively rare mineral. Found in metamorphic rocks; frequently with kyanite. Occurs in considerable amount in Guilford and Orange counties, North Carolina.

Use. Quarried in North Carolina and used for the same purpose as talc but does not command as high a price as the best grades of talc. A considerable part of the so-called *agalmatolite*, from which the Chinese carve small images, is this species.

Name. From the Greek meaning *fire* and a *leaf*, because it exfoliates on heating.

Mica Group

The micas, composed of *t-o-t* layers with interlayer cations and little or no exchangeable water, crystallize in the monoclinic system but with their crystallographic angle β close to 90°, so that their monoclinic symmetry is not easily seen. The crystals are usually tabular with prominent basal planes and have either a diamond-shaped or hexagonal outline with angles of approximately 60° and 120°. Crystals, as a rule, therefore, appear to be either orthorhombic or hexagonal. They are characterized by a highly perfect {001} cleavage. A blow with a somewhat dull-pointed instrument on a cleavage plate develops in all the species a six-rayed *percussion figure* (Fig. 13.100), two lines of which are nearly parallel to the prism edges and the third, which is most strongly developed, parallel to the mirror plane {010}. There is limited ionic substitution between various members of the group and two members may crystallize together in parallel

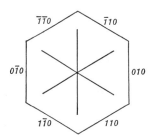

FIG. 13.100. Percussion figure in mica.

position in the same crystal plate with the cleavage extending through both.

MUSCOVITE—$KAl_2(AlSi_3O_{10})(OH)_2$

Crystallography. Monoclinic; $2/m$. Distinct crystals rare; usually tabular with prominent {001}. The presence of prism faces {110} at angles of nearly 60° gives some plates a diamond-shaped outline, making them simulate orthorhombic symmetry (Fig. 13.101). If {010} is also present, the crystals have a hexagonal appearance. The prism faces are roughened by horizontal striations and frequently taper. Penetration twins with [310] the twin axis (see Fig. 13.102). Foliated in large to small sheets; in scales that are, in some cases, aggregated into plumose or globular forms. Also cryptocrystalline and compact massive.

$C2/c$; a = 5.19, b = 9.04, c = 20.08 Å; β = 95°30'; Z = 4. d's: 9.95(10), 3.37(10), 2.66(8), 2.45(8), 2.18(8).

Physical Properties. *Cleavage* {001} perfect, allowing the mineral to be split into very thin sheets; folia flexible and elastic. **H** 2–2½. **G** 2.76–2.88. *Luster* vitreous to silky or pearly. *Color* colorless and trans-

FIG. 13.101. Muscovite. Diamond-shaped crystals. (Harvard Mineralogical Museum.)

FIG. 13.102. Muscovite twins, twin axis [310], producing a penetration twin, with (001) of the individual crystals coplanar. Methuen Township, Ontario, Canada. (Harvard Mineralogical Museum.)

parent in thin sheets. In thicker blocks translucent, with light shades of yellow, brown, green, red. Some crystals allow more light to pass parallel to the cleavage than perpendicular to it. *Optics:* $(-)$; $\alpha = 1.560-1.572$, $\beta = 1.593-1.611$, $\gamma = 1.599-1.615$; $2V = 30°-47°$; $Z = Y$, $X \wedge c = 0°-5°$; $r > v$.

Composition and Structure. Essentially $KAl_2(AlSi_3O_{10})(OH)_2$. Little solid solution occurs among the dioctahedral mica group or between members of the dioctahedral and trioctahedral groups. Minor substitutions may be Na, Rb, Cs for K; Mg, Fe^{2+}, Fe^{3+}, Li, Mn, Ti, Cr for Al; F for OH. *Paragonite*, $NaAl_2(AlSi_3O_{10})(OH)_2$, which is isostructural with muscovite, is practically indistinguishable from it without X-ray methods. At low to moderate temperatures a large miscibility gap exists between muscovite and paragonite (Fig. 13.103). The structure of muscovite is illustrated in Figs. 13.83 and 13.84. It consists of *t* sheets of composition $(Si,Al)_2O_5$ linked to octahedral gibbsite-like sheets to form *t-o-t* layers. These layers have a net negative charge which is balanced by K or Na ions (in paragonite) between them (see discussion on page 501). Although the muscovite structure may show several polytypic forms due to various ways of stacking succeeding Si_2O_5 sheets (see page 501 and Figs. 13.91 and 13.92*b*), the most commonly observed polytype is $2M_1$ with space group $C2/c$.

Diagnostic Features. Characterized by its highly perfect cleavage and light color.

Occurrence. Muscovite is a widespread and common rock-forming mineral. Characteristic of granites and granite pegmatites. In pegmatites, muscovite associated with quartz and feldspar may be in large crystals, called *books*, which in some localities are several feet across. It is also very common in meta-

morphic rocks, forming the chief constituent of certain mica schists. In the chlorite zone of metamorphism, muscovite is a characteristic constituent of albite-chlorite-muscovite schists (see Fig. 13.104 for a *P-T* stability diagram). In some schistose rocks it occurs as fibrous aggregates of minute scales with a silky luster and is known as *sericite*. The development of sericite from feldspar and other minerals, such as topaz, kyanite, spodumene, and andalusite, is a common feature of retrograde metamorphism. Sericite also forms as an alteration of the wall rock of hydrothermal ore veins. As *illite* it is a constituent of some shales, soils, and recent sediments.

Large and important deposits of muscovite occur in the CIS and India. The United States produces only a small amount of sheet mica, but is the largest producer of scrap (flake) mica with major production coming from North and South Carolina, Connecticut,

FIG. 13.103. Temperature-composition diagram for the system $KAl_2(AlSi_3O_{10})(OH)_2$-$NaAl_2(AlSi_3O_{10})(OH)_2$. Below about 700°C this shows a large miscibility gap, and above about 700°C it shows the reaction relations of (1) muscovite to K-feldspar and (2) paragonite to albite. At that temperature there is complete solid solution between K-feldspar and Na-feldspar (see also Fig. 13.121). Corundum is a reaction product in this reaction:

$$KAl(Al_2Si_3O_{10})(OH)_2 \rightarrow \underset{\text{K-feldspar}}{KAlSi_3O_8} + \underset{\text{corundum}}{Al_2O_3} + H_2O$$
$$\underset{\text{muscovite}}{}$$

(From Eugster, H. P. and Yoder, H. S., 1955. The join muscovite-paragonite. Carnegie Institute, Washington, Annual Report of the Director of the Geophysical Laboratory, 1954–55, p. 124.)

FIG. 13.104. Pressure-temperature stability diagram for muscovite, according to the reaction:

$$KAl(Al_2Si_3O_{10})(OH)_2 \rightarrow KAlSi_3O_8 + Al_2O_3 + H_2O$$
$$\text{muscovite} \qquad \text{K-feldspar} \quad \text{corundum}$$

(From Hewitt, D. A. and Wones, D. R., 1984, Experimental phase relations of the micas, in *Micas, Reviews in Mineralogy*, v. 13. Mineralogical Soc. America, Washington, D.C., pp. 201–256.)

Georgia, New Mexico, and South Dakota. Crystals measuring 7 to 9 feet across have been mined in Mattawan Township, Ontario, Canada. Good specimen material is obtained from pegmatites worldwide.

Use. Because of its high dielectric and heat-resisting properties, *sheet mica*, single cleavage plates, is used as an insulating material in the manufacture of electrical apparatus. The *isinglass* used in furnace and stove doors is sheet mica. Many small parts used for electrical insulation are built up of thin sheets of mica cemented together. They may thus be pressed into shape before the cement hardens. India is the largest supplier of mica used in this way. Ground mica is used in many ways: in the manufacture of wallpapers to give them a shiny luster; as a lubricant when mixed with oils; a filler; and as a fireproofing material.

Name. *Muscovite* was so called from the popular name of the mineral, *Muscovy-glass*, because of its use as a substitute for glass in Old Russia (Muscovy). *Mica* was probably derived from the Latin *micare*, meaning to *shine*.

Phlogopite—KMg$_3$(AlSi$_3$O$_{10}$)(OH)$_2$

Crystallography. Monoclinic; $2/m$. Usually in six-sided plates or in tapering prismatic crystals. Crystals frequently large and coarse. Found also in foliated masses.

$C2/m$; $a = 5.33$, $b = 9.23$, $c = 10.26$ Å; $\beta = 100°12'$; $Z = 2$. d's: 10.13(10), 3.53(4), 3.36(10), 3.28(4), 2.62(10).

Physical Properties. *Cleavage* {001} perfect. Folia flexible and elastic. **H** $2\frac{1}{2}$–3. **G** 2.86. *Luster* vitreous to pearly. *Color* yellowish-brown, green, white, often with copperlike reflections from the cleavage surface. Transparent in thin sheets. When viewed in transmitted light, some phlogopite shows asterism because of tiny oriented inclusions of rutile. *Optics:* $(-)$; $\alpha = 1.53$–1.59, $\beta = 1.56$–1.64, $\gamma = 1.56$–1.64; $2V = 0°$–15°; $Y = b$, $X \wedge c = 0°$–15'; $r < v$.

Composition and Structure. Small amount of Na and lesser amounts of Rb, Cs, and Ba may substitute for K. Fe^{2+} substitutes for Mg, forming a series toward biotite. F may substitute in part for OH. Only limited solid solution occurs between muscovite and the phlogopite-biotite series (see Fig. 13.105). The trioctahedral structure of phlogopite is schematically illustrated in Fig. 13.84. It consists of *t-o-t* layers bonded by interlayer K$^+$ ions; it is very similar to the muscovite structure except for the difference in the octahedral layer. It occurs most commonly in the 1M (space group $C2/m$) polytype, but 2M_1 and 3T polytypes occur occasionally (see Figs. 13.91 and 13.92b).

Diagnostic Features. Characterized by its micaceous cleavage and yellowish-brown color. It is, however, impossible to draw a sharp distinction between biotite and phlogopite.

Occurrence. Phlogopite is found in metamorphosed Mg-rich limestones, dolomites, and ultrabasic rocks. It is a common mineral in kimberlite. In the United States found chiefly in Jefferson and St. Lawrence counties, New York. Found abundantly in Canada in Ontario at North and South Burgess, and in various other localities in Ontario and Quebec.

Use. Same as for muscovite; chiefly as electrical insulator.

Name. Name from the Greek word *phlogos* meaning *fire*, an allusion to its color.

BIOTITE—K(Mg,Fe)$_3$(AlSi$_3$O$_{10}$)(OH)$_2$

Crystallography. Monoclinic; $2/m$. Rarely in tabular or short prismatic crystals with prominent {001} and pseudohexagonal outline. Usually in irregular foliated masses; often in disseminated scales or in scale aggregates.

$C2/m$; $a = 5.31$, $b = 9.23$, $c = 10.18$ Å; $\beta = 99°18'$; $Z = 2$. d's: 10.1(10), 3.37(10), 2.66(8), 2.54(8), 2.18(8).

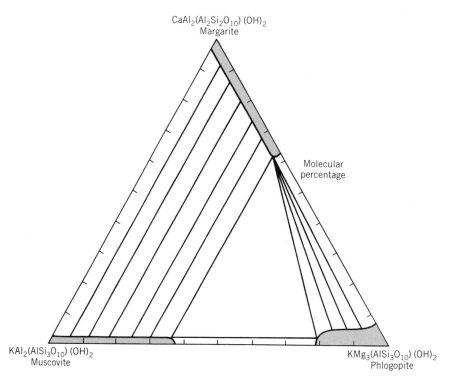

FIG. 13.105. The extent of solid solution for several micas in rocks that have been metamorphosed to a temperature of about 450°C (garnet zone metamorphism). Schematic tielines are drawn between the various solid solution fields, to indicate possible mica pairs, as well as a three-mica triangle (muscovite-phlogopite-margarite). (From Guidotti, C. V., 1984, Micas in metamorphic rocks, in *Micas, Reviews in Mineralogy*, v. 13, pp. 357–468, Mineralogical Society America, Washington, D.C.)

Physical Properties. *Cleavage* {001} perfect. Folia flexible and elastic. **H** $2\frac{1}{2}$–3. **G** 2.8–3.2. *Luster* splendent. *Color* usually dark green, brown to black. More rarely light yellow. Thin sheets usually have a smoky color (differing from the almost colorless muscovite). *Optics:* $(-)$; $\alpha = 1.57$–1.63, $\beta = 1.61$–1.70; $\gamma = 1.61$–1.70; $2V = 0°$–25°; $Y = b$, $Z \wedge a = 0°$–9°; $r < v$.

Composition and Structure. The composition is similar to phlogopite but with considerable substitution of Fe^{2+} for Mg. There is also substitution by Fe^{3+} and Al for Mg and by Al for Si. In addition Na, Ca, Ba, Rb, and Cs may substitute for K. A complete solid solution series exists between phlogopite and biotite. The trioctahedral biotite structure is the same as that of phlogopite (see Fig. 13.84). The 1*M* polytype with space group *C*2/*m* is most common but $2M_1$ and 3*T* also occur (see Figs. 13.91 and 13.92*b*).

Diagnostic Features. Characterized by its micaceous cleavage and dark color.

Occurrence. Biotite forms under a very great variety of geological environments. It is found in igneous rocks varying from granite pegmatites, to granites, to diorites, to gabbros and peridotites. It occurs also in felsite lavas and porphyries. In metamorphic rocks it is formed over a wide range of temperature and pressure conditions, and it occurs in regionally as well as contact metamorphosed rocks. The crystallization of biotite in argillaceous rocks is taken as the onset of the temperature and pressure conditions of the biotite zone. Typical associations at this metamorphic grade are: biotite-chlorite and biotite-muscovite. Occurs in fine crystals in blocks included in the lavas of Vesuvius.

Name. In honor of French physicist J. B. Biot.

Similar Species. *Glauconite*, similar in composition to biotite, is an authigenic mineral found in green pellets in sedimentary rocks. *Vermiculite* is a platy mineral formed chiefly as an alteration of biotite. The structure is essentially that of talc with interlayered water molecules (see Fig. 13.86). The name comes from the Latin word meaning *to breed worms*, for when heated, the mineral expands into wormlike forms. In the expanded condition it is used extensively in heat and sound insulating. Vermiculite is mined at Libby, Montana, and Macon, North Carolina.

Stilpnomelane, with approximate composition $K_{0.6}(Mg,Fe^{2+},Fe^{3+})_6Si_8Al(O,OH)_{27} \cdot 2$–$4H_2O$, can be an important constituent of low-grade, regionally metamorphosed schists, of glaucophane metamorphic facies rocks, and of essentially unmetamorphosed Precambrian banded iron-formations. It is impossible to distinguish stilpnomelane in hand specimen from biotite mica. Optical microscopic or chemical techniques are necessary to distinguish stilpnomelane from biotite.

LEPIDOLITE—K(Li,Al)$_{2-3}$(AlSi$_3$O$_{10}$)(O,OH,F)$_2$

Crystallography. Monoclinic; $2/m$. Crystals usually in small plates or prisms with hexagonal outline. Commonly in coarse- to fine-grained scaly aggregates.

$C2/m$; $a = 5.21$, $b = 8.97$, $c = 20.16$ Å; $\beta = 100°48'$; $Z = 4$. d's: 10.0(6), 5.00(5), 4.50(5), 2.58(10), 1.989(8).

Physical Properties. *Cleavage* {001} perfect. **H** $2\frac{1}{2}$–4. **G** 2.8–2.9. *Luster* pearly. *Color* pink and lilac to grayish-white. Translucent. *Optics:* (−); $\alpha = 1.53$–1.55, $\beta = 1.55$–1.59, $\gamma = 1.55$–1.59; 2V = 0°–60°; $Y = b$, $Z \wedge a = 0°$–7°; $r > v$.

Composition and Structure. The composition of lepidolite varies depending chiefly on the relative amounts of Al and Li in octahedral coordination. Analyses show a range of Li$_2$O from 3.3 to 7 weight percent. In addition Na, Rb, and Cs may substitute for K. It is possible that there is a continuous chemical series from dioctahedral muscovite to trioctahedral lepidolite, in which all of the octahedral sites are occupied. Lepidolite can occur as one of three polytypes (1M, 2M_2, and 3T) (see Figs. 13.91 and 13.92b).

Diagnostic Features. Characterized chiefly by its micaceous cleavage and usually by its lilac to pink color. Muscovite may be pink and lepidolite white, and X-ray powder diffraction techniques may be necessary to distinguish them.

Occurrence. Lepidolite is a comparatively rare mineral, found in pegmatites, usually associated with other lithium-bearing minerals such as pink and green tourmaline, amblygonite, and spodumene. Often intergrown with muscovite in parallel position. Notable foreign localities for its occurrence are Rozna, Moravia, Czechoslovakia; Bikita, Zimbabwe; and the Malagasy Republic. In the United States it is found in several localities in Maine; near Middletown, Connecticut; Pala, California; Dixon, New Mexico; and Black Hills, South Dakota.

Use. A source of lithium. Used in the manufacture of heat-resistant glass.

Name. Derived from the Greek word *lepidos* meaning *scale*.

Margarite—CaAl$_2$(Al$_2$Si$_2$O$_{10}$)(OH)$_2$

Crystallography. Monoclinic; $2/m$. Seldom in distinct crystals. Usually in foliated aggregates with micaceous habit.

$C2/c$; $a = 5.13$, $b = 8.92$, $c = 19.50$ Å; $\beta = 100°48'$; $Z = 4$. d's: 4.40(8), 3.39(8), 3.20(9), 2.51(10), 2.42(8).

Physical Properties. *Cleavage* {001} perfect.

H $3\frac{1}{2}$–5 (harder than the true micas). **G** 3.0–3.1. *Luster* vitreous to pearly. *Color* pink, white, and gray. Translucent. Folia somewhat brittle; because of this brittleness margarite is known as a *brittle mica. Optics:* (−); $\alpha = 1.632$–1.638; $\beta = 1.643$–1.648; $\gamma = 1.645$–1.650; 2V = 40°–67°; $Z = b$, $Y \wedge a = 7°$.

Composition and Structure. Most analyses are close to the above end member composition with CaO 14.0, Al$_2$O$_3$ 51.3, SiO$_2$ 30.2 and H$_2$O 4.5%. A small amount of Na may replace Ca. The dioctahedral structure of margarite is very similar to that of muscovite (see Fig. 13.106). In margarite, however, the tetrahedral layer has the composition (Si$_2$Al$_2$)O$_{10}$ instead of (Si$_3$Al)O$_{10}$ as in muscovite. Because of the greater electrical charge on the (Si$_2$Al$_2$)O$_{10}$ sheet the structure can be balanced by incorporated divalent Ca^{2+} ions instead of monovalent K$^+$. The bond strength between the layers is therefore greater; this is expressed in the brittle nature of margarite.

Diagnostic Features. Characterized by its micaceous cleavage, brittleness, and association with corundum.

Occurrence. Margarite occurs usually with corundum and diaspore and apparently as an alteration product. It is found in this way with the emery deposits of Asia Minor and on the islands of Naxos and Nicaria, of the Grecian Archipelago. In the United States associated with emery at Chester, Massachusetts; Chester County, Pennsylvania; and with corundum deposits in North Carolina.

Name. From the Greek *margarites* meaning *pearl*.

Similar Species. Other *brittle micas* are *clintonite* and *xanthophyllite* which may be regarded as the Ca analogues of phlogopite.

Chlorite Group

A number of minerals are included in the chlorite group all of which have similar chemical, crystallographic, and physical properties. Without quantitative chemical analyses or careful study of the optical and X-ray properties, it is extremely difficult to distinguish between the members. The following is a composite description of the principal members of the group.

CHLORITE— (Mg,Fe)$_3$(Si,Al)$_4$O$_{10}$(OH)$_2$·(Mg,Fe)$_3$(OH)$_6$

Crystallography. Monoclinic; $2/m$, some polymorphs of chlorite are triclinic. In pseudohexagonal tabular crystals, with prominent {001}. Similar in habit to crystals of the mica group, but distinct crystals rare.

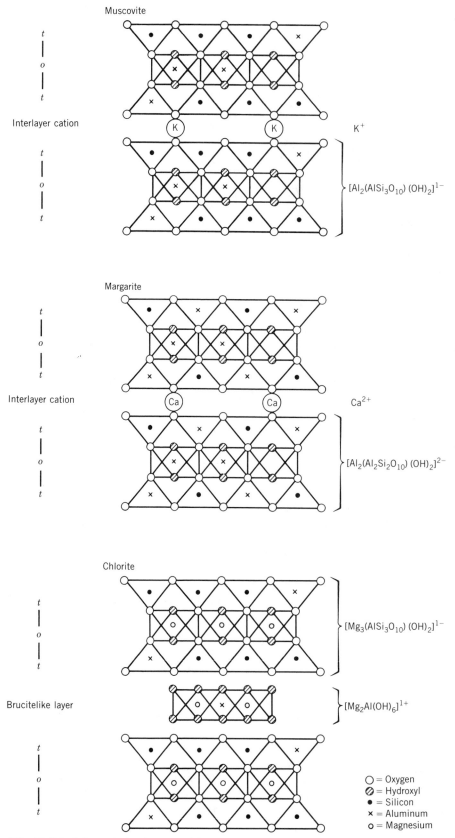

FIG. 13.106. Schematic development of the structures and compositions of muscovite, margarite, and chlorite (compare with Fig. 13.84).

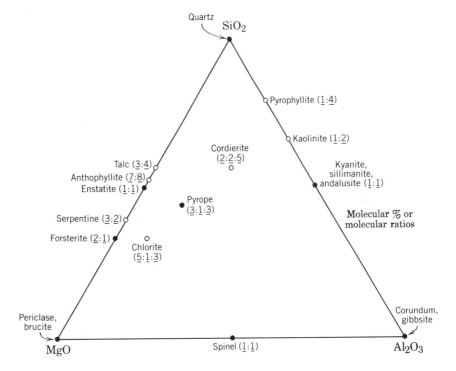

Pyrophyllite—$Al_2Si_4O_{10}(OH)_2$
Kaolinite—$Al_2Si_2O_5(OH)_4$
Kyanite
Sillimanite } —Al_2SiO_5
Andalusite
Talc—$Mg_3Si_4O_{10}(OH)_2$
Anthophyllite—$Mg_7Si_8O_{22}(OH)_2$
Enstatite—$MgSiO_3$
Serpentine—$Mg_3Si_2O_5(OH)_4$
Forsterite—Mg_2SiO_4
Spinel—$MgAl_2O_4$
Cordierite—$Mg_2Al_4Si_5O_{18} \cdot nH_2O$
Pyrope—$Mg_3Al_2Si_3O_{12}$
Chlorite—$(Mg_5Al)(Si_3Al)O_{10}(OH)_8$

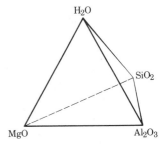

FIG. 13.107. The composition of an idealized chlorite, and those of other end member compositions of rock-forming minerals in the system MgO-Al_2O_3-SiO_2-H_2O. In this diagram, which is
water-free, and which may be considered as a projection from an H_2O corner (in a tetrahedron,
see insert), the hydrous formulas are plotted on the basis of their molecular (MgO, Al_2O_3 and
SiO_2) ratios, without reference to the OH groups or H_2O content. The opaque dots are anhydrous
compositions, and the open circles hydrous compositions. The molecular ratios (MgO:SiO_2;
Al_2O_3:SiO_2; MgO:Al_2O_3:SiO_2, and MgO:Al_2O_3) are noted next to the mineral names.

Usually foliated massive or in aggregates of minute
scales; also in finely disseminated particles.

Cell parameters vary with composition. For clinochlore: $C2/m$; $a = 5.2$–5.3, $b = 9.2$–9.3, $c = 28.6$
Å; $\beta = 96°50'$; $Z = 4$. d's: 3.54(10), 2.53(6), 2.00(6),
1.562(4), 1.534(7).

Physical Properties. *Cleavage* {001} perfect.
Folia flexible but not elastic. **H** 2–$2\frac{1}{2}$. **G** 2.6–3.3. *Luster* vitreous to pearly. *Color* green of various shades.
Rarely yellow, white, rose-red. Transparent to translucent. *Optics:* Most $(+)$, some $(-)$; in all Bxa nearly
$\perp$ {001}. $\alpha = 1.57$–1.66, $\beta = 1.57$–1.67, $\gamma =
1.57$–$1.67$; $2V = 20°$–$60°$. Pleochroism in green $(+)$,
$X, Y > Z$; $(-)$ $X < Y, Z$. The indices increase with
increasing iron content.

Composition and Structure. The composition
of chlorite can be regarded as made up of *t-o-t* layers
of composition $[Mg_3(AlSi_3O_{10})(OH)_2]^{1-}$ interleaved
with a "brucitelike" layer in which one out of three
Mg^{2+} ions is replaced by one Al^{3+}, resulting in a

composition of $[Mg_2Al(OH)_6]^{1+}$. This is illustrated
schematically in Fig. 13.106. When the composition
of the *t-o-t* layer and the "brucitelike" layer are
added, the composition $(Mg,Al)_6(Si,Al)_4O_{10}(OH)_8$ results, which is a general formula for magnesian chlorite. In most chlorites there is considerable deviation
from this formula because of Fe^{2+}, Fe^{3+}, and additional Al substitution. A diagrammatic sketch of the
chlorite structure is given in Fig. 13.85.

The general formula of chlorite may be represented as follows: $A_{5-6}Z_4O_{10}(OH)_8$, where $A =$ Al,
Fe^{2+}, Fe^{3+}, Li, Mg, Mn, Ni, and $Z =$ Al, Si, Fe^{3+}.
Because of the extensive solid solution, many varietal
names have been given to members of the chlorite
group. Examples are: *chamosite, clinochlore, pennan-
tite,* and *sudoite.* The composition of an idealized Mg
end member chlorite, $(Mg_5Al)(Si_3Al)O_{10}(OH)_8$, is
plotted on an anhydrous basis in Fig. 13.107. Such a
diagram, which is frequently used in petrologic discussions, allows for the comparison of, for example,

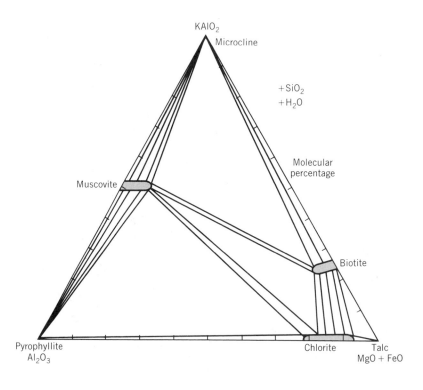

FIG. 13.108. Common assemblages involving chlorite, and other layer silicates, as well as microcline in metamorphic rocks of the biotite zone of the greenschist facies. The chemistry of this system is: $KAlO_2$-Al_2O_3-$(Mg+Fe)O$-(SiO_2)-(H_2O). The right-hand corner of the triangle shows two components, allowing for the representation of some solid solution extent in natural minerals. SiO_2 and H_2O are assumed to be present for the representation of the composition of anhydrous and hydrous silicates. Tielines connect coexisting mineral pairs. Triangles represent three-mineral assemblages. (Adapted from *Petrologic Phase Equilibria*, 21st edition, by W. G. Ernst. Copyright © 1976 by W. H. Freeman and Company. Reprinted by permission.)

the composition of chlorite with that of other rock-forming minerals. Figure 13.108 illustrates the common occurrence of chlorite, in association with other layer silicates, as well as microcline, in metamorphic rocks of the biotite zone of the greenschist facies (see also Chapter 14, section III, "Metamorphic Rocks").

The great range in composition of chlorite is reflected in variations in physical and optical properties as well as X-ray parameters (e.g., interplanar spacings and variation in unit cell size).

Diagnostic Features. Characterized by its green color, micaceous habit and cleavage, and by the fact that the folia are not elastic.

Occurrence. Chlorite is a common mineral in metamorphic rocks and it is the diagnostic mineral of the greenschist facies. In pelitic schists it occurs in quartz-albite-chlorite-sericite-garnet assemblages. It is also commonly found with actinolite and epidote. Chlorite is also a common constituent of igneous rocks where it has formed as an alteration of Mg-Fe silicates such as pyroxenes, amphiboles, biotite, and garnet. The green color of many igneous rocks is due to the chlorite to which the ferromagnesian silicates have altered; and the green color of many schists and slates is due to finely disseminated particles of the mineral.

Name. Chlorite is derived from the Greek word *chloros* meaning *green*, in allusion to the common color of the mineral.

Apophyllite—$KCa_4(Si_4O_{10})_2F \cdot 8H_2O$

Crystallography. Tetragonal; $4/m2/m2/m$. Usually in crystals showing a combination of {110}, {011}, and {001} (Fig. 13.109). Crystals may resemble a combination of cube and octahedron (Fig. 13.110), but are shown to be tetragonal by difference in luster between faces of prism and base.

$P4mnc$; $a = 9.02$, $c = 15.8$ Å; $Z = 2$. d's: 4.52(2), 3.94(10), 3.57(1), 2.98(7), 2.48(3).

Physical Properties. *Cleavage* {001} perfect. **H** $4\frac{1}{2}$–5. **G** 2.3–2.4. *Luster* of base pearly, other faces vitreous. *Color* colorless, white, or grayish; may show pale shades of green, yellow, rose. Transparent to translucent. *Optics:* (+); $\omega = 1.537$, $\epsilon = 1.535$. May be optically (−).

Composition and Structure. The structure of apophyllite differs from that of the other phyllosilicates in that the sheets are composed of 4-fold and

FIG. 13.109. Apophyllite.

FIG. 13.110. Apophyllite crystals with stilbite, India.

8-fold rings. These sheets are linked to each other by Ca, K, and F ions. Water, which is bonded to the Si_2O_5 sheet by hydrogen-bonding, is lost at about 250°C. This indicates that the water is more strongly held than adsorbed water, but less strongly than structural (OH) groups.

Diagnostic Features. Recognized usually by its crystals, color, luster, and basal cleavage.

Occurrence. Apophyllite occurs as a secondary mineral lining cavities in basalt and related rocks, associated with zeolites, calcite, datolite, and pectolite.

It is found in fine crystals at Andreasberg, Harz

Mountains, Germany; near Bombay, India; Iceland; and Guanajuato, Mexico. In the United States at Bergen Hill and Paterson, New Jersey; and Lake Superior copper district. Found in fine crystals in Nova Scotia.

Name. Apophyllite is derived from two Greek words meaning *from* and *a leaf*, because of its tendency to exfoliate when ignited.

PREHNITE—$Ca_2Al(AlSi_3O_{10})(OH)_2$

Crystallography. Orthorhombic; 2mm. Distinct crystals are rare, commonly tabular parallel to {001}. Usually reniform, stalactitic (Fig. 13.111), and in rounded groups of tabular crystals.

$P2cm$; $a = 4.65$, $b = 5.48$, $c = 18.49$ Å; $Z = 2$. d's: 3.48(9), 3.28(6), 3.08(10), 2.55(10), 1.77(7).

Physical Properties. H $6-6\frac{1}{2}$. G 2.8–2.95. *Luster* vitreous. *Color* usually light green, passing into white. Translucent. *Optics:* (+); $\alpha = 1.616$, $\beta = 1.626$, $\gamma = 1.649$; $2V = 66°$; $X = a$, $Z = c$; $r > v$.

Composition and Structure. CaO 27.1, Al_2O_3 24.8, SiO_2 43.7, H_2O 4.4%. Some Fe^{3+} may replace Al. The structure of prehnite contains layers of Al and Si tetrahedra parallel to {001}. Ca is in 7-coordination and lies between Si tetrahedra in adjoining layers.

Diagnostic Features. Characterized by its green color and crystalline aggregates forming reniform surfaces. Resembles hemimorphite but is of lower specific gravity.

Occurrence. Prehnite occurs as a secondary mineral lining cavities in basalt and related rocks. Associated with zeolites, datolite, pectolite, and calcite. In the United States, it occurs notably at Paterson and Bergen Hill, New Jersey. In Canada, at Asbestos, Quebec.

Use. Sometimes used as an ornamental and gem material.

Name. In honor of Colonel Prehn, who brought the mineral from the Cape of Good Hope.

CHRYSOCOLLA ≈ $Cu_4H_4Si_4O_{10}(OH)_8$

Crystallography and Structure. Generally amorphous and thus in the strict sense cannot be considered a mineral. Partially crystalline material shows the presence of Si_4O_{10} layers in a very much defect structure. Massive, compact; in some cases earthy. Individual specimens inhomogeneous.

Physical Properties. Fracture conchoidal. H 2–4. G 2.0–2.4. *Luster* vitreous to earthy. *Color* green to greenish-blue; brown to black when impure. Refractive index variable, usually about 1.50.

Composition. Chrysocolla is a hydrogel or ge-

FIG. 13.111. Prehnite, Paterson, New Jersey.

latinous precipitate and shows a wide range in composition. Chemical analyses report: CuO 32.4–42.2, SiO_2 37.9–42.5, H_2O 12.2–18.8%. In addition Al_2O_3 and Fe_2O_3 are usually present in small amounts.

Diagnostic Features. Characterized by its green or blue color and conchoidal fracture. Distinguished from turquoise by inferior hardness.

Occurrence. Chrysocolla forms in the oxidized zones of copper deposits associated with malachite, azurite, cuprite, or native copper, for example. It is commonly found in the oxidized portions of porphyry copper ores. Found in the copper districts of Arizona and New Mexico; at Chuquicamata, Chile; and Zaire.

Use. A minor ore of copper. Sometimes cut as a gemstone, but most chrysocolla used in jewelry is intergrown with chalcedony and has the refractive index of chalcedony.

Name. Chrysocolla, derived from two Greek words meaning *gold* and *glue*, which was the name of a similar-appearing material used to solder gold.

Similar Species. *Dioptase*, $Cu_6(Si_6O_{18})\cdot6H_2O$, is a rhombohedral cyclosilicate occurring in well-defined green rhombohedral crystals. It is a minor gem mineral. *Plancheite*, $Cu_8(Si_4O_{11})(OH)_2\cdot H_2O$, and *shattuckite*, $Cu_5(SiO_3)_4(OH)_2$, have inosilicate structures.

TECTOSILICATES

Approximately 64% of the rocky crust of the Earth is made up of minerals built about a three-dimensional framework of linked SiO_4 tetrahedra (see Fig. 13.1). These minerals belong to the *tectosilicate* class in which all the oxygen ions in each SiO_4 tetrahedron are shared with neighboring tetrahedra. This results in a stable, strongly bonded structure in which the ratio of Si : O is 1 : 2 (Fig. 13.3). We will discuss in detail the following groups and species:

TECTOSILICATES

SiO_2 Group

Quartz ⎤	
Tridymite ⎬	SiO_2
Cristobalite ⎦	
Opal	$SiO_2\cdot nH_2O$

Feldspar Group

K-Feldspars	
Microcline ⎤	
Orthoclase ⎬	$KAlSi_3O_8$
Sanidine ⎦	
Plagioclase Feldspars	
Albite	$NaAlSi_3O_8$
Anorthite	$CaAl_2Si_2O_8$

TECTOSILICATES (*continued*)

Feldspathoid Group

Leucite	$KAlSi_2O_6$
Nepheline	$(Na,K)AlSiO_4$
Sodalite	$Na_8(AlSiO_4)_6Cl_2$
Lazurite	$(Na,Ca)_8(AlSiO_4)_6(SO_4,S,Cl)_2$
Petalite	$LiAlSi_4O_{10}$

Scapolite Series

Marialite	$Na_4(AlSi_3O_8)_3(Cl_2,CO_3,SO_4)$
Meionite	$Ca_4(Al_2Si_2O_8)_3(Cl_2,CO_3,SO_4)$
Analcime	$NaAlSi_2O_6\cdot H_2O$

Zeolite Group

Natrolite	$Na_2Al_2Si_3O_{10}\cdot2H_2O$
Chabazite	$CaAl_2Si_4O_{12}\cdot6H_2O$
Heulandite	$CaAl_2Si_7O_{18}\cdot6H_2O$
Stilbite	$NaCa_2Al_5Si_{13}O_{36}\cdot14H_2O$

SiO_2 Group

An SiO_2 framework that does not contain other structural units is electrically neutral. There are at least nine different ways in which such a framework can be built. These modes of geometrical arrangement correspond to nine known polymorphs of SiO_2, one of which is synthetic (see Table 13.4). Each of these polymorphs has its own space group, cell dimensions, characteristic morphology, and lattice energy. Which polymorph is stable is determined chiefly by energy considerations; the higher-temperature forms with great lattice energy posses the more expanded structures which are reflected in lower specific gravity and refractive index. In addition to the nine polymorphs of SiO_2 there are two related and essentially amorphous substances, *lechatelierite*, a high silica glass of variable composition and *opal*, $SiO_2\cdot nH_2O$, with a locally ordered structure of silica spheres and highly variable H_2O content.

The principal naturally occurring SiO_2 polymorphs fall into three structural categories: *low quartz*, with the lowest symmetry and the most compact structure; *low tridymite*, with higher symmetry and a more open structure; and *low cristobalite*, with the highest symmetry and the most expanded structure of the three polymorphs. These polymorphs are related to each other by reconstructive transformations, a process that requires considerable energy. The sluggishness and energy requirements of reconstructive transformations allow the phases to exist metastably for long periods of time. The temperatures of reconstructive inversions vary widely, depending chiefly on the rate and direction of temperature change. Each of the three above structure types has also a high–low inversion, as shown by the existence of high and low quartz, high and low tridymite, and high and low cristobalite (see Table 13.4). These

Table 13.4
POLYMORPHS OF SiO₂

Name	Symmetry	Space Group	Specific Gravity	Refractive Index (Mean)
Stishovite*	Tetragonal	$P4_2/mnm$	4.35	1.81
Coesite	Monoclinic	$C2/c$	3.01	1.59
Low (α) quartz	Hexagonal	$P3_221$ (or $P3_121$)	2.65	1.55
High (β) quartz	Hexagonal	$P6_222$ (or $P6_422$)	2.53	1.54
Keatite (synth.)	Tetragonal	$P4_12_12$ (or $P4_32_12$)	2.50	1.52
Low (α) tridymite	Monoclinic or Orthorhombic	$C2/c$ (or Cc) $C222_1$	2.26	1.47
High (β) tridymite	Hexagonal	$P6_3/mmc$	2.22	1.47
Low (α) cristobalite	Tetragonal	$P4_12_12$ (or $P4_32_12$)	2.32	1.48
High (β) cristobalite	Isometric	$Fd3m$	2.20	1.48

*Only polymorph with Si in octahedral coordination with oxygen.

Table 13.5
INVERSION TEMPERATURES FOR DISPLACIVE TRANSFORMATIONS IN SOME SiO₂ POLYMORPHS

High T Polymorph	Minimum Crystallization T for Stable Form at 1 Atmosphere P	Inversion to Low T Form at 1 Atmosphere P
High cristobalite	1470°C	~268°C
High tridymite	870°C	~120°–140°C
High quartz	574°C	573°C

transformations are displacive; they take place quickly and reversibly at a fairly constant and sharply defined temperature of inversion and may be repeated over and over again without physical disintegration of the crystal. The nearly instant high → low inversions take place with the release of a fairly constant amount of energy; very near the same temperature the low → high inversions take place with the absorption of energy (see Table 13.5). The structures of several SiO₂ polymorphs are illustrated in Fig. 13.112 (see Fig. 3.42 for high and low quartz structures).

The low-temperature form of each of the displacive polymorphic pairs has a lower symmetry than the higher-temperature form (see Table 13.4), but this symmetry change is less than in the reconstructive transformations. The effect of increased pressure is to raise all inversion temperatures and for any temperature to favor the crystallization of the polymorph most economical of space (Fig. 13.113).

The most dense of the silica polymorphs are *coesite* and *stishovite*. *Coesite* was synthesized in 1953 and *stishovite*, which is isostructural with rutile, TiO_2, in 1961 and it was not until later that they were found in nature. They have both been identified in very small amounts at Meteor Crater, Arizona. Their formation is attributed to the high pressure and high temperature resulting from the impact of a meteorite. *Coesite* has recently been found in a sanidine-coesite eclogite from a kimberlite pipe in South Africa (see

FIG. 13.112. Structures of some polymorphs of SiO₂ (see Fig. 3.42 for illustration of low and high quartz structures). *(a)* Tetrahedral layers in high tridymite projected onto (0001). *(b)* Portion of the high cristobalite structure projected onto (111).

(a)

(b)

(continued)

(c)

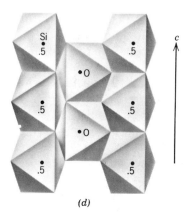

(d)

FIG. 13.112. *(continued) (c)* Coesite structure showing four-membered tetrahedral rings that lie parallel to (001). *(d)* Structure of stishovite, with Si in octahedral coordination with oxygen, projected on (100) (*a, b, c,* and *d* after Papike, J. J. and Cameron, M., 1976, Crystal chemistry of silicate minerals of geophysical interest, *Reviews of Geophysics and Space Physics,* v. 14, pp. 37–80.)

Fig. 4.52; also Fig. 9.4 and related text). *Keatite* has not been found in nature.

QUARTZ—SiO₂

Crystallography. Quartz, hexagonal; 32. High-quartz, hexagonal; 622. Crystals commonly prismatic, with prism faces horizontally striated. Terminated usually by a combination of positive and negative rhombohedrons, which often are so equally developed as to give the effect of a hexagonal dipyramid

(Fig. 13.114*a*). In some crystals one rhombohedron predominates or occurs alone (Fig. 13.114*b*). It is chosen as *r* {10$\bar{1}$1}. The prism faces may be wanting, and the combination of the two rhombohedrons gives what appears to be a hexagonal dipyramid (a *quartzoid*) (Fig. 13.114*c*). Some crystals are malformed, but the recognition of the prism faces by their horizontal striations assists in the orientation. The trigonal trapezohedral faces *x* are occasionally observed and reveal the true symmetry. They occur at the upper right of alternate prism faces in right-hand quartz and to the

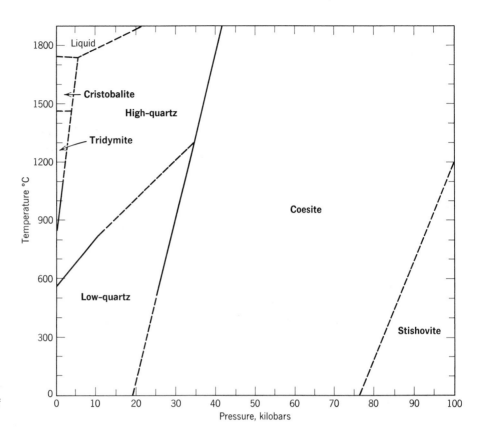

FIG. 13.113. Stability relations of the SiO₂ polymorphs.

upper left of alternate prism faces in left-hand quartz (Figs. 13.114d and e). The right- and left-hand trigonal trapezohedrons are enantimorphous forms and reflect the arrangement of the SiO₄ tetrahedra, either in the form of a right- or left-hand screw (see space groups).

FIG. 13.114. Quartz crystals.

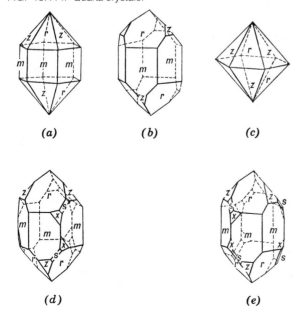

In the absence of x faces, the "hand" can be recognized by observing whether plane polarized light passing parallel to c is rotated to the left or right.

Crystals may be elongated in tapering and sharply pointed forms, and some appear twisted or bent. Equant crystals with apparent 6-fold symmetry (Figs. 13.114a and c) are characteristic of high-temperature quartz, but the same habit is found in quartz that crystallized as the low-temperature form. Most quartz is twinned according to one or both of two laws (see Figs. 2.118c and d). These are *Dauphiné*, c the twin axis; and *Brazil*, {11$\bar{2}$0} the twin plane. Both types are penetration twins and external evidence of them is rarely seen.

The size of crystals varies from individuals weighing several tons to finely crystalline coatings, forming "drusy" surfaces. Also common in massive forms of great variety. From coarse- to fine-grained crystalline to flintlike or cryptocrystalline, giving rise to many variety names (see below). May form in concretionary masses.

$P3_221$ or $P3_121$, $a = 4.91$, $c = 5.41$ Å; $Z = 3$. d's: 4.26(8), 3.34(10), 1.818(6), 1.541(4), 1.081(5) (see Fig. 7.19).

Physical Properties. **H** 7. **G** 2.65. Fracture conchoidal. *Luster* vitreous, in some specimens greasy, splendent. *Color* usually colorless or white, but frequently colored by impurities and may then be any color. Transparent to translucent. Strongly piezoelectric and pyroelectric. *Optics:* (+); $\omega = 1.544$, $\epsilon = 1.553$.

Composition and Structure. Of all the minerals, quartz is most nearly a pure chemical compound with constant physical properties. Si 46.7, O 53.3%. The structure of low (α) quartz with one of two enantiomorphic space groups, $P3_121$ or $P3_221$, is discussed on page 155 and illustrated in Fig. 3.42b. At 573°C, and at atmospheric pressure, this structure transforms instantaneously to high (β) quartz, with hexagonal symmetry and one of two enantiomorphic space groups, $P6_222$ or $P6_422$. (This structure is shown in Fig. 3.42a). The displacive transformation from low to high quartz involves only minor atomic adjustments without breaking of Si-O bonds. Upon cooling high quartz, through the inversion point at 573°C, Dauphiné twinning may be produced (see page 168).

Diagnostic Features. Characterized by its glassy luster, conchoidal fracture, and crystal form. Distinguished from calcite by its hardness and from white varieties of beryl by its inferior hardness.

Varieties. A great many different forms of quartz exist, to which varietal names have been given. The more important varieties, with a brief description of each, follow.

Coarsely Crystalline Varieties

Rock Crystal. Colorless quartz, commonly in distinct crystals.

Amethyst. Quartz colored various shades of violet, often in crystals. The color results from the presence of trace amounts of iron as $[FeO_4]^{4-}$ color centers (see Table 6.4).

Rose Quartz. Coarsely crystalline but usually without crystal form, color a rose-red or pink. Often fades on exposure to light. Small amounts of Ti^{4+} appear to be the coloring agent.

Smoky Quartz; Cairngorm Stone. Frequently in crystals; smoky yellow to brown to almost black. Named *cairngorm* for the locality of Cairngorm in Scotland. The dark color is attributed to the presence of trace amounts of Al^{3+} ions which produce $[AlO_4]^{4-}$ color centers upon irradiation of originally colorless quartz (see Table 6.4 and related discussion).

Citrine. Light yellow resembling topaz in color (see Plate II, nos. 1 to 4, Chapter 15 for illustrations of colored varieties of quartz).

Milky Quartz. Milky white owing to minute fluid inclusions. Some specimens have a greasy luster.

Quartz may contain parallel fibrous inclusions which give the mineral a chatoyancy. When stones are cut *en cabochon* they are called *quartz cat's eye* (see Fig. 6.24). *Tiger's eye* is a yellow fibrous quartz pseudomorphic after the fibrous amphibole crocidolite. It is also chatoyant.

With Inclusions. Many other minerals occur as inclusions in quartz and thus give rise to variety names. *Rutilated quartz* has fine needles of rutile penetrating it (Fig. 13.115). Tourmaline and other minerals are found in quartz in the same way. *Aventurine* quartz includes brilliant scales of colored minerals such as hematite (red) or chromium mica (green) and is used as a gem material. Liquids and gases may occur as inclusions; both liquid and gaseous carbon dioxide exist in some quartz.

Microcrystalline Varieties

Depending on their structure, the microcrystalline varieties of quartz may be divided into two types: *fibrous* and *granular*. It is difficult to distinguish between them macroscopically.

A. Fibrous Varieties

Chalcedony is the general term applied to fibrous varieties. More specifically it is a brown to gray, translucent variety, with a waxy luster, often mammillary and in other imitative shapes. Chalcedony has been de-

FIG. 13.115. Rutilated quartz, Brazil. (Harvard Mineralogical Museum.)

FIG. 13.116. Agate cut and polished, Brazil. (Harvard Mineralogical Museum.)

posited from aqueous solutions and is frequently found lining or filling cavities in rocks. Color and banding give rise to the following varieties:

Carnelian, a red chalcedony that grades into brown *sard*.

Chrysoprase is an apple-green chalcedony colored by nickel oxide.

Agate is a variety with alternating layers of chalcedony having different colors and porosity. The colors are usually in delicate, fine parallel bands which are commonly curved, in some specimens concentric (Fig. 13.116). Most agate used for commercial purposes is colored by artificial means. Some agates have the different colors not arranged in bands but irregularly distributed. *Moss agate* is a variety in which the variation in color is caused by visible impurities, commonly manganese oxide in mosslike patterns (see Plate II, nos. 5 and 6, Chapter 15).

Wood that has been petrified by replacement by clouded agate is known as *silicified* or *agatized wood*.

Onyx, like agate, is a layered chalcedony, with layers arranged in parallel planes. *Sardonyx* is an onyx with sard alternating with white or black layers.

Heliotrope or *bloodstone* is a green chalcedony with small red spots of jasper in it.

B. Granular Varieties

Flint and *chert* resemble each other and there is no sharp distinction between them. Dark, siliceous nodules, usually found in chalk, are called flint, whereas lighter-colored bedded deposits are called chert.

Jasper is a granular microcrystalline quartz with dull luster usually colored red by included hematite.

Prase has a dull green color; otherwise it is similar to jasper, and occurs with it.

Occurrence. Quartz is a common and abundant mineral occurring in a great variety of geological environments. It is present in many igneous and metamorphic rocks and is a major constituent of granite pegmatites. It is the most common gangue mineral in hydrothermal and metal-bearing veins and in many veins is essentially the only mineral present. In the form of flint and chert, quartz is deposited on the sea floor contemporaneously with the enclosing rock; or solutions carrying silica may replace limestone to form chert horizons. On the breakdown of quartz-bearing rocks, the quartz, because of its mechanical and chemical stability, persists as detrital grains to accumulate as sand. Quartz-rich sandstone and its metamorphic equivalent, quartzite, may be composed mainly of quartz.

Rock crystal is found widely distributed, some of the more notable localities being the Alps; Minas Gerais, Brazil; the Malagasy Republic; and Japan. The best quartz crystals from the United States are found near Hot Springs, Arkansas, and Little Falls, Herkimer County, New York. Important occurrences of amethyst are in the Ural Mountains, Czechoslovakia, Uruguay, Zambia, and Brazil. Found at Thunder Bay on the north shore of Lake Superior. In the United States found in Delaware and Chester counties, Pennsylvania; Oxford County, Maine. Smoky quartz is found in large and fine crystals in Switzerland; and in the United States at Pikes Peak, Colorado; Alexander County, North Carolina; and Oxford County, Maine.

The chief source of agates at present is in southern Brazil and northern Uruguay. Most of these agates are cut at Idar-Oberstein, Germany, itself a famous agate locality. In the United States agate is found in numerous places, notably in Oregon and Wyoming. The chalk cliffs at Dover, England, are famous for the flint nodules that weather from them. Similar nodules are found on the French coast of the English channel and on islands off the coast of Denmark. Massive quartz, occurring in veins or with feldspar in pegmatite dikes, is mined in Connecticut, New York, Maryland, and Wisconsin for its various commercial uses.

Use. Quartz has many and varied uses. It is widely used as gemstones or ornamental material, as amethyst, rose quartz, smoky quartz, tiger's eye, aventurine, carnelian, agate, and onyx. As sand, quartz is used in mortar, in concrete, as a flux, as an abrasive, and in the manufacture of glass and silica brick. In powdered form it is used in porcelain, paints, sandpaper, scouring soaps, and as a wood filler. In the form of quartzite and sandstone it is used as a building stone and for paving purposes.

Quartz has many uses in scientific equipment. Because of its transparency in both the infrared and ultraviolet portions of the spectrum, quartz is made into lenses and prisms for optical instruments. The *optical activity* of quartz (the ability to rotate the plane of polarization of light) is utilized in the manufacture of an instrument to produce monochromatic light of differing wavelengths. Quartz wedges, cut from transparent crystals, are used as an accessory to the polarizing microscope. Because of its piezoelectric property, quartz has specialized uses. It is cut into small oriented plates and used as radio oscillators to permit both transmission and reception on a fixed frequency. The tiny quartz plate used in digital quartz watches serves the same function as quartz oscillators used to control radio frequencies (see page 270). This property also renders it useful in the measurement of instantaneous high pressures such as result from firing a gun or an atomic explosion.

Artificial. Since 1947 much of the quartz used for optical and piezoelectrical purposes has been manufactured by hydrothermal methods. More recently yellow, brown, blue, and violet quartz has been synthesized for use as gem material.

Name. The name *quartz* is a German word of ancient derivation.

Similar Species. *Lechatelierite*, SiO_2, is fused silica or silica glass. Found in fulgurites, tubes of fused sand formed by lightning, and in cavities in some lavas. Lechatelierite is also found at Meteor Crater, Arizona, where sandstone has been fused by the heat generated by the impact of a meteorite.

Tridymite—SiO_2

Crystallography. Low (α) tridymite: monoclinic or orthorhombic; $2/m$, m, or 222. High (β) tridymite: hexagonal; $6/m2/m2/m$. Crystals are small and commonly twinned and at room temperature are pseudomorphs after high tridymite.

Low tridymite, for $C2/c$ or Cc; $a = 18.54$, $b = 5.01$, $c = 25.79$ Å; $\beta = 117°40'$; $Z = 48$; for $C222_1$; $a = 8.74$, $b = 5.04$, $c = 8.24$ Å; $Z = 8$. d's: (low tridymite): 4.30(10), 4.08(9), 3.81(9), 2.96(6), 2.47(6). High tridymite, $P6_3/mmc$; $a = 5.04$, $c = 8.24$ Å; $Z = 4$.

Physical Properties. **H** 7. **G** 2.26. *Luster* vitreous. *Color* colorless to white. Transparent to translucent. *Optics:* (+); $\alpha = 1.468-1.479$, $\beta = 1.470-1.480$, $\gamma = 1.475-1.483$; $2V = 40°-90°$.

Composition and Structure. Ideally SiO_2. However, small amounts of Na and Al may be in solid solution. The crystal structure of low (α) tridymite has not been investigated in detail but it is undoubtedly closely related to that of high tridymite. The high (β) tridymite structure consists of sheets of tetrahedra that lie parallel to {0001}; the tetrahedra within each sheet share corners to form six-membered rings (see Fig. 13.112a) and in these rings tetrahedra alternatively point up or down, providing linkage between the sheets. Tridymite is the stable form of SiO_2 at temperatures between 870° and 1470°C, at atmospheric pressure (see Fig. 13.113). At higher temperatures it transforms to cristobalite, at lower temperatures to high quartz. These transformations are reconstructive and extremely sluggish.

Diagnostic Features. It is impossible to identify tridymite by macroscopic means, but under the microscope its crystalline outline and refractive index distinguish it from the other silica minerals.

Occurrence. Tridymite occurs commonly in certain siliceous volcanic rocks such as rhyolite, obsidian, and andesite, and for this reason may be considered an abundant mineral. Commonly associated with sanidine and cristobalite. It is found in large amounts in the lavas of the San Juan district of Colorado. It is also found in stony meteorites and the lunar basalts.

Name. From the Greek meaning *threefold*, in allusion to its common occurrence in trillings.

Cristobalite—SiO_2

Crystallography. Low (α) cristobalite: tetragonal; 422. High (β) cristobalite: isometric; $4/m\bar{3}2/m$. Crystals small octahedrons; this form is retained on inversion from high to low cristobalite. Also in spherical aggregates.

Low cristobalite, $P4_12_12_1$ (or $P4_32_12$): $a = 4.97$, $c = 6.93$ Å; $Z = 4$. d's (low cristobalite): 4.05(10), 2.84(1), 2.48(2), 1.929(1), 1.870(1). High cristobalite, $Fd3m$: $a = 7.13$ Å; $Z = 8$.

Physical Properties. **H** $6\frac{1}{2}$. **G** 2.32. *Luster* vitreous. Colorless. Translucent. *Optics:* (+); $\omega = 1.484$, $\epsilon = 1.487$.

Composition and Structure. Ideally SiO_2, but most natural material contains some Na and Al in solid solution. The low cristobalite structure is tetragonal, whereas high cristobalite is isometric. In high cristobalite six-membered tetrahedral rings are stacked parallel to {111}; see Fig. 13.112b. High cristobalite is stable from 1470°C to the melting point, 1728°C, at atmospheric pressure (see Fig. 13.113).

The transformation at 1470°C to tridymite is of the reconstructive type.

Diagnostic Features. The occurrence in small lava cavities as spherical aggregates is characteristic, but it cannot be determined with certainty without optical or X-ray measurements.

Occurrence. Cristobalite is present in many siliceous volcanic rocks, both as the lining of cavities and as an important constituent in the fine-grained groundmass. It is, therefore, an abundant mineral. Associated with tridymite in the lavas of the San Juan district, Colorado.

Name. From the Cerro San Cristobal near Pachuca, Mexico.

OPAL—$SiO_2 \cdot nH_2O$

Crystallography. Generally amorphous. Massive; often botryoidal, stalactitic. Although X-ray studies indicate that much opal is essentially amorphous, precious opals contain silica spheres in an ordered packing (see Structure, below, and also Figs. 6.21 and 6.22).

Physical Properties. *Fracture* conchoidal. **H** 5–6. **G** 2.0–2.25. *Luster* vitreous; often somewhat resinous. *Color* colorless, white, pale shades of yellow, red, brown, green, gray, and blue. The darker colors result from impurities. Often has a milky or "opalescent" effect and may show a fine play of colors. Transparent to translucent. Some opal, especially *hyalite*, shows a greenish-yellow fluorescence in ultraviolet light. *Optics:* Refractive index 1.44–1.46.

Composition and Structure. $SiO_2 \cdot nH_2O$. The water content, usually between 4 and 9%, may be as high as 20%. The specific gravity and refractive index decrease with increasing water content.

Although opal is essentially amorphous, it has been shown to have an ordered structure. It is not a crystal structure with atoms in a regular three-dimensional array but is made up of closely packed spheres of silica in hexagonal and/or cubic closest packing (see Figs. 6.21 and 6.22). Air or water occupies the voids between the spheres. In common opal the domains of equal-size spheres with uniform packing are small or nonexistent, but in precious opal large domains are made up of regularly packed spheres of the same size. The sphere diameters vary from one opal to another and range from 1500 Å to 3000 Å. When white light passes through the essentially colorless opal and strikes planes of voids between spheres, certain wavelengths are diffracted and flash out of the stone as nearly pure spectral colors. This phenome-

non has been described as analogous to the diffraction of X-rays by crystals. In X-ray diffraction the interplanar spacings (*d*) are of the same order of magnitude as the wavelengths of X-rays. In precious opal the spacings, determined by sphere diameters, are far greater but so are the wavelengths of visible light (4000–7000 Å). The different wavelengths that satisfy the Bragg equation are diffracted with change in the angle of incident light (θ). Because light is refracted when it enters precious opal, the equation must include the refractive index, μ (= 1.45), and the equation is written $n\lambda = \mu 2d \sin \theta$ (see Fig. 6.22).

Diagnostic Features. Distinguished from microcrystalline varieties of quartz by lesser hardness and specific gravity and by the presence of water.

Varieties. *Precious opal* is characterized by a brilliant internal play of colors that may be red, orange, green, or blue (see Plate II, no. 7, Chapter 15). The body color is white, milky-blue, yellow, or black (*black opal*). *Fire opal* is a variety with intense orange to red reflections.

Common Opal. Milk-white, yellow, green, red, etc., without internal reflections.

Hyalite. Clear and colorless opal with a globular or botryoidal surface.

Geyserite or *Siliceous Sinter.* Opal deposited by hot springs and geysers. Found about the geysers in Yellowstone National Park.

Wood Opal. Fossil wood with opal as the petrifying material.

Diatomite. Fine-grained deposits, resembling chalk in appearance. Formed by sinking from near the surface and the accumulation on the sea floor of the siliceous tests of diatoms. Also known as *diatomaceous earth* or *infusorial earth*.

Occurrence. Opal may be deposited by hot springs at shallow depths, by meteoric waters, or by low-temperature hypogene solutions. It is found lining and filling cavities in rocks and may replace wood buried in volcanic tuff. The largest accumulations of opal are as siliceous tests of silica-secreting organisms.

Precious opals are found at Caernowitza, Hungary; in Querétaro, Mexico; Queensland and New South Wales, Australia; and Brazil. Black opal has been found in the United States in Virgin Valley, Nevada. Diatomite is mined in several western states, principally at Lompoc, California.

Artificial. Pierre Gilson in Switzerland has synthesized precious opal that is identical to natural material in chemical and physical properties, including a beautiful play of colors.

Use. As a gem. Opal is usually cut in round

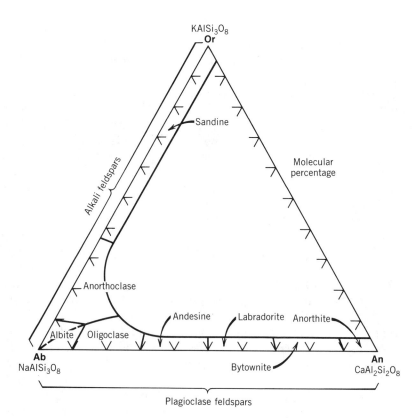

FIG. 13.117. Nomenclature for the plagio-
clase feldspar series and the high-tempera-
ture alkali feldspars. (After Deer, W. A.,
Howie, R. A., and Zussman, J. 1963, *Rock-
Forming Minerals*, v. 4. John Wiley and Sons,
New York, p. 2.)

shapes, *en cabochon*. Stones of large size and exceptional quality are very highly prized. Diatomite is used
extensively as an abrasive, filler, filtration powder,
and in insulation products.

   Name.   The name *opal* originated in the San-
skrit, *upala*, meaning stone or precious stone.

Feldspar Group

The compositions of the majority of common feldspars can be expressed in terms of the system
$KAlSi_3O_8$ (orthoclase; Or)–$NaAlSi_3O_8$ (albite; Ab)–
$CaAl_2Si_2O_8$ (anorthite; An). The members of the series
between $KAlSi_3O_8$ and $NaAlSi_3O_8$ are known as the
alkali feldspars and the members in the series between
$NaAlSi_3O_8$ and $CaAl_2Si_2O_8$ as the *plagioclase feldspars*. Members of both of these feldspar groups are
given specific names as shown in Fig. 13.117. The
chemical compositions of feldspars in this ternary
system are generally expressed in terms of molecular
percentages or Or, Ab, and An; for example,
$Or_{20}Ab_{75}An_5$. Barium feldspars such as *celsian*,
$BaAl_2Si_2O_8$, and *hyalophane*, $(K,Ba)(Al,Si)_2Si_2O_8$, are
relatively rare. All feldspars show good cleavages in
two directions which make an angle of 90°, or close
to 90°, with each other. Their hardness is about 6 and
specific gravity ranges from 2.55 to 2.76 (excluding
the Ba feldspars).

The unambiguous characterization of a feldspar
requires a knowledge not only of the chemical composition but also of the structural state of the species.
The structural state, which refers to the Al and Si distribution in tetrahedral sites of the framework structure, is a function of the crystallization temperature
and subsequent thermal history of a feldspar. In general, feldspars that cooled rapidly after crystallization
at high temperature show a disordered Al-Si distribution (high structural state). Those that cooled very
slowly from high temperatures or those that crystallized at low temperatures generally show an ordered
Al-Si distribution (low structural state).

Structure

The feldspar structure, similar to the structures of the
various polymorphs of SiO_2, consists of an infinite
network of SiO_4 as well as AlO_4 tetrahedra. The feldspar structure can be considered a "stuffed" derivative of the SiO_2 structures, by incorporation of Al into
the tetrahedral network, and concomitant housing of
Na^+ (or K^+ or Ca^{2+}) in available voids. When only
one Si^{4+} (per feldspar formula unit) is substituted by
Al^{3+}, the structure can be neutralized by incorporation of one K^+ or one Na^+. Similarly when two Si^{4+}
(per feldspar formula unit) are substituted for by Al^{3+},
the electrostatic charge of the network can be bal-

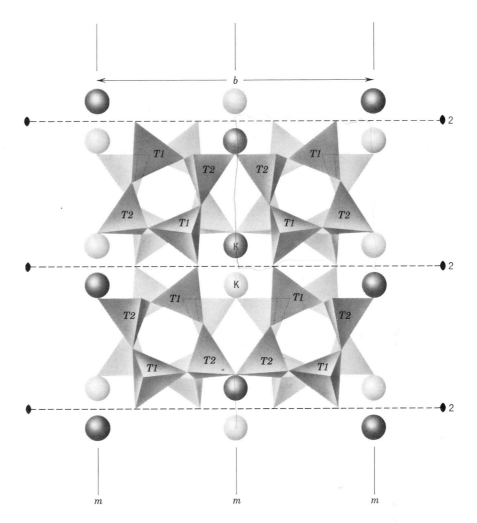

FIG. 13.118. The structure of high sanidine, KAlSi$_3$O$_8$, projected on ($\bar{2}$01). Mirror planes (*m*) and 2-fold rotation axes (2) are shown. Other symmetry elements such as glide planes and 2-fold screw axes are also present but their location is not shown here. (After Papike, J. J. and Cameron, M. 1976, Crystal chemistry of silicate minerals of geophysical interest. *Reviews of Geophysics and Space Physics*, v. 14, p. 66.)

anced by a divalent cation such as Ca^{2+}. We may state this as follows:

$$Si_4O_8 \text{ framework} \rightarrow Na(AlSi_3O_8) \rightarrow Ca(Al_2Si_2O_8)$$

In the plagioclase structures the amount of tetrahedral Al varies in proportion to the relative amounts of Ca^{2+} and Na$^+$ so as to maintain electrical neutrality; the more Ca^{2+}, the greater the amount of Al^{3+}.

The general architecture of the feldspar structure can be illustrated with the aid of a drawing of the high-temperature polymorph of KAlSi$_3$O$_8$, *sanidine*, with space group *C2/m* (see Fig. 13.118). In this structure the Al-Si distribution is completely disordered, meaning that the Al and Si ions are randomly distributed among the two crystallographically distinct tetrahedral sites, *T1* and *T2*. The K$^+$ ions, bonded to nine nearest oxygens in large interstices, occupy special positions on mirror planes perpendicular to the *b* axis. The Si-Al tetrahedral framework consists of four-membered rings of tetrahedra that are linked into chains (of a double crankshaft type) parallel to the *a*

axis (see Fig. 13.119). The square, blocky outline of these chains, imparted by the four-membered rings, finds outward expression in the right-angled cleavage and pseudotetragonal habit characteristic of feldspar. The structure of *microcline*, a low-temperature polymorph of KAlSi$_3$O$_8$, has triclinic symmetry (space group *C$\bar{1}$*) and lacks the mirror planes and rotation axes of sanidine as shown in Fig. 13.118. In other words, its structure is less symmetric and the K$^+$ ions are no longer located in special positions. The Al-Si distribution is completely ordered in what is known as low-temperature or maximum microcline (maximum refers to maximum "triclinicity" which results from the complete order). The tetrahedra that contain Al in this structure can be unambiguously located, whereas in sanidine the Al-Si distribution is completely random. *Orthoclase* represents a polymorph of KAlSi$_3$O$_8$ in which the Al-Si distribution is between the total randomness found in sanidine, and the total order of microcline. Orthoclase, with space group *C2/m*, crystallizes at intermediate temperatures (see Fig. 13.121). Although unambiguous distinction be-

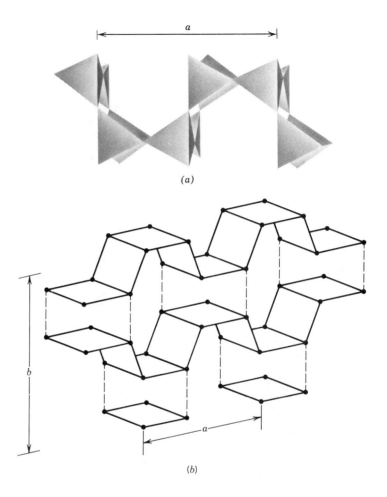

FIG. 13.119. (a) The four-membered rings in Fig. 13.118 are linked to form crankshaft-like chains that run parallel to the a axis. (After Papike, J. J. and Cameron, M. 1976, Crystal chemistry of silicate minerals of geophysical interest. *Reviews of Geophysics and Space Physics*, v. 14, p. 67.) (b) Schematic representation of the location and orientation of the four-membered crankshaft-like chains in the feldspar structure. Black dots are locations of Si. Dashed lines locate bonds between adjoining (Si, Al)O$_4$ tetrahedra. (Modified with permission after Ribbe, P., 1987, Feldspar. *McGraw-Hill Encyclopedia of Science and Technology*, 6th ed., v. 7, pp. 38–47, McGraw-Hill Book Co., New York; reprinted with permission of McGraw-Hill.)

tween the three structurally different K-feldspars, sanidine, orthoclase, and microcline, is based on careful measurements of unit cell dimensions and/or optical parameters, such as 2V and/or extinction angle $b \wedge Z$, we will use the terms orthoclase and microcline more broadly. In the following descriptions the terms will be based on generally recognizable characteristics in hand specimens. It must be remembered, however, that the definitions of the three types of K-feldspar as used in the recent literature are based on parameters that can be obtained only by X-ray and optical techniques (see Smith and Brown, 1988, *Feldspar Minerals*, or Smith, 1974, *Feldspar Minerals*, Springer-Verlag, New York; complete references are given at the end of this chapter). Such measurements allow for the definition of maximum microcline, and high, intermediate, and low orthoclase, and provide the investigator with information about the state of order or disorder of the Al-Si distribution in the feldspar structure.

The question of whether a specific, originally high-temperature feldspar retains its high-temperature

FIG. 13.120. The various possible temperature-cooling rate paths that an originally high-temperature K-feldspar (sanidine) can follow. The path marked "quenched" depicts a path in which the original high-temperature sanidine is cooled so rapidly that all characteristics of the high-temperature state are preserved in the final cooled product. (Modified with permission from Putnis, A. and McConnell, J. D. C., 1980, *Principles of Mineral Behavior*, Fig. 7.8, Blackwell Scientific Publications, Oxford, England, 257 pp; reprinted with permission.)

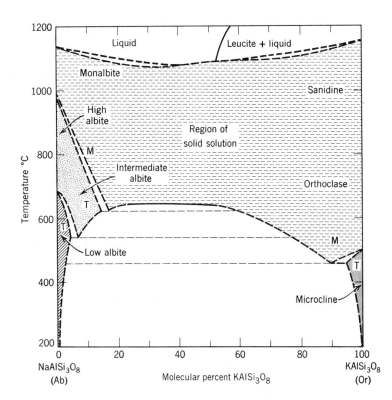

FIG. 13.121. Schematic phase diagram for the system $NaAlSi_3O_8$ (Ab)–$KAlSi_3O_8$ (K-spar) showing a large miscibility gap at temperatures below approximately 650°C. M and T denote mean monoclinic and triclinic, respectively. Compare with Fig. 9.7*b*. (Modified with permission after Smith, J. V. and Brown, W. L., 1988, *Feldspar Minerals*, v. 1, Fig. 1.2, Springer Verlag, New York, 828 pp.)

(disordered) structure type or whether it will transform (upon cooling) to a lower temperature (more ordered) structural state is much influenced by the cooling rate of the process. Figure 13.120 is a schematic illustration of the various cooling paths for a potassium feldspar, as a function of temperature and cooling rate.

Microcline is particularly characteristic of deep-seated rocks and pegmatites, orthoclase of intrusive rocks formed at intermediate temperatures and sanidine of extrusive, high-temperature lavas.

The general structure of members of the plagioclase series is very similar to that of microcline. The Na end member, *albite*, is generally triclinic (space group $C\bar{1}$) with a low albite form that shows a highly ordered Al-Si distribution and a high albite form with a highly disordered Al-Si distribution. A monoclinic variety of albite occurs at very high temperature and is known as *monalbite*. The Ca end member, *anorthite*, is also triclinic with space group $P\bar{1}$ at room temperature, and perfect Al-Si ordering in the structure. At elevated temperatures the structure of anorthite becomes body-centered with space group $I\bar{1}$. The general stability fields of the various forms of feldspar are shown in Figs. 13.121, 13.124, and 13.126.

Composition

The alkali feldspar series ($NaAlSi_3O_8$ to $KAlSi_3O_8$) shows complete solid solution only at high tempera-

tures (see Fig. 13.121). For example, members of the sanidine–high albite series are stable at elevated temperatures but at lower temperature two separate phases, low albite and microcline, become stable. As can be seen from Fig. 13.121, the compositional ranges of low albite and microcline are very small. If a homogeneous feldspar, of composition $Or_{50}Ab_{50}$, in which the Na^+ and K^+ ions are randomly distributed, is allowed to cool slowly, segregation of Na^+ and K^+ ions will result because the size requirements of the surrounding structure become more stringent. The Na^+ will diffuse to form Na-rich regions and the K^+ will segregate into K-rich regions in the structure, causing the originally homogeneous feldspar to become a heterogeneous intergrowth. The separation most commonly results in thin layers of albite in a host crystal of K feldspar. Such intergrowths are known as *perthites*, and are the result of *exsolution* (see Fig. 5.21 and discussion on page 236). In the alkali feldspar series the orientation of the exsolution lamellae is roughly parallel to {100}. When these intergrowths are visible to the naked eye they are known as *macroperthite*, when visible only by optical microscope they are referred to as *microperthite*, and when detectable only by X-ray or electron microscope techniques, they are called *cryptoperthite* (see Fig. 13.122). More rarely the host mineral is a plagioclase feldspar and the lamellae are of $KAlSi_3O_8$ composition; this is called *antiperthite*.

FIG. 13.122. Microstructure in an alkali feldspar of composition 57.3 weight percent Or. The Na-rich lamellae are twinned according to the albite law. This very high magnification photograph was taken with a transmission electron microscope. (From Champness, P. E. and Lorimer, G. W. Exsolution in Silicates, chapter 4.1 in *Electron Microscopy in Mineralogy*, H. R. Wenk, ed. Springer-Verlag, New York.)

Only very limited solid solution occurs between $KAlSi_3O_8$ and $CaAl_2Si_2O_8$ (see Fig. 13.123). Essentially complete solid solution, however, exists at elevated temperatures in the plagioclase series ($NaAl-Si_3O_8$ to $CaAl_2Si_2O_8$) (see Fig. 13.124). The general formula of a feldspar in this series may be written as: $Na_{1-x}Ca_x(Si_{3-x}Al_{1+x})O_8$, where x ranges from 0 to 1. The structural interpretation of the region of essen-

tially complete solid solution is complicated because of the varying ratio of Al/Si from albite, $NaAlSi_3O_8$, to anorthite, $CaAl_2Si_2O_8$. Three types of exsolution textures found in the plagioclase series are not visible to the naked eye, but may be detected because of iridescence. *Peristerite* intergrowths occur in the range An_2 to An_{15} (see Fig. 13.125). *Bøggild intergrowths* occur in some plagioclase with composition between An_{47} and An_{58}; their presence is indicated by the play of colors in labradorite. A third intergrowth occurs in the An_{60} to An_{85} region and is known as *Huttenlocher intergrowths*. These discontinuities in the solid solution series between Ab and An are on a very fine scale, however, and most properties, such as specific gravity or refractive index, show a generally linear change with chemical composition. Thus determination of a suitable property with sufficient precision permits a close approximation of the chemical composition in the plagioclase series (see Fig. 13.136). Figure 13.126 is a simplified representation of the three-component feldspar system ($KAlSi_3O_8$—K-spar; $NaAlSi_3O_8$—albite; $CaAl_2Si_2O_8$—anorthite) showing the major miscibility gaps, and melting temperatures at water pressure of about 5 kilobars (compare with Fig. 9.7).

FIG. 13.123. Experimentally determined extent of solid solution in the system Or-Ab-An at $P_{H_2O} = 1$ kilobar. (After Ribbe, P. H., 1975, *Feldspar Mineralogy*, *Reviews in Mineralogy*, v. 2, Fig. R-1, Mineralogical Soc. of America, Washington, D.C.)

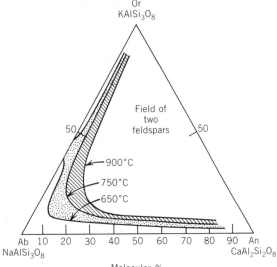

K-Feldspars

MICROCLINE—$KAlSi_3O_8$

Crystallography. Triclinic; $\overline{1}$. The habit and crystal forms are similar to those of orthoclase, and microcline may be twinned according to the same

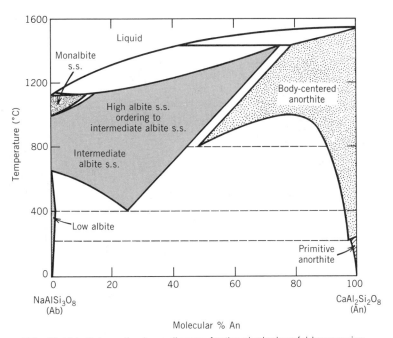

FIG. 13.124. Schematic phase diagram for the plagioclase feldspar series showing a range of almost complete solid solution at high temperatures, and three miscibility regions at lower temperatures. (Simplified with permission after Smith, J. V. and Brown, W. L., 1988, *Feldspar Minerals*, v. 1, Fig. 1.4, Springer Verlag, New York, 828 pp.) See Fig. 9.6*a* for the approximate location of the three miscibility gaps.

FIG. 13.125. Microstructure in the peristerite region of the plagioclase series (between An_2 and An_{15}), showing sharply defined, very fine lamellae approximately parallel to $(0\bar{4}1)$. These lamellae act as a diffraction grating for white light, producing the delicate bluish sheen of moonstone. This photograph was taken with a transmission electron microscope. (From Champness, P. E. and Lorimer, G. W., 1976, Exsolution in silicates, chapter 4.1, in *Electron Microscopy in Mineralogy*, H. R. Wenk, ed., Springer-Verlag, New York.)

0·5 μm.

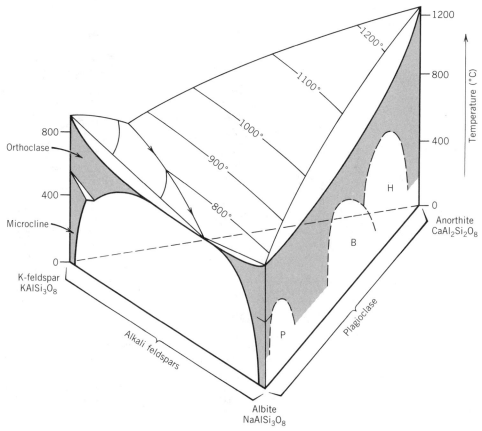

FIG. 13.126. Highly schematic temperature-composition diagram for the three-component system KAlSi$_3$O$_8$ (K-feldspar)–NaAlSi$_3$O$_8$ (albite)–CaAl$_2$Si$_2$O$_8$ (anorthite), at water pressure of about 5 kilobars. Details of the interior are complex and have been omitted. The upper, contoured surface of the diagram is the liquidus surface. The three intergrowth regions, as a result of miscibility gaps, in the lower temperature range of the plagioclase series are: P = peristerite, B = Bøggild intergrowth, H = Huttenlocher intergrowth. Solid solution is shown by shading. Compare the two vertical sides of this diagram with more detailed information given in Figs. 13.121 and 13.124. (Adapted with permission from Ribbe, P. H., 1987, Feldspars, in *McGraw-Hill Encyclopedia of Science and Technology*, 6th ed., v. 7, p. 45. McGraw-Hill Book Co., New York; reprinted with permission of McGraw-Hill.)

laws as orthoclase; *Carlsbad* twins are common, but *Baveno* and *Manebach* twins are rare (see Fig. 13.127). It is also twinned according to the *albite law*, with {010} the twin plane, and the *pericline law*, with {010} the twin axis. These two types of twinning are usually present in microcline and on {001} the lamellae cross

FIG. 13.127. Twinning in feldspar. *(a)* Carlsbad twin. *(b)* Manebach twin. *(c)* Baveno twin.

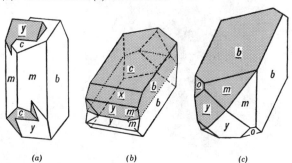

(a) *(b)* *(c)*

at nearly 90° giving a characteristic "tartan" structure (see Fig. 2.114). Microcline is found in cleavable masses, in crystals, and as a rock constituent in irregular grains. Microcline probably forms the largest known crystals. In a pegmatite in Karelia, CIS, a mass weighing over 2000 tons showed the continuity of a single crystal. Also in pegmatites microcline may be intimately intergrown with quartz, forming *graphic granite* (Fig. 13.128).

Microcline frequently has irregular and discontinuous bands crossing {001} and {010} that result from the exsolution of albite. The intergrowth as a whole is called *perthite*, or, if fine, *microperthite* (see page 535 and Fig. 13.129).

C$\bar{1}$; for low microcline $a = 8.59$, $b = 12.97$, $c = 7.22$ Å; $\alpha = 90°41'$, $\beta = 115°59'$, $\gamma = 87°39'$; $Z = 4$. d's: 4.21(5), 3.83(3), 3.48(3), 3.37(5), 3.29(5), 3.25(10).

Physical Properties. *Cleavage* {001} perfect, {010} good at an angle of 89°30'. **H** 6. **G** 2.54–2.57.

FIG. 13.128. Graphic granite, Hybla, Ontario. Quartz dark, microcline light.

Luster vitreous. *Color* white to pale yellow, more rarely red or green. Green microcline is known as *amazonite*. Translucent to transparent. *Optics*: (−); $\alpha = 1.522$, $\beta = 1.526$, $\gamma = 1.530$; $2V = 83°$; $r > v$.

Composition and Structure. For $KAlSi_3O_8$, K_2O 16.9, Al_2O_3 18.4, SiO_2 64.7%; there is some substitution of Na for K (see Fig. 13.121). The structure of microcline is triclinic and therefore less symmetrical than that of sanidine, which is shown in Fig. 13.118. Maximum low microcline shows perfect Al-Si order in the tetrahedral framework. With increasing disorder (increasing T) the terms low, intermediate, and high microcline are applied. At considerably higher T the microcline structure transforms to orthoclase or sanidine (see Fig. 13.121 and also Fig. 3.45 and related text).

Diagnostic Features. Distinguished from orthoclase only by determining the presence of microcline ("tartan") twinning, which can rarely be determined without the aid of the microscope. With only minor exceptions, deep green feldspar is microcline.

Occurrence. Microcline is a prominent constituent of igneous rocks such as granites and syenites that cooled slowly and at considerable depth. In sedimentary rocks it is present in arkose and conglomerate. In metamorphic rocks in gneisses. Microcline is the common K-feldspar of pegmatites and was quarried extensively in North Carolina, North and South Dakota, Colorado, Virginia, Wyoming, Maine, and Connecticut. *Amazonite*, a green variety of microcline, is found in the Ural Mountains, CIS, and in various places in Norway and the Malagasy Republic. In the United States it is found at Pikes Peak, Colorado (see Fig. 13.130), and Amelia Court House, Virginia.

Use. Feldspar is used chiefly in the manufacture of porcelain. It is ground very fine and mixed with kaolin or clay, and quartz. When heated to high temperature the feldspar fuses and acts as a cement to bind the material together. Fused feldspar also furnishes the major part of the glaze on porcelain ware. A small amount of feldspar is used in the manufacture of glass to contribute alumina to the batch. Amazonite is polished and used as an ornamental material.

Name. *Microcline* is derived from two Greek words meaning *little* and *inclined*, referring to the

FIG. 13.129. Microperthite as seen in polarized light under the microscope.

0.1 mm

FIG. 13.130. Microcline and smoky quartz, Crystal Peak, Colorado. (Harvard Mineralogical Museum.)

slight variation of the cleavage angle from 90°. *Feldspar* is derived from the German word *feld*, field.

ORTHOCLASE—KAISi₃O₈

Crystallography. Monoclinic; 2/m. Crystals are usually short prismatic elongated parallel to *a*, or elongated parallel to *c* and flattened on {010}. The prominent forms are *b*{010}, *c*{001}, and *m*{110}, often with smaller {0kl} and {hkl} prisms (Fig. 13.131). Frequently twinned according to the following laws: *Carlsbad* penetration twin with *c* as twin axis; *Baveno* with {021} as twin and composition plane; *Manebach* with {001} as twin and composition plane (see Fig. 13.127). Commonly in crystals or in coarsely cleavable to granular masses; more rarely fine-grained, massive, and cryptocrystalline. Most abundantly in rocks as formless grains.

$C2/m$; $a = 8.56$, $b = 1.96$, $c = 7.21$ Å; $\beta = 115°50'$; $Z = 4$. d's 4.22(6), 3.77(7), 3.46(5), 3.31(10), 3.23(8).

Physical Properties. *Cleavage* {001} perfect, {010} good, {110} imperfect. **H** 6. **G** 2.57. *Luster* vitreous. *Color* colorless, white, gray, flesh-red, rarely yellow or green. *Streak* white. *Optics*: (−); $\alpha = 1.518$, $\beta = 1.524$, $\gamma = 1.526$; $2V = 10°–70°$; $Z = b$, $X \wedge a = 5°$; $r > v$.

Composition and Structure. Most analyses contain small amounts of Na, but a complete solid solution series is possible at high temperature (see Fig. 13.121). Orthoclase represents a K-feldspar structure that crystallizes at intermediate temperature and has

a partially ordered Al-Si distribution. At higher temperatures the Al-Si distribution in the structure becomes less ordered and is known as sanidine (see Fig. 3.45 and related text).

Diagnostic Features. Is usually recognized by its color, hardness, and cleavage. Distinguished from the other feldspars by its right-angle cleavage and the lack of twin striations on the best cleavage surface.

Occurrence. Orthoclase is a major constituent of granites, granodiorites, and syenites, which have cooled at moderate depth and at reasonably fast rates. In more slowly cooled granites and syenites microcline will be the characteristic K-feldspar.

Name. The name *orthoclase* refers to the right-angle cleavage possessed by the mineral.

Similar Species. *Adularia*, KAISi₃O₈, is a colorless, translucent to transparent variety of K-feldspar which is commonly found in pseudo-orthorhombic crystals (Fig. 13.132). It occurs mainly in low-temperature veins in gneisses and schists. Some adularia shows an opalescent play of colors and is called *moonstone*.

FIG. 13.131. Orthoclase and sanidine.

(a)   (b)   (c)

FIG. 13.132. Adularia.

SANIDINE—(K,Na)AlSi$_3$O$_8$

Crystallography. Monoclinic; $2/m$. Crystals are often tabular parallel to {010}; also elongated on a with square cross section as in Fig. 13.131a. Carlsbad twins common.

$C2/m$; for high sanidine, $a = 8.56$, $b = 13.03$, $c = 7.17$ Å; $\beta = 115°59'$; $Z = 4$. d's: 4.22(6), 3.78(8), 3.31(10), 3.278(6), 3.225(8).

Physical Properties. Cleavage {001} perfect, {010} good. **H** = 6. **G** = 2.56–2.62. *Luster* vitreous. *Color* colorless and commonly transparent. *Streak* white. *Optics*: (−); α = 1.518–1.525, β = 1.523–1.530, γ = 1.525–1.531. Occurs in two orientations with optic plane parallel with {010} and 2V = 0°–60°, and with optic plane normal to {010} and 2V = 0°–25°.

Composition and Structure. A complete solid solution exists at high temperature between sanidine and high albite; part of the intermediate region is known as *anorthoclase* (see Fig. 13.117). The structure of sanidine shows a disordered (random) distribution of Al and Si in the tetrahedral framework (see Fig. 13.118). The Al and Si distribution in orthoclase is more ordered (see Fig. 3.45 and related text).

Diagnostic Features. Can be characterized with confidence only by optical or X-ray techniques. Optic orientation and 2V differ from orthoclase. Microcline and plagioclase show different types of twinning.

Occurrence. As phenocrysts (see Fig. 13.133) in extrusive igneous rocks such as rhyolites and trachytes. Sanidine is characteristic of rocks that cooled quickly from an initial high temperature of eruption. Most sanidines are cryptoperthitic.

Name. Sanidine is derived from the Greek *sanis*, tablet, and *idos*, appearance, in allusion to its typical tabular habit.

Plagioclase Feldspar Series

The plagioclase feldspars form, at elevated temperatures, an essentially complete solid solution series

FIG. 13.133. Polished slab of basalt with sanidine phenocrysts.

from pure *albite* (Ab), NaAlSi$_3$O$_8$, to pure *anorthite* (An), CaAl$_2$Si$_2$O$_8$ (see Fig. 13.124). Discontinuities in this series, as shown by the existence of peristerite and other intergrowths (see page 536), can be detected only by careful electron optical and X-ray studies. The nomenclature for the series, as based on six arbitrary divisions, is given in Fig. 13.117. If species names are not given, the composition is expressed by Ab$_{20}$An$_{80}$, for example, which may be shortened to An$_{80}$. Most of the properties of the various species in this series vary in a uniform manner with the change in chemical composition. For this reason the series can be more easily understood if one comprehensive description is given, rather than six individual descriptions, and the dissimilarities between members indicated. The distinction between high- and low-temperature forms of the series can be made only by X-ray and optical means.

ALBITE—NaAlSi$_3$O$_8$
ANORTHITE—CaAl$_2$Si$_2$O$_8$

Crystallography. Triclinic; $\bar{1}$. Crystals commonly tabular parallel to {010}; occasionally elon-

FIG. 13.134. Albite crystals.

gated on *b* (Fig. 13.134). In anorthite crystals may be prismatic elongated on *c*.

Crystals frequently twinned according to the various laws governing the twins of orthoclase, that is *Carlsbad, Baveno,* and *Manebach.* In addition, they are nearly always twinned according to the *pericline law.* Albite twinning with twin plane {010} is commonly polysynthetic and because (010) ∧ (001) ≈ 86°, {001} either as a crystal face or cleavage is crossed by parallel groovings or striations (Figs. 13.135 and 2.113*a*). Often these striations are so fine as to be invisible to the unaided eye, but on some specimens they are coarse and easily seen. The presence of the striations

FIG. 13.135. *(a)* Albite twinning (see also Fig. 2.113). *(b)* Labradorite feldspar showing albite twinning striations on the basal pinacoid {001}. (Harvard Mineralogical Museum.)

(a)

(b)

on the basal cleavage is one of the best proofs that a feldspar belongs to the plagioclase series. Pericline twinning with *b* as the twin axis is also polysynthetic. Striations resulting from it can be seen on {010}. Their direction on {010} is not constant but varies with the composition.

Distinct crystals are rare. Usually in twinned, cleavable masses; as irregular grains in igneous rocks.

For albite, $C\overline{1}$: $a = 8.14$, $b = 12.8$, $c = 7.16$ Å; $\alpha = 94°20'$ $\beta, = 116°34'$, $\gamma = 87°39'$; $Z = 4$. For anorthite, $I\overline{1}$ or $P\overline{1}$: $a = 8.18$, $b = 12.88$, $c = 14.17$ Å; $\alpha = 93°10'$, $\beta = 115°51'$, $\gamma = 91°13'$; $Z = 8$. *d*'s (Ab-An): 4.02–4.03(7), 3.77–3.74(5), 3.66–3.61(6), 3.21(7)–3.20(10), 3.18(10)–3.16(7).

Physical Properties. Cleavage {001} perfect, {010} good. **H** 6. **G** 2.62 in albite to 2.76 in anorthite (Fig. 13.136). *Color* colorless, white, gray; less frequently greenish, yellowish, flesh-red. *Luster* vitreous to pearly. Transparent to translucent. A beautiful play of colors is frequently seen, especially in labradorite and andesine. *Optics*: Albite: (+); $\alpha = 1.527$, $\beta = 1.532$, $\gamma = 1.538$; 2V = 74°. Anorthite: (−); $\alpha = 1.577$, $\beta = 1.585$, $\gamma = 1.590$, 2V = 77° (see Fig. 13.136).

Composition and Structure. An essentially complete solid solution series extends from $NaAlSi_3O_8$ to $CaAl_2Si_2O_8$ at elevated temperatures (see Fig. 13.124). Considerable K may be present, especially toward the albite end of the series (see Figs. 13.123 and 13.124). Very fine scale discontinuities in the series, which can be detected only by electron optical and X-ray means, are reflected in the occurrence of very fine lamellar intergrowths: peristerite, Bøggild and Huttenlocher intergrowths (see page 536). The structure of albite is triclinic ($C\overline{1}$) at low to moderate temperatures, and shows a highly ordered Al-Si dis-

FIG. 13.136. Variation of refractive indices and specific gravity with composition in the plagioclase feldspar series.

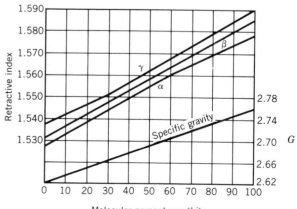

tribution; low albite is structurally analogous to low microcline. High albite is also triclinic and has a highly disordered Al-Si distribution. At temperatures above 980°C a monoclinic variety of albite, known as *monalbite*, with a totally disordered Al-Si distribution exists. Anorthite has space group $P\overline{1}$ at low temperatures. At elevated temperatures the structure of anorthite inverts to $I\overline{1}$ (see Fig. 13.124).

Diagnostic Features. The plagioclase feldspars can be distinguished from other feldspars by the presence of albite twin striations {001}. They can be placed accurately in their proper places in the series only by quantitative chemical analysis or optical tests, but they can be roughly distinguished from one another by specific gravity.

Occurrence. The plagioclase feldspars, as rock-forming minerals, are even more widely distributed and more abundant than the potash feldspars. They are found in igneous, metamorphic, and, more rarely, sedimentary rocks.

The classification of igneous rocks is based largely on the kind and amount of feldspar present (see Table 14.2). As a rule, the greater the percentage of SiO_2 in a rock the fewer the dark minerals, the greater amount of potash feldspar, and the more sodic the plagioclase; and conversely, the lower the percentage of SiO_2 the greater the percentage of dark minerals and the more calcic the plagioclase.

In an igneous crystallization sequence the more refractory minerals crystallize before the less refractory ones. The An end member of the plagioclase series has a much higher melting point than the Ab end member (see Fig. 13.124). For this reason plagioclase feldspars that crystallize early from a magma are generally more Ca-rich than those that crystallize later. Continuous chemical zoning occurs commonly in phenocrysts with Ca-rich centers and more Na-rich rims.

Albite is included with orthoclase and microcline in what are known as the *alkali* feldspars, all of which have a similar occurrence. They are commonly found in granites, syenites, rhyolites, and trachytes. Albite is common in pegmatites where it may replace earlier microcline; as the platy variety, *cleavelandite*. Some albite and oligoclase shows an opalescent play of colors and is known as *moonstone*. The name *albite* is derived from the Latin *albus*, meaning *white*, in allusion to the color.

Oligoclase is characteristic of granodiorites and monzonites. In some localities, notably at Tvedestrand, Norway, it contains inclusions of hematite, which give the mineral a golden shimmer and sparkle Such feldspar is called *aventurine* oligoclase, or *sunstone*. The name is derived from two Greek words

meaning *little* and *fracture*, because it was believed to have less perfect cleavage than albite.

Andesine is rarely found except as grains in andesites and diorites. Named from the Andes Mountains where it is the chief feldspar in the andesite lavas.

Labradorite is the common feldspar in gabbros and basalts and in anorthosite it is the only important constituent. Found on the coast of Labrador in large cleavable masses that show a fine iridescent play of colors. The name is derived from this locality.

Bytownite is rarely found except as grains in gabbros. Named from Bytown, Canada (now the city of Ottawa).

Anorthite is rarer than the more sodic plagioclases. Found in rocks rich in dark minerals and in druses of ejected volcanic blocks and in granular limestones of contact metamorphic deposits. Name derived from the Greek word meaning *oblique*, because its crystals are triclinic.

Use. Plagioclase feldspars are less widely used than potash feldspars. Albite, or *soda spar*, as it is called commercially, is used in ceramics in a manner similar to microcline. Labradorite that shows a play of colors is polished and used as an ornamental stone. Those varieties that show opalescence are cut and sold under the name of *moonstone*.

Name. The name *plagioclase* is derived from the Greek word *plagios* meaning *oblique*, an allusion to the oblique angle between the cleavages. (See under "Occurrence" for names of specific species.)

Feldspathoid Group

The feldspathoids are anhydrous framework silicates that are chemically similar to the feldspars. The chief chemical difference between feldspathoids and feldspars lies in the SiO_2 content. The feldspathoids contain only about two-thirds as much silica as alkali feldspar and hence tend to form from melts rich in alkalis (Na and K) and poor in SiO_2 (see Fig. 13.137). The structures of the feldspathoids are closely related to those of the feldspars and silica minerals. However, several of the feldspathoids show the development of somewhat larger structural cavities (than feldspar or silica minerals) as a result of four- and six-membered tetrahedral linkages. This general increase in openness of the feldspathoid structures, as compared with those of the feldspars, is expressed in their ranges of specific gravity:

G range for feldspars: 2.54–2.75

G range for feldspathoids: 2.15–2.5

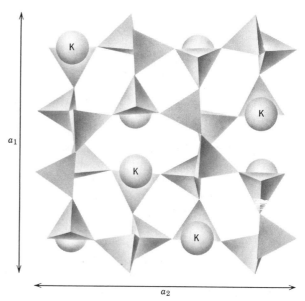

FIG. 13.137. Feldspathoid compositions, as compared to those of the feldspars, in the system SiO_2 (quartz)–$NaAlSiO_4$ (nepheline)–$KAlSiO_4$ (kalsilite)–$CaAl_2O_4$. Shaded areas represent average extents of solid solution (compare with Figs. 9.10, 9.11, and 9.14).

FIG. 13.138. Portion of the leucite structure projected down the *c* axis. (After Papike, J. J. and Cameron, M., 1976, Crystal chemistry of silicate minerals of geophysical interest. *Reviews of Geophysics and Space Physics*, v. 14, p. 74.)

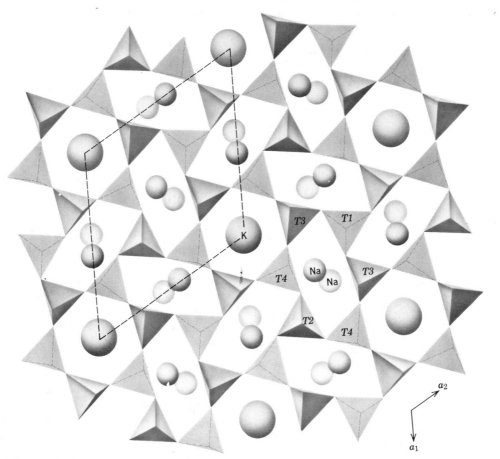

FIG. 13.139. Projection of the nepheline structure onto (0001). Dashed lines outline unit cell. *T1, T2, T3* and *T4* refer to crystallographically distinct tetrahedral sites. (After Papike, J. J. and Cameron, M., 1976, Crystal chemistry of silicate minerals of geophysical interest. *Reviews of Geophysics and Space Physics*, v. 14, p. 56.)

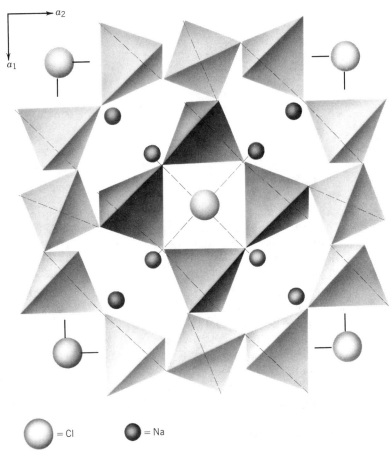

FIG. 13.140. The structure of sodalite, $Na_8(AlSiO_4)_6Cl_2$. This can be visualized as a cubic unit cell with a four-membered tetrahedral ring on each cube face, and a six-membered tetrahedral ring around each corner of the cube. (From Papike, J. J., 1988, Chemistry of the rock-forming silicates: Multiple chain, sheet, and framework structures. *Reviews of Geophysics*, v. 26, pp. 407–444.)

Examples of the aluminosilicate frameworks in the feldspathoids are given in Figs. 13.138 to 13.140. The structure of *leucite*, $KAlSi_2O_6$, has tetragonal symmetry (space group $I4_1/a$) at low to intermediate temperatures; at approximately 605°C it inverts to a cubic structure with space group $Ia3d$. In the low-temperature form the Al and Si tetrahedra share corners to form four- and six-membered rings (see Fig. 13.138). The K^+ ions are in 12-coordination with oxygen in large cavities in the structure. The structure of *nepheline*, $(Na,K)AlSiO_4$ (space group $P6_3$), can be considered as a derivative of the high tridymite structure with Al replacing Si in half the tetrahedra and Na and K in interstitial voids. The Si and Al are ordered in specific tetrahedral sites of the structure; the *T1* and *T4* sites are Al-rich, the *T2* and *T3* sites Si-rich. One-fourth of the interstitial sites, which are filled by K, are nearly hexagonal in geometry (see Fig. 13.139). The other three-fourths of the sites are irregular in configuration and are occupied by Na. The structure of *sodalite*, $Na_8(AlSiO_4)_6$ (see Fig. 13.140), has large cavities occupied by Na^+ and Cl^-. The structural framework consists of corner-sharing, alternating SiO_4 and AlO_4 tetrahedra. The cagelike cavities result from the linking of six four-membered and eight six-membered tetrahedral rings. The six-membered rings form channelways parallel to the cube diagonals of the structure. The large central cavities are occupied by Cl^-, which are tetrahedrally coordinated by Na^+.

Some of the members of the feldspathoid group contain unusual anions. Sodalite contains Cl and cancrinite incorporates CO_3. Noselite houses SO_4 and lazurite SO_4, S, and Cl ions. These large anionic groups and anions are located in large interstices of the structure.

LEUCITE—$KAlSi_2O_6$

Crystallography. Tetragonal; $4/m$ below 605°C. Isometric, $4/m\bar{3}2/m$ above 605°C. Usually in trapezohedral crystals which at low temperature are pseudomorphs of the high-temperature form (Fig. 13.141).

Below 605°C: $I4_1/a$; $a = 13.04$, $c = 13.85$ Å; $Z = 16$. Above 605°C: $Ia3d$; $a = 13.43$ Å; $Z = 16$. d's (low T form): 5.39(8), 3.44(9), 2.27(10), 2.92(7), 2.84(7).

Physical Properties. **H** $5\frac{1}{2}$–6. **G** 2.47. *Luster* vit-

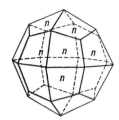

FIG. 13.141. Leucite.

reous to dull. *Color* white to gray. Translucent. *Optics*: (+); $\omega = 1.508$, $\epsilon = 1.509$.

Composition and Structure. Most leucites are close to $KAlSi_2O_6$ with K_2O 21.5, Al_2O_3 23.5, SiO_2 55.0%. Small amounts of Na may replace K. The structure of the low-temperature polymorph is shown in Fig. 13.138.

Diagnostic Features. Characterized by its trapezohedral form. Leucite is usually embedded in a fine-grained matrix (see Fig. 13.142), whereas analcime is usually in cavities in free-growing crystals.

Occurrence. Although leucite is a rather rare mineral it is abundant in certain recent lavas; rarely observed in deep-seated rocks. It is found only in silica-deficient rocks and thus never in rocks containing quartz. Its most notable occurrence is as phenocrysts in the lavas of Mount Vesuvius. In the United States it is found in rocks of the Leucite Hills, Wyoming, and in certain of the rocks in the Highwood and Bear Paw Mountains, Montana. Pseudomorphs of a mixture of nepheline, orthoclase, and analcime after leucite, *pseudoleucite*, are found in syenites of Arkansas, Montana, and Brazil.

Name. From the Greek word *leukos* meaning *white*.

Similar Species. *Pollucite*, $CsAlSi_2O_6 \cdot H_2O$, isostructural with leucite, is a rare isometric mineral usually occurring in pegmatites, notably at Bernie Lake, Manitoba.

NEPHELINE—(Na,K)AlSiO₄

Crystallography. Hexagonal; 6. Rarely in small prismatic crystals with base. Almost invariably massive, compact, and in embedded grains.

$P6_3$; $a = 10.01$, $c = 8.41$ Å; $Z = 8$. *d*'s: 4.18(7), 3.27(7), 3.00(10), 2.88(7), 2.34(6).

Physical Properties. Cleavage $\{10\bar{1}0\}$ distinct. **H** $5\frac{1}{2}$–6. **G** 2.60–2.65. *Luster* vitreous in clear crystals; greasy in the massive variety. *Color* colorless, white, or yellowish. In the massive variety gray, greenish, and reddish. Transparent to translucent. *Optics*: (−); $\omega = 1.529$–1.546, $\epsilon = 1.526$–1.542.

Composition and Structure. The end member composition $NaAlSiO_4$ contains Na_2O 21.8, Al_2O_3

FIG. 13.142. Polished slab of basalt showing large leucite phenocrysts.

35.9, SiO_2 42.3%. The amount of K in natural nephelines may range from 3 to 12 weight percent K_2O. A large miscibility gap exists at low to moderate temperatures between $NaAlSiO_4$ and *kalsilite*, $KAlSiO_4$. However, at temperatures above about 1000°C a complete solid solution series exists between Na and K end members. Figure 13.143 illustrates the large miscibility gap in this chemical system. As a function of decreasing temperature, the Na^+ and K^+ ions in an originally homogeneous solid solution become segregated, forming finely intergrown exsolution lamellae of differing compositions. The process of exsolution in this chemical system is analogous to that discussed for the alkali feldspars (see Fig. 13.121 and page 535). The structure of nepheline is illustrated in Fig. 13.139 and discussed on page 545.

Diagnostic Features. Characterized in massive varieties by its greasy luster. Distinguished from quartz by inferior hardness.

Occurrence. Nepheline is a rock-forming mineral found in silica-deficient intrusive and extrusive rocks. Crystals are present in the lavas of Mount Vesuvius. The largest known mass of intrusive nepheline rocks is on the Kola Peninsula, CIS, where, locally, nepheline is associated with apatite. Extensive masses of nepheline rocks are found in Norway and South Africa. In the United States nepheline, both massive and in crystals, is found at Litchfield, Maine, associated with cancrinite. Found near Magnet Cove, Arkansas, and Beemerville, New Jersey. Common in the syenites of the Bancroft region of Ontario, Canada, where the associated pegmatites contain large masses of nearly pure nepheline.

Use. Iron-free nepheline, because of its high Al_2O_3 content, has been used in place of feldspar in the glass industry. Most nepheline for this purpose comes from Ontario. Nepheline produced as a by-product of apatite mining on the Kola Peninsula, CIS, is used by the Russians in several industries including ceramics, leather, textile, wood, rubber, and oil.

Name. *Nepheline* is derived from the Greek word *nephele* meaning a *cloud*, because when immersed in acid the mineral becomes cloudy.

Similar Species. *Cancrinite*, $Na_6Ca(CO_3)(AlSiO_4)_6 \cdot 2H_2O$, is a rare mineral similar to nepheline in occurrence and associations.

SODALITE—$Na_8(AlSiO_4)_6Cl_2$

Crystallography. Isometric; $\bar{4}3m$. Crystals rare, usually dodecahedrons. Commonly massive, in embedded grains.

$P\bar{4}3n$; $a = 8.83$–8.91 Å; $Z = 2$. d's: 6.33(8), 3.64(10), 2.58(5), 2.10(2), 1.75(3).

Physical Properties. *Cleavage* {011} poor. **H** $5\frac{1}{2}$–6. **G** 2.15–2.3. *Luster* vitreous. *Color* usually blue, also white, gray, green. Transparent to translucent. *Optics*: refractive index 1.483.

Composition and Structure. Na_2O 25.6, Al_2O_3 31.6, SiO_2 37.2, Cl 7.3%. Only small substitution of K for Na. In the structure of sodalite the $AlSiO_4$ framework is linked such as to produce cagelike, cubooctahedral cavities in which the Cl is housed (see Fig. 13.140).

Diagnostic Features. Usually distinguished by its blue color, and told from lazurite by the different occurrence and absence of associated pyrite. If color is not blue, a test for chlorine is necessary to distinguish it from analcime, leucite, and haüynite.

Occurrence. Sodalite is a comparatively rare rock-forming mineral associated with nepheline, cancrinite, and other feldspathoids in nepheline syenites, trachytes, phonolites, and so forth. Found in transparent crystals in the lavas of Vesuvius. The massive blue variety is found at Bancroft, Ontario; Ice River, British Columbia; and Namibia.

Use. May be carved as an ornamental material.

Name. Named in allusion to its sodium content.

Similar Species. Other rare feldspathoids are *haüynite*, $(Na,Ca)_{4-8}(AlSiO_4)_6(SO_4)_{1-2}$ and *noselite*, $Na_8(AlSiO_4)_6SO_4$.

Lazurite—$(Na,Ca)_8(AlSiO_4)_6(\underline{SO_4},S,Cl)_2$

Crystallography. Isometric; $\bar{4}3m$. Crystals rare, usually dodecahedral. Commonly massive, compact.

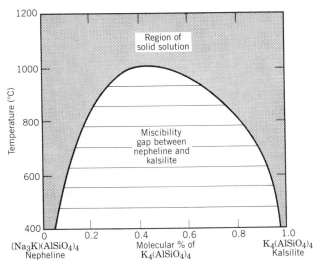

FIG. 13.143. Temperature-composition section for the join nepheline, $Na_3K(AlSiO_4)_4$–kalsilite, $K_4(AlSiO_4)_4$. Compare with Fig. 13.121, which is a similar diagram for the alkali feldspar system. (From Ferry, J. and Blencoe, J. G., 1978, Subsolidus phase relations in the nepheline-kalsilite system at 0.5, 2.0, and 5.0 Kbar, *American Mineralogist*, v. 63, pp. 1225–1240.)

$P\bar{4}3m; a = 9.08$ Å; $Z = 1$. d's: 3.74(10), 2.99(10), 2.53(9), 1.545(9), 1.422(7).

Physical Properties. Cleavage {011} imperfect. **H** 5–5½. **G** 2.4–2.45. *Luster* vitreous. *Color* deep azure-blue, greenish-blue. Translucent. *Optics*: refractive index 1.50±.

Composition and Structure. Small amounts of Rb, Cs, Sr, and Ba may substitute for Na. There is considerable variation in the amounts of SO_4, S, and Cl. Lazurite appears isostructural with sodalite (see Fig. 13.140 and discussion on page 545). The large cage-like cavities in the structure house the Cl and S anions and the (SO_4) anionic groups.

Diagnostic Features. Characterized by its blue color and the presence of associated pyrite.

Occurrence. Lazurite is a rare mineral, occurring usually in crystalline limestones as a product of contact metamorphism. *Lapis lazuli* is a mixture of lazurite with small amounts of calcite, pyroxene, and other silicates, and commonly contains small disseminated particles of pyrite. The best quality of lapis lazuli comes from northeastern Afghanistan. Also found at Lake Baikal, Siberia, CIS, and in Chile.

Use. Lapis lazuli is highly prized as an ornamental stone, for carvings, and so on (see Plate III, no. 5, Chapter 15). As a powder it was formerly used as the paint pigment ultramarine. Now ultramarine is produced synthetically.

Name. Lazurite is an obsolete synonym for azurite, and hence the mineral is named because of its color resemblance to azurite.

Petalite—Li(AlSi₄O₁₀)

Crystallography. Monoclinic; $2/m$. Crystals rare, flattened on {010} or elongated on [100]. Usually massive or in foliated cleavable masses.

$P2/a$; $a = 11.76$, $b = 5.14$, $c = 7.62$ Å; $\beta = 112°24'$; $Z = 2$. d's: 3.73(10), 3.67(9), 3.52(3), 2.99(1), 2.57(2).

Physical Properties. *Cleavage* {001} perfect, {201} good. Brittle. **H** 6–6½. **G** 2.4. *Luster* vitreous, pearly on {001}. *Color* colorless, white, gray. Transparent to translucent. *Optics*: (+); $\alpha = 1.505$, $\beta = 1.511$, $\gamma = 1.518$; 2V = 83°; $Z = b$. $X \wedge a = 2°$–8°. $r > v$.

Composition and Structure. LiO 4.9, Al_2O_3 16.7, SiO_2 78.4%. The structure of petalite consists of a framework of AlO_4 and SiO_4 tetrahedra that contains Si_4O_{10} layers linked by AlO_4 tetrahedra. The Li is in tetrahedral coordination with oxygen.

Diagnostic Features. Characterized by its platy habit. Distinguished from spodumene by its cleavage and lesser specific gravity.

Occurrence. Petalite is found in pegmatites where it is associated with other lithium-bearing minerals such as spodumene, tourmaline, and lepidolite. Petalite was considered a rare mineral until it was found to be abundant at the Varutrask pegmatite, Sweden. More recently it has been found abundantly in Bikita, Zimbabwe, and in Namibia. At these localities, associated with lepidolite and eucryptite, it is mined extensively.

Use. An important ore of lithium. (See spodumene, page 483.) Colorless petalite may be faceted as a gem.

Name. From the Greek word *petalos* meaning *leaf*, alluding to the cleavage.

Scapolite Series

The *scapolites* are metamorphic minerals with compositions suggestive of the feldspars. There is a complete solid solution series between *marialite*, $3NaAlSi_3O_8 \cdot NaCl$, and *meionite*, $3CaAl_2Si_2O_8 \cdot CaSO_4$ (or $CaCO_3$). When the formulas are written in this way, it is clear that the composition of marialite is the equivalent of three formula weights of albite, $NaAlSi_3O_8$, plus one formula weight NaCl, and that the meionite composition is equivalent to three formula weights of anorthite, $CaAl_2Si_2O_8$, plus one formula weight of $CaCO_3$ or $CaSO_4$. In this series there is complete substitution of Ca for Na with charge compensation effected as in the feldspars by concomitant substitution of Al for Si. There is also complete substitution of CO_3, SO_4, and Cl_2 for each other. The name *wernerite* has been used for members intermediate in composition between marialite and meionite.

The structure of scapolite consists of a framework of SiO_4 and AlO_4 tetrahedra with large cavities containing the (Ca,Na) ions and (CO_3,Cl_2,SO_4) anionic groups (see Fig. 13.144).

Marialite—Meionite

Crystallography. Tetragonal; $4/m$. Crystals usually prismatic. Prominent forms are prisms {010} and {110} and dipyramid {011}; rarely shows the faces of the tetragonal dipyramid z (Fig. 13.145). Crystals are usually coarse, or with faint fibrous appearance.

$I4/m$; $a = 12.2$, $c = 7.6$ Å; $Z = 2$. a varies slightly with composition. d's: (marialite), 4.24(7), 3.78(9), 3.44(10), 3.03(10), 2.68(9).

Physical Properties. *Cleavage* {100} and {110} imperfect but distinct. **H** 5–6. **G** 2.55–2.74. The specific gravity and refractive indices increase with increasing Ca content. *Luster* vitreous when fresh. *Color* white, gray, pale green; more rarely yellow, bluish, or

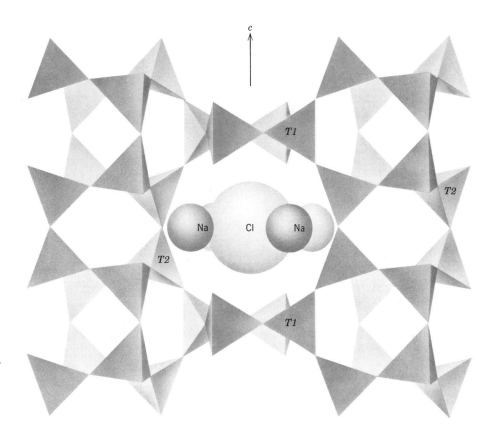

FIG. 13.144. The structure of scapolite projected on (100). (After Papike, J. J. and Cameron, M., 1976, Crystal chemistry of silicate minerals of geophysical interest. *Reviews of Geophysics and Space Physics*, v. 14, p. 75.)

reddish. Transparent to translucent. *Optics*: $(-)$; $\omega = 1.55–1.60$, $\epsilon = 1.54–1.56$.

Composition and Structure. The extent of solid solution between *marialite*, $Na_4(AlSi_3O_8)_3(Cl_2,CO_3,SO_4)$, and *meionite*, $Ca_4(Al_2Si_2O_8)_3(Cl_2,CO_3,SO_4)$, is discussed on page 548. The structure, illustrated in Fig. 13.144, contains two crystallographically distinct tetrahedra, $T1$ and $T2$. In marialite compositions Al is confined to the $T2$ tetrahedra (46% Al, 54% Si) and only Si occupies the $T1$ sites. In meionite compositions Al is housed in both tetrahedral sites.

Diagnostic Features. Characterized by its crystals with square cross section and four cleavage directions at 45°. When massive, resembles feldspar but has a characteristic fibrous appearance on the cleavage surfaces.

Occurrence. Scapolite occurs in crystalline

FIG. 13.145. Scapolite.

schists, gneisses, amphibolites, and granulite facies rocks. In many cases it is derived by alteration from plagioclase feldspars. It is also characteristic in crystalline limestones as a contact metamorphic mineral. Associated with diopside, amphibole, garnet, apatite, titanite, and zircon.

Crystals of gem quality with a yellow color occur in the Malagasy Republic. In the United States found in various places in Massachusetts, notably at Bolton; and in Orange, Lewis, and St. Lawrence counties, New York. Also found at various points in Ontario, Canada.

Use. Colorless crystals may be cut as faceted gemstones.

Name. From the Greek word *skapos* meaning a *shaft*, in allusion to the prismatic habit of the crystals.

ANALCIME—NaAlSi$_2$O$_6$·H$_2$O[1]

Crystallography. Isometric; $4/m\bar{3}2/m$. Usually in trapezohedrons (Figs. 13.146a and 13.147). Cubes with trapezohedral truncations also known (Fig. 13.146b). Usually in crystals, also massive granular.

[1]Analcime may be classified as a member of the zeolite group; however, its structure, chemistry, and occurrence are very similar to those of the feldspathoids.

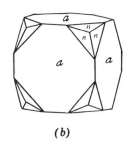

FIG. 13.146. Analcime crystals.

Ia3d; a = 13.71 Å; *Z* = 16. *d*'s: 5.61(8), 4.85(4), 3.43(10), 2.93(7), 2.51(5).

Physical Properties. **H** $5-5\frac{1}{2}$. **G** 2.27. *Luster* vitreous. *Color* colorless or white. Transparent to translucent. *Optics*: refractive index 1.48–1.49.

Composition and Structure. The chemical composition of most analcimes is fairly constant with minor amounts of K or Ca substituting for Na; also some Al substitution for Si. $NaAlSi_2O_6 \cdot H_2O$ contains Na_2O 14.1, Al_2O_3 23.2, SiO_2 54.5, H_2O 8.2%. The structure is made of a framework of AlO_4 and SiO_4 tetrahedra with four- and six-membered tetrahedral rings. This framework contains continuous channels along the 3-fold axes of the structure which are occupied by H_2O molecules. The octahedrally coordinated Na is housed in somewhat smaller voids in the structure.

Diagnostic Features. Usually recognized by its free-growing crystals and its vitreous luster. Crystals resemble leucite, but leucite is always embedded in rock matrix.

Occurrence. Analcime occurs as a primary mineral in some igneous rocks and is also the product

FIG. 13.147. Analcime from Ischia, Italy. Scanning electron microscope (SEM) photograph (From Gottardi, G. and Galli, E., 1985, *Natural Zeolites*. Springer-Verlag, New York; with permission.)

of hydrothermal action in the filling of basaltic cavities. It is an original constituent of analcime basalts and may occur in alkaline rocks such as aegirine-analcime-nepheline syenites. It is also found in vesicles of igneous rocks in association with prehnite, calcite, and zeolites. Fine crystals are found in the Cyclopean Islands near Sicily; in the Val di Fassa and on the Seiser Alpe, Trentino, Italy; in Victoria, Australia; and Kerguelen Island in the Indian Ocean. In the United States found at Bergen Hill, New Jersey; in the Lake Superior copper district; and at Table Mountain, near Golden, Colorado. Also found at Cape Blomidon, Nova Scotia.

Name. Derived from the Greek word *analkidos* meaning *weak*, in allusion to its weak electrical property when heated or rubbed.

Zeolite Group

The zeolites form a large group of hydrous silicates that show close similarities in composition, association, and mode of occurrence. They are framework aluminosilicates with Na, Ca, and K, and highly variable amounts of H_2O in the generally large voids of the framework. Many of them fuse readily with marked intumescence (a swelling up), hence the

FIG. 13.148. *(a)* Chains of AlO_4 and SiO_4 tetrahedra as found in natrolite and other fibrous zeolites.

(a)

name *zeolite*, from two Greek words meaning *to boil* and *stone*. Traditionally, the zeolites have been thought of as well-crystallized minerals found in cavities and veins in basic igneous rocks. More recently, large deposits of zeolites have been found in the western United States and in Tanzania as alterations of volcanic tuff and volcanic glass. Such zeolites are the diagenetic alteration products of silicic tuffs in Cenozoic lacustrine deposits, especially those that were deposited in highly saline and alkaline waters. The occurrence of zeolites in various rock types is used to define the low-grade regional metamorphic zone known as the "zeolite facies."

Physical properties, such as hardness (**H**) and specific gravity (**G**), of members of the zeolite group show a pronounced decrease, when compared to those of the silica minerals, feldspars, and feldspathoids:

	H	**G**
Quartz	7	2.65
Feldspars	6	2.54–2.75
Feldspathoids	5–6	2.15–2.5
Zeolites	$3\frac{1}{2}–5\frac{1}{2}$	2.0–2.4

The zeolites, like the feldspars and feldspathoids, are built of frameworks of AlO_4 and SiO_4 tetrahedra; the zeolite frameworks, however, are very open with large interconnecting spaces or channels. The zeolites can be divided into those with a fibrous habit and an underlying chain structure (Fig. 13.148), those with a platy habit and an underlying sheet structure, and

FIG. 13.148. *(continued) (b)* The structure of natrolite projected on (001).

(b)

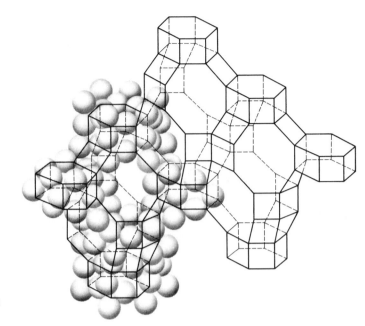

FIG. 13.149. Schematic representation of the structure of chabazite. Si and Al (not shown) occupy the corners of the framework outlined by the lines. The balls represent oxygen in close packing. Each framework unit contains a cavity which is connected to adjacent cavities by channels. The channel-diameter in chabazite is 3.9 Å. (From Breck, D. W. and Smith, J. V., 1959, Molecular sieves. *Scientific American*, v. 200, no. 1, p. 88. All rights reserved.)

those with an equant habit and an underlying framework structure (Fig. 13.149).

Much of the interest in zeolites derives from the presence of the spacious channels and the water molecules and variable amounts of Na, Ca, and K that are housed in the channels and intraframe cavities. The water molecules are weakly tied by hydrogen bonding to anionic framework atoms. When a zeolite is heated, the water in the channelways is given off easily and continuously as the temperature rises, leaving the structure intact. This phenomenon of continuous water loss as a function of increasing temperature is illustrated in Fig. 13.150 for several common, natural zeolites. The formulas in Fig. 13.150 show the large and variable numbers of water molecules per formula unit. The thermal gravimetric analysis (TGA) curves show a continuous loss of water with increasing temperature and also illustrate the fact that 80 to 90% of all the water is lost from the structure below about 350°C. This relatively low-temperature water loss occurs via the channels of the structures without collapse of the framework. Collapse and overall destruction of the framework (known as decrepitation) commonly occurs at temperatures above 600°C. Such water loss behavior is in sharp contrast to that observed in silicates with (OH) groups in which (OH) is an integral part of their structures (e.g., amphiboles and micas). Such (OH) groups are generally lost over a very small temperature range at high temperatures. As the result of such loss, new, commonly anhydrous minerals form (see the schematic curve at the bottom of Fig. 13.150). Dehydration behavior similar to that

of zeolites, but at much lower temperature than for (OH)-containing silicates, is shown by gypsum, $CaSO_4 \cdot 2H_2O$ (see Fig. 12.40), which converts initially (at about 65°C) to $CaSO_4 \cdot \frac{1}{2}H_2O$ and at about 100°C to the γ polymorph of $CaSO_4$.

The dehydrated zeolite structure can be completely rehydrated when it is immersed in water. In industrial applications, the property of rehydration allows zeolites to be used as desiccants, such as in the removal of water from gaseous hydrocarbons and petroleum. Zeolites in their dehydrated state can absorb other molecules as well, as long as the overall size of the zeolite channels is large enough to accommodate such molecules. Molecules too large to pass through the channels are excluded, giving rise to the "molecular sieving" property of most zeolites. For example, a specific zeolite with channel diameters of about 4.5 Å is able to absorb normal hydrocarbons, such as octane and pentane (with effective cross-sectional diameters of about 4.3 Å), but will exclude branch chain hydrocarbons, such as isooctane and isopentane (with effective cross-sectional diameters of 5.0 Å or larger). This sieving property is illustrated in Fig. 13.151.

Zeolites, in addition to their commercial use as "molecular sieves," are also much exploited on account of their cation exchange properties. The cations (such as Na^+, K^+, and Ca^{2+}) are only loosely bonded to the tetrahedral framework and can be removed or exchanged easily by washing with a strong solution of another ion. This ability for cation exchange is the basis for water softeners, in which "hard" water (with high Ca^{2+} concentrations) is made "soft" by ex-

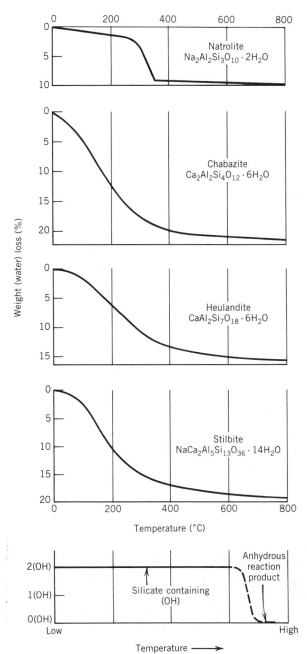

FIG. 13.150. Thermal gravimetric analysis (TGA) curves for four selected zeolites. In each of these four curves there is a continuous water loss with increasing temperature. In every case about 80 to 90% of the water in these structures is lost by about 350°C. (From Gottardi, G. and Galli, E., 1985, *Natural Zeolites*. Springer-Verlag, New York, with permission.) The schematic curve at the bottom illustrates the behavior of silicates with (OH) groups (e.g., amphiboles, micas) and the abrupt loss of this (OH) at some high temperature, at which the original (OH)-containing structure is destroyed and converts to an anhydrous reaction product + H_2O.

changing the Ca^{2+} (in the water) with Na^{2+} (supplied by a natural or synthetic zeolite, e.g. $Na_2Al_2SiO_{10} \cdot 2H_2O$). "Hard" water, that is, water containing many calcium ions in solution, is passed through a tank filled with zeolite grains. The Ca^{2+} ions replace the Na^+ ions in the zeolite, forming $CaAl_2Si_3O_{10} \cdot 2H_2O$, contributing Na^+ ions to the solution. Water containing sodium does not form scum and is said to be "soft." When the zeolite in the tank has become saturated with calcium, a strong NaCl brine is passed through the tank. The high concentration of sodium ions forces the reaction to go in the reverse direction, and the $Na_2Al_2Si_3O_{10} \cdot 2H_2O$ is reconstituted, calcium going into solution. By base exchange many ions, including silver, may be substituted for the alkali-metal cations in the zeolite structure. These cation exchange properties are used in removing harmful ions from radioactive waste, or ammonia from sewage and agricultural waste.

Because of the great demand for various zeolites used in commercial applications such as in catalytic cracking (cracking is the molecular weight reduction process by which the heavier components of crude oil are converted to lighter, more volatile materials such as those used in gasoline), there is a large industrial production (through laboratory synthesis) of zeolites. The first synthetic zeolites were produced in the 1950s by the Linde Division of the Union Carbide Corporation. Such syntheses have produced many structures and compositions that have no natural counterpart. The number of naturally occurring zeolites is about 41, and only a few of these are found in sufficient quantity and purity to be used as a commercial product. Synthetic zeolite production now occurs on a large scale with a production of over 12,000 tons per year. Very specific zeolite structures (with unique channel diameters and/or cation exchange properties) are being produced for unique industrial applications.

NATROLITE—$Na_2Al_2Si_3O_{10} \cdot 2H_2O$

Crystallography. Orthorhombic; *mm*2. Prismatic, often acicular with prism zone vertically striated. Usually in radiating crystal groups (Fig. 13.152); also fibrous, massive, granular, or compact.

*Fdd*2; $a = 18.29$, $b = 18.64$, $c = 6.59$ Å; $Z = 8$. *d*'s: 6.6(9), 5.89(8), 3.19(9), 2.86(10), 2.42(6).

Physical Properties. *Cleavage* {110} perfect. **H** $5-5\frac{1}{2}$. **G** 2.25. *Luster* vitreous. *Color* colorless or white, rarely tinted yellow to red. Transparent to translucent. *Optics*: (+); $\alpha = 1.480$, $\beta = 1.482$, $\gamma = 1.493$; $2V = 63°$; $X = a$; $Y = b$; $r < v$.

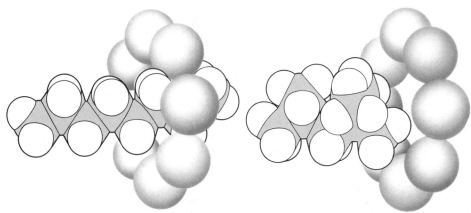

FIG. 13.151. Schematic illustration of how a zeolite sieve separates a straight chain hydrocarbon (e.g., octane) from a branching chain hydrocarbon (e.g., isooctane). These two organic compounds have almost identical properties and thus are very hard to separate by other methods. Isooctane has the higher "antiknock" rating. (From Flanigen, E. M. and Mumpton, F. A., 1977, Commercial properties of natural zeolites, in *Mineralogy and Geology of Natural Zeolites, Reviews in Mineralogy*, v. 4, Mineralogical Soc. America, pp. 165–175.)

Composition and Structure. Na_2O 16.3, Al_2O_3 26.8, SiO_2 47.4, H_2O 9.5%. Some K and Ca may replace Na. Natrolite is one of the group of fibrous zeolites. Its structure consists of an Si-Al-O framework in which chains parallel to the *c* axis are prominent (see Fig. 13.148). The chains are linked laterally by the sharing of some oxygens of opposing tetrahedra. Na is coordinated between the chains to six oxygens, four of which are tetrahedral oxygens and two of which are from water molecules. The perfect {110} cleavage in natrolite is due to the relatively small number of bonds between chains as compared to the large number of bonds within the chainlike structure.

Diagnostic Features. Light color and commonly fibrous habit.

Occurrence. Natrolite is characteristically found lining cavities in basalt associated with other zeolites and calcite. Notable localities are Aussig and Salesel, Bohemia, Czechoslovakia; Puy-de-Dôme, France; and Val di Fassa, Trentino, Italy. In the United States found at Bergen Hill, New Jersey. Also found in various places in Nova Scotia.

FIG. 13.152. Natrolite from Altavilla, Vicenza, Italy. (From Gottardi, G. and Galli, E., 1985, *Natural Zeolites*. Springer-Verlag, New York; with permission).

FIG. 13.153. Chabazite.

Name. From the Latin *natrium*, meaning *sodium*, in allusion to its composition.

Similar Species. *Scolecite*, $CaAl_2Si_3O_{10} \cdot 3H_2O$, is another fibrous zeolite, similar in structure to natrolite but monoclinic in symmetry.

CHABAZITE—$Ca_2Al_2Si_4O_{12} \cdot 6H_2O$

Crystallography. Hexagonal; $\bar{3}2/m$. Usually in rhombohedral crystals, $\{10\bar{1}1\}$, with nearly cubic angles. May show several different rhombohedrons (Fig. 13.153). Often in penetration twins.

$P\bar{1}$; $a = 9.41, b = 9.42, c = 9.42$ Å; $\alpha = 94°11'$, $\beta = 94°16'$, $\gamma = 94°21'$; $Z = 6$. *d*'s: 9.35(5), 5.03(3), 4.33(7), 3.87(3), 2.925(10).

Physical Properties. *Cleavage* $\{10\bar{1}1\}$ poor. **H** 4–5. **G** 2.05–2.15. *Luster* vitreous. *Color* white, yellow, pink, red. Transparent to translucent. *Optics*: (+) or (−); $\alpha = 1.4848$, $\beta = 1.4852$, $\gamma = 1.4858$, 2V ~ 70°.

Composition and Structure. The ideal composition is $Ca_2Al_2Si_4O_{12} \cdot 6H_2O$, but there is considerable replacement of Ca by Na and K as well as (Na,K)Si for CaAl. The structure of chabazite consists of an Al-Si-O framework with large cagelike openings bounded by rings of tetrahedra (see Fig. 13.149). The cages are connected to each other by channels which allow for the diffusion of molecules through the structure of a size comparable to the diameter (3.9 Å) of the channels. Argon (3.84 Å in diameter) is rapidly absorbed by the chabazite structure, but *iso*-butane (5.6 Å in diameter) cannot enter the structure. In this way chabazite can act as a molecular sieve.

Diagnostic Features. Recognized usually by its rhombohedral-appearing crystals (Fig. 13.154), and distinguished from calcite by its poorer cleavage and lack of effervescence in HCl.

Occurrence. Chabazite is found, usually with other zeolites, lining cavities in basalt. Notable localities are the Faeroe Islands; the Giant's Causeway, Ireland; Aussig, Bohemia, Czechoslovakia; Seiser Alpe, Trentino, Italy; Oberstein, Germany; and India. In the United States found at West Paterson, New Jersey, and

FIG. 13.154. Chabazite from Heinekeim, Germany. (From Gottardi, G. and Galli, E., 1985, *Natural Zeolites*. Springer-Verlag, New York; with permission.)

Goble Station, Oregon. Also found in Nova Scotia and there known as *acadialite*.

Name. Chabazite is derived from a Greek word which was an ancient name for *a stone*.

HEULANDITE—$CaAl_2Si_7O_{18} \cdot 6H_2O$

Crystallography. Monoclinic; $2/m$ but crystals often simulate orthorhombic symmetry (Figs. 13.155 and 13.156); often diamond-shaped with $\{010\}$ prominent.

$C2/m$; $a = 17.71, b = 17.94, c = 7.46$ Å; $\beta = 116°20'$; $Z = 4$. *d*'s: 8.9(7), 5.10(4), 3.97(10), 3.42(5), 2.97(7).

Physical Properties. *Cleavage* $\{010\}$ perfect. **H** $3\frac{1}{2}$–4. **G** 2.18–2.2. *Luster* vitreous, pearly on $\{010\}$. *Color* colorless, white, yellow, red. Transparent to translucent. *Optics*: (+); $\alpha = 1.482$, $\beta = 1.485$, $\gamma = 1.489$; 2V = 50°; $X = b$. $Y \wedge c = 35°$. $r > v$.

Composition and Structure. There is considerable variation in the Si/Al ratio with concomitant variation in the proportions of Ca and Na. The structure

FIG. 13.155. Heulandite.

FIG. 13.156. Heulandite from Val di Fassa, Italy. (From Gottardi, G. and Galli, E., 1985, *Natural Zeolites.* Springer-Verlag, New York; with permission.)

of heulandite consists of a very open Si-Al-O framework in which two-thirds of the $(Si,Al)O_4$ tetrahedra are linked to form networks of six-membered rings parallel to {010}; this is responsible for the perfect {010} cleavage. The framework contains several sets of channels which house the water molecules as well as Ca.

Diagnostic Features. Characterized by its crystal form and one direction of perfect cleavage with pearly luster.

Occurrence. Heulandite is usually found in cavities of basic igneous rocks associated with other zeolites and calcite. Found in notable quality in Iceland; the Faeroe Islands; Andreasberg, Harz Mountains, Germany; Tyrol, Austria; and India, near Bombay. In the United States found at West Paterson, New Jersey. Also found in Nova Scotia.

Name. In honor of the English mineral collector, H. Heuland.

STILBITE—$NaCa_2Al_5Si_{13}O_{36}\cdot14H_2O$

Crystallography. Monoclinic; $2/m$. Crystals usually tabular on {010} or in sheaflike aggregates (Figs. 13.157 and 13.158). They may also form cruciform penetration twins.

$C2/m$; $a = 13.61$, $b = 18.24$, $c = 11.27$ Å; $\beta = 127°51'$; $Z = 4$. d's: 9.1(9), 4.68(7), 4.08(10), 3.41(5), 3.03(7).

Physical Properties. *Cleavage* {010} perfect.

H $3\frac{1}{2}$–4. G 2.1–2.2. *Luster* vitreous; pearly on {010}. *Color* white, more rarely yellow, brown, red. Translucent. *Optics*: $(-)$; $\alpha = 1.494$, $\beta = 1.498$, $\gamma = 1.500$; $2V = 33°$; $Y = b$, $X \wedge a = 5°$.

Composition and Structure. Na and K are usually present, substituting for Ca by the coupled substitution mechanisms $Na^+Si^{4+} \leftrightarrows Ca^{2+}Al^{3+}$ and $(Na,K)^+Al^{3+} \leftrightarrows Si^{4+}$. The structure of stilbite is similar to that of heulandite with sheets of six-membered $(Si,Al)O_4$ tetrahedra parallel to {010}. This accounts for the tabular habit and excellent {010} cleavage.

Diagnostic Features. Characterized chiefly by its cleavage, pearly luster on the cleavage face, and common sheaflike groups of crystals.

Occurrence. Stilbite is found in cavities in basalts and related rocks associated with other zeolites

FIG. 13.157. Stilbite.

FIG. 13.158. Stilbite, Jewel Tunnel, Bombay, India.

and calcite. Notable localities are Poona, India; Isle of Skye; Faeroe Islands; Kilpatrick, Scotland; and Iceland. In the United States it is found in northeastern New Jersey. Also found in Nova Scotia.

Name. Derived from the Greek word *stilbo* meaning *luster*, in allusion to the pearly luster.

Similar Species. Other zeolites of lesser importance than those described are:

Phillipsite, $KCa(Al_3Si_5O_{16}) \cdot 6H_2O$, monoclinic.

Harmotome, $Ba(Al_2Si_6O_{16}) \cdot 6H_2O$, monoclinic.

Gmelinite, $(Na_2,Ca)(Al_2Si_4O_{12}) \cdot 6H_2O$, hexagonal.

Laumontite, $Ca(Al_2Si_4O_{12}) \cdot 4H_2O$, monoclinic.

Scolecite, $Ca(Al_2Si_3O_{10}) \cdot 3H_2O$, monoclinic.

Thomsonite, $NaCa_2(Al_5Si_5O_{20}) \cdot 6H_2O$, orthorhombic. A compact, massive type is used as a gem. When polished, red, yellow, or green "eyes" are seen on a rounded surface.

REFERENCES AND SUGGESTED READING

Abelson, P. H., 1990, The asbestos removal fiasco (editorial). *Science*, v. 247, p. 1017.

The Al₂SiO₅ Polymorphs. Reviews in Mineralogy, 1990, v. 22, 406 pp.

Amphiboles. Reviews in Mineralogy, 1981, v. 9A, 372 pp.; 1982, v. 9B, 390 pp.

Breck, D. W., 1974, *Zeolite Molecular Sieves*. John Wiley & Sons, New York, 771 pp.

Buseck, P. R., 1983, Electron microscopy of minerals. *American Scientist*, v. 71, pp. 175–185.

Buseck, P. R. and Iijima, S., 1974, High resolution electron microscopy of silicates. *American Mineralogist*, v. 59, pp. 1–22.

Darragh, P. J., Gaskin, A. J. and Sanders, J. V., 1976, Opals. *Scientific American*, v. 234, no. 4, pp. 84–95.

Deer, W. A., Howie, R. A. and Zussman, J., 1962 and 1963, *Rock-Forming Minerals*, v. 1–4. John Wiley & Sons, New York.

——, 1978, *Single-Chain Silicates*, v. 2A; 1982, *Orthosilicates*, v. 1A; 1986, *Disilicates and Ring Silicates*, v. 1B. John Wiley & Sons, New York, 668 pp., 919 pp., and 629 pp., respectively.

——, 1992, *An Introduction to the Rock-Forming Minerals*, 2nd ed. John Wiley & Sons, New York, 691 pp.

Feldspar Mineralogy. Reviews in Mineralogy, 1983, 2nd ed., v. 2, 362 pp.

Frondel, C., 1962, *The System of Mineralogy*, v. 3, *Silica Minerals*. John Wiley & Sons, New York, 334 pp.

Gottardi, G. and Galli, E., 1985, *Natural Zeolites*. Springer-Verlag, New York, 409 pp.

Grim, R. E., 1968, *Clay Mineralogy*, 2nd ed. McGraw-Hill Book Co., New York, 596 pp.

Hawthorne, F. C., 1983, The crystal chemistry of amphiboles. *Canadian Mineralogist*, v. 21, pp. 173–480.

Hydrous Phyllosilicates. Reviews in Mineralogy, 1988, v. 19, 725 pp.

Liebau, F., 1985, *Structural Chemistry of Silicates: Structure, Bonding, and Classification*. Springer-Verlag, New York, 347 pp.

Micas. Reviews in Mineralogy, 1984, v. 13, 584 pp.

Mineralogy and Geology of Natural Zeolites. Reviews in Mineralogy, 1977, v. 4, 233 pp.

Morimoto, N., Fabries, J., Ferguson, A. K., Ginzburg, I. V., Ross, M., Siefert, F. A., Zussman, J., Aoki, K. and Gottardi, G., 1988, Nomenclature of pyroxenes. *American Mineralogist*, v. 73, pp. 1123–1133.

Mossman, B. T., Bignon, J., Corn, M., Seaton, A. and Gee, J. B. L., 1990, Asbestos: Scientific developments and implications for public policy. *Science*, v. 247, pp. 294–301.

Mossman, B. T. and Gee, J. B. L., 1989, Asbestos-related diseases. *New England Journal of Medicine*, v. 320, pp. 1721–1729.

Nesse, W. D., 1991, *Introduction to Optical Mineralogy*, 2nd ed. Oxford University Press, New York, 335 pp.

Orthosilicates. Reviews in Mineralogy, 1980, v. 5, 381 pp.

Papike, J. J. and Cameron, M., 1976, Crystal chemistry of silicate minerals of geophysical interest. *Reviews of Geophysics and Space Physics*, v. 14, pp. 37–80.

Pyroxenes. Reviews in Mineralogy, 1980, v. 7, 525 pp.

Ross, M., 1981, The geologic occurrences and health hazards of amphiboles and serpentine asbestos, in *Amphiboles. Reviews in Mineralogy*, v. 9A, pp. 279–325.

——, 1984, A survey of asbestos related disease in trades and mining occupations and in factory and mining communities as a means of predicting health risks of nonoccupational exposure to fibrous minerals. American Society for Testing and Materials, Special Publication no. 834, pp. 51–104.

Smith, J. V., 1974, *Feldspar Minerals*, v. 1, 2. Springer-Verlag, New York, 627 pp. and 690 pp., respectively.

Smith, J. V. and Brown, W. L., 1988, *Feldspar Minerals. vol. 1, Crystal Structures, Physical, Chemical, and Microtextural Properties*. Springer-Verlag, New York, 828 pp.

Smyth, J. R. and Bish, D. L., 1988, *Crystal Structures and Cation Sites for the Rock-Forming Minerals*. Allen and Unwin, Boston, 332 pp.

Strunz, H., 1970, *Mineralogische Tabellen*, 5th ed. Akademische Verlaggeselschaft, Geest und Portig K.G., Leipzig, 621 pp.

Wenk, H. R., ed., 1976, *Electron Microscopy in Mineralogy*. Springer-Verlag, New York, 564 pp.

Zoltai, T., 1960, Classification of silicates and other minerals, with tetrahedral structures. *American Mineralogist*, v. 45, pp. 960–973.

CHAPTER 14

MINERAL ASSEMBLAGES: INTRODUCTION TO ROCK TYPES AND VEIN MINERALIZATION

In the previous four chapters we have discussed specific mineral species and mineral groups. Under the heading of "Occurrence" we have mentioned typical assemblages or rock types in which a specific mineral is found, and we have often illustrated the stability of a mineral or mineral group with pressure-temperature (P-T) or temperature-composition (T-X) diagrams (see also Chapter 9). In our previous treatment we have not, however, addressed ourselves to the systematics of mineral occurrences as reflected in naturally occurring rock types. Because rocks are generally made up of a variety of minerals, the study of rocks, known as *petrology*, concerns itself to a large degree with the identification of the individual minerals in a rock, their textures, abundances, and grain size; this type of information is fundamental to the understanding of the origin of a rock and to its classification. A petrologist must have a strong background in mineralogy and mineral identification, but in addition he or she must be conversant with the theories of origin of rocks and with experimental studies that elucidate their possible origins. Within the context of this book we cannot cover the many aspects of petrology that are treated extensively in books on the subject (see references at end of chapter); we will, however, attempt to provide a link between the study of minerals and the study of rocks.

All rocks can be classified in terms of three major groups: *igneous, sedimentary*, and *metamorphic*.

IGNEOUS ROCKS

Igneous rocks make up approximately 95 percent of the upper 10 miles (16 km) of the Earth's crust, but their great abundance is hidden on the Earth's surface by a relatively thin but widespread layer of sedimentary and metamorphic rocks. Igneous rocks have crystallized from a silicate melt, known as a *magma*. The high temperatures (about 900° to 1600°C) necessary to generate magmas, as concluded from experimental liquidus diagrams obtained in the laboratory and from the similarly high temperatures measured in the boiling magmas of active volcanoes, suggest that their source lies deep in the Earth's crust. A magma is principally made up of O, Si, Al, Fe, Ca, Mg, Na, and K, but it also contains considerable amounts of H_2O and CO_2 as well as lesser gaseous components such as H_2S, HCl, CH_4, and CO. As a magma cools there is generally a definite order of crystallization of the various mineral constituents. In a magma consisting mainly of O, Si, Mg, and Fe, for example, the mineral with the highest melting point, Mg-rich olivine, Mg_2SiO_4, would crystallize first, followed by more

Fe-rich olivines approaching fayalite, Fe_2SiO_4 (see Figs. 9.5*a* and 13.9). In a magma of the appropriate composition for pyroxene crystallization, the Mg-rich ortho- and clinopyroxenes (see Fig. 13.47), such as enstatite and diopsidic augite, would crystallize before hedenbergite and Fe-rich orthopyroxenes that have lower relative melting points. In magmas rich in the plagioclase component (see Figs. 9.5*b* and 13.124) the An-rich plagioclases will crystallize before the Ab-rich members. All of the above crystallization sequences are the result of *continuous reactions* that take place under equilibrium conditions between the melt and the precipitated crystals as a function of falling temperature. If chemical equilibrium is not maintained between melt and crystals during cooling, the resulting crystals may show compositional zoning. This is especially common in the plagioclase feldspars and in members of the pyroxene series. Igneous rocks, however, also illustrate *discontinuous reactions* that occur at fairly definite temperatures and do not take place over a range of temperatures as do the continuous reactions. In Fig. 9.9*b* we have shown the peritectic reaction of early-formed forsterite transforming to enstatite + melt at lower temperatures. In igneous rocks of appropriate bulk composition it is not uncommon to find early-formed olivine crystals rimmed by later orthopyroxenes. Such reactions (see Fig. 9.9*b*) force the composition of the melt toward the eutectic point at which crystallization of an SiO_2 phase (e.g., cristobalite or tridymite in high-temperature volcanic rocks) occurs.

Early-formed crystals produced in a cooling melt may separate from the liquid by settling out because of gravitational forces, or may be removed from the melt by tectonic deformation. Therefore, the crystals may not remain in equilibrium with the melt from which they crystallized and systematic changes will occur in the bulk composition of the remaining (residual) magma.

The above continuous and discontinuous reactions as well as separation of magma and crystals lead to what is known as *magmatic differentiation*. This concept was first developed by N. L. Bowen, an American petrologist, on the basis of his studies of the textures and mineralogical make-up of many igneous rock types and his correlative experimental studies. Mg-Fe-rich igneous silicates constitute a series of mineral groups that are related to each other by discontinuous reactions. For example, early-formed olivine may be rimmed by pyroxene; amphiboles may form at the expense of the pyroxene rims; and biotite may form as a reaction product of the earlier crystallized amphiboles. On the other hand, members of the plagioclase feldspar series represent a continuous reaction series in which An-rich plagioclase crystallizes early from the melt, enriching the residual melt in alkalies (Na and K). Figure 14.1 illustrates schematically what is known as *Bowen's reaction series*. As a magma of basaltic composition cools, olivine and An-rich plagioclase may crystallize first. If these minerals remain in contact with the magma they will tend to form pyroxene and more Ab-rich plagioclase and the resulting rock will be a gabbro or a basalt. If, however, the bulk of the early olivine and An-rich plagioclase is removed, as by crystal settling, the bulk composition of the remaining melt will tend to become enriched in Si, Al, Fe^{2+}, alkalies and H_2O and CO_2. Such a melt could produce a mineral assemblage consisting mainly of amphiboles, some mica, alkali feldspar, and SiO_2. It should be noted in Fig. 14.1 that amphiboles,

FIG. 14.1. Bowen's reaction series, relating compositional and mineral variation during the crystallization of a cooling magma.

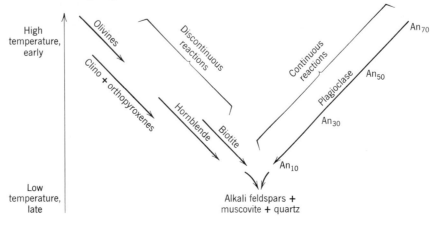

micas, alkali feldspars, and quartz are relatively low-temperature, late-stage crystallization products. The various processes that cause magmatic differentiation can produce a large variety of igneous rocks from a common parent magma.

General Occurrence and Texture

There are two major types of igneous rocks, *extrusive* and *intrusive*. The first group includes those igneous rocks that reached the Earth's surface in a molten or partly molten state. Modern volcanoes produce lava flows that pour from a vent or fracture in the Earth's crust. Such extrusive or *volcanic* rocks tend to cool and crystallize rapidly with the result that their grain size is generally small. If the cooling has been so rapid as to prevent the formation of even small crystals of the mineral constituents, the resulting rock may be a *glass*. Ordinarily the mineral constituents of fine-grained extrusive rocks can be determined only by microscopic examination of thin sections of the rocks. *Intrusive* or *plutonic* rocks are the result of crystallization from a magma that did not reach the Earth's surface. Magmatic intrusions that are discordant with the surrounding country rock are referred to as *batholiths* or *plutons*, depending on their size and shape; when the intrusive is tabular and concordant it is known as a *sill*. A magma that is deeply buried in the Earth's crust generally cools slowly and the mineral constituents crystallizing from it have time to grow to considerable size, giving the rock a medium- to coarse-grained texture. The mineral grains in such rocks can generally be identified with the naked eye. When a magma intrudes in *dikes* (discordant tabular bodies) the textures are usually finer grained than those of plutonic rocks but coarser than those of volcanic rocks; these rocks of intermediate grain size are known as *hypabyssal*.

Some igneous rocks show distinct crystals of some minerals embedded in a much finer-grained matrix. The larger crystals are *phenocrysts*, and the finer-grained material is the *groundmass* (see Fig.13.133). Such rocks are known as *porphyries*. The phenocrysts may vary in size from crystals a centimeter or more across down to very small individuals. The groundmass may also be composed of fairly coarse-grained material, or its grains may be microscopic. The difference in size between the phenocrysts and the particles of the groundmass is the distinguishing feature of a porphyry. The porphyritic texture develops when some of the crystals grow to considerable size before the main mass of the magma consolidates into the finer and uniform-grained material. Any of the types of igneous rocks described below may have a porphyritic variety, such as *granite porphyry, diorite porphyry, rhyolite porphyry*. Porphyritic varieties are most common in volcanic rocks, especially in the more siliceous types.

Chemical Composition

The chemical bulk compositions of igneous rocks exhibit a fairly limited range. The largest oxide component, SiO_2, ranges from about 40 to 75 weight percent in common igneous rock types (see Table 14.1 and Fig. 14.2). Al_2O_3 ranges generally from about 10 to 20 weight percent (except for peridotite and dunite, see analyses in Table 14.1) and each of the other major components generally does not exceed 10 weight percent (except MgO in peridotite and dunite, see Table 14.1).

When the magma is fairly low in SiO_2 the resulting rocks will contain mainly relatively silica-poor minerals such as olivine, pyroxene, hornblende, or biotite and little or no free SiO_2 (i.e., quartz, cristobalite, tridymite; see Figs. 14.2 and 14.3). These rocks,

				Table 14.1 AVERAGE CHEMICAL COMPOSITION OF SOME IGNEOUS ROCKS*					
Oxide	Nepheline syenite	Syenite	Granite	Tonalite	Diorite	Gabbro	Peridotite	Dunite
SiO_2	54.83	59.41	72.08	66.15	51.86	48.36	43.54	40.16
TiO_2	0.39	0.83	0.37	0.62	1.50	1.32	0.81	0.20
Al_2O_3	22.63	17.12	13.86	15.56	16.40	16.84	3.99	0.84
Fe_2O_3	1.56	2.19	0.86	1.36	2.73	2.55	2.51	1.88
FeO	3.45	2.83	1.67	3.42	6.97	7.92	9.84	11.87
MnO	trace	0.08	0.06	0.08	0.18	0.18	0.21	0.21
MgO	trace	2.02	0.52	1.94	6.12	8.06	34.02	43.16
CaO	1.94	4.06	1.33	4.65	8.40	11.07	3.46	0.75
Na_2O	10.63	3.92	3.08	3.90	3.36	2.26	0.56	0.31
K_2O	4.16	6.53	5.46	1.42	1.33	0.56	0.25	0.14
H_2O	0.18	0.63	0.53	0.69	0.80	0.64	0.76	0.44
P_2O_5	—	0.38	0.18	0.21	0.35	0.24	0.05	0.04
Total	99.77	100.00	100.00	100.00	100.00	100.00	100.00	100.00

*All analyses except the nepheline syenite from Nockolds, S. R., 1954, *Geological Society of America Bulletin*, v. 65, pp. 1007–1032.

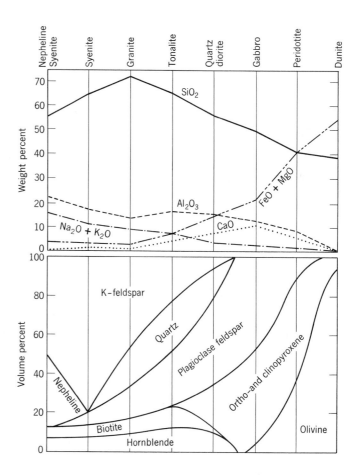

FIG. 14.2. Relationships of variation in chemical and mineral compositions in igneous rocks.

which tend to be dark because of their high percentage of ferromagnesian minerals, are known as *mafic* rock types. When the melt is poor in SiO_2 (*subsiliceous*, or silica undersaturated) and high in alkalies and Al_2O_3 (as in a nepheline syenite composition, see Table 14.1) the resultant crystallization products will contain SiO_2-poor minerals such as feldspathoids and will lack free SiO_2 as quartz (see Figs. 14.2 and 14.4). Crystallization of a melt high in SiO_2 (silica over-saturated) results in rocks with abundant quartz and alkali feldspars, with or without muscovite, and only minor amounts of ferromagnesian minerals. Such rock types are referred to as *felsic* (alkali feldspar-rich) or *silicic*, and are lighter in overall color than mafic rocks. In general, the darker the rock the greater the abundance of ferromagnesian minerals and the lighter the rock the greater the abundance of quartz, or feldspars, or feldspathoids.

Classification

Because of considerable variation in magmas both in chemistry and in conditions of crystallization, igneous rock types show a wide variation in mineralogy and texture. There is a complete gradation from one rock type into another, so the names of igneous rocks, and the boundaries between types, are largely arbitrary (see Figs. 14.2 and 14.5).

Many schemes have been proposed for the classification of igneous rocks, but the most practical for the introductory student is based on mineralogy and texture. In general, four criteria are to be considered in classifying an igneous rock. (1) The relative amount of silica; quartz (or tridymite, or cristobalite) indicates an excess of silica; feldspathoids indicate a deficiency of silica. (2) The kinds of feldspar (alkali feldspar versus plagioclase) and the relative amount of each kind. (3) The relative amount and type of dark minerals. (4) The texture or size of the grains. Is the rock coarse- or fine-grained; that is, is it plutonic or volcanic?

It is obvious that the exact determination of the kind of feldspar or a correct estimate of the amount of each kind is impossible in the field or in the hand specimen. It is also impossible in many fine-grained rocks to recognize individual minerals. Such precise work must be left for the laboratory and carried out by the microscopic examination of thin sections of rocks. Nevertheless, it is important that the basis for the gen-

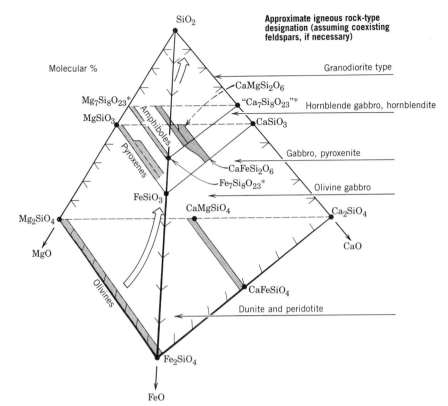

FIG. 14.3. End member compositions, and approximate extents of solid solution in major mafic mineral groups, in terms of molecular percentages of SiO_2, MgO, FeO, and CaO. The triangular diagrams for olivines, pyroxenes, and amphiboles are equivalent to those given in Figs. 13.8, 13.47, and 13.65, respectively. In this representation all compositions are given as anhydrous; in the amphiboles $2(OH)^{1-}$ are replaced by one oxygen, O^{2-} for reasons of charge balance.

The large arrow inside the tetrahedron shows the general, discontinuous progression in mafic silicate groups from subsiliceous to more SiO_2-rich (compare with Fig. 14.1). Some igneous rock types, on the right of the tetrahedron, are correlated with approximate bulk compositions at various heights above the base in the tetrahedron on the left.

*amphiboles, in this anhydrous system, are represented by anhydrous formulas, in which 2OH are replaced by one oxygen, e.g. $Mg_7Si_8O_{22}(OH)_2$ becomes $Mg_7Si_8O_{23}$

eral classification be understood in order that a simplified field classification may have more meaning.

Three major divisions may be made on the basis of the silica content (see Fig. 14.4). (1) Quartz present in amounts greater than 5% (silica oversaturated). (2) No quartz and no feldspathoids present (silica saturated). (3) Feldspathoids in amounts greater than 5% (silica undersaturated). The above divisions made on the basis of silica content are further subdivided ac-

cording to the kind and amount (or the absence) of feldspar. Most of the rocks thus classified have a coarse- and fine-grained variety, which receive different names. Figure 14.5 illustrates the classification of the principal plutonic and volcanic rock types. Table 14.2 gives examples of the principal rock types according to such a classification. Although these rock names are the most important, more than 600 have been proposed to indicate specific types.

FIG. 14.4. Feldspar and feldspathoid compositions in the system SiO_2 (quartz)-$NaAlSiO_4$ (nepheline)-$KAlSiO_4$ (kalsilite)-$CaAl_2O_4$. Bars and fields indicate average extent of solid solution. A melt of composition x_1 (*silica oversaturated*) would produce a quartz-feldspar assemblage; a melt of x_2 composition (*silica saturated*) and lying in the feldspar plane would produce only feldspars, and a melt of composition x_3 (*silica undersaturated*) would produce feldspar-feldspathoid assemblages.

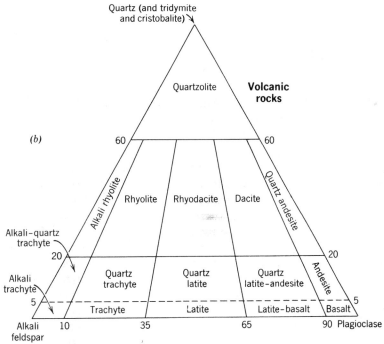

FIG. 14.5. General classification and nomenclature of some common plutonic rock types *(a)* and some common volcanic rock types *(b)*. This classification is based on the relative percentages of quartz, alkali feldspar, and plagioclase, measured in volume percent. (Adapted from Subcommission on the Systematics of Igneous Rocks, *Geotimes*, 1973, v. 18, no. 10, pp. 26–30, and Hyndman, D. W., 1972, *Petrology of Igneous and Metamorphic Rocks.* McGraw-Hill Book Co., New York, p. 35.)

Table 14.2

SIMPLIFIED CLASSIFICATION OF THE IGNEOUS ROCKS

Feldspar	Quartz > 5%		No Quartz; No Feldspathoids		Nepheline or Leucite > 5%	
	Coarse	Fine	Coarse	Fine	Coarse	Fine
K-feldspar* > plagioclase	Granite	Rhyolite	Syenite	Trachyte	Nepheline syenite Leucite syenite	Phonolite Leucite phonolite
Plagioclase > K-feldspar	Granodiorite	Dacite	Monzonite	Latite	Nepheline monzonite	
Plagioclase (oligoclase or andesine)	Tonalite	Quartz andesite	Monzodiorite	Latite basalt	Nepheline diorite	
Plagioclase (labradorite to anorthite)	Quartz diorite	Andesite	Gabbro	Basalt	Nepheline gabbro	Tephrite (− olivine) Basanite (+ olivine)
No feldspar			Peridotite (olivine dominant) Pyroxenite (pyroxene dominant) Hornblendite (hornblende dominant)		Ijolite	Nephelinite (− olivine) Nepheline basalt (+ olivine)

*K-feldspar includes orthoclase, microcline and microperthite; in high *T* volcanic rocks it can be sanidine or anorthoclase.

Mineralogical Composition

Many minerals are found in the igneous rocks, but those that can be called rock-forming minerals are comparatively few. Table 14.3, which lists the major mineral constituents of igneous rocks, is divided into two parts: (1) the common rock-forming minerals of igneous rocks, and (2) the accessory minerals of igneous rocks. See Table 14.4 for a listing of common mineral assemblages in some plutonic rocks and Table 14.5 for a volume percentage listing of mineral constituents in some common plutonic rock types.

Plutonic Rocks

Granite-Granodiorite. Granite is a granular rock of light color and even texture consisting chiefly of feldspar and quartz. Usually both K-feldspar and oligoclase (or albite) are present; K-feldspar may be flesh-colored or red, whereas oligoclase (or albite) is commonly white and can be recognized by the presence of albite twinning striations. The quartz can be recognized by its glassy luster and lack of cleavage. Granites usually carry a small amount (about 8%) of mica or hornblende. The mica is commonly biotite, but muscovite may also be present. The minor accessory minerals are zircon, titanite, apatite, magnetite, ilmenite. Figure 14.6 illustrates the liquidus diagram for the hydrous granite system, $NaAlSi_3O_8$-$KAlSi_3O_8$-

Table 14.3

MINERALOGY OF THE IGNEOUS ROCKS

Common Rock-Forming Minerals	Common Accessory Minerals
1. Quartz, tridymite, cristobalite	1. Zircon

2. Feldspars
   Orthoclase
   Microcline
   Sanidine
   Plagioclase
	2. Titanite
	3. Magnetite
	4. Ilmenite
	5. Hematite

3. Nepheline	6. Apatite
4. Sodalite	7. Pyrite
5. Leucite	8. Rutile

6. Micas
   Muscovite
   Biotite
   Phlogopite
	9. Corundum
	10. Garnet

7. Pyroxenes
   Augite
   Orthopyroxene
   Aegirine	

8. Amphiboles
   Hornblende
   Arfvedsonite
   Riebeckite	

9. Olivine	

SiO_2-H_2O, with a temperature minimum of about 770°C in the central part of a low-temperature trough between the liquidus fields of K-Na feldspar solid so-

Table 14.4
TYPICAL MINERAL ASSEMBLAGES OF SOME PLUTONIC ROCKS
(See also Table 14.5)

	Major Constituents (>10%)	Minor Constituents (each <10%)
Granite	mic (often perthitic) + Ab (or olig) + qtz	bio, musc, hbld, mg, zirc, ap, ti
Granodiorite	and + K-spar + qtz + hbld	bio, ti, ap, mag
Syenite	mic + Ab (or olig)	qtz, bio, hbld, ti, ap, zirc
Nepheline syenite	orth + neph	Ab, aeg, arf, sod, ap, zirc, ti
Monzonite	plag + orth + aug	hbld, qtz, bio, mag, ap, ti
Diorite	olig (or and) + hbld	bio, orth, qtz, hbld, pyx, ap, zirc, mag
Gabbro	plag (An_{50}-An_{90}) + aug + opx + oliv	bio, mag, sp, ilm, hbld
Norite	lab (or byt) + opx + some oliv	aug, hbld, bio, ap, mag
Anorthosite	lab (~90%)	aug, hbld, bio
Peridotite	oliv (forsteritic) + opx	hbld, chrom, plag
Dunite	oliv (forsteritic) ~90%	chrom, mag, ilm, pyrrh

Abbreviations: Ab = albite, aeg = aegirine, and = andesine, ap = apatite, arf = arfvedsonite, aug = augite, bio = biotite, byt = bytownite, chrom = chromite, hbld = hornblende, ilm = ilmenite, K-spar = K-feldspar, lab = labradorite, mag = magnetite, mic = microcline, musc = muscovite, neph = nepheline, olig = oligoclase, oliv = olivine, opx = orthopyroxene, orth = orthoclase, plag = plagioclase, pyrrh = pyrrhotite, pyx = pyroxene, qtz = quartz, sp = spinel, sod = sodalite, ti = titanite, zirc = zircon.

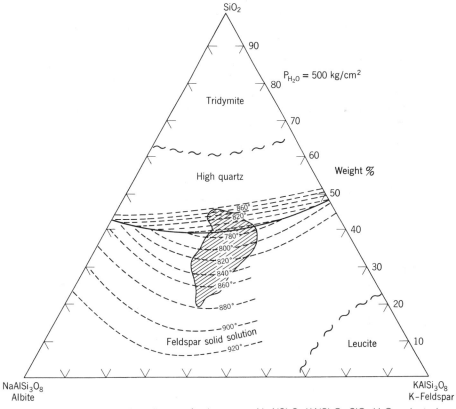

FIG. 14.6. Isobaric equilibrium diagram for the system $NaAlSi_3O_8$-$KAlSi_3O_8$-SiO_2-H_2O projected onto the anhydrous base of the $NaAlSi_3O_8$-$KAlSi_3O_8$-SiO_2-H_2O tetrahedron. Contour lines indicate melting temperatures. Stability fields are approximately located. The shaded area in the center represents percentages of quartz, albite, and K-feldspar, as calculated for hundreds of chemically analyzed granites. (From Tuttle, O. F. and Bowen, N. L., 1958, Origin of granite in the light of experimental studies in the system $NaAlSi_3O_8$-$KAlSi_3O_8$-SiO_2-H_2O. *Geological Society of America Memoir*, no. 74, 153 pp.)

lutions and high quartz. When the mineral compositions of hundreds of granites are expressed in terms of percentages of albite, K-feldspar, and quartz, they can be plotted on Fig. 14.6. The distribution of the resulting data points represents the residual melt compositions from which the granites formed during their crystallization history. These melt compositions coincide broadly with the experimentally determined region of minimum temperature, between 860° and 770°C. Many granites, therefore, are concluded to be the result of crystallization from melts at relatively low temperatures.

There is a complete series grading from granite, with feldspar almost entirely the potash varieties, to granodiorite with feldspar mostly plagioclase and only slightly more than 10% K-feldspar. The boundary between the two types is arbitrarily set. Granites are those rocks in which K-feldspar generally exceeds plagioclase; granodiorites are those in which plagioclase exceeds K-feldspar. In most instances it so happens that as the plagioclase increases in amount the percentage of dark minerals also increases, and thus, in general, granodiorites are darker than granites. However, in the field or in a hand specimen, it is usually impossible to distinguish between the two rock types with certainty.

Syenite-Monzonite. A syenite is a granular rock of light color and even texture composed essentially of K-feldspar and oligoclase, with lesser amounts of hornblende, biotite, and pyroxene. It thus resembles granite but differs from granite in that it contains less than 5% quartz. Accessory minerals are apatite, titanite, zircon, and magnetite.

A series exists between syenite and monzonite, with increasing plagioclase content in the monzonite (see Fig. 14.5). Monzonites are usually darker than syenites, for an increase in dark minerals frequently accompanies an increase in plagioclase. However, without microscopic aid it is rarely possible to distinguish between the two types.

Nepheline is present in some syenites; if the amount exceeds 5%, the rock is called a *nepheline syenite*. The nepheline has a greasy luster and may be mistaken for quartz, but can be distinguished by its hardness ($5\frac{1}{2}$–6). Some nepheline syenites may contain sodalite; others, corundum.

Syenites in which leucite is present in amounts greater than 5% are called *leucite syenites*. The leucite can be recognized by its trapezohedral form. Such rocks are extremely rare.

Tonalite-Quartz Diorite. A tonalite is composed essentially of plagioclase feldspar and quartz with only minor amounts of K-feldspar (less than 5%). The plagioclase is oligoclase or andesine. Dark minerals, especially biotite and hornblende, are plentiful; pyroxene is more rarely present. Apatite, magnetite, and titanite are common accessory minerals. Although not essential to the classification, dark minerals are usually abundant, and thus, in general, tonalites are darker in color than granites; see color index at the bottom of Table 14.5.

As the plagioclase becomes more An-rich and

Table 14.5
APPROXIMATE MINERAL COMPOSITIONS OF SOME PLUTONIC ROCK TYPES (IN VOLUME PERCENT)*

	Granite	Syenite	Granodiorite	Quartz Diorite	Diorite	Gabbro	Olivine Diabase	Diabase	Dunite
Quartz	25		21	20	2				
Orthoclase and microperthite	40	72	15	6	3				
Oligoclase	26	12							
Andesine			46	56	64				
Labradorite						65	63	62	
Biotite	5	2	3	4	5	1		1	
Amphibole	1	7	13	8	12	3		1	
Orthopyroxene				1	3	6			2
Clinopyroxene		4		3	8	14	21	29	
Olivine						7	12	3	95
Magnetite	2	2	1	2	2	2	2	2	3
Ilmenite	1	1				2	2	2	
Apatite	Trace	Trace	Trace	Trace	Trace				
Titanite	Trace	Trace	1	Trace	Trace				
Color Index†	9	16	18	18	30	35	37	38	98–100

*After Larsen, E. S., 1942, in *Handbook of Physical Constants*, Birch, F., Schairer, Y. F., and Spicer, H. C. (eds.), *Geological Society of America, Special Paper* 36, Table 1.2, p. 3. The percentage values are based on grain counts of minerals in a thin section using a polarizing microscope. This is known as *modal analysis*.
†Color index—a number that represents the percentage, by volume, of dark-colored (i.e., mafic) minerals in a rock.

with a lessening of the amount of quartz, tonalite grades into a *quartz-diorite* and finally into a *diorite*.

Diorite-Gabbro. A diorite is a granular rock characterized by plagioclase feldspar (oligoclase to andesine) but lacking quartz and K-feldspar in appreciable amounts. Hornblende is the principal dark mineral, but biotite is usually present. Pyroxenes are minor constituents. Magnetite, ilmenite, apatite, and, less commonly, titanite and zircon are accessory minerals. Normally dark minerals are present in sufficient amount to give the rock a dark appearance (see color index at the bottom of Table 14.5).

If the plagioclase is more calcic in composition than andesine (labradorite to anorthite), the rock is a gabbro. Although the distinction is made on this criterion alone, it so happens that rocks carrying labradorite or more calcic plagioclase usually have pyroxene as the chief dark constituent, whereas the diorites with more sodic plagioclase usually have amphiboles as dark minerals. Olivine and orthopyroxene are also present in most gabbros. The association of pyroxene-olivine-An-rich plagioclase is diagnostic of the relatively high temperatures of crystallization of mafic rock types. In Fig. 9.11 forsterite and protoenstatite are shown crystallizing together from 1557°C to about 1300°C (along the crystallization path between the two fields). These temperatures are much higher than those determined for granites, for example, in Fig. 14.6. See also Fig. 14.1 which expresses these relative temperature differences qualitatively.

The name *norite* is given to a gabbro in which the pyroxene is essentially orthopyroxene (hypersthene); it is usually impossible to make this distinction without microscopic aid. A type of igneous rock known as *anorthosite* is composed almost entirely of plagioclase feldspar and may therefore be light in color.

If amounts of nepheline in diorites and gabbros exceed 5%, the rocks are called respectively *nepheline diorite* and *nepheline gabbro*. These rocks are rare and unimportant.

The term *diabase* is sometimes used to indicate a fine-grained gabbro characterized by a certain texture. This "diabasic" texture is shown microscopically to have augite filling the interstices of tabular plagioclase grains.

Peridotite. A peridotite is a granular rock composed of dark minerals; feldspar is negligible (less than 5%). The dark minerals are chiefly pyroxene and olivine in varying proportions, but hornblende may be present. If the rock is composed almost wholly of pyroxene, it is called a *pyroxenite*; if it is composed almost wholly of olivine, it is called a *dunite*. The name *hornblendite* is given to a rare type of rock composed almost wholly of hornblende. Magnetite, chromite, il-

menite, and garnet are frequently associated with peridotites. Platinum is associated with chromite in some peridotites, usually dunites, whereas diamond is found in a variety of altered peridotite known as *kimberlite*.

The olivine in peridotites is usually altered in whole or in part to the mineral serpentine. If the entire rock is thus altered, the name *serpentinite* is given to it (see also page 589).

Volcanic Rocks

Because of their fine-grained texture it is much more difficult to distinguish between the different types of volcanic rocks than between their plutonic equivalents. In the field only an approximate classification, based chiefly on whether the rock is light or dark in color, can be made. The term *felsite* is thus used to include the dense, fine-grained rocks of all colors except dark gray, dark green, or black. Felsite thus embraces the following types described below: rhyolite, trachyte, quartz latite, latite, dacite, and andesite. Experienced petrologists may be able, by the aid of a hand lens, to discern differences in texture or mineral composition that enable them to classify these rocks, but to the untrained observer they frequently appear much the same.

Fine-grained rocks that are a very dark green or black are called *traps*. This term is applied to dark, fine-grained rocks of indefinite mineral composition irrespective of whether they have been intruded as dikes or extruded as lava flows. It so happens that most rocks thus classified as traps in the field or hand specimen are basalts and satisfy the more rigorous classification based on microscopic examination.

Rhyolite is a dense, fine-grained rock, the volcanic equivalent of a granite. It is thus composed essentially of alkali feldspar and quartz, but much of the silica may be present as tridymite or cristobalite. Phenocrysts of quartz, sanidine, and oligoclase are common. Dark minerals are never abundant, but dark brown biotite is most common. Augite and hornblende are found in some rhyolites.

Rhyolites may be very uniform in appearance or may show a flow structure, giving a banded or streaked appearance to the rock. The groundmass may be partly or wholly glassy. When the rock is completely glassy and compact, it is known as *obsidian* and is usually black. Similar glassy rocks of a brown, pitchy appearance are called *pitchstones*. *Pumice* is rhyolite glass in which expanding gas bubbles have distended the magma to form a highly vesicular material. In pumice, therefore, cavities are so numerous as to make up the bulk of the rock and give it an apparent low specific gravity.

Trachyte is the volcanic equivalent of syenite. It is thus composed chiefly of alkali feldspar with some dark minerals but lacks quartz. Small amounts of tridymite and cristobalite are often found in gas cavities. Phenocrysts of sanidine are frequently present and characteristically show Carlsbad twinning; phenocrysts of oligoclase, biotite, hornblende, and pyroxene are less common. Olivine may be present.

Banding or streaking, due to flow, is common in the trachytes. Unlike the rhyolites, glass is seldom found in the groundmass, and there are thus few glassy or vesicular types. As a result of flow the tabular feldspar frequently shows a subparallel orientation which is so common in trachytes that it is called *trachytic texture*.

Phonolite is the volcanic equivalent of nepheline syenite and is thus poorer in silica than trachyte. This is expressed mineralogically by the presence of feldspathoids. Orthoclase or sanidine is the common feldspar; albite is rarely present. Nepheline occurs in the groundmass as minute hexagonal crystals and can be observed only by microscopic aid. Sodalite and other feldspathoids may be present, usually altered to zeolites. Leucite is present in *leucite phonolite.* It is in well-formed crystals that range from microscopic sizes to a centimeter in diameter. Aegirine is the common dark mineral and normally occurs as phenocrysts, but biotite may be abundant in the leucite-rich rocks. The phonolites are completely crystalline, and there are thus no glassy varieties.

Latite and **quartz latite** are the volcanic equivalents of monzonite and quartz monzonite, respectively. They, therefore, contain about equal amounts of plagioclase and alkali feldspar. The dark minerals are chiefly biotite and hornblende. The distinction between them rests on the amount of quartz present; quartz latites would contain more than 5% quartz, latites less than 5% quartz. Both of these rock types occur infrequently.

Dacite is the dense volcanic equivalent of granodiorite. It contains plagioclase feldspar and quartz, both of which may occur as phenocrysts. The dark mineral is usually hornblende, but biotite is found in some varieties. Some glass may be present in the groundmass, but glassy equivalents of dacites are rare.

Andesite is the volcanic equivalent of quartz diorite and thus is composed chiefly of oligoclase or andesine feldspar. K-feldspar and quartz are absent or present in amounts of less than 10%. Hornblende, biotite, augite, or orthopyroxene may be present, frequently as phenocrysts. Andesites are usually named according to the dark mineral present, such as *hornblende andesite* or *hypersthene andesite.* In some andesites the groundmass is partly glassy and in rarer types completely so.

Andesites are abundant in certain localities, notably in the Andes Mountains of South America, from which locality the rock receives its name.

Basalt is a dark-colored, fine-grained rock, the volcanic equivalent of gabbro. Labradorite feldspar is the chief constituent of the groundmass, whereas more calcic plagioclase (bytownite or anorthite) may be present as phenocrysts. Augite and olivine are usually present; the augite is frequently found both as phenocrysts and in the groundmass, but olivine, as a rule, is only in phenocrysts. Brown hornblende and brown biotite are present in some basalts.

The groundmass of some basalts contains small amounts of interstitial glass and in rare instances is wholly glassy. Gas cavities near the top of basalt flows may be abundant enough to make the rock vesicular.

The presence of nepheline or leucite in basalt gives rise to the rare rock types *tephrite* and *leucite tephrite.*

Basalts are the most abundant of the volcanic rocks and form extensive lava flows in many regions; the most noted are the Columbia River flows in the western United States and the Deccan "traps" of western India. The oceanic basins are underlain by extensive flows of basaltic composition. Many of the great volcanoes, such as form the Hawaiian Islands, are built up of basaltic material. In addition to forming extrusive rock masses, basalt is widely found forming many small dikes and other intrusives.

Fragmental Igneous Rocks

During periods of igneous activity volcanoes eject much fragmental material which accumulates and forms the *fragmental igneous rocks,* or *pyroclastic rocks.* The ejecta vary greatly in size. Rock composed of finer particles of *volcanic ash* and *volcanic dust* is called *tuff;* that composed of coarser volcanic bombs is called *agglomerate,* or *volcanic breccia.* Such rocks are frequently waterlaid and bedded, and thus form a transition between the igneous and sedimentary rocks.

Pegmatites

Pegmatites are extremely coarse-grained bodies that are commonly closely related genetically and in space to large masses of plutonic rocks. They may be found as veins or dikes traversing the granular igneous rock but more commonly extend out from it into the surrounding country rock. Granites more frequently than any other rock have pegmatites genetically associated with them; consequently, unless modified by other terms, pegmatite refers to *granite pegmatite.* The minerals in most pegmatites, therefore, are the common minerals in granite (quartz, feldspar, mica) but of extremely large size. Crystals of these minerals measuring a foot across are common, and in some localities they reach gigantic sizes. Probably the largest crystals ever found were of feldspar in pegmatites in Karelia, CIS, where material weighing thousands of tons was mined from single crystals. Quartz crystals weighing thousands of pounds and mica crystals over 10 feet across have been found. A common characteristic of pegmatites is the simultaneous and interpenetrating crystallization of quartz and K-feldspar (usually microcline) to form *graphic granite* (see Fig. 13.128).

Although most pegmatites are composed entirely of the minerals found abundantly in granite, those of greatest interest contain other, and rarer, minerals. In these pegmatites there has apparently been a definite sequence in deposition. The earliest minerals are microcline and quartz, with smaller amounts of garnet and black tourmaline. These are followed by and partly replaced by albite, lepidolite, gem tourmaline, beryl, spodumene, amblygonite, topaz, apatite, and fluorite. A host of rarer minerals such as triphylite, columbite, monazite, molybdenite, and uranium minerals may be present. In places some of the above minerals are abundant and form large crystals that are mined for their rare constituent elements. Thus spodumene crystals over 40 feet long have been found in the Black Hills of South Dakota, and beryl crystals from Albany, Maine, have measured as much as 27 feet long and 6 feet in diameter.

The formation of pegmatite dikes is believed to be directly connected with the crystallization of the larger masses of associated plutonic rock. The process of crystallization brings about a concentration of the volatile constituents in the residual melt of the magma. The presence of these volatiles (H_2O, B, F, Cl, and P) decreases the viscosity and thus facilitates crystallization. Such an end product of magmatic crystallization is also enriched in the rare elements originally disseminated through the magma. When this residual liquid is injected into the cooler surrounding rock, it crystallizes from the borders inward, frequently giving a zonal distribution of minerals with massive quartz at the center.

Nepheline syenite pegmatites have been found in a number of localities. They are commonly rich in unusual constituents and contain numerous zirconium, titanium, and rare-earth minerals.

SEDIMENTARY ROCKS

Sedimentary rocks cover approximately 80% of the Earth's land surface, but their total contribution to the upper 10 miles (16 km) of the Earth's crust is estimated to be only about 5%. As such, the sedimentary sequences we see represent only a veneer over a crust, consisting mainly of igneous and metamorphic rocks.

The materials of which sedimentary rocks are composed have been derived from the weathering of previously existing rock masses that were elevated above sea level. *Chemical weathering* decomposes minerals in the rocks, and *mechanical weathering* is responsible for the physical destruction of the original rock. The decomposition and disintegration products are transported to and deposited in areas of accumulation by the action of water or, less frequently, by glacial or wind action. Such loose deposits are converted into rocks by the processes of *diagenesis* and *lithification*, which include compaction and cementation of the loose materials.

The products of *chemical* decomposition may be transported in solution by water into lakes and seas, where chemical changes (such as due to evaporation) or organisms, may cause precipitation. These chemical (or biochemical) precipitates result, upon induration, diagenesis and lithification, in *chemical sedimentary rocks* (see Fig. 14.7). Such chemically deposited sediments are represented by carbonate (such as travertine) and evaporite sequences and finely banded sedimentary iron-formations. These truly chemical sediments are also known as *orthochemical* (from the Greek, meaning *correct* or *true*) sedimentary rocks. If organisms have caused the precipitation of the major sedimentary mineral components, or if the precipitated minerals have undergone substantial movement (and redeposition) after their crystallization, the resulting sediment is referred to as *allochemical* (from the Greek, meaning *different*). Examples of such allochemical sedimentary rocks are oolitic and fossiliferous limestone.

More generally, weathering includes both chemical decomposition and mechanical disintegration and thus the end-products consist of sedimentary materials that are the products of chemical as well as mechanical action. Both processes produce solid fragments and particles known as *detritus*, or clastic material. Mechanically deposited loose sediments include gravel and sand which, upon lithification (compaction and cementation) form conglomerates and sandstones (see Fig. 14.7). These sediments, with clastic textures, are known as *terrigenous* sediments. Detrital materials consist most commonly of chemically inert minerals such as quartz, K-feldspar, garnet, zircon, rutile, and magnetite, and of rock fragments made of these minerals.

The two categories, *chemical* and *detrital*, are not exclusive because most chemical sediments contain some detrital material and most clastic rocks also carry some chemical sediment.

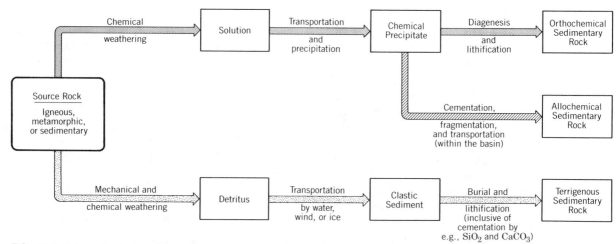

FIG. 14.7. Schematic diagram for the sequence: source rock→weathering→sedimentary rock. The arrows represent processes; the boxes represent products. (Modified after Suttner, L. J. and Meyers, J., 1991, Field study of the petrology of sedimentary rocks, in *Manual for Geologic Field Study of Northern Rocky Mountains*. Indiana University, Bloomington, pp. 305–326.)

Chemical Composition

The range in chemical compositions of igneous rocks (see Table 14.1) is generally relatively small because the crystallization sequence of a magma is governed by physicochemical and chemical principles that control the sequence of products crystallizing from the melt. Sedimentary rocks, on the other hand, present a much larger compositional range (see Table 14.6). For example, a sandstone consisting essentially of quartz grains may contain as much as 99% SiO_2; $FeO + Fe_2O_3$ may be as high as 58% in banded iron-formations rich in hematite and magnetite; CaO may reach 55% in pure limestones. This large range in compositions, which is illustrated in Fig. 14.8, is caused by the weathering cycle that tends to produce

mechanical sediments that are compositionally very distinct from chemical sediments.

Mineralogical Composition

The minerals of sedimentary rocks can be divided into two major groups: minerals that are resistant to the mechanical and chemical breakdown of the weathering cycle and minerals that are newly formed from the products of chemical weathering. The relative stabilities of minerals to weathering are shown in Table 14.7. This listing means that quartz is one of the chemically and mechanically most resistant minerals, whereas olivine is easily altered chemically. The positions of minerals listed between quartz and olivine

Table 14.6 CHEMICAL COMPOSITION OF AVERAGE IGNEOUS ROCK COMPARED WITH AVERAGE COMPOSITIONS OF SOME SEDIMENTARY ROCK TYPES*	Average Continental Igneous Rock	Average Sandstone	Average Shale	Average Limestone
SiO_2	59.14	78.33	58.10	5.19
TiO_2	1.05	0.25	0.65	0.06
Al_2O_3	15.34	4.77	15.40	0.81
Fe_2O_3	3.08	1.07	4.02	0.54
FeO	3.80	0.30	2.45	—
MgO	3.49	1.16	2.44	7.89
CaO	5.08	5.50	3.11	42.57
Na_2O	3.84	0.45	1.30	0.05
K_2O	3.13	1.31	3.24	0.33
H_2O	1.15	1.63	5.00	0.77
P_2O_5	0.30	0.08	0.17	0.04
CO_2	0.10	5.03	2.63	41.54
SO_3	—	0.07	0.64	0.05
C(elemental)	—	—	0.80	—
Total	99.50	99.95	99.95	99.84

*After Clarke, F. W., 1924, Data of Geochemistry. *U.S. Geological Survey Bulletin*, no. 770.

represent intermediate stabilities ("survival rates") during the weathering process. It is of interest to note that this sequence is very similar to Bowen's reaction series in reverse (see Fig. 14.1). This similarity indicates that the minerals formed at the lowest temperature in the crystallization of a melt are also the most stable at ordinary temperatures and pressures (atmospheric conditions).

The most important chemical breakdown reactions involving high-temperature igneous (or metamorphic) minerals, as the reactants in a hydrous (weathering) environment at low temperature and pressure (atmospheric conditions), have reaction products that fall into three categories: (1) layer silicates such as kaolinite and montmorillonite, (2) silica in solution as H_4SiO_4, and (3) Na^+, K^+, Ca^{2+}, and Mg^{2+} ions in solution. Fe^{2+} released from the breakdown of mafic minerals, is instantly oxidized to Fe^{3+} and is precipitated as $Fe(OH)_3$, as a precursor to goethite, or hematite. Examples of some of these reactions are as follows:

$$(Ca,Na)(Mg,Fe^{2+},Al)(Si,Al)_2O_6 + H_2O + H^+ \rightarrow$$
$$\text{augite}$$

$$\rightarrow \text{Na- and Ca-montmorillonite} + H_2SiO_4 + Na^+$$
$$+ Ca^{2+} + Fe(OH)_3,$$

$$CaAl_2Si_2O_8 + H_2O + H^+ \rightarrow$$
$$\text{anorthite}$$

$$\rightarrow \text{Ca-montmorillonite} + H_2SiO_4 + Ca^{2+},$$

$$3KAlSi_3O_8 + 12H_2O + 2H^+ \rightarrow$$
$$\text{orthoclase}$$

$$\rightarrow KAl_3Si_3O_{10}(OH)_2 + 6H_4SiO_4 + 2K^+,$$
$$\text{muscovite}$$

in which the muscovite continues its alteration to kaolinite by the following reaction:

$$2KAl_3Si_3O_{10}(OH)_2 + 3H_2O + 2H^+ \rightarrow$$
$$\text{muscovite}$$

$$\rightarrow 3Al_2Si_2O_5(OH)_4 + 2K^+$$
$$\text{kaolinite}$$

Detrital sedimentary rocks consist mainly of the most resistant rock-forming minerals, quartz, K-feldspar, mica and lesser plagioclase, as well as small amounts of garnet, zircon, and spinel (magnetite). The detrital rock types may be regarded as accidental, mechanical mixtures of genetically unrelated resistant minerals. For example, a feldspar-rich sandstone may contain orthoclase and microcline, as well as several members of the plagioclase series. Such a random va-

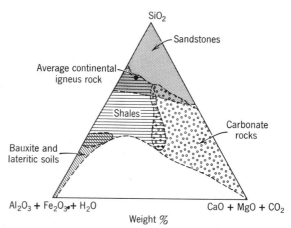

FIG. 14.8. Compositional variations of common sediments. (After Mason, B. and Moore, C. B., 1982, *Principles of Geochemistry*, 4th ed. John Wiley & Sons, New York.)

riety of feldspar compositions is not found in igneous assemblages because of physical-chemical controls in the crystallization sequence.

Chemical sedimentary rocks that result from the inorganic or organic precipitation of minerals can be interpreted in large part in terms of chemical and physico-chemical principles that apply at low temperatures (25°C) and atmospheric pressure (see, for example, Figs. 9.20 and 9.21). Such chemical sedimentary assemblages, therefore, are not random or accidental but reflect the concentrations of ions in solution, as well as conditions such as temperature, pressure, and salinity of the sedimentary basin. For example, the sequence of minerals in evaporite beds can be related to the concentration of ions in solution in the brine from which the sequence precipitated (see page 577). Examples of common chemical precipitates are calcite, aragonite, gypsum, anhydrite, and halite. In sedimentary iron-formations hematite, magnetite, siderite and ankerite as well as chert are considered products of chemical sedimentation.

Table 14.7
RELATIVE STABILITIES OF SOME ROCK-FORMING MINERALS IN THE WEATHERING CYCLE

High stability		Quartz
		Muscovite
		K-feldspar
		Biotite
↑ Increasing stability	Albite { intermediate { plagioclases Anorthite	Hornblende
		Augite
Low stability		Olivine

Table 14.8

SEDIMENTARY ROCKS SUBDIVIDED INTO THREE MAJOR CATEGORIES

I Terrigenous Sedimentary Rocks (Clastic Texture)	II Allochemical Sedimentary Rocks (Biochemical/biogenic; with Clastic Texture)	III Orthochemical Sedimentary Rocks
conglomerates, breccias, sandstones, and mudstones	limestones, dolostones, phosphorites, chert, and coal	evaporites, chert, travertine, and iron-formations
agglomerates and volcaniclastic sandstones		

Classification

Sedimentary rocks are, in general, *stratified*, that is, they are characterized by layers or beds distinguished from each other by differences in grain size, mineral composition, color, or internal structure. Other features that are uniquely diagnostic of a sedimentary origin are: primary sedimentary structures; the presence of fossils; the occurrence of grains whose shape is the result of transportation (referred to as *clasts* or *detrital grains*); and the presence of a mineral that is invariably of sedimentary origin, such as glauconite.

All sedimentary rocks can be grouped in three broad categories, terrigenous, allochemical, and orthochemical, as shown in Table 14.8. The identification of a sedimentary rock type within each of these three broad categories is based upon its mineralogical composition and its texture (inclusive of grain size and grain shape).

Terrigenous Sedimentary Rocks

Terrigenous sedimentary rocks consist of *detrital grains* which form the framework of the rock, and which are joined together by *cement*; these detrital grains (or clasts) are known as the *framework grains*. Variable amounts of *matrix*, which consists of fragmental material substantially smaller than the mean size of the framework grains, may also be present. Because of the normally very fine grain size of matrix material, it may be impossible to determine whether this is indeed of detrital rather than diagenetic origin. The textural expression of rocks dominated by detrital material is a composite of the grain size, grain shapes,

Table 14.9

TERMS AND SIZES FOR CLASTIC SEDIMENTS AND CLASTIC SEDIMENTARY ROCK TYPES*

	Name	Millimeters	Micrometers	Φ
GRAVEL		4,096		−12
	Boulder	256		−8
	Cobble	64		−6
	Pebble	4		−2
	Granule			
		2		−1
SAND	Very coarse sand	1		0
	Coarse sand	0.5	500	1
	Medium sand	0.25	250	2
	Fine sand	0.125	125	3
	Very fine sand			
		0.062	62	4
MUD	Coarse silt	0.031	31	5
	Medium silt	0.016	16	6
	Fine silt	0.008	8	7
	Very fine silt			
		0.004	4	8
	Clay			

*After J. A. Udden (1898) and C. K. Wentworth (1924). The Φ scale, devised by W. C. Krumbein (1934), is based on a logarithmic transformation, $\Phi = -\log_2 S$, where S is grain size in millimeters. The Φ scale is commonly used in sedimentological studies because it is more convenient in presenting data than if values are given in millimeters.

Very well sorted | Well sorted | Moderately sorted | Poorly sorted | Very poorly sorted

FIG. 14.9. Degrees of sorting in sandstones and conglomerates. (From Compton, R. R., 1962, *Manual of Field Geology*. John Wiley & Sons, New York.)

sorting, and angularity of the framework grains. A universally adopted grain size scale for *detrital* (or *clastic*) sediments is given in Table 14.9. Examples of various degrees of sorting are given in Fig. 14.9.

Clastic rocks are dominated by detrital (framework) grains and include conglomerates, breccias, sandstones, and mudrocks. *Conglomerates* and *breccias* consist of large clasts (boulders, cobbles, pebbles, and granules) with or without a sandy matrix; conglomerates show pronounced rounding of the clasts and breccias contain more angular clasts. Volcanic debris may be a predominant component of some sedimentary rocks; such are commonly referred to as *pyroclastic rocks*. Coarse-grained pyroclastics with a grain size over 32 mm are known as *agglomerates* or *volcanic breccias*. *Sandstones* are finer-grained (grain size between 2 and 0.062 mm) and most easily classi-

fied by recognizing the amounts of clastic (framework) grains composed of quartz (and chert), feldspar, and lithic fragments. A commonly accepted scheme of sandstone classification, as based on the population of these three components in the clastic grains, is given in Fig. 14.10. Such a classification scheme is most applicable to the study of sandstones in thin section under the microscope, because the name assignment is based on a modal (volume percent) analysis of the constituent clastic grains. However, with close inspection in the field with a hand lens, a reasonably correct name can often be assigned. The matrix material of sandstones is commonly made of clay minerals and very fine-grained quartz. When the matrix of a sandstone constitutes more than about 10 volume percent of the rock, it is classified as a *wacke*. *Greywackes* are mostly dark gray sandstones with abun-

FIG. 14.10. Classification of common sandstones. (From Tucker, M. E., 1982, *Field Description of Sedimentary Rocks*. John Wiley & Sons, New York.)

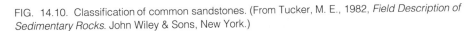

dant matrix. *Volcaniclastic* sandstones are made chiefly of lava fragments, volcanic glass, and crystals. Many tend to be green through chlorite replacement. Common cements in sandstones are quartz, calcite, and clay minerals. Diagenetic hematite may stain sandstones red.

Mudstone is a general term for sediments composed mainly of silt-sized (0.062 to 0.004 mm) and clay-sized (<0.004 mm) particles. Mudstones are essentially impossible to study in hand specimen because of their fine grain size. Modern laboratory methods of study for these very fine-grained rock types include X-ray diffraction, electron microprobe, and scanning electron microscope techniques. *Siltstones* and *claystones* are rock types made up mainly of silt and clay particle size materials, respectively. *Shale* is characterized by *fissility*, the ability to split into thin sheets, generally parallel to the bedding. Mudstones are nonfissile, commonly with a massive or blocky texture. *Slate* refers to a mudstone with a well-developed cleavage (which may or may not be parallel to the bedding and is commonly the result of metamorphism).

A schematic classification of common terrigenous sedimentary rocks is given in Table 14.10.

Allochemical Carbonate Rocks

Allochemical carbonate rocks show clastic (fragmental) textures analogous to those seen in terrigenous rocks, but the textural interpretation in these mineralogically simple rocks is not always straightforward. Because calcite recrystallizes easily and because secondary dolomite (replacing original calcium carbonates; this process is known as *dolomitization*) often destroys the texture of the original carbonate, textural interpretations of such "crystalline limestones" or "crystalline dolostones" may be impossible.

Limestones consist of two classes of constituents, orthochemical and allochemical. These constituents originate from within the basin of deposition of the limestone, and are referred to as *intrabasinal*. The orthochemical components are of two types: (1) *microcrystalline calcite ooze*, which is a very fine-grained carbonate precipitate that has settled to the bottom of the basin, and (2) *sparry calcite cement*, which is coarser in grain size than the ooze, and tends to be clear or translucent. This coarser-grained type of calcite is a pore-filling cement that was precipitated in place. The allochemical components of limestone are of four types: *intraclasts*, *oolites*, *fossils*, and *pellets*. *Intraclasts* represent fragments of weakly con-

Table 14.10
CLASSIFICATION OF TERRIGENOUS ROCKS*
(compare with Fig. 14.10 for sandstone classification)

GRAIN SIZE			COMPOSITION			
		Conglomerate (rounded), breccia (angular)	Lithic fragments (e.g., chert, limestone, volcanic, granite)		Quartz	Feldspar
	Cobble		Cobble conglomerate (or breccia) (e.g., granite cobble conglomerate)		Quartz cobble conglomerate (or breccia)	
	Pebble		Pebble conglomerate (or breccia) (e.g., chert pebble conglomerate)		Quartz pebble conglomerate (or breccia)	
	Granule		Granule conglomerate (or breccia) (e.g., limestone granule conglomerate)		Quartz granule conglomerate (or breccia)	Feldspar granule conglomerate (or breccia)
	Sand	Sandstone	**Wacke (>10% matrix)**	Lithic† wacke	Quartz wacke	Feldspathic‡ wacke
			Arenite (<10% matrix)	Lithic arenite	Quartz arenite	Feldspathic‡ arenite
	Silt		*(composition cannot be evaluated because of fine grain size)*			
			Siltstone			
	Mud		Mudstone—lacking fissility Shale—showing fissility			

*Modified after Suttner, L. J. and Meyers, J., 1991, Field study of the petrology of sedimentary rocks, in *Manual for Geologic Field Study of Northern Rocky Mountains.* Indiana University, Bloomington, pp. 305–326.
†Dark, highly indurated lithic wackes are also referred to as greywackes.
‡Red- or pink-colored feldspathic wackes and arenites can be referred to as arkoses.

solidated carbonate sediment that has been torn up, transported, and redeposited by currents within the basin of deposition. They consist of various types of limestone and can range in size from very fine to pebble or boulder size. *Oolites*, in a size range of 0.1 to 1.0 mm in diameter, are spherical and show radial and concentric structures, and resemble fish roe. They are commonly formed around nuclei such as shell fragments, pellets, or quartz-sand grains. They develop by chemical accretion under the rolling influence of waves in shallow marine environments. *Fossils* of many types are common constituents of limestones. *Pellets* are well-rounded, homogeneous aggregates of microcrystalline calcite in a size range of 0.03 to 0.2 mm in diameter. They are mainly the feces of mollusks, worms, and crustaceans.

Because carbonate rocks tend to be mixtures of (1) the allochemical components noted above, (2) microcrystalline ooze, and (3) sparry calcite cement, Folk (1959) notes that, as a first-order approach, limestones can be classified on the basis of the volumetric abundance of these three types of materials. Such a classification of limestones, which ignores any of the possibly present terrigenous components (e.g., detrital quartz-sand grains), is given in Fig. 14.11. In this figure the *allochems* (intraclasts, oolites, fossils, and pellets) represent the framework material of the rock. The microcrystalline matrix is equivalent to a clay-rich matrix in a poorly washed sandstone. The sparry cement in the filling of pore spaces, as in quartz-rich sandstones, is a chemical precipitate. A rock composed only of microcrystalline limestone is referred to as a *micrite*. Folk (1959) recommends that after limestones have been divided into type I, type II, and type III (see caption of Fig. 14.11) it is essential to note which allochemical components (intraclasts, oolites, fossils, or pellets) predominate. Once this is known, it can be incorporated into a scheme of nomenclature by using parts of the allochem names ("intra" from intraclast; "oo" from oolite; "pel" from pellet; and "bio" for biogenic, in place of fossil) as prefixes. This leads, for example, to biosparite and biomicrite, both consisting of fossil fragments but in different matrix types (*biosparite* = a limestone with more than 10% allochems in a dominantly sparry calcite matrix; *biomicrite* = a limestone with more than 10% allochems set in a micritic matrix). Other such terms are shown at the upper part of the triangle in Fig. 14.11.

There is another commonly used classification of limestones, as outlined by R. J. Dunham (1962), with emphasis on the depositional texture of limestones, instead of Folk's emphasis on the micritic content of

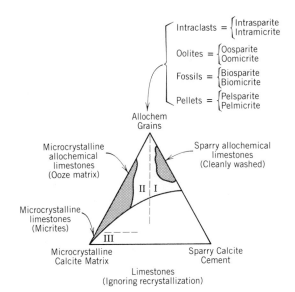

FIG. 14.11. Classification of limestones in terms of volume percentage of allochem grains, microcrystalline calcite matrix, and sparry calcite cement. (After Folk, R. L., 1959, Practical petrographic classification of limestones. *Bulletin of the American Association of Petroleum Geologists*, v. 43, pp. 1–38.) The field of common limestones is shaded. Type I limestones consist of a mixture of allochemical constituents and sparry calcite cement; type II of a variable percentage of allochemical grains in a microcrystalline matrix; and type III of microcrystalline calcite matrix only (referred to as *micrite*).

the rock. Dunham's classification stresses the question of "were the framework grains (Folk's allochems) in close contact with each other, that is, were they well packed when they were deposited?" Because both classifications are almost equally popular, both are combined in a graphic representation in Fig. 14.12. The terms *mudstone, wackestone, packstone, grainstone,* and *boundstone* were introduced by Dunham (1962) so as to reflect, at one extreme, less than 10% grains (mudstone) and, at the other extreme, the predominance of original components that are closely bound together (boundstone). The term *boundstone* is equivalent to *biolithite*, a limestone made up of organic structures that grew *in situ* (in place), forming a coherent rock mass during growth. *Wackestone* reflects more than 10% grains in a microcrystalline ooze (micrite), *packstone* is a grain-supported limestone with micrite matrix and sparry calcite cement, and *grainstone* is a grain-supported limestone with very little micrite, if any. Either of the two schemes of Fig. 14.12 can be used to classify limestones on the basis of careful observation by hand lens in the field, or binocular microscope in the laboratory.

Dolostones are Ca-Mg-rich carbonate rocks consisting mainly of the mineral dolomite. The term *do-*

FIG. 14.12. Classification of carbonate rocks after Folk (1959) and Dunham (1962). The headings at the bottom of the chart represent the classification of R. L. Folk; those at the top of the chart, the classification of R. J. Dunham. (From Suttner, L. J. and Meyers, J., 1991, *Manual for Geologic Study of Northern Rocky Mountains.* Indiana University, Bloomington, pp. 305–326.)

lomite is commonly used for both the rock type and the mineral. Dolostones in limestone-dolostone sequences may show irregular or cross-cutting relationships with the limestones, indicating that the dolostone has formed by replacement of early calcite by later dolomite. Such observations form the basis for the interpretation of the process of *dolomitization*, which involves the replacement of original calcium carbonate in limestone by Ca-Mg-carbonate (dolomite) at any time during or after deposition. Sometimes it is possible with a hand lens to see rhombic outlines of dolomite grains cutting across fossil fragments; however, such observations are best made in thin section with a petrographic microscope. Because of the secondary nature of dolostones and their commonly coarsely recrystallized grain size, little can be learned about their formation in hand specimen.

Orthochemical Sedimentary Rocks

Orthochemical sedimentary rocks are the result of direct precipitation by chemical action in the depositional basin as a result of environmental (e.g., climatic) changes. Three sedimentary rock occurrences that fall into this category are: evaporites, banded iron-formation, and travertine. Bedded cherts may also be of direct chemical origin.

Evaporites. When a restricted body of seawater or the waters of saline lakes evaporate, the elements in solution (see Table 14.11) are precipitated in what are known as evaporites. More than 80 minerals (excluding clastic material) have been recorded in evaporites and most of these are chlorides, sulfates, carbonates, and borates; only about 11 rank as major constituents (see Table 14.11). On evaporation the general sequence of precipitation is: some calcite (when the original volume of seawater is reduced by evaporation to about one-half), gypsum or anhydrite (with the volume reduced to one-fifth of the original),

halite (with the volume reduced to one-tenth of the original), and finally sulfates and chlorides of Mg and K. If all the salt in a 1000 foot (305 m) column of seawater were precipitated, it would form 0.5 feet (0.15 m) of calcium sulfate, 11.8 feet (3.6 m) of NaCl, and 2.6 feet (0.8 m) of K- and Mg-bearing salts, producing a total salt column of 15 feet (4.6 m) thick.

In natural deposits, however, the minerals that precipitate early in the sequence tend to show increased abundance. Hence, gypsum and anhydrite are by far the most abundant evaporate minerals and commonly form massive beds. The deposition of gypsum or anhydrite depends on the temperature and salinity of the brine; anhydrite is formed at higher salt concentrations and at higher temperatures than gypsum. Halite forms about 95% of the chloride minerals in an evaporite sequence; thick and extensive beds of rock salt overlie the gypsum-anhydrite zone of marine evaporites. Deposits of the more soluble salts such as sylvite, carnallite, and polyhalite are rare because they are not deposited until nearly complete dryness has been reached. Nevertheless, in places large deposits (particularly of sylvite) have formed and are mined extensively as the chief source of potassium.

Iron-Formation. Sedimentary, banded iron-formations are most common in rock sequences of late Precambrian (Proterozoic) age. These iron-rich formations, which commonly weather to a dark, reddish-brown color as a result of the formation of red hematite and yellow-brown goethite (and/or limonite), may show well-developed sedimentary stratification (such as banding on a scale of millimeters, centimeters, or meters) as well as oolitic textures. The overall bulk chemistry of these iron-formations reflects the highly unusual chemistry of the original waters from which these sedimentary sequences precipitated. Table 14.12 gives the major oxide components of a sedimentary carbonate-silicate-chert-magnetite iron-

Table 14.11 **MAJOR IONIC CONSTITUENTS OF SEAWATER AND MOST COMMON MINERALS IN EVAPORITE SEQUENCES***	Normal Seawater (Ion Concentration as Parts per million)	Common Minerals in Marine Evaporites
K^+	380	Halite—NaCl
Na^+	10,556	Sylvite—KCl
Ca^{2+}	400	Carnallite—$KMgCl_3 \cdot 6H_2O$
Mg^{2+}	1,272	Anhydrite—$CaSO_4$
Cl^{-1}	18,980	Gypsum—$CaSO_4 \cdot 2H_2O$
SO_4^{-2}	2,649	Langbeinite—$K_2Mg_2(SO_4)_3$
HCO_3^{-1}	140	Polyhalite—$K_2Ca_2Mg(SO_4)_4 \cdot 2H_2O$
Total	34,387	Kieserite—$MgSO_4 \cdot H_2O$
		Calcite—$CaCO_3$
		Magnesite—$MgCO_3$
		Dolomite—$CaMg(CO_3)_2$

*After Stewart, F. H., 1963, *Marine Evaporites*. U.S. Geological Survey Professional Paper no. 440-Y.

Table 14.12

CHEMICAL COMPOSITION OF A CARBONATE-SILICATE-CHERT-MAGNETITE IRON-FORMATION FROM THE LABRADOR TROUGH AREA, CANADA, AND COMMON MINERALS IN IRON FORMATION*

	Weight Percent	Common Minerals
SiO_2	29.6	Chert—SiO_2
Al_2O_3	0.78	Magnetite—Fe_3O_4
Fe_2O_3	11.7	Hematite—Fe_2O_3
FeO	32.0	Siderite—$FeCO_3$
MnO	0.85	Ankerite—$CaFe(CO_3)_2$
MgO	3.62	Calcite—$CaCO_3$
CaO	2.73	Greenalite—$(Fe,Mg)_3Si_2O_5(OH)_4$
Na_2O	0.01	Minnesotaite—$(Fe,Mg)_3Si_4O_{10}(OH)_2$
K_2O	0.25	Stilpnomelane~$K_{0.6}(Mg,Fe)_6Si_8Al(O,OH)_{27} \cdot 2-4H_2O$
H_2O	0.83	Riebeckite—$Na_2Fe_3^{2+}Fe_2^{3+}Si_8O_{22}(OH)_2$†
CO_2	18.1	
Total	100.47	

*From Klein, C. and Fink, R. P., 1976, Petrology of the Sokoman Iron Formation in the Howells River area, at the western edge of the Labrador Trough. *Economic Geology*, v. 71, pp. 453–487.
†Common as crocidolite in Western Australian and South African iron-formations.

formation. Its major components are SiO_2, FeO, Fe_2O_3, MgO, CaO, and CO_2, with only small amounts of Al_2O_3, MnO, Na_2O, and K_2O. When the average chemical compositions of many well-studied iron-formations are recalculated on an H_2O- and CO_2-free basis, the curves in Fig. 14.13 result. The chemical data plotted on this graph represent the bulk compositions of sedimentary (and thus unmetamorphosed) iron-formations as well as metamorphosed iron-formation sequences. During metamorphism of

FIG. 14.13. Plot of the average chemistry of several Precambrian banded iron-formations. The analyses are recalculated to 100% on a CO_2- and H_2O-free basis; *n* refers to the number analyses represented by each set of data; BIF means banded iron formation. (From Klein, C., 1983, Diagenesis and metamorphism of Precambrian banded iron-formations, in *Iron-Formation: Facts and Problems*, A. C. Trendall and R. C. Morris, eds. Elsevier, New York, pp. 417–470.)

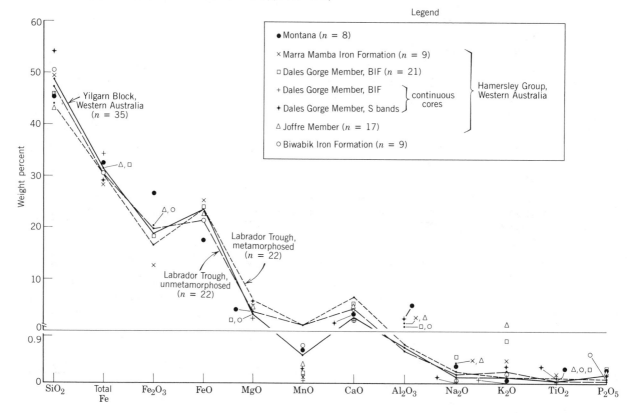

Legend

- ● Montana (*n* = 8)
- × Marra Mamba Iron Formation (*n* = 9)
- □ Dales Gorge Member, BIF (*n* = 21)
- + Dales Gorge Member, BIF
- ✦ Dales Gorge Member, S bands
- △ Joffre Member (*n* = 17)
- ○ Biwabik Iron Formation (*n* = 9)

continuous cores

Hamersley Group, Western Australia

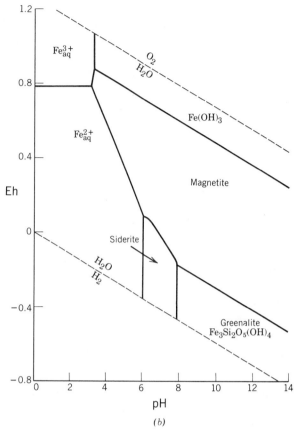

FIG. 14.14. Graphic representation of the mineral compositions in unmetamorphosed iron-formations and of the low-temperature stability fields of some of these minerals. *(a)* Carbonate and silicate minerals in iron-formations as part of the system CaO-MgO-(FeO + MnO). In this representation the CO_2 content of the carbonates and the SiO_2 and H_2O contents of the silicates are ignored (see Fig. 14.17 and pages 582 and 584 for a discussion of this type of representation). *(b)* Stability diagram for some common minerals in iron-formation at 25°C and 1 atmosphere total pressure. The $Fe(OH)_3$ phase is considered the sedimentary precursor to hematite. Compare this figure with Figs. 9.20, 9.21, and 14.22. (*a* and *b* are from Klein, C., 1983, Diagenesis and metamorphism of Precambrian banded iron-formations, in *Iron-Formation: Facts and Problems*, A. F. Trendall and R. C. Morris, eds. Elsevier, New York, pp. 417–470.)

iron-formations CO_2 and H_2O are lost in decarbonation and dehydration reactions. Therefore, in order to compare the chemistry of unmetamorphosed and metamorphosed materials, it is best to compare them without reference to CO_2 and H_2O. Figure 14.13 shows well the unique average chemistry of these sedimentary sequences.

Common minerals in Precambrian iron-formations are listed in Table 14.12 and plotted in Fig. 14.14. Generally, chert, magnetite, and Fe-rich carbonates (siderite and ankerite) are the most common constituents. Hematite and various Fe-rich silicates (greenalite, minnesotaite, and stilpnomelane) can be locally abundant as well. Greenalite, $(Fe,Mg)_3Si_2O_5(OH)_4$, is the Fe-rich analogue of antigorite; minnesotaite, $(Fe,Mg)_3Si_4O_{10}(OH)_2$, is the Fe-rich equivalent of talc; and stilpnomelane,

$K_{0.6}(Mg,Fe)_6Si_8Al(O,OH)_{27}.2-4H_2O$, is a complex layer-silicate structure closely related to that of biotite. Riebeckite, $Na_2Fe_3^{2+}Fe_2^{3+}Si_8O_{22}(OH)_2$, is abundant, in its fibrous variety *crocidolite*, in the Proterozoic, banded iron-formations of Western Australia and South Africa. This sodium-rich amphibole is considered diagenetic in origin in these sedimentary sequences.

Regions of vast Proterozoic iron-formations are found in the Lake Superior region of the United States, the Labrador Trough of Canada, the Hamersley Range of Western Australia, the Transvaal region of South Africa, and in Brazil.

Travertine. This is a calcareous material deposited from spring waters (frequently thermal springs) under atmospheric conditions. If the deposit is porous, it is known as *calcareous tufa*. Such deposits are prevalent in limestone regions where circulating groundwater containing CO_2 has incorporated considerable calcium carbonate in solution. When the groundwater reaches the surface as springs, some of the CO_2 is given off, resulting in the precipitation of some of the calcium carbonate as travertine.

Other Rock Types

Conglomerate. Conglomerates may be considered consolidated gravels. They are composed of coarse clasts, ranging from boulder to granule size, that have been rounded by transportation. The individual clasts may be composed entirely of quartz (e.g., *quartz pebble conglomerate,* or *quartzose conglomerate*), or may be rock fragments that have not been decomposed (e.g., *granite cobble conglomerate,* or *limestone granule conglomerate*). Fine conglomerates grade into coarse sandstones.

Agglomerate. This is a pyroclastic rock containing abundant subangular or rounded fragments (bombs) in a fine-grained volcanic matrix. The rounding of the boulders and pebbles is due to the erosion by running water. Agglomerates are common near volcanic vents. A *volcanic breccia* is of similar composition, but the fragments are more angular.

Sandstone. A cemented or indurated sedimentary rock with grain sizes between 0.062 and 2 mm. The constituent grains are usually rounded and waterworn but may be more or less angular. The detrital particles (clasts) may be quartz, rock fragments, volcanic debris, organic material, or any other clastic material. Sandstones are subdivided texturally into *arenites* and *wackes* on the basis of detrital matrix content (see Fig. 14.10). The term *arenite* or *wacke* is prefixed with a compositional modifier as in *lithic arenite, arkosic arenite, feldspathic wacke,* and *quartz wacke.* The cement that binds the sand grains together may be silica, a carbonate (usually calcite), an iron oxide (hematite or goethite), or clay minerals. The color of the rock depends in large measure on the character of the cement. The rocks with silica or calcite as their binding materials are light in color, usually white to gray, pale yellow, or buff; those that contain an iron oxide are red to reddish-brown. It is to be noted that when a sandstone breaks it is commonly the cement that is fractured, the individual grains remaining unbroken, so that the fresh surfaces of the rock may have a granular appearance and feeling. *Arkose* is an arenite with more than 25% feldspar among the clastic grains. Arkoses are often pink or red because of the large content of K-feldspar, and they may resemble a granite in appearance. A *greywacke* is a type of sandstone with grayish-green color, clay matrix, poor sorting of sand grains, and abundant lithic fragments.

Siltstone, Mudstone, and Shale. These rock names are based on the fine grain size ($<\frac{1}{16}$ mm) of the sedimentary particles. It is not possible in these rock types to describe the composition because of the fine grain size. *Siltstones* lack fissility, that is, they lack ease of splitting, as along bedding planes. *Mudstone* is a nonfissile rock consisting of mud-sized detritus. A *shale* is a fissile sedimentary silicate-rich rock with particles mostly in the silt and clay size. The color of these rock is commonly some tone of gray, although they may be brown, red, or green to black. They are composed chiefly of clay minerals with quartz and mica but are too fine-grained to permit the recognition of their mineral constituents by the eye alone. By the introduction of quartz and an increase in their grain size they grade into greywackes and arenites, and with the presence of calcite they grade into limestones.

Limestone. Limestone composed essentially of calcite, $CaCO_3$, is the most abundant chemically precipitated sedimentary rock. Although the calcite may be precipitated directly from seawater, most limestone is the result of organic precipitation. Many organisms living in the sea extract calcium carbonate from the water to build hard protective shells. On the death of the organisms the hard calcareous parts accumulate on the sea floor. When marine life is abundant, great thicknesses of shells and other hard parts may build up and, when consolidated, become limestone. There are many varieties of limestone. *Chalk* is a soft, micritic limestone composed for the most part of foraminiferal shells. *Coquina* is a loosely cemented fragmental limestone made of fossil shells and fragments; it is found along the coast of Florida. *Lithographic limestone* is a compact, very fine-grained limestone, formerly used in printing. It is predominantly a micrite, with the most famous example occurring in Solenhofen, Germany. A *mudstone* is a mud-supported limestone that contains less than 10% sand- or gravel-sized grains; this is equivalent to a *micrite*. The term *micrite*, for a limestone of microcrystalline calcium carbonate, is strongly preferred over mudstone, which can be confused with terrigenous mudstone. A *wackestone* is a mud-supported limestone with more than 10% sand- or gravel-sized particles. Subdivisions of wackestone (according to R. J. Dunham, 1962, and R. L. Folk, 1959, respectively; see Fig. 14.12) are: pellet wackestone (pelmicrite), biowackestone (biomicrite), oolitic wackestone (oomicrite), and intraclast wackestone (intramicrite). *Packstone* is a clastic limestone in which the grains rest on each other, but some micrite is present. *Grainstone* is carbonate rock in which the grains are in contact with each other and which contains only a minor amount of mud. Grainstones have been divided by R. J. Dunham (1962) into pellet grainstone, biograinstone, oolitic grainstone, and intraclast grainstone (see Fig. 14.12). Equivalent terms (according to R. L. Folk, 1959) are: pelsparite, biosparite, oosparite, and intrasparite. A *boundstone* is a carbonate rock in which the skeletal parts were cemented in place during their formation and remain in their growth position. This is equivalent to *biolithite*, which is typical of reef frameworks (organic structures) in which skeletal elements and fragments are cemented by biogenic carbonate.

Dolostone. This name was introduced to distinguish the rock, dolomite, from the mineral, dolomite. Dolostone resembles limestone so closely in its appearance that it is usually impossible to distinguish the two rock types without a chemical test. Dolostones have generally not formed as original chemical precipitates but are the result of alteration of limestone in which part of the calcium is replaced by magnesium. This process of *dolomitization* is believed to have taken place either by the action of seawater shortly after original deposition, or by the action of circulating groundwater after the rock has been consolidated and raised above sea level.

Siliceous Sinter. In certain volcanic regions hot springs deposit an opaline material known as siliceous sinter or *geyserite*. The deposit apparently results from both evaporation and secretion of silica by algae.

Diatomaceous Earth. This is a light-colored, soft, friable, and porous siliceous deposit made of diatoms. Diatoms are minute one-celled organisms that live in both fresh and seawater and have the power of secreting tests of opaline material. When the organisms die, their tiny shells accumulate to build up a chalk-like deposit of *diatomaceous earth*.

Evaporites, iron-formation, and **travertine** are described on page 577.

METAMORPHIC ROCKS

Metamorphic rocks are derived from preexisting rocks (igneous, sedimentary, or metamorphic) by mineralogical, textural, and structural changes. Such changes may be the result of marked variations in temperature, pressure, and shearing stress at considerable depth in the Earth's crust. Weathering effects, at atmospheric conditions, are not considered part of metamorphism, nor are chemical reactions involving partial melting as these are part of igneous processes. Metamorphic changes such as recrystallization and chemical reactions among mineral constituents take place essentially in the solid state, although the solids may exchange chemical species with small amounts of a liquid phase consisting mainly of H_2O (as water, steam, or supercritical fluid, depending on the tem-

perature and pressure at which the reactions took place). The general conditions of formation of metamorphic rocks lie between those of sedimentary rocks that form at essentially atmospheric conditions of T and P and those of igneous rocks that are the result of crystallization from a melt at high T. Metamorphic rocks may be the result also of very large changes in pressure in conjunction with increasing metamorphic temperature; mineral assemblages in the Earth's mantle have formed in response to very high confining pressures (see page 216). Excluding gain or loss of H_2O, many metamorphic reactions are generally considered to be essentially *isochemical*; this implies that during recrystallization and the processes of chemical reactions the bulk chemistry of rocks has remained essentially constant. If this is not the case, and if additional elements have been introduced into the rock by circulating fluids, for example, it is said to have undergone *metasomastism*.

The most obvious textural feature of most metamorphic rocks (except those of contact metamorphic origin, see below) is the alignment of minerals along planar surfaces. For example, a shale that has undergone only slight metamorphic changes may show well-developed cleavage along planar surfaces, producing a *slate*. With increased temperature of metamorphism recrystallization of originally very fine-grained minerals produces coarser-grained *schists* which exhibit minerals aligned in parallel layers (known as *schistosity*). The coarsest-grained metamorphic rocks that show distinct mineralogical banding are known as *gneisses* (exhibiting *gneissosity*).

In general, metamorphic rocks can be divided into two groups: (1) those formed by contact metamorphism and (2) those formed by regional metamorphism. *Contact metamorphic rocks* occur as concentric zones (*aureoles*) around hot igneous intrusive bodies. Such metamorphic rocks lack schistosity, and the relatively large temperature gradient from the intrusive contact to the unaffected country rock gives rise to zones that may differ greatly in mineral assemblages. Sandstones are converted to *quartzites*, and shales are changed to *hornfels*, a fine-grained dense rock. *Regional metamorphic rocks* are the result of increases in T or P, or both, on a regional scale (areas a few hundred to thousands of miles in extent) in response to mountain building, or deep burial of rocks. Contact and regionally metamorphosed rocks generally reflect an increase in temperature (progressive metamorphism) in their assemblages. However, when assemblages of high-temperature origin (e.g., igneous or high-temperature metamorphic rocks) fail to survive conditions of lower-temperature metamorphism, the process is referred to as *retrograde* (or retrogressive) metamorphism.

Chemical Composition

As metamorphic rocks are the recrystallized and generally foliated (from the Latin *folia*, meaning leaves) products of rocks of igneous or sedimentary bulk compositions, the general range of their chemical compositions is as large as that of igneous and sedimentary rocks combined. Although metamorphic reactions take place without addition of chemical species to the rock, some chemical components may be lost, especially during the higher-temperature ranges of metamorphism. For example, a shale, which is largely composed of hydrous minerals such as clays, may convert into a slate and subsequently into a schist at increasing metamorphic temperatures, during which processes the mineral assemblage becomes less and less hydrous, due to progressive *dehydration*. In other words, with an increase in metamorphic temperatures water is generally lost from the rock. Similarly, carbonate-rich sedimentary rocks while undergoing increasing temperatures of metamorphism will tend to lose CO_2 which is known as *decarbonation*. Such processes can be illustrated as follows:

$$Al_4Si_4O_{10}(OH)_8 \rightarrow Al_4Si_{4-x}O_{10} + xSiO_2 + H_2O \nearrow$$
$$\text{kaolinite} \qquad \text{``meta-kaolinite''}$$

and

$$CaCO_3 + SiO_2 \rightarrow CaSiO_3 + CO_2 \nearrow$$
$$\text{calcite} \quad \text{chert} \quad \text{wollastonite}$$

Although such losses of H_2O and CO_2 to a thin layer of fluid phase surrounding the mineral grains are common during the conditions of progressive metamorphism, additions of chemical species from the fluid phase to the rock are considered part of metasomatic processes.

Mineralogical Composition

The mineralogical composition of a rock that has not undergone metamorphic conditions will generally be very different from that of its metamorphosed equivalent. The extent of the differences is largely controlled by the marked changes of T and P during metamorphism; the relative degrees of metamorphism can be expressed in terms of *grades*, such as very low, low, medium, and high grade (see Fig. 14.15). *Changes in metamorphic grade are reflected in changes in the mineral assemblages of the rocks.* Petrologists, in their description of the mineralogical changes in rocks as a function of increasing metamorphic temperature, commonly represent such changes graphically. Such a graphic approach is shown in Fig. 14.16 for a sedimentary cherty dolo-

mitic limestone that has been subjected to various metamorphic temperature conditions. The cation components of the mineral assemblage in the original sedimentary rock (Fig. 14.16*A*) are correctly accounted for by the components of the corners of the triangle. However, CO_2, although present in the dolomite and calcite composition, is not depicted. Similarly, the cation proportions of any hydrous minerals, as in Fig. 14.16*C* (e.g., talc and tremolite) are accounted for, but (OH) is not shown. This is a relatively common practice in metamorphic petrology, so as to allow for the representation of anhydrous, hydrous, and carbonate minerals on the same composition diagrams. The theoretical justification for this is that H_2O (or OH) and CO_2 are major constituents of the metamorphic fluid which, in most cases, is considered to have been present, and in equilibrium with the various minerals throughout the metamorphic process; such components are often referred to as *perfectly mobile components*.

In graphic terms this means that all CO_2-containing and hydrous minerals in the system CaO-MgO-SiO_2-CO_2-H_2O have been projected from their chemical location in a 4-component system (e.g., CaO-MgO-SiO_2-CO_2 for the carbonates; and CaO-MgO-SiO_2-H_2O for the hydrous silicates) onto a carbonate-free and anhydrous plane, respectively. Indeed, the CaO-MgO-SiO_2-CO_2 or CaO-MgO-SiO_2-H_2O systems can be regarded as tetrahedra in which the apex (above the plane of the page) is the CO_2 or H_2O component. The triangles shown in Fig. 14.16 are the triangular bases of these tetrahedra onto which the carbonates or hydrous cmpositions have been projected.

In going from Fig. 14.16*A* to *B*, the original sediment has undergone very low-grade metamorphic

FIG. 14.15. Schematic *P-T* diagram outlining approximate fields for various metamorphic grades. The shaded area marked "diagenetic conditions" represents the general conditions of lithification of a sediment at low temperature. (After Winkler, H. G. F., 1974, *Petrogenesis of Metamorphic Rocks.* Springer-Verlag, New York, p. 5, with some modifications.)

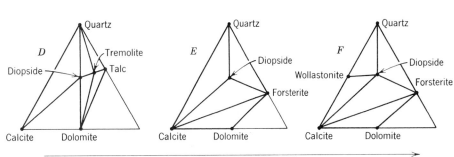

FIG. 14.16. Some of the assemblage changes as a result of increasing temperature in the system CaO-MgO-SiO₂-CO₂-H₂O. Triangle *A* represents the original, sedimentary assemblage in this system (the composition of a naturally occurring cherty, dolomitic limestone is only partly represented in this triangle because it also contains small amounts of FeO). Triangle *F* represents the highest-temperature (contact metamorphic) assemblage, and the other triangles represent intermediate grades of metamorphism.

conditions (T between about 150° and 250°C; see Fig. 14.15) as shown by the formation of talc according to:

$$3CaMg(CO_3)_2 + 4SiO_2 + H_2O \rightarrow$$
$$\text{dolomite} \qquad \text{chert}$$

$$Mg_3Si_4O_{10}(OH)_2 + 3CaCO_3 + 3CO_2\nearrow$$
$$\text{talc} \qquad \text{calcite}$$

At low-grade metamorphic conditions (T between about 250° and 450°C) the assemblage on the right of the above equation may react as follows:

$$2Mg_3Si_4O_{10}(OH)_2 + 3CaCO_3 \rightarrow$$
$$\text{talc} \qquad \text{calcite}$$

$$Ca_2Mg_5Si_8O_{22}(OH)_2 + CaMg(CO_3)_2 + CO_2\nearrow + H_2O\nearrow$$
$$\text{tremolite} \qquad \text{dolomite}$$

The tremolite product is plotted in Fig. 14.16C.

At medium-grade metamorphic conditions (about 450° to 600°C) the tremolite may give way to diopside according to:

$$Ca_2Mg_5Si_8O_{22}(OH)_2 + 3CaCO_3 + 2SiO_2 \rightarrow$$
$$\text{tremolite} \qquad \text{calcite} \qquad \text{quartz}$$

$$5CaMgSi_2O_6 + 3CO_3\nearrow + H_2O\nearrow$$
$$\text{diopside}$$

As a result, diopside is now part of the assemblage, as shown in Fig. 14.16D.

At high-grade metamorphic conditions (above about 600°C) forsterite may form as follows:

$$CaMgSi_2O_6 + 3CaMg(CO_3)_2 \rightarrow$$
$$\text{diopside} \qquad \text{dolomite}$$

$$2Mg_2SiO_4 + 4CaCO_3 + 2CO_2\nearrow$$
$$\text{forsterite} \qquad \text{calcite}$$

The occurrence of forsterite is noted in Fig. 14.16E.

At very high temperatures, such as in the immediate border zone of an igneous intrusive (contact metamorphism) wollastonite may be formed by the reaction:

$$CaCO_3 + SiO_2 \rightarrow CaSiO_3 + CO_2\nearrow$$
$$\text{calcite} \quad \text{quartz} \quad \text{wollastonite}$$

The wollastonite reaction product is shown in Fig. 14.16F.

The above types of reactions can be studied experimentally, providing us with quantitative information in terms of temperature and pressure ranges at which reactants give way to products (see Figs. 9.18 and 9.19). This provides us with reasonably close estimates of the conditions of metamorphism of rocks that are similar in bulk chemistry to those studied in the laboratory.

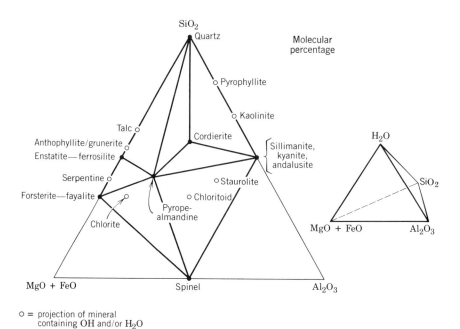

FIG. 14.17. The composition of common diagenetic and metamorphic silicates in the system MgO-FeO-Al$_2$O$_3$-SiO$_2$-H$_2$O. The triangle, which is water-free, can be considered as a projection from an H$_2$O corner in a tetrahedron (see insert). All minerals containing OH and/or H$_2$O are shown by open circles; all anhydrous minerals with opaque dots. With the triangle, tielines are shown which connect the compositions of minerals that commonly form at very high temperatures of metamorphism (between about 600° and 850°C). These tielines represent coexistences (i.e., compatibilities). For the graphic location of mineral compositions in this triangle see Fig. 13.107.

o = projection of mineral containing OH and/or H$_2$O

In order to depict graphically the mineralogical and assemblage changes that take place in rocks of more complex composition than the example shown in Fig. 14.16, the investigator may be forced to combine (lump) some chemical components by which the chemical system is defined. The cation (or cation-oxide) components most commonly combined in graphic representations, are those that substitute for each other in the same cation site in a crystal structure. For this reason (MgO + FeO + MnO), (K$_2$O + Na$_2$O), and (Al$_2$O$_3$ + SiO$_2$) are commonly selected as combined components. The triangular composition diagram in Fig. 14.17 allows for the graphic representation of a relatively large number of common metamorphic silicate compositions as a result of the combination of (MgO + FeO) in the lower left corner; the other corners represent single oxide components. In this diagram hydrous minerals are plotted by ignoring the (OH) or H$_2$O contents in their formulas. This is equivalent to stating that the hydrous mineral compositions have been projected onto the (MgO + FeO)-Al$_2$O$_3$-SiO$_2$ triangular base from an H$_2$O apex in a tetrahedral configuration. Such composition diagrams are very useful in a visual overview of all possible minerals that can occur in a specific chemical system. They may also be used to indicate, in a very general sense, what minerals are found together at a specific metamorphic grade. In Fig. 14.17 tielines connect only those minerals that may be found to coexist at high metamorphic temperatures (over an estimated range of about 600° to 850°C).

A major drawback in this diagram is that binary Fe-Mg series (e.g., Mg$_2$SiO$_4$ to Fe$_2$SiO$_4$) appear only as points instead of line segments (or compositional bars). In order to avoid the need for combining components in triangular composition diagrams, various other graphic projection schemes have been devised in metamorphic petrology.

One projection scheme that is very commonly used in the study of metamorphism of argillaceous (clay-rich) rocks such as shales is due to J. B. Thompson, Jr. (1957). It allows for the graphic representation of the changes that take place among the mineral constituents of argillaceous rocks as a function of increasing temperature. In the succession from shale (clay minerals + chlorite + muscovite + feldspar + quartz) to slate, to staurolite-almandine-biotite-muscovite-quartz schist, and finally to sillimanite-almandine-cordierite-K-feldspar-quartz gneiss (at the highest temperatures of regional metamorphism), many reactions have taken place among the mineral constituents. As in the case of the reactions in carbonate rocks (see Fig. 14.16) the sequence ranges from fine-grained, H$_2$O-rich mineral assemblages at low temperature, to coarse-grained anhydrous assemblages at the highest temperatures (see Table 14.14).

A projection scheme that allows for a graphic display of the changing mineral assemblages in argillaceous rocks is shown in Fig. 14.18. The compositions of the most common minerals in these Al-rich rocks (i.e., clay-rich in the unmetamorphosed, sedimentary assemblage) can be approximated in the chemical

(a)

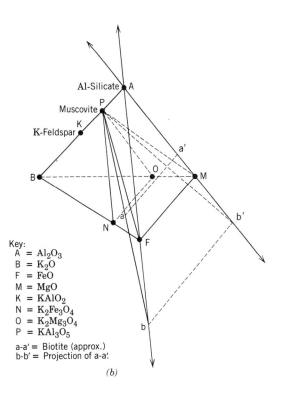

Key:
A = Al_2O_3
B = K_2O
F = FeO
M = MgO
K = $KAlO_2$
N = $K_2Fe_3O_4$
O = $K_2Mg_3O_4$
P = KAl_3O_5

a-a' = Biotite (approx.)
b-b' = Projection of a-a'.

(b)

(c)

FIG. 14.18. *(a)* The system SiO_2-Al_2O_3-MgO-FeO-K_2O-H_2O showing minerals stable with quartz, as plotted in the tetrahedron Al_2O_3-MgO-FeO-K_2O. *(b)* Projection of compositions in the tetrahedron from an idealized muscovite composition, onto the plane Al_2O_3-FeO-MgO. *(c)* The resultant projection on the plane Al_2O_3-FeO-MgO. The extents of solid solution for various silicate minerals are shown schematically. (From Thompson, J. B., Jr., 1957, The graphical analysis of mineral assemblages in pelitic schists. *American Mineralogist*, v. 42, pp. 842–858.)

system: SiO_2-Al_2O_3-MgO-FeO-K_2O-H_2O. In order to portray this six-oxide component system graphically, two oxide components, at least, will need to be eliminated (or "ignored"). If the investigator restricts the graphic analysis to mineral assemblages *that always contain quartz*, the SiO_2 component can be ignored. That is so because any increase or decrease in the amount of SiO_2 in a quartz-containing rock will be reflected simply in an increase or decrease of the amount of quartz in the assemblage; such a change in SiO_2 content does not affect the compositions of the other silicates present. The other component that is commonly "ignored" in a graphic representation of this kind is H_2O. It is well known that argillaceous

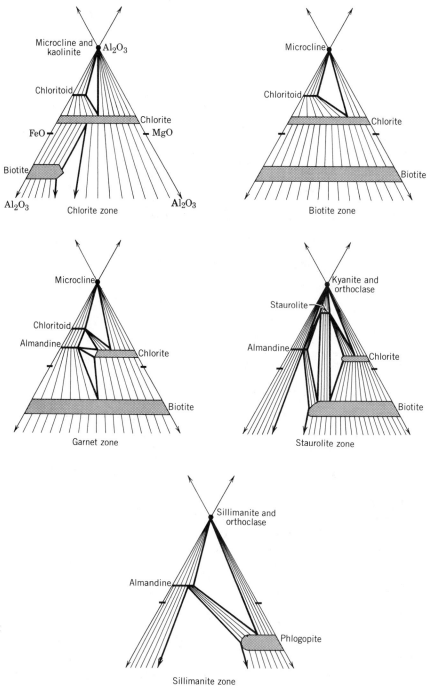

FIG. 14.19. Schematic projections of assemblages with muscovite and quartz in the system SiO_2-Al_2O_3-MgO-FeO-K_2O-H_2O for various metamorphic zones. Tielines and bundles of tielines connect possible coexisting minerals.

sediments lose H_2O during increasing grades of metamorphism. In other words, water is lost as a function of increasing temperature; however, components such as Al_2O_3, MgO, FeO, and K_2O are not lost. It is, therefore, commonly stated that the chemical system is "open" to H_2O but "closed" to the relative amounts of Al_2O_3, MgO, FeO, and K_2O. H_2O is considered a "mobile" component, an externally controlled component. For further discussion of this see J. B. Thomp-

son (1957). The resulting tetrahedral space for the system Al_2O_3-MgO-FeO-K_2O is shown in Fig. 14.18a. The silicates and their compositional ranges, as can be approximated in this tetrahedron, are shown. It should be noted that this tetrahedron contains many of the minerals depicted in Fig. 14.17; however, K-containing minerals such as muscovite, biotite, and K-feldspar can be shown as well. Because muscovite is present in almost all metamorphosed argillaceous assemblages, J. B. Thompson recommends projection of all of the mineral compositions in the tetrahedron from the point of muscovite composition (KAl_3O_5 in Fig. 14.18a). If one imagines one's eye at this compositional point (marked P in Fig. 14.18b), then all other minerals in the tetrahedron can be viewed as projected from that point onto the Al_2O_3-MgO-FeO plane on the right of the tetrahedron. This type of projection is specifically shown for the biotite compositional range (marked as a-a') with the resultant projected composition on the lower extension of the triangle, at position b-b'. It can be said that all mineral compositions in the volume A-F-M-N-O-P (Fig. 14.18b) project in the *positive* sense onto the Al_2O_3-FeO-MgO plane, and that all compositions in the volume B-N-O-P project in the *negative* sense onto that same plane. As such, the feldspar composition is projected in the negative sense onto the Al_2O_3 corner. Figure 14.18c shows the resultant projected mineral compositions and their extents of solid solution. This type of projection is especially useful in the graphic representation of mineral compositions, and the coexistences of minerals, as a function of increasing temperature in argillaceous rock types. Figure 14.19 gives examples of the common mineral compositions and coexistences in various metamorphic zones as a result of the regional metamorphism of argillaceous rocks. The names as applied to the various metamorphic zones in Fig. 14.19 are explained below. It should be noted that all of the assemblages shown contain quartz and muscovite, as a result of the projection scheme outlined above. The various assemblage configurations in Fig. 14.19 are the result of generally complex reactions between the various silicates. Examples are the appearance of almandine and staurolite, and the disappearance of chloritoid, chlorite, and staurolite. The reaction from sillimanite to kyanite is a polymorphic reaction in the system Al_2SiO_5.

In a systematic study of shales and their metamorphic equivalents in the Scottish Highlands, published in 1912, G. Barrow delineated various metamorphic zones on the basis of the occurrence of *index minerals.*[1] At successively higher grades of metamorphism, he noted that the argillaceous rocks showed the development of the following index minerals: first chlorite, then biotite, next almandine, sub-

sequently staurolite, then kyanite, and at the highest temperatures sillimanite. Similar progressions of minerals are found in other metamorphic regions containing argillaceous rocks such as in the New England region. When areas of specific index mineral occurrence are outlined on a geological map, such as a chlorite-rich region or a biotite-rich region, the line that marks the first appearance of an index mineral is known as an *isograd*. The isograds reflect positions of similar metamorphic grade in terms of T and P. Isograds are labeled with the name of the appropriate index mineral, such as biotite isograd, garnet isograd, staurolite isograd, and so on.

In regions where very high pressures have been generated during metamorphism the final mineral assemblages will be very different, for a specific bulk rock composition, from those where high temperatures have been most prevalent. Argillaceous rocks metamorphosed at high temperature commonly contain sillimanite, whereas those subjected to high pressures (with some increase in temperature, see Fig. 9.19) contain kyanite. Similarly, high-temperature basalt assemblages occur as coarse-grained eclogites (see Fig. 9.17) in the high-pressure regions of the lower crust of the Earth.

In addition to the subdivision of P and T fields in terms of very low, low, medium, and high grades of metamorphism (Fig. 14.15) metamorphic rocks are often classified in terms of metamorphic facies. A *metamorphic facies* is *a set of metamorphic mineral assemblages, repeatedly associated in space and time, such that there is a constant and therefore predictable relation between mineral composition and chemical composition.* A metamorphic facies, therefore, is defined not in terms of a single index mineral but by an association of mineral assemblages; some examples of metamorphic facies follow. *Zeolite facies* represents the lowest grade of metamorphism. The mineral assemblages include zeolites, chlorite, muscovite, and quartz. *Greenschist facies* is the low-grade metamorphic facies of many regionally metamorphosed terranes. The mineral assemblages may include chlorite, epidote, muscovite, albite, and quartz. *Amphibolite facies* occurs in medium- to high-grade metamorphic terranes. The mineral constituents include hornblende, plagioclase, and almandine. This facies occurs where staurolite and sillimanite-grade metamorphic conditions have prevailed. *Glaucophane-lawsonite schist* (or *blueschist*) *facies* is represented by relatively low temperatures but elevated pressures of metamorphism in young orogenic zones, such as California and Japan. Characteristic constituents are lawsonite, jadeite, albite, glaucophane, muscovite, and garnet. *Granulite facies* reflects the maximum temper-

[1]An index mineral is a mineral that developed under a particular set of temperature and pressure conditions, thus characterizing a particular degree of metamorphism. When dealing with progressive (or prograde) metamorphism, it is a mineral whose first appearance (in going from low to higher grades of metamorphism) marks the outer limit of the lower grade.

ature conditions of regional metamorphism such as have been commonly attained in Archean terranes. Characteristic mineral constituents are plagioclase, orthopyroxene, garnet, and diopside. *Eclogite facies* represents the most deep-seated conditions of metamorphism. Characteristic mineral constituents are pyrope-rich garnet and omphacite. Such assemblages are common in *kimberlite* pipes, many of which carry associated diamond. A diagram outlining the approximate fields of the various metamorphic facies in terms of *P* and *T* is given in Fig. 14.20.

Minerals that are especially common in metamorphic rocks are listed in Table 14.13. Examples of mineral assemblages in carbonate-rich and argillaceous rocks as a function of increasing metamorphic grade are given in Table 14.14.

Rock Types
Some of the most common metamorphic rocks are briefly described below.

Slate. Slates are exceedingly fine-grained rocks that have a remarkable property known as *slaty cleavage* which permits them to be split into thin, broad sheets. Their color is commonly gray to black but may be green, yellow, brown, and red. Slates are usually the result of the metamorphism of shales. Their characteristic slaty cleavage may or may not be parallel to the bedding planes of the original shales. Slate is rather common in occurrence.

Marble. A marble is a metamorphosed limestone. It is a crystalline rock composed of grains of calcite or, more rarely, dolomite. The individual grains may be so small that they cannot be distinguished by the eye, and again they may be coarse and show clearly the characteristic calcite cleavage. Like limestone, a marble is characterized by its softness and its effervescence with acids. When pure, marble is white, but various impurities may create a wide range of color. It is a rock that is found in many localities and may be in thick and extensive beds. Commercially *marble* is used to indicate any Ca-carbonate rock capable of taking a polish and thus includes some limestones.

Schist. Schists are metamorphic rocks that are distinguished by the presence of well-developed foliation or schistosity, along which the rock may be easily broken. A common example is a *mica schist*, which consists essentially of quartz and mica, usually muscovite or biotite, or both. Mica is the prominent mineral, occurring in irregular leaves and in foliated masses. The mica plates all lie with their cleavage planes parallel to each other and give to the rock a striking laminated appearance. Mica schists frequently carry characteristic accessory minerals, such as garnet, staurolite, kyanite, sillimanite, andalusite, epidote, and hornblende; thus the rock may be called *garnet-mica schist* or *staurolite-mica schist*. Other varieties of schists that are derived chiefly by the metamorphism of igneous rocks rich in ferromagnesian minerals are *talc schist, chlorite schist, hornblende schist,* and *amphibolite*. They are characterized, as their names indicate, by the preponderance of some metamorphic ferromagnesian mineral.

Gneiss. When the word *gneiss* is used alone it

FIG. 14.20. Tentative scheme of metamorphic facies in relation to pressure (P_{H_2O}) and temperature. All boundaries are gradational. (Modified from Turner, F. J., 1968, *Metamorphic Petrology*. McGraw-Hill Book Co., New York, p. 366. Copyright 1968 by McGraw-Hill, Inc. Used with permission of McGraw-Hill Book Co.) Compare with Fig. 14.15.

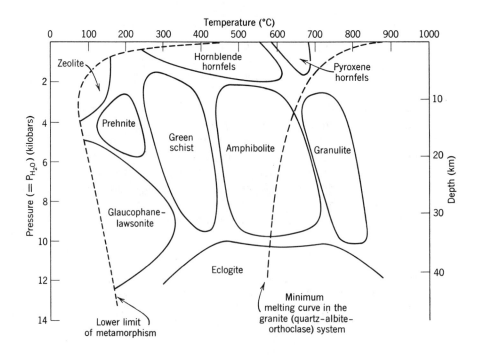

Table 14.13

COMMON SILICATES OF METAMORPHIC ROCKS

Phyllosilicates	talc serpentine chlorite muscovite biotite
Inosilicates	anthophyllite cummingtonite-grunerite tremolite-actinolite hornblende glaucophane clinopyroxene with jadeite component (high pressure) diopside orthopyroxene wollastonite*
Tectosilicates	quartz plagioclase, except for very An-rich compositions microline and orthoclase
Nesosilicates	garnet (pyrope at high pressures) epidote kyanite, sillimanite, andalusite vesuvianite* forsterite staurolite chloritoid

*Especially common in contact metamorphic rocks.

refers to a coarsely foliated metamorphic rock. The banding is caused by the segregation of quartz and feldspar into layers alternating with layers of dark minerals. Because the metamorphism of many igneous or sedimentary rocks may result in a gneiss, there are many varieties, with varied mineral associations. Thus they are given such names as *plagioclase-biotite gneiss, hornblende gneiss,* or *pyroxene-garnet gneiss.* When it is certain that a gneiss is the result of metamorphism of an earlier-formed igneous rock, the igneous rock name is used in the metamorphic terminology, such as *granite gneiss* or *syenite gneiss.* Granite gneisses, that is, rocks derived from the metamorphism of granites, are common, especially in Archean terranes.

Quartzite. As its name indicates, a quartzite is a rock composed essentially of quartz. It has been derived from a quartz-rich sandstone by high-grade metamorphism. It is a common and widely distributed rock in which solution and redeposition of silica have yielded a compact rock of interlocking quartz grains. It is distinguished from an unmetamorphosed quartz arenite by noting the fracture, which in a quartzite passes through the grains but in a sandstone passes around them.

Serpentinite. A serpentinite is a rock composed essentially of the mineral serpentine, derived by retrograde metamorphism from high-temperature igneous rocks such as dunite or peridotite. The primary forsteritic olivine gives way, in a lower temperature (ret-

Table 14.14

EXAMPLES OF METAMORPHIC MINERAL ASSEMBLAGES PRODUCED DURING PROGRADE METAMORPHISM IN CARBONATE-RICH ROCKS AND ARGILLACEOUS SHALE

	Ca carbonate-rich rock (see Fig. 14.16)	Argillaceous shale
Very low grade	calcite-dolomite-talc and calcite-quartz-talc	muscovite-chlorite-quartz-feldspar
Low grade	calcite-dolomite-tremolite and calcite-tremolite-quartz	biotite-muscovite-chlorite-quartz-feldspar
Medium grade	calcite-dolomite-diopside and calcite-diopside-quartz	staurolite-garnet-biotite-muscovite-quartz-feldspar
High grade	calcite-dolomite-forsterite and calcite-diopside-quartz	sillimanite-garnet-biotite-muscovite-quartz-feldspar

rograde) and hydrous metamorphic environment, to serpentine and brucite according to:

$$4Mg_2SiO_4 + 6H_2O \rightarrow Mg_6Si_4O_{10}(OH)_8 + 2Mg(OH)_2$$

forsterite $\qquad$ serpentine $\qquad$ brucite

Serpentinites are compact, of a green to greenish-yellow color, and may have a slightly greasy feel. Serpentinites may be a source for associated chromite and platinum, as in the Ural Mountains, CIS, or a source of nickel from associated garnierite, as in New Caledonia.

Soapstone. A metamorphic rock with a massive or schistose texture and soft feel, composed essentially of fine-grained talc. It is generally the metamorphic alteration product of an originally ferromagnesian silicate-rich rock.

Amphibolite. An unfoliated or somewhat foliated metamorphic rock composed essentially of amphibole and plagioclase. Some varieties may consist almost completely of hornblende. Mineral qualifiers are used as prefixes to indicate the presence of specific minerals, for example, *garnet-amphibolite, biotite-amphibolite*. Amphibolites may be the metamorphic products of basic igneous rocks or of argillaceous carbonate rocks.

Greenschist. A well-foliated greenish rock consisting largely of chlorite, actinolite (or hornblende), epidote, and albite. *Greenstone* is a field term applied to somewhat metamorphosed basaltic or doleritic rocks with a characteristic green color caused by the presence of chlorite, hornblende, and epidote.

Granulite. A high-grade, relatively coarse-grained metamorphic rock with a strong layering due to the occurrence of well-oriented, flattened lenses of quartz and feldspar. Other major constituents may be pyroxene, garnet, kyanite, or sillimanite. It generally contains little or no mica, and lacks prismatic minerals. The term *granulite* can be prefaced by a mineral name, or a series of mineral names, such as *garnet-granulite* or *garnet-kyanite-granulite*. Similarly, the mineralogical and chemical composition of the rock may be expressed by the use of a prefix, for example, *syenite-granulite*.

Eclogite. A dense, granulose rock which has resulted from high-pressure and high-temperature metamorphism. It is composed chiefly of omphacite (a pyroxene composition along the diopside-jadeite series) and pyrope-rich garnet. It has the bulk composition of a gabbro (or basalt; see Fig. 9.17). Other minerals that may be present are: quartz, kyanite, enstatite, olivine, and rutile. Commonly mineral names are used as qualifiers, as in *kyanite-eclogite*.

Skarn Rock. A contact metamorphic and meta-somatic rock lying within the aureole of an intrusive igneous body. It is the result of considerable metasomatic replacement of the original rock type, which commonly is a limestone. Mineral prefixes are used to indicate the dominant mineral, for example, *andradite skarn*.

Hornfels. An old German name for a compact, horny-looking rock. Its use is now restricted to a fairly massive, contact-metamorphosed rock. It is commonly made up of quartz, mica, feldspar, garnet, andalusite and/or cordierite, and minor amounts of amphibole and pyroxene. The name may be modified by noting the most important constituents, for example, *cordierite-andalusite-hornfels*.

VEINS AND VEIN MINERALS

Although ore and mineral deposits can be of igneous, metamorphic, or sedimentary origin a large number of mineral deposits exist as tabular or lenticular bodies known as *veins*. The veins have been formed by the filling with minerals of a preexisting fracture or fissure.

The shape and general physical character of a vein depends on the type of fissure in which its minerals have been deposited, and the type of fissure in turn depends on the character of the rock in which it occurs. In a firm, homogeneous rock, like a granite, fissures may be fairly regular and clean-cut. They are likely to be comparatively narrow with respect to their horizontal and vertical extent and reasonably straight in their course. On the other hand, if a rock that is easily fractured and splintered, like a slate or a schist, is subjected to a breaking stress, a zone of narrow and interlacing fissures is more likely to have formed than one straight crack. In an easily soluble rock like a limestone, a fissure will often be extremely irregular, owing to the differential solution of its walls by the solutions that have moved through it.

A typical vein consists of minerals that have filled a fissure solidly from wall to wall and shows sharply defined boundaries. There are, however, many variations from this type. Frequently, irregular openings termed *vugs* may occur along the center of the vein. It is from these vugs that many well-crystallized mineral specimens are obtained. Again, the walls of a vein may not be sharply defined. The mineralizing solutions that filled the fissure may have acted on the wall rocks and partially replaced them with the vein minerals. Consequently, there may be an almost complete gradation from the unaltered rock to the vein filling with no sharp line of division between. Some deposits have been largely formed by the deposition of vein minerals in the surrounding rocks (wall rocks) and are known as *replacement deposits*. There is every gradation possible, from a vein formed by filling of an open fissure with sharply defined walls to a replacement deposit with indefinite boundaries.

Vein mineralization is generally the result of what is known as deposition from *hydrothermal solutions*. This term refers to heated or hot magmatic emanations rich in water as well as heated aqueous solutions that have no demonstrable magmatic affiliations. An example of relatively high-temperature vein mineralization in and adjacent to granitic intrusions occurs in Cornwall, England. The mineral assemblage in such veins consists of quartz-mica-tourmaline-wolframite-stannite-cassiterite-molybdenite. In the Tintic district, Utah, veins peripheral to granitic intrusions in limestone consist of pyrite, enargite, tetrahedrite, and galena. In many mining districts, however, there is no apparent relationship between the ore veins and possible igneous activity.

Associated with the economically useful minerals (*ore minerals*) are minerals of no commercial value (*gangue*). Careful assemblage studies, involving both ore and gangue mineralizations, together with the study of very small inclusions of remaining hydrothermal fluid (*fluid inclusions*) in mineral grains allow for the division of hydrothermal ore deposits in terms of temperature of origin: low (50°–150°C), intermediate (150°–400°C), and high temperature (400°–600°C).

The *primary* or *hypogene* ore minerals may alter near the surface to *secondary* or *supergene* minerals. The primary sulfide minerals such as pyrite, chalcopyrite, galena, and sphalerite are particularly subject to such alteration. Under the influences of low-

FIG. 14.22. Stability relations of some copper minerals commonly found in supergene enrichment deposits in the system Cu-H_2O-O_2-S-CO_2 at 25°C and 1 atmosphere total pressure. (Adapted from Anderson, J., in Garrels, R. M. and Christ, C. L., 1965, *Solutions, Minerals, and Equilibria*, Freeman, Cooper, and Company, San Francisco, Cal., Fig. 7.27*b* on p. 240; see same book for construction of such diagrams.) Compare with Figs. 9.20, 9.21 and 14.14*b*.

temperature, oxygenated meteoric waters, sphalerite alters to hemimorphite and smithsonite; galena to anglesite and cerussite; copper sulfides such as chalcopyrite and bornite to covellite, chalcocite, native copper, cuprite, malachite, and azurite. The circulating meteoric waters dissolve primary minerals at near-surface conditions, producing a barren and leached zone (see Fig. 14.21), and frequently redeposit part of the soluble species below and close to the groundwater table, producing *secondary enrichment zones*. This process of *supergene enrichment* is especially important in ore formation because metals leached from the oxidized upper parts of mineral deposits may be redeposited at depth. Because all such chemical processes take place at essentially atmospheric conditions (about 25°C and 1 atmosphere pressure) Eh-pH stability diagrams for appropriate chemical systems can be applied to the assemblages in supergene deposits. Figure 14.22 shows the stability fields of malachite, cuprite, native copper, chalcocite, and covellite. Malachite and cuprite stability fields occupy the high Eh

FIG. 14.21. Diagrammatic representation of weathering and enrichment of a vein rich in primary Cu sulfides.

(highly oxygenated) part of the diagram; covellite and chalcocite are found in more reducing conditions, directly below the water table.

Because pyrite is generally a very abundant sulfide in veins as well as other mineral deposits, it reacts, upon oxidation, to form insoluble compounds such as goethite or limonite, which remain at the surface. The upper portion of a vein is thus frequently converted to a cellular rusty mass of goethite called *gossan* (see Fig. 14.21). The yellow-brown, rusty appearance of such outcrops is so characteristic as to be a guide to ore.

REFERENCES AND SUGGESTED READING

Best, M. G., 1982, *Igneous and Metamorphic Petrology*. W.H. Freeman and Co., San Francisco, 630 pp.

Blatt, H., 1992, *Sedimentary Petrology*, 2nd ed. W.H. Freeman and Co., San Francisco, 514 pp.

Blatt, H., Middleton, G. and Murray, R., 1980, *Origin of Sedimentary Rocks*, 2nd ed. Prentice Hall, Englewood Cliffs, N.J., 782 pp.

Boggs, S., 1992, *Petrology of Sedimentary Rocks*. Macmillan Publishing Co., New York, 707 pp.

Burns, R. G., ed., 1979, *Marine Minerals. Reviews in Mineralogy*, v. 6, 380 pp.

Dietrich, R. V. and Skinner, B. J., 1979, *Rocks and Rock Minerals*. John Wiley & Sons, New York, 319 pp.

Dunham, R. J., 1962, Classification of carbonate rocks according to depositional texture. *American Association of Petroleum Geologists Memoir*, no. 1, pp. 108–121.

Ehlers, E. G. and Blatt, H., 1980, *Petrology: Igneous, Sedimentary and Metamorphic*. W.H. Freeman and Co., San Francisco, 732 pp.

Folk, R. L., 1959, Practical petrographic classification of limestones. *American Association of Petroleum Geologists Bulletin*, no. 43, pp. 1–38.

Fry, N., 1983, *The Field Description of Metamorphic Rocks*. John Wiley & Sons, New York, 170 pp.

Granitic Pegmatites, 1992, Martin R.F., and Černý, P., eds., *Canadian Mineralogist*, v. 30, pp. 497–954.

Guilbert, J. M. and Park, C. F., Jr., 1986, *The Geology of Ore Deposits*. Freeman and Co., New York, 984 pp.

Hess, P. C., 1989, *Origins of Igneous Rocks*. Harvard University Press, Cambridge, Mass., 336 pp.

Hyndman, D. W., 1972, *Petrology of Igneous and Metamorphic Rocks*. McGraw-Hill Book Co., New York, 533 pp.

Jahns, R. H., 1952, The study of pegmatites. *Economic Geology*, 50th anniversary volume, pp. 1025–1130.

Klein, C., 1989, *Minerals and Rocks: Exercises in Mineralogy, Petrology, and Hand Specimen Petrology*. John Wiley & Sons, New York, 402 pp.

Mason, B. and Moore, C. B., 1982, *Principles of Geochemistry*, 4th ed. John Wiley & Sons, New York, 344 pp.

Philpotts, A. R., 1990, *Principles of Igneous and Metamorphic Petrology*, Prentice Hall, Englewood Cliffs, N.J., 498 pp.

Skinner, B. J. and Barton, P. B., Jr., 1973, Genesis of mineral deposits. *Annual Review of Earth and Planetary Sciences*, v. 1, pp. 183–211.

Stewart, F. H., 1963, *Marine Evaporites*. U.S. Geological Survey Professional Paper no. 440-Y, 53 pp.

Thompson, J. B., Jr., 1957, The graphical analysis of mineral assemblages in pelitic schists. *American Mineralogist*, v. 42, pp. 842–858.

Tucker, M. E., 1982, *The Field Description of Sedimentary Rocks*. John Wiley & Sons, New York, 112 pp.

Turner, F. J., 1980, *Metamorphic Petrology*, 2nd ed. McGraw-Hill Book Co., New York, 512 pp.

Winkler, H. G. F., 1979, *Petrogenesis of Metamorphic Rocks*, 5th ed. Springer-Verlag, New York, 348 pp.

Yardley, B. W. D., 1989, *An Introduction to Metamorphic Petrology*, John Wiley & Sons, New York, 248 pp.

CHAPTER 15

GEM MINERALS

Because most gems are minerals, it is fitting to devote a brief section to gemology, the science of gemstones, in a textbook of mineralogy. In fact, the use of the term *gem* is legally restricted in the American gem trade to refer to stones of natural origin. A gem is defined as follows:

a gem is a mineral which, by cutting and polishing, possesses sufficient beauty to be used in jewelry or for personal adornment.

Although beauty is the prime requirement, the more desirable and *more valuable gems are both rare and durable.* Included in the term gem are several organic gem materials, pearl, amber, coral, and jet, which, although products of nature, are not strictly minerals. They will, therefore, be omitted from our discussion. The reason for the restricted use of the term *gem* is to exclude all manufactured simulants; but in spite of the legislation, they are frequently referred to as *synthetic gems*. In a study of gems these synthetics cannot be ignored, for one of the major tasks of the gemologist, and frequently a difficult one, is to determine whether a stone is natural or synthetic. Such determinations are important because the rare natural gem may have a value several hundred times greater than a similar-appearing synthetic substitute. It is also important to know whether the color of a gemstone is natural or has been artificially induced. Thus, in the latter part of this chapter we will discuss synthetics briefly, how they are manufactured, how they can be distinguished from the gems they simulate, and the manner in which the color of gems can be enhanced by treatment.

GEM MINERALS

Because a mineral must have certain qualifications to be placed in the special category of gem minerals, the number is limited. Of the approximately 3500 known mineral species, approximately 70 meet the requirements, and of these about 15 can be considered as important gem minerals.

Chapters 10 to 13 on "Systematic Mineralogy" include all the minerals commonly used as gems as well as many others less frequently encountered as cut stones. In the following list these minerals are arranged by chemical groups, the order in which they are described on earlier pages, with the names of the more important gem minerals in boldface type. However, it should be pointed out that the list is not all-inclusive and that other rarer minerals occasionally appear as gemstones. The gemologist is frequently called upon to identify the minerals from which ornamental objects have been carved. Therefore, included in Table 15.1 are some more commonly used for carvings than for cut stones.

Table 15.1
GEM MINERALS*

NATIVE ELEMENTS	TUNGSTATES	SILICATES (continued)
Diamond	Scheelite	Axinite
SULFIDES	PHOSPHATES	**Beryl**
Sphalerite	Beryllonite	Cordierite
Pyrite	Apatite	**Tourmaline**
OXIDES	Amblygonite	Enstatite-
Zincite	Brazilianite	hypersthene
Corundum	**Turquoise**	Diopside
Hematite	Variscite	**Jadeite** (jade)
Rutile	SILICATES	Spodumene
Anatase	Phenacite	Rhodonite
Cassiterite	Willemite	**Tremolite-**
Spinel	**Olivine**	**actinolite**
Gahnite	**Garnet**	(nephrite jade)
Chrysoberyl	**Zircon**	Serpentine
HALIDES	Euclase	Talc
Fluorite	Andalusite	Prehnite
CARBONATES	Sillimanite	Chrysocolla
Calcite	Kyanite	Dioptase
Rhodochrosite	**Topaz**	**Quartz**
Smithsonite	Staurolite	**Opal**
Aragonite	Datolite	Feldspar
Malachite	Titanite	Danburite
Azurite	Benitoite	Sodalite
SULFATES	Zoisite	Lazurite
Gypsum	Epidote	Petalite
	Vesuvianite	Scapolite
		Thomsonite

*Colored photographs of many of these minerals and gems cut from them are given in Plates I–IV.

GEM QUALIFICATIONS

Of the several attributes of gemstones, beauty is most important. Factors contributing to beauty are color, luster, transparency, and, through skillful cutting, brilliance and fire. Most gems possess two or more of these qualities, but in some nontransparent stones, such as turquoise, beauty lies in color alone. Opal, one of the most beautiful gems, owes its attractiveness to flashes of spectral colors diffracted from the stone's interior, a phenomenon known as play of color.

Because most gemstones are used for personal adornment, they should be able to resist scratching and abrasion, which would dull their luster and mar their beauty. Thus durability, which depends on hardness and toughness, is the second requisite of a gem mineral. It is generally considered that a gemstone hard enough to resist abrasion should have a minimum hardness of 7, that of quartz. Only 10 or 12 gems satisfy this requirement, but others fashioned from softer minerals will retain their luster for many years if worn with care. This does not mean that care should not be exercised in wearing hard gemstones, for some

are brittle and a sharp blow may cause them to fracture or cleave. A few minerals with a hardness less than 7 may be extremely tough. This is true of nephrite jade. A blow that would shatter a diamond would have little effect on the nephrite.

In addition to the two intrinsic properties of beauty and durability, other factors influence the desirability of a gem. Chief of these is *rarity*, long an attribute of the highly prized gems—diamond, emerald, ruby, and sapphire. As the supply diminishes and they become rarer, their value increases. If, on the other hand, a gem through new discoveries becomes abundant, it decreases in value and loses its appeal to the few that formerly could afford it. The desirability of gems is also subject to the vagaries of *fashion*. For example, the dark red pyrope garnet was much in vogue during the nineteenth century but is in little demand today. A final factor that can be added to the desirable properties of gems is *portability*. Gemstones are unique in combining high value with small volume and weight and can be transported easily from one country to another during periods of political unrest and economic instability.

TYPES OF GEM CUTS

Space does not permit an account of the manner in which gem rough is fashioned into gemstones. Yet it is important to point out the two basic types of cuts because mention is frequently made of them on the following pages. There are the cabochon cut and the faceted cut. The cabochon has a smooth-domed top and most commonly a flat base (Fig. 15.1). Faceted gemstones are bounded by polished plane surfaces (facets) to which different names are given depending on their position. When viewed from above, the outline of faceted stones may be round, oval, rectangular, square, or other shapes. The brilliant cut, most common for diamond, is shown in Fig. 15.2. The upper part of the stone is called the *bezel, crown*, or *top*; the lower part is called the *pavilion, base*, or *back*. The central top facet is the *table* and the small facet parallel to it is the *culet*. The edge between bezel and pavilion is the *girdle*.

FIG. 15.1. Cross sections of cabochon cut stones. *(a)* Simple cabochon with domed top and flat base. *(b)* Double cabochon. *(c)* Hollow cabochon.

| (a) | (b) | (c) |

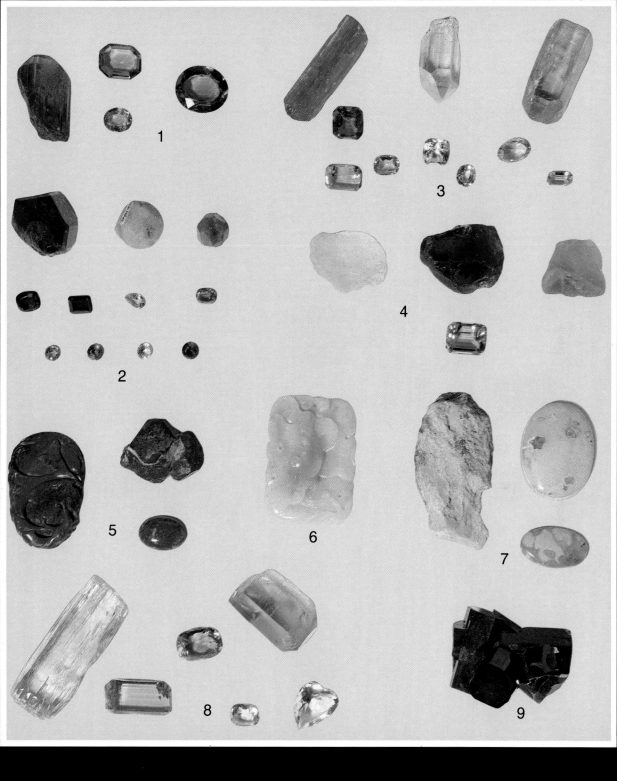

Plate III.
1. Olivine (peridot).
2. Garnet.
3. Topaz, natural colors.
4. Topaz. From left to right: colorless stream pebble;
colored by irradiation; heat treated after irradiation;
below: gem cut from treated material.

5. Lapis lazuli.
6. Jadeite.
7. Turquoise.
8. Beryl.
9. Synthetic emerald crystals.

(Harvard Mineralogical Museum.)

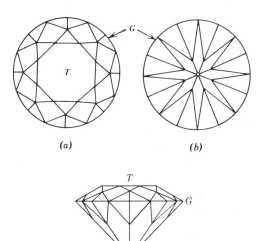

FIG. 15.2. A faceted gemstone (brilliant cut) with 58 facets. *(a)* The bezel showing the top facets. *(b)* The pavilion showing the bottom facets. *(c)* Front view showing the table (T), culet (C), and girdle (G).

THE EARLY USES OF GEMS

Although gemology as a science is modern, interest in gems is ancient and extends back beyond the dawn of history. Archaeological finds present clear evidence that our remote ancestors collected and treasured gem minerals as objects of beauty. By the beginning of history, numerous gems were not only known but were shaped to enhance their beauty and even pierced, permitting them to be strung together and worn as necklaces or bracelets. Four thousand years before the Christian Era, gems of lapis lazuli, chalcedony, amazonite, and jasper were cut and carved in Babylonia. The presence of lapis lazuli, undoubtedly from Afghanistan, shows that gem trade routes existed at that time. Later trade extended as far as Egypt, for lapis lazuli has been found in Egyptian tombs of the Twelfth Dynasty erected 4400 years ago. But far earlier in the predynastic period (5000–3000 B.C.) ornaments were fashioned in Egypt from a large number of minerals. They included several varieties of quartz—rock crystal, chalcedony, agate, carnelian, chrysoprase, and jasper—as well as turquoise, chrysocolla, amazonite, olivine, fluorite, and malachite. Emerald was also a gem well known to the early Egyptians. But unlike many of the other gems found in Egyptian tombs, whose source is unknown, emeralds were obtained from the Zabara Mountains in upper Egypt near the shore of the Red Sea. Even today, one can see abundant evidence of these ancient mining activities.

By the time of the ancient Greek and Roman civilizations, many of the gem minerals were known and the art of carving them was well advanced. The people of these civilizations valued gems as things of beauty and used them, as they are used today, for personal adornment. But of equal or perhaps greater importance were the presumed supernatural powers that they bestowed on the wearer. The belief in these mystical properties of gems was already old by Roman times, and each successive culture or civilization has added to the lore or independently created its own. Wearing gems as talismans or amulets was widespread, with different virtues attributed to different stones. There was virtually no disaster or human ailment that some gem would not guard against. They protected the wearer from disease, poison, lightning, fire, and intoxication; others rendered him invisible or invulnerable or endowed him with strength and wisdom.

Astrology played a part in the superstitions surrounding gems, remnants of which persist to the present day. A particular stone was believed to possess special virtues if worn under zodiacal control, that is, during a given month. A definite gem was assigned to each sign of the zodiac and thus, for most effective influence, a different stone should be worn each month. Because this required 12 gems, which few could possess, and because the most potent stone corresponded to the sign in which the birthday of the wearer fell, the use of month stones or birthstones gradually evolved.

The astrologers who assigned gems to zodiacal signs were obviously influenced by earlier religious symbolism of gems, because the early lists of birthstones bear a striking resemblance of gems mentioned in the Bible. In Exodus 28 instructions are given for preparing the breastplate of the high priest of the 12 tribes of Israel as follows: "And thou shall set it in settings of stones, even four rows of stones; the first row shall be a sardius, a topaz, and a carbuncle. And the second row shall be an emerald, a sapphire, and a diamond; and the third row a ligure, an agate, and an amethyst; and the fourth row a beryl, and an onyx, and a jasper." With the exception of *ligure* (possibly zircon), all the names of the various stones are still in use, but certainly many do not refer to stones so known today. In ancient times minerals were classified chiefly by color, with a single name given to what we recognize as several species. Carbuncle was red and could refer to ruby or spinel but most likely to garnet. Topaz was a green stone, possibly olivine, and a sapphire a blue one, perhaps lapis lazuli. Diamond could not have been our present gem, for one large enough for the breastplate was rare, if indeed known, and was too hard to engrave with the name of one of the tribes. It is possible that emerald, amethyst, beryl, sardius (sard), and onyx refer to gems as we know them.

Although the relation of some early described gems to modern gem and mineral names is uncertain, it seems probable that the gem minerals listed as "important" (Table 15.1) were, with few exceptions, known 2000 years ago. The esteem in which individual gems are held, and hence their value, has changed from time to time through the centuries and is still subject to change. Even at the present time there is no unanimity among gemologists as to the ordering of gems according to value.

IMPORTANT GEMS— YESTERDAY AND TODAY

Diamond

For most of historic time diamond has held the preeminent position as the most coveted gem, but it was rare among the ancients. Pliny writing in 100 A.D. stated it is "the most valuable of gemstones, but known only to kings." Its great value results from its high hardness, brilliant luster, and high dispersion, giving rise to flashes of spectral colors. Moreover, it is quite insoluble in acids or alkalis. Because of these resistant properties it was called *adamas*, from the Greek meaning "the invincible," from which the name *diamond* is derived.

India was the chief source of early diamonds. There were three separate areas from which diamonds were won from alluvial diggings, but the most important was Golconda. This was not a mine but a trading center whose name was given to a district in the southern part of the country that yielded many famous stones. Some of these gems are known today, whereas for others, although they undoubtedly exist, their whereabouts are unknown. But each in its time has left a trail of treachery, intrigue, murder, and wars. The most celebrated of these ancient diamonds is the Koh-i-nor, which, after recutting weighs 108.83 carats,[1] and is included in the British crown jewels. The Great Mogul (180 carats), described in detail by the French gem merchant Tavernier in the middle of the seventeenth century, vanished after the sack of Delhi in 1739. The blue Hope diamond (45.52 carats), with a long history of tragedy and disaster, now occupies a central position in the gem display of the Smithsonian Institution, Washington, D.C.

[1]One carat equals 0.2 gram.

With the discovery of diamonds in river gravels in Brazil in 1725, that country became and remained for 140 years the world's leading diamond producer. In 1866 diamonds were discovered in South Africa, first in stream gravels but shortly thereafter in diamond "pipes," the rock in which they formed. During the following 100 years numerous finds of diamonds, both alluvial and in situ, were made in a dozen African countries, but South Africa remains the largest producer of African gem diamonds. It was at the Premier Mine in that country that the Cullinan (3106 carats), the world's largest gem diamond, was found in 1905. In recent years diamonds have been coming from the Commonwealth of Independent States. Although the figures are unreliable, it is estimated that the CIS accounts for about 15% of the world production and is expanding its capacity. In 1988 the world diamond production totaled about 97 million carats, but the supply is increasing. Each year brings reports of new finds and expanded production from old mines. For example, Botswana produces about 17% and Australia about 37% of total world diamond production, which includes gem as well as industrial diamonds.

About 75% of the diamonds produced are of industrial grade, not suitable for gems. Stones cut from the remaining 25% have great variation in value, which depends on what is sometimes called the four C's; these are color, clarity, cut, and carat weight. Diamonds are graded into many categories on the basis of color and clarity (imperfections). At the top of the scale, the flawless stone of best color has a value at least 10 times that of a stone of equal weight at the bottom of the scale. To give the optimum brilliance and fire, a diamond must be properly proportioned. A well-cut stone may be worth 50% more than one poorly cut but of equal weight, color, and clarity. For this reason many old diamonds are recut to increase their value even though they lose weight in the process. The price of a diamond is obviously related to its weight but not in proportion to it. If a 1-carat stone is worth $6500, a 2-carat stone of like quality may be worth $8400 per carat, and one of 5 carats $12,000 per carat. Examples of several diamond cuts are shown in Fig. 15.3.

Beryl

As a gem mineral, beryl occurs in transparent crystals in several color varieties to which special names have been given. These are: *goshenite*, colorless; *morganite*, pink; *aquamarine*, blue-green; and *golden beryl*, yellow. These beryls make handsome gems but it is only the deep green variety, *emerald*, that has ranked

FIG. 15.3. Diamond, octahedral crystal in kimberlite matrix with faceted gems. The types of cuts, clockwise from upper left are: brilliant, marquise, heart-shaped, emerald cut, pendoloque (pear-shaped), oval. The brilliant cut stone (upper left) is approximately 4 carats. (Courtesy of the Diamond Information Center, New York, N.Y.)

as a "precious" stone. Since ancient times a high value has been placed on emerald, and today its value may equal, or even exceed, that of diamond and ruby.

Although emeralds were mined in ancient Egypt, nearly 4000 years ago, it was not until the Spanish conquest in the sixteenth century that large, high-quality stones reached Europe and the rest of the world. When the Spanish arrived in what is now Colombia, Ecuador, and Peru, they found the Indians there had large stones of emeralds. These they seized, but it was not until later that they learned that the source was Colombia. Other emerald deposits have since been found in Russia (1830), Australia (about 1900), South Africa (1927), and Zimbabwe (1956). But Colombia remains the unquestioned source of fine stones, as it has been for over 400 years. There are two separate mining districts: Chivor, 75 km northeast of Bogata, and Muzo, 100 km north of Bogata. The mines at both localities have been at various times under private and government ownership. But whoever the operator, stealing, bribery, and even murder have accompanied the mining operation. Today at nationalized mines, the work is carried out under armed guard. Still the same problems exist and a very high percentage of Colombian emerald production is mined and sold illegally.

Ruby and Sapphire

These gem varieties of corundum have since ancient times been held in high esteem, and with diamond and emerald they are regarded as the most valuable of gemstones. Ruby receives its name from the Latin *ruber*, meaning red. Sapphire comes from the Latin *sapphirus*, meaning blue, and was first used to denote any blue gemstone. Today all gem corundum, except red, is called sapphire but the color is specified as: yellow sapphire, purple sapphire, and so on. In the gem trade, however, *sapphire* is assumed to be blue.

Rubies may be of various shades of red, but the deep red, known as pigeon-blood, is of greatest value. Flawless stones of this color are seldom larger than three carats, and those that exceed ten carats are extremely rare. Yet in the past, large rubies have been reported but the whereabouts of most of them are unknown. Travernier in the seventeenth century described a ruby owned by the King of Bijapur as weighing over 50 carats. The German Emperor Rudolph II is reported to have possessed a ruby the size of a hen's egg. Because rubies have been confused with other red stones, particularly spinel and tourmaline, it may be that some of the earlier reported large rubies "disappeared" on the discovery that they were other gems. Blue sapphires are far more abundant than rubies of similar quality and thus are of less value. Also large sapphire crystals are not uncommon and numerous stones exceeding 100 carats have been described, many of which are known today. Transparent rubies and sapphires are commonly faceted, but nontransparent stones of fine color may be cut cabochon. This is particularly true if there is indication of asterism in the rough. Care must be taken in cutting such a stone to have the c axis perpendicular to the base of the cabochon so that the star appearing on the upper curved surface will be centered. Since the advent of synthetic star rubies and star sapphires, the popularity of natural "stars" has decreased, and they are of no greater value than cabochons of similar size and color that lack asterism.

Since the fifteenth century, Burma has been the source of the finest rubies. They come from three separate areas, but the most important is near Mogok. The rubies originated in a metamorphic limestone, and mining of the parent rock has yielded some fine stones. However, most of the gems are recovered, as they have been since ancient times, from the soil of the mountain slopes and from the gravels of the stream valleys. Stones of poorer quality than the Burmese have long come from alluvial deposits in Cambodia, Thailand, and Sri Lanka. The most recent major ruby discovery is in the Tsavo National Park, Kenya. Mining, which began there in 1980, has produced stones of high quality said to rival those from Burma.

Because ruby and sapphire are both varieties of corundum, it is not surprising that they form under similar conditions and have similar occurrences. Excellent sapphires of a cornflower blue—the most desirable color—come from alluvial deposits near Batambang in Thailand and adjacent areas of Cambodia. In Sri Lanka sapphires of good quality are recovered from the gem gravels. In 1881 in Kashmir in the Himalaya Mountains, a landslide revealed the presence of sapphires. These "Kashmir" blue stones are highly prized but are somewhat lighter in color than those from Cambodia and Thailand.

Opal

Precious opal lacks the high hardness, clarity, and brilliance of most of the other important gems. But even without these properties, it is highly prized. The great value of precious opal lies in a subtle beauty resulting from an internal display of flashing colors. Various names have been given to precious opal, such as black opal, from which a brilliant play of color is emitted from a dark body color; and fire opal, in which a translucent background gives forth flashes of red to orange colors. Handsome carvings have been executed from opal but as a gem it is always cut cabochon. Many opal gemstones are doublets; that is, a thin slice of precious opal is cemented to a slice of nongem opal, or other material, and the assembly cut and polished. Because most opals are nontransparent and the color flashes a surface phenomenon, the doublet resembles a stone cut from a single piece and the deception, if such is intended, is difficult to detect. Some doublets are capped with colorless quartz, which protects the softer opal from damage, yet does not detract from the beauty of the stone.

Opal was well known and greatly valued in Roman times. The high esteem in which it was held is indicated by Pliny in the story of Nonius, a Roman senator, who possessed an opal the size of a hazelnut. The emperor, Marcus Antonius, demanded the stone, but rather than part with it, Nonius abandoned his other possessions and fled the country. With only minor fluctuations, opals maintained their high value until early in the nineteenth century. At that time a decrease in popularity and a consequent lessening of value may have resulted from the superstition that opal was an unlucky stone. Today, with the superstition either forgotten or ignored, opal commands a high place among important gems.

Ancient opals of fine quality are believed to have come from India and those of poorer quality from Egypt. From Roman times until late in the nineteenth century, most precious opal came from mines near Czerevenitza, Hungary (now in Czechoslovakia). Today Australia is the principal source, with production from several widely separated localities. There are a number of occurrences of opal in Mexico, making that country the second major producer.

Jade

The name *jade* is given to two different minerals: jadeite, a pyroxene, and tremolite-actinolite, an amphibole (called nephrite). The two types have in common an extreme toughness and generally a green color. Although nephrite is composed of a felted mass of fibers and jadeite is an aggregate of interlocking grains, the textural differences are obscured on polished pieces, making identification difficult. Distinction is best made by specific gravity: nephrite $G = 3.0\pm$; jadeite $G = 3.3$.

When the Spanish conquistadores arrived in Mexico and Central America, they found carved jadeite that was greatly valued by the natives for the important role it played in their social and religious life. It is reported that Montezuma considered a gift of two pieces of jade to Cortez equal to two cartloads of gold. Recent finds of jadeite in Central America have been made, but it is uncertain as to whether they are the source of the ancient jade of the Aztecs and Mayas.

On the discovery of New Zealand in 1769, Captain Cook found that the native Maoris also placed a high value on jade and carved it into ornaments, utensils, and weapons. This was nephrite found in stream boulders and along the west coast of the south island. The source of this jade has been located in the mountains to the east and is actively worked today.

Many centuries before the West knew of jade in America and New Zealand, it was valued in China above all other substances, and the art of working it into elaborate carvings was ancient. The early Chinese jade was nephrite from Chinese Turkestan; it was much later that jadeite, the rarer and more valued jade, was brought from Burma to be carved by Chinese artisans. Jadeite is still being recovered as stream boulders from the Uru river in Upper Burma.

Chrysoberyl

The most common chrysoberyl occurs in shades of yellow, green, or brown. Its high hardness ($8\frac{1}{2}$), transparency, and pleasing color make it a desirable gemstone. However, it is for the far less common varieties of *cat's eye* and *alexandrite* that the gem is best known and most valued. A cabochon cut cat's eye shows a marked chatoyancy, a sharp band of light crossing a yellow, green, or brown background. Numerous other gems show a chatoyancy, but the term *cat's eye* should be reserved for chatoyant chrysoberyl. These rare and highly prized stones come mostly from Sri Lanka.

Alexandrite is a dark-colored chrysoberyl with the remarkable property of appearing green in daylight and red in artificial (incandescent) light. Because of this color change, it has been described as "an emerald by day and a ruby by night." Alexandrite was discovered in the Ural Mountains, CIS, in 1830 on the day the Czarevitch, later Czar Alexander II of Russia, reached his majority, and was named for him. Alexandrites are also found in the gem gravels of Sri Lanka and, although their color change is less dramatic than in the Russian stones, they command a high price.

Topaz

For centuries the name *topaz* has been subject to much confusion, a situation that still exists with both jeweler and layperson. The ancients undoubtedly used the word for gem olivine, *peridot*, that came from Zebirget, an island in the Red Sea. The first use of the name for the mineral we know today as *topaz* was by Henckel in 1737 for yellow to sherry-colored gems from Saxony. Since then it has been used for stones of this color, sometimes called *precious topaz*, to set it apart from other similar-appearing gems that may be called topaz. The most common confusion is with yellow quartz, citrine, for which various names have been used to imply the stone is topaz or a close relative.

The confusion regarding topaz is compounded by the fact that gem-quality crystals of the mineral are not always wine-yellow. In fact, they are most commonly colorless, but may be pink, red, orange, greenish, pale blue, or brown. Furthermore, certain colorless topaz can be turned a deep aquamarine-blue by irradiation followed by heat treatment. Most of this treated topaz comes from Brazil, which is also the source of most of the precious topaz.

Tourmaline

Tourmaline is unique among gem minerals in occurring in shades of all colors as well as in multicolored crystals. Names have been given to the various colored varieties, but it is more meaningful to call them tourmaline with the appropriate color modifier. Tourmaline is a widespread mineral and undoubtedly was early used as a gem, but it was not until early in the eighteenth century that its identity became known. At that time it received its name from the Singhalese word, *turamali*, used in Ceylon (now Sri Lanka) for yellow zircon. It is reported that when a parcel of turamali sent to Holland turned out to be yellow tourmaline, the name was given to that mineral, and, although a misnomer, it has persisted.

There are numerous world-wide occurrences of gem tourmaline. Some of the more important are in Minas Gerais, Brazil; Malagasy Republic; Mozambique; and Namibia. In the United States the first significant gem find was of tourmaline at Mount Mica, Maine; and in 1972 another major deposit was discovered not far away in Newry, Maine. Other important localities are in California, mostly in San Diego County. Mining of gem tourmaline began there 100 years ago and, with reopening of old mines and the discovery of new deposits, is still producing superb specimens.

Quartz

Quartz is listed with the "important" gem minerals not for its rarity or high value but because of its many common and abundant gem varieties. In all likelihood, quartz was the first mineral used by early man for personal adornment and has occupied a prominent place among gems ever since. Of the 12 stones in the breastplate of the high priest, at least half appear to have been a quartz variety. Theophrastus in his treatise *On Stones*, written about 300 B.C., mentions quartz and its varieties more than any other mineral, and states "among the ancients, there was no precious stone in more common use." It is still in common use and all but two or three of its many coarsely crystalline and fine-grained varieties are used as gem materials. One should refer to "Quartz" (Chapter 13), for the names and properties of the gem varieties.

Turquoise

Turquoise, valued chiefly for its color, is an ancient gem material. The oldest mines are at Sarabit Elkhadem on the Sinai Peninsula and date as far back as 4000 B.C. It appears certain that carved turquoise in bracelets of Queen Zer of Egypt's First Dynasty came from this locality. Nishapur, Iran (Persia), is the source of the finest turquoise today just as it has been since mining began there over 2000 years ago. The only other deposits of importance today are in China, Tibet, and in southwestern United States.

Turquoise came from many localities in the American Southwest, but the most famous is in the Los Cerillos Mountains, New Mexico, where the American Indians may have begun mining as early as 1000 years ago. Indian jewelry set with turquoise has become very popular in recent years. The result has been not only a large increase in price but the use of substitutes and the manufacture of synthetic turquoise.

Garnet

The great range in the chemistry of garnet, greater than in any other gem material, is reflected in the diversity of color and other physical properties of its several varieties. For a description of the various species and the extensive solid solution between them, one should refer to the discussion of the garnet group in Chapter 13. Since Biblical times the garnet used as gems is dark red and thus to many is the only color associated with the mineral. Yet garnet occurs in all colors except blue. Mention is made here of only the more unusual colors that may be encountered in garnet gemstones.

Pyrope and *almandine*, and solid solutions between them, are various shades of red and violet. These are the garnets used in early times as gems. *Grossular* may also be red, but it is the orange-yellow (hessonite) to orange-brown (essonite) material that has been most widely used as gemstones. Recently a transparent, emerald-green variety, called tsavorite, has been found in the Tsavo National Park, Kenya. Stones of tsavorite are small but highly valued. A jade-green, massive variety of grossular from South Africa has been used as a jade substitute. *Andradite* may be yellow-green to black but it is only the transparent green variety, demantoid, that is important as a gem. It is the garnet of greatest value and occupies a high position among all gems. *Spessartine* of gem quality is rare, but yellow, yellow-brown, and orange-brown crystals are cut into lovely stones. *Uvarovite*, the chromium garnet, occurs in brilliant, deep green crystals. The crystals are too small for cutting but were they larger, they would make striking gems.

Zircon

Zircon is an Eastern gem which for centuries has come from the gem gravels of Sri Lanka and Indochina. It is recovered as water-worn crystals, usually in shades of red or brown, but it may be green, gray, or colorless. For a long time, the reddish-brown crystals have been heat treated to produce more attractive colors. Heating in air usually yields a yellow stone, but blue or colorless stones result when heated in a reducing environment. The most popular colored stone is blue, which is sold under the name of *starlite*.

Zircon has a high refractive index and a high dispersion, giving a brilliance and fire to a cut stone approaching those of a diamond. Because of the resemblance, colorless zircons from Matura, Sri Lanka, were called "Matura diamonds."

There are two types of zircon used as gems: a crystalline variety and a metamict type. The chemical composition is given as $ZrSiO_4$, but hafnium is always present and sometimes small amounts of thorium and uranium. Originally all zircon was crystalline, but irradiation by these radioactive elements destroys the structure, making the crystals amorphous. Metamict zircons are usually green and have lower refractive indices and density than the crystalline variety.

Olivine

The olive-green gem variety of olivine is today known as *peridot*, but until recently was called chrysolite (a synonym for olivine) and "evening emerald." The gem was known to the ancients as topazian, a name presumably derived from Topazias, an island in the Red Sea. It was called topaz until the eighteenth century, when the name was transferred to another gem, our present-day topaz. Although the early stones undoubtedly came from the Red Sea island, for centuries their source remained a mystery until it was rediscovered early in the twentieth century. Since rediscovery, the island, known as Zebirget (formerly Saint John's Island), has yielded a large quantity of fine peridot crystals. In 1958 the deposit was nationalized by the

Egyptian government. Although peridot has come from Burma, Australia, Norway, and the United States, Zebirget remains the outstanding locality for large stones of high quality.

GEM PROPERTIES AND INSTRUMENTS FOR THEIR DETERMINATION

Because gems are basically minerals, the student who has digested the material presented in the first 13 chapters of this book has in essence learned of the properties of gems and how to determine them. Many of the methods developed by the mineralogist for mineral determination are applied equally effectively to gemstone identification. In the study of rough gem material, the procedures are identical. In general, however, there is a fundamental difference in approach. Whereas the mineralogist can scratch, powder, or dissolve the mineral under investigation, the gemologist is restricted to nondestructive tests when working with cut and polished gemstones. As a result, specialized instruments and techniques have been developed for their study. In the following pages are discussed the most important of these methods used in determining the properties of gems. Because the characterizing properties of gems are obviously the same as for the minerals from which they were fashioned, they are not given here; for them, one should refer to descriptions of the individual minerals. In the Appendix, however, are two tables helpful in gem identification: (1) a list of minerals in order of increasing specific gravity (Table A.2), and (2) a list of nonopaque minerals in order of increasing refractive index (Table A.3). In most cases an unknown gem can be identified by determination of these two properties.

Observation

The first step in identification of a gem is to look at it with the unaided eye, for several properties are seen as surface phenomena. These include luster, color, and, in a faceted stone, fire caused by dispersion of the spectral colors. Other surface features, observed best in cabochon stones, are (with examples): play of color, opal; opalescence, moonstone; iridescence or labradorescence, labradorite; asterism, star ruby; and chatoyancy, cat's eye, (pages 267–270). Such observations may yield sufficient information for the experienced gemologist to identify a gem.

Hand Lens

Perhaps the most important gemological instrument is a simple hand lens; one with a 10X magnification is most commonly used. With it one can determine the quality of the cut. Are the facets well polished, are they symmetrically disposed, do they meet at common points, and are the edges between them sharp or rounded? The cut of an imitation is usually poorer than that of the gem it simulates. The hand lens will also reveal major flaws or inclusions that detract from the value of a stone.

The Microscope

For a critical examination, one that will reveal imperfections not seen with a hand lens, it is necessary to use a microscope. The type most useful in gemology is a low-power binocular microscope with magnification ranging from 10X to 60X. This instrument differs from the polarizing microscope (page 293) in that it gives true stereoscopic vision with an erect rather than an inverted image. Provision is made to view the stone in both incident and transmitted light. With the light source above the microscope stage, the external features are observed in light reflected from the surface. The interior of the stone is best seen in transmitted light, that is, in light passing through it from a substage illuminator. However, it is difficult to inspect the complete interior of a gemstone in normal transmitted light because of reflection and refraction by the facets. This problem is largely overcome in the gemological microscope (Fig. 15.4) by a substage adapter, giving what is called dark field illumination. Using this lighting technique, the gemstone is illuminated by a hollow cone of light that does not fall directly on the microscope objective. The gemstone is held at the apex of this cone and any flaws or inclusions within it appear bright against a dark background. Analysis of these internal imperfections aids both in identification and in quality determination. Also it frequently provides a means of determining whether a stone is of natural origin or is a synthetic imitation.

The Polariscope

The most informative tests are based on the optical properties and thus are usually the first made in gemstone identification. A determination as to whether a stone is isotropic or anisotropic can be made quickly using a polarizing microscope (Fig. 8.10). Equally effective for the purpose is a simple polariscope used by most gemologists. It is composed of two polarizing plates, one held above the other with a light source below, see Fig. 15.5. The plates are adjusted to the crossed position by turning the upper polar until a minimum of light passes through it. A transparent gemstone placed on the lower plate and slowly turned will remain dark if it is isotropic (isometric or noncrystalline), but will become alternately light and dark if anisotropic (page 291).

FIG. 15.4. A low-power stereoscopic wide-field microscope, designed to provide several different methods for illuminating gemstones. (Courtesy of Gemological Institute of America, GEM Instruments Corp., Santa Monica, Calif.)

Refractive Index and the Refractometer

Of the several measurable properties of gems, refractive index is most informative and determined most easily. The **RI** (the abbreviation used by gemologists for refractive index) can be determined on both set and unset gemstones, using especially designed refractometers. Several makes of refractometers are available, but all operate on the same principle: total reflection and the critical angle (page 290). But they are so constructed that, instead of measuring the angle, the refractive index is read directly from a scale. Central to all the instruments is a lead glass of high refractive index cut in the form of a hemisphere, hemicylinder, or prism with a flat, polished upper surface. The refractometer developed by the Gemological Institute of America (Fig. 15.6) uses a hemicylinder.

To make an **RI** determination, a facet of the gemstone, usually the table, is placed on the polished surface but separated from it by a film of liquid. Light entering through a ground glass at the back of the instrument passes through the hemicylinder, striking the

gemstone at varying angles of incidence. When the light rays are at angles greater than the critical angle, they are totally reflected back through the hemicylinder to fall on a transparent scale. The image of the scale is reflected by a mirror and is observed through a telescope (Fig. 15.7a). The position of the boundary separating light from dark portions (the shadow edge) is determined by the refractive index, which can be read from the scale (Fig. 15.7b).

The contact liquid usually used with a standard jeweler's refractometer is a mixture of methelyne iodide, sulfur, and tetraiodethylene with an **RI** of 1.81. Because the liquid must have an **RI** greater than that of the gemstone, the maximum refractive index determination is about 1.80. *Caution*: Refractive index liquids available in most mineralogical laboratories with **RI** greater than 1.78 should not be used. They are corrosive and will etch the lead glass of the refractometer. However, they may be used in a refractometer in which the glass is replaced by chemically inert cubic zirconia (**RI** 2.16). With this instrument, using a liquid of **RI** 2.10, measurements of refractive index as high as 2.08 can be made.

FIG. 15.5. Polariscope used for distinguishing isotropic from anisotropic materials, determining pleochroism, and obtaining interference figures. (Courtesy of Gemological Institute of America, GEM Instruments Corp., Santa Monica, Calif.)

FIG. 15.6. Jeweler's refractometer. (Courtesy of Gemological Institute of America, GEM Instruments Corp., Santa Monica, Calif.)

Dispersion

Refractive index varies with the wavelength of light and is less for red than for violet light, the phenomenon called *dispersion* (pages 290 and 306). Dispersion is a characterizing feature of gemstones and is given as a number representing the difference of **RI** determined in red light and in violet light. For diamond it is 0.044 and for quartz 0.013. The high dispersion of diamond gives it its fire, the flashes of color from a cut stone. If white light is used in determining the **RI** of a gemstone, the shadow edge on the scale is not a line but, because of dispersion, is a band of spectral colors.

For precise measurements it is necessary to use a monochromatic source such as sodium light. The refractive index usually reported is that determined in sodium light.

Refractive Indices of Anisotropic Gems

Although, basically, the method of refractive index determination outlined above is applicable to all gemstones, it strictly applies only to those that are isotropic with a single refractive index. Anisotropic crystals have two refractive indices if uniaxial and three if biaxial.

FIG. 15.7. *(a)* Schematic diagram of light rays entering a refractometer and striking a gemstone of spinel at varying angles of incidence. Rays at angles less than the critical angle (CA) pass upward through the stone. Rays at angles greater than CA are totally reflected through the hemicylinder. *(b)* Image of scale as seen through the optical system with reading at 1.720, the RI of spinel. The faint edge at 1.81 is the RI of the contact liquid.

(a)   *(b)*

Uniaxial stones, in general, give rise to two shadow edges on the refractometer, one resulting from vibrations of the O ray, the other from vibrations of the E ray. If the stone is turned on the hemicylinder, one edge remains constant; this is ω, the **RI** of the O ray. The other edge, resulting from the E ray, varies as the stone is turned giving values of ϵ'. The true value of ϵ is read when the two edges are farthest apart. The difference between ω and ϵ is the birefringence. The stone is optically positive if $\omega < \epsilon$ and negative if $\omega > \epsilon$. If the polished surface of the gem in contact with the refractometer is perpendicular to the optic axis (c crystal axis), both shadow edges remain constant as the stone is turned. In this situation, measurements should be made with the stone lying on another facet. One can then determine whether the high or low index is variable and thus the optic sign.

Biaxial gemstones normally show two shadow edges on the refractometer but, unlike uniaxial, both edges move as the stone is turned. The maximum reading of the higher edge is γ, the minimum reading of the lower edge is α, and their difference is the birefringence. To determine the optic sign, it is necessary to know the third index, β. For $(\gamma - \beta) > (\beta - \alpha)$ is positive and $(\gamma - \beta) < (\beta - \alpha)$ is negative. As the stone is rotated, the value of β is observed as *either*: (1) the lowest reading of the higher edge or (2) the highest reading of the lower edge. If (1) is less than the midpoint between α and γ, β is the nearer to α than to γ and the stone is positive; if (2) is greater than this midpoint, β is nearer to γ than to α and the stone is negative. If neither shadow edge passes the midpoint the gemstone should be placed on a different face and readings (1) and (2), the two possibilities for β, again determined. The value common to both sets of readings is β.

Pleochroism

Light passing through an anisotropic crystal may be absorbed differently in different vibration directions. The phenomenon is called dichroism in uniaxial crystals (page 296) and pleochroism in biaxial crystals (page 307) although the term *pleochroism* is commonly used for both. It may be evidenced only in a difference in intensity of light of a given wavelength, but frequently different wavelengths are absorbed in each vibration direction, resulting in color variations characteristic of a given gem.

Uniaxial gemstones with only two rays, O and E, can show only two pleochroic colors, which vary as do the refractive indices. When light moves parallel to the optic axis, only one color is seen, that of the O

ray. The color of the E ray is most pronounced when light moves perpendicular to the optic axis. In some uniaxial gems the color difference is only in intensity. For example, in green tourmaline the absorption of the O ray is much greater than that of the E ray and is expressed as O > E or $\omega > \epsilon$. When absorption results in different pleochroic colors, the colors are given. For example in emerald: O (ω) yellow-green, E (ϵ) blue-green. In biaxial crystals light absorption may be different for light vibrating in each of the principal crystallographic directions, X, Y, Z. For example, the variety of chrysoberyl, alexandrite, has strong pleochroism expressed as X (α) red, Y (β) orange, and Z (γ) green.

Because pleochroic colors result from rays vibrating at right angles to each other, it is possible to observe one by eliminating the other. This can be done with a polariscope by rotating the stone to an extinction position. The upper polar is then removed and the color noted in plane polarized light. If the stone is pleochroic, a different color or color intensity will be observed on rotating it 90°. A gemstone either uniaxial or biaxial may be pleochroic, yet no color difference will be seen if light passes through it parallel to an optic axis. Thus in testing for pleochroism, it is advisable to make observations with light passing through the stone in several directions.

The dichroscope is an instrument long used by gemologists for detection of pleochroism. The instrument is a metal tube with a square or rectangular opening at one end and a lens at the other. Within the tube is an elongated cleavage piece of optical calcite (Iceland spar). Because of the strong double refraction of calcite, two images of the opening are seen through the instrument (Fig. 15.8). If a pleochroic stone is held over a bright light source and viewed through the dichroscope, the two images have different colors. Although the same colors are seen with a polariscope, they are observed simultaneously with the dichroscope and compared more easily.

Fluorescence

Some gemstones fluoresce under ultraviolet light but for only a few, as for their parent minerals, is the fluorescent color constant and therefore diagnostic. Of this small number that are occasionally cut as gems, willemite fluoresces a yellow-green, scheelite a pale blue, and benitoite a bright blue in shortwave ultraviolet. Gems cut from some minerals are always nonfluorescent, but those cut from other minerals may or may not fluoresce. Thus in most cases fluorescence cannot be used as a definitive test.

FIG. 15.8. Dichroscope, diagrammatic section. *R* is a calcite cleavage rhombohedron held in a metal tube by a cork setting, *C*. Glass prisms, *G*, are cemented to the calcite to aid in light transmission. To the right of the drawing is a double image of the window, *W*, as seen through lens, *L*.

Filtering of Color

A colored gemstone absorbs certain wavelengths of white light, and the color results from a combination of the unabsorbed wavelengths. To the eye two gems may have nearly identical colors but the color of one may result from quite a different mixture of wavelengths than the color of the other. Color filters offer a means of resolving the basic difference and thus aid in gem identification.

The most commonly used filter is the "Chelsea," sometimes called the "emerald filter" because its principal use is to distinguish emerald from other green gems and imitations. The filter transmits only the long red wavelengths and a band in the yellow-green portion of the spectrum. Emerald transmits red light but absorbs some of the yellow-green. It therefore appears red when illuminated by a bright incandescent light and viewed through the filter. Most other green gems as well as glass imitations absorb red light and appear green. Although this is a helpful test in emerald identification, it must be used with caution, because a few other green gems as well as synthetic emerald may appear red and the color of emerald from some localities remains unchanged.

The Spectroscope

In Chapter 5 the spectroscope is briefly discussed as a tool used in mineralogical research to determine chemical composition. Using the instrument described there, the mineral is vaporized by an electrical arc. In this excited (vapor) state the atoms of the elements present emit light of definite wavelengths, pro-

ducing a characteristic spectrum for each element. This method, known as emission spectroscopy, finds little use in gemology because the test sample is consumed. However, gems are studied spectroscopically by analyzing the light transmitted through them or reflected from their surfaces.

The instrument used by gemologists is a small, direct vision spectroscope. It is essentially a metal tube with an adjustable slit at one end and an eyecap window at the other. Between the ends are arranged a lens to focus on the slit and a train of prisms, as shown in Fig. 15.9. When the slit is directed toward a source of white light, one observes a ribbon of the complete visible spectrum, red at one end, violet at the other. Different portions of the spectrum are brought into sharp focus by pulling out or pushing in the drawtube carrying the slit.

When a gemstone is interposed between the bright light and the spectroscope, the continuous spectrum is visible but it may be crossed by dark bands or lines where certain wavelengths have been eliminated (Fig. 15.10). This is an absorption spectrum. The dark bands or lines fall in positions identical to the bright bands or lines of the emission spectrum but are not as sharp or as well defined. Nevertheless, their distribution depends on the elements present and yields information that aids in the identification of certain gems. Bands or lines resulting from both major elements and minor elements, acting as coloring agents, may be present. In some cases with data obtained with the spectroscope, one can differentiate between a natural or synthetic gem or tell whether the color is natural or has been artificially induced.

FIG. 15.9. Section through a direct vision prism spectroscope of the type manufactured by R. and J. Beck, London.

FIG. 15.10. Absorption spectrum of a pale yellow diamond. Seen through the spectroscope the background would be a continuous spectrum of violet through red, crossed by dark lines. (Courtesy of Gemological Institute of America, Santa Monica, Calif.)

Physical Properties

Cleavage and Fracture. These properties that frequently enable one to determine a mineral specimen are far less important in gem identification. Evidence of them is usually obliterated in a well-cut and polished gemstone. However, careful inspection of a cut stone with a hand lens or microscope may reveal interior reflecting surfaces, indicating incipient cleavage cracks. A mounted stone should also be examined at the points of contact with the prongs where pressure of mounting may have produced tiny fractures or cleavages.

Hardness. A scratch test for hardness of a gemstone should be made only when other identifying tests have failed, and then with extreme caution. To determine hardness the gemologist uses a set of *hardness points*. These are metal tubes set with sharp-pointed mineral fragments usually having hardnesses of 10, 9, $8\frac{1}{2}$, 8, $7\frac{1}{2}$, 7, and 6. When it is necessary to check hardness on a faceted stone, the scratch, as short as possible, should be attempted on a back facet near the girdle or on the girdle itself. To avoid scratching an unset stone, the gem girdle can be used as the "scratcher" with attempts to scratch the scale minerals. Begin with a mineral low in the scale and work toward those of higher hardness until no scratch is produced. A short scratch usually can be made on the back of a cabochon without injury to the stone.

Specific Gravity. The specific gravity is as characteristic a property of gems as it is of minerals. Its determination, coupled with refractive index, in many cases suffices for identification. But unlike the refractive index, it cannot be determined for a mounted stone. The most reliable methods of determining specific gravity are by hydrostatic weighing or by suspension in a heavy liquid (page 258). The definite values obtained by these methods are more helpful in pinpointing a gem than approximate specific gravities obtained in the manner outlined below.

It is common practice for gemologists to maintain a number of bottles containing liquids of known densities. Densities found most convenient are 2.67 (bromoform + acetone), in which quartz floats and beryl sinks; 2.89 (bromoform), in which beryl floats and tourmaline sinks; 3.10 (methylene iodide + acetone), in which gem tourmaline just floats and fluorite sinks; 3.33 (methylene iodide), in which fluorite sinks and jadeite is suspended or sinks slowly. By making successive tests, the specific gravity of a gem can be bracketed or determined as greater than 3.33 or less than 2.67. For example, a gemstone that sinks in the 3.10 liquid but floats in the 3.33 liquid has a specific gravity between these limits. To avoid contamination, the stone should be well cleaned before transferring from one liquid to another.

X-ray Diffraction

Of the several methods used today for mineral identification, the one yielding the most definitive information is an X-ray powder photograph or powder diffractogram (pages 283 and 288). Thus in studying a mineral, this is frequently the first, and perhaps the only, test made even though simple ones may have been adequate. In studying a gem, the reverse is true. Only when all other tests have failed to identify a gem is resort made to an X-ray powder photograph. This is due in part to the fact that X-ray equipment is unavailable and in part to the difficulty in obtaining the requisite sample. However, with care and without injury to the stone, sufficient material can be ground from the back of a cabochon or scraped from the girdle of a faceted gemstone to make a powder mount.

SYNTHESIS OF GEM MATERIALS

During the twentieth century mineralogical research has resulted in the laboratory synthesis of many minerals. Most of these are fine-grained aggregates or in small crystals, but a few have been grown in large, single crystals. These latter are our immediate concern and particularly those that duplicate gem minerals. Because duplication is not only in appearance but in all respects—chemistry, crystal structure, and physical properties—detection is a difficult problem.

The method of synthesis varies depending on the desired product. Some techniques have been developed to produce only a single gem (e.g., diamond and opal), whereas others are used to synthesize several. The methods of more general application, outlined below, are followed by a consideration of the most important synthetic gem materials. Included are not only synthetic minerals but also synthetic products without natural equivalents.

Verneuil Process

We can consider that gem synthesis on a commercial scale had its beginning in 1902. In that year Auguste Verneuil announced that by a process of flame fusion he had synthesized ruby in sizes large enough and clear enough to be cut as gemstones. An industry based on the Verneuil technique grew rapidly and within a decade millions of carats, not only of ruby, but of sapphires both colorless and in a variety of colors, were being produced annually. Later (ca. 1920) spinel and still later (ca. 1947) rutile were manufactured by the same process. The Verneuil method with only minor changes remains today the principal means of gem synthesis.

In the Verneuil process a powder with the chemical composition of the desired crystal (for ruby, Al_2O_3 plus a coloring agent) is melted as it passes through an oxygen-hydrogen flame. Droplets of the fused powder fall on a ceramic plate and, on cooling, form a carrot-shaped single crystal of corundum called a boule. Most crystals grown by the Verneuil method contain small gas bubbles, usually spherical but sometimes elongated. Also present are curved striae resulting from successive layers of molten material spreading over the curved surface of the growing boule. Detection in a gemstone of these bubbles and growth striations signifies a synthetic origin.

The Czochralski Process

By this technique, also called *crystal pulling*, large, high-quality crystals of several compounds including corundum can be grown. A melt of the composition of the desired crystal is contained in an iridium crucible. A "seed" crystal, held on the end of a rotating rod, is touched to the surface of the melt and then slowly withdrawn. On withdrawal, material contin-

ually crystallizes, forming a rod-shaped crystal. Ruby crystals 40 cm long and 5 cm in diameter are grown in this manner.

Flux Growth

In this method a powder of the composition of the desired crystal is mixed with a flux, a material having a relatively low melting point. As the mixture is heated in an inert crucible, the flux melts and, in the molten state, dissolves the other material. After the melt is thoroughly mixed, it is allowed to cool slowly. On reaching a critical temperature, crystal nuclei of the desired material form and gradually increase in size as the temperature is further lowered. When cool, the crystals are recovered by dissolving the flux. Several gem materials, including emerald and ruby, are grown by this method.

Hydrothermal Growth

In nature many minerals crystallize from hot, water-rich solutions. The constituent elements, held in solution at high temperatures, are precipitated as the solutions move toward the surface into regions of lower temperature. Using equipment designed to withstand high temperature and high pressure, hydrothermal growth in the laboratory closely duplicates the natural process. The apparatus, called an autoclave or "bomb," is a heavy-walled steel cylinder closed at one end. To prepare an autoclave for crystal growth of quartz, for example, pure quartz fragments as source material are placed in the bottom and seed plates of thin slices of single quartz crystals are held in the upper part. The bomb is then 85% filled with water to which a crystallization catalyst has been added to increase the solubility of the quartz fragments. After sealing, the bomb is placed in a furnace and heated to about 400°C in the lower part and to about 340°C in the upper part. As the temperature is raised, the water expands to fill the bomb and the pressure reaches 20,000 to 25,000 psi. The convection current generated by the temperature gradient causes the solution to rise, carrying silica dissolved from the feed material. When it reaches the zone of lower temperature, the silica is deposited on the seed plates. Although ruby and emerald are grown hydrothermally, quartz is by far the most important product grown by this method.

TREATMENT OF GEMSTONES

The color of a number of gemstones can be changed or enhanced by "treatment." The change in some is temporary, lasting only a few days or weeks, but in others it is permanent, resulting in a color identical to that of a naturally colored stone. The principal methods of treatment are dyeing, heating, and irradiation, or, for some gems, a combination of heating and irradiation.

Dyeing

An old and an obvious way to color a gemstone is to stain it with an appropriate dye. Because the material must be sufficiently porous to permit penetration of the pigmenting solution, the method is restricted to relatively few gems. Dyeing has been used most effectively in coloring chalcedony, jade, and turquoise.

Heat Treatment

It has long been known that heating certain gemstones will result in a color change or will render an off-color stone colorless. It is very likely that most, if not all, gem materials have been subjected to heat treatment in both oxidizing and reducing environments with the hope that they too could be similarly enhanced. However, only a few have responded favorably, and some have been downgraded by heating. The gems for which heat treatment is most successful are zircon, quartz, beryl, topaz, and zoisite (tanzanite).

Irradiation

Compared with dyeing and heat treatment, irradiation of gemstones is a new process, a twentieth-century development. Irradiation has produced desirable color changes in numerous gems but in only a few is the color permanent. There are several irradiation sources: X-rays, alpha particles from a cyclotron, neutrons from atomic piles, and gamma rays from cobalt 60. Irradiation is most effective in producing desirable color changes in diamond, topaz, and quartz.

SYNTHETIC AND TREATED GEMS

Gems that have been synthesized and those that have responded most favorably to treatment are listed below, designated s for synthesis and t for treatment.

Beryl (s, t)	Quartz (s, t)
Chrysoberyl (s)	Rutile (s)
Corundum (s, t)	Spinel (s)
Diamond (s, t)	Topaz (t)
Jade (t)	Turquoise (s, t)
Opal (s, t)	Zoisite (t)

Below are brief discussions of the above gems, giving for each the method by which it is synthesized and/or treated. Also mentioned, under "Imitation," are the materials, both natural and synthetic, that are sometimes used to imitate the natural gem. Glass is not included with the materials used as imitations, but it could be in each case. Therefore, the gemologist should be ever alert to the possibility that a stone is glass or "paste" as it is sometimes called. Although glass may correspond in color to the gem it simulates, usually the deception can be quickly detected by determination of other properties.

Beryl

Synthesis. Beryl synthesis has been directed chiefly to emerald because of its high value compared with other gem beryls. Efforts have been highly successful and emerald has been synthesized by both flux-growth and hydrothermal methods. Also, a special hydrothermal method (the Lechtleitner process) has been developed by which an overgrowth of emerald can be produced on gemstones of colorless beryl.

Most natural emeralds are nonfluorescent but synthetics usually fluoresce a dull red under longwave ultraviolet. Also natural emeralds usually appear red when viewed through the Chelsea filter, whereas most synthetics and imitations appear green. Unfortunately neither test is unequivocal. Flux-grown emeralds contain wisplike inclusions that signify a synthetic origin. The refractive indices and specific gravity of synthetics made by the different processes differ slightly from one another as well as from natural emerald as shown below.

Emerald type	ω	ε	G
Natural Colombian	1.583	1.577	2.71–2.77
Hydrothermal	1.573	1.568	2.66–2.70
Flux grown (some)	1.579	1.571	2.65–2.68
Flux grown (normal)	1.564	1.561	2.65–2.66

Treatment. Gems of green, yellow, and pink beryl are usually of natural colors, but the deep blue of many aquamarines has been produced by heat treatment of greenish-yellow beryl. Emeralds with surface cracks may be "oiled," that is, immersed in an oil of the same refractive index as emerald. The cracks, filled with oil, are thus obscured and the color of the stone deepened.

Imitation. A composite stone simulating emerald, called a triplet, is made with a crown and pavilion of colorless beryl (or colorless synthetic spinel) between which there is a layer of green-colored cement.

Chrysoberyl

Synthesis. Alexandrite, the variety of chrysoberyl that appears green in daylight and red in incandescent light, has been synthesized by flux growth. The properties of these synthetics, including the color change, are very similar to those of natural crystals.

Imitation. Synthetic corundum and, to a lesser extent, synthetic spinel, are made to resemble alexandrite. However, the color change in these synthetics is less marked than in alexandrite, being more violet in artificial light and a grayish-green in daylight.

Corundum (Ruby and Sapphire)

Synthesis. Synthetic corundum is produced most extensively by the Verneuil process. It is synthesized as ruby (red) and as sapphire both colorless and in a wide variety of colors, some of which do not occur in nature. Synthetic sapphire in colors usually displayed by other gems may be sold erroneously under the names of those gems; for example, purple as "amethyst" and yellow as "topaz." Since 1947 star rubies and blue star sapphires have been manufactured by the Verneuil method. As cabochon stones they rival in beauty the natural stones they imitate. Tiny bubbles and curved growth striae are characteristic of corundum grown by the Verneuil process.

Corundum is also synthesized hydrothermally and by crystal pulling and flux-growth methods. Rubies made by Kashan, Inc. of Austin, Texas, by flux growth lack the characterizing features of Verneuil-grown crystals and detection as a synthetic is difficult, and in some cases impossible. Moreover, the color of "Kashan rubies" is strikingly similar to Burmese rubies and, for this reason they command a relatively high price.

Treatment. The color of some rubies and blue sapphires can be enhanced by heat treatment. Gray sapphires heated to a high temperature in the proper gaseous environment may become a deep blue. Also a surface color can be induced by diffusion of impurities (e.g., TiO_2) into a stone during heat treatment.

Diamond

Synthesis. Using apparatus designed to maintain both high temperature and high pressure, diamonds were synthesized in 1955 by the General Electric Company (see also pages 346–349). Although these were small crystals, not suitable for gems, they signaled the beginning of an industry to manufacture abrasive diamond. By 1980 diamond grit was being made in several countries, with a total annual production of over 100 million carats. In 1970 the General Electric Company announced the synthesis of gem quality diamonds large enough (about 1 carat each) to be cut into gemstones. However, the process is so expensive that synthetic diamonds do not compete in the gem market with natural stones.

Treatment. Color can be produced or changed in diamonds by irradiation and heat treatment. Early in the twentieth century diamonds were colored green by exposure to radiation from radium compounds. This treatment induced not only color but a radioactivity that remains dangerously high. Today diamonds are colored green, without induced radioactivity, in a nuclear reactor and with a cyclotron. Heat treatment of these green stones changes the color to yellow or brown. Diamonds have been colored blue by high-energy electron bombardment. Distinguishing between treated and naturally colored stones presents a difficult problem for the gemologist.

Imitation. For a long time various colorless gems have been used to imitate diamond, but the properties of most of them are so different from those of diamond as to permit easy recognition. Zircon, with **RI** beyond the range of the jeweler's refractometer and relatively high dispersion, is most deceptive, but unlike diamond, it is birefringent. With the manufacture of gem materials that have no natural counterparts, detection of diamond simulants became more difficult. These substances are briefly described at the end of this chapter under "Manufactured Gem Materials Without Natural Counterparts." They and the minerals used to imitate diamond are listed in Table 15.2 with several diagnostic properties, in order of decreasing refractive index. For comparison, diamond and its properties are given first.

Jade

Jade is not a single mineral, but includes the pyroxene, jadeite, and the amphibole, tremolite-actinolite, known as nephrite. Although the two types of jade are quite distinct mineralogically, they are frequently difficult to distinguish from one another.

Treatment. Jadeite occurs in all colors but the most highly prized has a deep emerald-green color. White jadeite has been dyed the color of this "imperial" jade. Under magnification, the color can be seen distributed along veinlets and grain boundaries.

Imitation. Several minerals have been used as jade substitutes, some of which closely resemble true jade. These include massive green vesuvianite, *californite*; massive grossular garnet; and particularly *bowenite*, a translucent yellow-green serpentine.

Table 15.2
**PROPERTIES OF
DIAMOND AND DIAMOND
IMITATIONS**

Material	RI*	Disp†	Opt.‡	G§	H
Diamond	2.42	0.044	I	3.52	10
Rutile (syn.)	2.75±	0.33	A	4.26	6
Strontium titanate‖	2.41	0.19	I	5.13	5½
Cubic zirconia‖	2.16	0.06	I	5.9±	8
GGG‖	1.97	0.038	I	7.02	7
Zircon	1.95±	0.039	A	4.7	7+
YAG‖	1.83	0.028	I	4.55	8+
Sapphire (nat., syn.)	1.77	0.018	A	4.02	9
Spinel (syn.)	1.73±	0.020	I	3.64	8
Topaz	1.62±	0.014	A	3.5±	8
Beryl	1.58±	0.014	A	2.7	8
Quartz	1.55	0.013	A	2.65	7

*RI is average refractive index; ± indicates a variation greater than 0.01.
†Disp. is dispersion
‡Opt. I, isotropic; A, anisotropic.
§G is average specific gravity; ± indicates a variation of 0.1 or greater.
‖Manufactured materials without natural counterparts.

Other substitutes, more easily distinguished because of low hardness, are *pseudophite*, a massive form of chlorite; *agalmatolite*, a compact talc or pyrophyllite; and *verdite*, a green rock containing the chrome mica, *fuchsite*.

Opal

Synthesis. The determination in 1964 that opal is made up of close-packed silica spheres (page 267) pointed the way to its eventual synthesis. By developing a process that would closely duplicate this "structure," Pierre Gilson produced synthetic opal in 1974. These synthetics have chemical and physical properties, including a play of color, that are essentially identical to those of natural precious opal. Detection of the synthetic product is thus difficult. Under magnification in transmitted light, however, synthetic opal shows a mosaic structure not seen in natural opal.

Treatment. A play of color can be induced in some opals by immersing them in water, which apparently fills voids between the spheres. But as the water slowly leaves, so does the color effect. In such an opal the play of color is retained if it is impregnated with paraffin or plastic.

Imitation. A material simulating opal, called Opal-Essence, is manufactured by John S. Slocum of Rochester, Michigan. It is basically a glass with included platelets from which emanate color flashes. It can be distinguished from opal by its greater refractive index (1.49–1.51) and specific gravity (2.41–2.50). A realistic opal imitation is made of plastic spheres embedded in a plastic matrix, but it has much lower hardness and specific gravity than does opal.

Quartz

Synthesis. The many efforts to synthesize quartz during and immediately following World War II met with success in 1947, when optical grade crystals large enough for scientific uses were grown by the hydrothermal method. Since then increasingly large amounts have been produced each year, with an estimated world production today of over 700 tons! Most of this quartz is colorless and used for technological purposes. But colored quartz is also synthesized, duplicating the natural colors of citrine, amethyst, and smoky quartz. Other synthetic crystals are grown in colors in which natural quartz does not occur, for example, dark green and deep blue. If not of an unusual color, it is difficult if not impossible to distinguish synthetic from natural quartz in a cut stone.

Because it is general practice in quartz synthesis to use seed plates cut from single, untwinned crystals, the resulting synthetic product is also in single, untwinned crystals. Although natural citrine and amethyst gems may be single crystals, they are commonly twinned according to the Brazil law (optical twinning). Thus the presence of this twinning in a gemstone points to a natural origin.

Treatment. Most colorless quartz, natural and synthetic, will turn smoky on irradiation but, on heating, again becomes colorless. With heat treatment, amethyst usually becomes an orange-brown citrine but some turns green. In either case the violet amethyst color returns on irradiation.

Dyed Microcrystalline Quartz. Although agate and other types of chalcedony occur in several natural colors, much of this material used for ornamental purposes has been dyed. Depending on the dye, a gray to white agate may be rendered black or colored red, yellow, blue, or brown. Chert dyed the blue color of lapis lazuli has been sold under the name of "Swiss lapis."

Rutile

Although natural rutile is too dark to be used as a gem, synthetic rutile of gem quality has been produced by the Verneuil process since 1947. The boules as grown are black, but on annealing in oxygen, become transparent and pale yellow. When first made, synthetic rutile was a popular diamond substitute, but it is little used today. With its high refractive index (average 2.75) and extremely high dispersion (0.33) it outshines diamond and is thus easily detected. Furthermore, its high birefringence (0.287) causes a pronounced doubling of the back facets, giving a "fuzzy" appearance to cut stones.

Spinel

Synthesis. Synthetic spinel is made by the Verneuil process in a variety of colors not found in the mineral and, as such, is used more to imitate other gems than to simulate natural spinel. Most commonly it is colorless or blue to imitate diamond, blue sapphire, and aquamarine. Different shades of green imitate tourmaline, peridot, emerald, and chrysoberyl. Alexandrite-colored synthetic spinel is used to simulate alexandrite, and a cloudy form makes a rather convincing imitation of moonstone.

Synthetic spinel has the curved striations and gas bubbles characteristic of Verneuil-grown crystals. Its refractive index (1.73) and specific gravity (3.64) are both slightly greater than of natural spinel, which are respectively 1.72 and 3.60.

Turquoise

Synthesis. Synthetic turquoise was introduced to the gem market in 1972 by Pierre Gilson as Gilson-created Turquoise. It is of a fine color and made both with and without "matrix," but the method of manufacture has not been revealed.

Treatment. Natural turquoise may be impregnated with oil, paraffin, or plastic to enhance its color and increase its hardness. Such treatment can usually be detected by touching the material with a red-hot needle.

Imitation. Glass, dyed chalcedony or chert, enamel, and plastic are used to imitate turquoise. Powdered turquoise, when compacted and bonded with plastic, forms a realistic imitation of the naturally

massive material. When scratched with a knife, such impregnated turquoise tends to cut rather than to powder.

MANUFACTURED GEM MATERIALS WITHOUT NATURAL COUNTERPARTS

These manufactured materials that do not duplicate gem minerals are used primarily as diamond substitutes, although colored varieties of some make attractive cut stones in their own right. Included with them is, paradoxically, the mineral name *garnet*. It seems appropriate to do this because manufactured "garnets" differ both chemically and physically from natural garnets.

Garnet

The chemical composition of garnet is expressed by the general structural formula, $A_3B_2(SiO_4)_3$. Several different ions may occupy both the A and B sites, giving rise to a wide range in chemical composition. In synthetic garnets the general garnet structure is preserved, but unusual elements substitute in the A and B sites, and the silicon site may be occupied by aluminum or gallium. The synthetic "garnets" used as diamond imitations are rare earth compounds and usually are designated by the abbreviations "YAG" and "GGG," reflecting the elements present (see below).

YAG is yttrium aluminum "garnet," $Y_3Al_2(AlO_4)_3$. It is isometric with the following properties: **H** 8+, **G** 4.55, **RI** 1.833, disp. 0.028. In addition to colorless material, YAG is produced in red, yellow, green, and blue colors.

GGG is gadolinium gallium "garnet," $Gd_3Ga_2(GaO_4)_3$. It is isometric with the following properties; **H** $6\frac{1}{2}$, **G** 7.05, **RI** 1.97, disp. 0.038. Because of its higher refractive index and dispersion, GGG is a more impressive diamond substitute than YAG. Both YAG and GGG are produced by flux-growth and Czochralski techniques.

Strontium Titanate, $SrTiO_3$

Strontium titanate has been produced since 1955 by the Verneuil process. The boules as they come from the furnace are black due to a deficiency of oxygen, but on annealing in an oxygen atmosphere, they become quite colorless. Strontium titanate is isometric, with **H** $5\frac{1}{2}$ and **G** 5.13. The **RI** is 2.41, only slightly below that of diamond (2.42); but the dispersion, 0.19, over four times that of diamond (0.044), gives the faceted stone an excessive fire. Furthermore, because of the low hardness, polishing scratches and rounded facet junctions are usually present.

Cubic Zirconia

Since its introduction to the gem market in 1976, cubic zirconia, sometimes abbreviated CZ, has become the most important diamond simulant. It is isometric and has the following properties: **H** 8, **G** $5.9\pm$, **RI** $2.16\pm$, disp. 0.06. On consideration of these properties, it becomes immediately apparent why CZ is the most convincing diamond substitute. Because of its high hardness, it takes a good polish and resists scratching. The high refractive index produces a brilliance approaching that of diamond and the dispersion, although high, is not excessive.

A unique type of melt growth called "skull melting" has been devised for the manufacture of cubic zirconia; the melting point of ZrO_2 (2750°C) is so high that conventional crucibles cannot be used. The "skull" is a container that is open at the top with a cylindrical wall composed of closely spaced, vertical fingers of water-cooled copper tubes. The material to be melted is placed in the "skull," which remains relatively cool while, by use of a radio frequency generator, the central material is brought to the necessary temperature and melts within a shell of its own powder. On cooling, crystals grow upward from the bottom as CZ, the isometric polymorph of zirconia. To prevent the crystals from inverting to the low-temperature monoclinic form on further cooling, a stabilizer of yttrium oxide or calcium oxide is added to the ZrO_2 powder to be melted. The refractive index and specific gravity vary with the kind and amount of stabilizer.

REFERENCES AND SUGGESTED READING

Anderson, B. W., 1990, *Gem Testing*, 10th ed. Butterworth & Co., London.

Arem, J. E., 1987, *Color Encyclopedia of Gemstones*, 2nd ed. Van Nostrand Reinhold, New York, 248 pp., 64 color plates.

Branson, O. T., 1976, *Turquoise, the Gem of the Centuries*. Treasure Chest Publications, Inc., Tucson, Ariz., 62 pp.

Gübelin, E. J. and Koivula, J. I., 1986, *Photoatlas of Inclusions in Gemstones*. ABC Edition, Zurich, 532 pp.

Hurlbut, C. S., Jr. and Kammerling, R. C., 1991, *Gemology*. John Wiley & Sons, New York, 336 pp.

Keller, P., 1990, *Gemstones and Their Origins*. Van Nostrand Reinhold, New York, 144 pp.

Liddicoat, R. T., 1989, *Handbook of Gem Identification*, 12th ed. Gemological Institute of America, Santa Monica, Calif., 450 pp.

Nassau, K., 1980, *Gems Made by Man*. Gemological Institute of America, Santa Monica, Calif., 382 pp.

Sinkankas, 1959, 1976, *Gemstones of North America*, v. I, II. Van Nostrand Reinhold Co., New York.

Webster, R., 1983, *Gems*, 4th ed. Butterworth & Co., London.

APPENDIX

DETERMINATIVE TABLES

The determinative Table A.1 of this appendix is based on the physical properties of minerals that can be easily and quickly determined. These properties should be used with the understanding that for many minerals there is a variation in physical properties from specimen to specimen. Color for some minerals is constant but for others is extremely variable. Hardness, although more definite, may vary slightly and, by change in the state of aggregation of a mineral, may appear to vary more widely. Cleavage may also be obscured in a fine-grained aggregate of a mineral. In using the tables it is often impossible to differentiate between two or three similar species. However, by consulting the descriptions of these possible minerals in the sections on "Systematic Mineralogy" and the specific tests given there, a definite decision can usually be made.

In the determinative tables only the common minerals or those which, though rarer, are of economic importance have been included. The chances of having to determine a mineral that is not included in these tables are small, but it must be borne in mind that there is such a possibility. The names of the minerals have been printed in three different styles of type, as **PYRITE**, CHALCOCITE, and Covellite, in order to indicate their relative importance and frequency of occurrence. Whenever there is ambiguity in placing a mineral, it has been included in the two or more possible divisions.

The general scheme of classification is outlined on page 614. The proper division in which to look for a mineral can be determined by means of the tests indicated there. The tables are divided into two main sections on the basis of luster: (1) metallic and submetallic and (2) nonmetallic. The metallic minerals are opaque and give black or dark-colored streaks, whereas nonmetallic minerals are nonopaque and give either colorless or light-colored streaks. The tables are next subdivided according to hardness. These divisions are easily determined. For nonmetallic minerals they are: $<2\frac{1}{2}$, can be scratched by the fingernail; $>2\frac{1}{2}-<3$, cannot be scratched by fingernail, can be scratched by copper cent; $>3-<5\frac{1}{2}$, cannot be scratched by copper cent, can be scratched by a knife; $>5\frac{1}{2}-<7$, cannot be scratched by a knife, can be scratched by quartz; >7, cannot be scratched by quartz. Metallic minerals have fewer divisions of hardness; they are $<2\frac{1}{2}$; $>2\frac{1}{2}-<5\frac{1}{2}$; $>5\frac{1}{2}$.

Nonmetallic minerals are further subdivided according to whether they show a *prominent* cleavage. If the cleavage is not prominent and imperfect or obscure, the mineral is included with those that have no cleavage. Minerals in which cleavage may not be observed easily, because of certain conditions in the state of aggregation, have been included in both divisions.

The minerals that fall in a given division of the tables have been arranged according to various methods. In some cases, those that possess similar cleavages have been grouped together; frequently color determines the order, and so on. The column farthest to the left will indicate the method of arrangement.

The figures given in the column headed **G** are specific gravity. This is an important diagnostic physical property but less easily determined than luster, hardness, color, and cleavage. For a discussion of specific gravity and methods for its accurate determination, see page 256. However, if a specimen is pure and of sufficient size, its approximate specific gravity can be determined by simply weighing it in the hand. Below is a list of common minerals over a wide range of specific gravity. By experimenting with specimens of these one can become expert in the approximate determination of specific gravity. Following the determinative tables (Table A.1) is a list of the common minerals arranged according to increasing specific gravity (Table A.2).

Gypsum	2.32	Pyrite	5.02
Orthoclase	2.57	Arsenopyrite	6.07
Quartz	2.65	Cassiterite	6.95
Calcite	2.71	Galena	7.50
Fluorite	3.18	Cinnabar	8.10
Topaz	3.53	Copper	8.9
Corundum	4.02	Silver	10.5
Barite	4.45		

Many minerals are polymorphic. The crystal system given in the tables is that of the most commonly observed form.

DETERMINATIVE TABLES

General Classification of the Tables
Luster–Metallic or Submetallic
  I. Hardness: $<2\frac{1}{2}$. (Will leave a mark on paper.) Page 615
 II. Hardness: $>2\frac{1}{2}$, $<5\frac{1}{2}$. (Can be scratched by knife; will not readily leave a mark on paper.) Page 616
III. Hardness: $>5\frac{1}{2}$. (Cannot be scratched by knife.) Page 620

Luster—Nonmetallic
 I. Streak definitely colored. Page 622
II. Streak colorless.
  A. Hardness: $<2\frac{1}{2}$. (Can be scratched by fingernail.) Page 625
  B. Hardness: $>2\frac{1}{2}$, <3. (Cannot be scratched by fingernail; can be scratched by cent.)
    1. Cleavage prominent. Page 627
    2. Cleavage not prominent.
      a. A small splinter is fusible in the candle flame. Page 628
      b. Infusible in candle flame. Page 629
  C. Hardness: >3, $<5\frac{1}{2}$. (Cannot be scratched by cent; can be scratched by knife.)
    1. Cleavage prominent. Page 630
    2. Cleavage not prominent. Page 635
  D. Hardness: $>5\frac{1}{2}$, <7. (Cannot be scratched by knife; can be scratched by quartz.)
    1. Cleavage prominent. Page 638
    2. Cleavage not prominent. Page 640
  E. Hardness: >7. (Cannot be scratched by quartz.)
    1. Cleavage prominent. Page 644
    2. Cleavage not prominent. Page 645

Table A.1
**LUSTER: METALLIC
OR SUBMETALLIC**

I. Hardness: $<2\frac{1}{2}$

(Will leave a mark
on paper)

Streak	Color	G	H	Remarks	Name, Composition, Crystal System
Black	Iron-black	4.7	1–2	Usually splintery or in radiating fibrous aggregates.	**PYROLUSITE** p. 382 MnO_2 Tetragonal
	Steel-gray to iron-black	2.23	$1–1\frac{1}{2}$	Cleavage perfect {0001}. May be in hexagonal-shaped plates. Greasy feel.	**GRAPHITE** p. 349 C Hexagonal
Black to greenish black	Blue-black	4.7	$1–1\frac{1}{2}$	Cleavage perfect {0001}. May be in hexagonal-shaped leaves. Greenish streak on glazed porcelain (graphite, black). Greasy feel.	MOLYBDENITE p. 367 MoS_2 Hexagonal
Gray-black	Blue-black to lead-gray	7.6	$2\frac{1}{2}$	Cleavage perfect cubic {100}. In cubic crystals. Massive granular.	**GALENA** p. 354 PbS Isometric
	Blue-black	4.5	2	Cleavage perfect {010}. Bladed with cross striations. Fuses in candle flame.	**STIBNITE** p. 363 Sb_2S_3 Orthorhombic
Bright red	Red to vermillion	8.1	$2–2\frac{1}{2}$	Cleavage perfect {$10\bar{1}0$}. Luster adamantine. Usually granular massive.	**CINNABAR** p. 361 HgS Hexagonal
Red-brown	Red to vermillion	5.2	1+	Earthy. Frequently as pigment in rocks. Crystalline hematite is harder and black.	**HEMATITE** p. 379 Fe_2O_3 Hexagonal
Black. May mark paper	Gray-black	7.3	$2–2\frac{1}{2}$	Usually massive or earthy. Easily sectile. Bright steel-gray on fresh surfaces; darkens on exposure.	ACANTHITE p. 352 Ag_2S Morph.: isometric
	Blue; may tarnish to blue-black	4.6	$1\frac{1}{2}–2$	Usually in platy masses or in thin 6-sided platy crystals. Moistened with water turns purple.	COVELLITE p. 361 CuS Hexagonal

Table A.1 (Cont.)

LUSTER: METALLIC OR SUBMETALLIC

II. Hardness: $>2\frac{1}{2}$, $<5\frac{1}{2}$

(Can be scratched by a knife; will not readily leave a mark on paper)

Streak	Color	G	H	Remarks	Name, Composition, Crystal System
Black	Iron-black	4.7	1–2	Usually splintery or in radiating fibrous aggregates.	**PYROLUSITE** p. 382 MnO_2 Tetragonal
	Gray-black	4.4	3	Cleavage {110}. Usually in bladed masses showing cleavage. Associated with other copper minerals.	ENARGITE p. 369 Cu_3AsS_4 Orthorhombic
	Pale copper-red. May be silver-white, pinkish	7.8	5–5$\frac{1}{2}$	Usually massive. May be coated with green nickel bloom. Associated with cobalt and nickel minerals.	NICKELINE p. 359 $NiAs$ Hexagonal
	Fresh surface brownish-bronze; purple tarnish	5.1	3	Usually massive. Associated with other copper minerals; chiefly chalcocite and chalcopyrite.	**BORNITE** p. 353 Cu_5FeS_4 Morph.: isometric
	Brownish-bronze	4.6	4	Small fragments magnetic. Usually massive. Often associated with chalcopyrite and pyrite.	**PYRRHOTITE** p. 358 $Fe_{1-x}S$ Morph.: hexagonal
		4.6 to 5.0	3$\frac{1}{2}$–4	Octahedral parting. Resembles pyrrhotite with which it usually is associated but nonmagnetic.	Pentlandite p. 360 $(Fe,Ni)_9S_8$ Isometric
	Brass-yellow	4.1 to 4.3	3$\frac{1}{2}$–4	Usually massive, but may be in crystals resembling tetrahedrons. Associated with other copper minerals and pyrite.	**CHALCOPYRITE** p. 357 $CuFeS_2$ Tetragonal
	Brass-yellow. Slender crystals greenish	5.5	3–3$\frac{1}{2}$	Cleavage {10$\bar{1}$1}, rarely seen. Usually in radiating groups of hairlike crystals.	MILLERITE p. 360 NiS Hexagonal
	Blue; may tarnish to blue-black	4.6	1$\frac{1}{2}$–2	Usually in platy masses or in thin 6-sided platy crystals. Moistened with water turns purple.	Covellite p. 361 CuS Hexagonal

Table A.1 (Cont.)

LUSTER: METALLIC OR SUBMETALLIC

II. Hardness: $>2\frac{1}{2}$, $<5\frac{1}{2}$

(Can be scratched by a knife; will not readily leave a mark on paper)
(Continued)

Streak	Color	G	H	Remarks	Name, Composition, Crystal System
Gray-black. Will mark paper	Lead-gray	4.6	2	Cleavage perfect {010}. Fuses easily in candle flame. In bladed crystal aggregates with cross striations.	**STIBNITE** p. 363 Sb_2S_3 Orthorhombic
		7.5	$2\frac{1}{2}$	Cleavage perfect {100}. In cubic crystals and granular masses. If held in the candle flame does not fuse but small globules of metallic lead collect on the surface.	**GALENA** p. 354 PbS Isometric
Black. May mark paper	Steel-gray on fresh surface. Tarnishes to dull gray	7.3	$2-2\frac{1}{2}$	Easily sectile. Usually massive or earthy. Rarely in cubic crystals.	Acanthite p. 352 Ag_2S Morph.: isometric
Usually black. May be brownish	Black	3.7 to 4.7	5–6	Massive botryoidal and stalactitic. Usually associated with pyrolusite.	ROMANECHITE p. 394 $BaMn^{2+}Mn^{4+}O_{16}(OH)_4$ Appears amorphous
Black; may have brown tinge	Steel-gray. May tarnish to dead black on exposure	4.7 to 5.0	$3-4\frac{1}{2}$	Massive or in tetrahedral crystals. Often associated with silver ores.	TETRAHEDRITE p. 370 $(Cu,Fe,Zn,Ag)_{12}Sb_4S_{13}$ Isometric
Gray-black		5.7	$2\frac{1}{2}-3$	Somewhat sectile. Usually compact massive. Associated with other copper minerals.	CHALCOCITE p. 353 Cu_2S Morph: orthorhombic
Dark brown to black	Iron-black to brownish-black	4.6	$5\frac{1}{2}$	Luster pitchy. May be accompanied by yellow or green oxidation products. Usually in masses in peridotites.	**CHROMITE** p. 389 $FeCr_2O_4$ Isometric
	Brown to black	7.0 to 7.5	$5-5\frac{1}{2}$	Cleavage perfect {010}. With greater amounts of Mn streak and color are darker.	WOLFRAMITE p. 430 $(Fe,Mn)WO_4$ Monoclinic

Table A.1 (Cont.)

LUSTER: METALLIC OR SUBMETALLIC

II. Hardness: $>2\frac{1}{2}$, $<5\frac{1}{2}$

(Can be scratched by a knife; will not readily leave a mark on paper) (*Continued*)

Streak	Color	G	H	Remarks	Name, Composition, Crystal System
Dark brown to black	Steel-gray to iron-black	4.3	4	In radiating fibrous or crystalline masses. Distinct prismatic crystals often grouped in bundles. Associated with pyrolusite.	MANGANITE p. 394 MnO(OH) Morph.: orthorhombic
Light to dark brown	Dark brown to coal-black. More rarely yellow or red	3.9 to 4.1	$3\frac{1}{2}$–4	Cleavage perfect {110} (6 directions). Usually cleavable, granular; may be in tetrahedral crystals. The streak is always of a lighter color than the specimen.	**SPHALERITE** p. 356 ZnS Isometric
Red-brown to brown-red	Dark brown to steel-gray to black	4.8 to 5.3	$5\frac{1}{2}$–$6\frac{1}{2}$	Usually harder than knife. Massive, radiating, reniform, micaceous.	**HEMATITE** p. 379 Fe_2O_3 Hexagonal
	Deep red to black	5.85	$2\frac{1}{2}$	Cleavage {10$\bar{1}$1}. Fusible in candle flame. Shows dark ruby-red color in thin splinters. Associated with other silver minerals.	Pyrargyrite p. 370 Ag_3SbS_3 Hexagonal
	Red-brown to deep red. Ruby-red if transparent	6.0	$3\frac{1}{2}$–4	Massive or in cubes or octahedrons. May be in very slender crystals. Associated with malachite, azurite, native copper.	CUPRITE p. 376 Cu_2O Isometric
Bright red	Ruby-red	5.55	2–$2\frac{1}{2}$	Cleavage {10$\bar{1}$1}. Fusible in candle flame. Associated with pyrargyrite.	Proustite p. 370 Ag_3AsS_3 Hexagonal
Yellow-brown. Yellow ocher	Dark brown to black	4.37	5–$5\frac{1}{2}$	Cleavage {010}. In radiating fibers, mammillary and stalactitic forms. Rarely in crystals.	**GOETHITE** p. 395 $\alpha FeO(OH)$ Orthorhombic
Dark red. (Some varieties mark paper)	Dark red to vermillion	8.10	$2\frac{1}{2}$	Cleavage {10$\bar{1}$1}. Usually granular or earthy. Commonly impure and dark red or brown. When pure, translucent or transparent and bright red.	**CINNABAR** p. 361 HgS Hexagonal

Table A.1 (Cont.)

LUSTER: METALLIC OR SUBMETALLIC

II. Hardness: $>2\frac{1}{2}$, $<5\frac{1}{2}$

(Can be scratched by a knife; will not readily leave a mark on paper) (*Continued*)

Streak	Color	G	H	Remarks	Name, Composition, Crystal System
Copper-red, shiny	Copper-red on fresh surface; black tarnish	8.9	$2\frac{1}{2}$–3	Malleable. Usually in irregular grains. May be in branching crystal groups or in rude isometric crystals.	COPPER p. 339 Cu Isometric
Silver-white, shiny	Silver-white on fresh surface. Black tarnish	10.5	$2\frac{1}{2}$–3	Malleable. Usually in irregular grains. May be in wire, plates, branching crystal groups.	SILVER p. 338 Ag Isometric
Gray, shiny	White or steel-gray	14 to 19	4–$4\frac{1}{2}$	Malleable. Irregular grains or nuggets. Unusually hard for a metal. Rare.	Platinum p. 340 Pt Isometric
Gold-yellow, shiny	Gold-yellow	15.0 to 19.3	$2\frac{1}{2}$–3	Malleable. Irregular grains, nuggets, leaves. Very heavy; specific gravity varies with silver content.	GOLD p. 336 Au Isometric

Table A.1 (Cont.)

LUSTER: METALLIC OR SUBMETALLIC

III. Hardness: $>5\frac{1}{2}$

(Cannot be scratched by a knife)

Streak	Color	G	H	Remarks	Name, Composition, Crystal System
Black	Silver- or tin-white	6.0 to 6.2	$5\frac{1}{2}$–6	Usually massive. Crystals pseudo-orthorhombic.	**ARSENOPYRITE** p. 368 FeAsS Monoclinic
		6.1 to 6.9	$5\frac{1}{2}$	Usually massive. Crystals pyritohedral. May be coated with pink nickel bloom.	Skutterudite-Nickel skutterudite p. 369 $(Co,Ni,Fe)As_3$-$(Ni,Co,Fe)As_3$ Isometric
		6.33	$5\frac{1}{2}$	Commonly in pyritohedral crystals with pinkish cast. Also massive.	Cobaltite-Gersdorffite p. 368 $(Co,Fe)AsS$-$NiAsS$ Pseudo-isometric
	Copper-red to silver-white with pink tone	7.5	5–$5\frac{1}{2}$	Usually massive. May be coated wtih green nickel bloom.	NICKELINE p. 359 NiAs Hexagonal
	Pale brass-yellow	5.0	6–$6\frac{1}{2}$	Often in pyritohedrons or striated cubes. Massive granular. Most common sulfide.	**PYRITE** p. 365 FeS_2 Isometric
	Pale yellow to almost white	4.9	6–$6\frac{1}{2}$	Frequently in "cock's comb" crystal groups and radiating fibrous masses.	**MARCASITE** p. 366 FeS_2 Isometric
	Black	5.18	6	Strongly magnetic. Crystals octahedral. May show octahedral parting.	**MAGNETITE** p. 387 Fe_3O_4 Isometric
Dark brown to black	Black	9.0 to 9.7	$5\frac{1}{2}$	Luster pitchy. Massive granular, botryoidal crystals.	Uraninite p. 384 UO_2 Isometric
		4.7	$5\frac{1}{2}$–6	May be slightly magnetic. Often associated with magnetite. Massive granular; platy crystals; as sand.	**ILMENITE** p. 380 $FeTiO_3$ Hexagonal
		3.7 to 4.7	5–6	Compact massive, stalactitic, botryoidal. Associated with other manganese minerals and told by greater hardness.	ROMANECHITE p. 394 $BaMn^{2+}Mn^{4+}O_{16}(OH)_4$ Orthorhombic

Table A.1 (Cont.)

LUSTER: METALLIC OR SUBMETALLIC

III. Hardness: $>5\frac{1}{2}$

(Cannot be scratched by a knife)
(*Continued*)

Streak	Color	G	H	Remarks	Name, Composition, Crystal System
Dark brown to black	Black	5.3 to 7.3	6	Luster black and shiny on fresh surface. May have slight bluish tarnish. Granular or in stout prismatic crystals	Columbite-Tantalite p. 390 $(Fe,Mn)(Nb,Ta)_2O_6$ Orthorhombic
Dark brown	Iron-brown to brownish-black	7.0 to 7.5	$5-5\frac{1}{2}$	Cleavage perfect {010}. With greater amounts of Mn streak and color are darker.	WOLFRAMITE p. 430 $(Fe,Mn)WO_4$ Monoclinic
		4.6	$5\frac{1}{2}$	Luster pitchy. Frequently accompanied by green oxidation products. Usually in granular masses in peridotites.	**CHROMITE** p. 389 $FeCr_2O_4$ Isometric
		5.15	6	Slightly magnetic. Granular or in octahedral crystals. Common only at Franklin, N.J., associated with zincite and willemite.	FRANKLINITE p. 388 $(Fe,Zn,Mn)(Fe,Mn)_2O_4$ Isometric
Red-brown, brown-red	Dark brown to steel-gray to black	4.8 to 5.3	$5\frac{1}{2}-6\frac{1}{2}$	Radiating, reniform, massive, micaceous. Rarely in steel-black rhombohedral crystals. Some varieties softer.	**HEMATITE** p. 379 Fe_2O_3 Hexagonal
Pale brown	Brown to black	4.18 to 4.25	$6-6\frac{1}{2}$	In prismatic crystals, vertically striated; often slender acicular. Crystals frequently twinned. Found in black sands.	**RUTILE** p. 381 TiO_2 Tetragonal
Yellow-brown to yellow-ocher	Dark brown to black	4.37	$5-5\frac{1}{2}$	Cleavage {010}. Radiating, colloform, stalactitic.	**GOETHITE** p. 395 $\alpha FeO(OH)$ Orthorhombic

Table A.1 (Cont.)
LUSTER: NONMETALLIC

I. Streak definitely colored

Streak	Color	G	H	Remarks	Name, Composition, Crystal System
Dark red	Dark red to vermillion	8.10	$2\frac{1}{2}$	Cleavage {10$\bar{1}$0}. Usually granular or earthy. Commonly impure and dark red or brown. When pure, translucent or transparent and bright red.	**CINNABAR** p. 361 HgS Hexagonal
	Red-brown. Ruby-red when transparent	6.0	$3\frac{1}{2}$–4	Massive or in cubes or octahedrons. May be in very slender crystals. Associated with malachite, azurite, native copper.	CUPRITE p. 376 Cu$_2$O Isometric
Red-brown	Dark brown to steel-gray to black	4.8 to 5.3	$5\frac{1}{2}$–$6\frac{1}{2}$	Radiating, reniform, massive, micaceous. Rarely in steel-black rhombohedral crystals. Some varieties softer.	**HEMATITE** p. 379 Fe$_2$O$_3$ Hexagonal
	Deep red to black	5.8	$2\frac{1}{2}$	Cleavage {10$\bar{1}$1}. Fusible in candle flame. Shows dark ruby-red color in thin splinters. Associated with other silver minerals.	Pyrargyrite p. 370 Ag$_3$SbS$_3$ Hexagonal
Bright red	Ruby-red	5.55	2–$2\frac{1}{2}$	Cleavage {10$\bar{1}$1}. Fusible in candle flame. Light "ruby silver." Associated with pyrargyrite.	Proustite p. 370 Ag$_3$AsS$_3$ Hexagonal
Pink	Red to pink	2.95	$1\frac{1}{2}$–$2\frac{1}{2}$	Cleavage perfect {010}. Usually reniform or as pulverulent or earthy crusts. Found as coatings on cobalt minerals.	Erythrite (cobalt bloom) p. 436 Co$_3$(AsO$_4$)$_2$·8H$_2$O Monoclinic
Yellow-brown to yellow-ocher	Dark brown to black	4.4	5–$5\frac{1}{2}$	Cleavage {010}. In radiating fibers, mammillary and stalactitic forms. Rarely in crystals.	**GOETHITE** p. 395 αFeO(OH) Orthorhombic

Table A.1 (Cont.)
LUSTER: NONMETALLIC

I. Streak definitely colored
(*Continued*)

Streak	Color	G	H	Remarks	Name, Composition, Crystal System
Brown	Dark brown	7.0 to 7.5	$5-5\frac{1}{2}$	Cleavage perfect {010}. With greater amounts of manganese the streak and color are darker.	WOLFRAMITE p. 430 $(Fe,Mn)WO_4$ Monoclinic
	Light to dark brown	3.83 to 3.88	$3\frac{1}{2}-4$	In cleavable masses or in small curved rhombohedral crystals. Becomes magnetic after heating in candle flame.	**SIDERITE** p. 409 $FeCO_3$ Hexagonal
Light brown	Light to dark brown	3.9 to 4.1	$3\frac{1}{2}-4$	Cleavage perfect {011} (6 directions). Usually cleavable, granular; may be in tetrahedral crystals. The streak is always of a lighter color than the specimen.	**SPHALERITE** p. 356 ZnS Isometric
	Brown to black	6.8 to 7.1	6–7	Occurs in twinned crystals. Fibrous, reniform and irregular masses; in rolled grains.	**CASSITERITE** p. 383 SnO_2 Tetragonal
	Reddish brown to black	4.18 to 4.25	$6-6\frac{1}{2}$	Crystals vertically striated; often acicular. Twinning common.	**RUTILE** p. 381 TiO_2 Tetragonal
Orange-yellow	Deep red to orange-yellow	5.68	$4-4\frac{1}{2}$	Cleavage {0001}. Found only at Franklin, N.J., associated with franklinite and willemite.	ZINCITE p. 377 ZnO Hexagonal
	Bright red	5.9 to 6.1	$2\frac{1}{2}-3$	Luster adamantine. In long slender crystals, often interlacing groups. Decrepitates in candle flame.	Crocoite p. 427 $PbCrO_4$ Monoclinic
	Deep red	3.48	$1\frac{1}{2}-2$	Frequently earthy. Associated with orpiment. Fusible in candle flame.	REALGAR p. 362 AsS Monoclinic

Table A.1 (Cont.)
LUSTER: NONMETALLIC

I. Streak definitely colored
(*Continued*)

Streak	Color	G	H	Remarks	Name, Composition, Crystal System
Pale yellow	Lemon-yellow	3.49	$1\frac{1}{2}$–2	Cleavage {010}. Luster resinous. Associated with realgar. Fusible in candle flame.	ORPIMENT p. 363 As_2S_3 Monoclinic
	Pale yellow	2.05 to 2.09	$1\frac{1}{2}$–$2\frac{1}{2}$	Burns with blue flame giving odor of SO_2. Crystallized, granular, earthy.	**SULFUR** p. 344 S Orthorhombic
Light green	Dark emerald-green	3.75 to 3.77	3–$3\frac{1}{2}$	One perfect cleavage {010}. In granular cleavable masses or small prismatic crystals.	Atacamite p. 402 $Cu_2Cl(OH)_3$ Orthorhombic
Light green	Dark emerald-green	3.9±	$3\frac{1}{2}$–4	One good cleavage {010}. In small prismatic crystals or granular masses.	Antlerite p. 429 $Cu_3(SO_4)(OH)_4$ Orthorhombic
	Bright green	3.9 to 4.03	$3\frac{1}{2}$–4	Radiating fibrous, mammillary. Associated with azurite and may alter to it. Effervesces in cold acid.	**MALACHITE** p. 417 $Cu_2CO_3(OH)_2$ Monoclinic
Light blue	Intense azure-blue	3.77	$3\frac{1}{2}$–4	In small crystals, often in groups. Radiating fibrous, usually as alteration from malachite. Effervesces in cold acid.	**AZURITE** p. 417 $Cu_3(CO_3)_2(OH)_2$ Monoclinic
Very light blue	Light green to turquoise-blue	2.0 to 2.4	2–4	Massive compact. Associated with oxidized copper minerals.	CHRYSOCOLLA p. 523 ~$Cu_4H_4Si_4O_{10}(OH)_8$ Amorphous

Table A.1 (Cont.)

LUSTER: NONMETALLIC

II. Streak colorless

A. Hardness: <2$\frac{1}{2}$

(Can be scratched by fingernail)

Cleavage, Fracture		Color	G	H	Remarks	Name, Composition, Crystal System
Perfect cleavage in one direction	The micas or related micaceous minerals, which possess such a perfect cleavage that they can be split into exceedingly thin sheets. They may occur as aggregates of minute scales, when the micaceous structure may not be readily apparent.	Pale brown, green, yellow, white	2.76 to 2.88	2–2$\frac{1}{2}$	In foliated masses and scales. Crystals tabular with hexagonal or diamond-shaped outline. Cleavage flakes elastic.	**MUSCOVITE** p. 515 $KAl_2(AlSi_3O_{10})(OH)_2$ Monoclinic
		Dark brown, green to black; may be yellow	2.95 to 3	2$\frac{1}{2}$–3	Usually in irregular foliated masses. Crystals have hexagonal outline, but rare. Cleavage flakes elastic.	**BIOTITE** p. 517 $K(Mg,Fe)_3$-$(AlSi_3O_{10})(OH)_2$ Monoclinic
		Yellowish-brown, green white	2.86	2$\frac{1}{2}$–3	Often in 6-sided tabular crystals; in irregular foliated masses. May show copperlike reflection from cleavage.	PHLOGOPITE p. 517 $KMg_3(AlSi_3O_{10})(OH)_2$ Monoclinic
		Green of various shades	2.6 to 2.9	2–2$\frac{1}{2}$	Usually in irregular foliated masses. May be in compact masses of minute scales. Thin sheets flexible but not elastic.	**CHLORITE** p. 519 $Mg,Fe_3(Si,Al)_4$-$O_{10}(OH)_2(Mg,$-$Fe)_3(OH)_6$ Monoclinic
		White, apple-green, gray. When impure, as in soapstone, dark gray, dark green to almost black	2.7 to 2.8	1	Greasy feel. Frequently distinctly foliated or micaceous. Cannot be positively identified by physical tests.	**TALC** p. 514 $Mg_3Si_4O_{10}(OH)_2$ Monoclinic
			2.8 to 2.9	1–2		Pyrophyllite p. 514 $Al_2Si_4O_{10}(OH)_2$ Monoclinic
		White, gray, green	2.39	2$\frac{1}{2}$	Pearly luster on cleavage face, elsewhere vitreous. Sectile. Commonly foliated massive, may be in broad tabular crystals. Thin sheets flexible but not elastic.	BRUCITE p. 393 $Mg(OH)_2$ Hexagonal
{001} perfect; seldom seen. Fracture earthy.		White; may be darker	2.6 to 2.63	2–2$\frac{1}{2}$	Compact, earthy. Breathed upon, gives argillaceous odor. Will adhere to the dry tongue.	**KAOLINITE** p. 512 $Al_2Si_2O_5(OH)_4$ Triclinic
Cubic {100}		Colorless or white	1.99	2	Soluble in water, bitter taste. Resembles halite but softer. In granular cleavable masses or cubic crystals.	Sylvite p. 400 KCl Isometric

Table A.1 (Cont.)

LUSTER: NONMETALLIC

II. Streak colorless

A. Hardness: $<2\frac{1}{2}$

(Can be scratched by fingernail) (*Continued*)

Cleavage, Fracture	Color	G	H	Remarks	Name, Composition, Crystal System
{010} perfect, {100}, {011} good	Colorless, white, gray. May be colored by impurities	2.32	2	Occurs in crystals, broad cleavage flakes. May be compact massive without cleavage, or fibrous with silky luster.	**GYPSUM** p. 427 $CaSO_4 \cdot 2H_2O$ Monoclinic
Rhomb. {10$\bar{1}$1} poor		2.29	1–2	Occurs in saline crusts. Readily soluble in water; cooling and salty taste. Fusible in candle flame.	NITRATITE p. 418 $NaNO_3$ Hexagonal
Prismatic {110}, seldom seen. Fracture conchoidal	Colorless or white	2.09 to 2.14	2	Usually in crusts, silky tufts and delicate acicular crystals. Readily soluble in water; cooling and salty taste. Fusible in the candle flame.	Niter, p. 418 KNO_3 Orthorhombic
Fracture uneven	Pearl-gray or colorless. Turns to pale brown on exposure to light	5.5±	2–3	Perfectly sectile. Translucent in thin plates. In irregular masses, rarely in crystals. Distinguished from other silver halides only by chemical tests.	Chlorargyrite p. 400 AgCl Isometric
	Pale yellow	2.05 to 2.09	1$\frac{1}{2}$–2$\frac{1}{2}$	Burns with blue flame, giving odor of SO_2. Crystallized, granular, earthy.	**SULFUR** p. 344 S Orthorhombic
	Yellow, brown, gray, white	2.0 to 2.55	1–3	In rounded grains, often earthy and claylike. Usually harder than 2$\frac{1}{2}$.	**BAUXITE** p. 397 A mixture of Al hydroxides
Cleavage seldom seen	White	1.95	1	Usually in rounded masses of fine fibers and acicular crystals.	ULEXITE p. 422 $NaCaB_5O_6(OH) \cdot 5H_2O$ Triclinic

Table A.1 (Cont.)
LUSTER: NONMETALLIC

II. Streak colorless

B. Hardness: $>2\frac{1}{2}$, <3

(Cannot be scratched by fingernail; can be scratched by cent)

1. Cleavage prominent

Cleavage, Fracture		Color	G	H	Remarks	Name, Composition, Crystal System
Perfect cleavage in one direction. See also the minerals of the mica group, p. 515 which may be harder than the fingernail.	{001}	Lilac, grayish white	2.8 to 3.0	$2\frac{1}{2}$–4	Crystals 6-sided prismatic. Usually in small irregular sheets and scales. A pegmatite mineral.	LEPIDOLITE p. 519 $K(Li,Al)_{2-3}(AlSi_3O_{10})$-$(O,OH,F)_2$ Monoclinic
	{001}	Pink, gray, white	3.0 to 3.1	$3\frac{1}{2}$–5	Usually in irregular foliated masses; folia brittle. Associated with emery.	MARGARITE p. 519 $CaAl_2(Al_2Si_2O_{10})(OH)_2$ Monoclinic
	{010}	Colorless or white	4.3	$3\frac{1}{2}$	Usually massive, with radiating habit. Effervesces in cold acid.	**WITHERITE** p. 413 $BaCO_3$ Orthorhombic
Cleavage in 2 directions {001} {100}		Colorless or white to gray	1.95	3	Occurs in cleavable crystalline aggregates.	KERNITE p. 419 $Na_2B_4O_6(OH)_2 \cdot 3H_2O$ Monoclinic
Cleavage in 3 directions at right angles	Cubic {100}	Colorless, white, red, blue	2.1 to 2.3	$2\frac{1}{2}$	Common salt. Soluble in water, taste salty, fusible in candle flame. In granular masses or in cubic crystals.	**HALITE** p. 399 NaCl Isometric
	Cubic {100}	Colorless or white	1.99	2	Resembles halite, but distinguished from it by more bitter taste and lesser hardness.	Sylvite p. 400 KCl Isometric
	{001} {010} {100}	Colorless, white, blue, gray, red	2.89 to 2.98	3–$3\frac{1}{2}$	Commonly in massive aggregates, not showing cleavage; then distinguished only by chemical tests.	**ANHYDRITE** p. 426 $CaSO_4$ Orthorhombic
Cleavage in 3 directions not at right angles. Rhombohedral {10$\bar{1}$1}		Colorless, white, and variously tinted	2.71	3	Effervesces in cold acid. Crystals show many forms. Occurs as limestone and marble. Clear varieties show strong double refraction.	**CALCITE** p. 405 $CaCO_3$ Hexagonal
Cleavage in 3 directions not at right angles. Rhombohedral {10$\bar{1}$1}.		Colorless, white, pink	2.85	$3\frac{1}{2}$–4	Usually harder than 3. Often in curved rhombohedral crystals with pearly luster; as dolomitic limestone and marble. Powdered mineral will effervesce in cold acid.	**DOLOMITE** p. 415 $CaMg(CO_3)_2$ Hexagonal

Table A.1 (Cont.)

LUSTER: NONMETALLIC

II. Streak colorless

B. Hardness: >2½, <3

(Cannot be scratched by fingernail; can be scratched by cent)

1. Cleavage prominent
(*Continued*)

Cleavage, Fracture	Color	G	H	Remarks	Name, Composition, Crystal System
Cleavage in 3 directions. Basal {001} at right angles to prismatic {210}	Colorless, white, blue, yellow red	4.5	3–3½	Frequently in aggregates of platy crystals. Pearly luster on basal cleavage. Characterized by high specific gravity and thus distinguished from celestite.	**BARITE** p. 424 $BaSO_4$ Orthorhombic
	Colorless, white, blue, red	3.95 to 3.97	3–3½	Similar to barite but lower specific gravity.	**CELESTITE** p. 425 $SrSO_4$ Orthorhombic
	Colorless or white. Gray-brown when impure	6.2 to 6.4	3	Adamantine luster. Usually massive but may be in small tabular crystals. Alteration of galena. When massive may need test for SO_4 to distinguish from cerrusite ($PbCO_3$).	**ANGLESITE** p. 426 $PbSO_4$ Orthorhombic

2. Cleavage not prominent

a. Small splinter is fusible in the candle flame

Color	G	H	Remarks	Name, Composition, Crystal System
Colorless or white	1.7±	2–2½	Soluble in water. One good cleavage, seldom seen. In crusts and prismatic crystals. In candle flame swells and then fuses. Sweetish alkaline taste.	**BORAX** p. 421 $Na_2B_4O_5(OH)_4 \cdot 8H_2O$ Monoclinic
	2.95 to 3.0	2½	Massive with peculiar translucent appearance. Fine powder becomes nearly invisible when placed in water. Ivigtut, Greenland, only important locality. Pseudocubic parting.	CRYOLITE p. 400 Na_3AlF_6 Monoclinic
	6.55	3–3½	Luster adamantine. In granular masses and platy crystals, usually associated with galena. Effervesces in cold nitric acid.	**CERUSSITE** p. 414 $PbCO_3$ Orthorhombic

Table A.1 (Cont.)
LUSTER: NONMETALLIC

II. Streak colorless

B. Hardness: $>2\frac{1}{2}$, <3

(Cannot be scratched by fingernail; can be scratched by cent)

2. Cleavage not prominent

b. Infusible in candle flame

Color	G	H	Remarks	Name, Composition, Crystal System
Colorless or white	4.3	$3\frac{1}{2}$	Often in radiating masses; granular; rarely in pseudohexagonal crystals. Effervesces in cold acid.	**WITHERITE** p. 413 $BaCO_3$ Orthorhombic
	2.6 to 2.63	$2-2\frac{1}{2}$	Usually compact, earthy. When breathed upon gives an argillaceous odor.	**KAOLINITE** p. 512 $Al_2Si_2O_5(OH)_4$ Triclinic
Colorless, white, blue, gray, red	2.89 to 2.98	$3-3\frac{1}{2}$	Commonly in massive fine aggregates, not showing cleavage, and can be distinguished only by chemical tests.	**ANHYDRITE** p. 426 $CaSO_4$ Orthorhombic
Yellow, brown, gray, white	2.0 to 2.55	$1-3$	Usually pisolitic; in rounded grains and earthy masses. Often impure.	**BAUXITE** p. 397 A mixture of aluminum hydroxides
Ruby-red, brown, yellow	6.7 to 7.1	3	Luster resinous. In slender prismatic and cavernous crystals; in barrel-shaped forms.	Vanadinite p. 436 $Pb_5(VO_4)_3Cl$ Hexagonal
Yellow, green, white, brown	2.33	$3\frac{1}{2}-4$	Characteristically in radiating hemispherical globular aggregates. Cleavage seldom seen.	Wavellite p. 437 $Al_3(PO_4)_2(OH)_3 \cdot 5H_2O$ Orthorhombic
Olive- to blackish-green, yellow-green, white	2.3 to 2.66	$2-5$	Massive. Fibrous in the asbestos variety, chrysotile. Frequently mottled green in the massive variety.	**SERPENTINE** p. 507 $Mg_3Si_2O_5(OH)_4$ Monoclinic

Table A.1 (Cont.)

LUSTER: NONMETALLIC

II. Streak colorless

C. Hardness: >3, <$5\frac{1}{2}$

(Cannot be scratched by cent; can be scratched by knife)

1. Cleavage prominent

One cleavage direction

Cleavage	Color	G	H	Remarks	Name, Composition, Crystal System
{100}	Blue, usually darker at center	3.56 to 3.66	5–7	In bladed aggregates with cleavage parallel to length. Can be scratched by knife parallel to the length of the crystal but not in a direction at right angles to this.	**KYANITE** p. 456 Al_2SiO_5 Triclinic
{010}	White, yellow, brown, red	2.1 to 2.2	$3\frac{1}{2}$–4	Characteristically in sheaflike crystal aggregates. May be in flat tabular crystals. Luster pearly on cleavage face.	**STILBITE** p. 556 $NaCa_2Al_5Si_{13}O_{36} \cdot 14H_2O$ Monoclinic
{001}	Colorless, white, pale green, yellow, rose	2.3 to 2.4	$4\frac{1}{2}$–5	In prismatic crystals vertically striated. Crystals often resemble cubes truncated by the octahedron. Luster pearly on base, elsewhere vitreous.	APOPHYLLITE p. 522 $KCa_4(Si_4O_{10})_2F \cdot 8H_2O$ Tetragonal
{010}	White, yellow, red	2.18 to 2.20	$3\frac{1}{2}$–4	Luster pearly on cleavage, elsewhere vitreous. Crystals often tabular parallel to cleavage plane. Found in cavities in igneous rocks.	HEULANDITE p. 555 $CaAl_2Si_7O_{18} \cdot 6H_2O$ Monoclinic
{010}	Colorless, white	2.42	4–$4\frac{1}{2}$	In crystals and in cleavable aggregates. Decrepitates violently in the candle flame.	COLEMANITE p. 422 $CaB_3O_4(OH)_3 \cdot H_2O$ Monoclinic
{010}	Colorless, white	4.3	$3\frac{1}{2}$	Often in radiating crystal aggregates; granular. Rarely in pseudohexagonal crystals.	WITHERITE p. 413 $BaCO_3$ Orthorhombic
{010} {110} poor	Colorless, white	2.95	$3\frac{1}{2}$–4	Effervesces in cold acid. Frequently in radiating groups of acicular crystals; in pseudohexagonal twins.	**ARAGONITE** p. 411 $CaCO_3$ Orthorhombic

Table A.1 (Cont.)
LUSTER: NONMETALLIC

II. Streak colorless

C. Hardness: >3, <5½

(Cannot be scratched by cent; can be scratched by knife)

1. Cleavage prominent
(*Continued*)

Two cleavage directions

Cleavage	Color	G	H	Remarks	Name, Composition, Crystal System
{001} {100}	Bluish gray, salmon to clove-brown	3.42 to 3.56	4½–5	Commonly cleavable, massive. Found in pegmatites with other lithium minerals.	Triphylite-lithiophilite p. 433 Li(Fe,Mn)PO$_4$ Orthorhombic
{001} {100}	Colorless, white, gray	2.8 to 2.9	5–5½	Usually cleavable, massive to fibrous. Also compact. Associated with crystalline limestone.	WOLLASTONITE p. 484 CaSiO$_3$ Triclinic
{001} {100}	Colorless, white, gray	2.7 to 2.8	5	In radiating aggregates of sharp acicular crystals. Associated with zeolites in cavities in igneous rocks.	Pectolite p. 487 Ca$_2$NaH(SiO$_3$)$_3$ Triclinic
{110}	Colorless, white	2.25	5–5½	Slender prismatic crystals, prism faces vertically striated. Often in radiating groups. Found lining cavities in igneous rocks.	**NATROLITE** p. 553 Na$_2$Al$_2$Si$_3$O$_{10}$·2H$_2$O Orthorhombic
{110}	Colorless, white	3.7	3½–4	In prismatic crystals and pseudohexagonal twins. Also fibrous and massive. Effervesces in cold acid.	**STRONTIANITE** p. 413 SrCO$_3$ Orthorhombic
{110}	White, pale green, blue	3.4 to 3.5	4½–5	Often in radiating crystal groups. Also stalactitic, mammillary. Prismatic cleavage seldom seen.	**HEMIMORPHITE** p. 464 Zn$_4$(Si$_2$O$_7$)(OH)$_2$·H$_2$O Orthorhombic
{110}	Brown, gray, green, yellow	3.4 to 3.55	5–5½	Luster adamantine to resinous. In thin wedge-shaped crystals with sharp edges. Prismatic cleavage seldom seen.	TITANITE p. 461 CaTiO(SiO$_4$) Monoclinic

Table A.1 (Cont.)

LUSTER: NONMETALLIC

II. Streak colorless

C. Hardness: >3, <5$\frac{1}{2}$

(Cannot be scratched by cent; can be scratched by knife)

1. Cleavage prominent
(*Continued*)

Cleavage	Color	G	H	Remarks	Name, Composition, Crystal System
Two cleavage directions — Prismatic at angles of 56° and 124°	White, green, black	3.0 to 3.3	5–6	Crystals usually slender, fibrous; may be asbestiform. Tremolite (white, gray, violet), actinolite (green) common in metamorphic rocks. Hornblende and arfvedsonite (dark green to black) common in igneous and metamorphic rocks.	**AMPHIBOLE GROUP** p. 488 Essentially hydrous Ca-Mg-Fe silicates Monoclinic
	Gray, clove-brown, green	2.85 to 3.2	5$\frac{1}{2}$–6	An amphibole. Distinct crystals rare. Commonly in aggregates and fibrous masses.	Anthophyllite p. 493 or grunerite p. 494 $(Mg,Fe)_7Si_8O_{22}(OH)_2$ Orthorhombic or monoclinic
Prismatic at nearly 90° angles.	White, green, black, brown	3.1 to 3.5	5–6	In stout prisms with rectangular cross section. Often in granular crystalline masses. Diopside (colorless, white, green), aegirine (brown, green), augite (dark green to black), members of the orthopyroxene series (light brown)	**PYROXENE GROUP** p. 475 Essentially Ca-Mg-Fe silicates Monoclinic and orthorhombic
Prismatic at nearly 90° angles	Rose-red, pink, brown	3.58 to 3.70	5$\frac{1}{2}$–6	Color diagnostic. Usually massive, cleavable to compact, in imbedded grains; in large rough crystals, with rounded edges.	RHODONITE p. 486 $MnSiO_3$ Triclinic

Table A.1 (Cont.)
LUSTER: NONMETALLIC

II. Streak colorless

C. Hardness: >3, <5½

(Cannot be scratched by
cent; can be scratched
by knife)

1. Cleavage prominent
(*Continued*)

Three cleavage directions — Three directions not at right angles. Rhombohedral {10$\bar{1}$1}

Cleavage	Color	G	H	Remarks	Name, Composition, Crystal System
	Colorless, white, and variously tinted	2.72	3	Effervesces in cold acid. Crystals show many forms. Occurs as limestone and marble. Clear varieties show strong double refraction.	**CALCITE** p. 405 $CaCO_3$ Hexagonal
	Colorless, white, pink	2.85	3½–4	Often in curved rhombohedral crystals with pearly luster. In coarse masses as dolomitic limestone and marble. Powdered mineral will effervesce in cold acid.	**DOLOMITE** p. 415 $CaMg(CO_3)_2$ Hexagonal
	White, yellow, gray, brown	3.0 to 3.2	3½–5	Commonly in dense compact masses; also in fine to coarse cleavable masses. Effervesces in hot hydrochloric acid.	**MAGNESITE** p. 408 $MgCO_3$ Hexagonal
	Light to dark brown	3.83 to 3.88	3½–4	In cleavable masses or in small curved rhombohedral crystals. Becomes magnetic after heating.	**SIDERITE** p. 409 $FeCO_3$ Hexagonal
	Pink, rose-red, brown	3.45 to 3.6	3½–4½	In cleavable masses or in small rhombohedral crystals. Characterized by its color.	RHODOCHROSITE p. 409 $MnCO_3$ Hexagonal
	Brown, green, blue, pink, white	4.35 to 4.40	5	In rounded botryoidal aggregates and honeycombed masses. Effervesces in cold hydrochloric acid. Cleavage rarely seen.	SMITHSONITE p. 410 $ZnCO_3$ Hexagonal

Table A.1 (Cont.)

LUSTER: NONMETALLIC

II. Streak colorless

C. Hardness: >3, <5½

(Cannot be scratched by cent; can be scratched by knife)

1. Cleavage prominent
(*Continued*)

Cleavage	Color	G	H	Remarks	Name, Composition, Crystal System
Three cleavage directions					
{10$\bar{1}$1}	White, yellow, flesh-red	2.05 to 2.15	4–5	In small rhombohedral crystals with nearly cubic angles. Found lining cavities in igneous rocks.	CHABAZITE p. 555 $Ca_2Al_2Si_4O_{12} \cdot 6H_2O$ Hexagonal
{001} {010} {100}	Colorless, white, blue, gray, red	2.89 to 2.98	3–3½	Commonly in massive fine aggregates, not showing cleavage, and can be distinguished only by chemical tests.	**ANHYDRITE** p. 426 $CaSO_4$ Orthorhombic
{001} at right angles to {110}	Colorless, white, blue, yellow, red	4.5	3–3½	Frequently in aggregates of platy crystals. Pearly luster on basal cleavage. Characterized by high specific gravity and thus distinguished from celestite.	**BARITE** p. 424 $BaSO_4$ Orthorhombic
	Colorless, white, blue, red	3.95 to 3.97	3–3½	Similar to barite but lower specific gravity.	**CELESTITE** p. 425 $SrSO_4$ Orthorhombic
Four cleavage directions					
{111} Octahedral	Colorless, violet, green, yellow, pink	3.18	4	In cubic crystals often in penetration twins. Characterized by cleavage.	**FLUORITE** p. 401 CaF_2 Isometric
{100} {110}	White, pink, gray, green, brown	2.65 to 2.74	5–6	In prismatic crystals, granular or massive. Commonly altered. Prismatic cleavage obscure.	SCAPOLITE p. 548 Essentially Na, Ca aluminum silicate Tetragonal
Six cleavage directions Dodecahedral {011}					
	Yellow, brown, white	3.9 to 4.1	3½–4	Luster resinous. Small tetrahedral crystals rare. Usually in cleavable masses. If massive, difficult to determine.	**SPHALERITE** p. 356 ZnS Isometric
	Blue, white, gray, green	2.15 to 2.3	5½–6	Massive or in embedded grains; rarely in crystals. A feldspathoid associated with nepheline, never with quartz.	SODALITE p. 547 $Na_6(AlSiO_4)_6Cl_2$ Isometric

Table A.1 (Cont.)
LUSTER: NONMETALLIC

II. Streak colorless

C. Hardness: >3, <$5\frac{1}{2}$

(Cannot be scratched by cent; can be scratched by knife)

2. Cleavage not prominent

Color	G	H	Remarks	Name, Composition, Crystal System
Colorless, white	2.25	5–$5\frac{1}{2}$	In slender prismatic crystals, prism faces vertically striated. Often in radiating groups. Found lining cavities in igneous rocks. Poor prismatic cleavage.	NATROLITE p. 553 $Na_2Al_2Si_3O_{10}\cdot 2H_2O$ Orthorhombic
	2.95	$3\frac{1}{2}$–4	Effervesces in cold acid. Falls to powder in the candle flame. Frequently in radiating groups of acicular crystals; in pseudohexagonal twins. Cleavage indistinct.	ARAGONITE p. 411 $CaCO_3$ Orthorhombic
	2.27	5–$5\frac{1}{2}$	Usually in trapezohedrons with vitreous luster. Found lining cavities in igneous rocks.	ANALCIME p. 549 $NaAlSi_2O_6\cdot H_2O$ Isometric
	3.7	$3\frac{1}{2}$–4	Occurs in prismatic crystals and pseudohexagonal twins. Also fibrous and massive. Effervesces in cold hydrochloric acid.	STRONTIANITE p. 413 $SrCO_3$ Orthorhombic
	3.0 to 3.2	$3\frac{1}{2}$–5	Commonly in dense compact masses showing no cleavage. Effervesces in hot hydrochloric acid.	MAGNESITE p. 408 $MgCO_3$ Hexagonal
	4.3	$3\frac{1}{2}$	Often in radiating masses; granular; rarely in pseudohexagonal crystals. Effervesces in cold hydrochloric acid.	WITHERITE p. 413 $BaCO_3$ Orthorhombic
	2.7 to 2.8	5	Commonly fibrous in radiating aggregates of sharp acicular crystals. Associated with zeolites in cavities in igneous rocks.	Pectolite p. 487 $Ca_2NaH(SiO_3)_3$ Triclinic
Colorless, pale green, yellow	2.8 to 3.0	5–$5\frac{1}{2}$	Usually in crystals with many brilliant faces. Occurs with zeolites lining cavities in igneous rocks.	DATOLITE p. 460 $CaB(SiO_4)(OH)$ Monoclinic
White, pale green, blue	3.4 to 3.5	$4\frac{1}{2}$–5	Often in radiating crystal groups. Also stalactitic, mammillary. Prismatic cleavage seldom seen.	HEMIMORPHITE p. 464 $Zn_4(Si_2O_7)(OH)_2\cdot H_2O$ Orthorhombic
White, pink, gray, green, brown	2.65 to 2.74	5–6	In prismatic crystals, granular or massive. Commonly altered. Prismatic cleavage obscure.	SCAPOLITE p. 548 Essentially Na, Ca-aluminum silicate Tetragonal

Table A.1 (Cont.)

LUSTER: NONMETALLIC

II. Streak colorless

C. Hardness: >3, <5½

(Cannot be scratched by cent; can be scratched by knife)

2. Cleavage not prominent (*Continued*)

Color	G	H	Remarks	Name, Composition, Crystal System
White, grayish, red	2.6 to 2.8	4	May be in rhombohedral crystals. Usually massive granular. Definitely determined only by chemical tests. Cleavage {0001}, poor.	ALUNITE p. 429 $KAl_3(SO_4)_2(OH)_6$ Hexagonal
Colorless, white, yellow, red, brown	1.9 to 2.2	5–6	Conchoidal fracture. Precious opal shows internal play of colors. Specific gravity and hardness less than fine-grained quartz.	**OPAL** p. 531 $SiO_2 \cdot nH_2O$ Essentially amorphous
Brown, green, blue, pink, white	4.35 to 4.40	5	Usually in rounded botryoidal aggregates and in honeycombed masses. Effervesces in cold hydrochloric acid. Cleavage rarely seen.	**SMITHSONITE** p. 410 $ZnCO_3$ Hexagonal
Brown, gray, green, yellow	3.4 to 3.55	5–5½	Adamantine to resinous luster. In thin wedge-shaped crystals with sharp edges. Prismatic cleavage seldom seen.	TITANITE p. 460 $CaTiO(SiO_4)$ Monoclinic
Colorless, white, yellow, red, brown	2.72	3	May be fibrous or fine granular, banded in Mexican onyx variety. Effervesces in cold hydrochloric acid.	**CALCITE** p. 405 $CaCO_3$ Hexagonal
Yellowish- to reddish-brown	5.0 to 5.3	5–5½	In small crystals or as rolled grains. Found in pegmatites.	Monazite p. 433 $(Ce,La,Y,Th)PO_4$ Monoclinic
Light to dark brown	3.83 to 3.88	3½–4	Usually cleavable but may be in compact concretions in clay or shale—clay ironstone variety. Becomes magnetic on heating.	**SIDERITE** p. 409 $FeCO_3$ Hexagonal
White, yellow, green, brown	5.9 to 6.1	4½–5	Luster vitreous to adamantine. Massive and in octahedral-like crystals. Frequently associated with quartz. Will fluoresce.	SCHEELITE p. 431 $CaWO_4$ Tetragonal
Yellow, orange, red, gray, green	6.8±	3	Luster adamantine. Usually in square tabular crystals. Also granular massive. Characterized by color and high specific gravity.	WULFENITE p. 431 $PbMoO_4$ Tetragonal
White, yellow, brown, gray	2.6 to 2.9	3–5	Occurs massive as rock phosphate. Difficult to identify without chemical tests.	**APATITE** p. 434 $Ca_5(PO_4)_3(F,Cl,OH)$ Appears amorphous
Yellow, brown, gray, white	2.0 to 2.55	1–3	Usually pisolitic; in rounded grains and earthy masses. Often impure.	**BAUXITE** p. 397 A mixture of aluminum hydroxides

Table A.1 (Cont.)
LUSTER: NONMETALLIC

II. Streak colorless

C. Hardness: >3, <5½

(Cannot be scratched by cent; can be scratched by knife)

2. Cleavage not prominent
(*Continued*)

Color	G	H	Remarks	Name, Composition, Crystal System
Green, blue, violet, brown, colorless	3.15 to 3.20	5	Usually in hexagonal prisms with pyramid. Also massive. Poor basal cleavage.	**APATITE** p. 434 $Ca_5(PO_4)_3(F,Cl,OH)$ Hexagonal
Green, brown, yellow, gray	6.5 to 7.1	$3\frac{1}{2}$–4	In small hexagonal crystals, often curved and barrel-shaped. Crystals may be cavernous. Often globular and botryoidal.	Pyromorphite p. 435 $Pb_5(PO_4)_3Cl$ Hexagonal
Yellow, green, white, brown	2.33	$3\frac{1}{2}$–4	Characteristically in radiating, hemispherical globular aggregates. Cleavage seldom seen.	Wavellite p. 437 $Al_3(PO_4)_2(OH)_3 \cdot 5H_2O$ Orthorhombic
Olive- to blackish-green, yellow-green, white	2.3 to 2.6	2–5	Massive. Fibrous in the asbestos variety, chrysotile. Frequently mottled green in the massive variety.	**SERPENTINE** p. 507 $Mg_3Si_2O_5(OH)_4$ Monoclinic
Yellow-green, white, blue, gray, brown	3.9 to 4.2	$5\frac{1}{2}$	Massive and in disseminated grains. Rarely in hexagonal prisms. Associated with red zincite and black franklinite at Franklin, N.J. Will fluoresce.	WILLEMITE p. 447 Zn_2SiO_4 Hexagonal
White, gray, blue, green	2.15 to 2.3	$5\frac{1}{2}$–6	Massive or in embedded grains; rarely in crystals. A feldspathoid associated with nepheline, never with quartz. Dodecahedral cleavage poor.	SODALITE p. 547 $Na_6(AlSiO_4)_6Cl_2$ Isometric
Deep azure-blue, greenish-blue	2.4 to 2.45	5–$5\frac{1}{2}$	Usually massive. Associated with feldspathoids and pyrite. Poor dodecahedral cleavage.	Lazurite p. 547 $(Na,Ca)_8(AlSiO_4)_6(SO_4,S,Cl)_2$ Isometric

Table A.1 (Cont.)
LUSTER: NONMETALLIC

II. Streak colorless

D. Hardness: $>5\frac{1}{2}$, <7

(Cannot be scratched by knife; can be scratched by quartz)

1. Cleavage prominent

	Cleavage	Color	G	H	Remarks	Name, Composition, Crystal System
One cleavage direction	{010}	White, gray, pale lavender, yellow-green	3.35 to 3.45	$6\frac{1}{2}$–7	In thin tabular crystals. Luster pearly on cleavage face. Associated with emery, margarite, chlorite.	Diaspore p. 395 $\alpha AlO(OH)$ Orthorhombic
	{010} perfect	Hair-brown, grayish-green	3.23	6–7	In long, prismatic crystals. May be in parallel groups—columnar or fibrous. Found in schists.	SILLIMANITE p. 456 Al_2SiO_5 Orthorhombic
	{001}	Yellowish to blackish green	3.35 to 3.45	6–7	In prismatic crystals striated parallel to length. Found in metamorphic rocks and crystalline limestones.	**EPIDOTE** p. 465 $Ca_2(Al,Fe)Al_2O(SiO_4)-(Si_2O_7)(OH)$ Monoclinic
	{100}	Blue. May be gray, or green	3.56 to 3.66	5–7	In bladed aggregates with cleavage parallel to length. H = 5 parallel to length of crystal.	**KYANITE** p. 456 Al_2SiO_5 Triclinic
Two cleavage directions	{100} good {110} poor	White, pale green, or blue	3.0 to 3.1	6	Usually cleavable, resembling feldspar. Found in pegmatites associated with other lithium minerals.	AMBLYGONITE p. 436 $LiAlFPO_4$ Triclinic
	{100} good {001}	Colorless, white, gray	2.8 to 2.9	5–$5\frac{1}{2}$	Usually cleavable, massive to fibrous. Also compact. Associated with crystalline limestone.	WOLLASTONITE p. 484 $CaSiO_3$ Triclinic
	{001} {100}	Grayish-white, green, pink	3.25 to 3.37	6–$6\frac{1}{2}$	In prismatic crystals, deeply striated. Also massive, columnar, compact. Luster pearly on cleavage, elsewhere vitreous.	Clinozoisite p. 465 $Ca_2Al_3O(SiO_4)(Si_2O_7)-(OH)$ Monoclinic
	{110}	Colorless, white	2.25	5–$5\frac{1}{2}$	In slender prismatic crystals, prism faces vertically striated. Often in radiating groups lining cavities in volcanic rocks.	**NATROLITE** p. 553 $Na_2Al_2Si_3O_{10}\cdot2H_2O$ Orthorhombic
	{110}	Brown, gray, green, yellow	3.4 to 3.55	5–$5\frac{1}{2}$	Luster adamantine to resinous. In thin wedge-shaped crystals with sharp edges. Prismatic cleavage seldom seen.	TITANITE p. 460 $CaTiO(SiO_4)$ Monoclinic

Table A.1 (Cont.)
LUSTER: NONMETALLIC

II. Streak colorless

D. Hardness: $>5\frac{1}{2}$, <7

(Cannot be scratched by knife; can be scratched by quartz)

1. Cleavage prominent
(*Continued*)

Cleavage	Color	G	H	Remarks	Name, Composition, Crystal System
{001} {010}	Colorless, white, gray, cream, red, green	2.54 to 2.56	6	In cleavable masses or in irregular grains as rock constituents. May be in crystals in pegmatites. Distinguished with certainty only with the microscope. Green amazonite is microcline.	**ORTHOCLASE** p. 540 Monoclinic **MICROCLINE** p. 536 Triclinic $KAlSi_3O_8$
{001} {010}	Colorless, white, gray, bluish. Often shows play of colors	2.62 (ab) to 2.76 (an)	6	In cleavable masses or in irregular grains as a rock constituent. On the better cleavage can be seen a series of fine parallel striations due to albite twinning; these distinguish it from orthoclase.	**PLAGIOCLASE** p. 541 Various proportions of albite, $NaAlSi_3O_8$, and anorthite, $CaAl_2Si_2O_8$ Triclinic
{110}	White, gray, pink, green	3.15 to 3.20	$6\frac{1}{2}$–7	In flattened prismatic crystals, vertically striated. Also massive, cleavable. Found in pegmatites. Frequently shows good {100} parting.	SPODUMENE p. 483 $LiAlSi_2O_6$ Monoclinic
{110}	White, green, black	3.1 to 3.5	5–6	In stout prisms with rectangular cross section. Often in granular crystalline masses. Diopside (colorless, white, green), aegirine (brown, green), augite (dark green to black). Characterized by cleavage.	**PYROXENE GROUP** p. 475 Essentially Ca-Mg-Fe silicates Monoclinic
{110}	Gray-brown, green, bronze-brown, black	3.2 to 3.5	$5\frac{1}{2}$	Crystals usually prismatic but rare. Commonly massive, fibrous, lamellar. Fe may replace Mg and mineral is darker.	ENSTATITE p. 478 $MgSiO_3$ Orthorhombic
{110}	Rose-red, pink, brown	3.58 to 3.70	$5\frac{1}{2}$–6	Color diagnostic. Usually massive; cleavable to compact, in embedded grains; in large rough crystals with rounded edges.	RHODONITE p. 486 $MnSiO_3$ Triclinic

Two cleavage directions at 90° or nearly 90°

Table A.1 (Cont.)

LUSTER: NONMETALLIC

II. Streak colorless

D. Hardness: $>5\frac{1}{2}$, <7

(Cannot be scratched by knife; can be scratched by quartz)

1. Cleavage prominent (*Continued*)

	Cleavage	Color	G	H	Remarks	Name, Composition, Crystal System
Two cleavage directions at 54° and 126° angles	{110}	White, green, black	3.0 to 3.3	5–6	Crystals usually slender fibrous; may be asbestiform. Tremolite (white, gray, violet) and actinolite (green) are common in metamorphic rocks. Hornblende and arfvedsonite (dark green to black) are common in igneous rocks. Characterized by cleavage angle.	**AMPHIBOLE GROUP** p. 488 Essentially hydrous Ca-Mg-Fe silicates Monoclinic
	{110}	Gray, clove-brown, green	2.85 to 3.2	$5\frac{1}{2}$–6	An amphibole. Distinct crystals rare. Commonly in aggregates and fibrous, massive.	Anthophyllite p. 493 and grunerite p. 494 $(Mg,Fe)_7(Si_8O_{22})(OH)_2$ Orthorhombic and monoclinic
Six cleavage directions	{110}	Blue, gray, white, green	2.15 to 2.3	$5\frac{1}{2}$–6	Massive or embedded grains; rarely in crystals. A feldspathoid associated with nepheline, never with quartz.	SODALITE p. 547 $Na_8(AlSiO_4)_6Cl_2$ Isometric

2. Cleavage not prominent

Color	G	H	Remarks	Name, Composition, Crystal System
Colorless	2.26	7	Occurs as small crystals in cavities in volcanic rocks. Difficult to determine without optical aid.	Tridymite p. 530 SiO_2 Pseudohexagonal
Colorless or white	2.27	5–$5\frac{1}{2}$	Usually in trapezohedrons with vitreous luster. Found lining cavities in igneous rocks.	ANALCIME p. 549 $NaAlSi_2O_6 \cdot H_2O$ Isometric
	2.32	7	Occurs in spherical aggregations in volcanic rocks. Difficult to determine without optical aid.	Cristobalite p. 530 SiO_2 Pseudoisometric
Colorless, yellow, red, brown, green, gray, blue	1.9 to 2.2	5–6	Conchoidal fracture. Precious opal shows internal play of colors. Gravity and hardness less than fine-grained quartz.	**OPAL** p. 531 $SiO_2 \cdot nH_2O$ Essentially amorphous

Table A.1 (Cont.)
LUSTER: NONMETALLIC

II. Streak colorless

D. Hardness: >$5\frac{1}{2}$, <7

(Cannot be scratched by
knife; can be scratched
by quartz)

**2. Cleavage not
prominent**
(*Continued*)

Color	G	H	Remarks	Name, Composition, Crystal System
Gray, white, colorless	2.45 to 2.50	$5\frac{1}{2}$–6	In trapezohedral crystals embedded in dark igneous rock. Does not line cavities as analcime does.	LEUCITE p. 545 $KAlSi_2O_6$ Pseudoisometric
Colorless, pale, green, yellow	2.8 to 3.0	5–$5\frac{1}{2}$	Usually in crystals with many brilliant faces. Occurs with zeolites lining cavities in igneous rocks.	DATOLITE p. 460 $CaB(SiO_4)(OH)$ Monoclinic
Colorless, white, smoky. Variously colored	2.65	7	Crystals usually show horizontally striated prism with rhombohedral terminations.	**QUARTZ** p. 526 SiO_2 Hexagonal
Colorless, gray, greenish, reddish	2.55 to 2.65	$5\frac{1}{2}$–6	Greasy luster. A rock constituent, usually massive; rarely in hexagonal prisms. Poor prismatic cleavage. A feldspathoid.	NEPHELINE p. 546 $(Na,K)AlSiO_4$ Hexagonal
White, gray, light to dark green, brown	2.65 to 2.74	5–6	In prismatic crystals, granular or massive. Commonly altered and prismatic cleavage obscure.	SCAPOLITE p. 548 Essentially Na-Ca aluminum silicate Tetragonal
Light yellow, brown, orange	3.1 to 3.2	6–$6\frac{1}{2}$	Occurs in disseminated crystals and grains. Commonly in crystalline limestones	Chondrodite p. 459 $Mg_5(SiO_4)_2(F,OH)_2$ Monoclinic
Light brown, yellow, red, green	2.6	7	Chalcedony, waxy luster, commonly colloform; as agate lining cavities. Dull luster, massive: jasper red, flint black, chert gray.	**MICROCRYSTALLINE QUARTZ** p. 528 SiO_2
Blue, bluish-green, green	2.6 to 2.8	6	Usually appears amorphous in reniform and stalactitic masses.	TURQUOISE p. 438 $CuAl_6(PO_4)_4(OH)_8 \cdot 5H_2O$ Triclinic
Apple-green, gray, white	2.8 to 2.95	6–$6\frac{1}{2}$	Reniform and stalactitic with crystalline surface. In subparallel groups of tabular crystals.	PREHNITE p. 523 $Ca_2Al(AlSi_3O_{10})(OH)_2$ Orthorhombic
Yellow-green, white, blue, gray, brown	3.9 to 4.2	$5\frac{1}{2}$	Massive and in disseminated grains. Rarely in hexagonal prisms. Associated with red zincite and black franklinite at Franklin, N.J. Will fluoresce.	WILLEMITE p. 447 Zn_2SiO_4 Hexagonal
Olive to grayish-green, brown	3.27 to 4.37	$6\frac{1}{2}$–7	Usually in disseminated grains in basic igneous rocks. May be massive, granular.	OLIVINE p. 449 $(Mg,Fe)_2SiO_4$ Orthorhombic

Table A.1 (Cont.)

LUSTER: NONMETALLIC

II. Streak colorless

D. Hardness: $>5\frac{1}{2}$, <7

(Cannot be scratched by knife; can be scratched by quartz)

2. Cleavage not prominent
(*Continued*)

Color	G	H	Remarks	Name, Composition, Crystal System
Black, green, brown, blue, red, pink, white	3.0 to 3.25	$7-7\frac{1}{2}$	In slender prismatic crystals with triangular cross section. Crystals may be in radiating groups. Found usually in pegmatites. Black most common; other colors associated with lithium minerals.	**TOURMALINE** p. 473 Complex boron silicate of Na-Ca-Al-Mg-Fe-Mn Hexagonal
Green, brown, yellow, blue, red	3.35 to 4.45	$6\frac{1}{2}$	In square prismatic crystals, vertically striated. Often columnar and granular massive. Found in crystalline limestones.	VESUVIANITE p. 466 Complex hydrous Ca-Mg-Fe-Al silicate Tetragonal
Clove-brown, gray, green, yellow	3.27 to 3.35	$6\frac{1}{2}-7$	In wedge-shaped crystals with sharp edges. Also lamellar.	AXINITE p. 471 $(Ca,Fe,Mn)_3Al_2(BO_3)(Si_4O_{12})(OH)$ Triclinic
Red-brown to brownish-black	3.65 to 3.75	$7-7\frac{1}{2}$	In prismatic crystals; commonly in cruciform penetration twins. Frequently altered on the surface and then soft. Found in schists.	**STAUROLITE** p. 458 $Fe_2Al_9O_6(SiO_4)_4(O,OH)_2$ Pseudo-orthorhombic
Reddish-brown, flesh-red, olive-green	3.16 to 3.20	$7\frac{1}{2}$	Prismatic crystals with nearly square cross section. Cross section may show black cross (chiastolite). May be altered to mica and then soft. Found in schists.	**ANDALUSITE** p. 455 Al_2SiO_5 Orthorhombic
Brown, gray, green, yellow	3.4 to 3.55	$5-5\frac{1}{2}$	Luster adamantine to resinous. In thin wedge-shaped crystals with sharp edges. Prismatic cleavage seldom seen.	**TITANITE** p. 460 $CaTiO(SiO_4)$ Monoclinic
Yellowish- to reddish-brown	5.0 to 5.3	$5-5\frac{1}{2}$	In isolated crystals, granular. Commonly found in pegmatites.	Monazite p. 433 $(Ce,La,Y,Th)PO_4$ Monoclinic
Brown to black	6.8 to 7.1	$6-7$	Rarely in prismatic crystals, twinned. Fibrous, giving reniform surface. Rolled grains. Usually gives light brown streak.	**CASSITERITE** p. 383 SnO_2 Tetragonal
Reddish-brown to black	4.18 to 4.25	$6-6\frac{1}{2}$	In prismatic crystals, vertically striated; often slender, acicular. Crystals frequently twinned. A constituent of black sands.	**RUTILE** p. 381 TiO_2 Tetragonal

Table A.1 (Cont.)
LUSTER: NONMETALLIC

II. Streak colorless

D. Hardness: >$5\frac{1}{2}$, <7

(Cannot be scratched by knife; can be scratched by quartz)

2. Cleavage not prominent
(*Continued*)

Color	G	H	Remarks	Name, Composition, Crystal System
Brown to pitch-black	3.5 to 4.2	$5\frac{1}{2}$–6	Crystals often tabular. Massive and in embedded grains. An accessory mineral in igneous rocks.	Allanite p. 466 Ce, La, Th epidote Monoclinic
Blue, rarely colorless	2.60 to 2.66	7–$7\frac{1}{2}$	In embedded grains and massive, resembling quartz. Commonly altered and foliated; then softer than a knife.	Cordierite p. 473 $(Mg,Fe)_2Al_4Si_5O_{18}\cdot nH_2O$ Orthorhombic (Pseudohexagonal)
Deep azure-blue, greenish-blue	2.4 to 2.45	5–$5\frac{1}{2}$	Usually massive. Associated with feldspathoids and pyrite. Poor dodecahedral cleavage.	LAZURITE p. 547 $(Na,Ca)_8(AlSiO_4)_6(SO_4,S,Cl)_2$ Isometric
Azure-blue	3.0 to 3.1	5–$5\frac{1}{2}$	Usually in pyramidal crystals, which distinguishes it from massive lazurite. A rare mineral.	LAZULITE p. 437 $(Mg,Fe)Al_2(PO_4)_2(OH)_2$ Monoclinic
Blue, green, white, gray	2.15 to 2.3	$5\frac{1}{2}$–6	Massive or in embedded grains; rarely in crystals. A feldspathoid associated with nepheline, never with quartz. Poor dodecahedral cleavage.	SODALITE p. 547 $Na_8(AlSiO_4)_6Cl_2$ Isometric

Table A.1 (Cont.)

LUSTER: NONMETALLIC

II. Streak colorless

E. Hardness: >7

(Cannot be scratched by quartz)

1. Cleavage prominent

Cleavage	Color	G	H	Remarks	Name, Composition, Crystal System
{001} *One cleavage direction*	Colorless, yellow, pink, bluish, greenish	3.4 to 3.6	8	Usually in crystals, also coarse to fine granular. Found in pegmatites.	**TOPAZ** p. 457 $Al_2SiO_4(F,OH)_2$ Orthorhombic
{010}	Brown, gray, greenish-gray	3.23	6–7	Commonly in long slender prismatic crystals. May be in parallel groups, columnar or fibrous. Found in schistose rocks.	**SILLIMANITE** p. 456 Al_2SiO_5 Orthorhombic
{110} *Two cleavage directions*	White, gray, pink, green	3.15 to 3.20	6½–7	In flattened prismatic crystals, vertically striated. Also massive, cleavable. Pink variety, kunzite; green, hiddenite. Found in pegmatites. Frequently shows good {100} parting.	SPODUMENE p. 483 $LiAlSi_2O_6$ Monoclinic
{010} {110} *Three cleavage directions*	Colorless pale blue, gray	3.09	8	Commonly in tabular or prismatic crystals in schists.	Lawsonite p. 464 $CaAl_2(Si_2O_7)(OH)_2 \cdot H_2O$ Orthorhombic
{111} *Four cleavage directions*	Colorless, yellow, red, blue, black	3.5	10	Adamantine luster. In octahedral crystals, frequently twinned. Faces may be curved.	Diamond p. 346 C Isometric
No cleavage. Rhombohedral and basal parting	Colorless, gray, blue, red, yellow, brown, green	3.95 to 4.1	9	Luster adamantine to vitreous. Parting fragments may appear nearly cubic. In rude barrel-shaped crystals.	**CORUNDUM** p. 377 Al_2O_3 Hexagonal

Table A.1 (Cont.)
LUSER: NONMETALLIC

II. Streak colorless

E. Hardness: >7

(Cannot be scratched
by quartz)

**2. Cleavage not
prominent**

Color	G	H	Remarks	Name, Composition, Crystal System
Colorless, white, smoky, variously colored	2.65	7	Crystals usually show horizontally striated prism with rhombohedral terminations.	**QUARTZ** p. 526 SiO_2 Hexagonal
White, colorless	2.97 to 3.0	$7\frac{1}{2}$–8	In small rhombohedral crystals. A rare mineral.	Phenacite p. 447 Be_2SiO_4 Hexagonal
White and almost any color	3.95 to 4.1	9	Luster adamantine to vitreous. Parting fragments may appear nearly cubic. In rude barrel-shaped crystals.	**CORUNDUM** p. 377 Al_2O_3 Hexagonal
Red, black, blue, green, brown	3.6 to 4.0	8	In octahedrons; twinning common. Associated with crystalline limestones.	**SPINEL** p. 385 $MgAl_2O_4$ Isometric
Bluish-green, yellow, pink, colorless	2.65 to 2.8	$7\frac{1}{2}$–8	Commonly in hexagonal prisms terminated by the base; pyramid faces are rare. Crystals large in places. Poor basal cleavage.	**BERYL** p. 471 $Be_3Al_2(Si_6O_{18})$ Hexagonal
Yellowish to emerald-green	3.65 to 3.8	$8\frac{1}{2}$	In tabular crystals; frequently in pseudohexagonal twins. Found in pegmatites.	CHRYSOBERYL p. 390 $BeAl_2O_4$ Orthorhombic
Green, brown, blue, red, pink, black	3.0 to 3.25	7–$7\frac{1}{2}$	In slender prismatic crystals with triangular cross section. Found usually in pegmatites. Black most common, other colors associated with lithium minerals.	TOURMALINE p. 473 Complex boron silicate of Na-Ca-Al-Mg-Fe-Mn Hexagonal
Green, gray, white	3.3 to 3.5	$6\frac{1}{2}$–7	Massive, closely compact. Poor prismatic cleavage at nearly 90° angles. A pyroxene.	Jadeite p. 482 $NaAlSi_2O_6$ Monoclinic
Olive to grayish-green, brown	3.27 to 4.37	$6\frac{1}{2}$–7	Usually in disseminated grains in basic igneous rocks. May be massive granular.	**OLIVINE** p. 449 $(Mg,Fe)_2SiO_4$ Orthorhombic

Table A.1 (Cont.)
LUSTER: NONMETALLIC

II. Streak colorless

E. Hardness: >7

(Cannot be scratched
by quartz)

**2. Cleavage not
prominent**
(*Continued*)

Color	G	H	Remarks	Name, Composition, Crystal System
Green, brown, yellow, blue, red	3.35 to 3.45	$6\frac{1}{2}$	In square prismatic crystals, vertically striated. Often columnar and granular, massive. Found in crystalline limestones.	VESUVIANITE p. 466 Complex hydrous Ca-Mg-Fe-Al silicate Tetragonal
Dark green	4.55	$7\frac{1}{2}$–8	Usually in octahedrons, characteristically striated. A zinc spinel.	Gahnite p. 387 $ZnAl_2O_4$ Isometric
Reddish-brown to black	6.8 to 7.1	6–7	Rarely in prismatic crystals, twinned. Fibrous, giving reniform surface. Rolled grains. Usually gives light brown streak.	**CASSITERITE** p. 383 SnO_2 Tetragonal
Reddish-brown, flesh-red, olive-green	3.16 to 3.20	$7\frac{1}{2}$	Prismatic crystals with nearly square cross section. Cross section may show black cross (chiastolite). May be altered to mica and then soft. Found in schists.	**ANDALUSITE** p. 455 Al_2SiO_5 Orthorhombic
Clove-brown, green, yellow, gray	3.27 to 3.35	$6\frac{1}{2}$–7	In wedge-shaped crystals with sharp edges. Also lamellar.	AXINITE p. 471 $Ca_2(Fe,Mn)$-$Al_2(BO_3)(Si_4O_{12})(OH)$ Triclinic
Red-brown to brownish-black	3.65 to 3.75	7–$7\frac{1}{2}$	In prismatic crystals; commonly in cruciform penetration twins. Frequently altered on the surface and then soft. Found in schists.	**STAUROLITE** p. 458 $Fe_2Al_9O_6(SiO_4)_4(O,OH)_2$ Pseudo-orthorhombic
Brown, red, gray, green, colorless	4.68	$7\frac{1}{2}$	Usually in small prisms truncated by the pyramid. An accessory mineral in igneous rocks. Found as rolled grains in sand.	**ZIRCON** p. 454 $ZrSiO_4$ Tetragonal
Usually brown to red. Also yellow, green, pink	3.5 to 4.3	$6\frac{1}{2}$–$7\frac{1}{2}$	Usually in dodecahedrons or trapezohedrons or in combinations of the two. An accessory mineral in igneous rocks and pegmatites. Commonly in metamorphic rocks. As sand.	**GARNET** p. 451 $A_3B_2(SiO_4)_3$ Isometric

Table A.2

MINERALS ARRANGED ACCORDING TO INCREASING SPECIFIC GRAVITY

G	Name	G	Name	G	Name
1.6	Carnallite	2.62–2.76	**Plagioclase**	3.35–3.45	**Epidote**
1.7	**Borax**	2.6–2.9	**Collophane**	3.35–3.45	**Vesuvianite**
1.95	**Kernite**	2.74	**Bytownite**	3.4–3.5	**Hemimorphite**
1.96	**Ulexite**	2.7–2.8	Pectolite	3.45	Arfvedsonite
1.99	Sylvite	2.7–2.8	**Talc**	3.40–3.55	Aegirine
		2.76	**Anorthite**	3.4–3.55	**Titanite**
2.0–2.19				3.48	**Realgar**
		2.8–2.99		3.42–3.56	Triphylite
2.0–2.55	**Bauxite**			3.49	**Orpiment**
2.0–2.4	**Chrysocolla**	2.6–2.9	**Collophane**	3.4–3.6	**Topaz**
2.05–2.09	**Sulfur**	2.8–2.9	Pyrophyllite	3.5	**Diamond**
2.05–2.15	**Chabazite**	2.8–2.9	**Wollastonite**	3.45–3.60	**Rhodochrosite**
1.9–2.2	**Opal**	2.85	**Dolomite**	3.5–4.3	Garnet
2.09–2.14	Niter	2.86	Phlogopite		
2.1–2.2	**Stilbite**	2.76–2.88	**Muscovite**	**3.6–3.79**	
2.16	**Halite**	2.8–2.95	**Prehnite**		
2.18–2.20	**Heulandite**	2.8–3.0	**Datolite**	3.27–4.37	Olivine
		2.8–3.0	**Lepidolite**	3.5–4.2	Allanite
2.2–2.39		2.89–2.98	**Anhydrite**	3.5–4.3	**Garnet**
		2.9–3.0	Boracite	3.6–4.0	**Spinel**
2.0–2.4	**Chrysocolla**	2.95	**Aragonite**	3.56–3.66	**Kyanite**
2.2–2.65	**Serpentine**	2.95	Erythrite	3.58–3.70	**Rhodonite**
2.23	**Graphite**	2.8–3.2	**Biotite**	3.65–3.75	**Staurolite**
2.25	**Natrolite**	2.95–3.0	**Cryolite**	3.7	**Strontianite**
2.26	Tridymite	2.97–3.00	Phenacite	3.65–3.8	**Chrysoberyl**
2.27	**Analcime**			3.75–3.77	Atacamite
2.29	Nitratite	**3.0–3.19**		3.77	**Azurite**
2.30	Cristobalite				
2.30	**Sodalite**	2.97–3.02	Danburite	**3.8–3.99**	
2.32	**Gypsum**	2.85–3.2	Anthophyllite		
2.33	Wavellite	3.0–3.1	**Amblygonite**	3.7–4.7	**Romanechite**
2.3–2.4	**Apophyllite**	3.0–3.1	**Lazulite**	3.6–4.0	**Spinel**
2.39	**Brucite**	3.0–3.2	**Magnesite**	3.6–4.0	Limonite
		3.0–3.1	Margarite	3.83–3.88	**Siderite**
2.4–2.59		3.0–3.25	**Tourmaline**	3.5–4.2	Allanite
		3.0–3.3	**Tremolite**	3.5–4.3	**Garnet**
2.0–2.55	**Bauxite**	3.09	Lawsonite	3.9	Antlerite
2.2–2.65	**Serpentine**	3.1–3.2	Autunite	3.9–4.03	**Malachite**
2.42	**Colemanite**	3.1–3.2	Chondrodite	3.95–3.97	**Celestite**
2.42	Petalite	3.15–3.20	**Apatite**		
2.4–2.45	**Lazurite**	3.15–3.20	**Spodumene**	**4.0–4.19**	
2.45–2.50	**Leucite**	3.16–3.20	**Andalusite**		
2.2–2.8	**Garnierite**	3.18	**Fluorite**	3.9–4.1	**Sphalerite**
2.54–2.57	**Microcline**			4.02	**Corundum**
2.57	**Orthoclase**	**3.2–3.39**		3.9–4.2	**Willemite**
2.6–2.79		3.1–3.3	Scorodite	**4.2–4.39**	
		3.2	**Hornblende**		
2.55–2.65	**Nepheline**	3.23	**Sillimanite**	4.1–4.3	**Chalcopyrite**
2.6–2.63	**Kaolinite**	3.2–3.3	**Diopside**	3.7–4.7	**Romanechite**
2.62	**Albite**	3.2–3.4	**Augite**	4.18–4.25	**Rutile**
2.60–2.66	Cordierite	3.25–3.37	Clinozoisite	4.3	**Manganite**
2.65	**Oligoclase**	3.26–3.36	Dumortierite	4.3	**Witherite**
2.65	**Quartz**	3.27–3.35	Axinite	4.37	**Goethite**
2.69	**Andesine**	3.27–4.37	**Olivine**	4.35–4.40	**Smithsonite**
2.6–2.8	Alunite	3.2–3.5	**Enstatite**		
2.6–2.8	Turquoise			**4.4–4.59**	
2.71	**Labradorite**	**3.4–3.59**			
2.65–2.74	**Scapolite**			4.43–4.45	**Enargite**
2.65–2.8	**Beryl**	3.27–4.27	Olivine	4.5	**Barite**
2.72	**Calcite**	3.3–3.5	Jadeite	4.55	Gahnite
2.6–3.3	**Chlorite**	3.35–3.45	**Diaspore**	4.52–4.62	**Stibnite**

Table A.2 (Cont.)

MINERALS ARRANGED ACCORDING TO INCREASING SPECIFIC GRAVITY

G	Name	G	Name	G	Name
4.6–4.79		**5.4–5.59**		**6.5–6.99**	
3.7–4.7	**Romanechite**	5.5	Millerite	6.5	**Skutterudite**
4.6	**Chromite**	5.5±	**Chlorargyrite**	6.55	**Cerussite**
4.58–4.65	**Pyrrhotite**	5.55	**Proustite**	6.78	Bismuthinite
4.7	**Ilmenite**			6.5–7.1	Pyromorphite
4.75	**Pyrolusite**	**5.6–5.79**		6.8	**Wulfenite**
4.6–4.76	Covellite			6.7–7.1	Vanadinite
4.62–4.73	**Molybdenite**	5.5–5.8	**Chalcocite**	6.8–7.1	**Cassiterite**
4.68	**Zircon**	5.68	**Zincite**		
		5.5–6.0	Jamesonite	**7.0–7.49**	
4.8–4.99		5.3–7.3	Columbite		
				7.0–7.5	**Wolframite**
4.6–5.0	Pentlandite	**5.8–5.99**		7.3	**Acanthite**
4.6–5.1	**Tetrahedrite-**				
	Tennantite	5.8–5.9	Bournonite	**7.5–7.99**	
4.89	**Marcasite**	5.85	**Pyrargyrite**		
				7.4–7.6	**Galena**
5.0–5.19		**6.0–6.49**		7.3–7.9	Iron
				7.78	**Nickeline**
5.02	**Pyrite**	5.9–6.1	Crocoite		
4.8–5.3	**Hematite**	5.9–6.1	**Scheelite**	**>8.0**	
5.06–5.08	**Bornite**	6.0	**Cuprite**		
5.15	**Franklinite**	6.07	**Arsenopyrite**	8.10	**Cinnabar**
5.0–5.3	Monazite	6.0–6.2	Polybasite	8.9	**Copper**
5.18	**Magnetite**	6.2–6.4	**Anglesite**	9.0–9.7	Uraninite
		5.3–7.3	Columbite	10.5	**Silver**
5.2–5.39		6.33	**Cobaltite**	15.0–19.3	**Gold**
				14–19	**Platinum**

Table A.3
**NONOPAQUE MINERALS
AND SOME SYNTHETIC
COMPOUNDS ARRANGED
ACCORDING TO
INCREASING
REFRACTIVE INDEX***

n, ω, β	ε, α	γ	Optic Sign	Name
1.338	1.338	1.339	B +	Cryolite
1.433			I	Fluorite
1.45			I	Opal
1.469	1.447	1.472	B −	Borax
1.472	1.454	1.485	B −	Kernite
1.475	1.473	1.479	B +	Tridymite
1.482	1.480		U −	Chabazite
1.482	1.480	1.493	B +	Natrolite
1.483			I	Sodalite
1.492	1.490	1.493	B +	Heulandite
1.487	1.484		U −	Cristobalite
1.487			I	Analcime
1.490			I	Sylvite
1.498	1.494	1.500	B −	Stilbite
1.50			I	Lazurite
1.50			≈I	Chrysocolla
1.504	1.332	1.504	B −	Niter
1.504	1.491	1.520	B +	Ulexite
1.508	1.509		U +	Leucite
1.511	1.505	1.518	B +	Petalite
1.526	1.522	1.530	B −	Microcline
1.523	1.520	1.530	B +	Gypsum
1.524	1.518	1.526	B −	Orthoclase
1.524–1.574	1.522–1.560	1.527–1.578	B ±	Cordierite
1.526	1.521	1.528	B −	Sanidine
1.532	1.527	1.538	B +	Albite
1.54–1.59	1.52–1.56		U −	Scapolite
1.537	1.550		U +	Chalcedony
1.537	1.534		U −	Nepheline
1.533	1.536		U +	Apophyllite
1.535	1.525	1.550	B +	Wavellite
1.540	1.540	1.550	B +	Chrysotile
1.542	1.540		U −	Apophyllite
1.543	1.539	1.547	B −	Oligoclase
1.544			I	Halite
1.544	1.531		U −	Marialite
1.544	1.553		U +	Quartz
1.547	1.540	1.550	B −	Cordierite
1.553	1.550	1.557	B +	Andesine
1.55			≈I	Antigorite
1.55–1.59	1.53–1.55	1.55–1.59	B −	Lepidolite
1.588	1.552	1.561	B −	Beryllonite
1.56–1.64	1.53–1.59	1.56–1.64	B −	Phlogopite
1.563	1.559	1.568	B +	Labradorite
1.564	1.559	1.565	B −	Kaolinite
1.57–1.60	1.56–1.60		U −	Beryl
1.570	1.590		U +	Brucite
1.57–1.67	1.57–1.66	1.57–1.67	B ±	Chlorite
1.572	1.567	1.576	B −	Bytownite
1.577	1.595		U +	Alunite
1.575	1.570	1.614	B +	Anhydrite
1.577	1.553		U −	Autunite
1.579	1.579	1.584	B +	Clinochlore
1.58			I	Variscite
1.581	1.572	1.586	B −	Anorthite
1.583	1.577		U −	Emerald
1.587	1.336		U −	Nitratite
1.587–1.610	1.575–1.595	1.590–1.622	B −	Amblygonite
1.588	1.544	1.598	B −	Pyrophyllite
1.589	1.539	1.589	B −	Talc

Table A.3 (Cont.)

NONOPAQUE MINERALS AND SOME SYNTHETIC COMPOUNDS ARRANGED ACCORDING TO INCREASING REFRACTIVE INDEX*

n, ω, β	ε, α	γ	Optic Sign	Name
1.59			I	Howlite
1.592	1.586	1.614	B+	Colemanite
1.598	1.551	1.598	B−	Phlogopite
1.594	1.585		U−	Beryl (gem)
1.602	1.556	1.603	B−	Muscovite
1.604	1.595	1.633	B+	Pectolite
1.61–1.70	1.57–1.63	1.61–1.70	B−	Biotite
1.609	1.602	1.621	B+	Brazilianite
1.609–1.631	1.606–1.630	1.616–1.638	B+	Topaz
1.61–1.71	1.60–1.69	1.62–1.72	B±	Anthophyllite
1.612	1.598	1.626	B±	Chondrodite
1.61–1.66	1.59–1.65	1.61–1.66	B−	Glaucophane
1.616	1.600	1.627	B−	Tremolite
1.617	1.614	1.636	B+	Hemimorphite
1.62–1.72	1.61–1.71	1.63–1.73	B−	Hornblende
1.62	1.61	1.65	B+	Turquoise
1.624	1.622	1.631	B+	Celestite
1.627	1.600	1.649	B+	Prehnite
1.630	1.619	1.640	B−	Anthophyllite
1.63–1.67	1.62–1.67		U−	Apatite
1.632	1.620	1.634	B−	Wollastonite
1.633	1.630	1.636	B−	Danburite
1.634	1.612	1.643	B−	Lazulite
1.640–1.60	1.61–1.66		U−	Tourmaline
1.637	1.622	1.649	B−	Actinolite
1.637	1.636	1.648	B+	Barite
1.638	1.632	1.643	B−	Andalusite
1.638	1.633	1.652	B+	Anthophyllite
1.64–1.68	1.63–1.66	1.65–1.70	B+	Cummingtonite
1.643	1.626	1.643	B−	Glaucophane
1.645	1.634	1.647	B−	Margarite
1.646	1.622	1.658	U−	Tourmaline (gem)
1.648	1.584		B−	Biotite
1.649	1.644		U−	Apatite
1.651	1.703		U+	Dioptase
1.65–1.88	1.63–1.88	1.67–1.87	B±	Olivine
1.651	1.635	1.670	B+	Forsterite
1.652	1.624	1.668	B−	Datolite
1.654	1.670		U+	Phenacite
1.658	1.486		U−	Calcite
1.658	1.657	1.677	B+	Sillimanite
1.659	1.654	1.667	B+	Jadeite
1.662	1.636		U−	Tourmaline
1.661	1.626	1.699	B±	Erythrite
1.665	1.650	1.679	B+	Cummingtonite
1.665	1.660	1.674	B+	Enstatite
1.666	1.654	1.670	B−	Hornblende
1.666	1.655	1.672	B+	Spodumene
1.667	1.520	1.669	B−	Strontianite
1.683	1.678	1.688	B−	Bronzite (orthopyroxene)
1.672	1.654	1.690	B+	Olivine (peridot)
1.69–1.72	1.68–1.74	1.71–1.75	B+	Augite
1.674	1.665	1.684	B+	Lawsonite
1.676	1.623	1.677	B−	Biotite
1.676	1.529	1.677	B−	Witherite
1.676	1.667	1.699	B+	Diopside
1.679	1.676	1.680	B+	Lithiophilite
1.680	1.530	1.685	B−	Aragonite
1.681	1.500		U−	Dolomite
1.688	1.681	1.692	B−	Axinite

Table A.3 (Cont.)
NONOPAQUE MINERALS AND SOME SYNTHETIC COMPOUNDS ARRANGED ACCORDING TO INCREASING REFRACTIVE INDEX*

n, ω, β	ε, α	γ	Optic Sign	Name
1.690	1.670	1.708	B−	Grunerite
1.691	1.719		U+	Willemite
1.692	1.690	1.694	B±	Triphylite
1.694	1.679	1.698	B−	Hornblende
1.694	1.693	1.702	B+	Zoisite
1.699	1.668	1.707	B−	Sinhalite
1.700	1.696	1.711	B−	Riebeckite
1.700	1.509		U−	Magnesite
1.701	1.680	1.720	B−	Olivine
1.702	1.703	1.728	B+	Pigeonite
1.703–1.752	1.700–1.746		U−	Vesuvianite
1.704	1.698	1.723	B+	Augite
1.704	1.694	1.707	B−	Hypersthene (orthopyroxene)
1.714–1.750			I	Pyrope
1.72–1.74	1.71–1.74	1.72–1.75	B+	Rhodonite
1.720–1.734	1.713–1.730	1.723–1.740	B−	Chloritoid
1.720	1.712	1.728	B−	Kyanite
1.722	1.702	1.850	B+	Diaspore
1.724			I	Spinel
1.72–1.78	1.71–1.75	1.73–1.80	B−	Epidote
1.725	1.718	1.748	B+	Hedenbergite
1.734			I	Grossular
1.737	1.733	1.747	B+	Rhodonite
1.738	1.726	1.789	B+	Antlerite
1.748	1.746	1.756	B+	Chrysoberyl
1.749	1.743	1.757	B+	Staurolite
1.755	1.733	1.765	B−	Epidote
1.757	1.804		U+	Benitoite
1.758	1.730	1.838	B+	Azurite
1.769	1.760		U−	Corundum
1.794	1.792	1.845	B+	Monazite
1.78–1.87	1.76–1.78	1.80–1.84	B−	Aegirine
1.80			I	Gahnite
1.80			I	Spessartine
1.816	1.579		U−	Rhodochrosite
1.819	1.776	1.836	B−	Aegirine
1.820			I	Almandine
1.833			I	YAG
1.861	1.831	1.880	B−	Atacamite
1.883	1.877	1.894	B+	Anglesite
1.850	1.623		U−	Smithsonite
1.875	1.633		U−	Siderite
1.875	1.655	1.909	B−	Malachite
1.877	1.835	1.886	B−	Fayalite
1.868			I	Uvarovite
1.887			I	Andradite
1.907	1.900	2.034	B+	Titanite
1.920	1.934		U+	Scheelite
1.920–1.960	1.967–2.015		U+	Zircon
1.93	1.75	1.95	B−	Carnotite
1.975			I	GGG
1.997	2.093		U+	Cassiterite
2.013	2.029		U+	Zincite
2.058	2.048		U−	Pyromorphite
2.077	1.804	2.079	B−	Cerussite
2.11			I	Chromite
2.16			I	Zirconia (cubic)
2.39	2.26	2.40	B−	Goethite
2.35	2.27		U−	Vanadinite
2.37			I	Sphalerite

Table A.3 (Cont.)

NONOPAQUE MINERALS AND SOME SYNTHETIC COMPOUNDS ARRANGED ACCORDING TO INCREASING REFRACTIVE INDEX*

n, ω, β	ϵ, α	γ	Optic Sign	Name
2.37	2.31	2.66	B+	Crocoite
2.404	2.283		U−	Wulfenite
2.409			I	Strontium titanate
2.417			I	Diamond
2.554	2.493		U−	Anatase
2.586	2.583	2.704	B+	Brookite
2.684	2.538	2.704	B−	Realgar
2.61	2.90		U+	Rutile
2.81	2.40	3.02	B−	Orpiment
2.85			I	Cuprite
2.85	3.20		U+	Cinnabar
2.98	2.71		U−	Proustite
3.08	2.88		U−	Pyrargyrite
3.22	2.94		U−	Hematite

*A continuous listing of minerals is given according to increasing refractive index: n (isotropic), ω (uniaxial), β (biaxial). Under optic sign: I = isotropic; U = uniaxial, + or −; B = biaxial, + or −. Although only a few minerals have constant refractive indices, a single listing is given for most. A range of refractive indices is given when the variation is appreciably greater than ±0.01.

MINERAL INDEX

In this index the mineral name is followed by the commonly sought information: composition, crystal system (XI Sys.), specific gravity (**G**), hardness (**H**), and index of refraction (*n*). For uniaxial crystals $n = \omega$, for biaxial crystals $n = \beta$. The refractive index is given as a single entry when the range is usually no greater than ±0.01. Mineral names in bold face (e.g., **Arsenopyrite**) represent species for which complete descriptions are given in the chapters on Systematic Mineralogy (Chapters 10 to 13); those in light face (e.g., Arfvedsonite) refer to minerals that are briefly treated or only mentioned in the text.

Name, Page	Composition	Xl Sys.	G	H	n	Remarks
A						
Acadialite, 555	....					See chabazite
Acanthite, 352	Ag_2S	{ Mon { Iso	7.3	$2-2\frac{1}{2}$	—	Low temp. Ag_2S
Achroite, 473	....					Colorless tourmaline
Acmite, 483	....					See aegirine
Actinolite, 495	$Ca_2(Mg,Fe)_5Si_8O_{22}(OH)_2$	Mon	3.1–3.3	5–6	1.64	Green amphibole
Adularia, 540	$KAlSi_3O_8$	Mon	....			Colorless, translucent K-feldspar
Aegirine, 483	$NaFe^{3+}Si_2O_6$	Mon	3.5	$6-6\frac{1}{2}$	1.82	Na-pyroxene
Aegirine-augite, 482	$(Na,Ca)(Fe^{3+},Fe^{2+},Mg,Al)Si_2O_6$	Mon	3.4–3.5	6	1.71–1.78	Pyroxene
Aenigmatite, 461	$Na_2Fe_5^{2+}TiO_2(Si_2O_6)_3$	Tric	3.75	$5\frac{1}{2}$	1.80	In prismatic crystals, black
Agalmatolite, 515, 610	....					Compact talc substitute for gem jade
Agate, 529	....					Concentric layers of chalcedony
Alabandite, 356, 383	MnS	Iso	4.0	$3\frac{1}{2}-4$	—	Black
Alabaster, 427	....					See gypsum
Albite, 541	$NaAlSi_3O_8(An_0\text{-}An_{10})$	Tric	2.62	6	1.53	Na end member of plagioclase
Alexandrite, 390, 599, 609	....					Gem chrysoberyl
Alkali feldspar, 536, 541	....					Na- or K-feldspar
Allanite, 466	$(Ca,Ce)_3(Fe^{2+},Fe^{3+})Al_2O(SiO_4)(Si_2O_7)(OH)$	Mon	3.5–4.2	$5\frac{1}{2}-6$	1.70–1.81	Brown-black, pitchy luster
Almandine, 452, 600	$Fe_3Al_2Si_3O_{12}$	Iso	4.32	7	1.83	A garnet
Altaite, 356	$PbTe$	Iso	8.16	3	—	Tin-white
Alumstone, 429	....					See alunite
Alunite, 429	$KAl_3(SO_4)_2(OH)_6$	Hex	2.6–2.8	4	1.57	Usually massive
Amalgam, 339	$Ag\text{-}Hg$		....			See silver
Amazonite, 539	....					Green microcline
Amblygonite, 436	$LiAlFPO_4$	Tric	3.0–3.1	6	1.60	Cleavable masses
Amethyst, 528	SiO_2		....			Purple quartz
Amosite, 494, 511	....					Asbestiform cummingtonite; "brown asbestos"
Amphibole, 488	....					A mineral group
Analcime, 549	$NaAlSi_2O_6\cdot H_2O$	Iso	2.27	$5-5\frac{1}{2}$	1.48–1.49	Usually in trapezohedrons
Anatase, 381	TiO_2	Tet	3.9	$5\frac{1}{2}-6$	2.6	Adamantine luster
Anauxite, 513	....	Mon	2.6	2	1.56	Si-rich kaolinite
Andalusite, 455	Al_2SiO_5	Orth	3.16–3.20	$7\frac{1}{2}$	1.64	Square cross sections
Andesine, 532, 543	$Ab_{70}An_{30}\text{-}Ab_{50}An_{50}$	Tric	2.69	6	1.55	Plagioclase feldspar
Andradite, 453, 600	$Ca_3Fe_2Si_3O_{12}$	Iso	3.86	7	1.89	A garnet
Anglesite, 426	$PbSO_4$	Orth	6.2–6.4	3	1.88	Cl{001} {210}
Anhydrite, 426	$CaSO_4$	Orth	2.89–2.98	$3-3\frac{1}{2}$	1.58	Cl{010} {100} {001}
Ankerite, 415	$CaFe(CO_3)_2$	Hex	2.95–3	$3\frac{1}{2}$	1.70–1.75	Cl{10$\bar{1}$1}
Annabergite, 359, 436	$Ni_3(AsO_4)_2\cdot 8H_2O$	Mon	3.0	$2\frac{1}{2}-3$	1.68	Nickel bloom; green
Anorthite, 541	$CaAl_2Si_2O_8(An_{90}\text{-}An_{100})$	Tric	2.76	6	1.58	Ca end member of plagioclase
Anorthoclase, 541	$(K,Na)AlSi_3O_8$	Tric	2.58	6	1.53	An alkali feldspar between K-spar and Ab
Anthophyllite, 493	$(Mg,Fe)_7Si_8O_{22}(OH)_2$	Orth	2.85–3.2	$5\frac{1}{2}-6$	1.61–1.71	An amphibole
Antigorite, 507	$Mg_3Si_2O_5(OH)_4$	Mon	2.5–2.6	~4	1.55	Platy serpentine
Antimony, 342	Sb	Hex	6.7	3	—	Cl{0001}
Antlerite, 429	$Cu_3SO_4(OH)_4$	Orth	3.9±	$3\frac{1}{2}-4$	1.74	Green
Apatite, 434	$Ca_5(PO_4)_3(F,Cl,OH)$	Hex	3.15–3.20	5	1.63	Cl{0001} poor
Apophyllite, 522	$KCa_4(Si_4O_{10})_2F\cdot 8H_2O$	Tet	2.3–2.4	$4\frac{1}{2}-5$	1.54	Cl{001}
Aquamarine, 472, 596	....					Greenish-blue gem beryl
Aragonite, 411	$CaCO_3$	Orth	2.95	$3\frac{1}{2}-4$	1.68	Cl{010} {110}

Name, Page	Composition	XI Sys.	G	H	n	Remarks
Arfvedsonite, 498	$Na_3Fe_4^{2+}Fe^{3+}Si_8O_{22}(OH)_2$	Mon	3.45	6	1.69	Deep green amphibole
Argentite, 353	Ag_2S	....				Now known as acanthite
Arsenopyrite, 368	FeAsS	Mon	6.07	$5\frac{1}{2}$–6	—	Pseudo-orthorhombic
Asbestos, 494, 497, 509, 511						See amphibole and serpentine
Astrophyllite, 461	$(K,Na)_3(Fe,Mn)_7(Ti,Zr)_2$-$Si_8(O,OH)_{31}$	Tric	3.35	3	1.71	Micaceous cleavage
Atacamite, 402	$Cu_2Cl(OH)_3$	Orth	3.75–3.77	3–$3\frac{1}{2}$	1.86	Green, Cl{010}
Augite, 481	$(Ca,Na)(Mg,Fe,Al)(Si,Al)_2O_6$	Mon	3.2–3.4	5–6	1.67–1.73	Dark green to black pyroxene
Aurichalcite, 418	$(Zn,Cu)_5(CO_3)_2(OH)_3$	Orth	3.64	2	1.74	Green to blue
Autunite, 438	$Ca(UO_2)_2(PO_4)_2 \cdot 10–12\,H_2O$	Tet	3.1–3.2	2–$2\frac{1}{2}$	1.58	Yellow-green
Aventurine, 528, 543	....					Quartz or oligoclase
Axinite, 471	$(Ca,Fe,Mn)_3Al_2(BO_3)_3$-$(Si_4O_{12})(OH)$	Tric	3.27–3.35	$6\frac{1}{2}$–7	1.69	Crystal angles acute
Azurite, 417	$Cu_3(CO_3)_2(OH)_2$	Mon	3.77	$3\frac{1}{2}$–4	1.76	Always blue

B

Name, Page	Composition	XI Sys.	G	H	n	Remarks
Balas ruby, 386	....					Red gem spinel
Barite, 424	$BaSO_4$	Orth	4.5	3–$3\frac{1}{2}$	1.64	Cl{001} {210}
Bauxite, 397	mixture of diaspore, gibbsite, boehmite	—	2.0–2.55	1–3	—	An earthy rock
Beidellite, 513	$(Ca,Na)_{0.3}Al_2(OH)_2$-$(Al,Si)_4O_{10} \cdot 4H_2O$	Mon	2–3	1–2	—	Member of montmorillonite group
Benitoite, 461	$BaTiSi_3O_9$	Hex	3.6	$6\frac{1}{2}$	1.76	Blue
Bentonite, 513	....					Montmorillonite alteration of volcanic ash
Beryl, 471, 596, 608	$Be_3Al_2(Si_6O_{18})$	Hex	2.65–2.8	$7\frac{1}{2}$–8	1.57–1.61	Usually green
Beryllonite, 437	$NaBePO_4$	Mon	2.81	$5\frac{1}{2}$	1.56	Rare gem mineral
Biotite, 517	$K(Mg,Fe)_3(AlSi_3O_{10})(OH)_2$	Mon	2.8–3.2	$2\frac{1}{2}$–3	1.61–1.70	Black mica
Bismuthinite, 365	Bi_2S_3	Orth	6.78 ± 0.03	2	—	Cl{010}
Black-band ore, 409	....					See siderite
Black jack, 357	....					See sphalerite
Bloodstone, 529	....					Green and red chalcedony
Boehmite, 397	$\gamma AlO(OH)$	Orth	3.01–3.06	$3\frac{1}{2}$–4	1.65	In bauxite
Bog-iron ore, 396	....					See limonite
Boracite, 422	$Mg_3ClB_7O_{13}$	Orth	2.9–3.0	7	1.66	Pseudoisometric
Borax, 421	$Na_2B_4O_5(OH)_4 \cdot 8H_2O$	Mon	1.7±	2–$2\frac{1}{2}$	1.47	Cl{100}
Bornite, 353	Cu_5FeS_4	{Tet Iso	5.06–5.08	3	—	Purple-blue tarnish
Bort, 346	C	Iso	3.5	10	2.4	Variety of diamond
Bowenite, 510, 609	....					Yellow-green variety of serpentine; substitute for jade
Bravoite, 365	$(Ni,Fe)S_2$	Iso	4.66	$5\frac{1}{2}$–6	—	Steel-gray
Brazilian emerald, 474	....					Green gem tourmaline
Brazilianite, 437	$NaAl_3(PO_4)_2(OH)_4$	Mon	2.98	$5\frac{1}{2}$	1.61	Yellow-green gem mineral
Breithauptite, 360	NiSb	Hex	8.23	$5\frac{1}{2}$	—	Copper-red
Brittle mica, 519	....					A type of mica
Brochantite, 429	$Cu_4SO_4(OH)_6$	Mon	3.9	$3\frac{1}{2}$–4	1.78	Cl{010}; green
Bromargyrite, 400	AgBr	Iso	5.9	1–$1\frac{1}{2}$	2.25	Sectile
Bronzite, 478	$(Mg,Fe)SiO_3$	Orth	3.3±	$5\frac{1}{2}$	1.68	Member of orthopyroxene series
Brookite, 381	TiO_2	Orth	3.9–4.1	$5\frac{1}{2}$–6	2.6	Adamantine luster
Brucite, 393	$Mg(OH)_2$	Hex	2.39	$2\frac{1}{2}$	1.57	Cl{0001}
Bustamite, 487	$(Mn,Ca,Fe)SiO_3$	Tric	3.3–3.4	$5\frac{1}{2}$–$6\frac{1}{2}$	1.67–1.70	Pink to brownish pyroxenoid
Bytownite, 532, 543	$Ab_{30}An_{70}$–$An_{10}An_{90}$	Tric	2.74	6	1.57	Plagioclase feldspar

Name, Page	Composition	Xl Sys.	G	H	n	Remarks
Citrine, 528	SiO_2					Yellow quartz
Clay ironstone, 409						See siderite
Clay minerals, 512						A mineral group
Cleavelandite, 543						White, platy albite
Clinochlore, 521					1.58	See chlorite
Clinoenstatite, 480	$MgSiO_3$	Mon	3.19	6	1.66	In meteorites
Clinoferrosilite, 480	$FeSiO_3$	Mon				Clinopyroxene end member
Clinohumite, 459	$Mg_9(SiO_4)(F,OH)_2$	Mon	3.1–3.2	6	1.64	See chondrodite
Clinohypersthene, 480	$(Mg,Fe)SiO_3$	Mon	3.4–3.5	5–6	1.68–1.72	In meteorites
Clinozoisite, 465	$Ca_2Al_3O(SiO_4)Si_2O_7(OH)$	Mon	3.35–3.37	$6–6\frac{1}{2}$	1.67–1.72	Crystals striated
Clintonite, 519	$Ca(Mg,Al)_{3-2}Al_2Si_2O_{10}(OH)_2$	Mon	3–3.1	$3\frac{1}{2}$	1.65	Brittle mica
Cobalt bloom, 436	$Co_3(AsO_4)_2\cdot 8H_2O$					See erythrite
Cobaltite, 368	$(Co,Fe)AsS$	Orth	6.33	$5\frac{1}{2}$	—	In pseudoisometric pyritohedrons
Coesite, 525	SiO_2	Mon	3.01	7	1.59	High-pressure form of SiO_2
Colemanite, 422	$CaB_3O_4(OH)_3\cdot H_2O$	Mon	2.42	$4–4\frac{1}{2}$	1.59	Cl{010} perfect
Collophane, 434						See apatite
Columbite, 390	$(Fe,Mn)Nb_2O_6$	Orth	5.2–7.3	6	—	Iron-black, submetallic
Common salt, 399						See halite
Copper, 339	Cu	Iso	8.9	$2\frac{1}{2}–3$		Malleable
Cordierite, 473	$(Mg,Fe)_2Al_4Si_5O_{18}\cdot nH_2O$	Orth	2.60–2.66	$7–7\frac{1}{2}$	1.53–1.57	Light bluish-gray
Corundum, 377	Al_2O_3	Hex	4.02	9	1.77	Rhombohedral parting
Cotton balls, 422						See ulexite
Covellite, 361	CuS	Hex	4.6–4.76	$1\frac{1}{2}–2$	—	Blue
Cristobalite, 530	SiO_2	Tet Iso	2.32	$6\frac{1}{2}$	1.48	In volcanic rocks
Crocidolite, 497, 511	$NaFe_3^{2+}Fe_2^{3+}Si_8O_{22}(OH)_2$	Mon	3.2–3.3	4	1.70	Blue amphibole asbestos
Crocoite, 427	$PbCrO_4$	Mon	5.9–6.1	$2\frac{1}{2}–3$	2.36	Orange-red
Crossite, 497						Amphibole between glaucophane and riebeckite
Cryolite, 400	Na_3AlF_6	Mon	2.95–3.0	$2\frac{1}{2}$	1.34	White
Cryptomelane, 395	KMn_8O_{16}					See romanechite
Cryptoperthite, 535						Extremely fine-grained intergrowth of K- and Na-feldspar
Cummingtonite, 494	$(Mg,Fe)_7Si_8O_{22}(OH)_2$	Mon	3.1–3.3	$5\frac{1}{2}–6$	1.66–1.68	Light beige, needlelike amphibole
Cuprite, 376	Cu_2O	Iso	6.1	$3\frac{1}{2}–4$	—	Ruby-red in transparent crystals
Cymophane, 390						Chatoyant chrysoberyl

D

Name, Page	Composition	Xl Sys.	G	H	n	Remarks
Danburite, 458	$Ca(B_2Si_2O_8)$	Orth	2.97–3.02	7	1.63	In crystals
Datolite, 460	$CaB(SiO_4)(OH)$	Mon	2.8–3.0	$5–5\frac{1}{2}$	1.65	Usually in crystals
Demantoid, 453						Green andradite garnet
Diallage, 481						Variety of diopside
Diamond, 346, 596, 609, 610	C	Iso	3.51	10	2.42	Adamantine luster
Diaspore, 395, 397	$\alpha AlO(OH)$	Orth	3.35–3.45	$6\frac{1}{2}–7$	1.72	Cl{010} perfect
Diatomaceous earth, 531						See opal
Diatomite, 531						See opal
Dichroite, 473						See cordierite
Dickite, 512	$Al_2Si_2O_5(OH)_4$	Mon	2.6	$2–2\frac{1}{2}$	1.56	Clay mineral
Digenite, 353	Cu_9S_5	Iso	5.6	$2\frac{1}{2}–3$	—	Similar to chalcocite
Diopside, 481	$CaMgSi_2O_6$	Mon	3.2	5–6	1.67	White to light green pyroxene

Name, Page	Composition	Xl Sys.	G	H	n	Remarks
Dioptase, 524	$Cu_6(Si_6O_{18}) \cdot 6H_2O$	Hex	3.3	5	1.65	Green; minor gem
Djurleite, 353	$Cu_{31}S_6$	Mon	5.75	2.5–3	—	Black; dark grey
Dolomite, 415	$CaMg(CO_3)_2$	Hex	2.85	$3\frac{1}{2}$–4	1.68	$Cl\{10\bar{1}1\}$
Dravite, 473						Brown tourmaline
Dry-bone ore, 410						See smithsonite
Dumortierite, 456	$Al_7O_3(BO_3)(SiO_4)_3$	Orth	3.26–3.36	7	1.69	Radiating

E

Name, Page	Composition	Xl Sys.	G	H	n	Remarks
Edenite, 497	$NaCa_2Mg_5AlSi_7O_{22}(OH)_2$	Mon	3.0	6	1.63	See hornblende
Elbaite, 473						See tourmaline
Electrum, 338						See gold
Emerald, 472, 596, 608						Deep green gem beryl
Emery, 378						Corundum with magnetite
Enargite, 369	Cu_3AsS_4	Orth	4.45	3	—	$Cl\{110\}$
Endlichite, 436						See vanadinite
Enstatite, 478	$MgSiO_3$	Orth	3.2–3.5	$5\frac{1}{2}$	1.65	$Cl\{210\} \sim 90°$
Epidote, 465	$Ca_2(Al,Fe)Al_2O(SiO_4)$-$(Si_2O_7)(OH)$	Mon	3.35–3.45	6–7	1.72–1.78	$Cl\{001\}$, $\{100\}$, green
Epsomite, 429	$MgSO_4 \cdot 7H_2O$	Orth	1.75	2–$2\frac{1}{2}$	1.46	Bitter taste
Epsom salt, 429						See epsomite
Erythrite, 436	$Co_3(AsO_4)_2 \cdot 8H_2O$	Mon	3.06	$1\frac{1}{2}$–$2\frac{1}{2}$	1.66	Pink; cobalt bloom
Essonite, 452, 600						See grossular; garnet
Euclase, 472	$BeAl(SiO_4)(OH)$	Mon	3.1	$7\frac{1}{2}$	1.66	$Cl\{010\}$
Eucryptite, 484	$LiAlSiO_4$	Hex	2.67	—	1.55	Spodumene alteration

F

Name, Page	Composition	Xl Sys.	G	H	n	Remarks
Fairy stone, 459						See staurolite
Famatinite, 370	Cu_3SbS_4	Tet	4.52	$3\frac{1}{2}$	—	Gray
Fayalite, 449	Fe_2SiO_4	Orth	4.39	$6\frac{1}{2}$	1.86	Olivine, Fe end member
Feldspar, 532						A mineral group
Feldspathoid, 543						A mineral group
Ferberite, 430	$FeWO_4$	Mon	7.0–7.5	5	—	See wolframite
Fergusonite, 391	$(Y,Er,Ce,Fe)NbO_4$	Tet	5.8	$5\frac{1}{2}$–6	—	Brown-black
Ferrimolybdite, 368	$Fe_2(MoO_4)_3 \cdot 8H_2O$	Orth?	2.99	~ 1	1.73–1.79	Canary-yellow, soft
Ferroactinolite, 495	$Ca_2Fe_5Si_8O_{22}(OH)_2$	Mon	3.2–3.3	5–6	1.68	Dark green amphibole
Ferropseudobrookite, 381	$FeTi_2O_5$					See ilmenite
Ferrosilite, 478	$FeSiO_3$	Orth	3.9	6	1.79	Orthopyroxene end member
Fersmannite, 461	$Na_4Ca_4Ti_4(SiO_4)_3(O,OH,F)_3$	Mon	3.44	$5\frac{1}{2}$	1.93	Brown; rare; in alkalic rocks
Fibrolite, 456						See sillimanite
Fischesserite, 338	Ag_3AuSe_2					Rare gold ore
Flint, 529	SiO_2	—	2.65	7	1.54	Microcryst. quartz
Flos ferri, 413						See aragonite
Fluorapatite, 434	$Ca_5(PO_4)_3F$					See apatite
Fluorite, 401	CaF_2	Iso	3.18	4	1.43	Cl octahedral
Forsterite, 449	Mg_2SiO_4	Orth	3.2	$6\frac{1}{2}$	1.63	Olivine; Mg end member
Fowlerite, 487						Zn-bearing rhodonite
Franklinite, 388	$(Zn,Fe,Mn)(Fe,Mn)_2O_4$	Iso	5.15	6	—	Spinel from Franklin, N.J.
Freibergite, 371						Ag-bearing tetrahedrite
Fuchsite, 515, 610						Chrome-rich muscovite

G

Name, Page	Composition	Xl Sys.	G	H	n	Remarks
Gadolinite, 472	$YFeBe_2(SiO_4)_2O_2$	Mon	4.0–4.5	$6\frac{1}{2}$–7	1.79	Black
Gahnite, 387	$ZnAl_2O_4$	Iso	4.55	$7\frac{1}{2}$–8	1.80	Spinel; dark green octahedrons
Galaxite, 387	$MnAl_2O_4$	Iso	4.03	$7\frac{1}{2}$–8	1.92	Mn spinel

Name, Page	Composition	XI Sys.	G	H	n	Remarks
Galena, 354	PbS	Iso	7.4–7.6	$2\frac{1}{2}$	—	Cl cubic
Garnet, 451, 600, 611	$A_3B_2(SiO_4)_3$	Iso	3.5–4.3	$6\frac{1}{2}$–$7\frac{1}{2}$	1.71–1.88	A mineral group; commonly in crystals
Garnierite, 511	$(Ni,Mg)_3Si_2O_5(OH)_4$	Mon	2.2–2.8	2–3	1.59	Green Ni serpentine
Gaylussite, 418	$Na_2Ca(CO_3)_2 \cdot 5H_2O$	Mon	1.99	2–3	1.52	Cl{110} perfect
Gedrite, 493	$\sim Na_{0.5}(Mg,Fe)_2(Mg,Fe)_{3.5}\text{-}(Al,Fe^{3+})_{1.5}Si_6Al_2O_{22}(OH)_2$	Orth	....			See anthophyllite
Geikielite, 380	$MgTiO_3$	Hex	4.05	$5\frac{1}{2}$–6	—	Cl{$10\bar{1}1$}
Gersdorffite, 368	NiAsS	Iso	5.9	$5\frac{1}{2}$	—	See cobaltite
Geyserite, 531			....			Opal in hot springs
Gibbsite, 394, 397	$Al(OH)_3$	Mon	2.3–2.4	$2\frac{1}{2}$–$3\frac{1}{2}$	1.57	Basal Cl{001}
Glaucodot, 368	$(Co,Fe)AsS$	Orth	6.04	5	—	Tin-white
Glauconite, 518	$(K,Na,Ca)_{0.5-1}(Fe^{3+},Al,\text{-}Fe^{2+},Mg)_2(Si,Al)_4O_{10}(OH)_2 \cdot nH_2O$	Mon	2.4±	2	1.62	In green sands
Glaucophane, 497	$Na_2Mg_3Al_2Si_8O_{22}(OH)_2$	Mon	3.1–3.3	$6.6\frac{1}{2}$	1.62–1.67	Blue to black amphibole
Gmelinite, 557	$(Na_2,Ca)(Al_2Si_4O_{12}) \cdot 6H_2O$	Hex	2.1±	$4\frac{1}{2}$	1.49	A zeolite
Goethite, 395	$\alpha FeO(OH)$	Orth	4.37	5–$5\frac{1}{2}$	2.39	Cl{010} perfect
Gold, 336	Au	Iso	15–19.3	$2\frac{1}{2}$–3	—	Yellow, soft
Goshenite, 471, 596			....			Colorless gem beryl
Graphite, 349	C	Hex	2.23	1–2	—	Black, platy
Greenalite, 511	$(Fe,Mg)_3Si_2O_5(OH)_4$	Mon	3.2	—	1.67	In iron-formations
Greenockite, 357	CdS	Hex	4.9	3–$3\frac{1}{2}$	—	Yellow-orange
Grossular, 452, 600	$Ca_3Al_2Si_3O_{12}$	Iso	3.59	$6\frac{1}{2}$	1.73	A garnet
Grunerite, 494	$Fe_7Si_8O_{22}(OH)_2$	Mon	3.6	6	1.71	Light brown, needlelike amphibole
Gypsum, 427	$CaSO_4 \cdot 2H_2O$	Mon	2.32	2	1.52	Cl{010} {100} {011}

H

Name, Page	Composition	XI Sys.	G	H	n	Remarks
Halite, 399	NaCl	Iso	2.16	$2\frac{1}{2}$	1.54	Cl cubic; salty
Halloysite, 513	$Al_2Si_2O_5(OH)_4$ and $Al_2Si_2O_5(OH)_4 \cdot 2H_2O$	Mon	2.0–2.2	1–2	1.54	Clay mineral
Harmotome, 557	$Ba(Al_2Si_6O_{16}) \cdot 6H_2O$	Mon	2.45	$4\frac{1}{2}$	1.51	A zeolite
Hastingsite, 497	$NaCa_2Fe_4(Al,Fe)Al_2Si_6\text{-}O_{22}(OH)_2$	Mon	3.2	6	1.66	See hornblende
Haüynite, 547	$(Na,Ca)_{4-8}(AlSiO_4)_6(SO_4)_{1-2}$	Iso	2.4–2.5	$5\frac{1}{2}$–6	1.50	A feldspathoid
Hectorite, 513	$(Mg,Li)_3Si_4O_{10}(OH)_2Na_{0.3} \cdot 4H_2O$	Mon	2.5	1–$1\frac{1}{2}$	1.52	Li montmorillonite
Hedenbergite, 481	$CaFeSi_2O_6$	Mon	3.55	5–6	1.73	Clinopyroxene end member
Heliotrope, 529			....			Green and red chalcedony
Hematite, 379	Fe_2O_3	Hex	5.26	$5\frac{1}{2}$–$6\frac{1}{2}$	—	Red streak
Hemimorphite, 464	$Zn_4(Si_2O_7)(OH)_2 \cdot H_2O$	Orth	3.4–3.5	$4\frac{1}{2}$–5	1.62	Cl{110}
Hercynite, 387	$FeAl_2O_4$	Iso	4.39	$7\frac{1}{2}$–8	1.80	Fe spinel
Heulandite, 555	$CaAl_2Si_7O_{18} \cdot 6H_2O$	Mon	2.18–2.20	$3\frac{1}{2}$–4	1.48	Cl{010} perfect
Hiddenite, 484			....			Green gem spodumene
Hollandite, 394	$Ba_2Mn_8O_{16}$	Tet / Mon	4.9	6	—	Silver-gray to black
Holmquistite, 489	$Li_2(Mg,Fe)_3(Al,Fe^{3+})_2\text{-}Si_8O_{22}(OH)_2$	Orth	3.06–3.13	5–6	1.64–1.66	Blue to violet-blue; near Li-pegmatites
Hornblende, 496	$(Ca,Na)_{2-3}(Mg,Fe,Al)_5\text{-}Si_6(Si,Al)_2O_{22}(OH)_2$	Mon	3.0–3.4	5–6	1.62–1.72	Green to black amphibole
Horn silver, 400			....			See chlorargyrite
Huebnerite, 430	$MnWO_4$	Mon	7.0	5	—	See wolframite
Humite, 459	$Mg_7(SiO_4)_3(F,OH)_2$	Orth	3.1–3.2	6	1.64	See chondrodite
Hyalite, 531			....			Colorless, globular opal
Hyalophane, 532	$(K,Ba)(Al,Si)_2Si_2O_8$	Mon	2.8	6	1.54	A feldspar
Hydroboracite, 422	$CaMgB_6O_8(OH)_6 \cdot 3H_2O$	Mon	2.17	2	1.53	Clear, colorless to white
Hydrogrossular, 452	$Ca_3Al_2Si_2O_8(SiO_4)_{1-m}(OH)_{4m}$	Iso	3.13–3.59	6–7	1.67–1.73	Hydrous garnet
Hydroxylapatite, 434	$Ca_5(PO_4)_3(OH)$		....			See apatite
Hydrozincite, 411	$Zn_5(CO_3)_2(OH)_6$	Mon	3.6–3.8	2–$2\frac{1}{2}$	1.74	Secondary mineral
Hypersthene, 478	$(Mg,Fe)SiO_3$	Orth	3.4–3.5	5–6	1.68–1.73	An intermediate member of the orthopyroxene series

Name, Page	Composition	XI Sys.	G	H	n	Remarks
I						
Iceland spar, 406	....					See calcite
Idocrase, 466	....					See vesuvianite
Illite, 513					1.57–1.61	Micalike clay mineral
Ilmenite, 380	$FeTiO_3$	Hex	4.7	$5\frac{1}{2}$–6	—	May be slightly magnetic
Ilvaite, 465	$CaFe_3^{2+}Fe^{3+}O(Si_2O_7)(OH)$	Orth	4.0	$5\frac{1}{2}$–6	1.91	Black
Indialite, 473	$(Mg,Fe)_2Al_4Si_5O_{18}\cdot nH_2O$	Hex	~2.6	7	—	High T form of cordierite
Indicolite, 473						Blue tourmaline
Inyoite, 422	$CaB_3O_3(OH)_5\cdot 4H_2O$	Mon	1.88	2	1.51	Colorless, transparent
Iodargyrite, 400	AgI	Hex	5.5–5.7	1–$1\frac{1}{2}$	2.18	Sectile
Iodobromite, 400	$Ag(Cl,Br,I)$	Iso	5.71	1–$1\frac{1}{2}$	2.20	Sectile; same as iodian bromargyrite
Iolite, 473	....					See cordierite
Iridium, 341	Ir	Iso	22.7	6–7	—	A platinum metal
Iridosmine, 341	Ir-Os	Hex	19.3–21.1	6–7	—	See platinum
Iron, 341	Fe	Iso	7.3–7.9	$4\frac{1}{2}$	—	Very rare
J						
Jacobsite, 388	$MnFe_2O_4$	Iso	5.1	$5\frac{1}{2}$–$6\frac{1}{2}$	2.3	A spinel
Jade, 483, 495, 598, 609	....					See nephrite and jadeite
Jadeite, 482, 598	$NaAlSi_2O_6$	Mon	3.3–3.5	$6\frac{1}{2}$–7	1.66	Green, compact
Jargon, 454	....					See zircon
Jarosite, 429	$KFe_3(SO_4)_2(OH)_6$	Hex	3.2±	3	1.82	Yellow-brown
Jasper, 529	....					Red, microcryst. quartz
Jimthompsonite, 491	$(Mg,Fe)_{10}Si_{12}O_{32}(OH)_4$	Orth				Microscopic alteration of anthophyllite; a "biopyribole"
Johannsenite, 481	$CaMnSi_2O_6$	Mon	3.4–3.5	6	1.71–1.73	Mn analogue of diopside
K						
Kainite, 400	$KMg(Cl,SO_4)\cdot 2\frac{3}{4}H_2O$	Mon	2.1	3	1.51	Cl{001}; salty, bitter
Kalsilite, 546	$KAlSiO_4$	Hex	2.61	6	1.54	Isostructural with nepheline
Kamacite, 341	Fe-Ni	Iso	7.3–7.9	4	—	In meteorites
Kaolin, 513	....					Mixture of clay minerals
Kaolinite, 512	$Al_2Si_2O_5(OH)_4$	Tric	2.6	2	1.55	Earthy
Keatite, 525	SiO_2	Tet	2.50	7	1.52	Synthetic
Kernite, 419	$Na_2B_4O_6(OH)_2\cdot 3H_2O$	Mon	1.9	3	1.47	Cl{001} {100}
K-feldspar, 536	....					See microcline, orthoclase, sanidine
Kieserite, 577	$MgSO_4\cdot H_2O$	Mon	2.57	$3\frac{1}{2}$	1.53	Massive, granular; whitish gray
Kostovite, 338	$CuAuTe_4$					Rare gold ore
Krennerite, 338	$AuTe_2$	Orth	8.62	2–3	—	Basal Cl{001}
Kunzite, 484	....					Pink gem spodumene
Kutnahorite, 409, 416	$CaMn(CO_3)_2$	Hex	3.12	$3\frac{1}{2}$–4	1.74	Mn-dolomite
Kyanite, 456	Al_2SiO_5	Tric	3.55–3.66	5–7	1.72	Blue, bladed
L						
Labradorite, 532, 543	$Ab_{50}An_{50}$-$Ab_{30}An_{70}$	Tric	2.71	6	1.56	A member of the plagioclase feldspar series
Lamprophyllite, 461	$Na_3Sr_2Ti_3(Si_2O_7)_2(O,OH,F)_2$	Orth	3.45	4	1.75	Platy
Langbeinite, 577	$K_2Mg_2(SO_4)_3$	Iso	2.83	$3\frac{1}{2}$–4	1.53	Colorless; in evaporites
Lapis lazuli, 548	....					See lazurite

Name, Page	Composition	Xl Sys.	G	H	n	Remarks
Perovskite, 381	$CaTiO_3$	Orth	4.03	$5\frac{1}{2}$	2.38	Pseudoisometric crystals
Perthite, 535						Coarse K-feldspar–albite intergrowth
Petalite, 548	$Li(AlSi_4O_{10})$	Mon	2.4	$6-6\frac{1}{2}$	1.51	Cl{001} {201}
Petzite, 338	$(Ag,Au)_2Te$	Iso	8.7–9.0	$2\frac{1}{2}-3$	—	Steel-gray to iron-black
Phenacite, 447	Be_2SiO_4	Hex	2.97–3.0	$7\frac{1}{2}-8$	1.65	In pegmatites
Phillipsite, 557	$KCa(Al_3Si_5O_{16})\cdot6H_2O$	Mon	2.2	$4\frac{1}{2}-5$	1.50	A zeolite
Phlogopite, 517	$KMg_3(AlSi_3O_{10})(OH)_2$	Mon	2.86	$2\frac{1}{2}-3$	1.56–1.64	Yellow-brown mica
Phosgenite, 415	$Pb_2CO_3Cl_2$	Tet	6.0–6.3	3	2.12	Adamantine luster
Phosphorite, 435						See apatite
Piemontite, 465	$Ca_2MnAl_2O(SiO_4)(Si_2O_7)(OH)$	Mon	3.4	$6\frac{1}{2}$	1.75–1.81	Reddish-brown
Pigeonite, 480	$\sim Ca_{0.25}(Mg,Fe)_{1.75}Si_2O_6$	Mon	3.30–3.46	6	1.64–1.72	High T pyroxene
Pitchblende, 384						Massive UO_2
Plagioclase, 541	$Ab_{100}An_0\text{-}Ab_0An_{100}$	Tric	2.62–2.76	6	1.53–1.59	The Na-Ca-feldspar series
Plancheite, 524	$Cu_8(Si_4O_{11})_2(OH)_2\cdot H_2O$	Orth	3.3	5.5	1.66	Fibrous, mammillary, blue
Platinum, 340	Pt	Iso	14–19	$4-4\frac{1}{2}$	—	Steel-gray with bright luster; malleable
Pleonaste, 385						Ferroan spinel
Pollucite, 546	$CsAlSi_2O_6\cdot H_2O$	Iso	2.9	$6\frac{1}{2}$	1.52	Colorless; in pegmatites
Polyhalite, 400, 577	$K_2Ca_2Mg(SO_4)_2\cdot2H_2O$	Tric	2.78	$2\frac{1}{2}-3$	1.56	Bitter taste
Potash feldspar, 536	$KAlSi_3O_8$					Microcline, orthoclase, sanidine
Powellite, 431	$CaMoO_4$	Tet	4.23	$3\frac{1}{2}-4$	1.97	Fluoresces yellow
Prase, 529						Dull green, microcrystalline quartz
Prehnite, 523	$Ca_2Al(AlSi_3O_{10})(OH)_2$	Orth	2.8–2.95	$6-6\frac{1}{2}$	1.63	Tabular crystals, green
Proustite, 370	Ag_3AsS_3	Hex	5.57	$2-2\frac{1}{2}$	3.09	Light ruby silver
Pseudobrookite, 381	Fe_2TiO_5					See ilmenite
Pseudoleucite, 546						See leucite
Pseudophite, 610						Massive form of chlorite; substitute for jade
Pseudowollastonite, 484	$CaSiO_3$	Tric				High T form of wollastonite
Psilomelane, 395						See romanechite
Pyrargyrite, 370	Ag_3SbS_2	Hex	5.85	$2-2\frac{1}{2}$	3.08	Dark ruby-silver
Pyrite, 365	FeS_2	Iso	5.02	$6-6\frac{1}{2}$	—	Crystals striated
Pyrochlore, 391	$(Ca,Na)_2(Nb,Ta)_2O_6(O,OH,F)$	Iso	4.3±	5	—	Usually metamict
Pyrolusite, 382	MnO_2	Tet	4.75	1–2	—	Sooty black
Pyromorphite, 435	$Pb_5(PO_4)_3Cl$	Hex	7.04	$3\frac{1}{2}-4$	2.06	Adamantine luster
Pyrope, 452, 600	$Mg_3Al_2Si_3O_{12}$	Iso	3.58	7	1.71	A garnet
Pyrophanite, 380	$MnTiO_3$	Hex	4.54	5–6	2.48	Cl{$02\bar{2}1$}
Pyrophyllite, 514	$Al_2Si_4O_{10}(OH)_2$	Mon	2.8	1–2	1.59	Smooth feel; micaceous habit
Pyroxene, 475						A mineral group
Pyroxenoid, 484						A mineral group
Pyroxferroite, 487	$Ca_{0.15}Fe_{0.85}SiO_3$	Tric	3.7	—	1.75	Pyroxenoid; in lunar rocks
Pyroxmangite, 487	$(Mn,Fe)SiO_3$	Tric	3.6–3.8	$5\frac{1}{2}-6$	1.72–1.75	From metamorphic Mn-rich rocks
Pyrrhotite, 358	$Fe_{1-x}S$	Mon Hex	4.58–4.65	4	—	Magnetic

Q

Name, Page	Composition	Xl Sys.	G	H	n	Remarks
Quartz, 526, 599, 610	SiO_2	Hex	2.65	7	1.54	Conchoidal fracture

R

Name, Page	Composition	Xl Sys.	G	H	n	Remarks
Ramsayite, 461	$Na_2Ti_2Si_2O_9$	Orth	3.43	6	2.01	Rare Ti-bearing silicate in alkalic rocks
Realgar, 362	AsS	Mon	3.48	$1\frac{1}{2}-2$	2.60	Cl{010}; red

SUBJECT INDEX

LOCATIONS OF SOME IMPORTANT TABLES AND ILLUSTRATIONS